Scientific Computing with MATLAB®

Second Edition

Scientific Computing with MATLAB®

Second Edition

Dingyü Xue
Northeastern University
Shenyang, China

YangQuan Chen
University of California
Merced, USA

CRC Press
Taylor & Francis Group
Boca Raton London New York

CRC Press is an imprint of the
Taylor & Francis Group, an **informa** business
A CHAPMAN & HALL BOOK

CRC Press
Taylor & Francis Group
6000 Broken Sound Parkway NW, Suite 300
Boca Raton, FL 33487-2742

First issued in paperback 2020

ISBN-13: 978-1-4987-5777-5 (hbk)
ISBN-13: 978-0-367-78313-6 (pbk)

Visit the Taylor & Francis Web site at
http://www.taylorandfrancis.com

and the CRC Press Web site at
http://www.crcpress.com

Contents

Preface

Since the first edition of this book published in 2008, computing landscape changed radically, from cloud computing to big data science, from wearable computing to internet of things, from deep learning to driverless cars... yet in more general sense, computing happens in all walks of life, from falling rocks to withering leaves, from climate change to extreme weather, from gene editing to digital matter... But one thing not changed is the scientific computing fundamentals that cover all college mathematics. Busy students, engineers and scientists need "fast-food" ways to compute and get problems solved reliably.

Oliver Heaviside once said "Mathematics is of two kinds, Rigorous and Physical. The former is Narrow: the latter Bold and Broad. To have to stop to formulate rigorous demonstrations would put a stop to most physico-mathematical inquiries. Am I to refuse to eat because I do not fully understand the mechanism of digestion?"[1] Today, we can ask a similar question: Am I to refuse to compute because I do not fully understand the mechanism of numerics? As we discussed in the Preface of the first edition, we need a new way of learning scientific computing so that we can focus more on "computational thinking."

With these goals in mind, this edition includes the following new features:

(1) A significant amount of new material is introduced, specifically: four-dimensional volume visualization, interval limit, infinite series convergence, numerical multiple integral, arbitrary matrix analysis, matrix power, difference equations, numerical integral transforms, Laurent series, matrix equation solutions, multi-objective optimizations, dynamic programming and shortest path problems, matrix differential equations, switching ODEs, delay ODEs, special functions, principal component analysis, Monte Carlo algorithm, outlier detection, radial basis network, particle swarm optimization, and a completely new section on fractional calculus.

(2) The three-phase solution procedure proposed by the authors has been followed throughout the book. Namely, to solve a problem, the physical explanation of the mathematical problem to be solved is given first, followed by the methodology of how to formulate the problem in MATLAB®-compatible framework, and finally, the third phase is to call MATLAB functions to solve the problem. The guideline is useful in real world problem solving with lots of illustrative examples.

(3) Mathematical branches are arranged more systematically. Using the traditional styles in mathematical presentation (as in typical mathematics courses), however, concentrations are made on how the problems are solved. If there are existing MATLAB functions, or third-party products, suggestions are made to use them directly. If there are not, or if existing ones are problematic, new MATLAB functions are written and easy-to-use calling syntaxes are designed and explained.

(4) Soon after the publication of the first edition, MATLAB R2008b was released, from which the symbolic engine is replaced, and some of the commands, especially those

[1] Edge A. Oliver Heaviside (1850–1927) - Physical mathematician. Teaching Mathematics and Its Applications 2: 55-61, 1983.

involving overload functions and Maple internal functions, cannot be used for the symbolic computation problems. In the new edition, compatibility with the new versions of MATLAB are supported.

(5) Enhanced examples and exercises are included to support the materials throughout the new edition. A complete set of teaching materials, composed of about 1500 PPT slides and a solutions manual, is provided with the book. The relevant materials can be downloaded from the authors-maintained web-site at

`https://mechatronics.ucmerced.edu/Scientific-Computing-with-MATLAB-2ndEd`

Financial support from the National Natural Science Foundation of China under Grant 61174145 is acknowledged. Thanks also go to Drs. Yanliang Zhang and Lynn Crisanti for arranging the first author's visit to MathWorks, Natick, MA for discussing a possible MOOC project for the book. A MOOC in Chinese is just made ready and will be released soon, thanks to the support from Liaoning Provincial Education Bureau and Northeastern University, China. Classroom videos in English are scheduled. New information and links on the MOOC progress will be anounced in the above web-site.

MATLAB and Simulink® are registered trademarks of The MathWorks, Inc. For product information, please contact:

The MathWorks, Inc.
3 Apple Hill Drive
Natick, MA, 01760-2098, USA
Tel: 508-647-7000
Fax: 508-647-7101
E-mail: `info@mathworks.com`
Web: `http://www.mathworks.com`

This new edition was suggested and supported by Sunil Nair, Publisher, CRC Press, Taylor and Francis Group. We are thankful for Sunil's patience and constructive comments. We wish to specially thank Michael Davidson, Project Editor, Taylor and Francis Group, LLC for helping us with an excellent copy-editing service. We would like to extend our appreciation to Professor Jian-Qiao Sun of University of California, Merced for adopting the first edition of the book for his ME021 "Engineering Computing" course and for motivating a new edition with a bigger format size. The new materials of the book have been used in a course entitled "MATLAB and Scientific Computing" for two semesters in Northeastern University, China. All the students of Lang Shijun Automation Experimental Classes 1309 and 1410, especially Mr. Weiming Mi and Mr. Huaijia Lin, are acknowledged for some new insights and the hard work of modifications of the PPT slides and solutions manual.

Last but not least, Dingyü Xue would like to thank his wife Jun Yang and his daughter Yang Xue; YangQuan Chen would like to thank his wife Huifang Dou and his sons Duyun, David and Daniel, for their patience, understanding and complete support throughout this book project.

Dingyü Xue, Shenyang, Liaoning, China
YangQuan Chen, Merced, California, USA

Preface of the First Edition

Computational Thinking[2], coined and promoted by Jeannette Wing of Carnegie Mellon University, is getting more and more attention. "It represents a universally applicable attitude and skill set everyone, not just computer scientists, would be eager to learn and use" as acknowledged by Dr. Wing, "Computational Thinking draws on math as its foundations." The present book responds to "Computational Thinking" by offering the readers enhanced math problem solving ability and therefore, the readers can focus more on "Computational Thinking" instead of "Computational Doing."

The breadth and depth of one's mathematical knowledge might not match his or her ability to solve mathematical problems. In today's applied science and applied engineering, one usually needs to get the mathematical problems at hand solved efficiently in a timely manner without complete understanding of the numerical techniques involved in the solution process. Therefore, today, arguably, it is a trend to focus more on how to formulate the problem in a form suitable for computer solution and on the interpretation of the results generated from the computer. We further argue that, even without a complete preparation of mathematics, it is possible to solve some advanced mathematical problems using a computer. We hope this book is useful for those who frequently feel that their level of math preparation is not high enough because they still can get their math problems at hand solved with the encouragement gained from reading this book.

Using computers to solve mathematical problems today is ubiquitous. MATLAB®/Simulink is considered as the dominant software platform for applied math related topics. Sometimes, one simply does not know one's problem could be solved in a much simpler way in MATLAB or Simulink. From what Confucius wrote, "The craftsman who wishes to work well has first to sharpen his implements,"[3] it is clear that MATLAB is the right, already sharpened "implement." However, a bothering practical problem is this: MATLAB documentation only shows "this function performs this," and what a user with a mathematical problem at hand wants is, "Given this math problem, through what reformulation, and then, use of what functions will get the problem solved." Frequently, it is very easy for one to get lost in thousands of functions offered in MATLAB plus the same amount, if not more, of functions contributed by the MATLAB users community. Therefore, the major contribution of this book is to bridge the gap between "problems" and "solutions" through well grouped topics and tightly yet smoothly glued MATLAB example scripts and reproducible MATLAB-generated plots.

A distinguishing feature of the book is the organization and presentation of the material. Based on our teaching, research and industrial experience, we have chosen to present the course materials following the sequence

- Computer Mathematics Languages — An Overview
- Fundamentals of MATLAB Programming
- Calculus Problems

[2]`http://www.cs.cmu.edu/afs/cs/usr/wing/www/Computational_Thinking.pdf`
[3]Confucius. `http://www.confucius.org/lunyu/ed1509.htm`.

- Linear Algebra Problems
- Integral Transforms and Complex Variable Functions
- Nonlinear Equations and Optimization Problems
- Differential Equations Problems
- Data Interpolation and Functional Approximation Problems
- Probability and Statistics Problems
- Nontraditional Methods

In particular, in the nontraditional mathematical problem solution methods, we choose to cover some interesting and practically important topics such as set theory and fuzzy inference system, neural networks, wavelet transform, evolutionary optimization methods including genetic algorithms and particle swarm optimization methods, rough set based data analysis problems, fractional-order calculus (derivative or integral of non-integer order) problems, etc., all with extensive problem solution examples. A dedicated CAI (computer aided instruction) kit including more than 1,300 interactive PowerPoint slides has been developed for this book for both instruction and self-learning purposes.

We hope that readers will enjoy playing with the scripts and changing them as they wish for a better understanding and deeper exploration with reduced efforts. Additionally, each chapter comes with a set of problems to strengthen the understanding of the chapter contents. It appears that the book is presenting in certain depth some mathematical problems. However, the ultimate objective of this book is to help the readers, after understanding *roughly* the mathematical background, to avoid the tedious and complex technical details of mathematics and find the reliable and accurate solutions to the interested mathematical problems with the use of MATLAB computer mathematics language. There is no doubt that the readers' ability to tackle mathematical problems can be significantly enhanced after reading this book.

This book can be used as a reference text for almost all college students, both undergraduates and graduates, in almost all disciplines which require certain levels of applied mathematics. The coverage of topics is practically broad yet with a balanced depth. The authors also believe that this book will be a good desktop reference for many who have graduated from college and are still involved in solving mathematical problems in their jobs.

Apart from the standard MATLAB, some of the commercial toolboxes may be needed. For instance, the Symbolic Math Toolbox is used throughout the book to provide alternative analytical solutions to certain problems. Optimization Toolbox, Partial Differential Equation Toolbox, Spline Toolbox, Statistics Toolbox, Fuzzy Logic Toolbox, Neural Network Toolbox, Wavelet Toolbox, and Genetic Algorithm and Direct Search Toolbox may be required in corresponding chapters or sections. A lot of MATLAB functions designed by the authors, plus some third-party free toolboxes, are also presented in the book. For more information on MATLAB and related products, please contact

The MathWorks, Inc.
3 Apple Hill Drive
Natick, MA, 01760-2098, USA
Tel: 508-647-7000
Fax: 508-647-7101
E-mail: `info@mathworks.com`
Web: `http://www.mathworks.com`

The writing of this book started more than 5 years ago, when a Chinese version[4] was

[4]Xue Dingyü and Chen YangQuan, *Advanced applied mathematical problem solutions using MATLAB*, Beijing: Tsinghua University Press, 2004

published in 2004. Many researchers, professors and students have provided useful feedback comments and input for the newly extended English version. In particular, we thank the following professors: Xinhe Xu, Fuli Wang of Northeastern University; Hengjun Zhu of Beijing Jiaotong University; Igor Podlubny of Technical University of Kosice, Slovakia; Shuzhi Sam Ge of National University of Singapore, Wen Chen of Hohai University, China. The writing of some parts of this book has been helped by Drs. Feng Pan, Daoxiang Gao, Chunna Zhao and Dali Chen, and some of the materials are motivated by the talks with colleagues at Northeastern University, especially Drs. Xuefeng Zhang and Haibin Shi. The computer aided instruction kit and solution manual were developed by our graduate students Wenbin Dong, Jun Peng, Yingying Liu, Dazhi E, Lingmin Zhang and Ying Luo.

Moreover, we are grateful to the Editors, LiMing Leong and Marsha Hecht, CRC Press, Taylor & Francis Group, for their creative suggestions and professional help. The "Book Program" from The MathWorks Inc., in particular, Hong Yang, MathWorks, Beijing, Courtney Esposito, Meg Vuliez and Dee Savageau, are acknowledged for the latest MATLAB software and technical problem support.

The authors are grateful to the following free toolbox authors, to allow the inclusion of their contributions in the companion CD:

Dr. Brian K Birge, for particle swarm optimization toolbox (PSOt)
John D'Errico, for `fminsearchbnd` Toolbox
Mr. Koert Kuipers for his BNB Toolbox
Dr. Johan Löfberg, University of Linköping, Sweden for YALMIP
Mr. Xuefeng Zhang, Northeastern University, China for RSDA Toolbox

Last but not least, Dingyü Xue would like to thank his wife Jun Yang and his daughter Yang Xue; YangQuan Chen would like to thank his wife Huifang Dou and his sons Duyun, David and Daniel, for their patience, understanding and complete support throughout this work.

Dingyü Xue
Northeastern University
Shenyang, China
`xuedingyu@mail.neu.edu.cn`

YangQuan Chen
Utah State University
Logan, Utah, USA
`yqchen@ieee.org`

Chapter 1

Computer Mathematics Languages — An Overview

Mathematical problems are essential in almost all aspects of scientific and engineering research. The mathematical models should normally be established first, then, the solutions of the models under investigation can be obtained. Specific knowledge is required for the establishment of mathematical models, which needs the expertise of the researchers, and with the established models, the numerical and analytical approaches presented in this book can be used to solve the problems. In this chapter, a brief introduction to computer mathematical problems is given, and it will be illustrated through simple examples why computer mathematical languages should be learned. A concise history of the development of computer mathematical languages and mathematical tools will be introduced. Finally the framework of the book is presented. Also, an overview of the mathematics branches involved is given in this chapter.

1.1 Computer Solutions to Mathematics Problems

1.1.1 Why should we study computer mathematics language?

We all know that manual derivation of solutions to mathematical problems is a useful skill when the problems are not so complicated. However, for a great variety of mathematical problems, manual solutions are laborious or even not possible. Therefore, computers must be employed for solving these problems. There are basically two ways of solving these problems by computers. One is to verbally implement the existing numerical algorithms using general purpose computer languages such as Fortran or C. The other way is to use specific computer languages with a good reputation. These languages include MATLAB, Mathematica and Maple. In this book, they are referred to as the *computer mathematics languages.* The numerical algorithms can only be used to handle computation problems by numbers, while for problems like to find the solutions to the symbolic equation $x^3+ax+c=d$, where a, c, d are not given numerical values but symbolic variables, the numerical algorithms cannot be used. The computer mathematics languages with symbolic computation capabilities should be used instead.

We shall use the term "mathematical computation" throughout the book, whereas the term really means both numerical and analytical computation of mathematical problems. Normally, analytical solutions are explored first, and if there are no analytical solutions, numerical solutions are obtained.

Before systematically introducing the contents of the book, the following examples are given such that the readers may understand and appreciate the necessity of using the computer mathematics languages.

Example 1.1 In calculus courses, the concepts and derivation methods are introduced with an emphasis on manual deduction and computation. If a function $f(x)$ is given by $f(x) = \dfrac{\sin x}{x^2+4x+3}$, how could one derive $\dfrac{\mathrm{d}^4 f(x)}{\mathrm{d}x^4}$ manually?

Solution *One can derive it using the methods taught in calculus courses. For instance, the first-order derivative* $\mathrm{d}f(x)/\mathrm{d}x$ *can be derived first, the second-order derivative, third-order derivative and finally fourth-order derivative of the function* $f(x)$ *can be evaluated in turn. In this way, even higher-order derivatives of the function can be derived manually, in theory. However, the procedure is more suitable to be carried out with computers. With suitable computer mathematics languages, the fourth-order derivative of the function* $f(x)$ *can be calculated using a single statement*

```
>> syms x; f=sin(x)/(x^2+4*x+3); y=diff(f,x,4)
```

and the result obtained is

$$y = \frac{\sin x}{x^2+4x+3} + 4\frac{(2x+4)\cos x}{(x^2+4x+3)^2} - 12\frac{(2x+4)^2\sin x}{(x^2+4x+3)^3} + \frac{12\sin x}{(x^2+4x+3)^2} + \frac{24\sin x}{(x^2+4x+3)^3}$$
$$+ 48\frac{(2x+4)\cos x}{(x^2+4x+3)^3} + 24\frac{(2x+4)^4\sin x}{(x^2+4x+3)^5} - 72\frac{(2x+4)^2\sin x}{(x^2+4x+3)^4} - 24\frac{(2x+4)^3\cos x}{(x^2+4x+3)^4}.$$

It is obvious that manual derivation could be a tedious and laborious work, and it could be quite complicated. Wrong results may be obtained even with a slightly careless manipulation of formulae. Therefore, even though the results can be obtained manually, after hours of hard work, the results may be suspicious and untrustworthy. If the computer mathematics languages are used, the tedious and unreliable work can be avoided. For example, by using MATLAB language, the accurate $\mathrm{d}^{100}f(x)/\mathrm{d}x^{100}$ *can be obtained within a second!*

Example 1.2 In many fields, the roots of polynomial equations are often needed. The well-known Abel–Ruffini Theorem states that there is no general solution in radicals to polynomial equations of degree five or higher. The problems can be solved numerically using the Lin–Bairstow algorithm. Solve a polynomial equation

$$s^6 + 9s^5 + \frac{135}{4}s^4 + \frac{135}{2}s^3 + \frac{1215}{16}s^2 + \frac{729}{16}s + \frac{729}{64} = 0.$$

Solution *Applying the Lin–Bairstrow method, under double-precision scheme, the roots of the equation obtained are*

$$s_{1,2} = -1.5056 \pm \mathrm{j}0.0032, \quad s_{3,4} = -1.5000 \pm \mathrm{j}0.0065, \quad s_{5,6} = -1.4944 \pm \mathrm{j}0.0032.$$

Substituting s_1 *back to the original equation, the error can be found to be* $-8.7041 \times 10^{-14} - \mathrm{j}1.8353 \times 10^{-15}$. *In fact, all the roots to the above equation are exactly* -1.5, *if the symbolic facilities of the computer mathematics languages are used. Below are the statements used in the example*

```
>> p=[1 9 135/4 135/2 1215/16 729/16 729/64]; roots(p) % numeric
   p1=poly2sym(p); solve(p1) % analytic solution of polynomial equations
```

Example 1.3 Do you remember how to find the determinant of an $n \times n$ matrix?

Solution *In linear algebra courses, the determinant of a matrix is suggested to be evaluated by* algebraic complements. *For instance, for an* $n\times n$ *matrix, its determinant can be evaluated from determinants of* n *matrices of size* $(n-1)\times(n-1)$. *Similarly, the determinant of* *each $(n-1)\times(n-1)$ matrix can be obtained from determinants of $n-1$ matrices of size $(n-2)\times(n-2)$. In other words, the determinant of an $n\times n$ matrix can eventually be obtained from the algebraic sum of many determinants of 1×1 matrices, i.e., the scalar itself. Therefore, it can be concluded that the analytical solution to the determinant of any given matrix exists.*

In fact, the above mathematical conclusion neglected the computability and feasibility issue. The computation load for such an evaluation task could be extremely tremendous, which requires $(n-1)(n+1)!+n$ operations. For instance, when $n=25$, the number of floating-point operations (flops) for the computation is 9.679×10^{27}, which amounts to 5580 years of computation on mainframe of 55 million billion (5.5×10^{16}) flops per second (the fastest mainframe in the world in 2014). Therefore, the algebraic complement method, although elegant and instructive, is not practically feasible. In real applications, the determinants of even larger sized matrices are usually needed $(n \gg 25)$, which is clearly not possible to directly apply the algebraic complement method mentioned above.

In numerical analysis courses, various algorithms have been devised. However, due to finite-precision numerical computation, these algorithms may have numerical problems when the matrix is close to being singular. For example, consider the Hilbert matrix given by

$$\boldsymbol{H}=\begin{bmatrix} 1 & 1/2 & 1/3 & \cdots & 1/n \\ 1/2 & 1/3 & 1/4 & \cdots & 1/(n+1) \\ \vdots & \vdots & \vdots & \ddots & \vdots \\ 1/n & 1/(n+1) & 1/(n+2) & \cdots & 1/(2n-1) \end{bmatrix}.$$

For $n=25$, an erroneous determinant $\det(\boldsymbol{H})=0$ could actually be obtained even if double-precision is used. On the other hand, if computer mathematics language is used, the analytical solution of the determinant of an 80×80 Hilbert matrix can be obtained within 2.5 seconds:

$$\det(\boldsymbol{H})=\frac{1}{\underbrace{99030101466993477878867678\cdots000000000000}_{3790 \text{ decimal digits, with some digits omitted}}}\approx 1.009794\times 10^{-3790}.$$

Below are the actual commands used in MATLAB for this problem

```
>> n=80; H=sym(hilb(n)); d=det(H) % build Hilbert matrix, find determinant
```

Example 1.4 Consider the well-known nonlinear *Van der Pol equation*

$$y''+\mu(y^2-1)y'+y=0,$$

and when μ is large, i.e., $\mu=1000$, how to numerically solve the equation.

Solution *The conventional numerical algorithms for solving differential equations such as the standard Runge–Kutta method may cause numerical problems. Specialized numerical algorithms for stiff ordinary differential equations (ODEs) should be used instead, rather than the standard Runge–Kutta methods in numerical analysis courses. Two lines of statements in MATLAB are adequate in solving numerically such an equation*

```
>> mu=1000; f=@(t,x)[x(2); -mu*(x(1)^2-1)*x(2)-x(1)]; % describe ODE
   [t,x]=ode15s(f,[0,3000],[-1;1]); plot(t,x) % solve with graphics
```

As another example, the first-order delay differential equation (DDE)

$$\frac{\mathrm{d}y(t)}{\mathrm{d}t}=-0.1y(t)+0.2\frac{y(t-30)}{1+y^{10}(t-30)}$$

cannot be solved using the commonly taught algorithms in numerical analysis courses. Specific MATLAB functions or block diagram modeling tool Simulink can be used instead. Details of the methods will be given later in the book.

Example 1.5 Solve the linear programming problem given below

$$\min_{\boldsymbol{x} \text{ s.t. } \begin{cases} 2x_2+x_3+4x_4+2x_5\leqslant 54 \\ 3x_1+4x_2+5x_3-x_4-x_5\leqslant 62 \\ x_1,x_2\geqslant 0,\ x_3\geqslant 3.32,\ x_4\geqslant 0.678,\ x_5\geqslant 2.57 \end{cases}} -2x_1 - x_2 - 4x_3 - 3x_4 - x_5.$$

Solution *Since the original problem is a linear constrained optimization problem, the analytical unconstrained method, i.e., setting the derivatives of the objective function with respect to each decision variable x_i to zeros, cannot be used. With linear programming tools in MATLAB, the following statements can be used*

```
>> P.f=[-2 -1 -4 -3 -1]; P.Aineq=[0 2 1 4 2; 3 4 5 -1 -1];
   P.Bineq=[54 62]; P.lb=[0;0;3.32;0.678;2.57]; P.solver='linprog';
   P.options=optimset; x=linprog(P) % solve linear programming problem
```

and the numerical solutions can be found easily as $x_1 = 19.7850$, $x_2 = 0$, $x_3 = 3.3200$, $x_4 = 11.3850$, $x_5 = 2.5700$.

Applying algorithms in numerical analysis or optimization courses, conventional constrained optimization problems can be solved. However, if other special constraints are introduced, for instance, the decision variables are constrained to be integers, the integer programming must be used. There are not so many books introducing softwares that can tackle the integer and mixed-integer programming problems. If we use MATLAB, the solutions to this example problem are easily found as $x_1=19, x_2=0, x_3=4, x_4=10, x_5=5$.

```
>> P.solver='intlinprog'; P.options=optimoptions('intlinprog');
   P.intcon=1:5; x=intlinprog(P)
```

Example 1.6 In many other courses of applied mathematics branches, such as integral transform, complex-valued functions, partial differential equations, data interpolation and fitting, probability and statistics, can you still remember how to solve the problems after the final exams?

Example 1.7 With the rapid development of modern science and technology, many new mathematics branches, such as fuzzy set, rough set, artificial neural network, evolutionary computing algorithms have emerged. It would be a hard and time consuming task to use the branches to solve particular problems, without using specific computer tools. If low-level programming is expected, the researcher needs to fully understand the technical contents of the branches, as well as how to implement the algorithms with computer languages. However, if the existing tools and frameworks are used instead, the problems can be solved in a much simpler manner.

In many subjects, such as electric circuits, electronics, motor drive, power electronics, automatic control theory, more sophisticated examples and problems are usually skipped due to the lack of introduction of high-level computer software tools. If computer mathematics languages are introduced routinely in the above courses, complicated practical problems can be solved and innovative solutions to the problems can be explored.

1.1.2 Analytical solutions versus numerical solutions

The development of modern sciences and engineering depends heavily on mathematics. However, the research interests of pure mathematicians are different from other scientists and engineers. Mathematicians are often more interested in finding the analytical or closed-form solutions to mathematical problems. They are in particular interested in proving the existence and uniqueness of the solutions, and do not usually care much about what the solutions are. Engineers and scientists are more interested in finding the exact or approximate solutions to the problems at hand and usually do not care too much about the details on how the results are obtained, as long as the results are reliable and meaningful. The most widely used approaches for finding the approximate solutions are the numerical techniques.

It is quite common to find that analytical solutions do not exist in reality in many different mathematics branches. For instance, it is well-known that the definite integral $\dfrac{2}{\sqrt{\pi}}\displaystyle\int_0^a \mathrm{e}^{-x^2}\,\mathrm{d}x$ has no analytical solution. To solve the problem, mathematicians introduce a special function $\mathrm{erf}(a)$ to denote it and do not care what in particular the numerical value is. In order to find an approximate value, scientists and engineers have to use numerical approaches.

Another example is that the irrational number π has no closed-form solution. The ancient Chinese astronomer and scientist Zu Chongzhi (429–500), also known as *Tsu Ch'ung-chih*, found that the value is between 3.1415926 and 3.1415927, in about A.D. 480. This value is accurate enough in most science and engineering practices. Even with the imprecise value 3.14 found by Archimedes (B.C. 287–B.C. 212) in about B.C. 250, the solutions to most engineering problems are often acceptable.

The above discussions hint that an approximate numerical solution is ubiquitous. In many cases, only showing existence and uniqueness of solutions is not enough. We need to compute the solution using computers.

The breadth and depth of one's mathematical knowledge might not match one's ability of getting mathematical problems solved. In today's applied science and engineering, one usually needs to get the mathematical problems at hand solved efficiently in a timely manner without complete understanding of the numerical techniques involved even in the solution process. Therefore, today, arguably, it is a trend to focus more on how to formulate the problem in a form suitable for computer solution and on the interpretation of the results generated from the computer.

Numerical techniques have already been used in many scientific and engineering areas. For instance, in mechanics, finite element methods (FEM) have been used in solving partial differential equations. In aerospace and control, numerical linear algebra and numerical solutions to ordinary differential equations have successfully been used for decades. For simulation experiments in engineering and non-engineering areas, numerical solutions to difference and differential equations are the core problems. In hi-tech developments, digital signal processing based on fast Fourier transform (FFT) has been regarded as a routine task. There is no doubt that if one masters one or more practical computation tools, significant enhancement of mathematical problem solving capability can be expected.

1.1.3 Mathematics software packages: an overview

The emerging digital computers fueled the developments of numerical as well as symbolic computation techniques. In the early stages of the development of numerical computation techniques, some well established packages, such as the eigenvalue-based package EISPACK [1,2], linear algebra package LINPACK [3] in the USA, the NAG package by the Numerical Algorithm Group in the UK, and the package in the well accepted book *Numerical Recipes* [4], appeared and were widely used with good user feedback.

The famous EISPACK and LINPACK packages are both specific packages for numerical linear algebra applications. Originally developed in the USA, EISPACK and LINPACK packages were written in Fortran. To have a flavor of how to use the packages, let us consider eigenvalues ($\boldsymbol{W}_{\rm R}$, $\boldsymbol{W}_{\rm I}$ for their real and imaginary parts) and eigenvectors $\boldsymbol{Z}$ of an $N \times N$ real matrix $\boldsymbol{A}$. As suggested by EISPACK, the standard solution method is by sequentially calling relevant subroutines provided in EISPACK as follows:

```
CALL BALANC(NM,N,A,IS1,IS2,FV1)
CALL ELMHES(NM,N,IS1,IS2,A,IV1)
CALL ELTRAN(NM,N,IS1,IS2,A,IV1,Z)
CALL HQR2(NM,N,IS1,IS2,A,WR,WI,Z,IERR)
IF (IERR.EQ.0) GOTO 99999
CALL BALBAK(NM,N,IS1,IS2,FV1,N,Z)
```

Apart from the main body of the program, the user should also write a few lines to input or initialize the matrix $\boldsymbol{A}$ to the above program and return or display the results obtained by adding some display or printing statements. Then, the whole program should be compiled and linked with the EISPACK library to generate an executable program. It can be seen that the procedure is quite complicated. Moreover, if another matrix is to be solved, the whole procedure might be repeated, which makes the solution process even more complicated.

It is good news that the mathematical software packages are continuously developing, implementing the leading-edge numerical algorithms, providing more efficient, more reliable, faster and more stable packages. For instance, in the area of numerical algebra, a new LaPACK is becoming the leading package. Unlike the original purposes of EISPACK or LINPACK, the objectives of LaPACK have been changed. LaPACK is no longer aiming at providing libraries or facilities for direct user applications. Instead, LaPACK provides support to mathematical software and languages. For example, MATLAB and a freeware Scilab have abandoned the packages of LINPACK and EISPACK, and adopted LaPACK as their low-level library support.

1.1.4 Limitations of conventional computer languages

Many people are using conventional computer languages, such as C and Fortran, in their research. Needless to say that these languages were very useful, and they were the low-level supporting languages of the computer mathematics languages such as MATLAB. However, for the modern scientific and engineering researchers, these languages are not adequate for solving their complicated computational problems. For instance, even for very experienced C programmers, they may not be able to write C code to find the indefinite integral of a given function, these involve knowledge of mathematical mechanization. Even for numerical computations, there are limitations. Here two examples are given to illustrate the problems.

Example 1.8 It is known that Fibonacci sequence can be generated with the following recursive formula, $a_1 = a_2 = 1$, and $a_k = a_{k-1} + a_{k-2}$, $k = 3, 4, \cdots$. Please compute its first 100 terms.

Solution *Data type for each variable is assigned in C programming language first. Since* *the terms in the sequence are integers, it is natural to select the data types* `int` *or* `long`. *If* `int` *is selected, the following C program can be written.*

```
main()
{  int a1, a2, a3, i;
   a1=1; a2=1; printf("%d  %d  ",a1,a2);
   for (i=3; i<=100; i++){a3=a1+a2; printf("%d  ",a3); a1=a2; a2=a3;
}}
```

It seems that the sequence can be obtained simply by using the program. But wait. Are the results obtained correct? If the program is executed, from the 24th term, the value of the sequence is negative, and from that term on, the terms are sometimes positive, sometimes negative. It is obvious that some peculiar things must have happened in the program. The problem is caused by the `int` *data type, since the range of it is* $(-32767, 32767)$. *If a term is beyond this range, wrong results are generated. Even if* `long` *data type is adopted instead, the correct answers may only last till about 10 more terms. To solve the problem of finding the first 100 terms, or even more, of Fibonacci sequence is certainly beyond the capabilities of average C users, or even experienced C programmers. Extremely slight carelessness may lead to misleading results.*

With the use of MATLAB, this kind of trivial thing should not be considered. The following code can be written directly

```
>> a=[1 1]; for i=3:100, a(i)=a(i-1)+a(i-2); end; a
```

Besides, for more accurate representation of the terms, symbolic data type can be used instead, by substituting the first statement with a=`sym([1,1])`. *In this case, the 100th term is* $a_{100} = 354224848179261915075$, *and even more, the 10000th term may be obtained, with about 32 seconds of computation, and all the 2089 decimal digits can be found, whose display may occupy more than half a page of the book.*

Example 1.9 Write a universal C program to compute the product of two matrices.

Solution *It is known from linear algebra courses that if matrix* $\boldsymbol{A}$ *is an* $n \times p$ *one, while* $\boldsymbol{B}$ *is a* $p \times m$ *one, the product of the two matrices can be obtained with*

$$c_{ij} = \sum_{k=1}^{p} a_{ik}b_{kj},\ i = 1, \cdots, n,\ j = 1, \cdots, m$$

From the formula, the kernel part of the program is the triple loop structure

```
for (i=0; i<n; i++){for (j=0; j<m; j++){
   c[i][j]=0; for (k=0; k<p; k++) c[i][j]+=a[i][k]*b[k][j]; }}
```

It seems again that the problem can be solved with these simple statements. Unfortunately there is still a serious problem in the short code, the multiplicability of the two matrices is not considered. Imprecisely speaking, when the number of columns of $\boldsymbol{A}$ *equals the number of rows of* $\boldsymbol{B}$, *the product can be found, otherwise, they are not multiplicable. To solve the problem, an extra* `if` *statement should be added*

`if` cols_of_$\boldsymbol{A}$ `!=` rows_of_$\boldsymbol{B}$, display an error message

Unfortunately by introducing such a statement, a new problem emerges. In mathematics, when $\boldsymbol{A}$ *or* $\boldsymbol{B}$ *is a scalar, the product of* $\boldsymbol{A}$ *and* $\boldsymbol{B}$ *can be found, however, this case is expelled by introducing the above* `if` *statement. To solve the problem, more* `if` *statements are expected to check the scalar cases.*

Although the above modifications are made, this program is not a universal one, since complex matrices were not considered ar all. More statements are needed to make the program universal.

It can be seen from the example that, if C or similar computer languages are used, the programmers must be very careful to consider all the possible cases. If one or more cases were not considered, wrong or misleading results may be obtained. In MATLAB, this kind of trivial thing need not to be considered at all. The command $\boldsymbol{C} = \boldsymbol{A}*\boldsymbol{B}$ *can be used directly. If the two matrices are multiplicable, the product can be obtained, otherwise, an error message will be displayed to indicate why the product cannot be found.*

Of course, in real-time control and similar areas, C or similar languages have their own advantages. Although some of the MATLAB code can be translated into C automatically, this is beyond the scope of this book.

1.2 Summary of Computer Mathematics Languages

1.2.1 A brief historic review of MATLAB

In the late 1970s, Professor Cleve Moler, the Chairman of the Department of Computer Science at the University of New Mexico found that the solutions to linear algebraic problems using the then most advanced EISPACK and LINPACK packages were too complicated. MATLAB (MATrix LABoratory) was then conceived and developed. The first release of MATLAB was freely distributed in late 1970s. Cleve Moler and Jack Little co-founded The MathWorks Inc. in 1984 to develop the MATLAB language [5]. At that time, state space-based control theory was rapidly developing, and a significant amount of numerical algebra problems needed to be solved. The appearance of MATLAB and its Control Systems Toolbox soon attracted the attention of the control community. More and more control oriented toolboxes were written by distinguished experts in different control disciplines, which added higher reputations to MATLAB. It is true that MATLAB was initiated by a numerical mathematician, but its impacts and innovations were first built by the control community. Soon it became the general purpose language of control scientists and engineers. With more and more new toolboxes in many other engineering disciplines, MATLAB is becoming the *de facto* standard language of science and engineering.

1.2.2 Three widely used computer mathematics languages

There are three leading computer mathematics languages in the world with high reputations. They are MATLAB of MathWorks Inc., Mathematica of Wolfram Research and Maple of Waterloo Maple. They each have their own distinguishing merits, for instance, MATLAB is good at numerical computation and easy in programming, while Mathematica and Maple are powerful in pure mathematics problems involving symbolics and derivations.

The numerical computation capability of MATLAB is much stronger than the other two languages. Besides, various nice toolboxes by experts can be used to tackle the problems with high efficiency. In addition, the symbolic computation engine in Maple was used to solve symbolic computation problems, and now with the MuPAD engine. Therefore, the symbolic computation capability of MATLAB is essentially as good as Mathematica and

Maple for most engineering mathematical problems. When the readers have mastered such a computer mathematics language like MATLAB, the ability of handling mathematical problems could be enhanced significantly.

1.2.3 Introduction to free scientific open-source softwares

Although many extremely powerful scientific computation facilities have been provided in the computer mathematics languages such as MATLAB, Maple and Mathematica, there are certain limitations in their applications in research and education, for example, they are expensive commercial softwares. Moreover, some of the core source codes are not accessible to the users. Therefore, the open-source softwares are welcome in scientific computation as well. Some influential softwares include:

(i) **Scilab** Scilab is developed and maintained by INRIA, France. The syntaxes are very similar to MATLAB. It is a free open-source software which concentrates in particular on control and signal processing. The Scicos in Scilab is a block diagram simulation environment similar to Simulink. The web-page of Scilab is `http://www.scilab.org/`.

(ii) **Octave** Octave was first released in 1993. It is a promising open-source software for numerical computation, initiated from numerical linear algebra. The earlier objective of the software was to provide support in education. The web-page of Octave is `http://www.gnu.org/software/octave/`.

(iii) **Others** Some other small-scale numerical matrix computation softwares such as Freemat and SpeQ are all attractive free softwares.

1.3 Outline of the Book

The book can be used as a reference text or even a textbook of a new course on scientific computation. The applications of all branches of college mathematics can be taught in such a course with broad coverage, which enables the students to view mathematics from a different angle. This will significantly increase the ability of the students for mathematical problem solutions. The book can also be used as a reference book for actual mathematical problem solutions.

1.3.1 The organization of the book

The contents of the book are summarized below:

Chapter 1, the current chapter, answers the question "Why MATLAB" in scientific computation and gives an overview of the development of MATLAB and other computer mathematics languages.

Chapter 2, "Fundamentals of MATLAB Programming," introduces briefly the programming essentials of MATLAB, including data structure, flow control structures and M-function programming. Two-dimensional, three-dimensional graphics, or even four-dimensional graphics, through volume visualization techniques, are also presented. This chapter is the basis for the materials in the book.

Chapter 3, "Calculus Problems," covers the problems in college calculus, from a different viewpoint. The subjects introduced in the chapter include limits, derivatives and integrals of univariate and multivariate functions. Series expansion problems such as Taylor series and Fourier series expansions as well as series sums and products are covered. Convergency tests of infinite series are explored. MATLAB solutions to path, line and surface integrals are illustrated. Finally, numerical differentiation and integral (or quadrature) are also introduced.

Chapter 4, "Linear Algebra Problems," studies linear algebra problems using both analytical and numerical methods. Special matrices in MATLAB are first discussed followed by basic matrix analysis, matrix transformation and matrix decomposition problems. Matrix equation solutions, including linear equations, Lyapunov equation and Riccati equations, are introduced. How to evaluate matrix functions is introduced for both the exponential function and the functions of arbitrary forms. Also, the integer power of matrices are also explored.

Chapter 5, "Integral Transforms and Complex-valued Functions," includes the solutions to Laplace transform problems and their inverse, Fourier transforms and their variations, z, Mellin and Hankel transforms. Numerical solutions of integral transform are illustrated, where numerical inverse Laplace transform and fast Fourier transform are in particular discussed. The analysis of complex-valued functions are also introduced, including poles, residues, partial fraction expansion, Laurent series, and closed-path integral problems, all with many illustrative solution examples. Difference equation solutions are explored also in the chapter.

Chapter 6, "Nonlinear Equations and Optimization Problems," explores the search methods for linear equations, nonlinear equations and nonlinear matrix equations. The unconstrained optimization, constrained optimization and mixed integer programming problems are demonstrated. Linear matrix inequality (LMIs) problems are also covered in the chapter. Multi-objective optimization problems and dynamical programming problems are also introduced. Dynamic programming problems, with particular applications in shortest path planning problems, are demonstrated.

Chapter 7, "Differential Equations Problems," mainly covers analytical as well as numerical solutions to ordinary differential equations. Different types of ordinary differential equations, including stiff equations, implicit equations, differential algebraic equations, delay differential equations and the boundary valued equations are illustrated. An introduction to partial differential equations is also given briefly through examples. In particular, block diagram-based modeling and numerical solutions of differential equations are explored with the sophisticated Simulink environment.

Chapter 8, "Data Interpolation and Functional Approximation Problems," studies the interpolation problems such as simple interpolation, cubic spline and B-spline problems. We show that numerical differentiation and integration problems can be solved with splines. Polynomial fitting, continued fraction expansion and Padé approximation as well as least squares curve fitting methods are all covered and illustrated. An introduction to signal filtering and de-noising problems is also presented briefly in this chapter.

Chapter 9, "Probability and Mathematical Statistics Problems," studies the probability distributions and pseudorandom number generators first. Statistical analysis to the measured random data is then illustrated. Hypothesis tests for a few common applications are presented, and the analysis of variance method is demonstrated briefly through examples. An introduction to principal component analysis problems is also given.

Chapter 10, "Nontraditional Solution Methods," covers a wide variety of interesting topics, such as traditional set theory, rough set theory, fuzzy set theory and fuzzy inference system, neural networks, wavelet transform, evolutionary optimization methods including genetic algorithms and particle swarm optimization methods. Most interestingly, fractional-order calculus (derivatives or integrals of non-integer order) problems are introduced with basic numerical computational techniques and examples.

1.3.2 How to learn and use MATLAB

The best way to learn a computer language like MATLAB is learning through extensive practice. MATLAB should be installed on your computer, and when invoked, the main interface is shown in Figure 1.1. The Command Window is the place where MATLAB commands are issued, and `>>` is the MATLAB prompt. For new users of MATLAB, it is advised to type **`demo`** after the MATLAB prompt, and experience the facilities provided in an easy-to-understand manner.

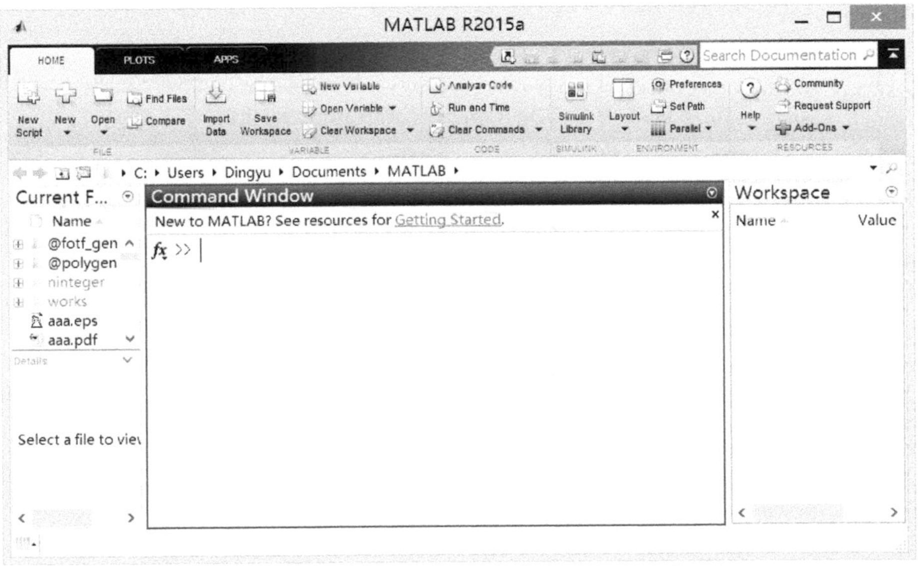

FIGURE 1.1: Main interface of MATLAB.

In this book, almost all the engineering mathematics branches are involved. On-line help facilities provided in MATLAB are useful for getting extra information of a particular function or a class of problems. The readers are recommended to get acquainted with such facilities. Help information can be obtained with the Help menu in the MATLAB command window, as shown in Figure 1.2(a). If the Using the Desktop menu item is selected, an on-line help window is opened, as shown in Figure 1.2(b).

Alternatively, the commands **`help`** or **`doc`** can be issued in the MATLAB command window, and the command **`lookfor`** can be used to search keywords.

1.3.3 The three-phase solution methodology

A *three-phase methodology* proposed by the authors is used throughout the book in presenting mathematical problem solutions [6]. The three phases are respectively "What," "How" and "Solve." In the "What" phase, the physical explanation of the mathematical problem to be solved is presented. Even though the students have not yet learned the

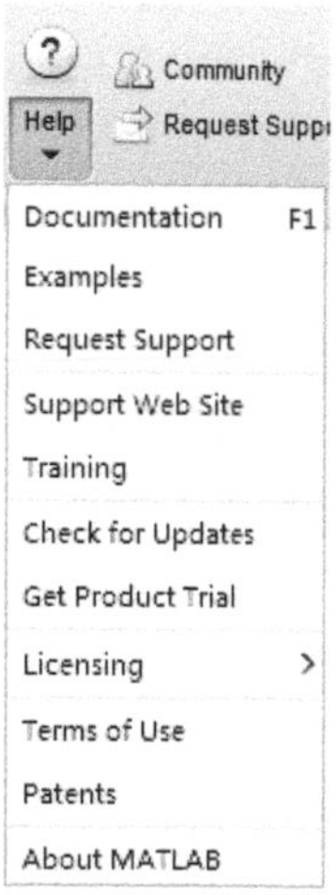

(a) the **Help** menu

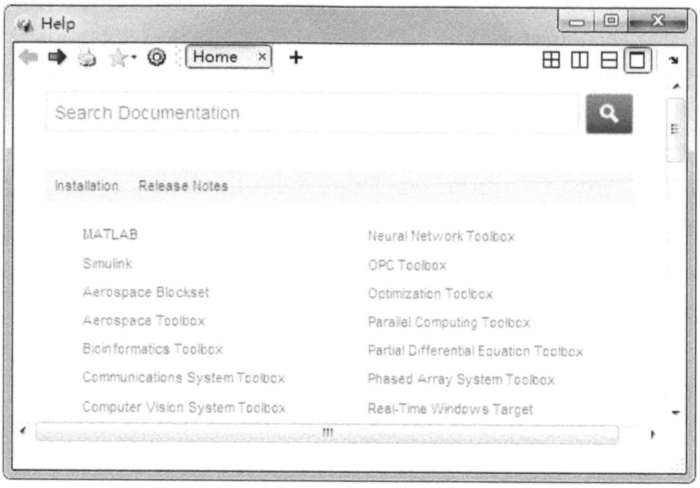

(b) on-line help window

FIGURE 1.2: On-line help information.

corresponding mathematics course, he can understand roughly what the problem is really about. In the "How" phase, the mathematical problem should be described in a manner understandable by MATLAB. In the final "Solve" phase, appropriate MATLAB functions should be called to solve the problem directly. If there is an existing MATLAB function, the syntax of the function is presented, and if there is not one, a universal function is written to solve the problem.

Example 1.10 Let us revisit the materials in Example 1.5, where the linear programming problem is involved, the philosophy used in the book is illustrated

$$\min_{\boldsymbol{x} \text{ s.t. } \begin{cases} 2x_2+x_3+4x_4+2x_5\leqslant 54 \\ 3x_1+4x_2+5x_3-x_4-x_5\leqslant 62 \\ x_1,x_2\geqslant 0,\ x_3\geqslant 3.32,\ x_4\geqslant 0.678,\ x_5\geqslant 2.57 \end{cases}} -2x_1 - x_2 - 4x_3 - 3x_4 - x_5.$$

Solution *To solve such a problem, even though the reader may have not learned optimization courses, the solution can be obtained following the three-phases.*

(i) ***"What" phase*** *In this book, we shall explain first the physical meaning of the mathematical problem. In this particular example, the mathematical formula means that under the simultaneous inequality constraints*

$$\begin{cases} 2x_2 + x_3 + 4x_4 + 2x_5 \leqslant 54 \\ 3x_1 + 4x_2 + 5x_3 - x_4 - x_5 \leqslant 62 \\ x_1, x_2 \geqslant 0,\ x_3 \geqslant 3.32,\ x_4 \geqslant 0.678,\ x_5 \geqslant 2.57, \end{cases}$$

how can we find a set of decision variables x_i, to minimize the objective function

$$f(\boldsymbol{x}) = -2x_1 - x_2 - 4x_3 - 3x_4 - x_5.$$

(ii) ***"How" phase*** *Illustrate the readers how to represent the problem in MATLAB. The code in Example 1.5 can be used to establish variable* `P`*. The mathematical problem should be expressed in a format understandable by MATLAB.*

```
>> P.f=[-2 -1 -4 -3 -1]; P.Aineq=[0 2 1 4 2; 3 4 5 -1 -1];
```

```
P.Bineq=[54 62]; P.lb=[0;0;3.32;0.678;2.57]; P.solver='linprog';
P.options=optimset; % express the linear programming problem in structure P
```

(iii) ***"Solve" phase*** *Call the* `linprog()` *function and get the result.*

```
>> x=linprog(P) % solve the problem with appropriate function linprog
```

It appears that the book is presenting in certain depth some mathematical problems. However, the ultimate objective of this book is to help the readers, after understanding roughly the mathematical background, to bypass the tedious and complex technical details of mathematics and find the reliable and accurate solutions to the interested mathematical problems with the use of MATLAB computer mathematics language. There is no doubt that the readers' capability in tackling mathematical problems can be significantly enhanced after reading this book.

Exercises

Exercise 1.1 *Install MATLAB on your machine, and issue the command* `demo`. *From the dialog boxes and menu items of the demonstration program, experience the powerful functions provided in MATLAB.*

Exercise 1.2 *In order to understand the three-phase learning stage, please revisit Example 1.5, and see which statements belong to the second and third phases. Also, please think about what should be done for the first phase.*

Exercise 1.3 *Solve the following Lyapunov equation by starting the command*

```
>> lookfor lyapunov
```

and see whether there is any function related to the keyword `lyapunov`. *If there is one, say, the* `lyap` *function is found, type* `doc lyap` *and see whether there is a way to solve this Lyapunov equation. Check the accuracy of the solution by back substitution.*

$$\begin{bmatrix} 8 & 1 & 6 \\ 3 & 5 & 7 \\ 4 & 9 & 2 \end{bmatrix} \boldsymbol{X} + \boldsymbol{X} \begin{bmatrix} 16 & 4 & 1 \\ 9 & 3 & 1 \\ 4 & 2 & 1 \end{bmatrix} = \begin{bmatrix} 1 & 2 & 3 \\ 4 & 5 & 6 \\ 7 & 8 & 0 \end{bmatrix}.$$

Bibliography

[1] Garbow B S, Boyle J M, Dongarra J J, et al. Matrix eigensystem routines — EISPACK guide extension, Lecture notes in computer sciences, volume 51. New York: Springer-Verlag, 1977

[2] Smith B T, Boyle J M, Dongarra J J, et al. Matrix eigensystem routines - EISPACK guide, Lecture notes in computer sciences, volume 6. New York: Springer-Verlag, second edition, 1976

[3] Dongarra J J, Bunsh J R, Molor C B. LINPACK user's guide. Philadelphia: Society of Industrial and Applied Mathematics, 1979

[4] Press W H, Flannery B P, Teukolsky S A, et al. Numerical recipes, the art of scientific computing. Cambridge: Cambridge University Press, 1986

[5] Moler C B. Evolution of MATLAB (Video of presentation at Tongji University, China, subtitled by Xue D Y). `http://v.youku.com/v_show/id_XNDcONTM4NzQw.html?tpa=dW5pb25faWQ9MjAwMDE0XzEwMDAwMV8wMV8wMQ`, 2012

[6] Xue D Y. Mathematics education made more practical with MATLAB. Presentation at the First MathWorks Asian Research Faculty Summit, Tokyo, Japan, November, 2014

Chapter 2

Fundamentals of MATLAB Programming and Scientific Visualization

MATLAB language is becoming a widely accepted scientific language, especially in the field of automatic control. In other engineering and non-engineering disciplines such as economics and even biology, MATLAB is also an attractive and promising computer mathematics language. In this book, we shall concentrate on the introduction to MATLAB with its applications in solving applied mathematics problems. A good working knowledge of MATLAB language will enable one not only to understand in depth the concepts and algorithms in research but also increase the ability to do creative research work and apply MATLAB to actively tackle the problems in other related research areas.

As a programming language, MATLAB has the following advantages:

(i) **Clarity and high efficiency** MATLAB language is a highly integrated language. A few MATLAB sentences may do the work of hundreds of lines of source code of other languages. Therefore, the MATLAB program is more reliable and easy to maintain.

(ii) **Scientific computation** The basic element in MATLAB is a complex matrix of double-precision. Matrix manipulations can be carried out directly. Numerical computation functions provided in MATLAB, such as the ones for solving optimization problems or other mathematical problems, can be used directly. Also, symbolic computation facilities are provided in MATLAB's Symbolic Math Toolbox to support formula derivation.

(iii) **Graphics facilities** MATLAB language can be used to visualize the experimental data in an easy manner. Implicit functions can be drawn with ease in the MATLAB environment. Moreover, the graphical user interface and object-oriented programming are also supported in MATLAB.

(iv) **Comprehensive toolboxes and blocksets** There are a huge amount of MATLAB toolboxes and Simulink blocksets contributed by experienced programmers and researchers.

(v) **Powerful system simulation facilities** The powerful block diagram-based modeling technique provided in Simulink can be used to analyze systems with almost any complexity. In particular, under Simulink, the control blocks, electronic blocks and mechanical blocks can be modeled together under the same framework, which is currently not possible in other computer mathematics languages.

In Section 2.1, the fundamental information about MATLAB programming, such as the data types, statement structures, colon expressions and sub-matrix extraction is introduced. In Section 2.2, the basic operations, including algebraic, logic and relationship operations, and simplification of symbolic formulae, and an introduction to number theory are presented. The flow control such as loop structures, conditional structures, switches and trial structures are introduced in Section 2.3. In Section 2.4, the most important programming structure, the M-function, is illustrated with useful hints on high-level programming. In Section 2.5, two-dimensional graphics facilities are presented, where two-dimensional sketching and implicit function expressions are illustrated in particular. Three-dimensional graphics are presented in Section 2.6, where mesh, surface and contour plots can be drawn. Viewpoint setting and other facilities are introduced. In Section 2.7, even more sophisticated four-dimensional volume visualization is also presented.

Only very essential introductory knowledge is presented in the chapter. More materials on MATLAB will be provided gradually through practical examples in later chapters.

2.1 Essentials in MATLAB Programming

2.1.1 Variables and constants in MATLAB

MATLAB variable names consist of a letter, followed by any number of letters, digits or underscores, with maximum length of 63. For instance, `MYvar12`, `MY_Var12` and `MyVar12_` are valid variable names, while `12MyVar` and `_MyVar12` are invalid ones, since the first character is not a letter. The variable names are case-sensitive, i.e., the variables `Abc` and `ABc` are different variables.

In MATLAB, some of the names are reserved for the constants. They can however be assigned to other values. It is suggested that these names should not be reassigned to other values if possible.

(i) `eps` — error tolerance for floating-point operation. The default value is `eps` $= 2.2204\times 10^{-16}$, and if the absolute value of a quantity is smaller than `eps`, it can be regarded as 0.

(ii) `i` and `j` — If `i` or `j` is not overwritten, they both represent $\sqrt{-1}$. If they are overwritten, they can be restored with the $i=$`sqrt(-1)` command.

(iii) `Inf` — the MATLAB representation of infinity quantity $+\infty$. It can also be written as `inf`. Similarly $-\infty$ can be written as `-Inf`. When 0 is used in the denominator, the value `Inf` can be generated, with a warning. This agrees with the IEEE standard. For mathematical computation, this definition has its advantages over C language.

(iv) `NaN` — not a number, which is often returned by the operations 0/0, `Inf/Inf` and others. Noted that `NaN` times `Inf` will return `NaN`.

(v) `pi` — double-precision representation of the circumference ratio π.

(vi) `lasterr` — returns the error message received last time. It can be a string variable, with empty string for no error message generated.

(vii) `lastwarn` — returns the last obtained warning message.

2.1.2 Data structures

I. Numeric data type

Numerical computation is the most widely used computation form in MATLAB. To ensure high-precision computations, double-precision floating-point data type is used, which is 8 bytes (64 bits). According to the IEEE standard, it is composed of 11 exponential bits, 52 fraction bits and a sign bit, representing the data range of $\pm 1.7\times 10^{308}$, with about 15~17 significant decimal digits. The MATLAB function for defining this data type is `double()`.

In other special applications, i.e., in image processing, an unsigned 8 bit integer can be used, whose function is `uint8()`, representing the value in $(0,255)$. Thus, significant memory space is saved. Also, the data types such as `int8()`, `int16()`, `int32()`, `uint16()` and `uint32()` can be used.

II. Symbolic data type

Symbolic variables are also defined in MATLAB in contrast to the conventional numerical variables. They can be used in formula derivation and analytical solutions of mathematical problems. Before finding analytical solutions, the related variables should be declared as symbolic ones, with the `syms` statement `syms var_list var_props`, where *var_list* is the list of variables to be declared, separated by spaces. If necessary, the types of the properties of the variables can be assigned by *var_props*, which can either be `real` or `positive`. For instance, if one wants to assume that a,b are symbolic variables, the statement `syms a b` can be used. Also, the statement `syms a real` can be used to indicate that a is a real variable.

Further assumptions can be made to specific symbolic variables. For instance, if symbolic variables $t \leqslant 4$ and $x \neq 2$ are to be assigned, they must first be specified as symbolic variables, then, `assume()` function can be used

```
>> syms t x; assume(t<=4); assume(x~=2); % declares t ⩽ 4 and x ≠ 2
```

Function `assumeAlso()` can also be used to pose more specifications on certain symbolic variables. For instance, if x is real and $-1 \leqslant x < 5$, the MATLAB declaration is

```
>> syms x real; assume(x>=-1); assumeAlso(x<5); % declare −1⩽x<5
```

The type of symbolic variables can be extracted with `assumptions()` function. For instance, if variable a is declared with the command `syms a real`, the function call `assumptions(a)` returns `R_`, while for x defined above, `assumptions()` function returns `[x<5,x>=1]`. Alternatively, the variable can be assigned with the command `assume(x>=-1&x<5)`.

Example 2.1 Please declare a positive symbolic integer k such that it is a multiple of 13, and it does not exceed 3000.

Solution *It can be found by simple calculation that 3000/13 is slightly larger than 230. Thus, the following MATLAB commands can be used to define such a* k

```
>> syms k1; assume(k1,'integer'); assumeAlso(k1<=230); % upper bound
   assumeAlso(k1>0); k=13*k1 % declare lower bound and multiples of 13
```

The variable precision arithmetic function `vpa()` can be used to display the symbolic variables in any precision. The syntax of the function is `vpa(A,n)` or `vpa(A)`, where A is the variable to be displayed, and n is the number of digits expected, with the default value of 32 decimal digits.

Example 2.2 Display the first 300 decimal digits of π.

Solution *The following statement can be used directly to display the exact value of* π*, with up to 300 terms*

```
>> vpa(pi,300) % display the first 300 decimal digits of π
```

and the result is shown as 3.14159265358979323846264338327950288419716939937510582097494459230781640628620899862803482534211706798214808651328230664709384460955058223172535940812848111745028410270193852110555964462294895493038196442881097566593344612847564823378678316527120190914564856692346034861045432664821339360726024914127.

One may also require a large number of digits to be displayed. Also, the result obtained with the statement `vpa(pi)` *is 3.1415926535897932384626433832795.*

III. Other data types

Apart from the commonly used numerical data types in MATLAB, the following data types are also provided such that

(i) **Strings** String variables are used to store messages. The string is quoted with single quotation marks, such as `'Hello world'`.

(ii) **Multi-dimensional arrays** Three-dimensional and even multi-dimensional arrays are the direct extension of matrices.

(iii) **Cell arrays** Cells are extension of matrices, whose elements are no longer values. The element, referred to as *cells*, of cell arrays can be of any data type. For instance, $A\{i,j\}$ can be used to represent the (i,j)th term of the cell array A.

(iv) **Structured variables** A structured variable A may have many *fields* represented as $A.b$, and field b can be of any data type.

(v) **Classes and objects** MATLAB allows the use of classes in the programming. For instance, the transfer function class in control can be used to represent a transfer function of a system in a single variable. An example of the creation and overload function programming of an object is given in Section 10.6.

2.1.3 Basic statement structures of MATLAB

Two types of MATLAB statements can be used:

(i) **Direct assignment** The basic structure of this type of statement is

`variable` = *expression*

and *expression* can be evaluated and assigned to the variable defined in the left-hand-side, and established in MATLAB workspace. If there is a semicolon used at the end of the statement, the result is not displayed. Thus, the semicolon can be used to suppress the display of intermediate results. If the left-hand-side variable is not given, the expression will be assigned to the reserved variable `ans`. Thus, the reserved variable `ans` always stores the latest statements without a left-hand-side variable.

Example 2.3 Specify the matrix $\boldsymbol{A}=\begin{bmatrix}1&2&3\\4&5&6\\7&8&0\end{bmatrix}$ into MATLAB workspace.

Solution *The matrix **A** can easily be entered into MATLAB workspace, with the following statement*

```
>> A=[1,2,3; 4 5,6; 7,8 0] % enter a matrix directly
```

where `>>` *is the MATLAB prompt, which is given automatically in MATLAB. Under the prompt, various MATLAB commands can be specified. For matrices, square brackets should be used to describe matrices, with the elements in the same row separated by commas, and the rows are separated by semicolons. The double-precision matrix variable **A** can then be established in MATLAB workspace. The matrix **A** can be displayed in MATLAB command window.*

A semicolon at the end of the statement suppresses the display of such a matrix. The size of a matrix can be expanded or reduced dynamically, with the following statements.

```
>> A=[1,2,3; 4 5,6; 7,8 0];       % assignment is made, however, no display
   A=[[A; [1 2 3]], [1;2;3;4]]; % dynamically define the size of matrix
```

Example 2.4 Enter complex matrix $\boldsymbol{B} = \begin{bmatrix} 1+\mathrm{j}9 & 2+\mathrm{j}8 & 3+\mathrm{j}7 \\ 4+\mathrm{j}6 & 5+\mathrm{j}5 & 6+\mathrm{j}4 \\ 7+\mathrm{j}3 & 8+\mathrm{j}2 & 0+\mathrm{j}1 \end{bmatrix}$ into MATLAB.

Solution *Specifying a complex matrix in MATLAB is as simple as with the case for real matrices. The notations* `i` *and* `j` *can be used to describe the imaginary unit. Thus, the following statement can be used to enter the complex matrix* $\boldsymbol{B}$

```
>> B=[1+9i,2+8i,3+7j; 4+6j 5+5i,6+4i; 7+3i,8+2j 1i] % complex matrix
```

(ii) **Function call statement** The basic function call statement is

[*returned_arguments*] = *function_name*(*input_arguments*)

where, the regulation for function names are the same as in variable names. Generally the function names are the file names in the MATLAB path. For instance, the function name `my_fun` corresponds to the file my_fun.m. Of course, some of the functions are built-in functions in MATLAB kernel, such as the `inv()` function.

More than one input argument and returned argument are allowed, in which case, commas should be used to separate the arguments. For instance, the function call [U S V] = `svd`($\boldsymbol{X}$) performs singular value decomposition to a given matrix $\boldsymbol{X}$, and the three arguments $\boldsymbol{U}, \boldsymbol{S}, \boldsymbol{V}$ will be returned.

2.1.4 Colon expressions and sub-matrices extraction

Colon expression is an effective way in defining row vectors. It is useful in generating vectors and in extracting sub-matrices. The typical form of colon expression is $\boldsymbol{v} = s_1 : s_2 : s_3$. Thus, a row vector $\boldsymbol{v}$ can be established in MATLAB workspace, with the initial value s_1, the increment s_2 and the final value s_3. If the term s_2 is omitted, a default increment of 1 is used instead. The examples given below illustrate the use of colon expressions.

Example 2.5 For different increments, establish vectors for $t \in [0, \pi]$.

Solution *One may select an increment 0.2. The following statement can be used to establish a row vector such that*

```
>> v1=0: 0.2: pi % generate a row vector
```

and the row vector is then established such that

$$\boldsymbol{v}_1 = [0, 0.2, 0.4, 0.6, 0.8, 1, 1.2, 1.4, 1.6, 1.8, 2, 2.2, 2.4, 2.6, 2.8, 3].$$

It is noted that the last term in $\boldsymbol{v}_1$ *is 3, rather than* π.

The following commands can be used to input row vectors using colon expressions

```
>> v2=0: -0.1: pi  % negative step here means no vector generated
   v3=0:pi         % with the default step-size of 1
   v4=pi:-1:0      % the new vector in the reversed order
```

thus, $\boldsymbol{v}_2$ *is a* 1×0 *(empty) matrix,* $\boldsymbol{v}_3 = [0, 1, 2, 3]$*, while* $\boldsymbol{v}_4 = [3.142, 2.142, 1.142, 0.142]$.

The sub-matrix of a given matrix $\boldsymbol{A}$ can be extracted with the MATLAB statement, and the matrix can be extracted with $\boldsymbol{B} = \boldsymbol{A}(\boldsymbol{v}_1, \boldsymbol{v}_2)$, where $\boldsymbol{v}_1$ vector contains the rows to retain, and $\boldsymbol{v}_2$ contains the numbers of columns. Therefore, the relevant columns and rows can be extracted from matrix $\boldsymbol{A}$. The sub-matrix can be returned in matrix $\boldsymbol{B}$. If $\boldsymbol{v}_1$ is assigned to :, all the elements in the $\boldsymbol{v}_2$ columns can be extracted. The keyword `end` can be used to indicate the last row or column.

Example 2.6 With the following statements, different sub-matrices can be extracted from the given matrix $\boldsymbol{A}$, such that

```
>> A=[1,2,3; 4,5,6; 7,8,0];
   B1=A(1:2:end, :)         % extract all the odd rows of matrix A
   B2=A([3,2,1],[1 1 1])  % copy the reversed first column to all columns
   B3=A(:,end:-1:1)         % flip left to right the given matrix A
   B4=A([2 2 2],:)           % copy the second row of A three times
```

and the sub-matrices extracted with the above statements are

$$\boldsymbol{B}_1=\begin{bmatrix}1&2&3\\7&8&0\end{bmatrix},\ \boldsymbol{B}_2=\begin{bmatrix}7&7&7\\4&4&4\\1&1&1\end{bmatrix},\ \boldsymbol{B}_3=\begin{bmatrix}3&2&1\\6&5&4\\0&8&7\end{bmatrix},\ \boldsymbol{B}_4=\begin{bmatrix}4&5&6\\4&5&6\\4&5&6\end{bmatrix}.$$

2.2 Fundamental Mathematical Calculations

2.2.1 Algebraic operations of matrices

Suppose matrix $\boldsymbol{A}$ has n rows and m columns, it is then referred to as an $n\times m$ *matrix*. If $n=m$, then, matrix $\boldsymbol{A}$ is also referred to as a *square matrix*. The following algebraic operations can be defined:

(i) **Matrix transpose** In mathematics textbooks, the transpose of matrices is often denoted as $\boldsymbol{A}^{\mathrm{T}}$. For an $n\times m$ matrix $\boldsymbol{A}$, the transpose matrix $\boldsymbol{B}$ can be defined as $b_{ji}=a_{ij}$, $i=1,\cdots,n$, $j=1,\cdots,m$, therefore, $\boldsymbol{B}$ is an $m\times n$ matrix. If matrix $\boldsymbol{A}$ contains complex elements, a special transpose can also be defined as $b_{ji}=a_{ij}^*$, $i=1,\cdots,n$, $j=1,\cdots,m$, i.e., the complex conjugate transpose matrix $\boldsymbol{B}$ is defined. This kind of transpose is referred to as the *Hermitian transpose*, denoted as $\boldsymbol{B}=\boldsymbol{A}^{\mathrm{H}}$. In MATLAB, `A'` can be used to evaluate the Hermitian matrix of $\boldsymbol{A}$. The simple transpose can be obtained with `A.'`. For a real matrix $\boldsymbol{A}$, `A'` is the same as `A.'`.

(ii) **Addition and subtraction** Assume that there are two matrices $\boldsymbol{A}$ and $\boldsymbol{B}$ in MATLAB workspace, the statements $\boldsymbol{C}=\boldsymbol{A}+\boldsymbol{B}$ and $\boldsymbol{C}=\boldsymbol{A}-\boldsymbol{B}$ can be used respectively to evaluate the addition and subtraction of these two matrices. If the matrices $\boldsymbol{A}$ and $\boldsymbol{B}$ are with the same size, the relevant results can be obtained. If one of the matrices is a scalar, it can be added to or subtracted from the other matrix. If the sizes of the two matrices are different, error messages can be displayed.

(iii) **Matrix multiplication** Assume that matrix $\boldsymbol{A}$ of size $n\times m$ and matrix $\boldsymbol{B}$ of size $m\times r$ are two variables in MATLAB workspace, and the columns of $\boldsymbol{A}$ equal the rows of $\boldsymbol{B}$, the two matrices are referred to as *compatible*. The product can be obtained from

$$c_{ij}=\sum_{k=1}^{m}a_{ik}b_{kj},\ \text{where } i=1,2,\cdots,n,\ j=1,2,\cdots,r. \tag{2-2-1}$$

If one of the matrices is a scalar, the product can also be obtained. In MATLAB, the multiplication of the two matrices can be obtained with $\boldsymbol{C}=\boldsymbol{A}*\boldsymbol{B}$. If the two matrices are not compatible, an error message will be given.

(iv) **Matrix left division** The left division of the matrices $\boldsymbol{A}\backslash\boldsymbol{B}$ can be used to solve the linear equations $\boldsymbol{AX}=\boldsymbol{B}$. If matrix $\boldsymbol{A}$ is nonsingular, then, $\boldsymbol{X}=\boldsymbol{A}^{-1}\boldsymbol{B}$. If $\boldsymbol{A}$ is not a square matrix, $\boldsymbol{A}\backslash\boldsymbol{B}$ can also be used to find the least squares solution to the equations $\boldsymbol{AX}=\boldsymbol{B}$.

(v) **Matrix right division** The statement $\boldsymbol{B}/\boldsymbol{A}$ can be used to solve the linear equations $\boldsymbol{XA}=\boldsymbol{B}$. More precisely, `B/A=(A'\B')'`.

(vi) **Matrix flip and rotation** The left-right flip and up-down flip of a given matrix $\boldsymbol{A}$ can be obtained with `B=fliplr(A)` and `C=flipud(A)` respectively, such that $b_{ij}=a_{i,n+1-j}$ and $c_{ij}=a_{m+1-i,j}$. The command `D=rot90(A)` rotates matrix $\boldsymbol{A}$ counterclockwise by 90°, such that $d_{ij}=a_{j,n+1-i}$. Command `D=rot90(A,k)` is also allowed to rotate matrix $\boldsymbol{A}$ by $90k°$ in a counterclockwise direction.

(vii) **Matrix power** $\boldsymbol{A}^x$ computes the matrix $\boldsymbol{A}$ to the power x when matrix $\boldsymbol{A}$ is square. In MATLAB, the power can be evaluated with `F=A^x`.

(viii) **Dot operation** A class of special operation is defined in MATLAB. The statement `C=A.*B` can be used to obtain element-by-element product of matrices $\boldsymbol{A}$ and $\boldsymbol{B}$, such that $c_{ij}=a_{ij}b_{ij}$. The dot product is also referred to as the *Hadamard product.*

Dot operation plays an important role in scientific computation. For instance, if a vector $\boldsymbol{x}$ is given, then, the vector $[x_i^5]$ cannot be obtained with `x^5`. Instead, the command `x.^5` should be used. In fact, some of the functions such as `sin()` can also be used in element-by-element operation.

Dot operation can be used to deal with other problems, for instance, the statement `A.^A` can be used, with the (i,j)th element then defined as $a_{ij}^{a_{ij}}$. Therefore, the matrix can be obtained

$$\begin{bmatrix}1^1 & 2^2 & 3^3\\ 4^4 & 5^5 & 6^6\\ 7^7 & 8^8 & 0^0\end{bmatrix}=\begin{bmatrix}1 & 4 & 27\\ 256 & 3125 & 46656\\ 823543 & 16777216 & 1\end{bmatrix}.$$

Example 2.7 Consider again the matrix $\boldsymbol{A}$ in Example 2.3. Find all the cubic roots of such a matrix and verify the results.

Solution *The cubic root of the matrix* $\boldsymbol{A}$ *can easily be found and validated with*

```
>> A=[1,2,3; 4,5,6; 7,8,0]; C=A^(1/3), e=norm(A-C^3) % cubic root & verify
```

and it can be found, with an error of $e=8.2375\times10^{-15}$, *that*

$$\boldsymbol{C}=\begin{bmatrix}0.7718+\mathrm{j}0.6538 & 0.4869-\mathrm{j}0.0159 & 0.1764-\mathrm{j}0.2887\\ 0.8885-\mathrm{j}0.0726 & 1.4473+\mathrm{j}0.4794 & 0.5233-\mathrm{j}0.4959\\ 0.4685-\mathrm{j}0.6465 & 0.66929-\mathrm{j}0.6748 & 1.3379+\mathrm{j}1.0488\end{bmatrix}.$$

In fact, the cubic root of matrix $\boldsymbol{A}$ *should have three solutions. The other two roots can be rotated as* $\boldsymbol{C}\mathrm{e}^{\mathrm{j}2\pi/3}$ *and* $\boldsymbol{C}\mathrm{e}^{\mathrm{j}4\pi/3}$, *with the following statements*

```
>> j1=exp(sqrt(-1)*2*pi/3); A1=C*j1, A2=C*j1^2 % rotate to find more roots
   norm(A-A1^3), norm(A-A1^3)                  % validate the results
```

and the other two roots, through verification, are

$$\boldsymbol{A}_1=\begin{bmatrix}-0.9521+\mathrm{j}0.3415 & -0.2297+\mathrm{j}0.4296 & 0.1618+\mathrm{j}0.2971\\ -0.3814+\mathrm{j}0.8058 & -1.1388+\mathrm{j}1.0137 & 0.1678+\mathrm{j}0.7011\\ 0.3256+\mathrm{j}0.7289 & 0.2497+\mathrm{j}0.9170 & -1.5772+\mathrm{j}0.6343\end{bmatrix},$$

and

$$\boldsymbol{A}_2=\begin{bmatrix}0.1803-\mathrm{j}0.9953 & -0.2572-\mathrm{j}0.4137 & -0.3382-\mathrm{j}0.0084\\ -0.5071-\mathrm{j}0.7332 & -0.3085-\mathrm{j}1.4931 & -0.6911-\mathrm{j}0.2052\\ -0.7941-\mathrm{j}0.0825 & -0.9190-\mathrm{j}0.2422 & 0.2393-\mathrm{j}1.6831\end{bmatrix}.$$

It should be pointed out that symbolic matrices like this cannot be used to directly find its cubit root, however, `vpa()` *can be used to increase the accuracy. If the following statements are used, the norm of the error matrix is around* 10^{-39}.

```
>> A=sym([1,2,3; 4,5,6; 7,8,0]); C=A^(sym(1/3)); C=vpa(C); norm(C^3-A)
```

2.2.2 Logic operations of matrices

Logical data was not implemented in earlier versions of MATLAB. The nonzero value is regarded as logic 1, while a zero value is defined as logic 0. In new versions of MATLAB, logical variables are defined and the above rules also apply.

Assume that the matrices $\boldsymbol{A}$ and $\boldsymbol{B}$ are both $n \times m$ matrices, the following logical operations are defined:

(i) **"And" operation** In MATLAB, the operator & is used to define element-by-element "and" operation. The statement $\boldsymbol{A}\,\&\,\boldsymbol{B}$ can then be defined.

(ii) **"Or" operation** In MATLAB, the operator | is used to define element-by-element "or" operation. The statement $\boldsymbol{A}\,|\,\boldsymbol{B}$ can then be defined.

(iii) **"Not" operation** In MATLAB, the operator ~ can be used to define the "not" operation such that $\boldsymbol{B}=\sim\boldsymbol{A}$.

(iv) **Exclusive or** The exclusive or operation of two matrices $\boldsymbol{A}$ and $\boldsymbol{B}$ can be evaluated from `xor(`$\boldsymbol{A}$`,`$\boldsymbol{B}$`)`.

2.2.3 Relationship operations of matrices

Various relationship operators are provided in MATLAB. For example, the command $\boldsymbol{C}=\boldsymbol{A}>\boldsymbol{B}$ performs element-by-element comparison between matrices $\boldsymbol{A}$ and $\boldsymbol{B}$, with the element $c_{ij}=1$ for $a_{ij}>b_{ij}$, and $c_{ij}=0$ otherwise. The equality relationship can be tested with $==$ operator, while the other operators $>=$, $\sim=$ can also be used.

The special functions such as `find()` and `all()` can also be used to perform relationship operations. For instance, the index of the elements in $\boldsymbol{C}$ equal to 1 can be obtained from `find(`$\boldsymbol{C}==1$`)`. The following commands

```
>> A=[1,2,3; 4 5,6; 7,8 0]; % enter a matrix
   i=find(A>=5)' % find all the indices in A whose value is larger than 5
```

can be used, and the indices can be found as $i=3,5,6,8$. It can be seen that the function arranges first the original matrix $\boldsymbol{A}$ in a single column, on a column-wise basis. The indices can then be returned.

The functions `all()` and `any()` can also be used to check the values in the given matrices. For instance

```
>> a1=all(A>=5) % check each column whether all larger than 5
   a2=any(A>=5) % check each column whether any larger than 5
```

and it can be found that $\boldsymbol{a}_1=[0,0,0]$, $\boldsymbol{a}_2=[1,1,1]$.

2.2.4 Simplifications and presentations of analytical results

The Symbolic Math Toolbox can be used to derive mathematical formulas. The results, however, are often not presented in their simplest form. The results should then be simplified. The easiest way of simplification is by the use of `simplify()` function, where different simplification methods are tested automatically until the simplest result can be obtained, with the syntax

f_1=`simplify(`f`)` % try various simplification methods automatically

where f is the original symbolic expression, and f_1 is the simplified result.

Apart from the easy-to-use `simplify()` function, the function `collect()` can be used to collect the coefficients, and function `expand()` can be used to expand a polynomial. The functions `factor()`, `lcm()` and `gcd()` can also be used in processing polynomials. The function `numden()` can be used to extract the numerator and denominator from a given expression.

Example 2.8 If a polynomial $P(s)$ is given by $P(s) = (s+3)^2(s^2+3s+2)(s^3+12s^2+48s+64)$, process it with various functions and understand the results converted.

Solution *A symbolic variable s should be declared first, then, the full polynomial can be expressed easily and the polynomial can then be established in MATLAB workspace. With the polynomial, one can first simplify it with the* `simplify()` *function*

```
>> syms s; P=(s+3)^2*(s^2+3*s+2)*(s^3+12*s^2+48*s+64) % build expression
   P1=simplify(P), P2=expand(P)                        % try different forms
```

and one finds that $P_1 = (s+3)^2(s+4)^3(s^2+3s+2)$, $P_2 = s^7+21s^6+185s^5+883s^4+2454s^3+3944s^2+3360s+1152$.

The new `factor()` *function extracts all the factors in a vector, thus, the factorized form can be obtained as* $P_4 = (s+1)(s+2)(s+3)^2(s+4)^3$.

```
>> P3=factor(P), P4=prod(P3) % find the factorized form
```

The function `subs()` provided in the Symbolic Math Toolbox can be used to perform variable substitution, and the syntaxes are

f_1=`subs(`f,x_1,x_1^*`)` % replace one variable

f_1=`subs(`f,$\{x_1,x_2,\cdots,x_n\}$,$\{x_1^*,x_2^*,\cdots,x_n^*\}$`)` % replace several together

where f is the original expression. With the statement, the variable x_1 in the original function can be substituted with a new variable or expression x_1^*. The result is given in the variable f_1. The latter syntax can be used to substitute many variables simultaneously.

The function `latex()` can be used to convert a symbolic expression into a LaTeX-readable string, which can be embedded into a LaTeX document.

Example 2.9 For a given function $f(t) = \cos(at+b) + \sin(ct)\sin(dt)$, evaluate its Taylor expression with the function `taylor()` and convert the results in LaTeX.

Solution *A full description on Taylor series expansion will be given in Section 3.4. Here the function* `taylor()` *can be used straightforwardly to get the results. Applying the function* `latex()` *to the results, the LaTeX can be obtained.*

```
>> syms a b c d t;                   % declare symbolic variables
   f=cos(a*t+b)+sin(c*t)*sin(d*t);   % define the function f(t)
   f1=taylor(f,'Order',5);           % find first 5 terms in Taylor series
   latex(f1)                         % can be converted to a LaTeX string
```

The results can be embedded into a LaTeX document, and through compilation, the following results can be obtained

$$f(t) \approx \cos b - at\sin b + \left(-\frac{a^2\cos b}{2} + cd\right)t^2 + \frac{a^3\sin b}{6}t^3 + \left(\frac{a^4\cos b}{24} - \frac{cd^3}{6} - \frac{c^3d}{6}\right)t^4.$$

Unfortunately, there are no directly usable converters to other word processing programs such as Microsoft Word.

2.2.5 Basic number theory computations

Basic data transformation and number theory functions are provided in MATLAB, as shown in Table 2.1. The following examples are used to illustrate the functions. Through the example, the readers can observe the results.

TABLE 2.1: Functions for data transformations.

function	syntax	function description
`floor()`	$\boldsymbol{n}$ = `floor(`$\boldsymbol{x}$`)`	round towards $-\infty$ for each value in variable $\boldsymbol{x}$, mathematically denoted as $\boldsymbol{n}=\lfloor\boldsymbol{x}\rfloor$
`ceil()`	$\boldsymbol{n}$ = `ceil(`$\boldsymbol{x}$`)`	round towards $+\infty$ for $\boldsymbol{x}$, mathematically denoted as $\boldsymbol{n}=\lceil\boldsymbol{x}\rceil$
`round()`	$\boldsymbol{n}$ = `round(`$\boldsymbol{x}$`)`	round to nearest integer for $\boldsymbol{x}$
`fix()`	$\boldsymbol{n}$ = `fix(`$\boldsymbol{x}$`)`	round towards zero for variable $\boldsymbol{x}$
`rat()`	$[\boldsymbol{n},\boldsymbol{d}]$ = `rat(`$\boldsymbol{x}$`)`	find rational approximation for variable $\boldsymbol{x}$, and the numerator and denominator are returned respectively in $\boldsymbol{n}$ and $\boldsymbol{d}$
`rem()`	$\boldsymbol{B}$ = `rem(`$\boldsymbol{A}$,$\boldsymbol{C}$`)`	find the remainder after division to variable $\boldsymbol{A}$
`gcd()`	k = `gcd(`n,m`)`	compute the greatest common divisor for n and m
`lcm()`	k = `lcm(`n,m`)`	compute the least common multiple for n and m
`factor()`	$\boldsymbol{f}$ = `factor(`n`)`	prime factorization
`perms()`	$\boldsymbol{V}$ = `perms(`$\boldsymbol{v}$`)`	for vector $\boldsymbol{v}$, list all the permutations of its elements to construct matrix $\boldsymbol{V}$. Limitations: maximum length of $\boldsymbol{v}$ is 10.
`isprime()`	$\boldsymbol{v}_1$ = `isprime(`$\boldsymbol{v}$`)`	check whether each component in $\boldsymbol{v}$ is prime or not. Set those values in $\boldsymbol{v}_1$ to 1 for prime numbers, otherwise, set to 0
`primes()`	$\boldsymbol{v}$ = `primes(`n`)`	find all the prime numbers less than n and return in vector $\boldsymbol{v}$

Example 2.10 For a given data set $-0.2765, 0.5772, 1.4597, 2.1091, 1.191, -1.6187$, observe the integers obtained using different rounding functions.

Solution *The following statements can be used to round the original vector*

```
>> A=[-0.2765,0.5772,1.4597,2.1091,1.191,-1.6187];
   v1=floor(A), v2=ceil(A) % round towards −∞ and +∞ respectively
   v3=round(A), v4=fix(A)  % round towards 0 and nearest integers
```

and the integer vectors obtained are $\boldsymbol{v}_1 = [-1, 0, 1, 2, 1, -2]$, $\boldsymbol{v}_2 = [0, 1, 2, 3, 2, -1]$, $\boldsymbol{v}_3 = [0, 1, 1, 2, 1, -2]$, $\boldsymbol{v}_4 = [0, 0, 1, 2, 1, -1]$.

Example 2.11 Assume that a 3×3 Hilbert matrix can be specified with the statement $\boldsymbol{A}$ = `hilb(3)`, perform the rational transformation.

Solution *The following statements can be used to find the rational approximation*

```
>> A=hilb(3); [n,d]=rat(A) % extract the numerator and denominator matrices
```

and the integer matrices obtained are $\boldsymbol{n}=\begin{bmatrix}1&1&1\\1&1&1\\1&1&1\end{bmatrix}$, $\boldsymbol{d}=\begin{bmatrix}1&2&3\\2&3&4\\3&4&5\end{bmatrix}$.

Example 2.12 Find the greatest common divisor and least common multiple to the numbers 1856120 and 1483720, and find the prime factorization to the least common multiplier obtained.

Solution *Since the values are very large, one should not use the double-precision representations. The symbolic representations must be used instead. The following statements can be used*

```
>> m=sym(1856120); n=sym(1483720); gcd(m,n), lcm(m,n), factor(lcm(n,m))
```

which yield the greatest common divisor of 1960 and the least common multiple of 1405082840, whose prime factorization is $(2)^3(5)(7)^2(757)(947)$, *while in new versions of MATLAB, the* `factor()` *function returns a vector* $[2,2,2,5,7,7,757,947]$.

Here the functions `gcd()` *and* `lcm()` *can only be used to deal with two variables. If more than two variables are expected, nested calls are allowed such that* `gcd(gcd(m,n),k)`.

Example 2.13 List all the prime numbers in the interval $[1,1000]$.

Solution *The prime numbers can easily be recognized by the function* `isprime(A)`. *All the prime numbers less than 1000 can be extracted and stored in vector* $\boldsymbol{b}$. *Alternatively, a simpler command* $\boldsymbol{b}$=`primes(1000)` *can be used.*

```
>> a=1:1000; b=a(isprime(a))  % extract all the prime numbers
```

Example 2.14 There are 5 persons to arrange to take group photographs. The persons are labeled 1 to 5. Please list all the possibilities of their locations.

Solution *This is a typical permutation problem, since we are not only interested in how many permutations, but also interested in what are all the possible permutations. The following commands can be issued, and all the permutations are returned in matrix* $\boldsymbol{P}$, *whose size is* 120×5, *where* $120=5!$.

```
>> P=perms(1:5), size(P)  % find the number of permutations
```

If the persons are labeled as `'a'`~`'e'`, *the command* $\boldsymbol{P}$=`perms('abcde')` *can be used instead, and again 120 permutations are represented in string matrix* $\boldsymbol{P}$.

2.3 Flow Control Structures of MATLAB Language

As a programming language, the loop structures, conditional control structures, switch structures and trial structures are provided in MATLAB. These structures are illustrated in this section.

2.3.1 Loop control structures

The loop structures can be introduced by the keywords `for` or `while`, and ended with the `end` command. The two kinds of loop structures are shown in Figures 2.1 (a) and (b), respectively.

(i) **The `for` loop structures** The syntax of the structure is

`for` $i=\boldsymbol{v}$, *loop statements body*, `end`

When using the `for` loop structure, a component in vector $\boldsymbol{v}$ is extracted and assigned to variable i each time, the *loop statements body* can be executed. Then, the control goes back to the `for` statement again to extract the next component in vector $\boldsymbol{v}$. This process goes on and on, until all the components in $\boldsymbol{v}$ are used.

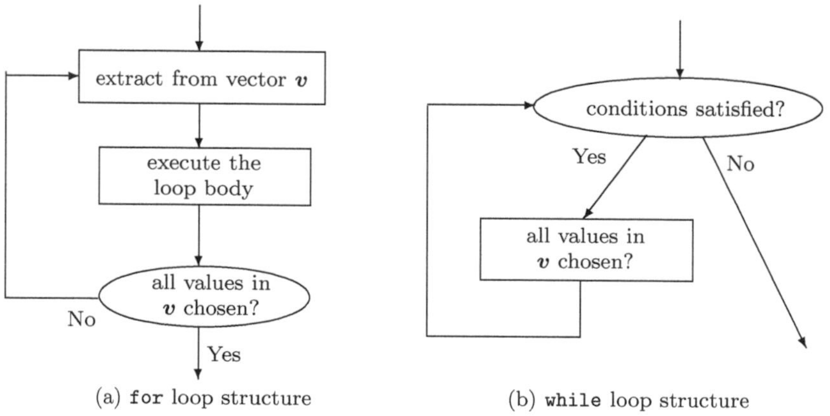

FIGURE 2.1: Illustration of loop structures.

If $\boldsymbol{v}$ is a matrix, then, each time a column of $\boldsymbol{v}$ is extracted to assign to the variable i, and the *loop statements body* is executed. This process goes on, until all the columns in matrix $\boldsymbol{v}$ are extracted.

(ii) **The `while` loop structures** The syntax of the structure is

`while` (*condition*), *loop statements body*, `end`

The *condition* expression is crucial in the `while` loop structure. If it is true, the *loop statements body* is executed, and the control returns back to the `while` command to evaluated *condition* again. The loop structure will be executed again and again, until *condition* becomes false.

There are differences between the two functions. Examples will be given below to show the advantages and disadvantages of these structures.

Example 2.15 Compute the sum of $S=\sum_{i=1}^{100} i$ using loop structures.

Solution *The `for` and `while` loop structures can be used with the following statements, and the same results can be obtained*

```
>> s1=0; for i=1:100, s1=s1+i; end
   s2=0; i=1; while (i<=100), s2=s2+i; i=i+1; end; s1, s2
```

where it can be seen that the `for` loop structure is simpler. In fact, the simplest statement *for this problem is* `sum(1:100)`. *In the function call, the built-in function* `sum()` *can be used to solve the problem.*

Example 2.16 Find the minimum value of m such that $S=\sum_{i=1}^{m} i>10000$.

Solution *It can be seen that it is not possible to solve such a problem with the* `for` *loop structure. However, the structure of* `while` *can be used easily to find* m

```
>> s=0; m=0; % set counter m and accumulate variable s to zeros
   while (s<=10000), m=m+1; s=s+m; end, [s,m] % the value of m is expected
```

with $m=141$ *and the sum is* $s=10011$.

The loop statements can be used in nested format. The statement `break` can be used to terminate the loop structure of the current level.

Example 2.17 Evaluate numerically the sum of the series $S=\sum_{i=1}^{100000}\left(\frac{1}{2^i}+\frac{1}{3^i}\right)$.

Solution *The execution time can be measured with the statements* `tic` *and* `toc`. *The time needed in vectorization is about 0.052 seconds, and the one needed in loops is 0.192 seconds. Therefore, the vectorization method is normally faster.*

```
>> tic, s=0; for i=1:100000, s=s+1/2^i+1/3^i; end; toc
   tic, i=1:100000; s=sum(1./2.^i+1./3.^i); toc % vectorized method
```

In order to demonstrate the performance of vectorization, the number of terms is exaggerated. Normally 20∼30 terms are sufficient for the exact solutions. The performance of loops was significantly speeded up in new versions of MATLAB.

The speed of loop is slightly slower, even in new versions of MATLAB, compared with other programming languages. Therefore, the loops may be avoided, and vectorized programming techniques are recommended.

2.3.2 Conditional control structures

Conditional control structures are the most widely used control structures in actual programming. The simplest `if` structure is

```
if condition, statements, end
```

In the structure, if the *condition* is satisfied, the *statements* will be executed, otherwise, the *statements* are bypassed. The control then returns to the point after `end`.

Complicated `if` structures with `else` and `elseif` branches are also supported. The structures can be shown in Figure 2.2.

```
if (condition 1) % If condition 1 is satisfied, statement group 1 is executed
     statement group 1 % other sub-level if can be nested
     statement group 1 % other sub-level if can be nested
elseif (condition 2) % Otherwise, if condition 2 is met, group 2 executed
     statement group 2
       ⋮   ⋮    % more conditional control statements
```

```
else    % if none of the above conditions are satisfied, define defaults
    statement group n + 1
end
```

In the structure, the ***condition 1*** is assessed first. If it is true, ***statement group 1*** is executed, and then, control is passed to the point after `end` statement; if ***condition 1*** is not true, the ***condition 2*** is evaluated, and if it is true, ***statement group 2*** is executed, then, control is passed to `end`; if none of the listed conditions are satisfied, ***statement group*** $n+1$ after `else` is executed.

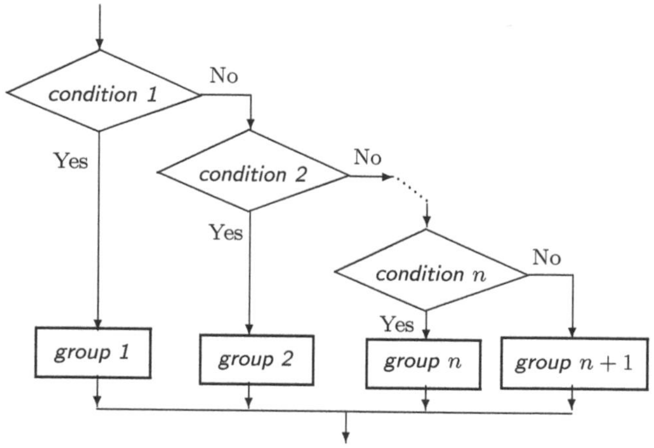

FIGURE 2.2: Illustrations of conditional control structures.

Example 2.18 Solve the problem in Example 2.16 again using the combinations of `for` and `if` structures.

Solution *It has been shown in Example 2.16 that the* `for` *loop structure is not suitable for finding the minimum* m *such that the sum is greater than 10000. The* `for` *loop can be used with* `if` *structure to solve the problem.*

```
>> s=0; for i=1:10000, s=s+i; if s>10000, break; end, end, s
```

The `break` *command is used to terminate current loop. Therefore, the structure of the program is more complicated than that of the* `while` *structure.*

2.3.3 Switch structure

The switch structure is illustrated in Figure 2.3, with the structure

```
switch switch expression
case expression 1, statements 1
case {expression 2, expression 3,···, expression m}, statements 2
    ⋮
```

```
otherwise, statements n
end
```

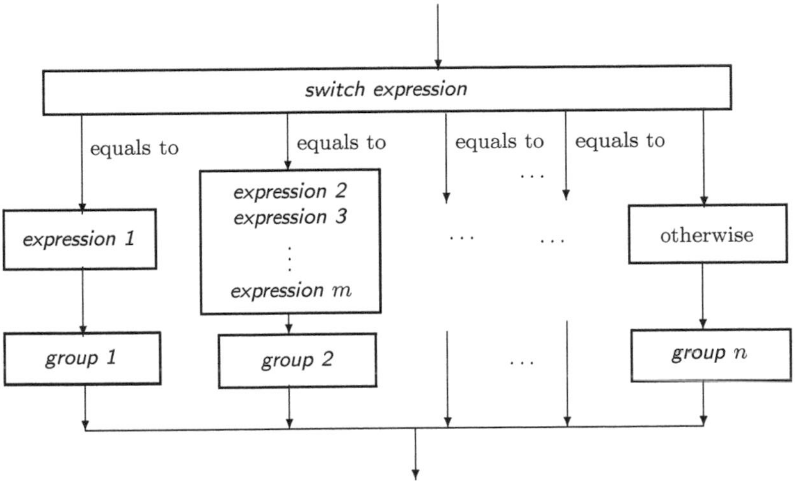

FIGURE 2.3: Illustrations of switch structures.

where the crucial part in switch structure is the evaluation of *switch expressions*. If it matches a value in a case statement, the statements after the case statement should be executed. Once completed, the switch structure is terminated.

There exist differences between the switch statements in MATLAB and in C languages. The following tips should be noted in programming with MATLAB:

(i) When the value of the switch expression equals *expression 1*, the *statement group 1* should be executed. After execution, the structure is completed. There is no need to introduce a `break` statement before the next `case`.

(ii) If one is checking whether one of several expressions is satisfied, the expressions must be given in cell format.

(iii) If none of the expressions are satisfied, the paragraph in `otherwise` should be executed. It is similar to the `default` statement in C language.

(iv) The execution results are independent of the orders of the `case` statement. When there exist two or more `case` statements having the same expressions, those listed behind may never be executed.

2.3.4 Trial structure

A trial structure is provided in MATLAB, whose syntax is

```
try,    statement group 1,  catch,    statement group 2,  end
```

Normally, only the *statement group 1* is executed. However, if an error occurs during execution of any of the statements, the error is captured into `lasterr`, and the *statement group 2* is executed. The trial structure is useful in practical programming.

2.4 Writing and Debugging MATLAB Functions

Two types of source programs are supported in MATLAB, both in ASCII format. One of the codes is the M-script program, which is a series of MATLAB statements to be evaluated in sequence, just as the batch files in DOS. The execution of this type of program is simple, one can simply key in the file name under the `>>` prompt. M-scripts process the data in MATLAB workspace, and the results are returned back to MATLAB workspace. M-scripts are suitable for dealing with small-scale computations.

Example 2.19 Consider again the problem in Example 2.16. The program can be used to find the smallest m such that the summation is greater than 10000. If one wants to find such m's for the summation greater than 20000 or 30000, the original program should be modified. This method is quite complicated and inconvenient. If a mechanism can be established such that the user may define 20000 or 30000 externally, without modifying the original program, the mechanism is quite reasonable. This kind of mechanism is often referred to as the *function*.

M-function is the major structure in MATLAB programming. In practical programming, M-script programming is not recommended. In this section, MATLAB functions and some tricks in programming are given.

2.4.1 Basic structure of MATLAB functions

MATLAB functions are led by the statement `function`, whose fundamental structure is

```
function [return argument list] = funname(input argument list)
   comments led by % sign
   input and output variables check
   main body of the function
```

The actual numbers of input and returned arguments can be extracted respectively by `nargin` and `nargout`. In the function call, the two variables are generated automatically.

If more than one input or returned argument is needed, they should be separated with commas in the lists. The comments led by `%` will not be executed. The messages in the leading comments can be displayed by the `help` command.

From the system view points, the MATLAB functions can be regarded as a variable processing unit, which receives variables from the calling function. Once the variables are processed, the results will be returned back to the calling function. Apart from the input and returned arguments, the other variables within the function are *local variables*, which vanish after function calls. Examples will be given to demonstrate the programming techniques.

Example 2.20 Rewrite the M-script in Example 2.19 in M-function.

Solution *Consider the requests in Example 2.19. One may choose the input argument as k, and returned arguments of m and s, where s is the sum of first m terms. The function can then be written as*

```
function [m,s]=findsum(k) % function framework
s=0; m=0; while (s<=k), m=m+1; s=s+m; end
```

The previous function can be saved in a file `findsum.m`*. One can then call such a function for different values of k, without modifying the function. For instance, if the targeted sum is 145323, the following statements can be used to find the smallest value of m, which returns* $m = 539$, $s_1 = 145530$.

```
>> [m1,s1]=findsum(145323) % find the terms make the sum larger than 145323
```

It can be seen that the calling format is quite flexible, and we may find the needed results without modifying the original program. Thus, this kind of method is recommended in programming.

Example 2.21 Assume that a MATLAB function is needed in obtaining an $n \times m$ Hilbert matrix, whose (i, j)th element is $h_{i,j} = 1/(i + j - 1)$. The following additional requests are also to be implemented:

(i) If only one input argument n is given in the calling command, a square matrix should be generated, such that $m = n$.

(ii) Certain help information to this function is required.

(iii) Check the formats of input and returned arguments.

Solution *In actual programming, it is better to write adequate comments, which are beneficial to the programmer as well as to the maintainer of the program. The required MATLAB function* `myhilb()` *can be written and stored as* `myhilb.m` *in the default MATLAB search path.*

```
function A=myhilb(n,m)
%MYHILB  is used to generate a rectangular Hilbert matrix
%        A=MYHILB(N,M) generates an NxM Hilbert matrix A
%        A=MYHILB(N) generates an NxN square Hilbert matrix A
%See also: HILB % the blank line below is deliberately used

% Designed by Professor Dingyu XUE, Northeastern University, PRC
%     5 April, 1995, Last modified by DYX at 30 July, 2001
if nargout>1, error('Too many output arguments.'); end
if nargin==1, m=n; % if one input argument used, generate square matrix
elseif nargin==0 | nargin>2, % check the number of input arguments
   error('Wrong number of input arguments.'); % display error message
end
for i=1:n, for j=1:m, A(i,j)=1/(i+j-1); end, end % Hilbert matrix
```

In the program, the comments are led by the % sign. To implement the requirement in item (i), one should check whether the number of input argument is 1, i.e., whether `nargin` *is 1. If so, the column number m is set to n, the row number, thus, a square matrix can be generated. If the numbers of input or returned arguments are not correct, the error messages can be given. The double* `for` *loops will generate the required Hilbert matrix.*

The on-line help command `help myhilb` *will display the following information*

```
MYHILB  The function is used to illustrate MATLAB functions.
   A=MYHILB(N, M) generates an NxM Hilbert matrix A;
   A=MYHILB(N) generates an NxN square Hilbert matrix A;
See also: HILB.
```

It should be noted that only the first few lines of information are displayed, while the author information is not displayed. This is because there is a blank line before the author information.

The following commands can be used to generate Hilbert matrices

```
>> A1=myhilb(4,3)  % two input arguments yield a rectangular matrix
   A2=myhilb(4)    % while one input argument yields a square matrix
```

and the two matrices can then be established as

$$A_1=\begin{bmatrix}1 & 0.5 & 0.33333\\ 0.5 & 0.33333 & 0.25\\ 0.33333 & 0.25 & 0.2\\ 0.25 & 0.2 & 0.16667\end{bmatrix},\ A_2=\begin{bmatrix}1 & 0.5 & 0.33333 & 0.25\\ 0.5 & 0.33333 & 0.25 & 0.2\\ 0.33333 & 0.25 & 0.2 & 0.16667\\ 0.25 & 0.2 & 0.16667 & 0.14286\end{bmatrix}.$$

There is in fact a bug in the function, since symbolic Hilbert matrix is not supported due to the use of the `for` *loop. The command such as* $\boldsymbol{H}$ `=myhilb(sym(3),7)` *cannot be used. To solve the problem, the last code line should be replaced by*

```
[i,j]=meshgrid(1:m,1:n); A=1./(j+i-1); % supporting symbolic arguments
```

Example 2.22 MATLAB functions can be called recursively, i.e., a function may call itself. Please write a recursive function to evaluate the factorial $n!$.

Solution *Consider the factorial* $n!$. *From the definition* $n! = n(n-1)!$, *it can be seen that the factorial of* n *can be evaluated from the factorial of* $n-1$, *while* $n-1$ *can be evaluated from* $n-2$, *and so on. The exits of the function call should be* $1! = 0! = 1$. *Therefore, the recursive function can be written as follows, with the comments omitted.*

```
function k=my_fact(n) % recursive function to compute factorial
if nargin~=1, error('Error: Only one input variable accepted'); end
if fix(n)~=n | n<0 % check whether n is a non-negative integer
   error('n should be a non-negative integer'); % if not, error message
end
if n>1, k=n*my_fact(n-1);           % if n > 1, evaluate recursively
elseif any([0 1]==n), k=1; end % if 0! = 1! = 1, assign the exits
```

In the function, `isinteger(`n`)` *cannot be used to check whether* n *is an integer or not, it is only applicable to check whether* n *is an integer data type or not. Some other functions started with* `is` *are,* `ischar`, `isreal`, `isfinite()`, *they are useful in actual programming activities.*

It can be seen that, in the function, the judgement whether n *is a non-negative integer is made. If not, an error message will be declared. If it is, the recursive function calls will be used such that when* $n = 1$ *or 0, the result is 1, which can be used as an exit to the function. For instance,* 11! *can be evaluated with* `my_fact(11)`, *and the result obtained is* 39916800.

In fact, the factorial for any non-negative integer can be evaluated directly with function `factorial(`n`)`, *and the kernel of such a function is* `prod(1:`n`)`. *Also,* `gamma(`$n+1$`)` *can be used. If* n *is substituted by* `sym(`n`)`, *an analytical solution can be found. For instance,* `factorial(sym(500))` *returns the exact value of 500!, with all the 1134 decimal digits.*

Example 2.23 Compare the advantages and disadvantages of a recursive algorithm with loop structure in constructing the Fibonacci arrays.

Solution *It is for sure that the recursive algorithm is an effective method for a class* *of problems. However, this method should not be misused. A counter-example is shown in this example. Consider the Fibonacci array, where* $a_1 = a_2 = 1$, *and the* k*th term can be evaluated from* $a_k = a_{k-1} + a_{k-2}$ *for* $k = 3, 4, \cdots$. *A MATLAB function can be written for the problem*

```
function a=my_fibo(k) % recursive function to generate Fibonacci sequence
if k==1 | k==2, a=1; else, a=my_fibo(k-1)+my_fibo(k-2); end
```

and for $k = 1, 2$, *the exit can be made such that it returns 1. If the 25th term is expected, the following statements can be used, and the time required is 7.6 seconds.*

```
>> tic, my_fibo(25), toc % never try more terms with recursive form
```

If one is expecting the term $k = 30$, *several hours of time might be required. If the loop structure is used, within 0.02 second, the whole array can be obtained for* $k = 100$.

```
>> tic, a=[1,1]; for k=3:100, a(k)=a(k-1)+a(k-2); end, toc
```

It can be seen that the ordinary loop structure only requires a very short execution time. Thus, the recursive function call should not be misused. Further observations show that since the values of the terms are too large, the above statements under double-precision scheme may not yield accurate results, symbolic data types should be used instead. For instance, if the first statement a = `[1 1]` *is replaced by* a = `sym([1 1])`, *an accurate sequence can be constructed, and* $a_{100} = 354224848179261915075$.

2.4.2 Programming of functions with variable numbers of arguments in inputs and outputs

In the following presentation, the variable number of input and returned arguments is introduced based on the cell data type. It should be mentioned that most of the MATLAB functions are implemented in this format.

Example 2.24 The product of two polynomials can be evaluated from the `conv()` function, based on the algorithm of finding the convolution of two arrays. Write a function to evaluate directly the multiplications of an arbitrary number of polynomials.

Solution *Cell data type can be used to write the function* `convs()`, *which can be used to evaluate the multiplication of an arbitrary number of polynomials.*

```
function a=convs(varargin) % with arbitrary number of input arguments
a=1; for i=1:nargin, a=conv(a,varargin{i}); end % use loop structure
```

The input argument list is passed to the function through the cell variable `varargin`. *Consequently, the returned arguments can be specified in* `varargout`, *if necessary. Under such a function, the multiplication of an arbitrary number of polynomials can be obtained. The following statements can be used to call the function*

```
>> P=[1 2 4 0 5]; Q=[1 2]; F=[1 2 3]; D=convs(P,Q,F) % multiply polynomials
   E=conv(conv(P,Q),F)  % nested calls are to be used with conv() function
   G=convs(P,Q,F,[1,1],[1,3],[1,1])   % direct use of convs()
```

where the obtained vectors are respectively

$$\boldsymbol{E} = [1, 6, 19, 36, 45, 44, 35, 30]^{\mathrm{T}},\ \boldsymbol{G} = [1, 11, 56, 176, 376, 578, 678, 648, 527, 315, 90]^{\mathrm{T}}.$$

2.4.3 Inline functions and anonymous functions

In order to describe simply the mathematics functions, *inline functions* can be used. The functions are equivalent to the M-functions. However, with inline function, it may no longer be necessary to save files. The format of inline function is *fun* = `inline(`*function expression*`,`*list of variables*`)`, where the *function expression* is the actual contents of the function to be expressed, and the *list of variables* contains all the input variables, with each variable given as a string. The inline function is useful in the descriptions in differential equations and objective functions given later. The function type accepts only one returned variable. The mathematical function $f(x, y) = \sin(x^2 + y^2)$ can be expressed as `f=inline('sin(x.^2+y.^2)','x','y')`.

Anonymous function is a better type of function definition, the structure of the function is similar to the inline function, but it is more concise and easy to use. The syntax of the function is

`f=@(`*list of variables*`)`*function_contents*, e.g., `f=@(x,y)sin(x.^2+y.^2)`

Note that the variable currently existing in MATLAB workspace can be used directly in the function. For instance, the variables a and b in MATLAB workspace can be used in the anonymous function `f=@(x,y)a*x.^2+b*y.^2` to describe the mathematical function $f(x, y) = ax^2 + by^2$. If such a function has been defined, while the variables a, b change after that, the values of those in the anonymous function will not change, unless it is defined again.

2.4.4 Pseudo code and source code protection

The aim of introducing pseudo code in MATLAB is two-fold. One of the aims is to increase the speed of code, since with pseudo code technique, MATLAB code can be converted to some extent executable code, and the conversion is no longer needed, such that the total execution time is reduced. The other aim is to convert ASCII MATLAB source code into binary code, so that other users can execute normally the code, but are unable to read the source code, so as to protect the source code.

The `pcode` command provided in MATLAB can be used to do the conversion, and the suffix of the converted files is .p. If one wants to convert a MATLAB function mytest.m to P code, the command to use is `pcode mytest`; if one wants the generated P code located in the same folder of the M code, the command `pcode mytest -inplace` can be used; if one wants to convert the .m files in a whole folder into P code, he can enter the folder with `cd` command, then, type the command `pcode *.m`. If there is no error in the source code, all the files can be converted, otherwise, the conversion will be aborted, and error messages will be displayed. The programmer can also use this method to check whether his source code has syntax errors or not. If both .m and .p exist in the same folder, the .p file has more advantage in execution.

Please note that, the user must save his original .m files in safe places. Make sure that the source files are not deleted, otherwise, they cannot be recovered from the .p code.

2.5 Two-dimensional Graphics

Graphics and visualization are the most significant advantages of MATLAB. A series of straightforward and simple functions are provided in MATLAB for two-dimensional and three-dimensional graphics. Experimental and simulation results can be easily interpreted in graphical form. In this section, the two-dimensional graphics functions will be illustrated.

2.5.1 Basic statements of two-dimensional plotting

Assume that a sequence of experimental data is acquired. For instance, at time instances $t = t_1, t_2, \cdots, t_n$, the function values are $y = y_1, y_2, \cdots, y_n$. The data can be entered to MATLAB workspace such that $\boldsymbol{t} = [t_1, t_2, \cdots, t_n]$ and $\boldsymbol{y} = [y_1, y_2, \cdots, y_n]$. The command `plot(t,y)` can be used to draw the curve for the data. The "curve" is in fact represented by poly-lines, joining the sample points.

It can be seen that the syntax of the function is quite straightforward. In actual applications, the `plot()` function can alternatively be called in other extended ways.

(i) $\boldsymbol{t}$ is still a vector, and $\boldsymbol{y}$ can be expressed by a matrix such that

$$\boldsymbol{y} = \begin{bmatrix} y_{11} & y_{12} & \cdots & y_{1n} \\ y_{21} & y_{22} & \cdots & y_{2n} \\ \vdots & \vdots & \ddots & \vdots \\ y_{m1} & y_{m2} & \cdots & y_{mn} \end{bmatrix}.$$

The same function can also be used to draw m curves, with each row of matrix $\boldsymbol{y}$ corresponding to a curve.

(ii) $\boldsymbol{t}$ and $\boldsymbol{y}$ are both matrices, and the sizes of the two matrices are the same. The plots between each row of $\boldsymbol{t}$ and $\boldsymbol{y}$ can be drawn.

(iii) Assume that there are many pairs of such vectors or matrices, $(\boldsymbol{t}_1, \boldsymbol{y}_1)$, $(\boldsymbol{t}_2, \boldsymbol{y}_2)$, $\cdots$, $(\boldsymbol{t}_m, \boldsymbol{y}_m)$, the following statement can be used directly to draw the corresponding curves.

plot($\boldsymbol{t}_1$,$\boldsymbol{y}_1$,$\boldsymbol{t}_2$,$\boldsymbol{y}_2$,$\cdots$,$\boldsymbol{t}_m$,$\boldsymbol{y}_m$)

(iv) The line types, line width and color information of the curves can separately be specified with the command

plot($\boldsymbol{t}_1$,$\boldsymbol{y}_1$,*option* 1,$\boldsymbol{t}_2$,$\boldsymbol{y}_2$,*option* 2,$\cdots$,$\boldsymbol{t}_m$,$\boldsymbol{y}_m$,*option* m)

where the available ***options*** are shown in Table 2.2. The combinations of the options are also allowed. For instance, the combination `'r-.pentagram'` indicates the red dash dot curve, with the sample points marked by pentagrams.

After the curves are drawn, the command **`grid on`** can be used to add grids to the curves, while the **`grid off`** command may remove the grids. Also, **`hold on`** command can reserve the current axis. Other `plot()` function can be used to superimpose curves on top of the existing ones. **`hold off`** command may remove the holding status.

Example 2.25 Draw the curve of $y = \sin(\tan x) - \tan(\sin x)$, $x \in [-\pi, \pi]$.

Solution *The curve of $f(x)$ can be drawn easily with the following statements*

```
>> x=[-pi : 0.05: pi];               % specify the vector with a step-size of 0.05
   y=sin(tan(x))-tan(sin(x));  % evaluate the function values
   plot(x,y)                         % draw the curve
```

TABLE 2.2: Options in MATLAB plotting commands.

line type		line color				markers			
opts	meaning	opts	meaning	opts	meaning	opts	meaning	opts	meaning
`'-'`	solid	`'b'`	blue	`'c'`	cyan	`'*'`	$*$	`'pentagram'`	☆
`'--'`	dash	`'g'`	green	`'k'`	black	`'.'`	dotted	`'o'`	○
`':'`	dotted	`'m'`	magenta	`'r'`	red	`'x'`	$\times$	`'square'`	□
`'-.'`	dash-dot	`'w'`	white	`'y'`	yellow	`'v'`	$\bigtriangledown$	`'diamond'`	◇
`'none'`	none					`'^'`	$\bigtriangleup$	`'hexagram'`	✡
						`'>'`	$\triangleright$	`'<'`	$\triangleleft$

and the curve in Figure 2.4 (a) can be obtained.

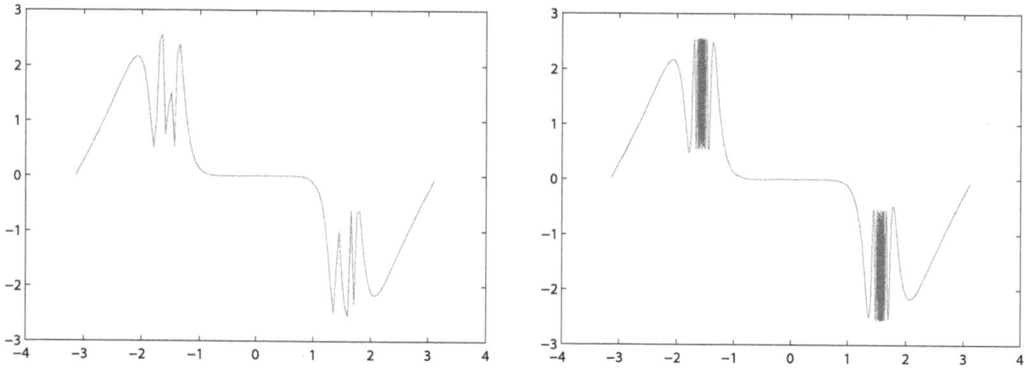

(a) with the default step-size of 0.05

(b) improved curve with variable-step-sizes

FIGURE 2.4: Two dimension curve for the given function.

It can be seen from the curve that it is rather sluggish over the intervals $x \in (-1.8, -1.2)$ *and* $x \in (1.2, 1.8)$*, since the step-size 0.05 is too large for these intervals. The step-size for these intervals should be selected smaller such that*

```
>> x=[-pi:0.05:-1.8,-1.801:.001:-1.2, -1.2:0.05:1.2,...
      1.201:0.001:1.8, 1.81:0.05:pi]; % with variable-step-size
   y=sin(tan(x))-tan(sin(x));          % evaluate the function
   plot(x,y)                           % draw the curve
```

The modified curve of the function is given in Figure 2.4 (b). It can be seen that the curve is significantly improved in the new plot. Alternatively, for the whole interval, a fixed step-size of 0.001 can be selected.

Example 2.26 Please draw the saturation function $y = \begin{cases} 1.1\mathrm{sign}(x), & |x| > 1.1 \\ x, & |x| \leqslant 1.1. \end{cases}$

Solution *It is obvious that one can create a vector of* $\boldsymbol{x}$*, then, for each point, construct an* `if` *clause to calculate the value of* $\boldsymbol{y}$*. An alternative way is to use vectorized format to evaluate the function values. With the following statements, the segmented function can be drawn as shown in Figure 2.5.*

```
>> x=[-2:0.02:2]; % generate an x vector with increment of 0.02
   y=1.1*sign(x).*(abs(x)>1.1)+x.*(abs(x)<=1.1); plot(x,y)
```

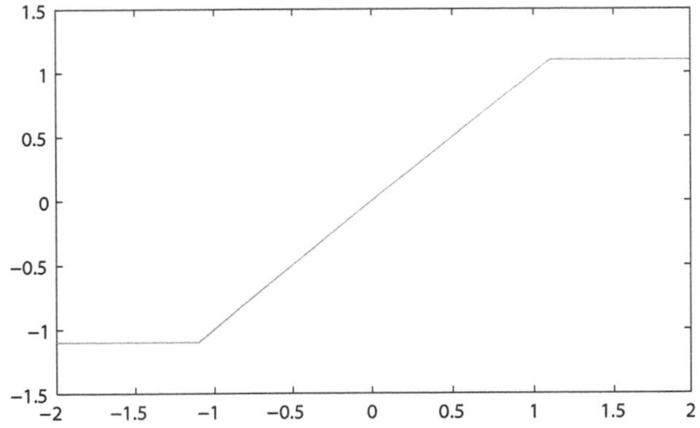

FIGURE 2.5: Segmented saturation function.

Even more simply, the command `plot([−3,−1.1,1.1,3],[−1.1,−1.1,1.1,1.1])` *can be used to draw the saturation poly-lines.*

In MATLAB graphics, each curve or the axis is an object, and the window is another object. The properties of the objects can be assigned by `set()` function, or extracted by `get()` function. The syntaxes of the functions are

```
set(handle,'prop_name 1',prop_value 1,'prop_name 2',prop_value 2,···)
v = get(object,prop_name')
```

where *prop_name* and *prop_value* are respectively the names and values of the corresponding properties. These two functions are very useful in graphical user interface programming.

2.5.2 Plotting with multiple horizontal or vertical axes

Suppose that we have two set of data, and the differences in the magnitude of them are huge, the readability of the one with smaller magnitude will be very poor, if they are plotted together with the same `plot()` function. In this case, the `plotyy()` function should be used, to have different y scales in the same plot, as will be demonstrated through the following example.

Example 2.27 Please draw together the two functions $y_1 = \sin x$, $y_2 = 0.01 \cos x$.

Solution *The two curves can be drawn together with the following statements, as shown in Figure 2.6(a). Since there are huge differences in magnitude, the curve of* y_2 *can hardly be read. Thus, the direct plotting is not suitable for this example.*

```
>> x=0:0.01:2*pi; y1=sin(x); y2=0.01*cos(x); plot(x,y1,x,y2,'--')
```

The function `plotyy()` *can be used instead to draw the plots with two* y *axes, as shown in Figure 2.6(b).*

```
>> plotyy(x,y1,x,y2) % draw curves in two y axes
```

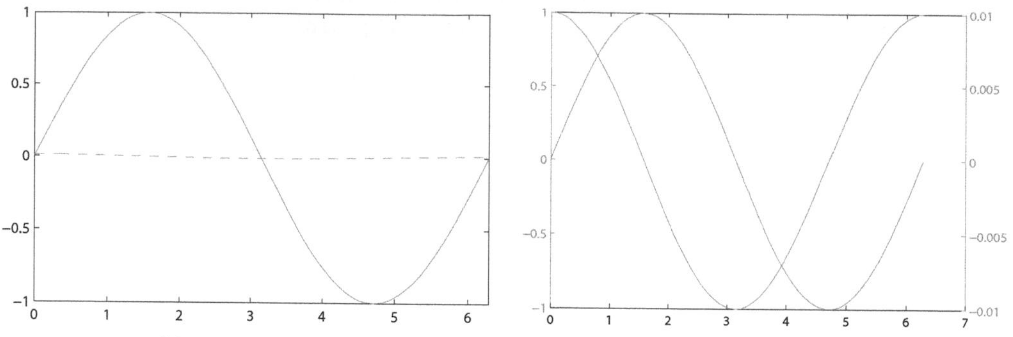

(a) improper plots (b) plots with two y axes

FIGURE 2.6: Plots of two curves with significant magnitude differences.

In some particular applications, three or even four vertical axes might be needed. This can be done by downloading the files `plotyyy()` [1], `plot4y()` [2] and others from MathWorks' File Exchange. The function `plotxx()` can also be used to draw plots with two horizontal axes [3].

2.5.3 Other two-dimensional plotting functions

Apart from the standard Descartes coordinate curves, MATLAB also provides other special two-dimensional graphical functions, and the common syntaxes of the functions are given in Table 2.3. In the functions, parameters $\boldsymbol{x}$,$\boldsymbol{y}$ are respectively the horizontal and vertical axis data, and `c` the color options. The parameters $\boldsymbol{y}_{\rm m}$,$\boldsymbol{y}_{\rm M}$ are the vectors of lower- and upper-boundaries in error plots. The functions are demonstrated through the following examples.

TABLE 2.3: Other two-dimensional plotting functions.

general syntax	explanation	general syntax	explanation
bar($\boldsymbol{x}$,$\boldsymbol{y}$)	two-dimensional bar chart	comet($\boldsymbol{x}$,$\boldsymbol{y}$)	comet trajectory
compass($\boldsymbol{x}$,$\boldsymbol{y}$)	compass plot	errorbar($\boldsymbol{x}$,$\boldsymbol{y}$,$\boldsymbol{y}_{\rm m}$,$\boldsymbol{y}_{\rm M}$)	errorbar plot
feather($\boldsymbol{x}$,$\boldsymbol{y}$)	feather plot	fill($\boldsymbol{x}$,$\boldsymbol{y}$,c)	filled plot
hist($\boldsymbol{y}$,n)	histogram	loglog($\boldsymbol{x}$,$\boldsymbol{y}$)	logarithmic plot
polar($\boldsymbol{x}$,$\boldsymbol{y}$)	polar plot	quiver($\boldsymbol{x}$,$\boldsymbol{y}$)	quiver graph
stairs($\boldsymbol{x}$,$\boldsymbol{y}$)	stairs plot	stem($\boldsymbol{x}$,$\boldsymbol{y}$)	stem plot
semilogx($\boldsymbol{x}$,$\boldsymbol{y}$)	x-semi-logarithmic plot	semilogy($\boldsymbol{x}$,$\boldsymbol{y}$)	y-semi-logarithmic plot

Example 2.28 Draw the polar plots for the function $\rho = 5\sin(4\theta/3)$.

Solution *A vector* $\boldsymbol{\theta}$ *can be constructed first, over the interval* $\theta \in (0, 2\pi)$, *then, the function value vector* $\boldsymbol{\rho}$ *can be calculated. With the* `polar()` *function, the polar plot can be drawn as shown in Figure 2.7 (a).*

```
>> theta=0:0.01:2*pi; rho=5*sin(4*theta/3); polar(theta,rho)
```

It seems that the polar plot thus obtained is incomplete, since the interval is selected too

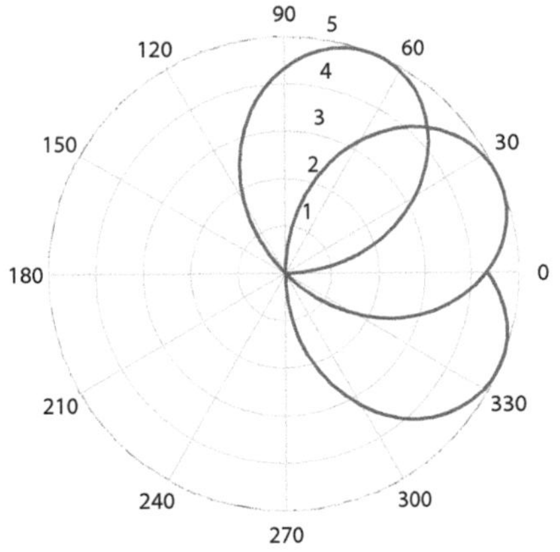

(a) $(0, 2\pi)$ interval

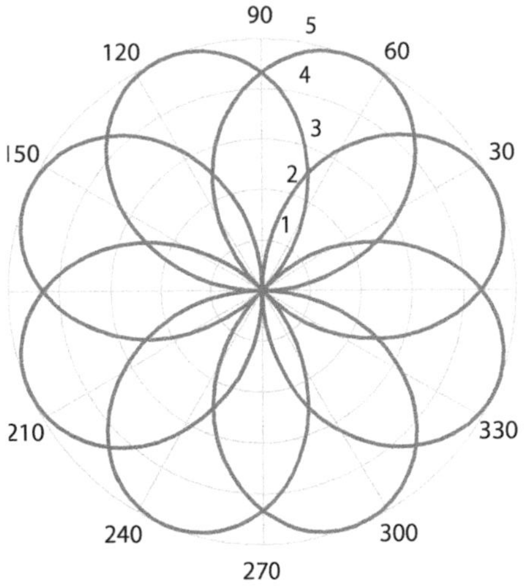

(b) $(0, 6\pi)$ interval

FIGURE 2.7: Polar plots.

small. To have a complete polar plot, the interval must be selected, at least, as $(0, 6\pi)$, *where* 6π *is the period of the polar function. The complete polar plot is shown in Figure 2.7 (b). If it is difficult to find the exact period, just select a large number, e.g.,* 100π.

Example 2.29 Draw the sinusoidal curve with different functions in different areas of the graphics window.

Solution *The following commands can be used to draw the expected curves as shown in Figure 2.8, where function* `subplot(n,m,k)` *can be used to divide the graphics window into several parts, with* n, m, *respectively the total numbers of rows and columns, and* k *indicates the serial of the area.*

```
>> t=0:.2:2*pi; y=sin(t);                % generate the data for plots
   subplot(2,2,1), stairs(t,y)           % partition the graphics window
   subplot(2,2,2), stem(t,y)             % stem plot in upper-right portion
   subplot(2,2,3), bar(t,y)              % bar chart in lower-left portion
   subplot(2,2,4), semilogx(t,y)         % semilogx in lower-right portion
```

2.5.4 Plots of implicit functions

For an implicit function $f(x, y) = 0$, the relationship between x and y cannot be explicitly formulated. Therefore, the conventional `plot()` function cannot be used. The MATLAB function `ezplot()` can be used to draw the implicit function curve, with default *range* of $[-2\pi, 2\pi]$

`ezplot(`*implicit function expression*`,`*range*`)`

The *implicit function expression* can be a string, a symbolic expression or an anonymous function. In the former two cases, the variables x, y can be regarded as symbols, thus, dot operation is not necessary.

Example 2.30 Draw the curve of the implicit function

$$f(x, y) = x^2 \sin(x + y^2) + y^2 e^{x+y} + 5\cos(x^2 + y) = 0.$$

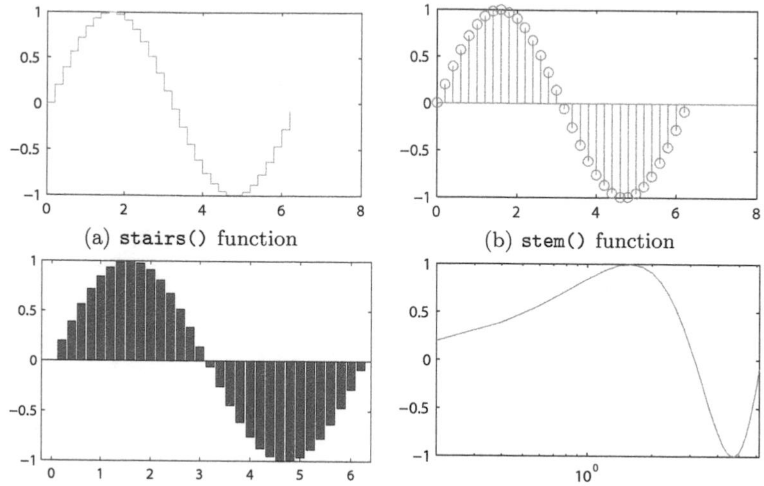

(a) `stairs()` function (b) `stem()` function

(c) `bar()` function (d) `semilogx()` function

FIGURE 2.8: Different representations of the same function.

Solution *From the given function, it can be seen that the analytical explicit solution of x-y relationship cannot be found. Thus, the* `plot()` *function cannot be used for such a function. The following MATLAB statements can be used to draw the implicit function as shown in Figure 2.9 (a).*

```
>> ezplot('x^2*sin(x+y^2)+y^2*exp(x+y)+5*cos(x^2+y)') % implicit function
```

The expression of the implicit function can also be expressed by a symbolic expression, and the same plots can be obtained

```
>> syms x y; f=x^2*sin(x+y^2)+y^2*exp(x+y)+5*cos(x^2+y); ezplot(f)
```

Alternatively, anonymous functions can be used to describe the implicit functions, however, dot operations can be used to speed up the function evaluation process.

```
>> f=@(x,y)x.^2.*sin(x+y.^2)+y.^2.*exp(x+y)+5*cos(x.^2+y); ezplot(f)
```

The above functions selected automatically the ranges of $x, y \in (-2\pi, 2\pi)$. The ranges can be specified to $(-10, 10)$ with the following statements, with the new implicit curve shown in Figure 2.9 (b).

```
>> ezplot('x^2*sin(x+y^2)+y^2*exp(x+y)+5*cos(x^2+y)',[-10 10])
```

2.5.5 Graphics decorations

The graphics window with editing tools is shown in Figure 2.10, by clicking the ▣ button in the toolbar. The user may choose to apply text and arrows to the plots.

In the graphics editing interface, there are three parts, with the left part corresponding to the View → Figure Palette menu item, where arrows and text can be added to the curve. 2D and 3D axes can also be added to the curve. The bottom part of the window corresponds

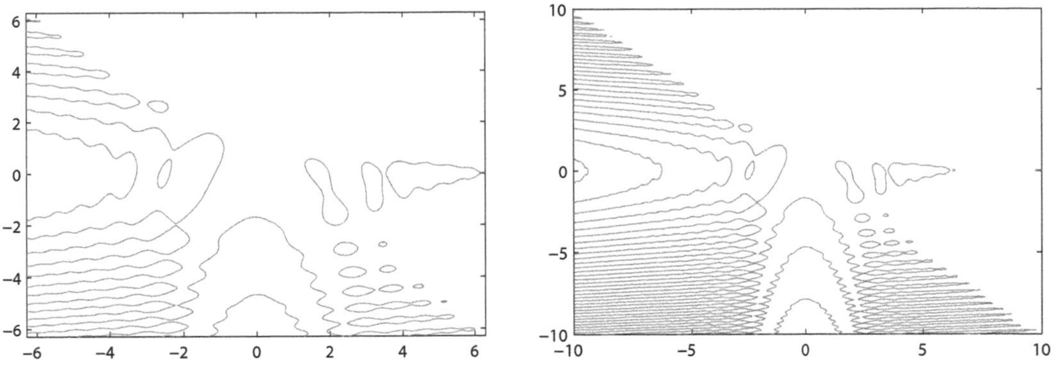

(a) default curves (b) curves within larger area

FIGURE 2.9: Curves of implicit functions.

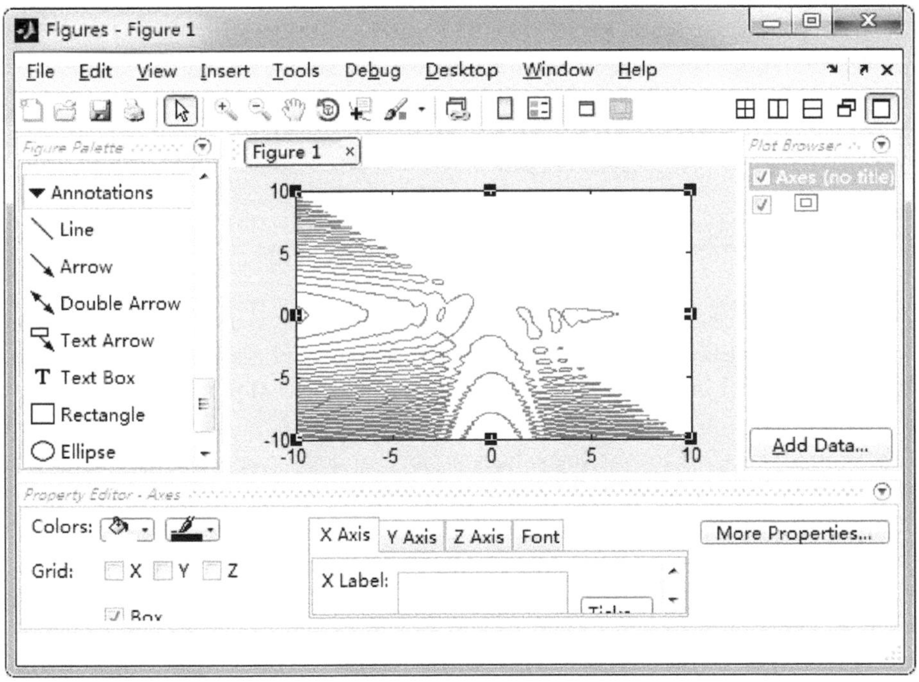

FIGURE 2.10: MATLAB graphics window with editing tools.

to the Property Editor menu item, which allows the selections of color, line styles or fonts to the selected objects. The right part of the window corresponds to the View → Plot Browser menu item, which allows the user to add new data or superimpose new curves.

Select the menu item View → Plot Edit Toolbar, an extra toolbar with be added to the figure window, as shown in Figure 2.11, also allowing to manipulate text and arrow objects on graphs.

Local zooming and 3D view point settings are also provided in the plots. The button in the toolbar can be used to read coordinate information of points on curves or surfaces, with simple mouse clicking. The icon can be used to change viewpoint of 3D plots. An

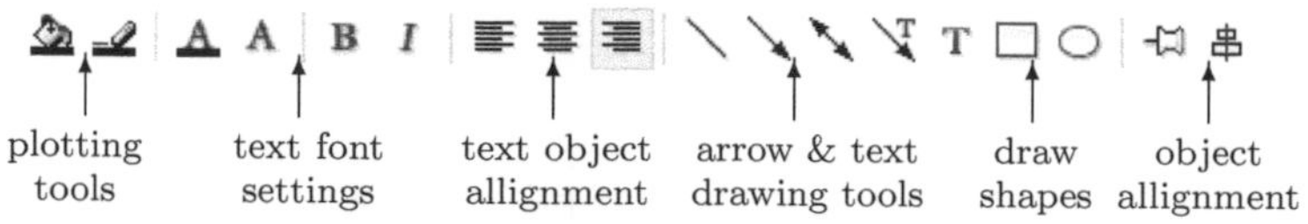

FIGURE 2.11: Figure editing toolbar.

example of a typical graphics display under the view-point change is shown in Figure 2.12, where a 2D curve is displayed under a 3D framework.

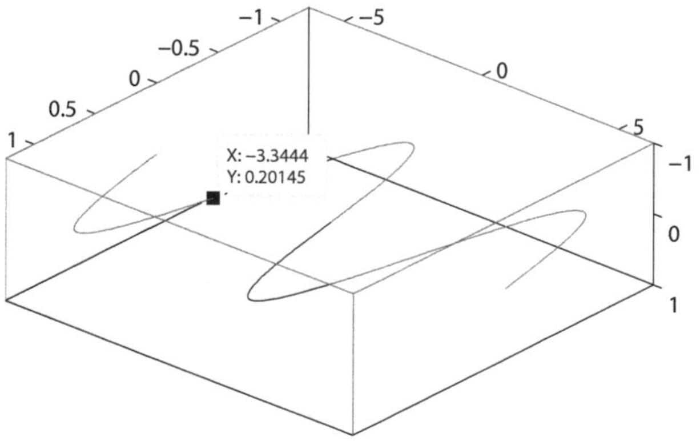

FIGURE 2.12: 3D representations of 2D curves.

A subset of LaTeX commands can be used to add mathematical formulas to the plots. LaTeX is a well established scientific type-setting system, and a subset of its mathematical symbols are supported in MATLAB. One may use them to superimpose formulas to the plots.

(i) The symbols are led by the backslash signs \, and the available symbols are listed in Table 2.4.

(ii) Superscripts and subscripts are represented by ^ and _, respectively. For instance, `a_2^2+b_2^2=c_2^2` represents $a_2^2 + b_2^2 = c_2^2$. If more than one symbol is used in the superscript, they should be written within the { and } signs. For instance `a^Abc` gives $a^A bc$, while `a^{Abc}` gives a^{Abc}.

LaTeX scientific type-setting system is widely used in the academic world. Interested readers may further refer to Reference [4].

2.5.6 Data file access with MATLAB

The command pair `save` and `load` can be used to access data files. If the `save` command is used alone, all the variables in MATLAB workspace will be saved to the file matlab.mat, in binary format. The contents of the binary files can only be retrieved with `load` command. If one wants to save the variables ***A***, ***B*** and ***C*** into a binary file mydat.mat, the command `save mydat` ***A B C*** can be used.

If the three variables are to be saved in the ASCII format into file mydat.dat, the command `save /ascii mydat.dat` ***A B C*** should be used instead.

TABLE 2.4: TEX compatible commands in MATLAB.

	c	TEX	*c*	TEX	*c*	TEX	*c*	TEX
	α	\alpha	β	\beta	γ	\gamma	δ	\delta
	ϵ	\epsilon	ε	\varepsilon	ζ	\zeta	η	\eta
lower-	θ	\theta	ϑ	\vartheta	ι	\iota	κ	\kappa
case	λ	\lambda	μ	\mu	ν	\nu	ξ	\xi
Greeks	o	o	π	\pi	ϖ	\varpi	ρ	\rho
	ι	\iota	κ	\kappa	ϱ	\varrho	σ	\sigma
	ς	\varsigma	τ	\tau	υ	\upsilon	ϕ	\phi
	φ	\varphi	χ	\chi	ψ	\psi	ω	\omega
upper-	Γ	\Gamma	Δ	\Delta	Θ	\Theta	Λ	\Lambda
case	Ξ	\Xi	Π	\Pi	Σ	\Sigma	Υ	\Upsilon
Greeks	Φ	\Phi	Ψ	\Psi	Ω	\Omega		
	$\aleph$	\aleph	$\prime$	\prime	$\forall$	\forall	$\exists$	\exists
common	$\wp$	\wp	$\Re$	\Re	$\Im$	\Im	∂	\partial
math	∞	\infty	∇	\nabla	$\surd$	\surd	$\angle$	\angle
symbols	$\neg$	\neg	$\int$	\int	$\clubsuit$	\clubsuit	$\diamondsuit$	\diamondsuit
	$\heartsuit$	\heartsuit	$\spadesuit$	\spadesuit				
binary	$\pm$	\pm	$\cdot$	\cdot	$\times$	\times	$\div$	\div
math	$\circ$	\circ	$\bullet$	\bullet	$\cup$	\cup	$\cap$	\cap
symbols	$\vee$	\vee	$\wedge$	\wedge	$\otimes$	\otimes	$\oplus$	\oplus
relat-	$\leq$	\leq	$\geq$	\geq	$\equiv$	\equiv	$\sim$	\sim
ional	$\subset$	\subset	$\supset$	\supset	$\approx$	\approx	$\subseteq$	\subseteq
math	$\supseteq$	\supseteq	$\in$	\in	$\ni$	\ni	$\propto$	\propto
symbols	$\mid$	\mid	$\perp$	\perp				
	$\leftarrow$	\leftarrow	$\uparrow$	\uparrow	$\Leftarrow$	\Leftarrow	$\Uparrow$	\Uparrow
arrows	$\rightarrow$	\rightarrow	$\downarrow$	\downarrow	$\Rightarrow$	\Rightarrow	$\Downarrow$	\Downarrow
	$\leftrightarrow$	\leftrightarrow	$\updownarrow$	\updownarrow				

If long or complicated path or file names are used, the functions `load()` should be used, with $\boldsymbol{X}$ = `load(`*filename*`)`, and a variable $\boldsymbol{X}$ will be loaded into MATLAB workspace.

Microsoft Excel files are also supported in MATLAB, with the functions `xlsread()` to load file into MATLAB workspace, in the syntax

$\boldsymbol{X}$ = `xlsread(`*filename*, *range*`)`

where, the *range* can be marked as a rectangular region such as `'B5:C67'`, meaning that the data in the Excel file, from column B to C, and from the 5th row to the 67th, all loaded into MATLAB workspace to form matrix $\boldsymbol{X}$. Another function, `xlswrite()`, can be used to write variables into Excel files.

Example 2.31 Suppose in the Excel file census.xls, the annual population records of a certain province provided, with column B the years, while column C the population record.

The effective data start from the 5th row and end at row 67. Please load the year information to vector $\boldsymbol{t}$ and population to vector $\boldsymbol{p}$.

Solution *It can be seen from the above description that the effective data are located in columns B and C, and from row 5 to row 67, the range specification can be represented as* `'B5:C67'`. *Therefore, the following statements can be used to load the year and population into MATLAB workspace, and then distribute them to the vectors* $\mathbf{t}$ *and* $\mathbf{p}$.

```
>> X=xlsread('census.xls','B5:C67'); t=X(:,1); p=X(:,2); plot(t,p)
```

2.6 Three-dimensional Graphics

2.6.1 Plotting of three-dimensional curves

The two-dimensional function `plot()` can be extended to a three-dimensional (3D) curve drawing with the new `plot3()` function, whose syntaxes are

`plot3(`$\boldsymbol{x}$,$\boldsymbol{y}$,$\boldsymbol{z}$`)`

`plot3(`$\boldsymbol{x}_1, \boldsymbol{y}_1, \boldsymbol{z}_1$, *option 1*, $\boldsymbol{x}_2, \boldsymbol{y}_2, \boldsymbol{z}_2$, *option 2*, $\cdots$, $\boldsymbol{x}_m, \boldsymbol{y}_m, \boldsymbol{z}_m$, *option m*`)`

where the *options* are the same as shown in Table 2.2.

Similar to other 2D curve drawing functions, the functions `stem3()`, `fill3()` and `bar3()` can also be applied to 3D curves.

Example 2.32 Suppose the position of a particle in 3D space is described by the parametric equations

$$x(t) = t^3 \mathrm{e}^{-t} \sin 3t, y(t) = t^3 \mathrm{e}^{-t} \cos 3t, z = t^2, \quad \text{where } t \in [0, 2\pi].$$

Please draw the trajectory of the particle.

Solution *A time vector* $\mathbf{t}$ *can be established first, then, the vectors* $\boldsymbol{x}, \boldsymbol{y}, \boldsymbol{z}$ *can be computed. The 3D curve can be drawn with the* `plot3()` *function, as shown in Figure 2.13 (a). It should be noted that dot operations are used in the evaluations.*

```
>> t=0:.1:2*pi;    % establish the t vector, with dot operation
   x=t.^3.*exp(-t).*sin(3*t); y=t.^3.*exp(-t).*cos(3*t); z=t.^2;
   plot3(x,y,z), grid  % 3D curve drawing
```

The `stem3()` *function can be used to obtained the plot in Figure 2.13 (b), superimposed by the 3D curve.*

```
>> stem3(x,y,z); hold on; plot3(x,y,z), grid
```

To display the trajectory of the 3D curve, `comet3()` *function is recommended.*

Alternatively function `ezplot3()` *can be used to draw 3D curves, and the latter statement can be used to draw dynamically the trajectory of the 3D curve.*

```
>> syms t; x=t^3*exp(-t)*sin(3*t); y=t^3*exp(-t)*cos(3*t); z=t^2;
   ezplot3(x,y,z,[0,2*pi]); figure; ezplot3(x,y,z,[0,2*pi],'animate');
```

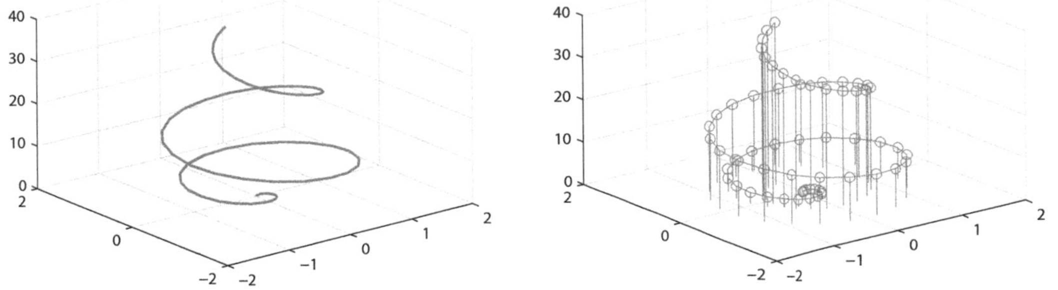

(a) 3D curve plots (b) superimposed with the plot with `stem3()`

FIGURE 2.13: Three-dimensional plots.

2.6.2 Plotting of three-dimensional surfaces

If function $z = f(x, y)$ is given, the 3D surface of the function can be drawn. One can generate mesh grid data in the x-y plane, with the `meshgrid()` function. The function values $\boldsymbol{z}$ for each grid can be evaluated. The functions `mesh()` and `surf()` can be used to draw the 3D mesh plots and surface plots. The syntaxes of the functions are

```
[x,y]=meshgrid(v1, v2)                  % mesh grid matrices generation
z=expression, for instance z=x.*y       % z matrix dot computation
surf(x,y,z)   or   mesh(x,y,z)          % mesh and surface plots
```

where $\boldsymbol{v}_1$ and $\boldsymbol{v}_2$ are the vectors with coordinate grids in the x and y axes, and two matrices $\boldsymbol{x}$, $\boldsymbol{y}$ can be generated. Dot operation is involved in the evaluation of the $\boldsymbol{z}$ matrix. The 3D surface can also be drawn with the `surfc()`, `surfl()` and `waterfall()` functions. Also, the `contour()` and `contour3()` functions can be used to draw 2D and 3D contour plots.

Other simple functions such as `ezsurf()`, `ezmesh()`, `ezcontour()` and `ezcontourf()` can also be used to draw 3D plots, with the mathematical explicit expression $z = f(x, y)$ specified.

Example 2.33 Consider the function $z = f(x, y) = (x^2 - 2x)\mathrm{e}^{-x^2-y^2-xy}$. Select in the x-y plane an area and draw the 3D plots.

Solution *One may use the* `meshgrid()` *function to specify the mesh grids on the x-y plane. The values of the function can be evaluated directly for the matrix $\boldsymbol{z}$. The mesh plot can be drawn as shown in Figure 2.14 (a).*

```
>> [x,y]=meshgrid(-3:0.1:3,-2:0.1:2); % generate mesh grid matrices
   z=(x.^2-2*x).*exp(-x.^2-y.^2-x.*y); mesh(x,y,z) % draw mesh plot
```

If one uses `surf()` *function to replace the* `mesh()` *function, the corresponding surface plot can be obtained as shown in Figure 2.14 (b).*

```
>> surf(x,y,z)   % surface plot
```

3D surface plots can be decorated by `shading` *command, and the options* `flat` *and* `interp` *can be used. The decorations are shown in Figures 2.15 (a) and (b), respectively.*

Other functions, such as `waterfall(x,y,z)` *and* `contour3(x,y,z,30)` *can be used to draw 3D plots as shown in Figures 2.16 (a) and (b).*

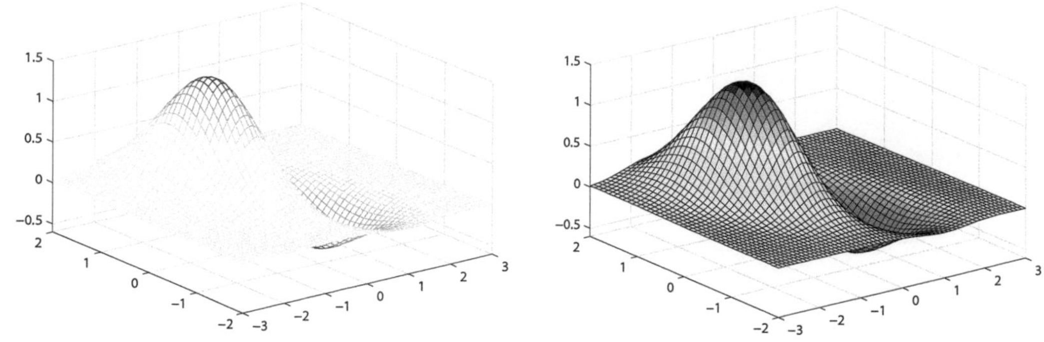

(a) `mesh()` plot

(b) `surf()` plot

FIGURE 2.14: Mesh and surface plots of a given function.

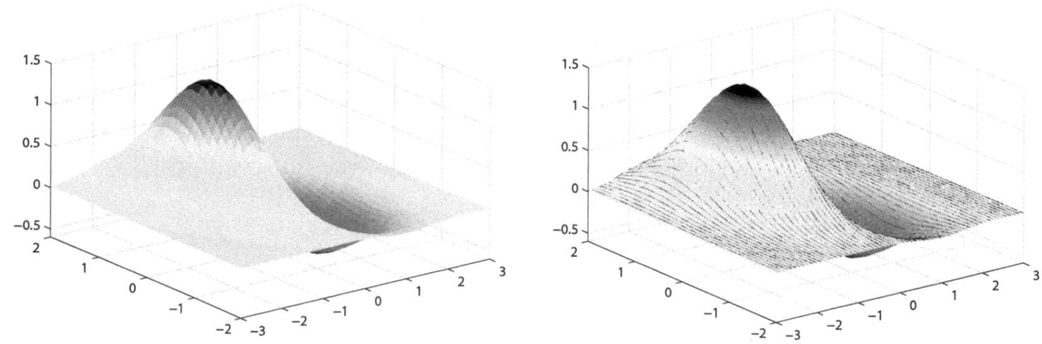

(a) `shading flat`

(b) `shading interp`

FIGURE 2.15: 3D surfaces decorated by the `shading` command.

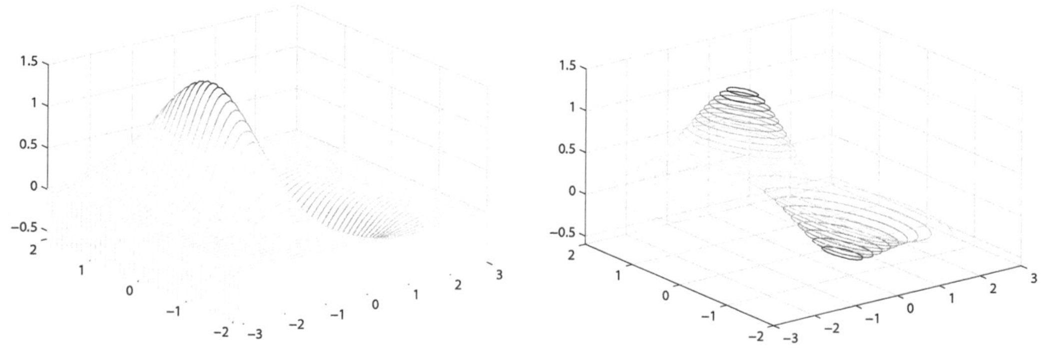

(a) `waterfall()` plots

(b) `contour3()` plots

FIGURE 2.16: Other 3D representations.

Example 2.34 Display graphically the surface of the following function

$$z = f(x, y) = \frac{1}{\sqrt{(1-x)^2 + y^2}} + \frac{1}{\sqrt{(1+x)^2 + y^2}}.$$

Solution *The following statements can be used to draw the 3D surface of the function, as shown in Figure 2.17 (a).*

```
>> [x,y]=meshgrid(-2:.1:2); % generate mesh grid data
   z=1./(sqrt((1-x).^2+y.^2))+1./(sqrt((1+x).^2+y.^2)); % evaluate function
   surf(x,y,z), shading flat % draw surface plot
```

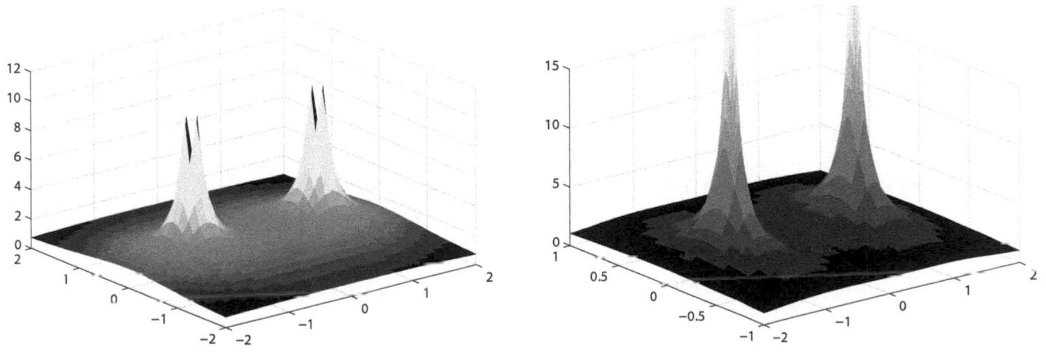

(a) fixed-step-size (b) variable-step-size

FIGURE 2.17: Three-dimensional surfaces under different grids.

In fact, there are problems around the $(\pm 1, 0)$ *points, where the function values tend to infinity. Thus, variable-step-size mesh grids can be constructed, and the new 3D surface can be obtained as shown in Figure 2.17 (b).*

```
>> xx=[-2:.1:-1.2,-1.1:0.02:-0.9,-0.8:0.1:0.8,0.9:0.02:1.1,1.2:0.1:2];
   yy=[-1:0.1:-0.2, -0.1:0.02:0.1, 0.2:.1:1];
   [x,y]=meshgrid(xx,yy); % generate a denser mesh grid data
   z=1./(sqrt((1-x).^2+y.^2))+1./(sqrt((1+x).^2+y.^2)); % evaluate function
   surf(x,y,z), shading flat; zlim([0,15]) % draw surface plot
```

Example 2.35 Assume that a piecewise function is described below [5]

$$p(x_1,x_2)=\begin{cases} 0.5457\exp(-0.75x_2^2-3.75x_1^2-1.5x_1), & x_1+x_2>1\\ 0.7575\exp(-x_2^2-6x_1^2), & -1<x_1+x_2\leqslant 1\\ 0.5457\exp(-0.75x_2^2-3.75x_1^2+1.5x_1), & x_1+x_2\leqslant -1.\end{cases}$$

Show the function in a three-dimensional surface.

Solution *It is obvious that the function value can be evaluated with loops and* `if` *statements, however, the process could be very complicated. Thus, the piecewise function configuration statements based on relational operations can be used instead to evaluate the functions as follows*

```
>> [x1,x2]=meshgrid(-1.5:.1:1.5,-2:.1:2); % evaluate piecewise function
   z= 0.5457*exp(-0.75*x2.^2-3.75*x1.^2-1.5*x1).*(x1+x2>1)+...
      0.7575*exp(-x2.^2-6*x1.^2).*((x1+x2>-1) & (x1+x2<=1))+...
      0.5457*exp(-0.75*x2.^2-3.75*x1.^2+1.5*x1).*(x1+x2<=-1);
   h=surf(x1,x2,z), xlim([-1.5 1.5]); shading flat % surface plot
```

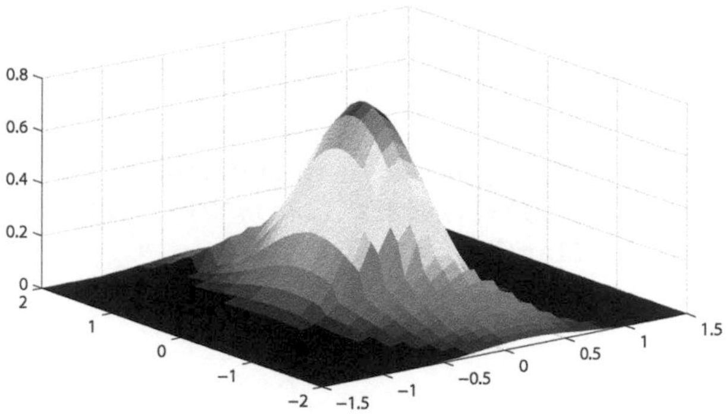

FIGURE 2.18: Surface of a piecewise function with two variables.

and the three-dimensional surface can be shown in Figure 2.18.

It is worth mentioning that all the logic conditions in the piecewise function must be mutually exclusive to avoid conflicts. Also, in the last sentence, the handle `h` *is deliberately returned. If one wants to remove the surface, a command* `delete(h)` *can be given.*

2.6.3 Viewpoint settings in 3D graphs

In the MATLAB 3D graphics facilities, viewpoint setting functions are provided, which allows the user to view the plot from any angle. Two ways are provided: one is the toolbar facility in the figure window, and the other is the `view()` function.

An illustration to the definition of the viewpoint is given in Figure 2.19 (a), where the two angles α and β can be used to define uniquely the viewpoint. The *azimuth angle* α is defined as the angle between the projection line in x–y plane with the negative y-axis, with a default value of $\alpha = -37.5°$. The elevation angle β is defined as the angle with the x-y plane, with a default value of $\beta = 30°$.

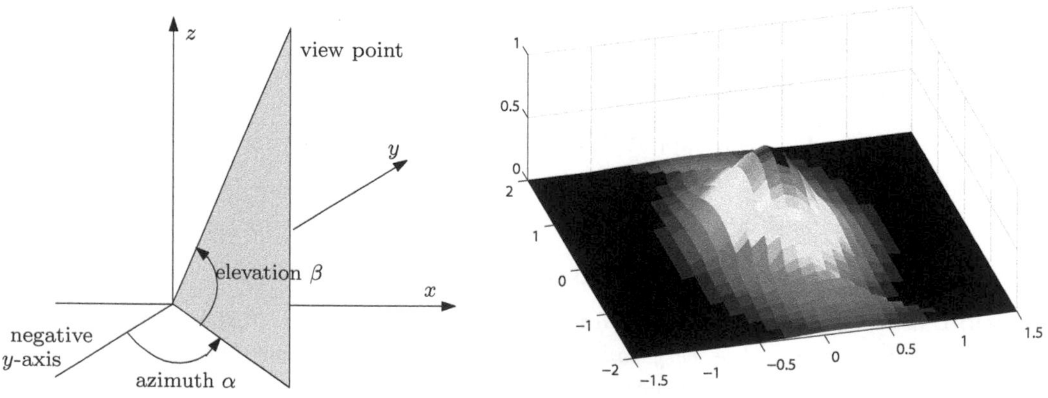

(a) definition of viewpoints (b) 3D surface after viewpoint change

FIGURE 2.19: Viewpoint settings of three-dimensional surfaces.

The function `view(`α`,`β`)` can be used to set the viewpoint, where the angles α and β are the azimuth and elevation angles, respectively. For instance, the setting `view(0,90)` shows the planform, while `view(0,0)` and `view(90,0)` show the front view and the side elevation, respectively.

For instance, one may change the viewpoint in the three-dimensional surface display shown in Figure 2.18. One may set $\alpha = 20°$, and $\beta = 50°$, the following statements can be used, and the results shown in Figure 2.19 (b) can be obtained.

```
>> view(20,50), xlim([-1.5 1.5]) % set the range of x-axis
```

Example 2.36 Consider again the surface plot in Example 2.33. View the surface from different angles.

Solution *The surface plots from different viewpoints can be obtained using the following statements, as shown in Figure 2.20.*

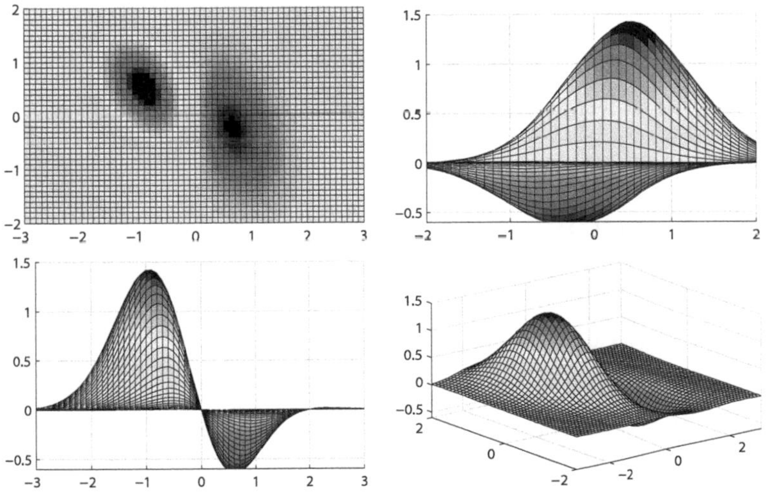

FIGURE 2.20: A surface with orthographic views.

```
>> [x,y]=meshgrid(-3:0.1:3,-2:0.1:2);        % generate mesh grid data
   z=(x.^2-2*x).*exp(-x.^2-y.^2-x.*y);       % evaluate function
   subplot(221), surf(x,y,z), view(0,90); % planform
   subplot(222), surf(x,y,z), view(90,0); % side elevation
   subplot(223), surf(x,y,z), view(0,0);  % front view
   subplot(224), surf(x,y,z),             % 3D surface plot
```

2.6.4 Surface plots of parametric equations

If the 3D function is given by parametric equations

$$x = f_x(u,v),\ y = f_y(u,v),\ z = f_z(u,v), \tag{2-6-1}$$

and $u_m \leqslant u \leqslant u_M$, $v_m \leqslant v \leqslant v_M$, the 3D surface plot can be obtained with `ezsurf(`f_x`,`f_y`,`f_z`,[`u_m`,`u_M`,`v_m`,`v_M`])`, with default boundaries of u, v in interval $(-2\pi, 2\pi)$.

Example 2.37 The well-known Möbius strip is mathematically expressed as

$$x = \cos u + v\cos\frac{u}{2}\cos u,\quad y = \sin u + v\cos\frac{u}{2}\sin u,\quad z = v\sin\frac{u}{2},$$

with, for instance, $0 \leqslant u \leqslant 2\pi$, $-0.5 \leqslant v \leqslant 0.5$. Please draw Möbius strip.

Solution *The variables u and v should be declared first as symbolic variables, and the parametric equations can be specified, and the Möbius strip can be obtained with the following statements, and the rotated version is as shown in Figure 2.21.*

```
>> syms u v; x=cos(u)+v*cos(u/2)*cos(u); y=sin(u)+v*cos(u/2)*sin(u);
   z=v*sin(u/2); ezsurf(x,y,z,[0,2*pi,-0.5,0.5])
```

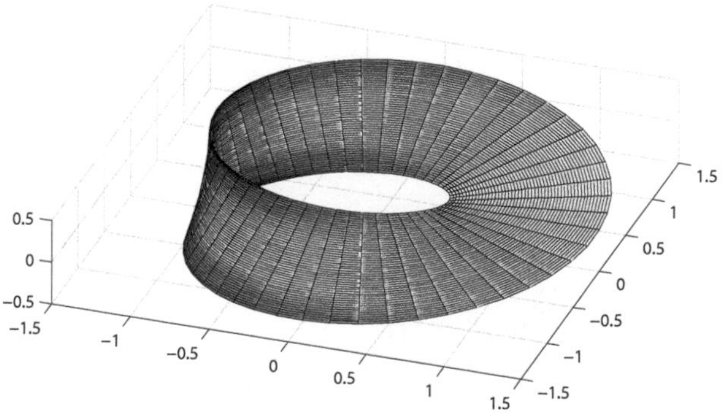

FIGURE 2.21: Rotated view of Möbius strip.

2.6.5 Spheres and cylinders

The data for a unit sphere, centered at the origin, can be generated with the function `[x,y,z]=sphere(n)`, and such a sphere can be drawn directly with `surf()`. If there is no returned argument in the function call, a sphere can be drawn directly. The argument n can be specified such that the sphere can have $n \times n$ faces.

Example 2.38 Please draw a unit sphere at the original, also on the same coordinates, draw another sphere with a radius of 0.3 and centered at $(0.9, -0.8, 0.6)$.

Solution *The data for the unit sphere can be generated first, and based on the data, the two spheres can be drawn as shown in Figure 2.22 (a).*

```
>> [x,y,z]=sphere(50); surf(x,y,z), hold on % draw the unit sphere first
   x1=0.3*x+0.9; y1=0.3*y-0.8; z1=0.3*z+0.6; surf(x1,y1,z1) % smaller one
```

Cylinders in MATLAB are created by rotating a curve around z-axis for 360°. If the curve is defined as a vector $\boldsymbol{r}$, standing for the radius, the data for the cylinder can be generated with `[x,y,z]=cylinder(r,n)`. If no returned arguments are specified, the cylinder can be drawn directly. Please note that the default interval of z is $z \in (0,1)$.

Example 2.39 Suppose the radius of the cylinder is $r(z) = \mathrm{e}^{-z^2/2}\sin z$, $z \in (-1,3)$, please draw the cylinder.

Solution *The radius can be calculated first, and the data for the standard cylinder can be obtained, and it can be mapped from $z \in (0,1)$ interval to $z \in (-1,3)$, with the following statements, as shown in Figure 2.22 (b).*

```
>> z0=-1:0.1:3; r=exp(-z0.^2/2).*sin(z0); [x,y,z]=cylinder(r);
   z=-1+4*z; surf(x,y,z) % map z axis from (0,1) to (-1,4)
```

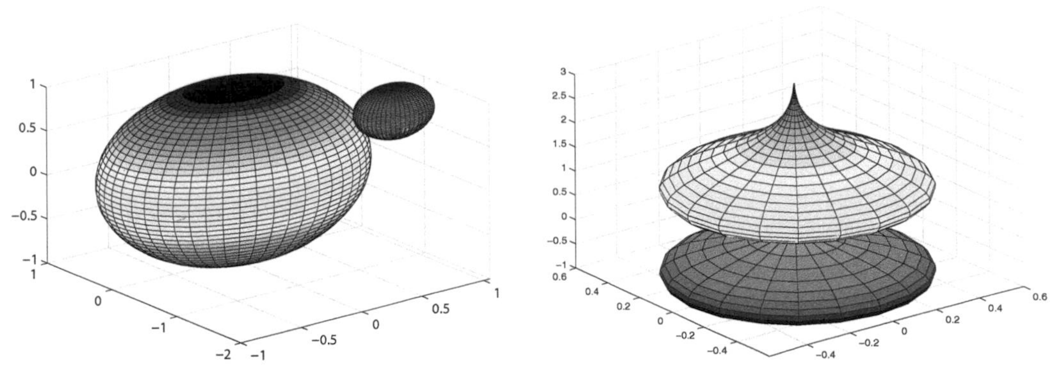

(a) two spheres (b) cylinder

FIGURE 2.22: Three-dimensional surfaces.

2.6.6 Drawing 2D and 3D contours

If the mesh grid data $\boldsymbol{x}$, $\boldsymbol{y}$, $\boldsymbol{z}$ for explicit function $z = f(x, y)$ are known, the function `contour()` can be used to draw 2D contour, with `contour(`$\boldsymbol{x}$`,`$\boldsymbol{y}$`,`$\boldsymbol{z}$`,`n`)`, where n is the expected number of contour lines and can be omitted. Alternatively, a pair of functions can be called

`[C,h]` $=$ `contour(`$\boldsymbol{x}$`,`$\boldsymbol{y}$`,`$\boldsymbol{z}$`,`n`), clabel(C,h)`

where `h` is the handle of all contour lines, and `C` returns all the height information, which can finally be superimposed on the contours with the last statement.

Functions `contourf()` and `contour3()` can be used to draw filled contour plots and 3D contours, respectively, with the syntaxes

`contourf(`$\boldsymbol{x}$`,`$\boldsymbol{y}$`,`$\boldsymbol{z}$`,`n`),` and `contour3(`$\boldsymbol{x}$`,`$\boldsymbol{y}$`,`$\boldsymbol{z}$`,`n`)`

Example 2.40 Consider the piecewise function used in Example 2.35, please draw its contour plots.

Solution *The mesh grid data can be generated first, as it was done in Example 2.35, and the ordinary contours can be drawn with* `contour()` *function, as shown in Figure 2.23(a). Also, the contours with superimposed height information can be obtained, as shown in Figure 2.23(b).*

```
>> [x1,x2]=meshgrid(-1.5:.1:1.5,-2:.1:2);
   z= 0.5457*exp(-0.75*x2.^2-3.75*x1.^2-1.5*x1).*(x1+x2>1)+...
      0.7575*exp(-x2.^2-6*x1.^2).*((x1+x2>-1) & (x1+x2<=1))+...
      0.5457*exp(-0.75*x2.^2-3.75*x1.^2+1.5*x1).*(x1+x2<=-1);
   [C,h]=contour(x1,x2,z); clabel(C,h)
```

The following statements can be used directly to draw filled and 3D contours, as shown in Figures 2.24(a) and (b). In the last statement, the number of contour lines is set to 30, otherwise, the contours are too sparsely distributed.

```
>> subplot(121), contourf(x1,x2,z); subplot(122), contour3(x1,x2,z,30)
```

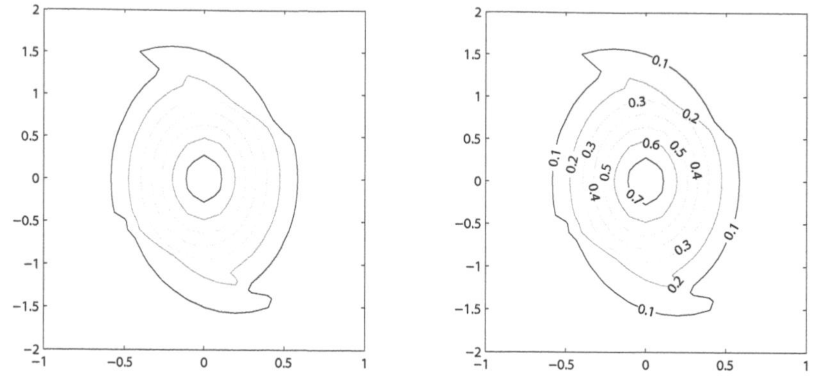

(a) ordinary contours (b) contours superimposed with heights

FIGURE 2.23: Contours of a piecewise function.

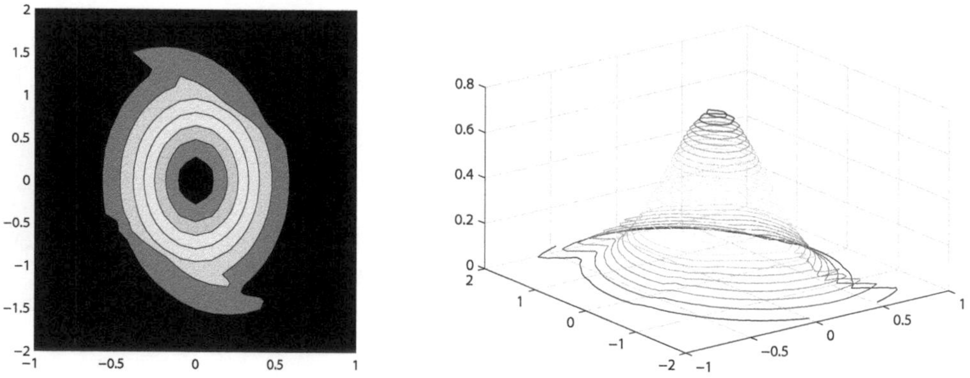

(a) filled contours (b) 3D contours with 30 lines

FIGURE 2.24: Filled and 3D contours.

2.6.7 Drawing 3D implicit functions

The functions `surf()` and `mesh()` discussed so far can only be used to draw 3D plots with known data, probably generated from known explicit function $z = f(x, y)$. If a three-dimensional implicit function $g(x, y, z) = 0$ is known, function `ezimplot3()` [6] can be used, with the syntax `ezimplot3(`*fun*`,[`x_m`,`x_M`,`y_m`,`y_M`,`z_m`,`z_M`])` , where, *fun* can be described by anonymous functions, string, M-functions or by symbolic expressions. The default axes ranges of $x_\mathrm{m}, x_\mathrm{M}, y_\mathrm{m}, y_\mathrm{M}, z_\mathrm{m}, z_\mathrm{M}$ are $\pm 2\pi$. If only one pair of $x_\mathrm{m}, x_\mathrm{M}$ is specified, all the three axes share the same settings.

Example 2.41 Assume that an implicit function

$$g(x, y, z) = x \sin\left(y+z^2\right) + y^2 \cos\left(x+z\right) + zx \cos\left(z+y^2\right) = 0$$

is known, and the interested ranges are $x, y, z \in (-1, 1)$, please draw the 3D surface.

Solution *Strings and anonymous functions can both be used to describe the original implicit function*

```
>> f=@(x,y,z)x*sin(y+z^2)+y^2*cos(x+z)+z*x*cos(z+y^2);
```

The surface of the function can be obtained with the following commands, as shown in Figure 2.25(a).

```
>> ezimplot3(f,[-1 1])  % draw 3D surface for the implicit function
```

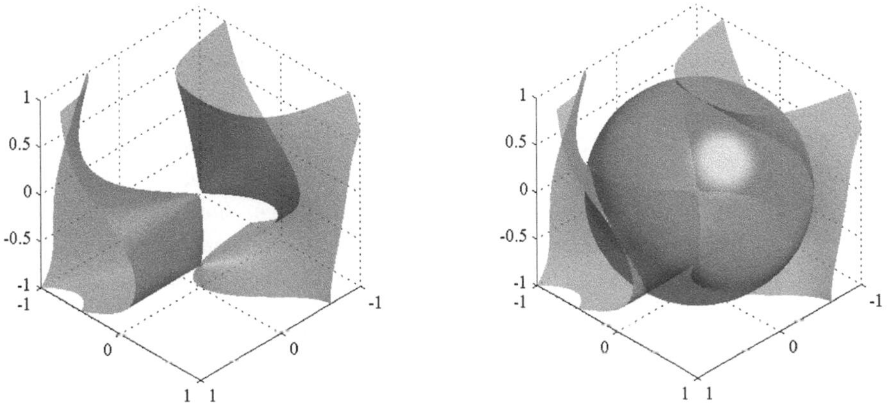

(a) the 3D surface (b) superimposed by a sphere

FIGURE 2.25: Surfaces of 3D implicit functions.

The following statements can be used to superimpose a unit sphere $x^2+y^2+z^2=1$ *on the surface, as shown in Figure 2.25(b).*

```
>> f1=@(x,y,z)x^2+y^2+z^2-1; ezimplot3(f1,[-1 1]); % no hold on needed
```

2.7 Four-dimensional Visualization

In the three-dimensional plotting facilities illustrated earlier, function $z = S(x, y)$ with two independent variables are involved. If a function has three independent variables, with its mathematical representation of $v = V(x, y, z)$, volume visualization should be introduced. There are quite a lot of practical examples for functions with three independent variables. For instance, the internal temperature of a three-dimensional solid object, or the flow rate or concentration of liquid. These kinds of functions are not suitable to be visualized with ordinary three-dimensional plots, and it is not possible to display four-dimensional plots on ordinary two-dimensional screens. Special three-dimensional plots must be introduced. In volume visualizations, slices are adopted to observe the values of the functions inside a three-dimensional function. Computerized tomography (CT) images are good examples in observing internal structures with slices. Of course, time-driven three-dimensional animation can also be regarded as four-dimensional plots.

Again the `meshgrid()` function can be used to generate three-dimensional mesh grids, and the three-dimensional arrays $\boldsymbol{x}$,$\boldsymbol{y}$,$\boldsymbol{z}$ can be established. The volume data $\boldsymbol{V}$ can be evaluated through dot operation. Then, the function `slice()` can be called to draw the slices, with the syntax `slice(`$\boldsymbol{x},\boldsymbol{y},\boldsymbol{z},\boldsymbol{V},\boldsymbol{x}_1,\boldsymbol{y}_1,\boldsymbol{z}_1$`)`, where, $\boldsymbol{x}$, $\boldsymbol{y}$, $\boldsymbol{z}$ and $\boldsymbol{V}$ are the data for volume visualization, while $\boldsymbol{x}_1$, $\boldsymbol{y}_1$ and $\boldsymbol{z}_1$ are the data for describing slices. If constant vectors are used, the slices are perpendicular to the corresponding axes. The slices can also be generated as rotated planes or even other types of surfaces, details will be given through examples.

Example 2.42 Assume that a function

$$V(x,y,z)=\sqrt{x^x+y^{(x+y)/2}+z^{(x+y+z)/3}}$$

is with three independent variables. Please observe the function through volume visualization techniques, and observe the properties through slices.

Solution *Since square roots are taken, the axes x, y and z are assigned within nonnegative ranges. The following statements can be used to visualize the data $\boldsymbol{V}$. The groups of slices are perpendicular to the axes can be specified. For instance, the first group of slices were fixed at $x=1$, $x=2$, perpendicular to x axis, and the second and third groups are at $y=1$, $y=2$ and $z=0$, $z=1$, respectively. The slices thus generated are shown in Figure 2.26(a).*

```
>> [x,y,z]=meshgrid(0:0.1:2); V=sqrt(x.^x+y.^((x+y)/2)+z.^((x+y+z)/3));
   slice(x,y,z,V,[1 2],[1 2],[0 1]); % volume visualization with slices
```

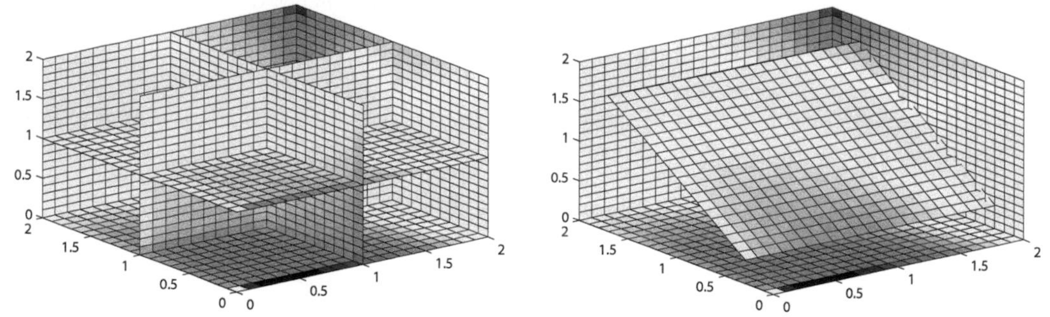

(a) slices perpendicular to the axes (b) a rotated slice

FIGURE 2.26: The effect of the slices.

As indicated earlier, an ordinary plane can be specified as $z=1$. Then, it can be rotated 45° along the x axis. In this case, the vectors $\boldsymbol{x}_1$, $\boldsymbol{y}_1$, $\boldsymbol{z}_1$ can be extracted from the rotated plane, and the volume visualization with such a slice can be obtained, as shown in Figure 2.26(b).

```
>> [x0,y0]=meshgrid(0:0.1:2); z0=ones(size(x0)); % generate z = 0 plane
   h=surf(x0,y0,z0); rotate(h,[1,0,0],45); % rotate plane 45° along x-axis
   x1=get(h,'XData'); y1=get(h,'YData'); z1=get(h,'ZData'); % get data
   slice(x,y,z,V,x1,y1,z1), hold on, slice(x,y,z,V,2,2,0) % draw slices
```

In order to observe the slices easily, a simple-to-use graphical user interface `vol_visual4d()` is designed. To use the interface, the volume data $\boldsymbol{x}$, $\boldsymbol{y}$, $\boldsymbol{z}$ and $\boldsymbol{V}$ should be generated first, and interface can then be called with `vol_visual4d(`$\boldsymbol{x}$`,`$\boldsymbol{y}$`,`$\boldsymbol{z}$`,`$\boldsymbol{V}$`)`. The controls in the interface can be used directly to adjust the positions of the slices.

Example 2.43 Function `vol_visual4d()` can be used to process the data generated in Example 2.42. The volume visualization with default slices is shown in Figure 2.27. The user

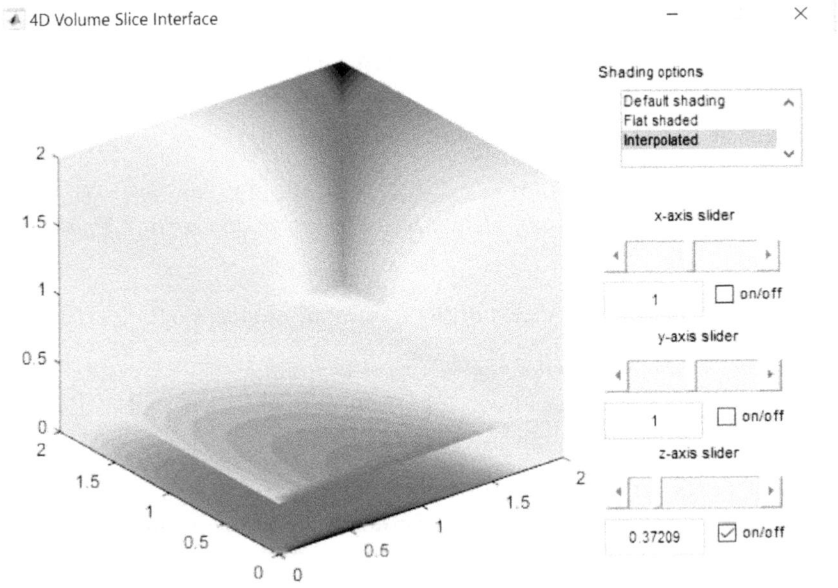

FIGURE 2.27: Default slices in volume visualization.

can now adjust the positions of the slices with mouse dragging actions. The check boxes can be turned on or off to control the slices.

```
>> [x,y,z]=meshgrid(0:0.1:2); V=sqrt(x.^x+y.^((x+y)/2)+z.^((x+y+z)/3));
   vol_visual4d(x,y,z,V); % volume visualization with GUI
```

Exercises

Exercise 2.1 *In MATLAB environment, the following statements can be given*

tic, $\boldsymbol{A}$=rand(500); $\boldsymbol{B}$=inv($\boldsymbol{A}$); norm($\boldsymbol{A}$*$\boldsymbol{B}$-eye(500)), toc

Run the statements and observe results. If you are not sure of the commands, just use the on-line help *facilities to display information on the related functions. Then, explain in detail the statement and the results.*

Exercise 2.2 *Suppose that a polynomial can be expressed by* $f(x) = x^5 + 3x^4 + 4x^3 + 2x^2 + 3x + 6$. *If one wants to substitute* x *by* $(s-1)/(s+1)$, *the function* $f(x)$ *can be changed into a function of* s. *Use the Symbolic Math Toolbox to do the substitution and get the simplest result.*

Exercise 2.3 *Please simplify* $\sin(k\pi + \pi/6)$, *for any integer* k.

Exercise 2.4 *Input the matrices* $\boldsymbol{A}$ *and* $\boldsymbol{B}$ *into MATLAB workspace where*

$$A = \begin{bmatrix} 1 & 2 & 3 & 4 \\ 4 & 3 & 2 & 1 \\ 2 & 3 & 4 & 1 \\ 3 & 2 & 4 & 1 \end{bmatrix}, \quad B = \begin{bmatrix} 1+\mathrm{j}4 & 2+\mathrm{j}3 & 3+\mathrm{j}2 & 4+\mathrm{j}1 \\ 4+\mathrm{j}1 & 3+\mathrm{j}2 & 2+\mathrm{j}3 & 1+\mathrm{j}4 \\ 2+\mathrm{j}3 & 3+\mathrm{j}2 & 4+\mathrm{j}1 & 1+\mathrm{j}4 \\ 3+\mathrm{j}2 & 2+\mathrm{j}3 & 4+\mathrm{j}1 & 1+\mathrm{j}4 \end{bmatrix}.$$

It is seen that $\boldsymbol{A}$ *is a* 4×4 *matrix. If a command* $\boldsymbol{A}(5,6)=5$ *is given, what will happen?*

Exercise 2.5 *Command* $\boldsymbol{A}$=`rand(3,4,5,6,7,8,9,10,11)` *can be used to generate a multi-dimensional array. How many elements are there in the array? Please find the sum of all its elements.*

Exercise 2.6 *Find the first 200 digits of the irrational numbers* $\sqrt{2}$, $\sqrt[6]{11}$, $\sin 1°$, e^2, $\ln 21$.

Exercise 2.7 *Please show the following identical equations*

$$(i)\ \mathrm{e}^{\mathrm{j}\pi}+1=0, \quad (ii)\ \frac{1-2\sin\alpha\cos\alpha}{\cos^2\alpha-\sin^2\alpha}=\frac{1-\tan\alpha}{1+\tan\alpha}.$$

Exercise 2.8 *If* $f(x)=x^2-x-1$, *please find* $f(f(f(f(f(f(f(f(f(f(x))))))))))$, *and also express the result in a polynomial. What is the degree of the polynomial?*

Exercise 2.9 *If the mathematical functions are known*

$$f(x)=\frac{x\sin x}{\sqrt{x^2+2(x+5)}},\ g(x)=\tan x,$$

please find out the functions $f(g(x))$ *and* $g(f(x))$.

Exercise 2.10 *Since double-precision scheme is quite limited in representing accurately large numbers, symbolic calculations are often used to find factorials of large numbers. Please use numerical and symbolic methods to calculate* C_{50}^{10}, *where,* $\mathrm{C}_m^n=m!/[n!(m-n)!]$. *Alternatively function* `nchoosek(sym(`m`),`n`)` *can be used.*

Exercise 2.11 *For a matrix* $\boldsymbol{A}$, *if one wants to extract all the even rows to form matrix* $\boldsymbol{B}$, *what command should be used? Suppose that matrix* $\boldsymbol{A}$ *is defined by* $\boldsymbol{A}$=`magic(8)`, *establish matrix* $\boldsymbol{B}$ *with suitable statements and see whether the results are correct or not.*

Exercise 2.12 *Please list all the positive integers such that it is a multiple of 11 and does not exceed 1000. Please also find all the integers which are multiples of 11 in the* $[3000,5000]$ *interval.*

Exercise 2.13 *How many prime numbers are there in the interval* $[1,1000000]$*? Please find the product of all the prime numbers in the interval. What is the number and how many digits are there? Measure the time elapsed in the evaluation.*

Exercise 2.14 *It is known in Example 2.12 that* `gcd()` *and* `lcm()` *functions can be used to find the greatest common divisor and least common multiple of two entities only. Please write functions* `gcds()` *and* `lcms()` *such that an arbitrary number of entities can be processed.*

Exercise 2.15 *Implement the following piecewise function where* $\boldsymbol{x}$ *can be given by scalar, vectors, matrices or even other multi-dimensional arrays, the returned argument* $\boldsymbol{y}$ *should be the same size as that of* $\boldsymbol{x}$. *The parameters* h *and* D *are scalars.*

$$\boldsymbol{y}=f(\boldsymbol{x})=\begin{cases} h, & \boldsymbol{x}>D \\ h/D\boldsymbol{x}, & |\,\boldsymbol{x}\,|\leqslant D \\ -h, & \boldsymbol{x}<-D. \end{cases}$$

Exercise 2.16 *A recursive formula is given by* $x_{n+1} = \dfrac{x_n}{2} + \dfrac{3}{2x_n}$, *with* $x_1 = 1$. *Please find a suitable number* n, *such that the terms after* x_n *approaches a certain constant. The accuracy requirement is* 10^{-14}. *Please find also the constant.*

Exercise 2.17 *Please calculate* $S = \displaystyle\prod_{n=1}^{\infty}\left(1+\frac{2}{n^2}\right)$, *under the precision requirement of* $\epsilon = 10^{-12}$.

Exercise 2.18 *It is known that*

$$\arctan(x) = x - \frac{x^3}{3} + \frac{x^5}{5} - \frac{x^7}{7} + \cdots.$$

Let $x = 1$, *the following formula can be derived*

$$\pi \approx 4\left(1 - \frac{1}{3} + \frac{1}{5} - \frac{1}{7} + \frac{1}{9} - \frac{1}{11} + \cdots\right).$$

Please calculate the approximate value of π, *with thc precision requirement* 10^{-6}.

Exercise 2.19 *Generate a* 100×100 *magic matrix, and find out the entities greater than 100, and substitute the entities by 0.*

Exercise 2.20 *Evaluate using numerical method the sum*

$$S = 1 + 2 + 4 + \cdots + 2^{62} + 2^{63} = \sum_{i=0}^{63} 2^i,$$

the use of vectorized form is suggested. Check whether accurate solutions can be found and why. Find the accurate sum using the symbolic computation methods. What happened in numerical and analytical solutions, if the number of terms is increased to 640.

Exercise 2.21 *Please use two algorithms to solve the equation* $f(x) = x^2\sin(0.1x+2)-3 = 0$.

(i) **Bisection method**. *If in an interval* (a, b), $f(a)f(b) < 0$, *there will be at least one solution. Take the middle point* $x_1 = (b-a)/2$, *and based on the relationship of* $f(x_1)$ *and* $f(a)$, $f(b)$, *determine in which half interval there exists solutions. Middle point in the new half interval can then be taken. Repeat the process until the size of the interval is smaller than the pre-specified error tolerance* ϵ. *Find the solution with bisection method in interval* $(-4, 0)$, *with* $\epsilon = 10^{-10}$.

(ii) **Newton–Raphson method**. *Select an initial guess of* x_n, *the next approximation can be obtained with* $x_{n+1} = x_n - f(x_n)/f'(x_n)$. *If the two points are close enough, i.e.,* $|x_{n+1} - x_n| < \epsilon$, *where* ϵ *is the error tolerance. Find the solution with* $x_0 = -4$, *and* $\epsilon = 10^{-12}$.

Exercise 2.22 *Write an M-function* `mat_add()`, *with the* $\boldsymbol{A}$=`mat_add(`$\boldsymbol{A}_1$,$\boldsymbol{A}_2$,$\boldsymbol{A}_3$,$\cdots$`)` *syntax. It is required that an arbitrary number of input arguments* $\boldsymbol{A}_i$ *are allowed.*

Exercise 2.23 *A MATLAB function can be written whose syntax is*

$\boldsymbol{v}$= `[`$h_1, h_2, h_m, h_{m+1}, \cdots, h_{2m-1}$`]` *and* $\boldsymbol{H}$=`myhankel(`$\boldsymbol{v}$`)`

where the vector $\boldsymbol{v}$ *is defined, and out of it, the output argument should be an* $m \times m$ *Hankel matrix.*

Exercise 2.24 *From matrix theory, it is known that if a matrix* $\boldsymbol{M}$ *is expressed as* $\boldsymbol{M} = \boldsymbol{A}+\boldsymbol{B}\boldsymbol{C}\boldsymbol{B}^{\rm T}$, *where* $\boldsymbol{A}$, $\boldsymbol{B}$ *and* $\boldsymbol{C}$ *are the matrices of relevant sizes, the inverse of* $\boldsymbol{M}$ *can be calculated by the following algorithm*

$$\boldsymbol{M}^{-1} = \left(\boldsymbol{A}+\boldsymbol{B}\boldsymbol{C}\boldsymbol{B}^{\rm T}\right)^{-1} = \boldsymbol{A}^{-1} - \boldsymbol{A}^{-1}\boldsymbol{B}\left(\boldsymbol{C}^{-1}+\boldsymbol{B}^{\rm T}\boldsymbol{A}^{-1}\boldsymbol{B}\right)^{-1}\boldsymbol{B}^{\rm T}\boldsymbol{A}^{-1}$$

The matrix inversion can be carried out using the formula easily. Suppose that there is a 5×5 *matrix* $\boldsymbol{M}$, *from which the three other matrices can be found.*

$$\boldsymbol{M} = \begin{bmatrix} -1 & -1 & -1 & 1 & 0 \\ -2 & 0 & 0 & -1 & 0 \\ -6 & -4 & -1 & -1 & -2 \\ -1 & -1 & 0 & 2 & 0 \\ -4 & -3 & -3 & -1 & 3 \end{bmatrix}, \quad \boldsymbol{A} = \begin{bmatrix} 1 & 0 & 0 & 0 & 0 \\ 0 & 3 & 0 & 0 & 0 \\ 0 & 0 & 4 & 0 & 0 \\ 0 & 0 & 0 & 2 & 0 \\ 0 & 0 & 0 & 0 & 4 \end{bmatrix},$$

$$\boldsymbol{B} = \begin{bmatrix} 0 & 1 & 1 & 1 & 1 \\ 0 & 2 & 1 & 0 & 1 \\ 1 & 1 & 1 & 2 & 1 \\ 0 & 1 & 0 & 0 & 1 \\ 1 & 1 & 1 & 1 & 1 \end{bmatrix}, \quad \boldsymbol{C} = \begin{bmatrix} 1 & -1 & 1 & -1 & -1 \\ 1 & -1 & 0 & 0 & -1 \\ 0 & 0 & 0 & 0 & 1 \\ 1 & 0 & -1 & -1 & 0 \\ 0 & 1 & -1 & 0 & 1 \end{bmatrix}.$$

Write the statement to evaluate the inverse matrix. Check the accuracy of the inversion. Compare the accuracy of the inversion method and the direct inversion method with `inv()` *function.*

Exercise 2.25 *Please generate the first 300 terms of the extended Fibonacci sequence* $T(n) = T(n-1)+T(n-2)+T(n-3), n = 4, 5, \cdots$, *with* $T(1) = T(2) = T(3) = 1$.

Exercise 2.26 *Consider the following iterative model*

$$\begin{cases} x_{k+1} = 1 + y_k - 1.4x_k^2 \\ y_{k+1} = 0.3x_k, \end{cases}$$

with initial conditions $x_0 = 0$, $y_0 = 0$. *Write an M-function to evaluate the sequence* x_i, y_i. *30000 points can be obtained by the function to construct the* $\boldsymbol{x}$ *and* $\boldsymbol{y}$ *vectors. The points can be expressed by a dot, rather than lines. In this case, the so-called* Hénon attractor *can be drawn.*

Exercise 2.27 *The well-known* Mittag–Leffler function *is defined as*

$$f_\alpha(z) = \sum_{k=0}^{\infty} \frac{z^k}{\Gamma(\alpha k+1)},$$

where $\Gamma(x)$ *is a Gamma function which can be evaluated with* `gamma(`x`)`. *Write an M-function with syntax* $\boldsymbol{f}$=`mymittag(`α,$\boldsymbol{z}$,ϵ`)`, *where* ϵ *is the error tolerance, with default value of* $\epsilon = 10^{-6}$. *Argument* $\boldsymbol{z}$ *is a numeric vector. Draw the curves for Mittag–Leffler functions with* $\alpha = 1$ *and* $\alpha = 0.5$.

Exercise 2.28 Chebyshev polynomials *are mathematically defined as*

$$T_1(x) = 1,\ T_2(x) = x,\ T_n(x) = 2xT_{n-1}(x) - T_{n-2}(x),\ n = 3, 4, 5, \cdots.$$

Please write a recursive function to generate a Chebyshev polynomial, and compute $T_{10}(x)$. *Please write a more efficient function as well to generate Chebyshev polynomials, and find* $T_{30}(x)$.

Exercise 2.29 *A regular triangle can be drawn by MATLAB statements easily. Use the loop structure to design an M-function that, in the same coordinates, a sequence of regular triangles can be drawn, each by rotating a small angle, for instance, 5°, from the previous one.*

Exercise 2.30 *Select suitable step-sizes and draw the function curve for* $\sin(1/t)$, $t \in (-1, 1)$.

Exercise 2.31 *For suitably assigned ranges of θ, draw polar plots for the following functions.*

(i) $\rho = 1.0013\theta^2$, *(ii)* $\rho = \cos(7\theta/2)$, *(iii)* $\rho = \sin(\theta)/\theta$, *(iv)* $\rho = 1 - \cos^3(7\theta)$.

Exercise 2.32 *Please draw the curves of $x \sin x + y \sin y = 0$ for $-50 \leqslant x, y \leqslant 50$.*

Exercise 2.33 *Find the solutions to the following simultaneous equations using graphical methods and verify the solutions.*

$$(i) \begin{cases} x^2 + y^2 = 3xy^2 \\ x^3 - x^2 = y^2 - y, \end{cases} \qquad (ii) \begin{cases} \mathrm{e}^{-(x+y)^2+\pi/2} \sin(5x + 2y) = 0 \\ (x^2 - y^2 + xy)\mathrm{e}^{-x^2-y^2-xy} = 0. \end{cases}$$

Exercise 2.34 *Please save a 100×100 magic matrix into an Excel file.*

Exercise 2.35 *Assume that the power series expansion of a function is*

$$f(x) = \lim_{N\to\infty} \sum_{n=1}^{N} (-1)^n \frac{x^{2n}}{(2n)!}.$$

If N is large enough, power series $f(x)$ converges to a certain function $\hat{f}(x)$. Please write a MATLAB program that plots the function $\hat{f}(x)$ in the interval $x \in (0, \pi)$. Observe and verify what function $\hat{f}(x)$ is.

Exercise 2.36 *Draw the 3D surface plots for the functions xy and $\sin xy$, respectively. Also, draw the contours of the functions. View the 3D surface plot from different angles, especially with orthographic views.*

Exercise 2.37 *Please draw 2D and 3D Lissajous figures under the parametric equations $x = \sin t$, $y = \sin at$ and $z = \sin bt$, for different parameters a and b, where, please try the following rational and irrational parameters and see what may happen.*

(i) $a = 1/2$, $b = 1/3$, *(ii)* $a = \sqrt[8]{2}$, $b = \sqrt{3}$.

Exercise 2.38 *For the parametric equations* [7], *please draw the surfaces*

(i) $x = 2\sin^2 u \cos^2 v, y = 2 \sin u \sin^2 v, z = 2 \cos u \sin^2 v, -\pi/2 \leqslant u, v \leqslant \pi/2$,

(ii) $x = u - \dfrac{u^3}{3} + uv^2, \ y = v - \dfrac{v^3}{3} + vu^2, \ z = u^2 - v^2, \ -2 \leqslant u, v \leqslant 2.$

Exercise 2.39 *A vertical cylinder can be described by parametric equation $x = r \sin u, y = r \cos u, z = v$, along z axis, with r the radius. If x and z are swopped, the cylinder can be represented along x axis. Please draw several cylinders with different radii and directions together in the same coordinate.*

Exercise 2.40 *Please draw a cone, whose top is at (0,0,2), and bottom at the plane $z = 0$, with a radius of 1.*

Exercise 2.41 *In graphics command, there is a trick in hiding certain parts of the plot. If the function values are assigned to* `NaN`*s, the point on the curve or the surface will not be shown. Draw first the surface plot of the function* $z = \sin xy$*. Then, cut off the region that satisfies* $x^2 + y^2 \leqslant 0.5^2$.

Exercise 2.42 *Please draw the 3D surface of* $f(x,y) = \dfrac{\sin\sqrt{x^2+y^2}}{\sqrt{x^2+y^2}}$ *for* $-8 \leqslant x, y \leqslant 8$.

Exercise 2.43 *Lambert W function is a commonly used special function, its mathematical form is* $W(z)\mathrm{e}^{W(z)} = z$*. Please draw the curve of the function.*

Exercise 2.44 *Draw the surface plot and contour plots for the following functions. Draw also with the functions* `surfc()`, `surfl()` *and* `waterfall()`*, and observe the results.*

$$(i)\;\; z = xy,\;\; (ii)\;\; z = \sin x^2 y^3,\;\; (iii)\;\; z = \frac{(x-1)^2 y^2}{(x-1)^2 + y^2},\;\; (iv)\;\; z = -xy\,\mathrm{e}^{-2(x^2+y^2)}.$$

Exercise 2.45 *Please draw the surface of the three-dimensional implicit function*

$$(x^2 + xy + xz)\mathrm{e}^{-z} + z^2 yx + \sin(x + y + z^2) = 0.$$

Exercise 2.46 *Please draw the two following two surfaces and observe the intersections*

$$x^2 + y^2 + z^2 = 64,\;\; y + z = 0.$$

Exercise 2.47 *Please draw the sliced volume visualization for the following functions*

$$(i)\; V(x,y,z) = \sqrt{\mathrm{e}^x + \mathrm{e}^{(x+y)-xy} + \mathrm{e}^{(x+y+z)/3-xyz}},\;\; (ii)\; V(x,y,z) = \mathrm{e}^{-x^2-y^2-z^2}.$$

Bibliography

[1] Gilbert D. Extended plotyy to three y-axes. MATLAB Central File ID: # 1017, 2001

[2] Bodin P. PLOTY4 support for four y axes. MATLAB Central File ID: # 4425, 2004

[3] Gilbert D. PLOTXX create graphs with two x axes. MATLAB Central File ID: # 317, 1999

[4] Lamport L. LaTeX: a document preparation system — user's guide and reference manual. Reading MA: Addision-Wesley Publishing Company, second edition, 1994

[5] Xue D. Analysis and computer aided design of nonlinear systems with Gaussian inputs. Ph.D. thesis, Sussex University, U.K., 1992

[6] Morales G. Ezimplot3: implicit 3D functions plotter. MATLAB Central File ID #23623

[7] Majewski M. MuPAD Pro computing essentials. Berlin: Springer, 2002

Chapter 3

Calculus Problems

The calculus established by Isaac Newton and Gottfried Wilhelm Leibniz is fundamental to many branches of sciences and engineering. In traditional calculus courses, limits, differentiations, integrals, series expansions such as Taylor series and Fourier series expansions for univariate and multivariate functions are the main topics discussed. The analytical solutions to these problems can be obtained by the direct use of the corresponding functions provided by the Symbolic Math Toolbox of MATLAB which will be discussed in Sections 3.1 to 3.3. The Taylor series expansions for univariate and multivariate functions as well as the Fourier series expansions are discussed in Section 3.4, moreover, the fitting quality and interval are assessed graphically, if finite-term series expansions are used. In Section 3.5, the sum and product of sequences problems are discussed. Convergency test for infinite series and convergent internals are explored. Sections 3.6 and 3.7 present methods for path integrals and surface integrals. Most of the materials presented in this chapter are symbolic-based, which cannot be solved using conventional computer programming languages such as C for average users. Computer mathematics languages such as MATLAB should be used instead.

In college mathematics courses, to find the limit or integral of a certain function required good experiences, and the experiences can only be acquired by doing a tremendous amount of exercises. In these sections, a unified three-phase solution pattern is introduced: (i) declare symbolic variables, (ii) express the functions as symbolic expressions, and (iii) use appropriate MATLAB function to find the answer. Equipped with the powerful computer tools, the readers are able to solve the calculus problems easily, without the need of mastering the skills in calculus problem solutions.

In many scientific and engineering researches, the analytical solutions to calculus problems may face difficulties, when the original functions are not given explicitly. For problems with measured data, numerical differentiations and integrals should be applied accordingly. They are illustrated in Sections 3.8 and 3.9, respectively. Alternative solutions to the same numerical calculus problems using spline interpolation will be given in Chapter 8. Solutions to differential equation problems will be presented in Chapter 7. As an extension to the traditional (integer-order) calculus, non-integer-order or fractional-order calculus, will be discussed in Chapter 10.

For readers who wish to check the detailed explanations of calculus, we recommend the free textbooks [1,2].

3.1 Analytical Solutions to Limit Problems

The Symbolic Math Toolbox of MATLAB can be used directly in solving the limit problems, the differentiation problems, and the integral problems. Using the methods presented in this and the following two sections, the readers will be equipped with the ability in solving ordinary calculus problems directly by computers.

3.1.1 Limits of univariate functions

Assume that the function to be analyzed is $f(x)$, the limit is denoted as

$$L = \lim_{x \to x_0} f(x), \tag{3-1-1}$$

meaning the value of the function when the independent variable x infinitely approaches to x_0, where x_0 can be either a given constant or infinity. For certain functions, the left or right limit can be denoted as

$$L_1 = \lim_{x \to x_0^-} f(x), \quad \text{or} \quad L_2 = \lim_{x \to x_0^+} f(x), \tag{3-1-2}$$

where the former means to approach the point x_0 from the left-hand side which is referred to as the *left limit* problem. The latter is referred to as the *right limit* problem. The limit problems summarized above can be solved by the use of the `limit()` function, where

```
L=limit(fun,x,x0)                      % calculate the limit
L=limit(fun,x,x0,'left' or 'right')    % the one-sided limit
```

To use the functions in Symbolic Math Toolbox, symbolic variables such as x should be declared first. Then, the limit function *fun* can be expressed. If x_0 is ∞, one can assign it to `inf`. If the one-sided limit is required, the `'left'` or `'right'` option should be specified. The following examples are used to demonstrate the use of the `limit()` function in MATLAB.

If there is only one symbolic variable used in first syntax, it can be omitted in the function call. The list of actual symbolic variables used in an expression f can be extracted with `symvar()` function, with `list=symvar(f)`.

Example 3.1 Find the limit $\displaystyle\lim_{x \to 0} \frac{\sin x}{x}$.

Solution *Everyone who has essential knowledge on calculus knows that the limit is 1. In order to let the computer solve the problem, the following three-step procedures should be used: (i) declare symbolic variables, (ii) write the symbolic expression of function f, and (iii) call* `limit()` *function to get the result. The three steps can be implemented in MATLAB with the following three statements*

```
>> syms x; f=sin(x)/x; limit(f,x,0) % the three-step procedures
```

Since in the second statement, x is a symbolic variable rather than a vector or a matrix, there is no need to describe the function with dot operations. The symbolic variables in expression f can be extracted with the `symvar()` *function, and it can be seen that x is the only symbolic variable in the expression, therefore, the argument x can be omitted in the function calls*

```
>> v=symvar(f), L=limit(f,0) % find the independent variable
```

Example 3.2 Solve the limit problem $\displaystyle\lim_{x \to \infty} x\left(1+\frac{a}{x}\right)^x \sin\frac{b}{x}$.

Solution *Similar to the example above, the three steps can also be used, and the obtained result is $L = \mathrm{e}^a b$. It can be seen that, for the user, the solution process of the complicated limit problem is not more difficult than the previous example.*

```
>> syms x a b; f=x*(1+a/x)^x*sin(b/x); L=limit(f,x,inf)
```

If $\boldsymbol{v}$ = `symvar(`*f*`)` *is used, the variable vector $\boldsymbol{v}$ returned is $[a,b,x]$. Therefore, x cannot be omitted in the function call.*

Example 3.3 Solve the one-sided limit problem $\displaystyle\lim_{x\to 0^+}\frac{\mathrm{e}^{x^3}-1}{1-\cos\sqrt{x-\sin x}}$.

Solution *With the* `limit()` *function, the one-sided limit can easily be solved, with the limit of 12.*

```
>> syms x; f=(exp(x^3)-1)/(1-cos(sqrt(x-sin(x))));
   L=limit(f,x,0,'right') % compute the right limit directly
```

One can further verify the above problem graphically over a proper range of interest. For instance, if the interval $(-0.01,0.01)$ is considered, the function over the interval can be drawn in Figure 3.1. It can be seen that the point at $x=0$ was deliberately excluded to avoid 0/0 computation, which yields `NaN`.

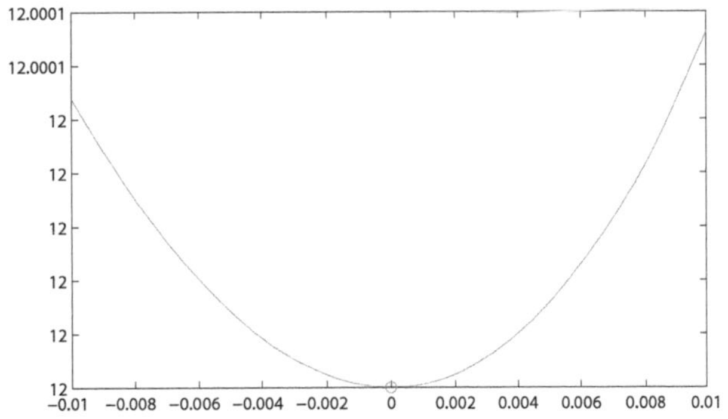

FIGURE 3.1: The curve of the function around $x=0$.

```
>> x0=-0.01:0.0005:0.01; x0=x0(x0~=0); % exclude 0 from the vector
   plot(x0,subs(f,x,x0),00,12,'o')     % draw the curve
   L2=limit(f,x,0), L3=limit(f,x,0,'left')
```

It can be seen that the limit of the original problem is also 12, no matter if approaching it from left or from right hand side.

Consider again the original problem. The aim of the original one-sided limit requirement ensures that the expression under the square root sign is nonnegative. In fact, for imaginary variables, one can still find from the Euler's formula that $\cos \mathrm{j}\alpha=(\mathrm{e}^{\alpha}+\mathrm{e}^{-\alpha})/2$. Thus, the one-sided limits for the function are the same for this example, which further verifies that the original function is continuous around $x=0$ as also seen from Figure 3.1.

For a certain point $x=a$, if the left and right limits of function $f(x)$ are equal, a is referred to as the *first-type discontinuity*, otherwise, it is referred to as the *second-type discontinuity*, as will be demonstrated by the following simple example.

Example 3.4 Please find the left and right limit of function $\tan t$ at point $t\to\pi/2$.

Solution *The two one-sided limits can be obtained easily, and the results are respectively $L_1=\infty$ and $L_2=-\infty$.*

```
>> syms t; f=tan(t); % declare symbolic variable and function
   L1=limit(f,t,pi/2,'left'), L2=limit(f,t,pi/2,'right') % compute limits
```

Example 3.5 Please find the limit of a sequence $\displaystyle\lim_{n\to\infty}\frac{\sqrt[3]{n^2}\sin n!}{n+1}$.

Solution *The procedures of finding the limit of a sequence are exactly the same as the limit of a function. The symbolic variables must be declared first, then, the sequence should be expressed as a symbolic expression. Finally the function* `limit()` *must be called to get the results. For this example, the following commands should be issued, and the result is 0.*

```
>> syms n; f=n^(2/3)*sin(factorial(n))/(n+1); F=limit(f,n,inf)
```

Example 3.6 Please find the limit of

$$\lim_{n\to\infty} n\arctan\left(\frac{1}{n(x^2+1)+x}\right)\tan^n\left(\frac{\pi}{4}+\frac{x}{2n}\right).$$

Solution *The expression is a function of x, and it is also a sequence. However, this does not cause extra difficulties when MATLAB is used. The following statements can be issued to find the limit $e^x/(x^2+1)$.*

```
>> syms x n;  % declare x and n as symbolic variables
   f=n*atan(1/(n*(x^2+1)+x))*tan(pi/4+x/2/n)^n; limit(f,n,inf)
```

Even more accurately, the variable n should be declared as an integer, and the first sentence should be rewritten as follows, and the result is the same.

```
>> syms x n; assume(n,'integer');  % declare n as an integer
```

3.1.2 Limits of interval functions

An example is used in this section, before the concept of interval function is introduced.

Example 3.7 Please find the limits $\displaystyle\lim_{n\to\infty} x^n$ and $\displaystyle\lim_{x\to\infty} x^n$.

Solution *The limits cannot be easily found with old versions of MATLAB. In the new versions, with MuPAD symbolic engine, piecewise functions are supported, thus, the limits can be found directly with the following statements*

```
>> syms x n real; f=x^n; L1=limit(f,n,inf), L2=limit(f,x,inf)
```

The results are piecewise functions, with the description of L_2 as follows

`piecewise([`$n==0$`,1],[`$0<n$`,Inf],[`$n<0$`,0])`

the two limits can be expressed mathematically as

$$L_1=\begin{cases}1, & x=1\\ \infty, & x>1\\ \textit{no limit}, & x<-1\\ 0, & 0<x<1 \textit{ or } -1<x<0,\end{cases}\qquad L_2=\begin{cases}1, & n=0\\ \infty, & n>0\\ 0, & n<0.\end{cases}$$

It seems that something imprecise in the last condition in L_1, where the point $x=0$ should be included, such that the condition should be $-1<x<1$.

For certain functions such as $\sin x$, the limit does not exist, when $x \to \infty$. With the low-level support of MuPAD, which is the actual symbolic engine in the new versions of the Symbolic Math Toolbox, the limit can be found, such that there are interval limits.

L = `feval(symengine,'limit',`f`,'x=infinity','Intervals')`

where, `feval()` function can be used to call directly the low-level function `limit()` under MuPAD, and the arguments are also passed to MuPAD.

Example 3.8 Assume that $a, b > 0$, please find the limit

$$\lim_{x\to\infty} a\sin 8x^2 + b\cos(2x-2).$$

Solution *With the low-level MuPAD command, it can be found that the interval limit is* $(-a-b, a+b)$.

```
>> syms a b positive, syms x; f=a*sin(8*x^2)+b*cos(2*x-2);
   L=feval(symengine,'limit',f,'x=infinity','Intervals')
```

Piecewise functions are supported in MuPAD, the new symbolic engine of MATLAB Symbolic Math Toolbox. Low-level MuPAD commands must be issued, and may not be convenient for ordinary MATLAB users. Thus, an interface function `piecewise()` is written to describe piecewise functions

```
function f=piecewise(varargin), str=[];
try
   for i=1:2:length(varargin),
      str=[str,'[',varargin{i},',',varargin{i+1},'],'];
   end
catch, error('Input arguments should be given in pairs.'), end
f=feval(symengine,'piecewise',str(1:end-1));
```

The syntax of the function is f = `piecewise(`*var1*`,`*var2*`,···)`, where the input arguments *var* must be provided in pairs, with the first describing the condition, and the last for the symbolic expression. They should both be declared as strings. In the condition string, the logic keywords such as `and`, `or` and `not` can be used.

The `try-catch` structure is used in the interface function to insure the arguments appear in pairs, otherwise, an error message will be displayed. Since there will be an extra comma by the end of the final string, the command `1:end-1` is used to discard it.

Example 3.9 Consider the saturation function in Example 2.26

$$y = \begin{cases} 1.1\,\mathrm{sign}(x), & |x| > 1.1 \\ x, & |x| \leqslant 1.1. \end{cases}$$

Describe it in piecewise function, and draw the curve.

Solution *With the use of the* `piecewise()` *interface, the saturation function can be expressed as a symbolic expression. Then, the curves can be drawn directly, and it is exactly the same as the one obtained in Example 2.26. It should be noted that, due to the limitations in symbolic calculation, the thus defined piecewise function cannot be drawn directly with* `ezplot()` *function.*

```
>> f=piecewise('abs(x)>1.1','1.1*sign(x)','abs(x)<=1.1','x');
   syms x; x0=-3:0.01:3; f1=subs(f,x,x0); plot(x0,f1)
```

The condition $|x| \leqslant 1.1$ can also be described mathematically as $-1.1 \leqslant x \leqslant 1.1$ and can also be understood as $x \geqslant -1.1$ and $x \leqslant 1.1$. Thus, the corresponding string can also be expressed as 'x>=-1.1 and x<=1.1'.

3.1.3 Limits of multivariate functions

Two kinds of limits are often encountered in multivariate functions, one is *sequential limit*, and the other is *multiple limit*. Assume for a function $f(x,y)$ with two independent variables, the sequential limits are defined as

$$L_1 = \lim_{x\to x_0}\left[\lim_{y\to y_0} f(x,y)\right], \quad \text{or} \quad L_2 = \lim_{y\to y_0}\left[\lim_{x\to x_0} f(x,y)\right], \tag{3-1-3}$$

where, x_0 and y_0 can either be values or functions. The sequential limits can also be evaluated with the nested calls to the `limit()` function

L_1=`limit(limit(`f,x,x_0`),`y,y_0`)`, or L_1=`limit(limit(`f,y,y_0`),`x,x_0`)`

Example 3.10 Please find the sequential limit

$$\lim_{y\to\infty}\left[\lim_{x\to 1/\sqrt{y}} \mathrm{e}^{-1/(y^2+x^2)}\frac{\sin^2 x}{x^2}\left(1+\frac{1}{y^2}\right)^{x+a^2y^2}\right].$$

Solution *Since $\sqrt{y}$ is involved, y should be declared as a positive symbolic variable (no need to do so in earlier versions). Therefore, the limit problem can be solved with the following statements, and the result is e^{a^2}.*

```
>> syms x a; syms y positive; % define y > 0. Not necessary in old versions
   f=exp(-1/(y^2+x^2))*sin(x)^2/x^2*(1+1/y^2)^(x+a^2*y^2);
   L=limit(limit(f,x,1/sqrt(y)),y,inf) % compute limit
```

Apart from sequential limits, there is also a multiple limit for multivariate functions defined as

$$L = \lim_{\substack{x\to x_0\\ y\to y_0}} f(x,y). \tag{3-1-4}$$

Normally speaking, if the two sequential limits both exist and are equal, the multiple limit may be equal to that value. It should be pointed out that in some special cases, even though the two sequential limits are the same, the multiple limit does not exist. In that case, approaching from different directions should also be considered.

Example 3.11 Try to solve the double limit problem

$$\lim_{\substack{x\to\infty\\ y\to\infty}}\left(\frac{xy}{x^2+y^2}\right)^{x^2}.$$

Solution *The two sequential limits can be obtained with the following statements, and they both are equal to 0. Moreover, two other directions, $x\to y^2$ and $y\to x^2$ can also be evaluated, and the same limits can be found.*

```
>> syms x y; f=(x*y/(x^2+y^2))^(x^2);
   L1=limit(limit(f,x,inf),y,inf), L2=limit(limit(f,y,inf),x,inf)
   L3=limit(limit(f,x,y^2),y,inf), L4=limit(limit(f,y,x^2),x,inf)
```

Example 3.12 Please check the existence of the multiple limit

$$\lim_{\substack{x \to 0 \\ y \to 0}} \frac{xy}{x^2+y^2}.$$

Solution *To definitely compute the multiple limit is difficult, since all the approaching directions should be considered. To indicate the nonexistence is comparatively easier. For instance, suppose $y = rx$, with r a symbolic variable, if the limit of $x \to 0$ is r dependent, that is adequate to indicate the nonexistence of the multiple limit. Other functions of y can also be tested.*

```
>> syms r x y; f=x*y/(x^2+y^2); L=limit(subs(f,y,r*x),x,0)
```

and the limit is $L = r/(r^2+1)$, therefore, the multiple limit is nonexistence.

3.2 Analytical Solutions to Derivative Problems

3.2.1 Derivatives and high-order derivatives

For a function described by $y = f(x)$, the first-order derivative of $y(x)$ with respect to x is defined as

$$y'(x) = \frac{\mathrm{d}y(x)}{\mathrm{d}x} = \lim_{\Delta x \to 0} \frac{f(x+\Delta x) - f(x)}{\Delta x}. \tag{3-2-1}$$

The second-order derivative is the first-order derivative of $y'(x)$ with respect to x. The high-order derivatives are also defined accordingly.

If the function $f(x)$ is described as a symbolic expression *fun*, the function `diff()` can be used to calculate its derivatives, with the syntaxes

```
y=diff(fun,x)      % find the derivative
y=diff(fun,x,n)    % evaluate the nth order derivative
```

If there is only one variable in the function, the independent variable name can be omitted. If n is omitted, the first-order derivative can be found. The three-step procedure discussed earlier can also be used to find the derivatives of given functions.

Example 3.13 Compute $\dfrac{\mathrm{d}^4 f(x)}{\mathrm{d}x^4}$ for a given function $f(x) = \dfrac{\sin x}{x^2+4x+3}$.

Solution *This was the very first example given at the beginning of the book. The derivatives can easily be obtained with the following MATLAB functions. The variable x should be declared as a symbolic variable first, then, the function `diff()` can be called to find the first-order derivative.*

```
>> syms x; f=sin(x)/(x^2+4*x+3); f1=diff(f) % three-step procedure
```

and the result obtained is

$$f_1 = \frac{\cos x}{x^2+4x+3} - \frac{(2x+4)\sin x}{(x^2+4x+3)^2}.$$

In recent versions, the above commands can alternatively be rewritten as

```
>> syms x; f(x)=sin(x)/(x^2+4*x+3); f1(x)=diff(f) % f(x) allowed
```

The function `ezplot()` *can be used to draw the original function and its first order derivative, as shown in Figure 3.2.*

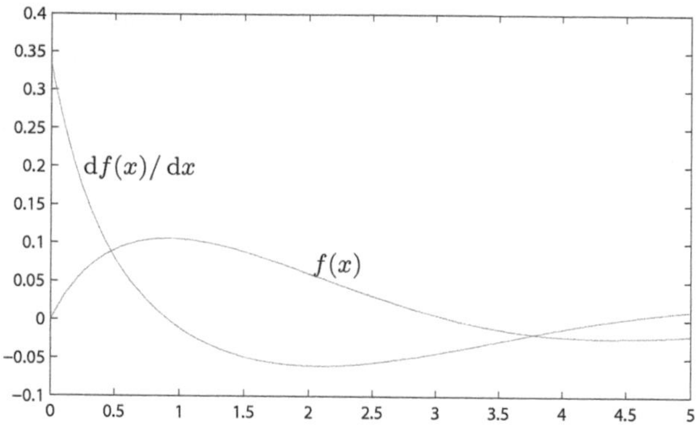

FIGURE 3.2: The curves of the original function and its derivative.

```
>> ezplot(f,[0,5]); hold on; ezplot(f1,[0,5]); % draw curves directly
```

The fourth-order derivative can be simply calculated from

```
>> f4=diff(f,x,4) % find the 4th order derivative
```

and the result obtained is

$$f_4 = \frac{\sin x}{x^2+4x+3} + \frac{12\sin x}{(x^2+4x+3)^2} + \frac{24\sin x}{(x^2+4x+3)^3} - \frac{24(2x+4)^3 \cos x}{(x^2+4x+3)^4}$$
$$-\frac{12(2x+4)^2 \sin x}{(x^2+4x+3)^3} - \frac{48(2x+4)^2 \sin x}{(x^2+4x+3)^4} + \frac{24(2x+4)^4 \sin x}{(x^2+4x+3)^5} + \frac{4\,(2x+4)\cos x}{(x^2+4x+3)^2}$$
$$+\frac{16\,(2x+4)\cos x}{(x^2+4x+3)^3} + \frac{8\,(8x+16)\cos x}{(x^2+4x+3)^3} - \frac{6\,(2x+4)\,(8x+16)\sin x}{(x^2+4x+3)^4}.$$

From the above simplified results, it is clear that the direct use of the function `simplify()` *is not sufficient for this example. For the given example, it can immediately be found that one may extract the terms* $\sin x$ *and* $\cos x$ *from the results and the coefficients for these terms can be simplified separately such that*

```
>> collect(simplify(f4),sin(x)), collect(simplify(f4),cos(x))
```

The even more concise results can be obtained shown as follows:

$$\frac{\mathrm{d}^4 f(x)}{\mathrm{d}x^4} = 8(x^5+10x^4+26x^3-4x^2-99x-102)\frac{\cos x}{(x^2+4x+3)^4}+$$
$$(x^8+16x^7+72x^6-32x^5-1094x^4-3120x^3-3120x^2+192x+1581)\frac{\sin x}{(x^2+4x+3)^5}.$$

The differentiation function `diff()` *can easily be used to find high-order derivatives. For instance, the 100th order derivative of the same function can be found within 15 seconds (in earlier versions, the results can be found within 1 second).*

```
>> tic, diff(f,x,100); toc % compute the 100th order derivative
```

Example 3.14 If the function is given by $F(t) = t^2 f(t)\sin t$, where $f(t)$ is another function, please derive the third order derivative formula. Also, if $f(t) = \mathrm{e}^{-t}$, please find the third order derivative of $F(t)$, and validate the result.

Solution *In recent versions, the command* `syms f(t)` *can be used to declare function $f(t)$. Therefore, the third order derivative of $F(t)$ can be found with*

```
>> syms t f(t); % define function f(t). In old versions, use f=sym('f(t)')
   F=f*t^2*sin(t); G=simplify(diff(F,t,3)) % evaluate 3rd order derivative
```

and the result is

$$\frac{\mathrm{d}^3F(t)}{\mathrm{d}t^3} = \left[\frac{\mathrm{d}^3f(t)}{\mathrm{d}t^3}\sin t + 3\frac{\mathrm{d}^2f(t)}{\mathrm{d}t^2}\cos t - 3\frac{\mathrm{d}f(t)}{\mathrm{d}t}\sin t - f(t)\cos t\right]t^2$$
$$+\left[6\frac{\mathrm{d}^2f(t)}{\mathrm{d}t^2}\sin t + 12\frac{\mathrm{d}f(t)}{\mathrm{d}t}\cos t - 6f(t)\sin t\right]t + 6\frac{\mathrm{d}f(t)}{\mathrm{d}t}\sin t + 6f(t)\cos t.$$

The following commands can be used to find the third order derivative of $F(t)$, by substituting $f(t)$ with e^{-t}, or by direct differentiation. It can be seen that the two results are exactly the same

$$y_1(t) = 2\mathrm{e}^{-t}\left(t^2\cos t + t^2\sin t - 6t\cos t + 3\cos t - 3\sin t\right).$$

```
>> y1=simplify(subs(G,f,exp(-t)))              % variable substitution
   simplify(diff(t^2*sin(t)*exp(-t),3)-y1) % validate the results
```

Please note that, in the last statement, `simplify()` *function is used to simplify the difference between the results of the two methods. It is a common way in MATLAB to show that the two results are identical, if the simplified difference is zero.*

Example 3.15 Assume a matrix function $\boldsymbol{H}(x)$ is given below, please find its third order derivative with respect to x.

$$\boldsymbol{H}(x) = \begin{bmatrix} 4\sin 5x & \mathrm{e}^{-4x^2} \\ 3x^2+4x+1 & \sqrt{4x^2+2} \end{bmatrix}.$$

Solution *Function* `diff()` *can also be used to find the high-order derivatives of a matrix function $\boldsymbol{H}(x)$. It takes derivatives to each matrix element $h_{i,j}(x)$ independently, in the same way as in dot operation. The new matrix function $\boldsymbol{N}(x)$ can be found*

```
>> syms x; H=[4*sin(5*x), exp(-4*x^2); 3*x^2+4*x+1, sqrt(4*x^2+2)]
   N=diff(H,x,3) % compute 3rd order derivative to each element individually
```

The resulting derivative matrix function is

$$\boldsymbol{N}(x) = \frac{\mathrm{d}}{\mathrm{d}x}H(x) = \begin{bmatrix} -500\cos 5x & 192x\,\mathrm{e}^{-4x^2} - 512x^3\,\mathrm{e}^{-4x^2} \\ 0 & \dfrac{24\sqrt{2}\,x^3}{(2x^2+1)^{5/2}} - \dfrac{12\sqrt{2}x}{(2x^2+1)^{3/2}} \end{bmatrix}.$$

3.2.2 Partial derivatives of multivariate functions

There is no direct function which can be used in finding the partial derivatives in MATLAB. The function `diff()` can actually be used instead. For instance, if a function

$f(x,y)$ with two variables is defined, the partial derivative $\partial^{m+n}f/(\partial x^m \partial y^n)$ can be evaluated by the nested use of the `diff()` function as follows:

```
f=diff(diff(fun,x,m),y,n), or f=diff(diff(fun,y,n),x,m)
```

Example 3.16 Find the partial derivatives of $z = f(x,y) = (x^2-2x)\mathrm{e}^{-x^2-y^2-xy}$ function and investigate the function further using graphical method.

Solution *The partial derivatives $\partial z/\partial x$ and $\partial z/\partial y$ can be evaluated easily using*

```
>> syms x y; z=(x^2-2*x)*exp(-x^2-y^2-x*y); % function expression
   zx=simplify(diff(z,x)), zy=simplify(diff(z,y)) % compute gradients
```

and the mathematical representations of the derivatives are

$$\frac{\partial z(x,y)}{\partial x} = -\mathrm{e}^{-x^2-y^2-xy}(-2x+2+2x^3+x^2y-4x^2-2xy)$$
$$\frac{\partial z(x,y)}{\partial y} = -x(x-2)(2y+x)\mathrm{e}^{-x^2-y^2-xy}.$$

Within the rectangular region where $x \in (-3,3), y \in (-2,2)$, mesh grids can be defined and the partial derivatives can be obtained numerically over the mesh grids. The three-dimensional surface of the original function is shown in Figure 3.3 (a).

```
>> [x0,y0]=meshgrid(-3:.2:3,-2:.2:2);  % generate mesh grid matrices
   z0=double(subs(z,{x,y},{x0,y0}));   % substituting the two variables
   surf(x0,y0,z0), axis([-3 3 -2 2 -0.7 1.5]) % three-dimensional surface
```

From the partial derivatives obtained, the numerical solutions at the mesh grids can be evaluated. The function `quiver()` *can then be used to draw attractive curves, and the curves can be superimposed over the contour of the original function with the following statements, as shown in Figure 3.3 (b).*

```
>> contour(x0,y0,z0,30), hold on    % contours of the function
   zx0=subs(zx,{x,y},{x0,y0}); zy0=subs(zy,{x,y},{x0,y0});
   quiver(x0,y0,-double(zx0),-double(zy0)) % draw the negative gradients
```

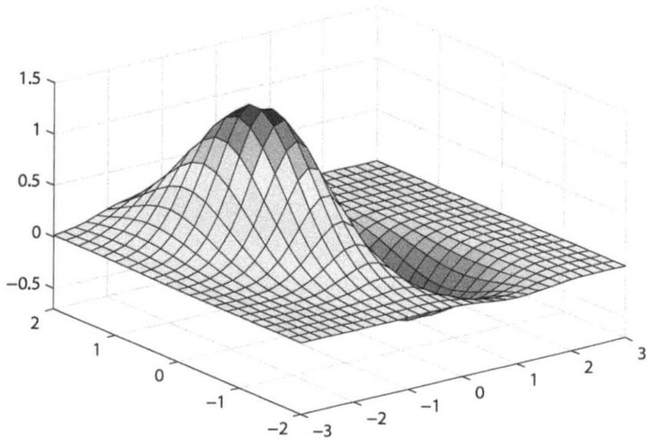

(a) three-dimensional surface

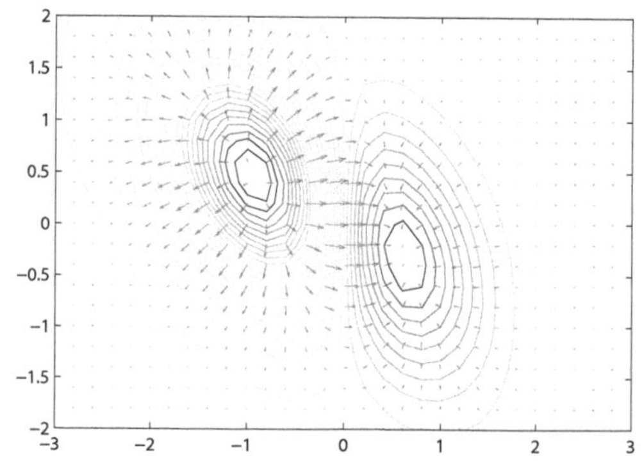

(b) contours with gradients

FIGURE 3.3: Graphical interpretation of the functions with two variables.

Example 3.17 For a given function with three independent variables x, y and z, $f(x,y,z)=\sin(x^2y)\mathrm{e}^{-x^2y-z^2}$, find the partial derivative $\partial^4 f(x,y,z)/(\partial x^2\partial y\partial z)$.

Solution *The following MATLAB statements can be given to solve this problem*

```
>> syms x y z; f=sin(x^2*y)*exp(-x^2*y-z^2);          % the original function
   df=diff(diff(diff(f,x,2),y),z); df=simplify(df) % compute directly
```

The results can be obtained as

$$-4z\mathrm{e}^{-x^2y-z^2}\left[\cos x^2y-10yx^2\cos x^2y+4x^4y^2\sin x^2y+4x^4y^2\cos x^2y-\sin x^2y\right].$$

3.2.3 Jacobian matrix of multivariate functions

Assume that there are n independent variables, and m functions defined as

$$\begin{cases} y_1=f_1(x_1,x_2,\cdots,x_n)\\ y_2=f_2(x_1,x_2,\cdots,x_n)\\ \quad\vdots \qquad \vdots\\ y_m=f_m(x_1,x_2,\cdots,x_n). \end{cases} \tag{3-2-2}$$

The partial derivative $\partial y_i/\partial x_j$ for each combination of i and j can be represented in the matrix form as

$$\boldsymbol{J}=\begin{bmatrix} \partial y_1/\partial x_1 & \partial y_1/\partial x_2 & \cdots & \partial y_1/\partial x_n\\ \partial y_2/\partial x_1 & \partial y_2/\partial x_2 & \cdots & \partial y_2/\partial x_n\\ \vdots & \vdots & \ddots & \vdots\\ \partial y_m/\partial x_1 & \partial y_m/\partial x_2 & \cdots & \partial y_m/\partial x_n \end{bmatrix} \tag{3-2-3}$$

and such a matrix is referred to as the *Jacobian matrix*. Jacobian matrices are quite useful in many research areas, such as robotics and image processing. Jacobian matrix can be obtained using the `jacobian()` function of the Symbolic Math Toolbox directly. The syntax of the function is

$\boldsymbol{J}=$`jacobian(`$[y_1,y_2,\cdots,y_m]$`,`$[x_1,x_2,\cdots,x_n]$`)`,

where $[x_1,x_2,\cdots,x_n]$ is the vector of independent variables, and the multivariate functions are represented in vector $[y_1,y_2,\cdots,y_m]$.

Example 3.18 Consider that the functions for coordinate transformation are defined as $x=r\sin\theta\cos\phi$, $y=r\sin\theta\sin\phi$ and $z=r\cos\theta$. Find the Jacobian matrix of these functions.

Solution *Three independent variables can be declared, and the three functions can then be expressed in a vector. The following statements can be used to find the Jacobian matrix*

```
>> syms r theta phi; x=r*sin(theta)*cos(phi); y=r*sin(theta)*sin(phi);
   z=r*cos(theta); J=jacobian([x; y; z],[r theta phi])
```

The Jacobian matrix is obtained as

$$\boldsymbol{J}=\begin{bmatrix} \sin\theta\cos\phi & r\cos\theta\cos\phi & -r\sin\theta\sin\phi\\ \sin\theta\sin\phi & r\cos\theta\sin\phi & r\sin\theta\cos\phi\\ \cos\theta & -r\sin\theta & 0 \end{bmatrix}.$$

3.2.4 Hessian partial derivative matrix

For a given scalar function $f(x_1, x_2, \cdots, x_n)$ with n independent variables, the definition of a Hessian matrix is

$$\boldsymbol{H} = \begin{bmatrix} \partial^2 f/\partial x_1^2 & \partial^2 f/\partial x_1 \partial x_2 & \cdots & \partial^2 f/\partial x_1 \partial x_n \\ \partial^2 f/\partial x_2 \partial x_1 & \partial^2 f/\partial x_2^2 & \cdots & \partial^2 f/\partial x_2 \partial x_n \\ \vdots & \vdots & \ddots & \vdots \\ \partial^2 f/\partial x_n \partial x_1 & \partial^2 f/\partial x_n \partial x_2 & \cdots & \partial^2 f/\partial x_n^2 \end{bmatrix}. \tag{3-2-4}$$

It can be seen that, a Hessian matrix is in fact a matrix second order derivatives of scalar function $f(x_1, x_2, \cdots, x_n)$, and the new MATLAB function `hessian()` can be used to find Hessian matrix, with $\boldsymbol{H}$ =`hessian(`f`,`$\boldsymbol{x}$`)`, where, $\boldsymbol{x} = [x_1, x_2, \cdots, x_n]$. In earlier versions of MATLAB, a Hessian matrix can be obtained with

$\boldsymbol{H}$ =`jacobian(jacobian(`f`,`$\boldsymbol{x}$`),`$\boldsymbol{x}$`)`

Example 3.19 Find the Hessian matrix for the function in Example 3.16.

Solution *The following statements can be used directly*

```
>> syms x y; f=(x^2-2*x)*exp(-x^2-y^2-x*y); % define original function
   H=simplify(hessian(f,[x,y]))              % compute Hessian matrix
```

and the result is

$$\boldsymbol{H} = \mathrm{e}^{-x^2-y^2-xy} \begin{bmatrix} 4x-2(2x-2)(2x+y)-2x^2-(2x-x^2)(2x+y)^2+2 & 2x-(2x-2)(x+2y)-x^2-(2x-x^2)(x+2y)(2x+y) \\ 2x-(2x-2)(x+2y)-x^2-(2x-x^2)(x+2y)(2x+y) & x(x-2)(x^2+4xy+4y^2-2) \end{bmatrix}.$$

3.2.5 Partial derivatives of implicit functions

Assume that an implicit function is defined as $f(x_1, x_2, \cdots, x_n) = 0$. The partial derivative $\partial x_i/\partial x_j$ among the independent variables can be obtained using the following formula

$$\frac{\partial x_i}{\partial x_j} = -\frac{\dfrac{\partial}{\partial x_j} f(x_1, x_2, \cdots, x_n)}{\dfrac{\partial}{\partial x_i} f(x_1, x_2, \cdots, x_n)}. \tag{3-2-5}$$

Since the derivatives of f with respect to x_i and x_j can easily be obtained separately with the function `diff()`, the partial derivative of $\partial x_i/\partial x_j$ can be obtained directly using the MATLAB functions F =`-diff(`f`,`x_j`)/diff(`f`,`x_i`)`.

For function $f(x, y)$, if $\partial y/\partial x = F_1(x, y)$ is found (here, partial derivative notation is still used, so that the formula can be used directly in multivariate functions), the second order derivative can easily be derived

$$F_2(x, y) = \frac{\partial^2 y}{\partial x^2} = \frac{\partial F_1(x, y)}{\partial x} + \frac{\partial F_1(x, y)}{\partial y} F_1(x, y). \tag{3-2-6}$$

Higher order derivatives can be formulated recursively from

$$F_n(x,y)=\frac{\partial^n y}{\partial x^n}=\frac{\partial F_{n-1}(x,y)}{\partial x}+\frac{\partial F_{n-1}(x,y)}{\partial y}F_1(x,y) \tag{3-2-7}$$

The above recursive formula can easily be implemented with MATLAB, and directly usable for multivariate functions. The nth order partial derivative $f_1=\partial^n y/\partial x^n$ can be evaluated directly with the syntax f_1=`impldiff(`f`,`x`,`y`,`n`)`. The listing of the function is

```
function dy=impldiff(f,x,y,n)
if mod(n,1)~=0 | n<0, error('n should positive integer')
else, F1=-simplify(diff(f,x)/diff(f,y)); dy=F1; % first-order
   for i=2:n, dy=simplify(diff(dy,x)+diff(dy,y)*F1);
end, end
```

Example 3.20 Consider the implicit function $f(x,y)=(x^2-2x)\mathrm{e}^{-x^2-y^2-xy}=0$. Please find $\partial y/\partial x$ and $\partial^3 y/\partial x^3$.

Solution *The first order derivative can be obtained directly, either by using Eqn* (3-2-5), *or by the recursive function*

```
>> syms x y; f=(x^2-2*x)*exp(-x^2-y^2-x*y); % special original function
   F1=impldiff(f,x,y,1)  % compute directly 1st order derivative function
```

with the result

$$\frac{\partial y}{\partial x}=F_1(x,y)=\frac{-3x^3+6x^2+4x-4}{2x\,(x+2y)\,(x-2)}-\frac{1}{2}.$$

Second and third order derivatives can be obtained with the recursive function

```
>> F2=impldiff(f,x,y,2), F3=impldiff(f,x,y,3),
   [n,d]=numden(F3), simplify(n) % extract the numerator and denominator
```

and the results are

$$F_2(x,y)=\frac{\partial^2 y}{\partial x^2}=-\frac{3x^4-12x^3+16x^2-8x+8}{2x^2\,(x+2y)\,(x-2)^2}-\frac{\left(-3x^3+6x^2+4x-4\right)^2}{2x^2(x+2y)^3(x-2)^2},$$

$$F_3(x,y)=\frac{\partial^3 y}{\partial x^3}=-\frac{\begin{array}{c}-54x^9+(324-54y)\,x^8+\left(-54y^2+324y-482\right)x^7\\+\left(324y^2-616y-408\right)x^6+\left(-552y^2+264y+1164\right)x^5\\+\left(128y^3+72y^2+432y+128\right)x^4-\left(256y^4+192y^2+96\right)\\+\left(64y^4-384y^3+816y^2-416y-888\right)x^3\\+\left(-192y^4+768y^3-672y^2+384y+96\right)x^2\\+\left(384y^4-512y^3+384y^2-192y+288\right)x\end{array}}{x^3(x+2y)^5(x-2)^3}.$$

Example 3.21 Please find the derivatives of the function $x^2+xy+y^2=3$ [3].

Solution *The following statements can be used to find the high order derivatives. Besides, since $x^2+xy+y^2=3$, the conditions can also be substituted back in the results to further simplify them.*

```
>> syms x y; f=x^2+x*y+y^2-3; F1=impldiff(f,x,y,1)
   f2=impldiff(f,x,y,2); F2=subs(f2,x^2+x*y+y^2,3)
   f3=impldiff(f,x,y,3); F3=subs(f3,x^2+x*y+y^2,3)
   f4=impldiff(f,x,y,4); F4=subs(f4,x^2+x*y+y^2,3)
```

and the following results can be obtained

$$F_1 = -\frac{2x+y}{x+2y},\ F_2 = -\frac{18}{(x+2y)^3},\ F_3 = -\frac{162x}{(x+2y)^5},\ F_4 = -\frac{648\left(4x^2+xy+y^2\right)}{(x+2y)^7},$$

where, sometimes the simplified results obtained with `subs()` *may still not be the simplest one. For instance,* F_4 *may still be simplified manually as* $F_4 = -\dfrac{1944(x^2+1)}{(x+2y)^7}$.

It should be noted that the result F_4 *in new versions of MATLAB is not thus simple, manual simplifications are needed.*

3.2.6 Derivatives of parametric equations

When the function $y(x)$ is given as parametric equations $y = f(t)$, $x = g(t)$, the nth order derivative of the function $\mathrm{d}^n y/\mathrm{d}x^n$ can be calculated recursively using the following formula

$$\begin{aligned}\frac{\mathrm{d}y}{\mathrm{d}x} &= \frac{f'(t)}{g'(t)}\\ \frac{\mathrm{d}^2y}{\mathrm{d}x^2} &= \frac{\mathrm{d}}{\mathrm{d}t}\left(\frac{f'(t)}{g'(t)}\right)\frac{1}{g'(t)} = \frac{\mathrm{d}}{\mathrm{d}t}\left(\frac{\mathrm{d}y}{\mathrm{d}x}\right)\frac{1}{g'(t)}\\ &\vdots\\ \frac{\mathrm{d}^ny}{\mathrm{d}x^n} &= \frac{\mathrm{d}}{\mathrm{d}t}\left(\frac{\mathrm{d}^{n-1}y}{\mathrm{d}x^{n-1}}\right)\frac{1}{g'(t)}.\end{aligned} \tag{3-2-8}$$

Using the recursive calling structure, the following MATLAB function can be written to implement the above algorithm directory

```
function result=paradiff(y,x,t,n)
if mod(n,1)~=0 | n<0, error('n should positive integer')
else, if n==1, result=diff(y,t)/diff(x,t);  % the exit
   else, result=diff(paradiff(y,x,t,n-1),t)/diff(x,t); % recursive call
end, end
```

Example 3.22 For the parametric equations $y = \dfrac{\sin t}{(t+1)^3}$, $x = \dfrac{\cos t}{(t+1)^3}$, find the third order derivative $\dfrac{\mathrm{d}^3y}{\mathrm{d}x^3}$.

Solution *From the above parametric equations, the derivative can be found by*

```
>> syms t; y=sin(t)/(t+1)^3; x=cos(t)/(t+1)^3; % input parametric equations
   f=paradiff(y,x,t,3); [n,d]=numden(f); F=simplify(n)/simplify(d)
```

The results can be simplified into the following form:

$$\frac{\mathrm{d}^3y}{\mathrm{d}x^3} = \frac{-3(t+1)^7[(t^4+4t^3+6t^2+4t-23)\cos t - (4t^3+12t^2+32t+24)\sin t]}{(t\sin t+\sin t+3\cos t)^5}.$$

3.2.7 Gradients, divergences and curls of fields

Fields are usually classified as *scalar fields* and *vector fields.* A scaler field is represented as a scalar function $\varphi(x, y, z)$, and a vector field is mathematically expressed as

$$\boldsymbol{v}(x,y,z) = [X(x,y,z), Y(x,y,z), Z(x,y,z)]. \tag{3-2-9}$$

Gradient of a scalar field is defined as

$$\operatorname{grad}\varphi(x,y,z)=\left[\frac{\partial\varphi(x,y,z)}{\partial x},\frac{\partial\varphi(x,y,z)}{\partial y},\frac{\partial\varphi(x,y,z)}{\partial z}\right], \tag{3-2-10}$$

and the gradient converts a scalar field into a vector field. The gradient of function $\varphi(x,y,z)$ can be evaluated with g=`jacobian(`φ`,[`x`,`y`,`z`])`.

The *divergence* and *curl* of vector field $\boldsymbol{v}(x,y,z)$ are defined respectively as

$$\operatorname{div}\boldsymbol{v}(x,y,z)=\frac{\partial X(x,y,z)}{\partial x}+\frac{\partial Y(x,y,z)}{\partial y}+\frac{\partial Z(x,y,z)}{\partial z}, \tag{3-2-11}$$

$$\operatorname{curl}\boldsymbol{v}(x,y,z)=\left[\left(\frac{\partial Z}{\partial y}-\frac{\partial Y}{\partial z}\right),\left(\frac{\partial X}{\partial z}-\frac{\partial Z}{\partial x}\right),\left(\frac{\partial Y}{\partial x}-\frac{\partial X}{\partial y}\right)\right]. \tag{3-2-12}$$

The divergence and curl of vector function $\boldsymbol{v}$ can be evaluated with symbolic functions, with d=`divergence(`$\boldsymbol{v}$`,[`x`,`y`,`z`])` and $\boldsymbol{c}$=`curl(`$\boldsymbol{v}$`,[`x`,`y`,`z`])`. The divergence of a vector field is a scalar function, and curl is a vector function. These functions can be nested to evaluate complicated mathematical operations.

Example 3.23 For the given vector function

$$X(x,y,z)=x^2\sin y,\ Y(x,y,z)=y^2\sin xz,\ Z(x,y,z)=xy\sin(\cos z),$$

please compute the divergence and curl.

Solution *The vector function can be expressed first, and the divergence and curl can be obtained directly with the following statements*

```
>> syms x y z; v=[(x^2)*sin(y), (y^2)*sin(x*z), x*y*sin(cos(z))];
   d=divergence(v,[x,y,z]), c=curl(v,[x,y,z]) % direct computation
```

and the results are

$$d=2y\sin xz+2x\sin y-xy\cos(\cos z)\sin z,$$

$$\boldsymbol{c}=\left[x\sin(\cos z)-xy^2\cos xz,\quad -y\sin(\cos z),\ y^2z\cos xz-x^2\cos y\right].$$

Example 3.24 Please show that $\operatorname{curl}[\operatorname{grad}u(x,y,z)]=\boldsymbol{0}$.

Solution *The above property can easily be proven, if the following statements are issued, where the results are a zero vector.*

```
>> syms x y z u(x,y,z); v=jacobian(u,[x,y,z]); simplify(curl(v,[x,y,z]))
```

3.3 Analytical Solutions to Integral Problems

In calculus, integral problems are the inverse problems of derivative problems. The integrals are often described mathematically as

$$\int f(x)\,\mathrm{d}x,\ \int_a^b f(x)\,\mathrm{d}x,\ \int\cdots\int f(x_1,x_2,\cdots,x_n)\,\mathrm{d}x_n\cdots\mathrm{d}x_2\,\mathrm{d}x_1 \tag{3-3-1}$$

where function $f(\cdot)$ is referred to as the *integrand.* The first integral is referred to as the *indefinite integral,* while $F(x)$ is referred to as the *primitive function.* The other two integrals are respectively referred to as the *definite integral* and *multiple integral.* To solve the integral problems, according to calculus courses, one has to select, largely by experience, the integration methods, such as integration by substitution, or integration by parts, or others. Therefore, solving integral problems could be a tedious task, and it may be totally dependent upon the one's experiences and skills. In this section, a skill and experience-independent approach is presented.

3.3.1 Indefinite integrals

The `int()` function provided in the Symbolic Math Toolbox of MATLAB can be used directly to evaluate the indefinite integrals to given functions. The syntax of the function is F = `int(fun,x)`, where the integrand can be described by the symbolic expression *fun*. If only one variable appears in the integrand, the argument x can be omitted. The returned argument is the primitive $F(x)$. In fact, the general solution to the indefinite integral problem is $F(x) + \mathrm{C}$, with C an arbitrary constant.

For any integrable functions, the use of the function `int()` can reduce the complicated work such that the primitive function can be obtained directly. However, for symbolically non-integrable functions, the `int()` function may not yield useful results either. In this case, numerical methods have to be used instead.

Example 3.25 Consider the function given in Example 3.13. The `diff()` function can be used to find the derivatives of $f(x)$. If the indefinite integrals are made upon the results, check whether the original function can be restored.

Solution *The original function can be defined and the integral can be taken on the first-order derivative such that*

```
>> syms x; y=sin(x)/(x^2+4*x+3); y1=diff(y); y0=int(y1)
```

the result is exactly the same as the original function $f(x)$. According to the rules of indefinite integrals, a group of functions $f(x) + \mathrm{C}$ are found.

Now consider taking the fourth-order derivative to the original function by applying `int()` *four times in a nested way as follows:*

```
>> y4=diff(y,4); F=int(int(int(int(y4)))); simplify(F)
```

and the result is still the same as the original function. In fact, not only $f(t)$, a cluster of functions can also be constructed manually

$$F(x) = \frac{\sin x}{x^2+4x+3} + \mathrm{C}_1 + \mathrm{C}_2 x + \mathrm{C}_3 x^2 + \mathrm{C}_4 x^3.$$

Example 3.26 Show that

$$\int x^3 \cos^2 ax \, \mathrm{d}x = \frac{x^4}{8} + \left(\frac{x^3}{4a} - \frac{3x}{8a^3}\right)\sin 2ax + \left(\frac{3x^2}{8a^2} - \frac{3}{16a^4}\right)\cos 2ax + \mathrm{C}.$$

Solution *The following MATLAB statements can be used:*

```
>> syms a x; f=simplify(int(x^3*cos(a*x)^2,x)) % evaluate the left-hand side
```

and the simplified results can be obtained as

$$\frac{1}{16a^4}\Big[4a^3x^3\sin(2ax)+2a^4x^4+6a^2x^2\cos(2ax)-6\,ax\sin(2ax)+3-3\cos(2ax)\Big].$$

It can be seen that the result is not the same as the one on the right-hand side. Let us check the difference. Using the following scripts

```
>> f1=x^4/8+(x^3/(4*a)-3*x/(8*a^3))*sin(2*a*x)+...
      (3*x^2/(8*a^2)-3/(16*a^4))*cos(2*a*x); % evaluate right-hand side
   simplify(f-f1)  % difference is taken and simplify to see whether it is zero
```

After simplification, the difference is $-3/(16a^4)$*, not zero. However, fortunately, since the difference between the two primitive functions is a constant, it can be included into the final constant* C*. Therefore, the original equation is proven.*

Example 3.27 Consider the two integrands

$$f(x)=\mathrm{e}^{-x^2/2},\quad \text{and}\quad g(x)=x\sin(ax^4)\mathrm{e}^{x^2/2}.$$

They are both known to be non-integrable. Compute the indefinite integral to the two functions.

Solution *Let us consider first the integral to the integrand* $f(x)=\mathrm{e}^{-x^2/2}$*. The following MATLAB functions can be used*

```
>> syms x; int(exp(-x^2/2)) % evaluate integral directly
```

and the result obtained is $\sqrt{2\pi}\,\mathrm{erf}(\sqrt{2}x)/2$*. Since the original integrand is not integrable, a special function* $\mathrm{erf}(x)=\dfrac{2}{\sqrt{\pi}}\displaystyle\int_0^x \mathrm{e}^{-t^2}\,\mathrm{d}t$ *is invented by mathematicians. Therefore, the "analytical" solution to the original problem can be obtained.*

The second integrand can be tested under the `int()` *function, with the following MATLAB statements*

```
>> syms a x; int(x*sin(a*x^4)*exp(x^2/2)) % try to compute directly
```

and the returned message shows the same statement as the original command, meaning that the explicit solutions cannot be obtained.

3.3.2 Computing definite, infinite and improper integrals

If the indefinite integral of $f(x)$ can be written as $F(x)+\mathrm{C}$, and $F(x)$ has no discontinuities in (a,b), the definite integral over the interval (a,b) can be obtained from $I=F(b)-F(a)$. In practical applications, the indefinite integrals may not exist, however, the definite or infinite integrals may be needed. For instance, the special function $\mathrm{erf}(x)$ in the previous examples cannot be directly solvable, while $\mathrm{erf}(1.5)$ is needed, numerical methods should be used instead.

The definite integrals and improper integrals are also part of calculus. For instance, although the function $\mathrm{erf}(x)$ is defined previously, the integral of a particular value of x cannot be obtained analytically. In this case, definite integrals, in cooperation with numerical methods, can be obtained. The function `int()` can be used to evaluate the definite and infinite integrals. The syntax of the function is $I=$ `int(`*fun*`,`x`,`a`,`b`)`, where x is the independent variable, (a,b) is the integral interval. For infinite integrals, the arguments

a and b can be assigned to `-Inf` or `Inf`. Also, if no exact value can be obtained directly, the `vpa()` function can be used to evaluate the solutions numerically. Alternatively, if x is not continuous in the interval $[a, b]$, the integral is referred to as *improper integral.* Suppose at $x = c$, $a \leqslant c \leqslant b$, the function $f(x)$ is not continuous, the improper integral should be evaluated mathematically with

$$\int_a^b f(x)\,\mathrm{d}x = \lim_{\epsilon\to 0+}\int_a^{c-\epsilon} f(x)\,\mathrm{d}x + \lim_{\epsilon\to 0+}\int_{c+\epsilon}^b f(x)\,\mathrm{d}x. \tag{3-3-2}$$

With MATLAB Symbolic Math Toolbox, the improper integral can be evaluated directly with I=`int(`f`,`x`,`a`,`b`)`, if the integral exists.

Example 3.28 Consider the integrands given previously in Example 3.27. When $a = 0$, $b = 1.5$ (or ∞), evaluate the values of the integral.

Solution *The following statements can be used in solving the definite and infinite integral problems*

```
>> syms x; I1=int(exp(-x^2/2),x,0,1.5), vpa(I1,70)
   I2=int(exp(-x^2/2),x,0,inf) % evaluate infinite integral
```

where $I_1 = \sqrt{2\pi}\,\mathrm{erf}(3\sqrt{2}/4)/2$, *and the high-precision numerical solution to the definite integral is* $I_1 = 1.085853317666016569702419076542265042534236293532156326729917229 30853$. *The analytical solution to the infinite integral is* $I_2 = \sqrt{\pi/2}$.

Example 3.29 Solve the definite integral problems for functional boundaries

$$I(t) = \int_{\cos t}^{\mathrm{e}^{-2t}} \frac{-2x^2+1}{(2x^2-3x+1)^2}\,\mathrm{d}x.$$

Solution *The function* `int()` *can be used in solving definite integrals, and the following statements can be used*

```
>> syms x t real; f=(-2*x^2+1)/(2*x^2-3*x+1)^2; % input integrand
   I=simplify(int(f,x,cos(t),exp(-2*t)))         % integral directly
```

and a piecewise result can be obtained. It is indicated in the result that, only when $t > 0$ *and* $(\cos t > 1/2$ *or* $\ln 2 < 2t)$, *the integral can be found*

$$I = -\frac{\left(\mathrm{e}^{-2t}-\cos t\right)\left(2\mathrm{e}^{-2t}\cos t-1\right)}{(\cos t-1)\,(2\cos t-1)\,(\mathrm{e}^{-2t}-1)\,(2\mathrm{e}^{-2t}-1)}.$$

With the alternative integral approach, the primitive function $F(x)$ *can be found first, and the integral seems to be computable with* $I = F(b) - F(a)$. *However, the assumption on* $F(x)$ *is continuous cannot be satisfied for all* t, *which means* $I = F(b) - F(a)$ *should be used conditionally. The conditions obtained above are the ones making* $I = F(b) - F(a)$ *usable, otherwise, the integral may be infinite or unsolvable.*

Example 3.30 Please evaluate the improper integral $\displaystyle\int_1^{\mathrm{e}} \frac{1}{x\sqrt{1-\ln^2 x}}\,\mathrm{d}x$.

Solution *It can be seen that at* $x = \mathrm{e}$, *the integrand is discontinuous, therefore, the integral is an improper integral. The problem can be solved directly with the following statements, and the result is* $\pi/2$.

```
>> syms x; f=1/x/sqrt(1-log(x)^2); I=int(f,x,1,exp(sym(1)))
```

3.3.3 Computing multiple integrals

Multiple integral problems can also be solved by using the same MATLAB function `int()`. Generally speaking, usually the inner integrals should be carried out first, and then, outer integrals. However, the sequence of integrals should be observed. In each integration step, the `int()` function can be used. Therefore, sometimes in certain integration steps, the inner integral may not yield a primitive function, which results in no analytical solution to the overall integral problem. If the sequence of integrals can be changed, analytical solutions may be obtained. Numerical solutions to multiple integral problems will be presented in Section 3.9.5.

Example 3.31 Compute the multiple integrals $\int\cdots\int F(x,y,z)\,\mathrm{d}x^2\,\mathrm{d}y\,\mathrm{d}z$ where the integrand $F(x,y,z)$ is defined as

$$-4z\mathrm{e}^{-x^2y-z^2}\left[\cos x^2y - 10yx^2\cos x^2y + 4x^4y^2\sin x^2y + 4x^4y^2\cos x^2y - \sin x^2y\right].$$

Solution *In fact, the above $F(x,y,z)$ function was obtained by taking partial derivatives to the function $f(x,y,z)$ defined in Example 3.17. Therefore, taking inverse operations in this example should restore the same primitive function.*

One may integrate once with respect to z, once to y and twice to x. The following results can be obtained through simplification

```
>> syms x y z; % evaluate integral in the sequential order z→y→x→x
   f0=-4*z*exp(-x^2*y-z^2)*(cos(x^2*y)-10*cos(x^2*y)*y*x^2+...
      4*sin(x^2*y)*x^4*y^2+4*cos(x^2*y)*x^4*y^2-sin(x^2*y));
   f1=int(f0,z); f1=int(f1,y); f1=int(f1,x); f1=simplify(int(f1,x))
```

with the primitive function $f_1 = \sin(x^2y)\mathrm{e}^{-x^2y-z^2}$, which is exactly the same as the one defined in Example 3.17.

Now if one alters the sequence of integrals, i.e., change the order to $z\to x\to x\to y$, the result is still the same. In earlier versions, different sequences of integration may yield different results.

```
>> f2=int(f0,z); f2=int(f2,x); f2=int(f2,x); f2=simplify(int(f2,y))
```

Example 3.32 Compute the definite integral $I = \displaystyle\int_0^2\int_0^\pi\int_0^\pi 4xz\mathrm{e}^{-x^2y-z^2}\,\mathrm{d}z\,\mathrm{d}y\,\mathrm{d}x$.

Solution *The following statements can be given to calculate the definite integral*

```
>> syms x y z % compute directly the triple integral with simple command
   I=int(int(int(4*x*z*exp(-x^2*y-z^2),z,0,pi),y,0,pi),x,0,2)
```

and the results obtained are

$$I = -\left(2\mathrm{e}^{-\pi^2}-2\right)\left(\frac{\gamma}{2}+\ln(2)+\frac{\ln(\pi)}{2}-\frac{\mathrm{Ei}(-4\pi)}{2}\right)$$

where **`eulergamma`** *is the Euler constant γ, $\mathrm{Ei}(x)=\displaystyle\int_{-\infty}^{x}\mathrm{e}^t/t\,\mathrm{d}t$ is an exponential integral. The integrand is not integrable analytically. However, numerical solutions can be found. Therefore, the accurate numerical solution to the original problem can be found from* `vpa(ans)` *command, and the integral value is 3.10807940208541272.*

3.4 Series Expansions and Finite-term Series Approximations

Taylor series expansions to univariate and multivariate functions will be discussed in this section. The Fourier series expansion to given functions are also to be discussed. In particular, with the powerful graphics facilities in MATLAB, the fitting quality and fitting interval can also be assessed using finite-term series approximations.

3.4.1 Taylor series expansion

I. Taylor series expansion of univariate functions

The Taylor series expansion about the point $x = 0$ can be written as

$$f(x) = a_1 + a_2x + a_3x^2 + \cdots + a_kx^{k-1} + \mathrm{o}(x^k) \tag{3-4-1}$$

where the coefficients a_i can be obtained from

$$a_i = \frac{1}{(i-1)!}\lim_{x\to 0}\frac{\mathrm{d}^{i-1}}{\mathrm{d}x^{i-1}}f(x),\quad i = 1, 2, 3, \cdots. \tag{3-4-2}$$

The expansion is also referred to as the *Maclaurin series.* If the Taylor series expansion is made about the $x = a$ point, the series can then be written as

$$f(x) = b_1 + b_2(x-a) + b_3(x-a)^2 + \cdots + b_k(x-a)^{k-1} + \mathrm{o}[(x-a)^k] \tag{3-4-3}$$

where the b_i coefficients can be obtained from

$$b_i = \frac{1}{(i-1)!}\lim_{x\to a}\frac{\mathrm{d}^{i-1}}{\mathrm{d}x^{i-1}}f(x),\quad i = 1, 2, 3, \cdots. \tag{3-4-4}$$

Taylor series expansion can be obtained by the use of the `taylor()` function, provided in the Symbolic Math Toolbox. The syntaxes of the function are

```
f1=taylor(fun,x,'Order',k)     % Taylor series about x = 0 point
f1=taylor(fun,x,'Order',k,a)   % expansion about the x = a point
```

where *fun* is a symbolic expression of the original function, and x is the independent variable. If there is only one independent variable in *fun*, x can be omitted. The argument k is the order required in the expansion, with a default number of terms of 6. If an extra argument a is given, the expansion is then made about the $x = a$ point. The Taylor series expansion solutions are demonstrated in the following examples. In old versions of Symbolic Math Toolbox, the argument `'Order'` should not be provided.

Example 3.33 Consider again the function $f(x) = \sin x/(x^2 + 4x + 3)$ given in Example 3.13. Find the first 9 terms of Taylor series expansion about $x = 0$ point. Consider also the series expansions about points $x = 2$ and $x = a$.

Solution *The following statements can be used to specify the given function. The first 9 terms of Taylor series expansion can be obtained with*

```
>> syms x; f=sin(x)/(x^2+4*x+3); f1=taylor(f,x,'Order',9)
```

and the result is

$$f_1 = -\frac{386459x^8}{918540} + \frac{515273x^7}{1224720} - \frac{3067x^6}{7290} + \frac{4087x^5}{9720} - \frac{34x^4}{81} + \frac{23x^3}{54} - \frac{4x^2}{9} + \frac{x}{3}.$$

In classical calculus courses, no analysis had been made upon the fitting quality of the finite number of terms approximation for a given function, since there were no ready tools available. With the use of MATLAB, the original function as well as the finite term Taylor series approximation can be compared graphically as shown in Figure 3.4(a).

```
>> ezplot(f,[-1,1]), hold on; ezplot(f1,[-1,1]) % compare the functions
```

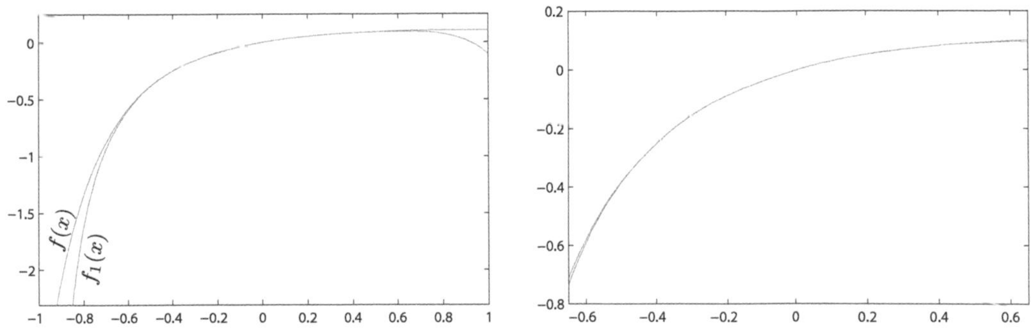

(a) $(-1,1)$ interval (b) $(-0.65, 0.65)$ interval

FIGURE 3.4: Finite term Taylor series approximation.

It can be seen that the 9th order Taylor series expansion for the original function is not good for interval $[-1,1]$. If the interval is reduced to $(-0.65, 0.65)$, the fitting quality is shown in Figure 3.4(b) which is good enough. Therefore, with the graphical facilities in MATLAB, the fitting qualities can be examined easily.

Now consider the Taylor series expansion about the point $x = 2$. The series can be derived using the following statement:

```
>> f2=taylor(f,x,'Order',9,2) % expand about x = 2
```

Since the expansion is lengthy, only first five terms are shown here

$$f_2 = \frac{\sin 2}{15} + \left(\frac{\cos 2}{15} - \frac{8\sin 2}{225}\right)(x-2) - \left(\frac{127\sin 2}{6750} + \frac{8\cos 2}{225}\right)(x-2)^2$$
$$+\left(\frac{23\cos 2}{6750} + \frac{628\sin 2}{50625}\right)(x-2)^3 + \left(-\frac{15697}{6075000}\sin(2) + \frac{28}{50625}\cos(2)\right)(x-2)^4.$$

If one wants to find the series expansion about the $x = a$ point, the Taylor series expansion can still be derived using similar statements

```
>> syms a; f3=taylor(f,x,'Order',9,a) % expand about x = a
```

Here only the first three terms are shown

$$f_3 = \frac{\sin a}{a^2+3+4a} + \left[\frac{\cos a}{a^2+3+4a} - \frac{(4+2a)\sin a}{(a^2+3+4a)^2}\right](x-a) + \left[-\frac{\sin a}{(a^2+3+4a)^2}\right.$$
$$\left.-\frac{\sin a}{2(a^2+3+4a)} - \frac{\left(a^2\cos a + 3\cos a + 4a\cos a - 4\sin a - 2a\sin a\right)(4+2a)}{\left(a^2+3+4a\right)^3}\right](x-a)^2.$$

Example 3.34 Expand the sinusoidal function $y = \sin x$ into Taylor series, and compare the approximation quality for different terms.

Solution *In order to find out the relationship between the fitting quality and the number of terms used, the loop structure should be used. The following statements can be issued to solve the problem, where the fitting curves shown in Figure 3.5 can be obtained.*

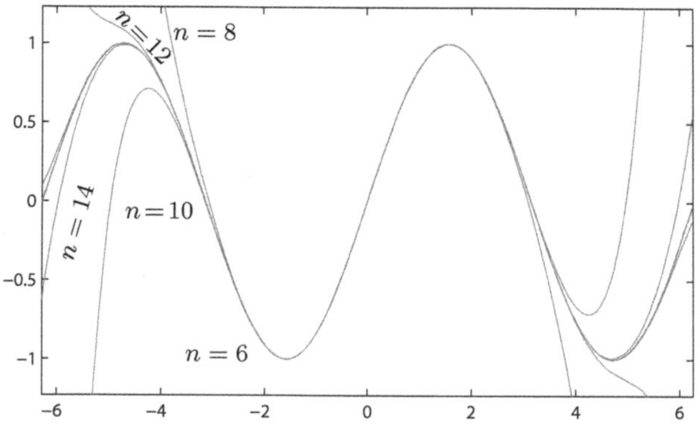

FIGURE 3.5: Taylor series approximation to given sinusoidal functions.

```
>> syms x; y=sin(x); ezplot(y); hold on % plot original function
   for n=[8:2:20], f1=taylor(y,x,'Order',n), ezplot(f1); end % different n
```

For fewer terms, the satisfactory fitting interval is small. If the number of terms is increased, the satisfactory fitting interval will also increase. For instance, if one selects $n = 16$, *the fitting is satisfactory over the interval* $(-2\pi, 2\pi)$. *The first 20 terms in the Taylor series expansion are obtained as*

$$\sin x \approx x - \frac{1}{6}x^3 + \frac{1}{120}x^5 - \frac{1}{5040}x^7 + \frac{1}{362880}x^9 - \frac{1}{39916800}x^{11} + \frac{1}{6227020800}x^{13} - \frac{1}{1307674368000}x^{15} + \frac{1}{355687428096000}x^{17} - \frac{1}{121645100408832000}x^{19}.$$

II. Taylor series expansion of multivariate functions

The Taylor series expansion of a multivariate function $f(x_1, x_2, \cdots, x_n)$ is

$$f(\boldsymbol{x}) = f(\boldsymbol{a}) + \left[(x_1 - a_1)\frac{\partial}{\partial x_1} + \cdots + (x_n - a_n)\frac{\partial}{\partial x_n}\right] f(\boldsymbol{x})\bigg|_{\boldsymbol{x}=\boldsymbol{a}} + \frac{1}{2!}\left[(x_1 - a_1)\frac{\partial}{\partial x_1} + \cdots + (x_n - a_n)\frac{\partial}{\partial x_n}\right]^2 f(\boldsymbol{x})\bigg|_{\boldsymbol{x}=\boldsymbol{a}} + \cdots + \frac{1}{k!}\left[(x_1 - a_1)\frac{\partial}{\partial x_1} + \cdots + (x_n - a_n)\frac{\partial}{\partial x_n}\right]^k f(\boldsymbol{x})\bigg|_{\boldsymbol{x}=\boldsymbol{a}} + \cdots \tag{3-4-5}$$

where $(a_1, \cdots, a_n)$ is the center point of Taylor series expansion. In order to avoid misunderstanding, the terms can be regarded as the derivatives of function *fun*. Then, the function evaluation can be made to the point $(a_1, a_2, \cdots, a_n)$. The Taylor series expansion to multivariate functions can be obtained from

F=`taylor(`f`,[`x_1`,`x_2`,`$\cdots$`,`x_n`],[`a_1`,`a_2`,`$\cdots$`,`a_n`],'Order',`k`)`

where k is the order of the expansion, and *fun* is the multivariate function.

Example 3.35 Consider again the function $z = f(x,y) = (x^2 - 2x)\mathrm{e}^{-x^2-y^2-xy}$ shown in Example 3.16. Find its Taylor series expansion.

Solution *The following statements can be used to get the Taylor series expansion about the origin*

```
>> syms x y; f=(x^2-2*x)*exp(-x^2-y^2-x*y); % original function
   F=taylor(f,[x,y],'Order',8) % 8th order Taylor series expansion
```

whose mathematical representation is

$$F = \frac{x^7}{3} + \left(y + \frac{1}{2}\right)x^6 + \left(2y^2 + y - 1\right)x^5 + \left(\frac{7y^3}{3} + \frac{3y^2}{2} - 2y - 1\right)x^4$$
$$+ \left(2y^4+y^3-3y^2-y+2\right)x^3 + \left(y^5+\frac{y^4}{2}-2y^3-y^2+2y+1\right)x^2 + \left(\frac{y^6}{3}-y^4+2y^2-2\right)x$$

If one wants to expand the original function about $x = 1, y = a$ point, the following statements can be used

```
>> syms a; F=taylor(f,[x,y],[1,a],'Order',3), F1=simplify(F)
```

and the expansion and its simplification can be found as

$$F(x,y) = -\mathrm{e}^{-a^2-a-1}\left\{\left[\left(\frac{a}{2}+1\right)(a+2)-2\right](x-1)^2 - (2a+1)(a-y)\right.$$
$$- (a-y)^2\left[(2a+1)\left(a+\frac{1}{2}\right)-1\right] + (a+2)(x-1) - 1$$
$$\left. + (a-y)(x-1)\left[(2a+1)\left(\frac{a}{2}+1\right) + (a+2)\left(a+\frac{1}{2}\right)-1\right]\right\}$$

$$F_1(x) = -\frac{1}{2}\,\mathrm{e}^{-a^2-a-1}\left(4a^4 - 4a^3x - 8a^3y + 8a^3 + a^2x^2 + 4a^2xy - 12a^2x\right.$$
$$+ 4a^2y^2 - 12a^2y + 14a^2 + 4ax^2 + 10axy - 12ax$$
$$\left. +4ay^2 - 12ay + 10a + 2xy - 4x - y^2 - 4y + 6\right)$$

3.4.2 Fourier series expansion

Consider a periodic function $f(x)$ defined over the interval $x \in [-L, L]$. The function is with a period of $T = 2L$. For the function defined on other intervals, it can be extended to periodic functions such that $f(x) = f(kT + x)$, where k is an arbitrary integer. A given function $f(x)$ can be approximated by an infinite series such that

$$F(x) = \frac{a_0}{2} + \sum_{n=1}^{\infty}\left(a_n \cos\frac{n\pi}{L}x + b_n \sin\frac{n\pi}{L}x\right) \tag{3-4-6}$$

where

$$\begin{cases} a_n = \dfrac{1}{L}\displaystyle\int_{-L}^{L} f(x)\cos\dfrac{n\pi x}{L}\,\mathrm{d}x, & n = 0, 1, 2, \cdots \\ b_n = \dfrac{1}{L}\displaystyle\int_{-L}^{L} f(x)\sin\dfrac{n\pi x}{L}\,\mathrm{d}x, & n = 1, 2, 3, \cdots. \end{cases} \tag{3-4-7}$$

Such a series is referred to as the *Fourier series* and a_n, b_n are referred to as *Fourier coefficients*. If the original function $f(x)$ is periodic over arbitrary interval $x \in (a, b)$, compute $L = (b - a)/2$, such that with variable substitution $x = \hat{x} + L + a$, the new function $f(\hat{x})$ is mapped in $\hat{x} \in (-L, L)$, and the above expansion can be carried out.

Unfortunately, there is no existing function for Fourier series expansion provided in MATLAB. Based on the above formula, the algorithm can be designed as follows

---

Require: Symbolic expression $f(x)$, independent variable x
  Set default variables $p \leftarrow 6$, $a \leftarrow -\pi$, $b \leftarrow \pi$
  Compute $L = (b - a)/2$, variable substitution $x = \hat{x} + L + a$ in $f(x)$
  Compute a_0, from (3-4-7), compute $F(\hat{x}) = a_0/2$
  for $n = 1$ To p **do**
    Compute and store a_n and b_n from (3-4-7), update $F(\hat{x})$ from (3-4-6)
  end for
  Variable substitution $\hat{x} = x - L - a$, map $F(\hat{x})$ back to $F(x)$.

The new function `fseries()` can be written to implement the algorithm, and it can be seen that the MATLAB statements are very concise implementations of the algorithm.

```
function [F,A,B]=fseries(f,x,varargin) % construct Fourier series
[p,a,b]=default_vals({6,-pi,pi},varargin{:});
L=(b-a)/2; f=subs(f,x,x+L+a); A=int(f,x,-L,L)/L; B=0; F=A/2;
for n=1:p % loop structure for the first p terms
   an=int(f*cos(n*pi*x/L),x,-L,L)/L; % compute the coefficients
   bn=int(f*sin(n*pi*x/L),x,-L,L)/L;
   A=[A,an]; B=[B,bn]; F=F+an*cos(n*pi*x/L)+bn*sin(n*pi*x/L);
end
F=subs(F,x,x-L-a); % variable substitution
```

A low-level supporting function to accept default argument is implemented in `default_vals()`, and it may be used by many other functions in the book

```
function varargout=default_vals(vals,varargin)
if nargout~=length(vals), error('number of arguments mismatch');
else, n=length(varargin)+1;
   varargout=varargin; for i=n:nargout, varargout{i}=vals{i};
end, end, end
```

The syntax of the function is `[F,A,B]=fseries(f,x,p,a,b)`, where f is the given function; x is the independent variable; p is number of the terms required in the expansion and (a, b) is the interval for x. If a, b arguments are omitted, the default interval $[-\pi, \pi]$ will be used. The returned arguments $\boldsymbol{A}$, $\boldsymbol{B}$ contain the Fourier coefficients, F is the symbolic expression of the Fourier series expansion.

Example 3.36 Find the Fourier series expansion to the function $y = x(x-\pi)(x-2\pi)$, where $x \in (0, 2\pi)$.

Solution *The Fourier series for the given function can easily be expressed*

```
>> syms x; f=x*(x-pi)*(x-2*pi); [F,A,B]=fseries(f,x,12,0,2*pi)
```

where the first 12 terms in the Fourier series are as follows:

$$F(x) = 12\sin x + \frac{3\sin 2x}{2} + \frac{4\sin 3x}{9} + \frac{3\sin 4x}{16} + \frac{12\sin 5x}{125} + \frac{\sin 6x}{18} + \frac{12\sin 7x}{343} + \frac{3\sin 8x}{128} + \frac{4\sin 9x}{243} + \frac{3\sin 10x}{250} + \frac{12\sin 11x}{1331} + \frac{\sin 12x}{144}.$$

From these results, the analytical form can be summarized as $f(x) = \sum_{n=1}^{\infty} \frac{12}{n^3}\sin nx$.

The first 12 terms in the Fourier series expansion and the original function can be graphically compared as shown in Figure 3.6 (a) with the following statements

```
>> ezplot(f,[0,2*pi]), hold on, ezplot(F,[0,2*pi]) % comparisons
```

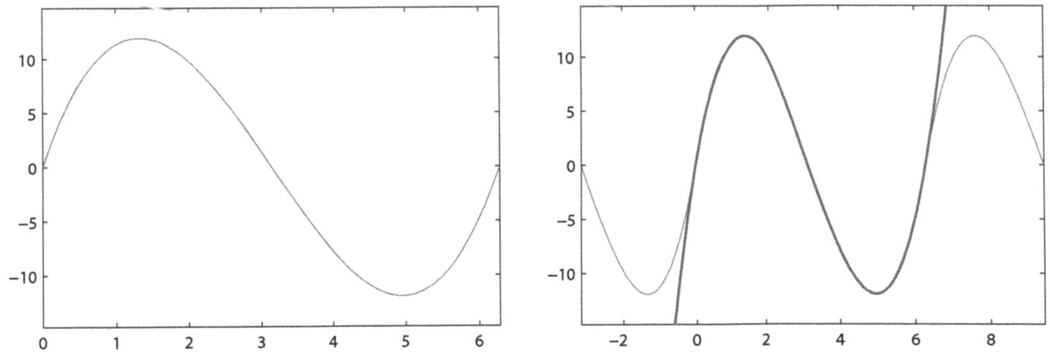

(a) over interval $(0, 2\pi)$ (b) a larger interval $(-\pi, 3\pi)$

FIGURE 3.6: Accuracy of finite term Fourier series approximation.

If one wants to further examine the approximation over a larger interval $x \in (-\pi, 3\pi)$, *the following statements should be used*

```
>> ezplot(f,[-pi,3*pi]), hold on, ezplot(F,[-pi,3*pi]) % larger interval
```

and the curves are shown in Figure 3.6 (b). It can be seen that over the $(0, 2\pi)$ *interval the fitting is quite good. In other regions, since the Fourier series is made upon the assumption that it is periodically extended, therefore, it cannot approximate the original function in other intervals at all.*

Example 3.37 Now consider a square wave defined over the interval $(-\pi, \pi)$, where $y = 1$ when $x \geqslant 0$, and $y = -1$ otherwise. Expand the function using Fourier series and observe how many terms in the function may give good approximation.

Solution *Since in symbolic expressions inequality cannot be used, the square wave can be expressed as* $f(x) = |x|/x$. *In this way, the numerical and analytical expressions in Fourier series can be obtained for different terms in the expression. The curves can be obtained as shown in Figure 3.7 (a).*

```
>> syms x; f=abs(x)/x;   % square wave definition
   xx=[-pi:pi/200:pi]; xx=xx(xx~=0); xx=sort([xx,-eps,eps]); % remove 0
   yy=subs(f,x,xx); plot(xx,yy), hold on  % draw the original function
   for n=1:20 % try different orders of Fourier series
      [f1,a,b]=fseries(f,x,n); y1=subs(f1,x,xx); plot(xx,y1)
   end
```

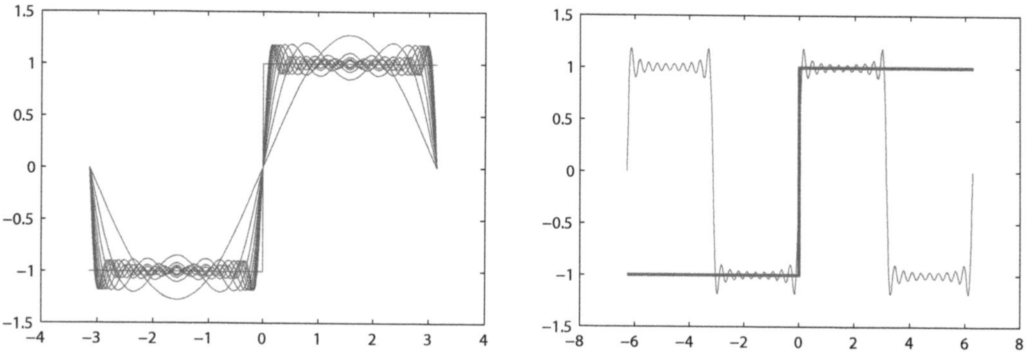

(a) over interval $(-\pi, \pi)$ (b) a larger interval $(-2\pi, 2\pi)$

FIGURE 3.7: Approximation of square wave by Fourier series.

It can be seen that when 10 terms are used, the approximation is satisfactory. Even if the number of terms increases, the fitting accuracy may not be improved significantly. A finite Fourier series of the original function can be obtained by

```
>> [f1,a,b]=fseries(f,x,14); f1 % 14th order Fourier series expansion
```

and the expansion can be written as

$$f_1(x) = 4\frac{\sin x}{\pi} + \frac{4\sin 3x}{3\pi} + \frac{4\sin 5x}{5\pi} + \frac{4\sin 7x}{7\pi} + \frac{4\sin 9x}{9\pi} + \frac{4\sin 11x}{11\pi} + \frac{4\sin 13x}{13\pi},$$

which can further be summarized as $f(x) = \dfrac{4}{\pi}\displaystyle\sum_{k=1}^{\infty}\frac{\sin(2k-1)x}{2k-1}$.

Again the Fourier series expansion is established upon the assumption that it is periodically extended over the original function, therefore, the fitting in other intervals may be incorrect, as shown in Figure 3.7 (b).

```
>> xx=[-2*pi:pi/200:2*pi]; xx=xx(xx~=0); xx=sort([xx,-eps,eps]);
   yy=subs(f,x,xx); plot(xx,yy), y1=subs(f1,x,xx); line(xx,y1)
```

3.5 Infinite Series and Products

Informally speaking, the term "series" is usually used to indicate the sum of sequences, and sometimes it is also used to indicate a sequence. The two meanings are not strictly distinguished in the book. In this section, the sum and product of sequences are introduced. Also, the convergent intervals of infinite series are also studied.

3.5.1 Series

The function `symsum()` provided in the Symbolic Math Toolbox can be used to evaluate the sum of finite and infinite series with known general terms. The syntax of the function is $S=$`symsum(`f_k`,`k`,`k_0`,`k_n`)`, where f_k is the general term of the series, k is the serial number, and k_0 and k_n are the initial and final terms of the series, respectively. They can be set to `inf` for infinite series. The sum of series can be written as

$$S = \sum_{k=k_0}^{k_n} f_k. \tag{3-5-1}$$

If there is only one symbolic variable defined in f_k, the variable k can be omitted in the function call.

Example 3.38 Compute the finite sum $S = 2^0 + 2^1 + 2^2 + \cdots + 2^{62} + 2^{63} = \sum_{i=0}^{63} 2^i$.

Solution *Numerical solution to the problem can be found from*

```
>> format long; s=sum(2.^[0:63]) % evaluate the sum of 64 terms numerically
```

with $s = 1.844674407370955{\times}10^{19}$. *Since the data type of* `double` *is used, only 16 digits can be reserved. Therefore, the exact result cannot be obtained under double-precision scheme. The function* `symsum()` *can be used to solve the problem*

```
>> syms k; symsum(2^k,0,63) % evaluate the sum of 64 terms symbolically
```

where $s_1 = 18446744073709551615$ *can be obtained. The problem can even be solved with a simpler command* $S=$`sum(sym(2).^[0:63])`, *and the same result can be obtained. The method can be extended to calculate for more terms, for instance, it is possible to calculate the sum to 201 terms*

```
>> s2=symsum(2^k,0,200) % computing series with more terms
```

and $s_2 = 3213876088517980551083924184682325205044405987565585670602751$. *The exact solution cannot possibly be obtained using the double-precision data type.*

Example 3.39 Compute the infinite series

$$S = \frac{1}{1\times 4} + \frac{1}{4\times 7} + \frac{1}{7\times 10} + \cdots + \frac{1}{(3n-2)(3n+1)} + \cdots .$$

Solution *With the use of the symbolic function*

```
>> syms n; s=symsum(1/((3*n-2)*(3*n+1)),n,1,inf) % compute infinite series
```

the sum result $s = 1/3$ *can be obtained. The same problem can be tried using numerical method with* `double` *data type. For instance, if 10,000,000 terms are selected to be added up, the following statements can be used directly*

```
>> m=1:10000000; s1=sum(1./((3*m-2).*(3*m+1))); format long; s1
```

and the sum is $s_1 = 0.33333332222165$. *It can be seen that although a very large number of terms are selected with a long time consumed, there still exists unavoidable difference and the error reaches* 10^{-6} *level. It can be seen that when* $m = 10^7$, *the value of the general term is around* 10^{-15}*, therefore, it seems that the additional error in the summation may not be very large. In fact, since double-precision data type is used, some of the terms may not be added to the* S *variable. Therefore, even though more terms are used in the summation, the accuracy cannot be further increased.*

Example 3.40 Evaluate the infinite series with an extra variable x.

$$J = 2\sum_{n=0}^{\infty} \frac{1}{(2n+1)(2x+1)^{2n+1}}.$$

Solution *In the examples studied earlier, numerical methods can be used to find the approximate solutions. If in the general term, extra independent variables are involved, numerical methods can no longer be used. Symbolic method has to be used to solve the problem. For instance, the sum can be evaluated with*

```
>> syms n x; s1=symsum(2/((2*n+1)*(2*x+1)^(2*n+1)),n,0,inf);
   s1=simplify(s1) % simplify the infinite series
```

and a piecewise function can be the result. It can be read as when $x > 0$ *or* $x < -1$*, the sum is* $s_1 = 2\arctan(1/(2x+1))$*. It can be seen that, not only the sum* s_1*, the convergent interval can also be obtained with only the new Symbolic Math Toolbox.*

Example 3.41 Solve the limit problem with the series

$$\lim_{n\to\infty}\left[\left(1+\frac{1}{2}+\frac{1}{3}+\frac{1}{4}+\cdots+\frac{1}{n}\right)-\ln n\right].$$

Solution *So far, the series and limit problems have been discussed and illustrated separately. For this mixed problem, the following MATLAB statements can be used to solve it, where the finite sum should be made first using* `symsum(1/m,m,1,n)`

```
>> syms m n; limit(symsum(1/m,m,1,n)-log(n),n,inf), vpa(ans)
```

and `eulergamma` *can be obtained, i.e., the Euler constant* γ *can be obtained whose value can be evaluated with* `vpa()` *function as* $\gamma = 0.57721566490153286060651209$.

It should be noted that in the computation, one should not evaluate the infinite sum before limit. Otherwise, the original problem cannot be correctly solved.

Example 3.42 Please find the sum

$$S = \lim_{n\to\infty}\left[\left(1+\frac{1}{n^2}\right)\sin\frac{\pi}{n^2}+\left(1+\frac{2}{n^2}\right)\sin\frac{2\pi}{n^2}+\cdots+\left(1+\frac{n-1}{n^2}\right)\sin\frac{(n-1)\pi}{n^2}\right].$$

Solution *To get correct results, the general term must be extracted correctly. In this example, the general term is* $a_k = (1+k/n^2)\sin(k\pi/n^2)$*, with* $k = 1, 2, \cdots, n-1$*. Thus, the sum can be obtained from the statement below, and the result is* $S = \pi/2$.

```
>> syms n k; % express the general term and find infinite series
   S=simplify(limit(symsum((1+k/n^2)*sin(k*pi/n^2),k,1,n-1),n,inf))
```

3.5.2 Product of sequences

The computation of sequence product $P = \prod_{n=a}^{b} f(n)$ can be directly obtained with the MATLAB function `symprod()`, with P = `symprod(`*fun*`,`n`,`a`,`b`))`.

Example 3.43 Calculate the sequence product $\prod_{k=2}^{\infty}\left(1-\dfrac{2}{k(k+1)}\right)$.

Solution *The general term can be written as* $p_k = 1-2/k/(k+1)$. *Therefore, with the following statements, the infinite product is* $P=1/3$.

```
>> syms k; p=1-2/k/(k+1); P=symprod(p,k,2,inf) % infinite product
```

Example 3.44 Please find the infinite sum of the series

$$S=1-\frac{1}{2}+\frac{1\times 3}{2\times 4}-\frac{1\times 3\times 5}{2\times 4\times 6}+\frac{1\times 3\times 5\times 7}{2\times 4\times 6\times 8}-\frac{1\times 3\times 5\times 7\times 9}{2\times 4\times 6\times 8\times 10}+\cdots.$$

Solution *This is the series problem, while from the second term on, the general term is the product of a finite series*

$$s_n=(-1)^n\prod_{k=1}^{n}\left[(2k-1)/(2k)\right],\ \textit{with}\ n=1,2,\cdots,\infty.$$

Therefore, the original problem can be rewritten as $S = 1+\sum_{n=1}^{\infty} s_n$, *which can be obtained with the following statements, and the sum is* $S=\sqrt{2}/2$.

```
>> syms k n, s=(-1)^n*symprod((2*k-1)/(2*k),k,1,n); % general term
   S=1+symsum(s,n,1,inf) % evaluate the infinite series
```

Example 3.45 Please find the product $P=\prod_{n=1}^{\infty}\left(1+\dfrac{x}{n}\right)\mathrm{e}^{-x/n}$.

Solution *The general term contains an independent variable* x, *however, this will not cause any difficulties for the user. The problem can be solved with*

```
>> syms n x; p=(1+x/n)*exp(-x/n); P=symprod(p,n,1,inf)
```

and the result is a piecewise function

$$P=\begin{cases} 0, & x\ \textit{is a negative integer}\\ \mathrm{e}^{-\gamma x}/\Gamma(x+1), & \textit{else, with}\ \gamma\ \textit{the Euler constant}\end{cases}$$

3.5.3 Convergence test of infinite series

There are different kinds of series in real applications, and sometimes, the closed-form solution of the infinite series cannot be obtained, even with function `symsum()` or other powerful tools. The convergence test is important in such a case. An infinite series

$$S=a_1+a_2+\cdots+a_n+\cdots=\sum_{k=1}^{\infty}a_n \tag{3-5-2}$$

is said to be *convergent*, if the sum S has a finite limit, when $n\to\infty$. If the limit is infinite, the series is *divergent*. If $a_n>0$ for all n, the series is referred to *positive series*.

There are several approaches in testing the convergence of a given series.

(i) If $\lim\limits_{n\to\infty} a_n\neq 0$, the series is divergent.

(ii) If the series $\sum\limits_{n=1}^{\infty}|a_n|$ is convergent, then, $\sum\limits_{n=1}^{\infty}a_n$ is also convergent. Moreover, it is referred to as *absolutely convergent.*

For positive series, the following tests can be made.

(iii) D'Alembert's test: Compute $\lim_{n\to\infty}\dfrac{a_{n+1}}{a_n}=\rho$. If $\rho<1$, the series is convergent; $\rho>1$, the series divergent; $\rho=1$, the convergency cannot be assessed directly.

(iv) Raabe's test: If $\rho=1$ in (iii), then, compute $\lim_{n\to\infty} n\left(\dfrac{a_n}{a_{n+1}}-1\right)=R$. If $R>1$, the series is convergent; if $R<1$, it is divergent; while if $R=1$, the convergency cannot be assessed.

For *alternating series*, defined as

$$S=b_1-b_2+b_3-b_4+\cdots+(-1)^{n-1}b_n+\cdots=\sum_{n=1}^{\infty}(-1)^{n-1}b_n, \tag{3-5-3}$$

the following tests can be made.

(v) Compute $\lim_{n\to\infty}\dfrac{b_{n+1}}{b_n}=\rho$. If $\rho<1$, the series is absolutely convergent; $\rho>1$, the series divergent; $\rho=1$, the convergency cannot be assessed directly.

(vi) If $b_{n+1}\leqslant b_n$, and the limit of b_n is 0, the series is convergent.

(vii) Compute $\rho=\lim_{n\to\infty} n\left(\dfrac{b_n}{b_{n+1}}-1\right)$, and for $b_n>0$, if $\rho>1$, the series is absolutely convergent; if $0<\rho\leqslant 1$, the alternating series is conditional convergent; otherwise, it is divergent.

Example 3.46 Please test the convergency of the following infinite series

$$S=\sum_{n=1}^{\infty}\frac{2^n}{1\times 3\times 5\times\cdots\times(2n-1)}=\sum_{n=1}^{\infty}\frac{2^n}{\prod\limits_{k=1}^{n}(2k-1)}.$$

Solution *For the positive series, the limit of a_{n+1}/a_n can be found easily with*

```
>> syms n k positive; assume(n,'integer'); a=2^n/symprod(2*k-1,k,1,n)
   F=simplify(subs(a,n,n+1)/a), L=simplify(limit(F,n,inf))
```

and it can be seen that the limit is 0, which means that the series is convergent. The term a_{n+1}/a_n can be obtained, however, the simplified form cannot be obtained by computer. Manual simplifications are made, with the ones after the $\Rightarrow$ sign.

$$\frac{a_{n+1}}{a_n}=\frac{4\,(2n)!\,(n+1)!}{(2n+2)!\,n!}\Rightarrow\frac{4(2n)!\,(n+1)n!}{(2n+2)(2n+1)(2n)!\,n!}=\frac{2}{2n+1}.$$

Example 3.47 Test the convergency of the following infinite series

$$\frac{1}{1}+\frac{1}{2}+\frac{1}{3}-\frac{1}{4}-\frac{1}{5}-\frac{1}{6}+\frac{1}{7}+\frac{1}{8}+\frac{1}{9}-\frac{1}{10}-\frac{1}{11}-\frac{1}{12}+\cdots$$

Solution *Since it is well-known that infinite series $\sum_{n=1}^{\infty}1/n$ is not convergent, the above series is not absolutely convergent. Therefore, the convergency cannot be tested in this way. We can express another alternating series with the entities in groups of 3, such that the general term can be written as*

$$b_n = \left(\frac{1}{3n-2}+\frac{1}{3n-1}+\frac{1}{3n}\right),\quad n=1,2,3,\cdots.$$

For the alternating series, (ii) is not applicable, since the limit is 1; with (v), manual processing is needed, although it is obvious that $b_{n+1} \leqslant b_n$. *Test (vii) can be tried with*

```
>> b=1/(3*n-1)+1/(3*n-2)+1/(3*n); L=limit(n*(b/subs(b,n,n+1)-1),n,inf)
```

and, since $L=1>0$, *the alternating series is conditional convergent.*

Example 3.48 A series of function is defined by

$$\sum_{n=1}^{\infty}\left[\frac{1\times 3\times 5\times\cdots\times(2n-1)}{2\times 4\times 6\times\cdots\times(2n)}\right]^p\left(\frac{x-1}{2}\right)^n,\quad p \text{ is real.}$$

Please find the interval of x, such that the infinite series is convergent.

Solution *When the symbolic variables are defined, the general term* a_n *can be expressed, and the limit of* a_{n+1}/a_n *can be taken, and the simplified result is* $L=(x-1)/2$. *To make the infinite series convergent,* $|L|<1$ *should be satisfied, and by solving* $|(x-1)/2|<1$, *it can be found that the convergent interval is* $x\in(-1,3)$.

```
>> syms n k positive; syms p real; assume(n,'integer');
   a=(symprod(2*k-1,k,1,n)/symprod(2*k,k,1,n))^p*((x-1)/2)^n;
   F=simplify(subs(a,n,n+1)/a), L=simplify(limit(F,n,inf))
```

Let $x=-1$, *the series is an alternating one, from (vii), it is found that* $L=p/2$, *which means when* $p>0$, $x=-1$ *is convergent. Therefore,* $x=-1$ *is referred to as a conditional convergent boundary.*

```
>> b=(symprod(2*k-1,k,1,n)/symprod(2*k,k,1,n))^p; % general term
   L=limit(n*(b/subs(b,n,n+1)-1),n,inf)           % with Raabe-like test
```

If $x=3$, *it is a positive series, test (ii) also have* $L=p/2$, *which means that the point is absolutely convergent when* $p>2$, *otherwise, it is divergent. If* $x=-1$, *the series is also absolutely convergent when* $p>2$.

3.6 Path Integrals and Line Integrals

Surprisingly, path integrals and line integrals cannot be solved by the existing MATLAB functions. In this section, the concepts and integration method for path and line integrals are summarized first and then, solutions to these problems will be demonstrated through examples.

3.6.1 Path integrals

Path integrals are originated from the evaluation of the total mass of a spatial wire with unevenly distributed density. Assume that the density of a path l is $f(x,y,z)$. Then, the total mass of the wire can be evaluated from the following equation

$$I_1=\int_l f(x,y,z)\,\mathrm{d}s \tag{3-6-1}$$

where ds is the arc length at a certain point. Thus, this kind of integral is also known as the *integral with respect to arc*. If $f(x, y, z) \equiv 1$, i.e., the density is evenly distributed and equals unity, the total length of the wire is calculated.

If the variables x, y and z are given respectively by parametric equations $x = x(t)$, $y = y(t)$, $z = z(t)$, they can be substituted into the $f(\cdot)$ function, and the differentiation of the arc ds can be written as

$$\mathrm{d}s = \sqrt{\left(\frac{\mathrm{d}x}{\mathrm{d}t}\right)^2 + \left(\frac{\mathrm{d}y}{\mathrm{d}t}\right)^2 + \left(\frac{\mathrm{d}z}{\mathrm{d}t}\right)^2}\,\mathrm{d}t, \text{ or } \mathrm{d}s = \sqrt{x_t^2 + y_t^2 + z_t^2}\,\mathrm{d}t. \tag{3-6-2}$$

Then, the path integral can be converted into an ordinary integral with respect to t

$$I = \int_{t_{\mathrm{m}}}^{t_{\mathrm{M}}} f[x(t), y(t), z(t)]\sqrt{x_t^2 + y_t^2 + z_t^2}\,\mathrm{d}t. \tag{3-6-3}$$

For the integrand with two variables, $f(x, y)$, it can also be converted into ordinary integrals. Therefore, the path integral problem can be solved with MATLAB using the previously described procedures. Especially if the two variables satisfy $y = y(x)$, the integral can be simplified to

$$I = \int_{x_{\mathrm{m}}}^{x_{\mathrm{M}}} f[x, y(x)]\sqrt{1 + y_x^2}\,\mathrm{d}x. \tag{3-6-4}$$

Based on the above formula, a MATLAB function can be written, where the integral can be evaluated in the first part of code, with the syntaxes

I=path_integral(f,[x,y],t,t_{m},t_{M}) % 2D integral
I=path_integral(f,[x,y,z],t,t_{m},t_{M}) % 3D integral

```
function I=path_integral(F,vars,t,a,b)
if length(F)==1, I=int(F*sqrt(sum(diff(vars,t).^2)),t,a,b);
else, F=F(:).'; vars=vars(:); I=int(F*diff(vars,t),t,a,b); end
```

In the syntaxes, [x, y] or [x,y,z] are the parametric equations of the curve, with two and three independent variables. If the curve is defined as $y = f(x)$, the vector can also be specified as [x,y].

Example 3.49 Compute $\displaystyle\int_l \frac{z^2}{x^2 + y^2}\,\mathrm{d}s$, where the path l is defined by parametric equations as $x = a\cos t, y = a\sin t, z = at$, with $0 \leqslant t \leqslant 2\pi$ and $a > 0$.

Solution *The following statements can be used for this path integral problem*

```
>> syms t; syms a positive; x=a*cos(t); y=a*sin(t); z=a*t;
   f=z^2/(x^2+y^2); I=path_integral(f,[x,y,z],t,0,2*pi)
```

and the result is $I = \dfrac{8\sqrt{2}}{3}\pi^3 a$.

Example 3.50 Compute $\displaystyle\int_l (x^2 + y^2)\,\mathrm{d}s$ where path l is defined as the positive direction curve encircled by the paths $y = x$ and $y = x^2$.

Solution *The following statements can be used to draw the two paths shown in Figure 3.8. The arrows, shown in counterclockwise direction are defined as the positive direction.*

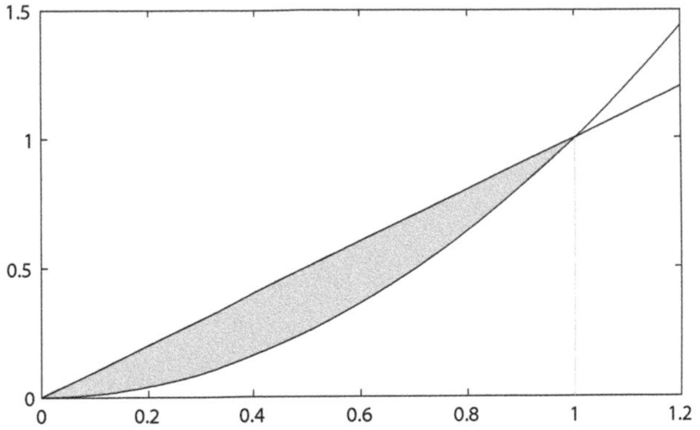

FIGURE 3.8: Illustration of the integration paths.

```
>> x=0:.001:1.2; y1=x; y2=x.^2; plot(x,y1,x,y2) % integral path
```

It can be seen that the original integration problem can be divided into two sub-integration problems. Therefore, the following statements can be used to add the two sub-integrals up to get the final solutions

```
>> syms x; y=x; f=(x^2+y^2); I1=path_integral(f,[x,y],x,1,0)
   y=x^2; f=(x^2+y^2); I2=path_integral(f,[x,y],x,0,1), I=I1+I2
```

and the result is

$$I = \frac{349}{768}\sqrt{5} - \frac{2}{3}\sqrt{2} - \frac{7\ln\left(\sqrt{5}+2\right)}{512}.$$

3.6.2 Line integrals

Line integral problems are originated from physics, where the total work is done by the force $\boldsymbol{f}(x,y,z)$ along a spatial curve l. This kind of integral problem can be expressed as

$$I_2 = \int_l \boldsymbol{f}(x,y,z)\,\mathrm{d}\boldsymbol{s} \tag{3-6-5}$$

where $\boldsymbol{f}(x,y,z) = [P(x,y,z), Q(x,y,z), R(x,y,z)]$ is a row vector. The differentiation of the line $\mathrm{d}\boldsymbol{s}$ is a column vector. If the line can be described by a parametric equation of t such as $x(t), y(t), z(t)$, with $t \in (a,b)$, the vector $\mathrm{d}\boldsymbol{s}$ can then be written as

$$\mathrm{d}\boldsymbol{s} = \left[\frac{\mathrm{d}x}{\mathrm{d}t}, \frac{\mathrm{d}y}{\mathrm{d}t}, \frac{\mathrm{d}z}{\mathrm{d}t}\right]^{\mathrm{T}} \mathrm{d}t. \tag{3-6-6}$$

The dot product of two vectors can be carried out directly, and the line integrals can be redefined as an ordinary integral as follows:

$$I_2 = \int_a^b [P(x,y,z), Q(x,y,z), R(x,y,z)] \left[\frac{\mathrm{d}x}{\mathrm{d}t}, \frac{\mathrm{d}y}{\mathrm{d}t}, \frac{\mathrm{d}z}{\mathrm{d}t}\right]^{\mathrm{T}} \mathrm{d}t \tag{3-6-7}$$

which can also be solved by using MATLAB function `path_integral()`, and the code is given in the second part of the function. The syntaxes of the function are

```
I=path_integral([P,Q],[x,y],t,a,b)          % 2D and 3D integrals
I=path_integral([P,Q,R],[x,y,z],t,a,b)  % with vector integrand
```

Example 3.51 Compute the integral $\int_l \frac{x+y}{x^2+y^2}\,dx - \frac{x-y}{x^2+y^2}\,dy$, where the line l is defined as the positive circle given by $x^2+y^2=a^2$, $a>0$.

Solution *If one wants to evaluate the line integral, the circle can be interpreted as the parametric equations $x=a\cos t$, $y=a\sin t$ for $0\leqslant t\leqslant 2\pi$. Thus, the following statements can be used to calculate the line integral, with the result $I=2\pi$.*

```
>> syms t; syms a positive; x=a*cos(t); y=a*sin(t);
   F=[(x+y)/(x^2+y^2),-(x-y)/(x^2+y^2)];
   I=path_integral(F,[x,y],t,2*pi,0) % integral
```

Example 3.52 Compute the line integral $\int_l (x^2-2xy)\,dx + (y^2-2xy)\,dy$, where the line l is defined as the parabolic curve $y=x^2$ $(-1\leqslant x\leqslant 1)$.

Solution *In fact, the equations given are already the parametric equations of x. The derivative of x with respective to x is 1. The following statements can be used to solve the line integral problem, with the result $I=-14/15$.*

```
>> syms x; y=x^2; F=[x^2-2*x*y,y^2-2*x*y]; % define 2D vector integrand
   I=path_integral(F,[x,y],x,-1,1)          % evaluate integral
```

3.7 Surface Integrals

Two types of surface integrals are considered in this section, the scalar type and the vector type. The definitions and solutions to the problems will be summarized first followed by the detailed solution procedures with MATLAB script-based examples.

3.7.1 Scalar surface integrals

The scalar-type surface integrals are defined as

$$I=\iint_S \phi(x,y,z)\,dS \tag{3-7-1}$$

where dS is the differentiated area. Thus, this kind of integral is also referred to as the *surface integrals with respect to area.* If $\phi(x,y,z)\equiv 1$, the area of the surface can be computed.

Let the surface S be defined by $z=f(x,y)$. The original surface integral can be converted into a double integral over the x-y plane, such that

$$I=\iint_{\sigma_{xy}} \phi[x,y,f(x,y)]\sqrt{1+f_x^2+f_y^2}\,dx\,dy \tag{3-7-2}$$

where σ_{xy} is the integration region, which is an ordinary double integral problem.

The MATLAB implementation of the above formula is given in the first part of the following function, with the syntax

$I=$ `surf_integral(`f`,`z`,[`x`,`y`],[`y_m`,`y_M`],[`x_m`,`x_M`])`

```
function I=surf_integral(f,xx,uu,um,vm)
if length(f)==1 % scalar surface integral
   if length(xx)==1 % surface by explicit function
      I=int(int(f*sqrt(1+diff(xx,uu(1))^2+diff(xx,uu(2))^2),...
          uu(2),um(1),um(2)),uu(1),vm(1),vm(2));
   else    % surface described by parametric equation
      xx=[xx(:).' 1]; x=xx(1); y=xx(2); z=xx(3); u=uu(1); v=uu(2);
      E=diff(x,u)^2+diff(y,u)^2+diff(z,u)^2;
      F=diff(x,u)*diff(x,v)+diff(y,u)*diff(y,v)+diff(z,u)*diff(z,v);
      G=diff(x,v)^2+diff(y,v)^2+diff(z,v)^2;
      I=int(int(f*sqrt(E*G-F^2),u,um(1),um(2)),v,vm(1),vm(2));
   end
else % vector surface integral
   if length(xx)==1 % surface by explicit function
      syms x y z; ua=sqrt(1+diff(xx,x)^2+diff(xx,y)^2);
      cA=-diff(xx,x)/ua; cB=-diff(xx,y)/ua; cC=1/ua;
      I=surf_integral(f(:).'*[cA; cB; cC],xx,uu,um,vm);
   else, x=xx(1); y=xx(2); z=xx(3); u=uu(1); v=uu(2);
      A=diff(y,u)*diff(z,v)-diff(z,u)*diff(y,v);
      B=diff(z,u)*diff(x,v)-diff(x,u)*diff(z,v);
      C=diff(x,u)*diff(y,v)-diff(y,u)*diff(x,v); % compute with (3-7-11)
      F=A*f(1)+B*f(2)+C*f(3);                    % integrand
      I=int(int(F,uu(1),um(1),um(2)),uu(2),vm(1),vm(2));
   end
end
```

Example 3.53 Compute $\displaystyle\iint_S xyz\,\mathrm{d}S$, where the integral surface S is defined as the region enclosed by the four planes $x=0$, $y=0$, $z=0$, $x+y+z=a$ and $a>0$.

Solution *Denote the four planes by* S_1,S_2,S_3 *and* S_4. *The original surface integral can be calculated using* $\displaystyle\iint_S=\iint_{S_1}+\iint_{S_2}+\iint_{S_3}+\iint_{S_4}$. *Considering the planes* S_1, S_2, S_3, *since the integrands are all 0, only the integral on the* S_4 *should be considered. The plane* S_4 *can mathematically be described as* $z=a-x-y$, *and the area of integral is then* $0\leqslant y\leqslant a-x$, $0\leqslant x\leqslant a$. *Then, the following statements can be used to evaluate the surface integral*

```
>> syms x y; syms a positive; z=a-x-y; f=x*y*z;
   I=surf_integral(f,z,[x,y],[0,a-x],[0,a])
```

which gives $I=\sqrt{3}a^5/120$.

If the parametric equations for the surface are given by

$$x=x(u,v),\ y=y(u,v),\ z=z(u,v), \tag{3-7-3}$$

the surface integral can then be obtained using the following formula

$$I = \iint_{\Sigma} \phi[x(u,v), y(u,v), z(u,v)]\sqrt{EG-F^2}\ \mathrm{d}u\,\mathrm{d}v \tag{3-7-4}$$

where

$$E = x_u^2 + y_u^2 + z_u^2,\ F = x_u x_v + y_u y_v + z_u z_v,\ G = x_v^2 + y_v^2 + z_v^2. \tag{3-7-5}$$

MATLAB implementation of the above formula are given in the second part of the function `surf_integral()`, with the syntax

$I=$`surf_integral(`f`,[`x`,`y`,`z`],[`u`,`v`],[`u_m`,`u_M`],[`v_m`,`v_M`])`

Example 3.54 Compute the surface integral $\displaystyle\iint_S (x^2y + zy^2)\,\mathrm{d}S$, where the surface S is defined as $x = u\cos v, y = u\sin v, z = v,\ 0 \leqslant u \leqslant a,\ 0 \leqslant v \leqslant 2\pi$.

Solution *The following statements can be used to calculate the integrals*

```
>> syms u v; syms a positive; x=u*cos(v); y=u*sin(v); z=v;
   f=x^2*y+z*y^2; I=surf_integral(f,[x,y,z],[u,v],[0,a],[0,2*pi])
```

and the result is $I = \dfrac{1}{8}\pi^2\left[2a^3\sqrt{a^2+1} - a\sqrt{a^2+1} - \mathrm{arcsinh}a\right]$.

3.7.2 Vector surface integrals

The second category of surface integral is also referred to as the *surface integrals in vector fields.* Suppose the integrand is given by a row vector $\boldsymbol{\Gamma} = [P,Q,R]$, while $\mathrm{d}\boldsymbol{v}$ is given by a column vector $\mathrm{d}\boldsymbol{v} = [\mathrm{d}y\,\mathrm{d}z, \mathrm{d}x\,\mathrm{d}z, \mathrm{d}x\,\mathrm{d}y]^\mathrm{T}$, the mathematical description to the problem is

$$I = \iint_{S^+} \boldsymbol{\Gamma}\,\mathrm{d}\boldsymbol{v} = \iint_{S^+} P(x,y,z)\,\mathrm{d}y\,\mathrm{d}z + Q(x,y,z)\,\mathrm{d}x\,\mathrm{d}z + R(x,y,z)\,\mathrm{d}x\,\mathrm{d}y, \tag{3-7-6}$$

where the positive surface S^+ is defined with $z = f(x,y)$. The surface integral problem can then be converted into the scalar surface integral problem

$$I = \iint_{S^+} [P(x,y,z)\cos\alpha + Q(x,y,z)\cos\beta + R(x,y,z)\cos\gamma]\,\mathrm{d}\boldsymbol{S} \tag{3-7-7}$$

where z is replaced by $f(x,y)$, and

$$\cos\alpha = \frac{-f_x}{\sqrt{1+f_x^2+f_y^2}}, \cos\beta = \frac{-f_y}{\sqrt{1+f_x^2+f_y^2}}, \cos\gamma = \frac{1}{\sqrt{1+f_x^2+f_y^2}}. \tag{3-7-8}$$

Therefore, the $\sqrt{1+f_x^2+f_y^2}$ term may cancel the relevant term in (3-7-2), and the surface integral can be written as

$$I = \iint_{\sigma_{xy}} -Pf_x\,\mathrm{d}y\,\mathrm{d}z - Qf_y\,\mathrm{d}x\,\mathrm{d}z + R\,\mathrm{d}x\,\mathrm{d}y. \tag{3-7-9}$$

If the surface is described by the parametric equations in (3-7-3), the following equations can be obtained

$$\cos\alpha = \frac{A}{\sqrt{A^2+B^2+C^2}},\quad \cos\beta = \frac{B}{\sqrt{A^2+B^2+C^2}},\quad \cos\gamma = \frac{C}{\sqrt{A^2+B^2+C^2}} \tag{3-7-10}$$

where

$$A = y_u z_v - z_u y_v,\ B = z_u x_v - x_u z_v,\ C = x_u y_v - y_u x_v. \tag{3-7-11}$$

Then, from the converted scalar surface integral (3-7-7), it can be found that the denominator in (3-7-10) cancels the $\sqrt{EG-F^2}$ term. Thus, the vector surface integral can be simplified as the following standard double integral

$$I = \int_{v_{\rm m}}^{v_{\rm M}} \int_{u_{\rm m}(v)}^{u_{\rm M}(v)} [AP(u,v) + BQ(u,v) + CR(u,v)]\,{\rm d}u\,{\rm d}v. \tag{3-7-12}$$

The above algorithms are implemented in `surf_integral()` function, in the final part of the code, with the syntaxes

```
I=surf_integral([P,Q,R],z,[u,v],[um,uM],[vm,vM])
I=surf_integral([P,Q,R],[x,y,z],[u,v],[um,uM],[vm,vM])
```

Example 3.55 Compute the surface integral $\iint x^3\,{\rm d}y\,{\rm d}z$, where the surface S is defined as the positive side of the ellipsoid surface $x^2/a^2 + y^2/b^2 + z^2/c^2 = 1$.

Solution *The parametric equations can be introduced such that* $x = a\sin u\cos v$, $y = b\sin u\sin v$, $z = c\cos u$ *and* $0 \leqslant u \leqslant \dfrac{\pi}{2}$, $0 \leqslant v \leqslant 2\pi$. *The following statements can be used to compute the surface integral, with the result* $I = 2\pi a^3 cb/5$.

```
>> syms u v; syms a b c positive;
   x=a*sin(u)*cos(v); y=b*sin(u)*sin(v); z=c*cos(u);
   I=surf_integral([x^3,0,0],[x,y,z],[u,v],[0,pi/2],[0,2*pi])
```

3.8 Numerical Differentiation

If the original function is symbolically given, the analytical solutions to the differentiation problem can be obtained directly with the MATLAB built-in function `diff()`. The 100th order derivative can be obtained within seconds. However, in some applications where the original function is not known, only experimental data are given, the analytical or symbolic methods cannot be used. In this case, numerical methods must be used to get the derivatives from the experimental data. There is no dedicated function available in solving numerical differentiation problems in MATLAB. Thus, simple numerical algorithms are presented in this section with detailed implementation of the algorithms together with examples on how to solve the numerical differentiation problems in MATLAB.

3.8.1 Numerical differentiation algorithms

Assume that there is a set of measured data (t_i, y_i) with evenly distributed time instances $t_i = i\Delta t$, $i = 1, \cdots, N$, and the sample time is Δt. Theoretically, the derivative is defined as

$$\frac{{\rm d}y(t)}{{\rm d}t} = \lim_{\Delta t\to 0} \frac{y(t+\Delta t) - y(t)}{\Delta t} \tag{3-8-1}$$

However, unfortunately, in practical applications, the condition $\Delta t \to 0$ cannot be satisfied. Therefore, only an approximated derivative can be obtained

$$y_i' \approx \frac{\Delta y_i}{\Delta t};\quad y_i' = \frac{y_{i+1}-y_i}{\Delta t} + \mathrm{o}(\Delta t). \tag{3-8-2}$$

where, $\mathrm{o}(\Delta t)$ is *infinitesimal quantity*, and it virtually means that the error is similar to the scale of Δt. This formula is also referred to as the *forward difference algorithm.*

Similarly, *backward difference formula* is defined as

$$y_i' \approx \frac{\Delta y_i}{\Delta t};\quad y_i' = \frac{y_i-y_{i-1}}{\Delta t} + \mathrm{o}(\Delta t). \tag{3-8-3}$$

When the value of Δt is large, the accuracy of the differentiation cannot be guaranteed. So other improved numerical differentiation algorithms should be considered. For instance, the *central-point algorithm* can be used. The first-order derivative can also be defined as

$$y_i' \approx \frac{y_{i+1}-y_{i-1}}{2\Delta t} + \mathrm{o}(\Delta t^2). \tag{3-8-4}$$

High-order differentiation formulae can be similarly derived as follows:

$$\begin{aligned} y_i'' &\approx \frac{y_{i+1}-2y_i+y_{i-1}}{\Delta t^2} \\ y_i''' &\approx \frac{y_{i+2}-2y_{i+1}+2y_{i-1}-y_{i-2}}{2\Delta t^3} \\ y_i^{(4)} &\approx \frac{y_{i+2}-4y_{i+1}+6y_i-4y_{i-1}+y_{i-2}}{\Delta t^4}. \end{aligned} \tag{3-8-5}$$

There is yet another set of central-point difference algorithms with even higher accuracy of $\mathrm{o}(\Delta t^4)$, defined as follows:

$$\begin{aligned} y_i' &\approx \frac{-y_{i+2}+8y_{i+1}-8y_{i-1}+y_{i-2}}{12\Delta t} \\ y_i'' &\approx \frac{-y_{i+2}+16y_{i+1}-30y_i+16y_{i-1}-y_{i-2}}{12\Delta t^2} \\ y_i''' &\approx \frac{-y_{i+3}+8y_{i+2}-13y_{i+1}+13y_{i-1}-8y_{i-2}+y_{i-3}}{8\Delta t^3} \\ y_i^{(4)} &\approx \frac{-y_{i+3}+12y_{i+2}-39y_{i+1}+56y_i-39y_{i-1}+12y_{i-2}-y_{i-3}}{6\Delta t^4}. \end{aligned} \tag{3-8-6}$$

3.8.2 Central-point difference algorithm with MATLAB implementation

The numerical differentiation algorithm given in (3-8-6) has the error level of $\mathrm{o}(\Delta t^4)$ which can be used to solve numerical differentiation problems with higher numerical accuracy. Even when Δt is not too small, good approximation can still be expected due to its error level. Based on the algorithm, a MATLAB function is prepared as follows:

```
function [dy,dx]=diff_ctr(y,Dt,n)
y1=[y 0 0 0 0 0 0]; y2=[0 y 0 0 0 0 0]; y3=[0 0 y 0 0 0 0];
y4=[0 0 0 y 0 0 0]; y5=[0 0 0 0 y 0 0]; y6=[0 0 0 0 0 y 0];
y7=[0 0 0 0 0 0 y];
switch n
```

```
case 1,
   dy=(-y1+8*y2-8*y4+y5)/12/Dt;
case 2,
   dy=(-y1+16*y2-30*y3+16*y4-y5)/12/Dt^2;
case 3,
   dy=(-y1+8*y2-13*y3+13*y5-8*y6+y7)/8/Dt^3;
case 4,
   dy=(-y1+12*y2-39*y3+56*y4-39*y5+12*y6-y7)/6/Dt^4;
end
dy=dy(5+2*(n>2):end-4-2*(n>2)); dx=([2:length(dy)+1]+(n>2))*Dt;
```

The syntax of the function is $[\boldsymbol{d}_y,\boldsymbol{d}_x]=$ `diff_ctr`$(\boldsymbol{y},\Delta t,n)$, where $\boldsymbol{y}$ is the vector containing measured data for evenly distributed points, and Δt is the sample time. The argument n specifies the order of derivatives. The returned arguments $\boldsymbol{d}_y$ is the derivative vector computed, while the argument $\boldsymbol{d}_x$ is the corresponding vector of independent variables. It should be noted that the two vectors are a few points shorter than the original $\boldsymbol{y}$ vector.

Example 3.56 The function defined in Example 3.13 is still used in the demonstration of the algorithm. Since the original function is known, the analytical solution can be obtained for comparison. Sample data of the function can be generated from the function, and with the help of the data, the derivatives of the first- up to the fourth-order can be calculated, and the results can be compared with the analytical solutions.

Solution *An evenly spaced vector* $\boldsymbol{x}$ *is generated first. Since the original function is known, the analytical solutions to derivatives can be obtained. Then, if one substitutes the vector* $\boldsymbol{x}$ *into the obtained analytical functions, the theoretical derivative vectors can be obtained for comparison.*

```
>> h=0.05; x=0:h:pi; syms x1; y=sin(x1)/(x1^2+4*x1+3);
   yy1=diff(y); f1=subs(yy1,x1,x); % get the contrast data analytically
   yy2=diff(yy1); f2=subs(yy2,x1,x); yy3=diff(yy2); f3=subs(yy3,x1,x);
   yy4=diff(yy3); f4=subs(yy4,x1,x);
```

From the data points y_i *generated above, the first-order up to the fourth-order derivatives from the data can be calculated easily with the function* `diff_ctr()`*, and the results are shown in Figure 3.9 together with the exact solutions. It can be seen that one may not observe the difference.*

```
>> y=sin(x)./(x.^2+4*x+3);       % generate the data to be used
   [y1,dx1]=diff_ctr(y,h,1); subplot(221), plot(x,f1,dx1,y1,':');
   [y2,dx2]=diff_ctr(y,h,2); subplot(222), plot(x,f2,dx2,y2,':')
   [y3,dx3]=diff_ctr(y,h,3); subplot(223), plot(x,f3,dx3,y3,':');
   [y4,dx4]=diff_ctr(y,h,4); subplot(224), plot(x,f4,dx4,y4,':')
```

Quantitative studies for the fourth-order derivative show that the maximum error between the exact results and the calculated results is as small as 3.5025×10^{-4}.

```
>> double(norm((y4-f4(4:60))./f4(4:60))) % find the norm of the error in f^(4)(x)
```

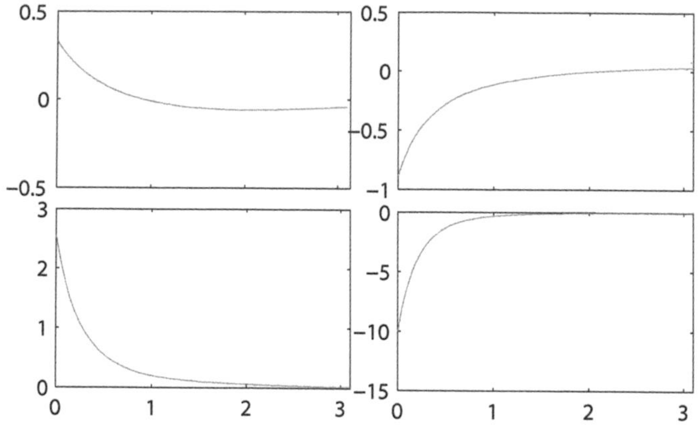

FIGURE 3.9: Comparisons of derivatives of different orders.

3.8.3 Gradient computations of functions with two variables

Consider the function $z(x, y)$ with two variables representing a 3D surface. The function `gradient()` can be used to calculate the gradients for the function. The syntax of the function is `[`f_x`,`f_y`]=gradient(`z`)`, where the "gradients" f_x and f_y thus calculated are not the actual gradients, since the coordinates x and y are not considered. If the matrix z is obtained, the gradients can be obtained using the following statements $f_x = f_x/\Delta x$, $f_y = f_y/\Delta y$, where Δx and Δy are respectively the step-sizes for x and y.

Example 3.57 Consider the function given in Example 3.16. Assume that the mesh grid data can be generated. Compute the gradients of the original function and analyze the error.

Solution *The data can be generated using the following statements. The gradients here are obtained from the data rather than from the analytical function. The 3D attractive curves can also be drawn as shown in Figure 3.10, and it should be the same as the one in Figure 3.3 (b).*

```
>> syms x y; z=(x^2-2*x)*exp(-x^2-y^2-x*y); % symbolic expression
   [x0,y0]=meshgrid(-3:.2:3,-2:.2:2); z0=subs(z,{x,y},{x0,y0}); % function
   [fx,fy]=gradient(double(z0)); fx=fx/0.2; fy=fy/0.2; % compute gradient
   contour(x0,y0,z0,30); hold on; quiver(x0,y0,fx,fy)   % draw quiver plot
```

The error surface is shown in Figure 3.10 where it can be seen that in most regions, the errors are relatively small. In other areas, the errors are large. This means that the spacing in the grid is too large to provide accurate gradient information. In order to reduce the error, the step-size should be reduced.

```
>> zx=diff(z,x); zx0=double(subs(zx,{x,y},{x0,y0}));
   zy=diff(z,y); zy0=double(subs(zy,{x,y},{x0,y0}));
   subplot(121), surf(x0,y0,abs(fx-zx0)); axis([-3 3 -2 2 0,0.08])
   subplot(122), surf(x0,y0,abs(fy-zy0)); axis([-3 3 -2 2 0,0.11])
```

If the spacing in grids is reduced both by half, the following statements can be used, and the new error surface can be calculated again as shown in Figure 3.11. It can be observed that the error is also reduced compared to Figure 3.10.

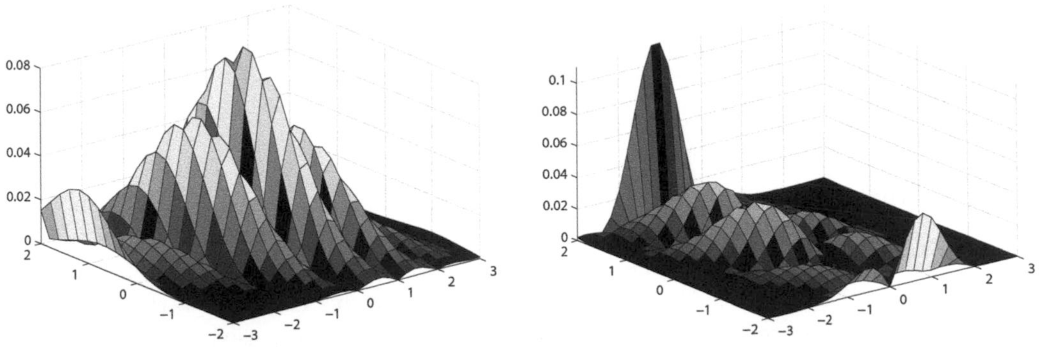

(a) $\partial z/\partial x$ (b) $\partial z/\partial y$

FIGURE 3.10: Error surface of the gradient of the function.

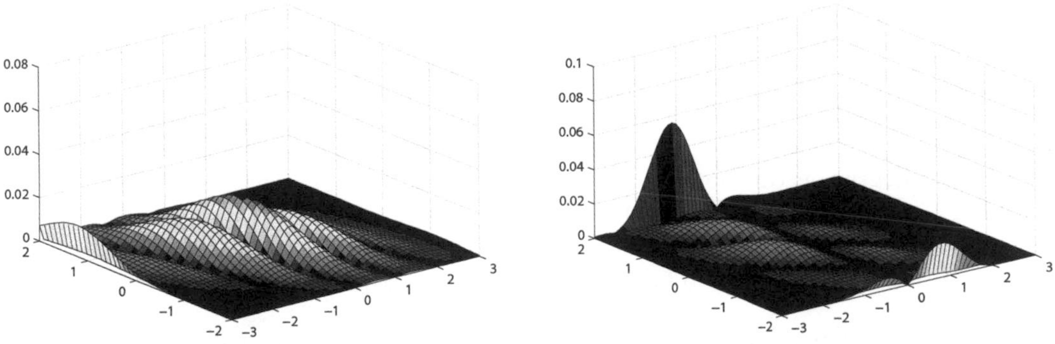

(a) $\partial z/\partial x$ (b) $\partial z/\partial y$

FIGURE 3.11: The error surface with reduced spacing in mesh grids.

```
>> [x1,y1]=meshgrid(-3:.1:3,-2:.1:2); z1=subs(z,{x,y},{x1,y1});
   [fx,fy]=gradient(double(z1)); fx=fx/0.1; fy=fy/0.1;
   z1=double(subs(zx,{x,y},{x1,y1})); z2=double(subs(zy,{x,y},{x1,y1}));
   subplot(121), surf(x1,y1,abs(fx-z1)); axis([-3 3 -2 2 0,0.08])
   subplot(122), surf(x1,y1,abs(fy-z2)); axis([-3 3 -2 2 0,0.1])
```

3.9 Numerical Integration Problems

Numerical integration problems are important in traditional numerical analysis courses. In this section, several cases are considered. If the integrand is not known, integral can be approximated based on a set of measured data. If the integrand is known but not integrable, various integrals can be approximated numerically.

3.9.1 Numerical integration from given data using trapezoidal method

Definite integral of a univariate function is defined as

$$I = \int_a^b f(x)\,\mathrm{d}x. \tag{3-9-1}$$

It is known that if the integrand $f(x)$ is theoretically not integrable, even with the powerful computer program, the analytical solutions to the problem cannot be obtained. Thus, numerical solutions to the problems should be pursued instead. Numerical computation of an integral of univariate function is also known as *quadrature*. There are various numerical quadrature algorithms to solve the integration problem. The widely used algorithms include the trapezoidal method, the Simpson's algorithm, the Romberg's algorithm, etc. The basic idea of the algorithms is to divide the whole interval $[a, b]$ into several sub-intervals $[x_i, x_{i+1}]$, $i = 1, 2, \cdots, N$, where $x_1 = a$ and $x_{N+1} = b$. Then, the integration problem can be converted to the summation problem as follows:

$$\int_a^b f(x)\,\mathrm{d}x = \sum_{i=1}^{N} \int_{x_i}^{x_{i+1}} f(x)\,\mathrm{d}x = \sum_{i=1}^{N} \Delta f_i. \tag{3-9-2}$$

The easiest method is to use trapezoidal approximation to each sub-interval. The numerical integration can be obtained by the use of `trapz()` function, whose syntax is S=`trapz(`$\boldsymbol{x}$`,`$\boldsymbol{y}$`)`, where $\boldsymbol{x}$ is a vector, and the number of rows of matrix $\boldsymbol{y}$ equals the number of the elements in vector $\boldsymbol{x}$. If the variable $\boldsymbol{y}$ is given as a multi-column matrix, the numerical integration to several functions can be evaluated simultaneously.

Example 3.58 Compute the definite integrals to the functions $\sin x, \cos x, \sin x/2$ within the interval $x \in (0, \pi)$ using the trapezoidal algorithm.

Solution *The vector for horizontal axis is generated first and from it, the values of different functions can be evaluated such that the numerical integration can be obtained*

```
>> x1=[0:pi/30:pi]'; y=[sin(x1) cos(x1) sin(x1/2)]; S=trapz(x1,y)
```

and the results are $S = [1.99817196134365, 0, 1.99954305299081]$.

Since the step-size is selected as $h = \pi/30 \approx 0.1$ *which is considered as quite large, there exist errors in the results. In Section 8.1.2, the algorithm will be used with the interpolation method to improve the quality of numerical integration results.*

Example 3.59 Compute $\displaystyle\int_0^{3\pi/2} \cos 15x\,\mathrm{d}x$ with various step-sizes.

Solution *Before solving the problem, the following statements can be used to draw the curves of the integrand as shown in Figure 3.12. It can be seen that there exists strong oscillation in the integrand.*

```
>> x=[0:0.01:3*pi/2, 3*pi/2]; y=cos(15*x); plot(x,y) % integrand curve
```

The theoretical solution to the problem is 1/15. For different step-sizes, $h = 0.1$*, 0.01, 0.001, 0.0001, 0.00001, 0.000001, the following statements can be used in approximately solving the integrals. The relevant results are given in Table 3.1.*

```
>> syms x, A=int(cos(15*x),0,3*pi/2) % analytical solution
   h0=pi./[30,300,3000,30000,300000,3000000]; v=[]; H=3*pi/2;
```

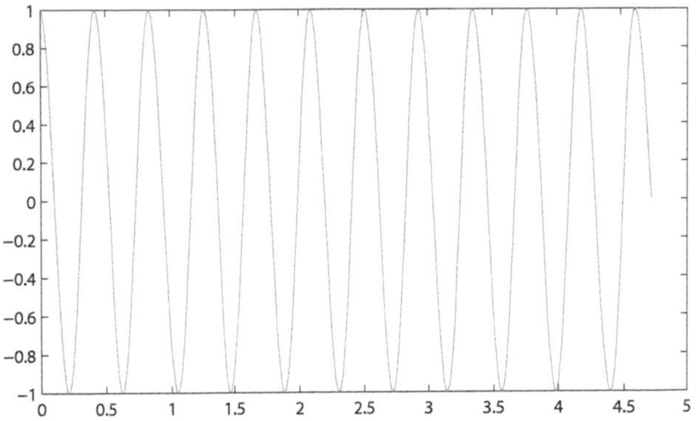

FIGURE 3.12: The plot of the integrand $f(x) = \cos 15x$.

TABLE 3.1: Step-size selection and computation results.

step-size	integral	error	time (s)	step-size	integral	error	time (s)
$\pi/30$	0.052359878	0.0143	0.000	$\pi/30000$	0.066666653	1.35×10^{-8}	0.0034
$\pi/300$	0.066529532	0.00014	0.0002	$\pi/300000$	0.066666667	1.35×10^{-10}	0.0356
$\pi/3000$	0.066665296	1.35×10^{-6}	0.005	$\pi/3000000$	0.066666667	1.35×10^{-12}	0.266

```
for h=h0, % evaluate integral with different step-sizes
   tic, x=[0:h:H]; y=cos(15*x); I=trapz(x,y); v=[v; h,I,1/15-I]; toc
end
```

It can be seen that when the step-size h reduces, the integral accuracy increases. For instance, if the step-size is selected as $h = 10^{-6}$, 11 digits can be preserved in the result. Thus, for this example, it takes as long as eight seconds for computation. If the step-size is further reduced, the computational effort demanded will be too high to be accepted.

3.9.2 Numerical integration of univariate functions

In traditional numerical analysis courses, several other numerical algorithms are usually explored for univariate functions.

A new powerful numerical integration function `integral()` is introduced in MATLAB 8.0, and the syntaxes are

I = `integral(`*fun*`,`a`,`b`)`, I = `integral(`*fun*`,`a`,`b`,`*property pairs*`)`

where *fun* can be used to specify the integrand. It can either be an M-file saved in `fun.m` file, or an anonymous function or an inline function. The syntax of such a function should be y = `fun(`x`)`. The arguments a and b are the lower- and upper-bounds in the definite integral, respectively. Other specifications can be described with *property pairs* and will be illustrated through examples later. In earlier versions, the functions `quad()`, `quadl()` and `quadgk()` should be used instead.

Example 3.60 For the integral $\mathrm{erf}(x) = \dfrac{2}{\sqrt{\pi}}\displaystyle\int_0^x \mathrm{e}^{-t^2}\,\mathrm{d}t$, which was shown not integrable, compute the integral using numerical methods.

Solution *Before finding the numerical integration of a given function, the integrand should be specified first. There are three ways for specifying the integrand.*

(i) **M-function** *The first method is to express the integrand using a MATLAB function, where the input argument is the variable x. Since many x values need to be processed simultaneously, $\boldsymbol{x}$ vector can finally be used as the input argument, and the computation within the function should be expressed in dot operations. An example for expressing such a function is shown as follows, saved in* `c3ffun.m` *file.*

```
function y=c3ffun(x)      % describe integrand with M-function
y=2/sqrt(pi)*exp(-x.^2); % dot operation should be used here
```

(ii) **Anonymous function** *Anonymous function expression is an effective way for describing the integrand. The format of the function is even more straightforward than the inline expression. The integrand can be expressed by the anonymous function as follows:*

```
>> f=@(x)2/sqrt(pi)*exp(-x.^2); % with anonymous function, no file needed
```

(iii) **Inline function** *The integrand can also be described by the old fashioned inline function, where the input argument $\boldsymbol{x}$ should be appended after the integrand expression.*

```
>> f=inline('2/sqrt(pi)*exp(-x.^2)','x');% inline function not recommended
```

It should be pointed out that the anonymous function expression is the fastest among the three. The drawbacks of the representation are that it can only return one argument, and function evaluations with intermediate computations are not allowed. Thus, the anonymous function is used throughout the book whenever possible. If anonymous function cannot be used, the M-function description will be used.

When the integrand has been declared by any of the above three methods, the `integral()` *function can be used to solve the definite integral problem*

```
>> f=@(x)2/sqrt(pi)*exp(-x.^2); % anonymous function expression
   I1=integral(f,0,1.5), I2=integral(@c3ffun,0,1.5)% same results
```

and $I_1 = I_2 = 0.966105146475311$. In fact, the high-precision solution to the same problem can be obtained with the use of Symbolic Math Toolbox

```
>> syms x, y0=vpa(int(2/sqrt(pi)*exp(-x^2),0,1.5),60)
```

where $y_0 = 0.966105146475310713936933729949905794996224943257461473285 75$.

Comparing the results obtained above, it can be found that the accuracy of the numerical method is the highest possible under double-precision scheme.

Example 3.61 Compute the integral of a piecewise function

$$I = \int_0^4 f(x)\,\mathrm{d}x, \text{ where } f(x) = \begin{cases} \mathrm{e}^{x^2}, & 0 \leqslant x \leqslant 2 \\ \dfrac{80}{4-\sin(16\pi x)}, & 2 < x \leqslant 4. \end{cases}$$

Solution *The piecewise function is displayed in filled curve in Figure 3.13. It can be seen that the curve is not continuous at $x = 2$ point.*

```
>> x=[0:0.01:2, 2+eps:0.01:4,4]; % x vector, with an extra value at 2+ε
   y=exp(x.^2).*(x<=2)+80./(4-sin(16*pi*x)).*(x>2); % piecewise function
   y(end)=0; x=[eps, x]; y=[0,y]; fill(x,y,'g')       % intrgral area in green
```

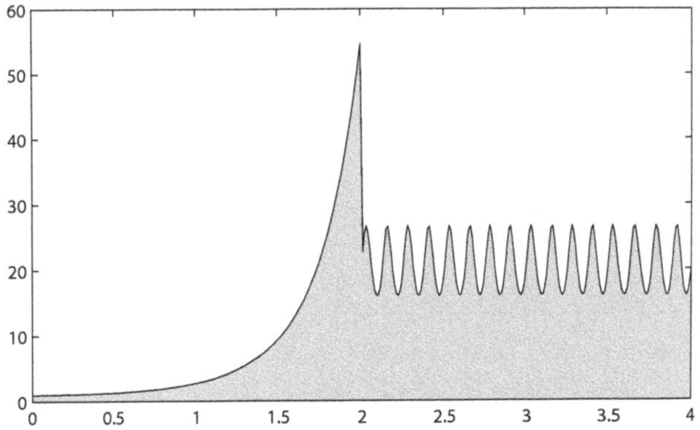

FIGURE 3.13: Filled plot of the integrand.

With the use of relationship expressions, the integrand can be described and the functions `integral()` *can be used respectively to solve the original problem*

```
>> f=@(x)exp(x.^2).*(x<=2)+80*(x>2)./(4-sin(16*pi*x));
   I1=integral(f,0,4) % evaluate integral with default setting
```

and it is found that $I_1 = 57.764450125048498$.

In fact, the original problem can also be divided into the integrals over $(0, 2)$ *and* $(2, 4)$ *intervals. The analytical solution function* `int()` *can then be used to find the analytical solutions to the original problem*

```
>> syms x; I=vpa(int(exp(x^2),0,2)+int(80/(4-sin(16*pi*x)),2,4))
```

with $I = 57.7644501250530103333152353851 82$.

Compared with the analytical solutions, the results obtained by the `integral()` *function is not quite accurate. The* property pair *argument should be used. In this case, it is better to the relative error tolerance, denoted by* `RelTol`*, to a very small number*

```
>> I2=integral(f,0,4,'RelTol',1e-20) % evaluate integral with error tolerance
```

the new result obtained is $I_2 = 57.764450125053017$, *again it is satisfactory under double-precision scheme.*

Example 3.62 Compute again the integral defined in Example 3.59.

Solution *From the fixed-step algorithm demonstrated in Example 3.59, it can be found that only when the step-size is selected to be a very small value, the high accuracy can be achieved. However, with the help of variable-step algorithms, the original problem can be solved within a much shorter time and with much higher accuracy.*

```
>> f=@(x)cos(15*x); tic, S=integral(f,0,3*pi/2), toc % numerical integral
```

It can be found that $S = 0.06666666666667$, *and the elapsed time is 0.005 seconds, much faster and more accurate than the fixed-step method.*

Thus, it can be concluded that the variable-step of integrals has much more advantages than the fixed-step method taught in numerical analysis courses. In real applications, this function is highly recommended.

Example 3.63 Please solve the integral in complex domain

$$\int_2^{6-\mathrm{j}5} \mathrm{e}^{-x^2-\mathrm{j}x}\sin(7+\mathrm{j}2)x\,\mathrm{d}x.$$

Solution *If there are complex quantities in the integrand or boundaries, there is no extra difficulties for the user to find the results. The commands below can be used to find accurate results of* $I = -0.924460417702 + 25.7920728107\mathrm{j}$.

```
>> f=@(x)exp(-x.^2-1i*x).*sin((7+2i)*x); % integrand description
   I=integral(f,2,6-5i,'RelTol',1e-20)    % numerical integral
   syms x; F=exp(-x^2-1i*x)*sin((7+2i)*x); I0=vpa(int(F,2,6-5i))
```

Example 3.64 Consider again the oscillatory problem in Example 3.59, with a much larger interval $[0, 100]$. Please find numerically the integral.

Solution *Since the interval is too large and the oscillation is too strong in the integrand, the numerical integral functions such as* `quadl()`, `quadgk()` *may all fail to get good results. With the new* `integral()` *function, the result obtained is* $I_1 = -0.066260130460300$, *while with analytical method,* $I = \sin(1500)/15 \approx -0.066260130460443564274$.

```
>> f=@(x)cos(15*x); I1=integral(f,0,100,'RelTol',1e-20) % numerical
   syms x; I=int(cos(15*x),x,0,100), vpa(I) % analytical solution
```

3.9.3 Numerical infinite integrals

The function `integral()` can be used to evaluate infinite integrals directly. In earlier versions, `quadgk()` function should be used [4]. For infinite integrals, just write `-inf` or `inf` in the appropriate intervals.

Example 3.65 Please evaluate the infinite integral $\int_0^\infty \mathrm{e}^{-x^2}\,\mathrm{d}x$.

Solution *With the following numerical function call, the results obtained are*

$$I = 0.886226925452758,\ I_1 = \sqrt{\pi}/2 \approx 0.88622692545275801365.$$

It can be seen that the error is around 10^{-16}.

```
>> f=@(x)exp(-x.^2); I=integral(f,0,inf,'RelTol',1e-20) % numerical
   syms x; I1=int(exp(-x^2),0,inf), vpa(I1) % analytical solution
```

Example 3.66 For a given function $I(\alpha) = \int_0^\infty \mathrm{e}^{-\alpha x^2}\sin(\alpha^2 x)\,\mathrm{d}x$, please draw the curve of $I(\alpha)$ versus α, where $\alpha \in (0, 4)$.

Solution *Previous discussed integrals are integrals of an individual function, while here integrals of a series of functions (for different* α*'s) are expected. With vectorized approach, the following statements can be used, and the expected curve is shown in Figure 3.14. In earlier versions, loop structures or* `quadv()` *function should be used.*

```
>> a=0:0.1:4; f=@(x)exp(-a*x.^2).*sin(a.^2*x);  % vector a is supported
   I=integral(f,0,inf,'RelTol',1e-20,'ArrayValued',true); plot(a,I)
```

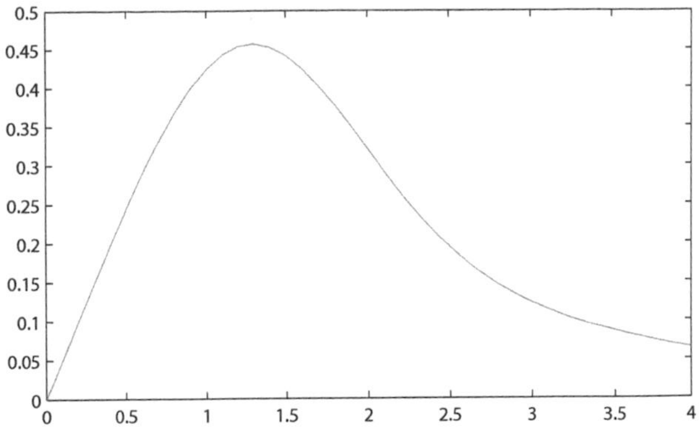

FIGURE 3.14: Integral $I(\alpha)$ versus α.

3.9.4 Evaluating integral functions

The numerical integrals discussed so far are definite integrals over the interval (a, b), and the integral function curves cannot be obtained. Here, the integral function evaluation problem is solved with the following function. The starting point of the integral function at $x = a$ is 0, since the definite integral in interval (a, a) is 0.

```
function [x,f1]=intfunc(f,a,b,n)
if nargin<=3, n=100; end; x=linspace(a,b,n); f1=0; f0=0;
for i=2:n,
   f2=f0+integral(f,x(i-1),x(i),'RelTol',1e-20); f1=[f1, f2]; f0=f2;
end
```

The syntax of the function is `[x,f1]=intfunc(f,a,b,n)`, with the default value of n as 100.

Example 3.67 Draw the integral curve of the piecewise function in Example 3.61.

Solution *Since the function* e^{x^2} *in not integrable analytically, the curve of the integral cannot be obtained with* `ezplot()` *function. Numerical methods should be used instead. The piecewise integrand should be defined first as an anonymous function, then, function* `intfunc()` *can be called to solve the original function and the curve is obtained as shown in Figure 3.15. It can also be seen that the definite integral in Example 3.61 is the right hand side point.*

```
>> f=@(x)exp(x.^2).*(x<=2)+80./(4-sin(16*pi*x)).*(x>2);
   [x1,f1]=intfunc(f,0,4,100); plot(x1,f1,x1(end),f1(end),'o'), f1(end)
```

Example 3.68 Evaluate numerically the Ei function $\text{Ei}(x) = \int_{-\infty}^{x} e^t/t\,dt$, and draw the curve in interval $x \in (1, 5)$.

Solution *Since the function* `intfunc()` *is written to evaluate integral functions in the interval* $(-\infty, x)$, *and the function expected in the example is in the interval (1,5), the original function can be written as the sum of integrals of two intervals,* $(-\infty, 1)$ *and* $(1, x)$. *The former one can be obtained with* `integral()` *function and the latter with* `intfunc()`. *The following statements can be issued, and the curve of Ei function is shown in Figure 3.16.*

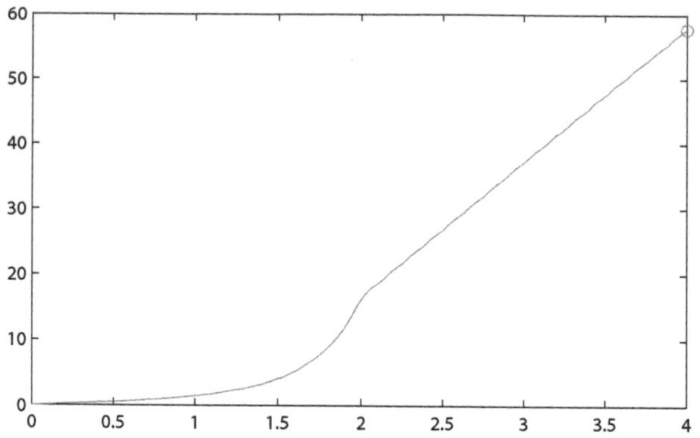

FIGURE 3.15: Integral function curve obtained.

```
>> f=@(t)exp(t)./t; I0=integral(f,-inf,1); % integrand description
   [x,I1]=intfunc(f,1,5); I=I1+I0; plot(x,I) % integral function plot
```

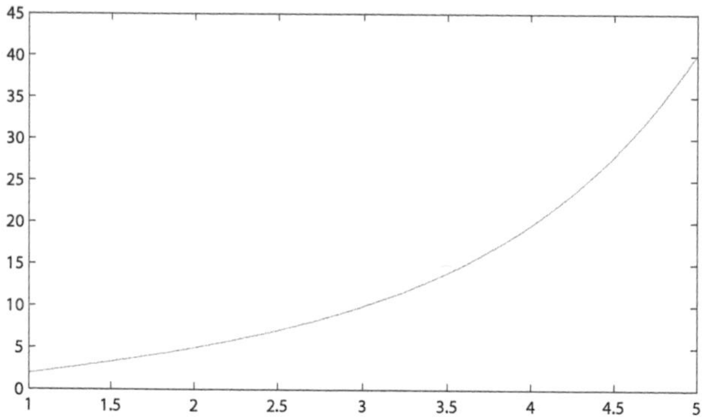

FIGURE 3.16: Plot of Ei function.

In fact, Ei function can be obtained directly with `ei()` *function, it can be seen that the curve shown in Figure 3.16 is exactly the same as the one by* `ei()` *function.*

```
>> I2=ei(x); line(x,I2); % draw the curve of Ei function directly
```

3.9.5 Numerical solutions to double integrals

Now consider the double integrals defined over a rectangular region

$$I = \int_{y_\mathrm{m}}^{y_\mathrm{M}} \int_{x_\mathrm{m}}^{x_\mathrm{M}} f(x, y)\, \mathrm{d}x\, \mathrm{d}y, \tag{3-9-3}$$

and the function `integral2()` can be used to solve this type of problem, with the syntaxes

$y=$`integral2(`*fun*`,`x_m`,`x_M`,`y_m`,`y_M`)` % double integral
$y=$`integral2(`*fun*`,`x_m`,`x_M`,`y_m`,`y_M`,`*property pairs*`)` % more properties

while in old versions, function `dblquad()` should be used instead.

Example 3.69 Compute the double definite integral

$$J = \int_{-1}^{1}\int_{-2}^{2} \mathrm{e}^{-x^2/2}\sin(x^2+y)\,\mathrm{d}x\,\mathrm{d}y.$$

Solution *With the anonymous function to describe the integrand, the double integral can be evaluated numerically from the following statements*

```
>> f=@(x,y)exp(-x.^2/2).*sin(x.^2+y);          % integrand description
   I1=integral2(f,-2,2,-1,1,'RelTol',1e-20) % numerical solutiom
   syms x y; f=exp(-x.^2/2).*sin(x.^2+y);    % analytical solution process
   I0=double(int(int(f,x,-2,2),y,-1,1))      % double precision display
```

and the result is $I_1 = 1.574498159218787$, *with* $I_0 = 1.57449815921736$. *It can be seen that the numerical result is quite accurate.*

Implementing the same idea in `intfunc()`, mesh grid data of the integral function can be obtained, and the surface of the integral can be obtained. Although sometimes the integrands may be non-integrable, the surface can still be drawn with ease. The syntax of the function is

$[x, y, f_1] =$ `intfunc2`$(f, x_m, x_M, y_m, y_M, n, m)$

where, f is the anonymous or M-function, (x_m, x_M) and (y_m, y_M) are integration rectangular region, and n, m are the numbers of cells in x, y axes, with default values of 50. The returned variable f_1`(end,end)` is the value of the definite integral.

```
function [yv,xv,f1]=intfunc2(f,xm,xM,varargin)
[ym,yM,n,m]=default_vals({xm,xM,50,50},varargin{:});
xv=linspace(xm,xM,n); yv=linspace(ym,yM,m); d=yv(2)-yv(1);
[x y]=meshgrid(xv,yv); f2=zeros(n,m);
for i=2:n, for j=2:m,
   f2(i,j)=integral2(f,xv(1),xv(i),yv(1),yv(j));
end, end
```

Example 3.70 Please draw the surface of the integral in Example 3.69.

Solution *The integrand is described by the anonymous function, and the surface of the integral can be obtained as shown in Figure 3.17.*

```
>> f=@(x,y)exp(-x.^2/2).*sin(x.^2+y); % integrand description
   [y,x,z]=intfunc2(f,-2,2,-1,1); surf(x,y,z), I=z(end,end)
```

The new MATLAB function `integral2()` also supports the numerical solution of the double integral defined as

$$I = \int_{x_m}^{x_M}\int_{y_m(x)}^{y_M(x)} f(x,y)\,\mathrm{d}y\,\mathrm{d}x. \tag{3-9-4}$$

In the syntax of the function `integral2()`, function handles can also be used for the boundaries y_m and y_M, i.e., M-functions or anonymous functions can be used to describe $y_m(x)$ and $y_M(x)$, while x_m and x_M must be fixed numbers. Please note the order of integrations is y first, then to x.

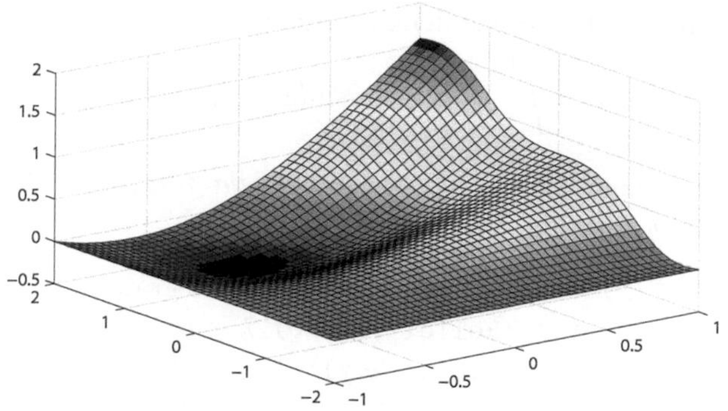

FIGURE 3.17: Surface of the integral.

Now consider the double integral, where the order of integration is x first, then to y.

$$I = \int_{y_\mathrm{m}}^{y_\mathrm{M}} \int_{x_\mathrm{m}(y)}^{x_\mathrm{M}(y)} f(x, y)\,\mathrm{d}x\,\mathrm{d}y. \tag{3-9-5}$$

The current version of the function `integral2()` does not support x_m described with function handles. Conversions must be made manually first. Let $\hat{x} = y$, and $\hat{y} = x$, after variable swopping, the original integral can be rewritten as

$$I = \int_{\hat{x}_\mathrm{m}}^{\hat{x}_\mathrm{M}} \int_{\hat{y}_\mathrm{m}(\hat{x})}^{\hat{y}_\mathrm{M}(\hat{x})} f(\hat{y}, \hat{x})\,\mathrm{d}\hat{y}\,\mathrm{d}\hat{x}. \tag{3-9-6}$$

Comparing (3-9-5) and (3-9-6), it can be seen that the integral can be directly obtained, by just swopping the orders of x and y in the integrand.

Example 3.71 Compute the double definite integral

$$J = \int_{-1/2}^{1} \int_{-\sqrt{1-x^2/2}}^{\sqrt{1-x^2/2}} \mathrm{e}^{-x^2/2} \sin(x^2 + y)\,\mathrm{d}y\,\mathrm{d}x.$$

Solution *This is the right integration order. Thus, the inner bounds $y_\mathrm{M}(x)$ and $y_\mathrm{m}(x)$ can be defined with anonymous functions, and then, the double integral can be evaluated directly with*

```
>> yM=@(x)sqrt(1-x.^2/2);  ym=@(x)-sqrt(1-x.^2/2); % inner bounds
   f=@(x,y)exp(-x.^2/2).*sin(x.^2+y);              % integrand
   I=integral2(f,-1/2,1,ym,yM,'RelTol',1e-20)      % numerical integration
```

and the value of integration can be found as $I = 0.411929546173382$. Now consider the analytical method

```
>> syms x y % analytical solution procedure
   i1=int(exp(-x^2/2)*sin(x^2+y),y,-sqrt(1-x^2/2),sqrt(1-x^2/2));
   int(i1,x,-1/2,1), I0=vpa(ans)  % warning message given
```

and it is prompted that the integral does not exist, and with `vpa()` *function, the high-precision numerical result can be written as 0.41192954617629511965175994017. It can be seen that the numerical solutions are very accurate.*

Example 3.72 If the original integral problem is changed to

$$J=\int_{-1}^{1}\int_{-\sqrt{1-y^2}}^{\sqrt{1-y^2}} e^{-x^2/2}\sin(x^2+y)\,dx\,dy$$

the analytical problem cannot be used to find the results, even with the help of `vpa()` function. Numerical method should be tried. Since the order of the integration is first to x, then to y, the anonymous function of the integrand should just be described as $f=$`@(y,x)`, and the solution can be found

```
>> fh=@(y)sqrt(1-y.^2); fl=@(y)-sqrt(1-y.^2); % inner bounds
   f=@(y,x)exp(-x.^2/2).*sin(x.^2+y);        % swopped order
   I=integral2(f,-1,1,fl,fh,'RelTol',eps)    % evaluate integral
```

which yields $I=0.536860382698816$, which is still very easy. For numerical methods, the numerical integration results will not be affected by whether the integrand is theoretically integrable or not.

3.9.6 Numerical solutions to triple integrals

The triple definite integral is described by

$$I=\int_{x_{\rm m}}^{x_{\rm M}}\int_{y_{\rm m}}^{y_{\rm M}}\int_{z_{\rm m}}^{z_{\rm M}} f(x,y,z)\,dz\,dy\,dx, \tag{3-9-7}$$

the problem can be solved with the `integral3()` function with the syntax

$I=$ `integral3(`*fun*`,`$x_{\rm m}$`,`$x_{\rm M}$`,`$y_{\rm m}$`,`$y_{\rm M}$`,`$z_{\rm m}$`,`$z_{\rm M}$`,`*property pairs*`)`

where *fun* describes the integrand. The arguments $y_{\rm m}$,$y_{\rm M}$,$z_{\rm m}$,$z_{\rm M}$ are all allowed to use constants or function handles. It is required that the handles of z must be functions of x and y, and handles of y must be functions of x. The boundaries of x must be constants. Therefore, the triple integrals over non-rectangular regions can also be evaluated easily with the function. Again, attention must be paid to the order of integral, which is $z\rightarrow y\rightarrow x$. If the order is otherwise given, rearrange them in the description of integrand, before carrying out numerical integral process.

Example 3.73 Compute the triple integral in Example 3.32

$$\int_0^2\int_0^{\pi}\int_0^{\pi} 4xze^{-x^2y-z^2}\,dz\,dy\,dx.$$

Solution *The anonymous function is used to specify the integrand. Therefore, the following statements can be used to compute the triple integral, and the result is* $I=3.108079402085465$.

```
>> f=@(x,y,z)4*x.*z.*exp(-x.*x.*y-z.*z); % integrand in dot operation
   tic, I=integral3(f,0,2,0,pi,0,pi,'RelTol',1e-20), toc % solve directly
```

Example 3.74 Evaluate numerically the triple integral

$$\int_0^2\int_{1-x}^{1+x}\int_0^{\sqrt{1-x^2-y^2}} 4xze^{-x^2y-z^2}\,dz\,dy\,dx.$$

Solution *Fortunately for this example, the high-precision results can be obtained with the statements as* $I_0=-177.14581094852176489995670313839$.

```
>> syms x y z; f=4*x*z*exp(-x^2*y-z^2);
   I0=vpa(int(int(int(f,z,0,sqrt(1-x^2-y^2)),y,1-x,1+x),x,0,2))
```

In fact, the numerical solution to the integral can also be found easily with the following statements, regardless whether the original problem is integrable or not. The numerical solution is -177.1458109485222*, and it is accurate enough for most practical applications.*

```
>> f=@(x,y,z)4*x.*z.*exp(-x.^2.*y-z.^2); % describe integrand
   zM=@(x,y)sqrt(1-x.^2-y.^2); ym=@(x)1-x; yM=@(x)1+x; % integral bounds
   I=integral3(f,0,2, ym,yM, 0,zM,'RelTol',1e-20) % numerical integral
```

3.9.7 Multiple integral evaluations

NIT Toolbox [5] can be used to solve multiple integral problems with other hyper-rectangular regions. For instance, the `quadndg()` function can be used for these problems.

Consider the standard multiple integral given by

$$I = \int_{x_{1\mathrm{m}}}^{x_{1\mathrm{M}}} \int_{x_{2\mathrm{m}}}^{x_{2\mathrm{M}}} \cdots \int_{x_{p\mathrm{m}}}^{x_{p\mathrm{M}}} f(x_1, x_2, \cdots, x_p)\, \mathrm{d}x_p \cdots \mathrm{d}x_2\, \mathrm{d}x_1, \tag{3-9-8}$$

the integral can be solved numerically with

$I=$ `quadndg(`f`,[`$x_{1\mathrm{m}}, x_{2\mathrm{m}}, \cdots, x_{p\mathrm{m}}$`],[`$x_{1\mathrm{M}}, x_{2\mathrm{M}}, \cdots, x_{p\mathrm{M}}$`],`$\epsilon$`)`

where, f is the M-function description of the integrand, and ϵ is the error tolerance.

Example 3.75 Solve again the triple integral problem in Example 3.73

$$\int_0^2 \int_0^{\pi} \int_0^{\pi} 4xz\mathrm{e}^{-x^2y-z^2}\, \mathrm{d}z\, \mathrm{d}y\, \mathrm{d}x.$$

Solution *Denote* $x_1 = x$, $x_2 = y$, $x_3 = z$, *the original integrand can be rewritten as* $f(\boldsymbol{x}) = 4x_1x_3\mathrm{e}^{-x_1^2x_2-x_3^2}$, *and anonymous or M-functions can be used to describe it, and the integral can be evaluated numerically, with the result* $I = 3.108079402085409$. *It can be seen that the result is almost the same as the one obtained in Example 3.73, however, the efficiency is much higher — for this example, just 1/10 of the time required.*

```
>> f=@(x)4*x(1)*x(3)*exp(-x(1)^2*x(2)-x(3)^2); % integrand description
   tic, I=quadndg(f,[0 0 0],[2,pi,pi]), toc     % evaluation of integral
```

Example 3.76 Solve the pentaple integral problem

$$I = \int_0^5 \int_0^4 \int_0^1 \int_0^2 \int_0^3 \sqrt[3]{v}\sqrt{w}x^2y^3z\, \mathrm{d}z\, \mathrm{d}y\, \mathrm{d}x\, \mathrm{d}w\, \mathrm{d}v.$$

Solution *For this example, its analytical solution can be found, as* $120\sqrt[3]{5}$.

```
>> syms x y z w v; F=v^(1/3)*sqrt(w)*x^2*y^3*z; % multiple integrand
   I=int(int(int(int(int(F,z,0,3),y,0,2),x,0,1),w,0,4),v,0,5)
```

In practical situations, there is no analytical solutions to pentaple integrals. Numerical approaches should be used instead. Denote $x_1 = v$, $x_2 = w$, $x_3 = x$, $x_4 = y$, $x_5 = z$, *the integrand can be rewritten as*

$$f(\boldsymbol{x}) = \sqrt[3]{x_1}\sqrt{x_2}x_3^2x_4^3x_5.$$

The integrand can be described with anonymous function, and the problem can be solved, with $I = 205.2205 \approx 120\sqrt[3]{5}$. The time elapsed for this example is about 48 seconds.

```
>> f=@(x)(x(1))^(1/3)*sqrt(x(2))*x(3)^2*x(4)^3*x(5); % describe integrand
   tic, I=quadndg(f,[0 0 0 0 0],[5,4,1,2,3]), toc    % numerical integral
```

Example 3.77 Evaluate the pentaple integral problem.

$$I = \int_0^5 \int_0^4 \int_0^1 \int_0^2 \int_0^3 \left(\mathrm{e}^{-\sqrt[3]{v}} \sin\sqrt{w} + \mathrm{e}^{-x^2y^3z} \right) \mathrm{d}z\, \mathrm{d}y\, \mathrm{d}x\, \mathrm{d}w\, \mathrm{d}v.$$

Solution *This integrand is not integrable, and numerical method should be used to solve the problem. Denote again $x_1 = v$, $x_2 = w$, $x_3 = x$, $x_4 = y$, $x_5 = z$, the integrand can be rewritten as*

$$f(\boldsymbol{x}) = \mathrm{e}^{-\sqrt[3]{x_1}} \sin\sqrt{x_2} + \mathrm{e}^{-x_3^2x_4^3x_5}.$$

The following statements can be used, and the result is $I = 113.60574122$. Although the integrand is much more complicated than the previous one, the time elapsed are almost the same.

```
>> f=@(x)exp(-(x(1))^(1/3))*sin(sqrt(x(2)))+exp(-x(3)^2*x(4)^3*x(5));
   tic, I=quadndg(f,[0 0 0 0 0],[5,4,1,2,3]), toc % numerical integral
```

Exercises

Exercise 3.1 *Compute the following limit problems:*

(i) $\lim\limits_{x\to\infty} (3^x + 9^x)^{1/x}$, (ii) $\lim\limits_{x\to\infty} \dfrac{(x+2)^{x+2}(x+3)^{x+3}}{(x+5)^{2x+5}}$, (iii) $\lim\limits_{x\to a} \left(\dfrac{\tan x}{\tan a}\right)^{\cot(x-a)}$,

(iv) $\lim\limits_{x\to 0} \left[\dfrac{1}{\ln\left(x+\sqrt{1+x^2}\right)} - \dfrac{1}{\ln(1+x)}\right]$,

(v) $\lim\limits_{x\to\infty} \left[\sqrt[3]{x^3+x^2+x+1} - \sqrt{x^2+x+1}\,\dfrac{\ln(\mathrm{e}^x+x)}{x}\right]$.

Exercise 3.2 *Please find sequential limits* $\lim\limits_{x\to a}\left[\lim\limits_{y\to b} f(x,y)\right]$ *and* $\lim\limits_{y\to b}\left[\lim\limits_{x\to a} f(x,y)\right]$

(i) $f(x,y) = \sin\dfrac{\pi x}{2x+y}$, $a=\infty, b=\infty$,

(ii) $f(x,y) = \dfrac{1}{xy}\tan\dfrac{xy}{1+xy}$, $a=0, b=\infty$.

Exercise 3.3 *Compute the following double limit problems:*

(i) $\lim\limits_{\substack{x\to -1\\ y\to 2}} \dfrac{x^2y+xy^3}{(x+y)^3}$, (ii) $\lim\limits_{\substack{x\to 0\\ y\to 0}} \dfrac{xy}{\sqrt{xy+1}-1}$, (iii) $\lim\limits_{\substack{x\to 0\\ y\to 0}} \dfrac{1-\cos\left(x^2+y^2\right)}{\left(x^2+y^2\right)\mathrm{e}^{x^2+y^2}}$.

Exercise 3.4 *Compute the derivatives of the following functions:*

$$\text{(i) } y(x)=\sqrt{x\sin x\sqrt{1-\mathrm{e}^x}},\quad \text{(ii) } y(t)=\sqrt{\frac{(x-1)(x-2)}{(x-3)(x-4)}},$$

$$\text{(iii) } \operatorname{atan}\frac{y}{x}=\ln(x^2+y^2),\quad \text{(iv) } y(x)=-\frac{1}{na}\ln\frac{x^n+a}{x^n},\quad n>0.$$

Exercise 3.5 *Please find the 4th order derivative* $y(t)=\sqrt{\dfrac{(x-1)(x-2)}{(x-3)(x-4)}}$.

Exercise 3.6 *Compute the 10th order derivative of the function* $y=\dfrac{1-\sqrt{\cos ax}}{x\left(1-\cos\sqrt{ax}\right)}$.

Exercise 3.7 *In calculus courses, when the limit of a ratio is required, where both the numerator and the denominator tend to 0 or* ∞*, simultaneously, L'Hôpital's law can be used, i.e., to evaluate the limits of derivatives of numerator and denominator. Verify the limit*

$$\lim_{x\to 0}\frac{\ln(1+x)\ln(1-x)-\ln(1-x^2)}{x^4}$$

by the consecutive use of L'Hôpital's law, and compare the result with the one directly obtained with `limit()` *function.*

Exercise 3.8 *For parametric equation* $\begin{cases}x=\ln\cos t\\ y=\cos t-t\sin t,\end{cases}$ *compute* $\dfrac{\mathrm{d}y}{\mathrm{d}x}$ *and* $\left.\dfrac{\mathrm{d}^2y}{\mathrm{d}x^2}\right|_{t=\pi/3}$.

Exercise 3.9 *Please find the first order derivatives for the parametric equations*

$$\text{(i) }\begin{cases}x(t)=a(\ln\tan t/2+\cos t-\sin t)\\ y(t)=a(\sin t+\cos t),\end{cases}\qquad \text{(ii) }\begin{cases}x(t)=2at/(1+t^3)\\ y=a(3at^2)/(1+t^3).\end{cases}$$

Exercise 3.10 *Assume that* $u=\cos^{-1}\sqrt{\dfrac{x}{y}}$. *Verify that* $\dfrac{\partial^2u}{\partial x\partial y}=\dfrac{\partial^2u}{\partial y\partial x}$.

Exercise 3.11 *For a given function* $\begin{cases}xu+yv=0\\ yu+xv=1,\end{cases}$ *compute* $\dfrac{\partial^2u}{\partial x\partial y}$.

Exercise 3.12 *Assume that* $f(x,y)=\displaystyle\int_0^{xy}\mathrm{e}^{-t^2}\,\mathrm{d}t$. *Compute* $\dfrac{x}{y}\dfrac{\partial^2f}{\partial x^2}-2\dfrac{\partial^2f}{\partial x\partial y}+\dfrac{\partial^2f}{\partial y^2}$.

Exercise 3.13 *Please find* $\mathrm{d}y/\mathrm{d}x$, $\mathrm{d}^2y/\mathrm{d}x^2$ *and* $\mathrm{d}^3y/\mathrm{d}x^3$ *from the parametric equation*

$$\text{(i) } x=\mathrm{e}^{2t}\cos^2t,\ y=\mathrm{e}^{2t}\sin^2t,\quad \text{(ii) } x=\arcsin\frac{t}{\sqrt{1+t^2}},\ y=\arccos\frac{t}{\sqrt{1+t^2}}.$$

Exercise 3.14 *If* $x^2-xy+2y^2+x-y-1=0$*, please calculate* $\mathrm{d}y/dx$, $\mathrm{d}^2y/\mathrm{d}x^2$ *and* $\mathrm{d}^3y/\mathrm{d}x^3$ *at* $x=0,y=1$.

Exercise 3.15 *Given a matrix* $\boldsymbol{f}(x,y,z)=\begin{bmatrix}3x+\mathrm{e}^yz\\ x^3+y^2\sin z\end{bmatrix}$*, compute its Jacobian matrix.*

Exercise 3.16 *If* $u = x - y + x^2 + 2xy + y^2 + x^3 - 3x^2y - y^3 + x^4 - 4x^2y^2 + y^4$, *please find* $\dfrac{\partial^4 u}{\partial x^4}$, $\dfrac{\partial^4 u}{\partial x^3 \partial y}$ *and* $\dfrac{\partial^4 u}{\partial x^2 \partial y^2}$.

Exercise 3.17 *If* $u = \ln\dfrac{1}{\sqrt{(x-\xi)^2 + (y-\eta)^2}}$, *please find* $\dfrac{\partial^4 u}{\partial x \partial y \partial \xi \partial \eta}$.

Exercise 3.18 *If* $z = \varphi\left(x^2 + y^2\right)$, *please find* $y\dfrac{\partial z}{\partial x} - x\dfrac{\partial z}{\partial y}$.

Exercise 3.19 *If* $u = x\phi(x+y) + y\varphi(x+y)$, *please find* $\dfrac{\partial^2 u}{\partial x^2} - 2\dfrac{\partial^2 u}{\partial x \partial y} + \dfrac{\partial^2 u}{\partial y^2}$.

Exercise 3.20 *Let* $z = F(r,\theta)$, *where* r *and* θ *are functions of* x *and* y, *defined by system of equations,* $x = r\cos\theta$, $y = r\sin\theta$. *Find* $\partial z/\partial x$ *and* $\partial z/\partial y$.

Exercise 3.21 *Please find the divergences and curls of the following vector function*

$$\text{(i) } \boldsymbol{v}(x,y) = \left[5x^2y - 4xy,\ 3x^2 - 2y\right], \quad \text{(ii) } \boldsymbol{v}(x,y,z) = \left[x^2y^2,\ 1,\ z\right],$$

$$\text{(iii) } \boldsymbol{v}(x,y,z) = \left[2xyz^2,\ x^2z^2 + z\cos yz,\ 2x^2yz + y\cos yz\right].$$

Exercise 3.22 *Compute the following indefinite integrals:*

$$\text{(i) } I(x) = -\int \frac{3x^2 + a}{x^2\left(x^2 + a\right)^2}\,\mathrm{d}x, \quad \text{(ii) } I(x) = \int \frac{\sqrt{x(x+1)}}{\sqrt{x} + \sqrt{1+x}}\,\mathrm{d}x,$$

$$\text{(iii) } I(x) = \int x\mathrm{e}^{ax}\cos bx\,\mathrm{d}x, \quad \text{(iv) } I(x) = \int \frac{\mathrm{e}^{\arctan x} + x\ln(1+x^2) + 1}{1+x^2}\,\mathrm{d}x,$$

$$\text{(v) } I(t) = \int \mathrm{e}^{ax}\sin bx \sin cx\,\mathrm{d}x, \quad \text{(vi) } I(x) = \int \frac{\mathrm{d}x}{\sqrt{(1+x^2)\ln(x+\sqrt{1+x^2})}}.$$

Exercise 3.23 *Compute the definite integrals and infinite integrals:*

$$\text{(i) } I = \int_0^\infty \frac{\cos x}{\sqrt{x}}\,\mathrm{d}x, \quad \text{(ii) } I = \int_0^1 \frac{1+x^2}{1+x^4}\,\mathrm{d}x, \quad \text{(iii) } \int_{\mathrm{e}^{-2\pi n}}^1 \left|\cos\left(\ln\frac{1}{x}\right)\right|\mathrm{d}x.$$

Exercise 3.24 *Please find the following definite integrals and improper integrals:*

$$\text{(i) } \int_0^{0.75} \frac{1}{(x+1)\sqrt{x^2+1}}\,\mathrm{d}x, \quad \text{(ii) } \int_0^1 \frac{\arcsin\sqrt{x}}{\sqrt{x(1-x)}}\,\mathrm{d}x,$$

$$\text{(iii) } \int_0^{\pi/4} \left(\frac{\sin x - \cos x}{\sin x + \cos x}\right)^{2n+1}\mathrm{d}x.$$

Exercise 3.25 *Please find the integral* $I(s) = \displaystyle\int_0^s \frac{\mathrm{e}^x\sqrt{\mathrm{e}^x - 1}}{\mathrm{e}^x + 3}\,\mathrm{d}x$.

Exercise 3.26 *Laplace transform of a given function* $f(t)$ *is defined as* $F(s) = \displaystyle\int_0^\infty \mathrm{e}^{-st}f(t)\,\mathrm{d}t$. *Please find the Laplace transforms to the following functions*

$$\text{(i) } f(t) = 1, \quad \text{(ii) } f(t) = \mathrm{e}^{\beta t}, \quad \text{(iii) } f(t) = \sin\alpha t, \quad \text{(iv) } f(t) = t^m.$$

Exercise 3.27 *Please find the following indefinite integrals:*

$$(i)\ \int \frac{\sin^2 x - 4\sin x\cos x + 3\cos^2 x}{\sin x + \cos x}\,\mathrm{d}x,\quad (ii)\ \int \frac{\sin^2 x - \sin x\cos x + 2\cos^2 x}{\sin x + 2\cos x}\,\mathrm{d}x.$$

Exercise 3.28 *For the function* $f(x) = \mathrm{e}^{-5x}\sin(3x+\pi/3)$*, compute* $\displaystyle\int_0^t f(x)f(t+x)\,\mathrm{d}x$.

Exercise 3.29 *Please find the triple integral* $\displaystyle\iiint_V x^3y^2z\,\mathrm{d}x\,\mathrm{d}y\,\mathrm{d}z$*, where* V *is given by* $0 \leqslant x \leqslant 1, 0 \leqslant y \leqslant x, 0 \leqslant z \leqslant xy$.

Exercise 3.30 *Please find the following multiple integrals:*

$$(i)\ \int_0^\pi\int_0^\pi |\cos(x+y)|\,\mathrm{d}x\,\mathrm{d}y,\quad (ii)\ \int_0^1\int_{-1}^{1-x} \arcsin(x+y)\,\mathrm{d}y\,\mathrm{d}x,$$

$$(iii)\ \iint_{|x|+|y|\leqslant 1} (|x|+|y|)\,\mathrm{d}x\,\mathrm{d}y,\quad (iv)\ \iint_{\pi^2\leqslant x^2+y^2\leqslant 4\pi^2} \sin\sqrt{x^2+y^2}\,\mathrm{d}x\,\mathrm{d}y.$$

Exercise 3.31 *For different values of a's, please find* $I = \displaystyle\int_0^\infty \frac{\cos ax}{1+x^2}\,\mathrm{d}x$.

Exercise 3.32 *Please show that for any function* $f(t)$*,* $\displaystyle\int_a^b f(t)\,\mathrm{d}t = -\int_b^a f(t)\,\mathrm{d}t$.

Exercise 3.33 *Please solve the multiple integral problems:*

$$(i)\ \int_0^2\int_0^{\sqrt{4-x^2}} \sqrt{4-x^2-y^2}\,\mathrm{d}y\,\mathrm{d}x,\quad (ii)\ \int_0^3\int_0^{3-x}\int_0^{3-x-y} xyz\,\mathrm{d}z\,\mathrm{d}y\,\mathrm{d}x,$$

$$(iii)\ \int_0^2\int_0^{\sqrt{4-x^2}}\int_0^{\sqrt{4-x^2-y^2}} z(x^2+y^2)\,\mathrm{d}z\,\mathrm{d}y\,\mathrm{d}x,$$

$$(iv)\ \int_0^1\int_0^x\int_0^y\int_0^z xyzu\mathrm{e}^{6-x^2-y^2-z^2-u^2}\,\mathrm{d}u\,\mathrm{d}z\,\mathrm{d}y\,\mathrm{d}x,$$

$$(v)\ \int_0^{7/10}\int_0^{4/5}\int_0^{9/10}\int_0^1\int_0^{11/10} \sqrt{6-x^2-y^2-z^2-w^2-u^2}\,\mathrm{d}w\,\mathrm{d}u\,\mathrm{d}z\,\mathrm{d}y\,\mathrm{d}x.$$

Exercise 3.34 *Compute the Fourier series expansions for the following functions, and compare graphically the approximation and exact results, using finite numbers of terms:*

$$(i)\ f(x) = (\pi - |x|)\sin x,\ -\pi \leqslant x < \pi,\qquad (ii)\ f(x) = \mathrm{e}^{|x|},\ -\pi \leqslant x < \pi,$$

$$(iii)\ f(x) = \begin{cases} 2x/l, & 0 < x < l/2 \\ 2(l-x)/l, & l/2 < x < l \end{cases},\ \text{where } l = \pi.$$

Exercise 3.35 *Obtain the Taylor series expansions for the following functions, and compare graphically the approximation and exact results with finite numbers of terms:*

$$(i)\ \int_0^x \frac{\sin t}{t}\,\mathrm{d}t,\quad (ii)\ \ln\left(\frac{1+x}{1-x}\right),\quad (iii)\ \ln\left(x+\sqrt{1+x^2}\right),\quad (iv)\ (1+4.2x^2)^{0.2},$$

(v) $\mathrm{e}^{-5x}\sin(3x+\pi/3)$ *expansions about* $x=0$ *and* $x=a$ *points, respectively.*

Exercise 3.36 *Find Taylor series expansions to the following multivariate functions:*

(i) $f(x, y) = e^x \cos y$, *about* $x = 0, y = 0$, *and* $x = a, y = b$,

(ii) $f(x, y) = \ln(1 + x)\ln(1 + y)$, *about* $x = 0, y = 0$ *and* $x = a, y = b$.

Exercise 3.37 *Find Taylor series expansion of the function* $f(x, y) = \dfrac{1 - \cos\left(x^2 + y^2\right)}{\left(x^2 + y^2\right) e^{x^2+y^2}}$ *about* $x = 1$, $y = 0$ *point.*

Exercise 3.38 *Compute the first n term finite sums and infinite sums:*

(i) $\dfrac{1}{1 \times 6} + \dfrac{1}{6 \times 11} + \cdots + \dfrac{1}{(5n-4)(5n+1)} + \cdots$

(ii) $\left(\dfrac{1}{2} + \dfrac{1}{3}\right) + \left(\dfrac{1}{2^2} + \dfrac{1}{3^2}\right) + \cdots + \left(\dfrac{1}{2^n} + \dfrac{1}{3^n}\right) + \cdots$

(iii) $\dfrac{1}{3}\left(\dfrac{x}{2}\right) + \dfrac{1 \times 4}{3 \times 6}\left(\dfrac{x}{2}\right)^2 + \dfrac{1 \times 4 \times 7}{3 \times 6 \times 9}\left(\dfrac{x}{2}\right)^3 + \dfrac{1 \times 4 \times 7 \times 10}{3 \times 6 \times 9 \times 12}\left(\dfrac{x}{2}\right)^4 + \cdots$.

Exercise 3.39 *Find the sum of the following infinite series:*

(i) $\displaystyle\sum_{n=1}^{\infty} \frac{\sin^2 n\alpha \sin nx}{n}$, $\left(0 < \alpha < \dfrac{\pi}{2}\right)$, (ii) $\displaystyle\sum_{n=0}^{\infty} \frac{(-1)^n n^3}{(n+1)!} x^n$, (iii) $\displaystyle\sum_{n=0}^{\infty} \frac{x^{4n+1}}{4n+1}$.

Exercise 3.40 *Compute the first n term finite sums and infinite sums of the following series*

(i) $\sqrt[3]{x} + (\sqrt[5]{x} - \sqrt[3]{x}) + (\sqrt[7]{x} - \sqrt[5]{x}) + \cdots + (\sqrt[2k+1]{x} - \sqrt[2k-1]{x}) + \cdots$,

(ii) $1 + \dfrac{m}{1!}x + \dfrac{m(m-1)}{2!}x^2 + \cdots + \dfrac{m(m-1)\cdots(m-n+1)}{n!}x^n + \cdots$.

Exercise 3.41 *If the general terms* a_n *are known, please find the infinite sum:*

(i) $a_n = \left(\sqrt{1+n} - \sqrt{n}\right)^p \ln\dfrac{n-1}{n+1}$, (ii) $a_k = \dfrac{1}{n^{1+k/\ln n}}$.

Exercise 3.42 *Find the sum of the sequences:*

(i) $\displaystyle\sum_{n=1}^{\infty} \frac{x^n}{(1+x)(1+x^2)\cdots(1+x^n)}$, (ii) $\displaystyle\sum_{n=2}^{\infty} \frac{(-1)^n}{n^2+n-2}$, (iii) $\displaystyle\sum_{n=2}^{\infty} \frac{1}{n^2(n+1)^2(n+2)^2}$.

Exercise 3.43 *Compute the following limits:*

(i) $\displaystyle\lim_{n\to\infty} \left[\frac{1}{2^2-1} + \frac{1}{4^2-1} + \frac{1}{6^2-1} + \cdots + \frac{1}{(2n)^2-1}\right]$,

(ii) $\displaystyle\lim_{n\to\infty} n\left(\frac{1}{n^2+\pi} + \frac{1}{n^2+2\pi} + \frac{1}{n^2+3\pi} + \cdots + \frac{1}{n^2+n\pi}\right)$.

Exercise 3.44 *Show that* $\cos\theta + \cos 2\theta + \cdots + \cos n\theta = \dfrac{\sin(n\theta/2)\cos[(n+1)\theta/2]}{\sin\theta/2}$.

Exercise 3.45 *Please find the products of the infinite sequences:*

(i) $\displaystyle\prod_{n=1}^{\infty} \frac{(2n+1)(2n+7)}{(2n+3)(2n+5)}$, (ii) $\displaystyle\prod_{n=1}^{\infty} \frac{9n^2}{(3n-1)(3n+1)}$, (iii) $\displaystyle\prod_{n=1}^{\infty} a^{(-1)^n/n}$, $a > 0$.

Exercise 3.46 *For* $a_n = \displaystyle\int_0^{\pi/4} \tan^n x \,\mathrm{d}x$*, compute* $S = \displaystyle\sum_{n=1}^{\infty} \frac{1}{n}(a_n + a_{n+2})$.

Exercise 3.47 *Please assess the convergency of the following infinite series:*

$$\text{(i)} \sum_{n=2}^{\infty} \left(\frac{n}{1+n^2}\right)^n, \quad \text{(ii)} \sum_{n=10}^{\infty} \frac{1}{\ln n \ln(\ln x)}, \quad \text{(iii)} \sum_{n=1}^{\infty} (-1)^n \frac{n+1}{(n+1)\sqrt{n+1}-1},$$

$$\text{(iv)} \frac{3}{2} - \frac{3\times 5}{2\times 5} + \frac{3\times 5\times 7}{2\times 5\times 8} + \cdots + (-1)^{n-1} \frac{3\times 5\times 7\times\cdots\times(2n+1)}{2\times 5\times 8\times\cdots\times(3n-1)} + \cdots.$$

Exercise 3.48 *Please find the interval of* x*, such that the infinite series converge.*

$$\text{(i)} \sum_{n=1}^{\infty} (-1)^n \left(\frac{2^n (n!)^2}{(2n+1)!}\right)^p x^n, \quad \text{(ii)} \sum_{n=1}^{\infty} \frac{3^{2n} n}{2^n} x^n (1-x)^n, \quad \text{(iii)} \sum_{n=1}^{\infty} \frac{1}{x^n} \sin\frac{\pi}{2^n}.$$

Exercise 3.49 *Compute the following path and line integrals:*

(i) $\displaystyle\int_l (x^2+y^2)\,\mathrm{d}s$, l: $x = a(\cos t + t\sin t), y = a(\sin t - t\cos t)$, *for* $0 \leqslant t \leqslant 2\pi$,

(ii) $\displaystyle\int_l (yx^3 + \mathrm{e}^y)\,\mathrm{d}x + (xy^3 + x\mathrm{e}^y - 2y)\,\mathrm{d}y$, *where* l *is given by the upper-semi-ellipsis of* $a^2x^2 + b^2y^2 = c^2$,

(iii) $\displaystyle\int_l y\,\mathrm{d}x - x\,\mathrm{d}y + (x^2+y^2)\,\mathrm{d}z$, l: $x = \mathrm{e}^t, y = \mathrm{e}^{-t}, z = at$, $0 \leqslant t \leqslant 1$, *for* $a > 0$,

(iv) $\displaystyle\int_l (\mathrm{e}^x \sin y - my)\,\mathrm{d}x + (\mathrm{e}^x \cos y - m)\,\mathrm{d}y$, *where* l *is defined as the closed path from* $(a,0)$ *to* $(0,0)$*, then, with the upper-semi-circle* $x^2 + y^2 = ax$.

Exercise 3.50 *Suppose a curve is described by polar function* $r = \rho(\theta)$*, with* $\theta \in (\theta_{\mathrm{m}}, \theta_{\mathrm{M}})$*, the length of the curve can be evaluated from*

$$L = \int_{\theta_{\mathrm{m}}}^{\theta_{\mathrm{M}}} \sqrt{\rho^2(\theta) + [\mathrm{d}\rho(\theta)/\,\mathrm{d}\theta]^2}\,\mathrm{d}\theta.$$

Please find the length of the curve $\rho = a\sin^2\theta/3$, $\theta \in (0, 3\pi)$.

Exercise 3.51 *Compute the surface integrals, where* S *is the bottom side of the semi-sphere* $z = \sqrt{R^2 - x^2 - y^2}$.

$$\text{(i)} \int_S xyz^3\,\mathrm{d}s, \quad \text{(ii)} \int_S (x + yz^3)\,\mathrm{d}x\,\mathrm{d}y.$$

Exercise 3.52 *For the measured data given in Table 3.2, evaluate numerically its derivatives and definite integral.*

TABLE 3.2: Computed data in Exercise 3.52.

x_i	0	0.1	0.2	0.3	0.4	0.5	0.6	0.7	0.8	0.9	1	1.1	1.2
y_i	0	2.2077	3.2058	3.4435	3.241	2.8164	2.311	1.8101	1.3602	0.9817	0.6791	0.4473	0.2768

Exercise 3.53 *Compute the gradient of the measured data in Table 3.3 for a function of two variables. Assume that the data were generated by the function* $f(x, y) = 4 - x^2 - y^2$. *Generate the data and verify the results of gradient with theoretical results.*

TABLE 3.3: Computed data in Exercise 3.53.

0	0	0.2	0.4	0.6	0.8	1	1.2	1.4	1.6	1.8	2
0	4	3.96	3.84	3.64	3.36	3	2.56	2.04	1.44	0.76	0
0.2	3.96	3.92	3.8	3.6	3.32	2.96	2.52	2	1.4	0.72	−0.04
0.4	3.84	3.8	3.68	3.48	3.2	2.84	2.4	1.88	1.28	0.6	−0.16
0.6	3.64	3.6	3.48	3.28	3	2.64	2.2	1.68	1.08	0.4	−0.36
0.8	3.36	3.32	3.2	3	2.72	2.36	1.92	1.4	0.8	0.12	−0.64
1	3	2.96	2.84	2.64	2.36	2	1.56	1.04	0.44	−0.24	−1
1.2	2.56	2.52	2.4	2.2	1.92	1.56	1.12	0.6	0	−0.68	−1.44
1.4	2.04	2	1.88	1.68	1.4	1.04	0.6	0.08	−0.52	−1.2	−1.96
1.6	1.44	1.4	1.28	1.08	0.8	0.44	0	−0.52	−1.12	−1.8	−2.56
1.8	0.76	0.72	0.6	0.4	0.12	−0.24	−0.68	−1.2	−1.8	−2.48	−3.24
2	0	−0.04	−0.16	−0.36	−0.64	−1	−1.44	−1.96	−2.56	−3.24	−4

Exercise 3.54 *Evaluate the definite integral* $\displaystyle\int_0^{\pi} (\pi - t)^{1/4} f(t)\,\mathrm{d}t,\ f(t) = \mathrm{e}^{-t}\sin(3t+1)$ *numerically. Also, evaluate the integration function* $\displaystyle F(t) = \int_0^t (t-\tau)^{1/4} f(\tau)\,\mathrm{d}\tau$ *numerically for different sample points of* t, *such that* $t = 0.1, 0.2, \cdots, \pi$, *and draw the* $F(t)$ *plot.*

Exercise 3.55 *Evaluate numerically the following multiple integral problems. It should be noted that there are no analytical solutions to these problems. Therefore, the obtained numerical results should be double-checked by varying step-sizes or default accuracies.*

(i) $\displaystyle\int_0^2 \int_0^{\mathrm{e}^{-x^2/2}} \sqrt{4 - x^2 - y^2}\,\mathrm{e}^{-x^2-y^2}\,\mathrm{d}y\,\mathrm{d}x$

(ii) $\displaystyle\int_0^2 \int_0^{\sqrt{4-x^2}} \int_0^{\sqrt{4-x^2-y^2}} z(x^2+y^2)\mathrm{e}^{-x^2-y^2-z^2-xz}\,\mathrm{d}z\,\mathrm{d}y\,\mathrm{d}x$

(iii) $\displaystyle\int_0^{7/10} \int_0^{4/5} \int_0^{9/10} \int_0^{1} \int_0^{11/10} \sqrt{6 - x^2 - y^2 - z^2 - w^2 - u^2}\,\mathrm{d}w\,\mathrm{d}u\,\mathrm{d}z\,\mathrm{d}y\,\mathrm{d}x.$

Bibliography

[1] Strang G. Calculus. Free textbook at `http://ocw.mit.edu/ans7870/resources/Strang/strangtext.htm`: Wellesley-Cambridge Press, 1991

[2] Dawkins P. Calculus I, II, & III. `http://tutorial.math.lamar.edu/pdf/CalcII/CalcI_Complete.pdf`; `http://tutorial.math.lamar.edu/pdf/CalcII/CalcII_Complete.pdf`; `http://tutorial.math.lamar.edu/pdf/CalcIII/CalcIII_Complete.pdf`, 2007

[3] Demidovich B P. Problems in mathematical analysis. Mowsco: MIR Publishers, 1970

[4] Shampine L F. Vectorized adaptive quadrature in MATLAB. Journal of Computational and Applied Mathematics, 2008, 211(2):131–140

[5] Wilson H, Gardner B. Numerical integration toolbox (NIT).

Chapter 4

Linear Algebra Problems

Linear algebra deals with vectors, vector spaces or linear spaces, linear maps or linear transformations and systems of linear equations. It is ubiquitous in modern applied mathematics and almost all engineering fields. Although nonlinear models are true models of the real world systems, in the natural sciences and the social sciences nonlinear models are usually approximated by linear ones for initial effective characterization. The importance of linear algebra and the ability to solve linear algebra problems is obvious.

Although many readers might not be able to quickly compute the determinant of a given 3×3 or even 4×4 matrix by hand, one should not feel bad about this inability. We leave this low-level computation to computer mathematics languages, such as MATLAB. In fact, many computer mathematics languages, such as MATLAB, originated from the early research of numerical linear algebra. For instance, the well-known EISPACK package [1] focused on the computation of eigen-systems of matrices. Another well-known package, LINPACK [2], was developed to solve general linear algebra problems using numerical algorithms. With the development of computer science, matrix computations are now no longer restricted to numerical computations. Analytical solutions can also be found for many linear algebra problems. Successful computer mathematics languages such as Mathematica, Maple and the Symbolic Math Toolbox of MATLAB can be used to analytically solve certain problems in linear algebra.

In Section 4.1, as a warming up, how to input some special matrices, such as identity matrix, companion matrix and Hankel matrix, and symbolic matrices is presented. In the section, arbitrary matrix input can also be entered in MATLAB workspace. In Section 4.2, basic concepts of matrix analysis are presented and illustrative MATLAB scripts are given for solving matrix determinant, trace, rank, norm, inverse matrix and eigen-system problems. Matrix decomposition methods such as similarity transformation, orthogonal decomposition, triangular factorization and singular value decomposition are explained and demonstrated in Section 4.3. These decomposition methods can be used to simplify matrix analysis problems among many other potential benefits. Section 4.4 presents solutions to various matrix equations, such as linear algebraic equations, Lyapunov equations, Sylvester equations as well as Riccati equations. An example in the solution of polynomial Diophantine equation is also presented. Both analytical solution algorithms and their MATLAB implementations are given. In Section 4.5, evaluations of matrix functions such as exponential functions and trigonometry functions will be discussed. In particular, this chapter ends with an introduction of a general method for computing matrix functions of arbitrary forms and power of matrix by using detailed illustrative examples with the respective MATLAB scripts.

For readers who wish to check the detailed explanations of linear algebra, we recommend the free textbooks [3–5]. In fact, the materials covered in the chapter are far beyond those in conventional linear algebra textbooks.

4.1 Inputting Special Matrices

4.1.1 Numerical matrix input

Although all the matrices can be entered into MATLAB workspace using the low-level statement discussed earlier, it might be complicated for some matrices with special structures. For instance, if one wants to enter an identity matrix, one should use the existing function `eye()` instead. In this section, the specifications of some special matrices will be presented.

I. Matrices of zeros, ones and identity matrices

In matrix theory, a matrix with all its elements 0 is referred to as a *zero matrix*, while a matrix with all its elements 1 is referred to as a *matrix of ones*. If the diagonal elements are 1 with the rest of the elements 0, the matrix is referred to as an *identity matrix*. This concept can be extended to $m \times n$ matrices. Matrix of zeros, matrix of ones and identity matrix can be entered into MATLAB using the following statements

```
A=zeros(n), B=ones(n), C=eye(n)    % n × n square matrix
A=zeros(m,n); B=ones(m,n); C=eye(m,n)   % m × n matrix
A=zeros(size(B))        % with the same size of B
```

Example 4.1 Please generate a 3×8 zero matrix $\boldsymbol{A}$ and an extended identity matrix $\boldsymbol{B}$, with the same size of $\boldsymbol{A}$.

Solution *The matrices can be entered into MATLAB environment using the following statements*

```
>> A=zeros(3,8), B=eye(size(A)) % the matrices can be entered easily
```

The two matrices can be established in MATLAB workspace

$$\boldsymbol{A} = \begin{bmatrix} 0 & 0 & 0 & 0 & 0 & 0 & 0 & 0 \\ 0 & 0 & 0 & 0 & 0 & 0 & 0 & 0 \\ 0 & 0 & 0 & 0 & 0 & 0 & 0 & 0 \end{bmatrix}, \ \boldsymbol{B} = \begin{bmatrix} 1 & 0 & 0 & 0 & 0 & 0 & 0 & 0 \\ 0 & 1 & 0 & 0 & 0 & 0 & 0 & 0 \\ 0 & 0 & 1 & 0 & 0 & 0 & 0 & 0 \end{bmatrix}.$$

Functions `zeros()` and `ones()` can also be used to define multi-dimensional arrays. For instance, `zeros(3,4,5)` can be used to define a 3×4×5 zero array.

II. Matrices with random elements

If all the elements in a matrix satisfy uniform distribution within the $[0, 1]$ interval, it can be defined using MATLAB function `rand()`. The syntaxes of such a function are as follows

```
A=rand(n)      % generates an n × n uniformly distributed random matrix
A=rand(n,m)    % generates an n × m random matrix
```

Function `rand()` can also be used to define multi-dimensional random arrays. Another function `randn()` can be used to define standard normal distributed random matrices. The statement $\boldsymbol{B}$=`rand(size(`$\boldsymbol{A}$`))` can be used to declare a random matrix of size $\boldsymbol{A}$.

Here, the random number is in fact pseudorandom numbers, and can be generated mathematically. It is easy to generate random numbers satisfying certain predefined distributions. Another advantage of pseudorandom numbers are that they can be generated repeatedly.

If one wants to obtain a uniformly distributed random number over (a,b) interval, one may generate uniformly distributed pseudorandom matrix $\boldsymbol{V}=$ `rand(`n`,`m`)` over $(0,1)$ interval, then, the expected matrix can be generated by $\boldsymbol{V}_1 = a+(b-a)*\boldsymbol{V}$ command.

If all the elements satisfy a standard normal distribution N(0,1), i.e., normal distribution with zero mean and unity variance, command $\boldsymbol{V}=$`randn(`n`,`m`)` can be used. If a normal distribution of $\mathrm{N}(\mu,\sigma^2)$ is satisfied, the sequence can be generated with $\boldsymbol{V}_1=\mu+\sigma*\boldsymbol{V}$ should be used.

III. Diagonal matrices

Mathematical description to a diagonal matrix is

$$\mathrm{diag}(\alpha_1,\alpha_2,\cdots,\alpha_n)=\begin{bmatrix}\alpha_1 & & & \\ & \alpha_2 & & \\ & & \ddots & \\ & & & \alpha_n\end{bmatrix}, \tag{4-1-1}$$

where all the non-diagonal elements are 0. A MATLAB function `diag()` can be used to deal with diagonal matrix related problems

```
A=diag(V)      % define a matrix from given vector
V=diag(A)      % extract diagonal vector from a given matrix
A=diag(V,k)    % define the kth diagonal elements V
```

Example 4.2 MATLAB function `diag()` is an interesting function. Different syntaxes are allowed for different tasks. For instance, the following statements can be used to define different matrices

```
>> C=[1 2 3]; V=diag(C), V1=diag(V)
   C=[1 2 3]; V2=diag(C,2), V3=diag(C,-1) % use of diag() function
```

and they yield

$$\boldsymbol{V}=\begin{bmatrix}1&0&0\\0&2&0\\0&0&3\end{bmatrix},\ \boldsymbol{V}_1=\begin{bmatrix}1\\2\\3\end{bmatrix},\ \boldsymbol{V}_2=\begin{bmatrix}0&0&1&0&0\\0&0&0&2&0\\0&0&0&0&3\\0&0&0&0&0\\0&0&0&0&0\end{bmatrix},\ \boldsymbol{V}_3=\begin{bmatrix}0&0&0&0\\1&0&0&0\\0&2&0&0\\0&0&3&0\end{bmatrix}.$$

In fact, k can be assigned to negative integers, indicating one wants to specify the kth lower-diagonal elements. With the use of such properties

```
>> V=diag([1 2 3 4])+diag([2 3 4],1)+diag([5 4 3],-1) % tri-diagonal
```

the tri-diagonal matrix $\boldsymbol{V}=\begin{bmatrix}1&2&0&0\\5&2&3&0\\0&4&3&4\\0&0&3&4\end{bmatrix}$ can then be established.

Assume that there exist the following matrices $\boldsymbol{A}_1$, $\boldsymbol{A}_2$, $\cdots, \boldsymbol{A}_n$, the MATLAB function `blkdiag()` can be used to generate block diagonal matrix $\boldsymbol{A}$, with $\boldsymbol{A}=$ `blkdiag(`$\boldsymbol{A}_1$`,`$\boldsymbol{A}_2$`,`$\cdots$`,`$\boldsymbol{A}_n$`)`, such that

$$\boldsymbol{A} = \begin{bmatrix} \boldsymbol{A}_1 & & & \\ & \boldsymbol{A}_2 & & \\ & & \ddots & \\ & & & \boldsymbol{A}_n \end{bmatrix}. \tag{4-1-2}$$

IV. Hankel matrices

The general form of a Hankel matrix is given below, with all the elements in each back-diagonal the same.

$$\boldsymbol{H} = \begin{bmatrix} c_1 & c_2 & \cdots & c_m \\ c_2 & c_3 & \cdots & c_{m+1} \\ \vdots & \vdots & \ddots & \vdots \\ c_n & c_{n+1} & \cdots & c_{n+m-1} \end{bmatrix}. \tag{4-1-3}$$

In MATLAB, the Hankel matrix based on two given vectors $\boldsymbol{c}$ and $\boldsymbol{r}$ can be constructed by $\boldsymbol{H}=$ `hankel(`$\boldsymbol{c}$`,`$\boldsymbol{r}$`)`, and the first column of matrix $\boldsymbol{H}$ can be assigned to vector $\boldsymbol{c}$, with the last row assigned to $\boldsymbol{r}$. Thus, using the properties of a Hankel matrix, the full Hankel matrix can then be established.

If there is only one vector $\boldsymbol{c}$ specified, the command $\boldsymbol{H}=$ `hankel(`$\boldsymbol{c}$`)` can be used to construct an upper-triangular Hankel matrix.

Example 4.3 Establish the following Hankel matrices using MATLAB statements

$$\boldsymbol{H}_1 = \begin{bmatrix} 1 & 2 & 3 & 4 & 5 & 6 & 7 \\ 2 & 3 & 4 & 5 & 6 & 7 & 8 \\ 3 & 4 & 5 & 6 & 7 & 8 & 9 \end{bmatrix}, \quad \boldsymbol{H}_2 = \begin{bmatrix} 1 & 2 & 3 \\ 2 & 3 & 0 \\ 3 & 0 & 0 \end{bmatrix}.$$

Solution *In order to construct the above Hankel matrix, the* $\boldsymbol{c}$ *and* $\boldsymbol{r}$ *vectors should be assigned to* $\boldsymbol{c}=$ `[1,2,3]`, $\boldsymbol{r}=$ `[3 4 5 6 7 8 9]`. *The Hankel matrices can then be established with the following statements.*

```
>> c=[1 2 3]; r=[3 4 5 6 7 8 9]; H1=hankel(c,r), H2=hankel(c)
```

V. Hilbert matrices and their inverses

Hilbert matrix is a special matrix whose (i,j)th element is defined as $h_{i,j} = 1/(i+j-1)$. An $n \times n$ square Hilbert can be written as

$$\boldsymbol{H} = \begin{bmatrix} 1 & 1/2 & 1/3 & \cdots & 1/n \\ 1/2 & 1/3 & 1/4 & \cdots & 1/(n+1) \\ \vdots & \vdots & \vdots & \ddots & \vdots \\ 1/n & 1/(n+1) & 1/(n+2) & \cdots & 1/(2n-1) \end{bmatrix}. \tag{4-1-4}$$

The syntax for generating the Hilbert matrix is $\boldsymbol{A}=$ `hilb(`n`)`.

Large-sized Hilbert matrices are bad-conditioned matrices. Overflow will often occur during inverting such a matrix. Thus, a direct inverse Hilbert matrix can be obtained with function $\boldsymbol{B}=$ `invhilb(`n`)`.

Since Hilbert matrices are very close to singular matrices, one must be very careful in dealing with such matrices. It is suggested here that symbolic computation be used. If numerical methods are used, do validate the results.

VI. Vandermonde matrices

For a given sequence $\boldsymbol{c} = \{c_1, c_2, \cdots, c_n\}$, a Vandermonde matrix can be established such that the (i, j)th element is defined as $v_{i,j} = c_i^{n-j}$, $i, j = 1, 2, \cdots, n$.

$$\boldsymbol{V} = \begin{bmatrix} c_1^{n-1} & c_1^{n-2} & \cdots & c_1 & 1 \\ c_2^{n-1} & c_2^{n-2} & \cdots & c_2 & 1 \\ \vdots & \vdots & \ddots & \vdots & \vdots \\ c_n^{n-1} & c_n^{n-2} & \cdots & c_n & 1 \end{bmatrix}. \tag{4-1-5}$$

A Vandermonde matrix can be established with $\boldsymbol{V}=$`vander(`$\boldsymbol{c}$`)` function in MATLAB for a given vector $\boldsymbol{c}$.

Example 4.4 Establish a Vandermonde matrix

$$\boldsymbol{A} = \begin{bmatrix} 1 & 1 & 1 & 1 & 1 \\ 16 & 8 & 4 & 2 & 1 \\ 81 & 27 & 9 & 3 & 1 \\ 256 & 64 & 16 & 4 & 1 \\ 625 & 125 & 25 & 5 & 1 \end{bmatrix}.$$

Solution *To generate such a matrix, one should select* `c= [1,2,3,4,5]`. *Thus, with the following statements, the corresponding Vandermonde can be constructed.*

```
>> c=[1, 2, 3, 4, 5];  V=vander(c) % create a Vandermonde matrix
```

VII. Companion matrices

Assume that there exists a *monic polynomial*, i.e., in the polynomial, the highest order term with coefficient 1

$$P(s) = s^n + p_1 s^{n-1} + p_2 s^{n-2} + \cdots + p_{n-1} s + p_n, \tag{4-1-6}$$

a companion matrix can be established such that

$$\boldsymbol{A}_{\rm c} = \begin{bmatrix} -p_1 & -p_2 & \cdots & -p_{n-1} & -p_n \\ 1 & 0 & \cdots & 0 & 0 \\ 0 & 1 & \cdots & 0 & 0 \\ \vdots & \vdots & \ddots & \vdots & \vdots \\ 0 & 0 & \cdots & 1 & 0 \end{bmatrix}. \tag{4-1-7}$$

A companion matrix can be established using $\boldsymbol{B}=$`compan(`$\boldsymbol{p}$`)`, where $\boldsymbol{p}$ is a polynomial coefficient vector, and `compan()` function will automatically transform it into monic form.

Example 4.5 For a polynomial $P(s) = 2s^4 + 4s^2 + 5s + 6$, find its companion matrix.

Solution *The characteristic polynomial can be entered first, it can be converted into a monic polynomial automatically, and then, the companion matrix* $\boldsymbol{A}$ *can be established using the following statements*

```
>> P=[2 0 4 5 6]; A=compan(P) % create a companion matrix
```

and the following companion matrix can be generated

$$A = \begin{bmatrix} 0 & -2 & -2.5 & -3 \\ 1 & 0 & 0 & 0 \\ 0 & 1 & 0 & 0 \\ 0 & 0 & 1 & 0 \end{bmatrix}.$$

VIII. Random integer matrices

Based on the `rand()` function, an integer matrix with random entities in the interval $[a, b]$ can be generated with A = `randintmat(`a`,`b`,`n`,`m`)`, and if m is omitted, an $n \times n$ matrix can be generated. The listing of the function is

```
function A=randintmat(a,b,n,m)  % generate a random integer matrix
if nargin==3, m=n; end          % generate a square matrix
a=floor(a); b=floor(b); A=floor(a+(b-a+1)*rand(n,m));
```

Example 4.6 Please generate a 10×10 nonsingular matrix with 0's and 1's.

Solution *An infinite loop can be used to generate such a matrix, and if the matrix generated is nonsingular,* `break` *command can be used to terminate the loop. The commands are*

```
>> while(1), A=randintmat(0,1,10); if det(A)~=0, break; end, end
```

4.1.2 Defining symbolic matrices

For a given numerical matrix A, one may transform it by B = `sym(`A`)` into a symbolic matrix. Thus, all the numerical matrices can be transformed into symbolic matrices so as to achieve higher accuracy. In some cases, it is even possible to find analytical solutions.

A symbolic matrix with all its elements numbers can be converted to a double-precision matrix with A_1 = `double(`A`)`.

A symbolic matrix with arbitrary elements a_{ij} can be generated with the `sym()` function, with A = `sym('a%d%d',[`n`,`m`])`. Arbitrary vectors with a single subscript can also be generated with

v = `sym('a%d',[1,`n`])`, or , v = `sym('a%d',[`n`,1])`

Example 4.7 Please generate the three arbitrary matrices and a column vector

$$A = \begin{bmatrix} a_{11} & a_{12} & a_{13} & a_{14} \\ a_{21} & a_{22} & a_{23} & a_{24} \\ a_{31} & a_{32} & a_{33} & a_{34} \\ a_{41} & a_{42} & a_{43} & a_{44} \end{bmatrix}, \quad B = \begin{bmatrix} a_{11} & a_{12} \\ a_{21} & a_{22} \\ a_{31} & a_{32} \\ a_{41} & a_{42} \end{bmatrix}, \quad C = \begin{bmatrix} f_{11} & f_{12} & f_{13} & f_{14} \\ f_{21} & f_{22} & f_{23} & f_{24} \\ f_{31} & f_{32} & f_{33} & f_{34} \\ f_{41} & f_{42} & f_{43} & f_{44} \end{bmatrix}, \quad v = \begin{bmatrix} v_1 \\ v_2 \\ v_3 \\ v_4 \end{bmatrix}.$$

Solution *The following commands can be used directly*

```
>> A=sym('a%d%d',4), B=sym('a%d%d',[4,2]), % create arbitrary matrices
   C=sym('f%d%d',4), v=sym('v%d',[4,1])
```

If one wants to declare the components of the matrices or vectors as real entities or entities having other properties, `assumeAlso()` *function can be used. For instance, the following commands can be used*

```
>> assumeAlso(A,'real'); assumeAlso(B,'integer') % set matrix properties
```

The arbitrary matrices constructed with `sym()` function are constant matrices. If an arbitrary matrix function is expected, for instance, $\boldsymbol{M} = \{m_{ij}(x,y)\}$, a MATLAB function `any_matrix()` can be written

```
function A=any_matrix(nn,a_str,varargin) % generate a matrix function
v=varargin; n=nn(1); if length(nn)==1, m=n; else, m=nn(2); end
s=''; k=length(v); K=0; if n==1 | m==1, K=1; end
if k>0, s='(';          % specify independent variables if needed
   for i=1:k, s=[s ',' char(v{i})]; end, s(2)=[]; s=[s ')'];
end
for i=1:n, for j=1:m, % processing each element of the matrix
   if K==0, str=[a_str int2str(i),int2str(j) s];
   else, str=[a_str int2str(i*j) s]; end % generate an arbitrary vector
   eval(['syms ' str]); eval(['A(i,j)=' str ';']); % assign elements
end, end
```

The vector or matrix functions can be established with commands such as

$\boldsymbol{A}$ = `any_matrix([5,1],'a',`x`,`y`)`, or, $\boldsymbol{v}$ = `any_matrix(5,'m',`t`)`

where a 5×1 vector $\boldsymbol{A}(x,y)$, with elements $a_i(x,y)$, and a 5×5 square matrix $\boldsymbol{v}(t)$, with elements $m_{ij}(t)$, can be established.

For other special matrices, such as Vandermonde matrix, Hankel matrix and companion matrix, the functions described above can also be used, in recent versions of MATLAB, to generate symbolic matrices.

Example 4.8 Establish a companion matrix from the following polynomial

$$P(\lambda) = a_1\lambda^9 + a_2\lambda^8 + a_3\lambda^7 + \cdots + a_8\lambda^2 + a_9\lambda + a_{10}.$$

Solution *With the use of the new support of* `compan()` *function, the required matrix can be established using the following statements*

```
>> a=sym('a%d',[1,10]); A=compan(a) % create symbolic companion matrix
```

the following matrix $\boldsymbol{A}$ *can be generated*

$$\begin{bmatrix} -a_2/a_1 & -a_3/a_1 & -a_4/a_1 & -a_5/a_1 & -a_6/a_1 & -a_7/a_1 & -a_8/a_1 & -a_9/a_1 & -a_{10}/a_1 \\ 1 & 0 & 0 & 0 & 0 & 0 & 0 & 0 & 0 \\ 0 & 1 & 0 & 0 & 0 & 0 & 0 & 0 & 0 \\ 0 & 0 & 1 & 0 & 0 & 0 & 0 & 0 & 0 \\ 0 & 0 & 0 & 1 & 0 & 0 & 0 & 0 & 0 \\ 0 & 0 & 0 & 0 & 1 & 0 & 0 & 0 & 0 \\ 0 & 0 & 0 & 0 & 0 & 1 & 0 & 0 & 0 \\ 0 & 0 & 0 & 0 & 0 & 0 & 1 & 0 & 0 \\ 0 & 0 & 0 & 0 & 0 & 0 & 0 & 1 & 0 \end{bmatrix}.$$

4.1.3 Sparse matrix input

In many applications, there might be extremely large scale matrices, with most of their elements zeros, with very few nonzero entities, this kind of matrix is also referred to as a *sparse matrix*. Some of the input and analysis functions in MATLAB support sparse matrices with high efficiencies.

Sparse matrices can be loaded into MATLAB workspace with function `sparse()`, with the syntax A=`sparse(p,q,w)`, where, p, q are the row and column indices of nonzero elements, while w is the vector composed of nonzero elements. The lengths of the three vectors must be the same, otherwise, an error message will be given.

With B=`full(A)`, a sparse matrix A can be converted into an ordinary matrix B, and with A=`sparse(B)`, an ordinary matrix can be converted to a sparse matrix. If a matrix with more than 2/3 of its elements zeros, it is more economic to store it in sparse form.

Example 4.9 Please generate a 10×10 identity matrix in sparse form.

Solution *To generate sparse matrices, the rows, columns and entities of the nonzero elements should be specified in vectors. Therefore, the matrix can be generated with*

```
>> i=1:10; j=1:10; v=ones(1,10); A=sparse(i,j,v) % sparse identity matrix
```

In fact, such a matrix can be generated with `speye()` *function directly.*

4.2 Fundamental Matrix Operations

4.2.1 Basic concepts and properties of matrices

I. Determinant

The determinant of matrix $A = \{a_{ij}\}$ is defined as

$$D = \mid A \mid = \det(A) = \sum (-1)^k a_{1k_1} a_{2k_2} \cdots a_{nk_n}, \tag{4-2-1}$$

where the sum is taken over all possible permutations on n elements, and the sign is positive if the permutation is even and negative if the permutation is odd.

There are many algorithms which can be used to compute the determinant of a matrix. A built-in function `det()` provided in MATLAB can be used to calculate the determinant d=`det(A)`, and this function applies both to symbolic and numerical matrices.

Example 4.10 Compute the determinant of a given matrix $A = \begin{bmatrix} 16 & 2 & 3 & 13 \\ 5 & 11 & 10 & 8 \\ 9 & 7 & 6 & 12 \\ 4 & 14 & 15 & 1 \end{bmatrix}$.

Solution *The determinant of matrix* A *can be obtained and it equals 0, which means that matrix* A *is singular.*

```
>> A=[16 2 3 13; 5 11 10 8; 9 7 6 12; 4 14 15 1]; det(A), det(sym(A))
```

Example 4.11 From the example given in Chapter 1, it is known that large-sized Hilbert matrix is very close to singular. Calculate analytically the determinant of an 80×80 Hilbert matrix.

Solution *The function* `hilb()` *can be used to declare a numerical* 80×80 *Hilbert matrix.*

It can then be transformed into a symbolic matrix. The `det()` *function in MATLAB can be used to calculate analytically the determinant of the matrix.*

```
>> A=sym(hilb(80)); det(A) % determinant of 80 × 80 Hilbert matrix
```

Therefore, the determinant can be obtained as

$$\det(\boldsymbol{H})=\frac{1}{\underbrace{99030101466993477878867678\cdots000000000000}_{3790 \textit{ digits, with some digits omitted}}}\approx 1.009794\times 10^{-3790}.$$

Example 4.12 Please derive the formula for the determinant of a 4×4 matrix.

Solution *A* 4×4 *arbitrary matrix can be generated first, and based on it, symbolic function* `det()` *can be used to find its determinant.*

```
>> A=sym('a%d%d',4); d=det(A) % determinant of an arbitrary 4 × 4 matrix
```

The determinant can be obtained as

$$\begin{aligned}d=&a_{11}a_{22}a_{33}a_{44}-a_{11}a_{22}a_{34}a_{43}-a_{11}a_{23}a_{32}a_{44}+a_{11}a_{23}a_{34}a_{42}+a_{11}a_{24}a_{32}a_{43}-a_{11}a_{24}a_{33}a_{42}\\&-a_{12}a_{21}a_{33}a_{44}+a_{12}a_{21}a_{34}a_{43}+a_{12}a_{23}a_{31}a_{44}-a_{12}a_{23}a_{34}a_{41}-a_{12}a_{24}a_{31}a_{43}+a_{12}a_{24}a_{33}a_{41}\\&+a_{13}a_{21}a_{32}a_{44}-a_{13}a_{21}a_{34}a_{42}-a_{13}a_{22}a_{31}a_{44}+a_{13}a_{22}a_{34}a_{41}+a_{13}a_{24}a_{31}a_{42}-a_{13}a_{24}a_{32}a_{41}\\&-a_{14}a_{21}a_{32}a_{43}+a_{14}a_{21}a_{33}a_{42}+a_{14}a_{22}a_{31}a_{43}-a_{14}a_{22}a_{33}a_{41}-a_{14}a_{23}a_{31}a_{42}+a_{14}a_{23}a_{32}a_{41},\end{aligned}$$

where the terms in the results are grouped in 6, deliberately by the author with square brackets. It can be seen that, the first group of terms are associated with a_{11}*, and the second group are associated with* a_{12}*, and so on. Thus, the determinant can be written as* $a_{11}A_{11}+a_{12}A_{12}+a_{13}A_{13}+a_{14}A_{14}$*, where,* A_{ij} *is referred to as the* algebraic complement *of* a_{ij}*, and it is in fact the determinant of the sub matrix, with the* i*th row and* j*th column elements removed, and then, multiplied by* $(-1)^{i+j}$.

If the algebraic complement of A_{23} *is expected, two approaches can be used. The following statements can be used to find it.*

$$A_{23}=-a_{11}a_{32}a_{44}+a_{11}a_{34}a_{42}+a_{12}a_{31}a_{44}-a_{12}a_{34}a_{41}-a_{14}a_{31}a_{42}+a_{14}a_{32}a_{41}$$

```
>> i=2; j=3; B=A; B(i,:)=[]; B(:,j)=[]; A23=(-1)^(i+j)*det(B)
```

Alternatively, delete all the terms which do not contain a_{23} *in d, then, divide the result by* a_{23}*. The results obtained are exactly the same.*

```
>> syms a23; A23_1=simplify((d-subs(d,a23,0))/a23)
```

Based on the concept of algebraic complement of a matrix, a new matrix

$$\operatorname{adj}(\boldsymbol{A})=\begin{bmatrix}A_{11}&A_{21}&\cdots&A_{n1}\\A_{12}&A_{22}&\cdots&A_{n2}\\\vdots&\vdots&\ddots&\vdots\\A_{1n}&A_{2n}&\cdots&A_{nn}\end{bmatrix}\tag{4-2-2}$$

can be constructed, and $\operatorname{adj}(\boldsymbol{A})$ is referred to as the *adjoint matrix* of the square matrix $\boldsymbol{A}$, and it is in fact the transpose of the matrix of algebraic complements of $\boldsymbol{A}$. The adjoint matrix of $\boldsymbol{A}$ can be obtained with $\boldsymbol{B}$= `adjoint(`$\boldsymbol{A}$`)`. The algebraic complement A_{23} in the previous example can be extracted with

```
>> B=adjoint(A).'; A23=B(2,3) % call function to get the whole adjoint matrix
```

II. Trace

For a square matrix $\boldsymbol{A} = \{a_{ij}\}$, $i, j = 1, 2, \cdots, n$, the trace of the $\boldsymbol{A}$ is defined as

$$\mathrm{tr}(\boldsymbol{A}) = \sum_{i=1}^{n} a_{ii}, \tag{4-2-3}$$

i.e., the trace of a matrix is defined as the sum of diagonal elements. From linear algebra theory, the trace of a matrix equals the sum of the eigenvalues. The trace of matrix $\boldsymbol{A}$ can be obtained using the MATLAB function `trace()`, such that t=`trace(A)`. The trace of the matrix in Example 4.10 can be obtained directly from trace($\boldsymbol{A}$) = 34.

III. Rank

If for a given $n \times m$ matrix, there exist maximum r_{c} linearly independent columns, the column rank of the matrix is r_{c}. If $r_{\mathrm{c}} = m$, the matrix is referred to as a *full column rank matrix*. Similarly, if there exist maximum r_{r} linearly independent rows, the row rank of the matrix is r_{r}. If $r_{\mathrm{r}} = n$, matrix $\boldsymbol{A}$ is referred to as a *full row rank matrix*. It can be shown that the column rank and row rank of the same matrix are identical, and they both can be called the *rank* of the matrix, i.e., $\mathrm{rank}(\boldsymbol{A}) = r_{\mathrm{c}} = r_{\mathrm{r}}$. The rank of matrix $\boldsymbol{A}$ is mathematically denoted as rank($\boldsymbol{A}$).

There are various algorithms for calculating the ranks of given matrices. Some of the algorithms may be numerically unstable. A built-in function `rank()` has been provided in MATLAB which applies to both numerical and symbolic matrices. The syntaxes of the function are

```
r=rank(A)     % symbolic or numerical
r=rank(A,ε)   % numerical rank with error tolerance of ε
```

where $\boldsymbol{A}$ is the given matrix, and ε is user specified error tolerance.

Example 4.13 Find the rank of matrix $\boldsymbol{A}$ in Example 4.10.

Solution *The MATLAB function* `rank(A)` *can be used to calculate the rank of matrix* $\boldsymbol{A}$ *and* rank($\boldsymbol{A}$) = 3, *which indicates that* $\boldsymbol{A}$ *is not a full rank matrix.*

Example 4.14 Now consider the 20×20 Hilbert matrix in Example 4.11. Find the rank of the matrix in numerical and analytic methods, respectively.

Solution *The following commands can be used to find the rank numerically*

```
>> H=hilb(20); rank(H) % numerical rank, may be wrong
```

and it can be seen that the rank of the matrix is 13, which means that the matrix is not of full rank. Now let us try the analytic method

```
>> H=sym(hilb(20)); rank(H) % analytical rank, correct
```

and it is concluded that the rank of the matrix is 20. Therefore, if the numerical method is adopted, one should be very careful, since misleading results can sometimes be obtained.

IV. Norms

The norms of a matrix can be considered as a measure of "size" of the matrix. Before introducing the concept of the norms of the matrices, the norms of the vectors are introduced first. For a vector $\boldsymbol{x}$ in linear space, if there exists a function $\rho(\boldsymbol{x})$ satisfying the following three conditions:

(i) $\rho(\boldsymbol{x}) \geqslant 0$ and $\rho(\boldsymbol{x}) = 0$ if and only if $\boldsymbol{x} = \boldsymbol{0}$;

(ii) $\rho(a\boldsymbol{x}) = |a|\rho(\boldsymbol{x})$, a is any given scalar;

(iii) for vectors $\boldsymbol{x}$ and $\boldsymbol{y}$, there exists $\rho(\boldsymbol{x}+\boldsymbol{y}) \leqslant \rho(\boldsymbol{x}) + \rho(\boldsymbol{y})$.

Then, $\rho(\boldsymbol{x})$ is referred to as the *norm* of the vector $\boldsymbol{x}$. There are various forms of the norms. It can be shown that a class of norms defined below satisfies all the above three conditions.

$$||\boldsymbol{x}||_p = \left(\sum_{i=1}^{n} |\ x_i\ |^p\right)^{1/p},\ p = 1, 2, \cdots,\ \text{and}\ ||\boldsymbol{x}||_\infty = \max_{1\leqslant i\leqslant n} |\ x_i\ |, \tag{4-2-4}$$

where the notation $||\boldsymbol{x}||_p$ is used to define the p-norm of the vector $\boldsymbol{x}$.

The definition of the norms of a matrix $\boldsymbol{A}$ is

$$||\boldsymbol{A}|| = \sup_{\boldsymbol{x}\neq 0} \frac{||\boldsymbol{A}\boldsymbol{x}||}{||\boldsymbol{x}||}, \tag{4-2-5}$$

for any non-zero vector $\boldsymbol{x}$. Similar to vector norms, the commonly used norms of a matrix are the following three

$$||\boldsymbol{A}||_1 = \max_{1\leqslant j\leqslant n} \sum_{i=1}^{n} |a_{ij}|,\ ||\boldsymbol{A}||_2 = \sqrt{s_{\max}(\boldsymbol{A}^{\mathrm{T}}\boldsymbol{A})},\ ||\boldsymbol{A}||_\infty = \max_{1\leqslant i\leqslant n} \sum_{j=1}^{n} |a_{ij}|, \tag{4-2-6}$$

where $s(\boldsymbol{X})$ is the eigenvalue of matrix $\boldsymbol{X}$, while $s_{\max}(\boldsymbol{A}^{\mathrm{T}}\boldsymbol{A})$ is the maximum eigenvalue of matrix $\boldsymbol{A}^{\mathrm{T}}\boldsymbol{A}$. In fact, $||\boldsymbol{A}||_2$ equals the maximum singular value of matrix $\boldsymbol{A}$.

The norms of a matrix can be evaluated using the MATLAB function `norm()`. Note that the `norm()` function applies only to numerical matrices. The syntaxes of the function are

```
N=norm(A)           % default for ||A||2
N=norm(A,options)   % options could be 1,2,inf, see Table 4.1
```

Different norms of matrix $\boldsymbol{A}$ in Example 4.10 can be evaluated using the following MATLAB statements

```
>> A=[16 2 3 13; 5 11 10 8; 9 7 6 12; 4 14 15 1];
   [norm(A), norm(A,2), norm(A,1), norm(A,Inf), norm(A,'fro')]
```

where $||\boldsymbol{A}||_1 = ||\boldsymbol{A}||_2 = ||\boldsymbol{A}||_\infty = 34$, $||\boldsymbol{A}||_{\mathrm{F}} = 38.6782$. The matrix here is a special matrix such that the norms $||\boldsymbol{A}||_1 = ||\boldsymbol{A}||_2 = ||\boldsymbol{A}||_\infty$ but in general cases, the specific values of the norms should not be the same.

It should be noted that the function `norm()` is not applicable to symbolic matrices containing variables.

V. Characteristic polynomials

When a symbolic variable s is introduced, the determinant of the matrix $s\boldsymbol{I} - \boldsymbol{A}$ can be constructed, and it is a polynomial of s, expressed as

$$C(s) = \det(s\boldsymbol{I} - \boldsymbol{A}) = s^n + c_1 s^{n-1} + \cdots + c_{n-1}s + c_n, \tag{4-2-7}$$

TABLE 4.1: The options for the `norm()` function.

options	definitions and algorithms
none	the maximum singular value, i.e., $\|\|\boldsymbol{A}\|\|_2$
2	the same as defaults, i.e., $\|\|\boldsymbol{A}\|\|_2$
1	1-norm of the matrix, i.e., $\|\|\boldsymbol{A}\|\|_1$
`Inf` or `'inf'`	infinite norm, i.e., $\|\|\boldsymbol{A}\|\|_\infty$
`'fro'`	Frobenius norm of the matrix, i.e., $\|\|\boldsymbol{A}\|\|_{\rm F}={\rm tr}(\boldsymbol{A}^{\rm T}\boldsymbol{A})$
integer p	applies only to vectors. For matrices, only 1,2, `inf` and `'fro'` are allowed
`-inf`	For vectors only, where $\|\|\boldsymbol{A}\|\|_{-\infty}=\min(\|\sum a_i\|)$

and polynomial $C(s)$ is referred to as the *characteristic polynomial* of matrix $\boldsymbol{A}$. In the formula, the coefficients c_i, $i=1,2,\cdots,n$ are referred to as the *coefficients of the characteristic polynomial.*

A MATLAB function `c=poly(`$\boldsymbol{A}$`)` can be used to evaluate the coefficients of the characteristic polynomial of matrix $\boldsymbol{A}$, where the returned vector $\boldsymbol{c}$ contains the coefficients in descending order of s of the characteristic polynomial. If the input argument $\boldsymbol{A}$ is a vector, the `poly()` function will return the polynomial coefficients vector whose roots are given in $\boldsymbol{A}$.

It should also be noted that this function is no longer usable to symbolic $\boldsymbol{A}$ matrix. An alternative function `charpoly()` with the same syntax can be used instead. To get characteristic polynomial expression with variable x, the function can be called with p`=charpoly(`$\boldsymbol{A}$`,`x`)`.

Example 4.15 Find the characteristic polynomial of matrix $\boldsymbol{A}$ in Example 4.10.

Solution *Using the* `poly()` *function, the following results can be obtained*

```
>> A=[16 2 3 13; 5 11 10 8; 9 7 6 12; 4 14 15 1]; c1=poly(A)
```

the coefficient vector is $\boldsymbol{c}_1=10^3\times[0.0010,-0.0340,-0.0800,2.7200,-0.0000]^{\rm T}$, *and it can be seen that there might exist minor errors in the results.*

Using symbolic matrix $\boldsymbol{A}$, *the function* `charpoly()` *will yield a characteristic polynomial such that* $\varphi(x)=x^4-34x^3-80x^2+2720x$.

```
>> A=sym(A); c2=charpoly(A) % coefficient vector is returned in new versions
```

In practical applications, there are other algorithms which can be used to calculate numerically the coefficients of the characteristic polynomial of a matrix. For instance, the Leverrier–Faddeev recursive algorithm is an accurate algorithm for solving such problems.

$$c_{k+1}=-\frac{1}{k}{\rm tr}(\boldsymbol{A}\boldsymbol{R}_k),\quad \boldsymbol{R}_{k+1}=\boldsymbol{A}\boldsymbol{R}_k+c_{k+1}\boldsymbol{I},\quad k=1,\cdots,n, \tag{4-2-8}$$

where $\boldsymbol{R}_1=\boldsymbol{I}, c_1=1$.

In the above algorithm, an identity matrix $\boldsymbol{I}$ is assigned to matrix $\boldsymbol{R}_1$. Then, for each value of k, the matrix $\boldsymbol{R}_k$ is recursively updated, and then, the value of c_k can be found. Based on such an algorithm, the following MATLAB function can be written

```
function c=poly1(A)
[nr,nc]=size(A); I=eye(nc);  R=I; c=[1 zeros(1,nc)];
for k=1:nc, c(k+1)=-1/k*trace(A*R); R=A*R+c(k+1)*I; end
```

Example 4.16 With the new `poly1()` function, accurate result can be achieved

```
>> c=poly1(A) % call the new function to get accurate result
```

such that $c = [1, -34, -80, 2720, 0]$, which is the accurate result.

Example 4.17 For a given vector $\boldsymbol{B} = [a_1, a_2, a_3, a_4, a_5]$, establish the corresponding Hankel matrix, then, find the characteristic polynomial.

Solution *The command* $\boldsymbol{A}$=`hankel(`$\boldsymbol{B}$`)` *can be used to construct a symbolic Hankel matrix. Thus, the function* `charpoly(`$\boldsymbol{A}$`)` *can then be used to calculate its characteristic polynomial coefficients.*

```
>> syms x; a=sym('a%d',[1,5]); A=hankel(a);  % construct a Hankel matrix
   p=charpoly(A,x); p=collect(p,x)
```

The mathematical representation of the characteristic equation can be written as

$$\begin{aligned}p(\boldsymbol{A}) &= x^5 + (-a_3 - a_1 - a_5)x^4 + (a_5a_1 + a_3a_1 + a_5a_3 - 2a_4^2 - 2a_5^2 - a_2^2 - a_3^2)x^3 \\ &+ (a_4^2a_1 + a_5^2a_3 + a_4^2a_5 + a_2^2a_5 + a_4^2a_3 - a_3a_1a_5 - 2a_2a_5a_4a_5^2a_1 - 2a_2a_4a_3 + 2a_5^3 + a_3^3)x^2 \\ &+ (-3a_4^2a_3a_5 + a_5^4 + a_4^2a_5^2 + a_4^4 - a_5^3a_1 + a_5^2a_3^2 + 2a_2a_5^2a_4 - a_5^3a_3)x - a_5^5.\end{aligned}$$

VI. Evaluation of polynomial matrices

Polynomial matrices take the following form

$$\boldsymbol{B} = a_1\boldsymbol{A}^n + a_2\boldsymbol{A}^{n-1} + \cdots + a_n\boldsymbol{A} + a_{n+1}\boldsymbol{I}, \tag{4-2-9}$$

where $\boldsymbol{A}$ is a given matrix, $\boldsymbol{I}$ is an identity matrix whose size is the same as matrix $\boldsymbol{A}$. The matrix $\boldsymbol{B}$ is then the polynomial matrix. In MATLAB, the polynomial matrix can be evaluated using the function `polyvalm()`, such that $\boldsymbol{B}$=`polyvalm(`$\boldsymbol{a}$`,`$\boldsymbol{A}$`)`, where $\boldsymbol{a}$= `[`$a_1, a_2, \cdots, a_n, a_{n+1}$`]` are the coefficients in descending order of s of the polynomial.

It should be noted that function `polyvalm()` can only be used when $\boldsymbol{a}$ is a double-precision vector. In order to extend it into symbolic computation, the following MATLAB function can be written.

```
function B=polyvalmsym(p,A)
E=eye(size(A)); B=zeros(size(A)); n=length(A);
for i=n+1:-1:1, B=B+p(i)*E; E=E*A; end
```

On the other hand, if polynomial operation is defined upon "dot operation" basis such that

$$\boldsymbol{C} = a_1\boldsymbol{x}\texttt{.\^{}}n + a_2\boldsymbol{x}\texttt{.\^{}}(n-1) + \cdots + a_{n+1}, \tag{4-2-10}$$

the matrix $\boldsymbol{C}$ can be evaluated from $\boldsymbol{C}$=`polyval(`$\boldsymbol{a}$`,`$\boldsymbol{x}$`)`.

Example 4.18 Cayley–Hamilton Theorem is a very important theorem in linear algebra. The theorem states that if the characteristic polynomial of matrix $\boldsymbol{A}$ is

$$f(s) = \det(s\boldsymbol{I} - \boldsymbol{A}) = a_1s^n + a_2s^{n-1} + \cdots + a_ns + a_{n+1}, \tag{4-2-11}$$

then, $f(\boldsymbol{A}) = 0$, i.e.,

$$a_1\boldsymbol{A}^n + a_2\boldsymbol{A}^{n-1} + \cdots + a_n\boldsymbol{A} + a_{n+1}\boldsymbol{I} = \boldsymbol{0}. \tag{4-2-12}$$

Assume that matrix $\boldsymbol{A}$ is a Vandermonde matrix. Verify that it satisfies the Cayley–Hamilton Theorem.

Solution *The following statements can be used to verify the theorem.*

```
>> A=vander([1 2 3 4 5 6 7]); p=poly(A); B=polyvalm(p,A); norm(B)
```

The norm of the error matrix is 2.1887×10^6, *which is too large because the* `poly()` *function is not accurate for the matrix. Therefore, something strange might have happened in the verification.*

It has been indicated that the function `poly()` *may cause errors, and for this particular matrix, the error is so large that misleading results are obtained. So for this matrix, the new function* `poly1()` *should be used instead. And it can be seen that with the new function, exact solutions can easily be obtained.*

```
>> p1=poly1(A); B1=polyvalm(p1,A); norm(B1)
```

It can be seen from the results that $\boldsymbol{B}$ *matrix obtained is a matrix of zeros, which means that for the given matrix, the Cayley–Hamilton Theorem holds.*

Example 4.19 Please show that an arbitrary 5×5 matrix satisfies Cayley–Hamilton Theorem.

Solution *An arbitrary matrix should be generated first, then,* `charpoly()` *function can be used to find the coefficients of the polynomial matrix, and finally* `polyvalm()` *function can be used to find the polynomial matrix. Simplification to the resulted matrix can be made and see whether it is a matrix of zeros. For the example, the elapsed time is around 5 seconds, and Cayley–Hamilton Theorem can be verified. To validate the theorem for an arbitrary* 6×6 *matrix, 2 minutes are needed.*

```
>> A=sym('a%d%d',5); p=charpoly(A); tic % generate an arbitrary matrix A
   E=simplify(polyvalmsym(p,A)), toc     % show Cayley–Hamilton Theorem
```

VII. Symbolic polynomials and coefficient extractions

A polynomial can either be represented in a numerical way or in a symbolic way. In the former case, a polynomial of x

$$p(x)=a_1x^n+a_2x^{n-1}+\cdots+a_nx+a_{n+1} \tag{4-2-13}$$

can be expressed by a coefficient vector such that $\boldsymbol{p}=$ `[`a_1`,` a_2`,` $\cdots$`,` a_{n+1}`]`. In the latter case, it can be expressed by symbolic polynomials. One may convert a numerical polynomial to a symbolic one using `poly2sym()` function, and the function `sym2poly()` can be used to convert in the other way. The syntaxes of the two functions are rather simple.

f `=poly2sym(`$\boldsymbol{p}$`)`, or f `=poly2sym(`$\boldsymbol{p}$`,`x`)`, or $\boldsymbol{p}$`=sym2poly(`f`)`

Example 4.20 Represent the $f=s^5+2s^4+3s^3+4s^2+5s+6$ in both numerical and symbolic forms.

Solution *For simplicity, a vector can be constructed from the coefficients of the polynomial, and then, the function* `poly2sym()` *can be used to find the symbolic representation*

```
>> P=[1 2 3 4 5 6];  % coefficients of the polynomial in descending order
   f=poly2sym(P,'v') % v is used as the operator
```

and the symbolic polynomial can be obtained as $f = v^5 + 2v^4 + 3v^3 + 4v^2 + 5v + 6$. *The numerical function can be obtained with* $\boldsymbol{P}$= `sym2poly(`f`)`.

Function $\boldsymbol{C}$= `coeffs(`P`,`x`)` can be used to extract coefficients, in ascending order of x, and if x is the only symbolic variable in P, it can be omitted.

Example 4.21 Please extract the coefficients of x from polynomial $(x + 2y)^8$.

Solution *It is obvious that the original polynomial is a polynomial of x and y. If the coefficients are to be extracted in ascending order of x, the following statements can be issued*

```
>> syms x y; P=(x+2*y)^8; p=coeffs(P,x) % extract coefficients of x in P
```

and the result is $\boldsymbol{p} = [256y^8, 1024y^7, 1792y^6, 1792y^5, 1120y^4, 448y^3, 112y^2, 16y, 1]$.

Unfortunately when there are missing terms in the polynomial, the behavior of `coeffs()` function may not be satisfactory, and `sym2poly()` function cannot handle polynomials with other symbolic parameters. A new MATLAB function is needed to extract the coefficients in descending order of x.

For a given polynomial in (4-2-13), it is easily found that

$$a_{n+1} = p(0), \text{ and } a_i = \frac{1}{(n-i+1)!} \left. \frac{\mathrm{d}^{n-i+1} p(x)}{\mathrm{d}x^{n-i+1}} \right|_{t=0}, \quad i = 1, 2, \cdots, n. \tag{4-2-14}$$

The above coefficient extraction algorithm can be implemented as

```
function c=polycoef(p,x)
c=[]; n=0; p1=p; n1=1; nn=1; if nargin==1, x=symvar(p); end
while (1), % loop for all the coefficients
   c=[c subs(p1,x,0)]; p1=diff(p1,x); n=n+1; n1=n1*n; nn=[nn,n1];
   if p1==0, c=c./nn(1:end-1); c=c(end:-1:1); break;
end, end
```

4.2.2 Matrix inversion

For an $n \times n$ nonsingular square matrix $\boldsymbol{A}$, if there exists a matrix $\boldsymbol{C}$ of the same size satisfying

$$\boldsymbol{AC} = \boldsymbol{CA} = \boldsymbol{I}, \tag{4-2-15}$$

where $\boldsymbol{I}$ is an identity matrix, then, matrix $\boldsymbol{C}$ is referred to as the *inverse matrix* of $\boldsymbol{A}$, denoted as $\boldsymbol{C} = \boldsymbol{A}^{-1}$.

A MATLAB function $\boldsymbol{C}$= `inv(`$\boldsymbol{A}$`)` is provided to calculate the inverse matrix $\boldsymbol{C}$, and this function is applicable for both numerical and symbolic matrices.

Example 4.22 Compute the inverse matrix for the given Hilbert matrix.

Solution *Let us consider first a 4×4 Hilbert matrix. The MATLAB function* `inv()` *can be used to find the inverse matrix*

```
>> format long; H=hilb(4); H1=inv(H), norm(H*H1-eye(4))
```

and the equation obtained below with the error is around 1.3931×10^{-12}*. The inverse matrix obtained is*

$$\begin{bmatrix} 15.999999999999 & -119.99999999999 & 239.99999999998 & -139.99999999999 \\ -119.99999999999 & 1199.9999999999 & -2699.9999999997 & 1679.9999999998 \\ 239.99999999998 & -2699.9999999997 & 6479.9999999994 & -4199.9999999996 \\ -139.99999999999 & 1679.9999999998 & -4199.9999999996 & 2799.9999999997 \end{bmatrix}.$$

Since a large-sized Hilbert matrix is close to a singular matrix, the use of numerical function `inv()` *is not recommended for Hilbert matrices. The function* `invhilb()` *can be used instead to find the accurate inverse matrix. For the* 4×4 *Hilbert matrix, the inverse matrix can be obtained*

```
>> H2=invhilb(4); norm(H*H2-eye(size(H)))
```

and the error is reduced to 5.6843×10^{-14}*, which means that the function* `invhilb()` *improves significantly for inverse matrices. Now consider a* 10×10 *Hilbert matrix, the inverse matrices by* `inv()` *and* `invhilb()` *can be obtained*

```
>> H=hilb(10); H1=inv(H); norm(H*H1-eye(size(H)))
   H2=invhilb(10); norm(H*H2-eye(size(H)))
```

and the errors by these approaches are respectively 0.0032 and 2.5249×10^{-5}*. The accuracy is very low. If the size of the matrix is further increased to 13, then, the commands*

```
>> H=hilb(13); H1=inv(H); norm(H*H1-eye(size(H)))
   H2=invhilb(13); norm(H*H2-eye(size(H)))
```

will detect the error norms by using the above two methods respectively as 81.1898, 11.7781. They are too high to be practically used.

Fortunately, the function `inv()` *is also provided in the Symbolic Math Toolbox, which can be used to evaluate the inverse matrix for symbolic matrices. Even for large-sized nonsingular matrices, the use of such a function can return error-free solutions. Using the following commands, the inverse matrix of a* 7×7 *Hilbert can be obtained and displayed*

```
>> H=sym(hilb(7)); inv(H) % exact computation of inverse matrix
```

and the exact inverse can then be found

$$\begin{bmatrix} 49 & -1176 & 8820 & -29400 & 48510 & -38808 & 12012 \\ -1176 & 37632 & -317520 & 1128960 & -1940400 & 1596672 & -504504 \\ 8820 & -317520 & 2857680 & -10584000 & 18711000 & -15717240 & 5045040 \\ -29400 & 1128960 & -10584000 & 40320000 & -72765000 & 62092800 & -20180160 \\ 48510 & -1940400 & 18711000 & -72765000 & 133402500 & -115259760 & 37837800 \\ -38808 & 1596672 & -15717240 & 62092800 & -115259760 & 100590336 & -33297264 \\ 12012 & -504504 & 5045040 & -20180160 & 37837800 & -33297264 & 11099088 \end{bmatrix}.$$

In fact, even for a 50×50 *Hilbert matrix, the exact inverse matrix can be obtained and it can be shown that there is zero error norm in the results. The elapsed time is around 9 seconds.*

```
>> tic, H=sym(hilb(50)); norm(H*inv(H)-eye(size(H))), toc % larger matrix
```

Example 4.23 Compute the inverse matrix for the matrix $\boldsymbol{A}$ in Example 4.10 and observe the differences using numerical and analytical methods.

Solution *One can enter the matrix, and then, use* `inv()` *function to find its numerical inverse*

```
>> A=[16 2 3 13; 5 11 10 8; 9 7 6 12; 4 14 15 1]; % input matrix
   B=inv(A), A*B   % compute and validate the inverse
```

and it prompted that "Warning: Matrix is close to singular or badly scaled. Results may be inaccurate. RCOND = 1.306145e-017," and the matrix $\boldsymbol{A}$ *and* $\boldsymbol{AB}$ *obtained are respectively*

$$\boldsymbol{B}=\begin{bmatrix}-2.6495 & -7.9484 & 7.9484 & 2.6495\\ -7.9484 & -23.845 & 23.845 & 7.9484\\ 7.9484 & 2.38455 & -23.845 & -7.9484\\ 2.6495 & 7.9484 & -7.9484 & -2.6495\end{bmatrix}\times 10^{14},\ \boldsymbol{AB}=\begin{bmatrix}1 & 0 & -1 & -0.25\\ -0.25 & 0 & 0 & 0.875\\ 0.25 & 0.5 & 0 & 0.25\\ 0.15625 & 0.125 & 0 & 1.7344\end{bmatrix}.$$

It can be seen that a warning message is displayed claiming that $\boldsymbol{A}$ *matrix is close to a singular matrix, thus, the results are useless. The product* $\boldsymbol{AB}$ *is no longer an identity matrix. The command* `norm(A*B-eye(size(A)))` *indicates that the norm of the error matrix is as big as* 1.6408.

In fact, for singular matrices, there is no inverse matrix satisfying (4-2-15). *For the same problem, from the Symbolic Math Toolbox function* $\boldsymbol{B}=$ `inv(sym(A))`, *the returned* $\boldsymbol{B}$ *is* `FAIL`.

Example 4.24 The symbolic function `inv()` can be applied also to matrices with symbolic variables. Find the inverse of Hankel matrix with variables.

Solution *For instance, the inverse matrix of a given Hankel matrix can easily be obtained with the direct use of* `inv()`

```
>> a=sym('a%d',[1,4]); H=hankel(a); inv(H) % direct inverse of Hankel matrix
```

the inverse matrix is

$$\boldsymbol{H}^{-1}=\begin{bmatrix}0 & 0 & 0 & 1/a_4\\ 0 & 0 & 1/a_4 & -1/a_4^2a_3\\ 0 & 1/a_4 & -1/a_4^2a_3 & -1/a_4^3(a_2a_4-a_3^2)\\ 1/a_4 & -1/a_4^2a_3 & -1/a_4^3(a_2a_4-a_3^2) & -(a_1a_4^2-2a_2a_3a_4+a_3^3)/a_4^4\end{bmatrix}.$$

In classical linear algebra textbooks, *reduced row echelon form* is useful in finding the inverse of a given matrix. An identity matrix can be appended to the right of the original matrix. After reduced row echelon conversion, the left half of the appended matrix is converted to an identity matrix, while the right half is the inverse. In MATLAB, $\boldsymbol{H}_1=$ `rref(H)` can be used to find the reduced row echelon form. This function can be used in the inverse matrix evaluation.

Example 4.25 To use reduced row echelon form to find the inverse in the previous example, the following statements can be issued

```
>> a=sym('a%d',[1,4]); H=hankel(a); H0=inv(H) % generate a Hankel matrix
   H1=[H eye(4)]; H2=rref(H1), H3=H2(:,5:8)   % compute inverse again
```

The inverse $\boldsymbol{H}_3$ is exactly the same as $\boldsymbol{H}_0$. The reduced row echelon form $\boldsymbol{H}_2$ is also shown, whose left half is an identity matrix, and the right is the inverse $\boldsymbol{H}_3$

$$\boldsymbol{H}_2=\begin{bmatrix}1 & 0 & 0 & 0 & 0 & 0 & 0 & 1/a_4\\ 0 & 1 & 0 & 0 & 0 & 0 & 1/a_4 & -1/a_4^2a_3\\ 0 & 0 & 1 & 0 & 0 & 1/a_4 & -1/a_4^2a_3 & -1/a_4^3(a_2a_4-a_3^2)\\ 0 & 0 & 0 & 1 & 1/a_4 & -1/a_4^2a_3 & -1/a_4^3(a_2a_4-a_3^2) & -(a_1a_4^2-2a_2a_3a_4+a_3^3)/a_4^4\end{bmatrix}.$$

Using the concept of adjoint matrix, the inverse of matrix $\boldsymbol{A}$ can also be obtained with $\boldsymbol{B}$=`adjoint(`$\boldsymbol{A}$`)/det(`$\boldsymbol{A}$`)`.

Example 4.26 Please show that for a 3×3 arbitrary matrix $\boldsymbol{A}(t)$, we have

$$\frac{\mathrm{d}\boldsymbol{A}^{-1}(t)}{\mathrm{d}t}=-\boldsymbol{A}^{-1}(t)\frac{\mathrm{d}\boldsymbol{A}(t)}{\mathrm{d}t}\boldsymbol{A}^{-1}(t).$$

Solution *In this example, arbitrary matrix function $\boldsymbol{A}(t)$ is involved. The function* `any_matrix()` *can be used to generate such a matrix, and the following commands can be used to validate the above equation. It can be seen that the simplified error is a zero matrix, hence, the equation is proved.*

```
>> syms t; A=any_matrix(3,'a',t); iA=inv(A); % generate matrix and inverse
   simplify(diff(iA,t)+iA*diff(A,t)*iA)       % validate the equation
```

4.2.3 Generalized matrix inverse

It can be seen that even with the Symbolic Math Toolbox, the inverse problems to a singular matrix cannot be handled. In fact, in a strict sense, no inverse matrix exists at all for singular matrices. In practical applications, one may need an "inverse" for singular or even rectangular matrices. Thus, a generalized matrix should be defined. For a given matrix $\boldsymbol{A}$, if there exists another matrix $\boldsymbol{N}$ such that

$$\boldsymbol{ANA}=\boldsymbol{A}, \tag{4-2-16}$$

the matrix $\boldsymbol{N}$ is referred to as the *generalized inverse matrix* of matrix $\boldsymbol{A}$, denoted by $\boldsymbol{N}=\boldsymbol{A}^{-}$. For an $n\times m$ rectangular matrix $\boldsymbol{A}$, the generalized inverse matrix $\boldsymbol{N}$ is an $m\times n$ matrix. It can be shown that there are an infinite number of $\boldsymbol{N}$'s satisfying such a condition.

One may introduce a norm criterion such that

$$\min_{\boldsymbol{N}}||\boldsymbol{AN}-\boldsymbol{I}|| \tag{4-2-17}$$

is minimized, and it can be shown that for any given matrix $\boldsymbol{A}$, there exists a unique matrix $\boldsymbol{M}$ such that the three conditions below are satisfied

(i) $\boldsymbol{AMA}=\boldsymbol{A}$,

(ii) $\boldsymbol{MAM}=\boldsymbol{M}$,

(iii) $\boldsymbol{AM}$ and $\boldsymbol{MA}$ are both Hermitian symmetrical matrices.

Such a matrix $\boldsymbol{M}$ is referred to as the *Moore–Penrose inverse* or *pseudoinverse* of $\boldsymbol{A}$, denoted by $\boldsymbol{M}=\boldsymbol{A}^{+}$.

A MATLAB function `pinv()` can be used to find the Moore–Penrose pseudoinverse of a given matrix. The syntaxes of the function are

```
M=pinv(A)     % evaluate the Moore-Penrose pseudoinverse
M=pinv(A,ε)   % evaluate the inverse numerically with precision of ε
```

where the variable ϵ is used to judge whether a value is zero or not. The returned matrix $\boldsymbol{M}$ is the Moore–Penrose pseudoinverse of the original matrix $\boldsymbol{A}$. If $\boldsymbol{A}$ is a nonsingular square matrix, the resulted pseudoinverse is in fact the inverse of the original matrix. However, the speed of the `pinv()` function is significantly lower than that of the `inv()` function. In new versions of MATLAB, `pinv()` also works for symbolic matrices.

Example 4.27 Find the pseudoinverse of the singular matrix $\boldsymbol{A}$ in Example 4.10.

Solution *For the singular matrix, the Moore–Penrose pseudoinverse of the matrix should be established instead, using the following statements.*

```
>> A=[16 2 3 13; 5 11 10 8; 9 7 6 12; 4 14 15 1]; B=pinv(A), A*B
```

The Moore–Penrose inverse of the matrix $\boldsymbol{B}$ *and* $\boldsymbol{AB}$ *are*

$$\boldsymbol{B}=\begin{bmatrix} 0.1011 & -0.0739 & -0.0614 & 0.0636 \\ -0.0364 & 0.0386 & 0.0261 & 0.001103 \\ 0.0136 & -0.0114 & -0.0239 & 0.0511 \\ -0.0489 & 0.0761 & 0.0886 & -0.0864 \end{bmatrix},\quad \boldsymbol{AB}=\begin{bmatrix} 0.95 & -0.15 & 0.15 & 0.05 \\ -0.15 & 0.55 & 0.45 & 0.15 \\ 0.15 & 0.45 & 0.55 & -0.15 \\ 0.05 & 0.15 & -0.15 & 0.95 \end{bmatrix}.$$

With symbolic call of the `pinv()` *function*

```
>> B1=pinv(sym(A)), B1*A, A*B1 % symbolic computation of pseudoinverse
```

the Moore–Penrose inverse can be written as

$$\boldsymbol{B}_1=\begin{bmatrix} 55/544 & -201/2720 & -167/2720 & 173/2720 \\ -99/2720 & 21/544 & 71/2720 & 3/2720 \\ 37/2720 & -31/2720 & -13/544 & 139/2720 \\ -133/2720 & 207/2720 & 241/2720 & -47/544 \end{bmatrix}.$$

With the result, the matrices $\boldsymbol{B}_1\boldsymbol{A}$ *and* $\boldsymbol{AB}_1$ *are the same*

$$\boldsymbol{AB}_1=\boldsymbol{B}_1\boldsymbol{A}=\begin{bmatrix} 19/20 & -3/20 & 3/20 & 1/20 \\ -3/20 & 11/20 & 9/20 & 3/20 \\ 3/20 & 9/20 & 11/20 & -3/20 \\ 1/20 & 3/20 & -3/20 & 19/20 \end{bmatrix}.$$

From these results obtained, it can be observed that the values in $\boldsymbol{A}^+$ *are meaningful, and* $\boldsymbol{AA}^+$ *is no longer an identify matrix. Now the three conditions for the Moore–Penrose pseudoinverse can be checked using the following statements:*

```
>> [norm(A*B*A-A), norm(B*A*B-B), norm(A*B-(A*B)'), norm(B*A-(B*A)')]
```

with errors respectively $2.2383\times10^{-14}, 7.6889\times10^{-17}, 1.0753\times10^{-15}, 9.3653\times10^{-16}$. *It can be seen that the above obtained matrix* $\boldsymbol{B}$ *is the Moore–Penrose inverse of the original matrix. Performing Moore–Penrose inverse to matrix* $\boldsymbol{B}$*, it can be seen that* $(\boldsymbol{A}^+)^+=\boldsymbol{A}$*, with error* 1.9278×10^{-14}.

```
>> pinv(B), norm(ans-A) % restore original matrix by taking pseudoinverse twice
```

Example 4.28 For the following rectangular matrix $\boldsymbol{A}$, find its rank and Moore–Penrose inverse. Check whether the pseudoinverse obtained is correct or not

$$\boldsymbol{A}=\begin{bmatrix} 6 & 1 & 4 & 2 & 1 \\ 3 & 0 & 1 & 4 & 2 \\ -3 & -2 & -5 & 8 & 4 \end{bmatrix}.$$

Solution *The following commands can be given to get the rank of the matrix. It can be seen that the rank is 2, rather than 3, which means that the matrix is not a full-rank matrix.*

```
>> A=[6,1,4,2,1; 3,0,1,4,2; -3,-2,-5,8,4]; rank(A)
```

Since matrix $\boldsymbol{A}$ is a singular rectangular matrix, the function `pinv()` *can be used to evaluate the Moore–Penrose pseudoinverse of the matrix. Then, each of the three conditions for Moore–Penrose pseudoinverse can be verified, and it can be shown that the resulted matrix is correct.*

```
>> A1=pinv(A), A2=pinv(sym(A)), A3=double(A2) % compute pseudoinverse
   [norm(A*A1*A-A), norm(A1*A-A'*A1'), ...    % validation
      norm(A1*A-A'*A1'), norm(A*A1-A1'*A')]
```

The pseudoinverse matrices are

$$\boldsymbol{A}_1=\begin{bmatrix}0.07303 & 0.041301 & -0.02215\\ 0.01077 & 0.001995 & -0.01556\\ 0.04589 & 0.017757 & -0.03851\\ 0.03272 & 0.043097 & 0.063847\\ 0.01636 & 0.021548 & 0.031923\end{bmatrix},\ \boldsymbol{A}_2=\begin{bmatrix}183/2506 & 207/5012 & -111/5012\\ 27/2506 & 5/2506 & -39/2506\\ 115/2506 & 89/5012 & -193/5012\\ 41/1253 & 54/1253 & 80/1253\\ 41/2506 & 27/1253 & 40/1253\end{bmatrix}$$

and $||\boldsymbol{A}^{+}\boldsymbol{A}\boldsymbol{A}^{+}-\boldsymbol{A}^{+}|| = 1.0263\times10^{-16}$, $||\boldsymbol{A}\boldsymbol{A}^{+}\boldsymbol{A}-\boldsymbol{A}|| = 8.1145\times10^{-15}$, $||\boldsymbol{A}^{+}\boldsymbol{A}-(\boldsymbol{A})^{\rm H}(\boldsymbol{A}^{+})^{\rm H}|| = 3.9098\times10^{-16}$, $||\boldsymbol{A}^{+}\boldsymbol{A}-(\boldsymbol{A})^{\rm H}(\boldsymbol{A}^{+})^{\rm H}|| = 1.6653\times10^{-16}$. *The double-precision representation of symbolic* $\boldsymbol{A}_2$ *is exactly the same as* $\boldsymbol{A}_1$.

4.2.4 Matrix eigenvalue problems

I. Eigenvalues and eigenvectors of a matrix

For the given matrix $\boldsymbol{A}$, if there exists a non-zero vector $\boldsymbol{x}$ and a scalar λ satisfying

$$\boldsymbol{A}\boldsymbol{x}=\lambda\boldsymbol{x}, \tag{4-2-18}$$

then, λ is referred to as an *eigenvalue* of matrix $\boldsymbol{A}$, while the vector $\boldsymbol{x}$ is referred to as the *eigenvector* of $\boldsymbol{A}$ corresponding to the eigenvalue λ. Strictly speaking, the eigenvector $\boldsymbol{x}$ should be referred to as the *right eigenvector*. If the eigenvalues are distinct, the eigenvectors are linearly independent. Thus, a nonsingular square eigen-matrix can be constructed. A diagonal matrix can be obtained if such a matrix is used to perform similar transformation. The eigenvalues and eigenvectors can easily be obtained using the `eig()` function. The syntaxes of the function are

```
d=eig(A)         % only eigenvalues are required
[V,D]=eig(A)     % if both eigenvalues and eigenvectors are expected
```

where, $\boldsymbol{d}$ is a vector containing all the eigenvalues, while $\boldsymbol{D}$ is a diagonal matrix whose diagonal elements are the eigenvalues of the matrix, and each column in matrix $\boldsymbol{V}$ contains the eigenvector to the corresponding eigenvalues. The following relationship $\boldsymbol{AV}=\boldsymbol{VD}$ is satisfied. This function applies also to complex $\boldsymbol{A}$ matrix as well as symbolic ones.

The definition of the roots of the characteristic polynomial discussed earlier is exactly the same as the eigenvalues. If the characteristic polynomial can be exactly known, the function `roots()` can also be used in evaluating the eigenvalues of the matrix.

Example 4.29 Compute the eigenvalues and eigenvectors of $\boldsymbol{A}$ in Example 4.10.

Solution *Using numerical method, the eigenvalues can be obtained by the direct use of the* `eig()` *function such that*

```
>> A=[16 2 3 13; 5 11 10 8; 9 7 6 12; 4 14 15 1]; eig(A)
```

and it can be seen that they are $34, \pm 8.9442719, -2.234826\times 10^{-15}$.

The `eig()` *function provided in the Symbolic Math Toolbox can also be used to evaluate the eigenvalues and eigenvectors of a given matrix. Even for large-sized matrices, results with very high accuracy can be obtained.*

```
>> eig(sym(A)) % symbolic computation of eigenvalues
```

Thus, the exact eigenvalues of the matrix are $0, 34, \pm 4\sqrt{5}$.

For the same matrix $\boldsymbol{A}$, *the eigenvalues and eigenvectors can be solved numerically such that*

```
>> [v,d]=eig(A) % find eigenvalues and eigenvector matrix
```

the numerical solutions to the eigenvector and eigenvalues are

$$\boldsymbol{v}=\begin{bmatrix} -0.5 & -0.8236 & 0.3764 & -0.2236 \\ -0.5 & 0.4236 & 0.02361 & -0.6708 \\ -0.5 & 0.02361 & 0.4236 & 0.6708 \\ 0.5 & 0.3764 & 0.8236 & 0.2236 \end{bmatrix},\quad \boldsymbol{d}=\begin{bmatrix} 34 & 0 & 0 & 0 \\ 0 & 8.9443 & 0 & 0 \\ 0 & 0 & -8.9443 & 0 \\ 0 & 0 & 0 & 9.416\times 10^{-16} \end{bmatrix}.$$

If the Symbolic Math Toolbox is used, the eigenvalues and eigenvectors can easily be found such that

```
>> [v,d]=eig(sym(A)) % symbolic computation
```

and it can be found

$$\boldsymbol{v}=\begin{bmatrix} -1 & 1 & -8\sqrt{5}-17 & 8\sqrt{5}-17 \\ -3 & 1 & 4\sqrt{5}+9 & -4\sqrt{5}+9 \\ 3 & 1 & 1 & 1 \\ 1 & 1 & 4\sqrt{5}+7 & -4\sqrt{5}+7 \end{bmatrix},\quad \boldsymbol{d}=\begin{bmatrix} 0 & 0 & 0 & 0 \\ 0 & 34 & 0 & 0 \\ 0 & 0 & 4\sqrt{5} & 0 \\ 0 & 0 & 0 & -4\sqrt{5} \end{bmatrix}.$$

If matrix $\boldsymbol{A}$ has repeated eigenvalues, the eigenvector matrix will be a singular matrix. When numerical algorithms are used, the obtained eigenvalues may not be exactly the same, due to numerical errors. Thus, the obtained $\boldsymbol{V}$ matrix may not be singular, although it is extremely close to a singular one.

II. Generalized eigenvalues and eigenvectors

Assume that there exists a scalar λ and a non-zero vector $\boldsymbol{x}$ such that

$$\boldsymbol{A}\boldsymbol{x}=\lambda\boldsymbol{B}\boldsymbol{x}, \tag{4-2-19}$$

where $\boldsymbol{B}$ is a symmetrical positive-definite matrix, λ is referred to as the *generalized eigenvalue*, while $\boldsymbol{x}$ is the *generalized eigenvector*. In fact, the ordinary eigenvalue problem is a special case of the generalized eigenvalue problem when $\boldsymbol{B}=\boldsymbol{I}$ is assumed.

If matrix $\boldsymbol{B}$ is a nonsingular square matrix, the generalized eigenvalue problem can be converted to the eigenvalue problem for matrix $\boldsymbol{B}^{-1}\boldsymbol{A}$.

$$\boldsymbol{B}^{-1}\boldsymbol{A}\boldsymbol{x}=\lambda\boldsymbol{x}, \tag{4-2-20}$$

i.e., λ and $\boldsymbol{x}$ are respectively the eigenvalues and eigenvectors of matrix $\boldsymbol{B}^{-1}\boldsymbol{A}$. In MATLAB, the function eig() can be used to compute directly the generalized eigenvalues and eigenvectors such that

```
d=eig(A,B)          % generalized eigenvalue evaluation
[V,D]=eig(A,B)      % generalized eigenvalues and eigenvectors
```

With the eig() function, the generalized eigenvalues and eigenvectors can be obtained in matrices $\boldsymbol{D}$ and $\boldsymbol{V}$, where $\boldsymbol{AV} = \boldsymbol{BVD}$. It should be noted that the matrix $\boldsymbol{B}$ is no longer restricted to positive-definite matrices.

Example 4.30 Now consider the matrices

$$\boldsymbol{A}=\begin{bmatrix}5 & 7 & 6 & 5\\ 7 & 10 & 8 & 7\\ 6 & 8 & 10 & 9\\ 5 & 7 & 9 & 10\end{bmatrix},\quad \boldsymbol{B}=\begin{bmatrix}2 & 6 & -1 & -2\\ 5 & -1 & 2 & 3\\ -3 & -4 & 1 & 10\\ 5 & -2 & -3 & 8\end{bmatrix}.$$

Compute the generalized eigenvalues and eigenvector matrices for the $(\boldsymbol{A},\boldsymbol{B})$ pair.

Solution *The following statements can be entered*

```
>> A=[5,7,6,5; 7,10,8,7; 6,8,10,9; 5,7,9,10];
   B=[2,6,-1,-2; 5,-1,2,3; -3,-4,1,10; 5,-2,-3,8];
   [V,D]=eig(A,B), norm(A*V-B*V*D) % generalized eigenvalues and validations
```

and the eigenvalues and eigenvector matrices can be obtained

$$\boldsymbol{V}=\begin{bmatrix}0.3697 & -0.37409+\mathrm{j}0.62591 & -0.37409-\mathrm{j}0.62591 & 1\\ 0.99484 & -0.067434-\mathrm{j}0.25314 & -0.067434+\mathrm{j}0.25314 & -0.60903\\ 0.79792 & 0.92389+\mathrm{j}0.026381 & 0.92389-\mathrm{j}0.026381 & -0.23164\\ 1 & -0.65986-\mathrm{j}0.32628 & -0.65986+\mathrm{j}0.32628 & 0.13186\end{bmatrix},$$

$$\boldsymbol{D}=\begin{bmatrix}4.7564 & 0 & 0 & 0\\ 0 & 0.047055+\mathrm{j}0.17497 & 0 & 0\\ 0 & 0 & 0.047055-\mathrm{j}0.17497 & 0\\ 0 & 0 & 0 & -0.003689\end{bmatrix}.$$

and the norm of the error matrix is 1.5783×10^{-14}.

4.3 Fundamental Matrix Transformations

4.3.1 Similarity transformations and orthogonal matrices

For a square matrix $\boldsymbol{A}$, if there exists a nonsingular matrix $\boldsymbol{B}$, then, the original $\boldsymbol{A}$ matrix can be transformed into the following form

$$\boldsymbol{X}=\boldsymbol{B}^{-1}\boldsymbol{AB}, \tag{4-3-1}$$

and this transformation is referred to as *similarity transformation*, and matrix $\boldsymbol{B}$ is referred to as the *similarity transformation matrix*. It can be shown that the determinant, rank, trace and eigenvalues of the transformed matrix are not changed. Through properly chosen transformation matrix $\boldsymbol{B}$, one may transform the original matrix $\boldsymbol{A}$ to other forms without changing important properties of matrix $\boldsymbol{A}$.

For a class of special transformation matrices $\boldsymbol{T}$, if it satisfies $\boldsymbol{T}^{-1}=\boldsymbol{T}^{\rm H}$, where $\boldsymbol{T}^{\rm H}$ is the Hermitian conjugate transpose of matrix $\boldsymbol{T}$, matrix $\boldsymbol{T}$ is then referred to as an *orthogonal matrix*, and it can be denoted that $\boldsymbol{Q}=\boldsymbol{T}$. Therefore, it can be seen that the orthogonal matrix $\boldsymbol{Q}$ satisfies

$$\boldsymbol{Q}^{\rm H}\boldsymbol{Q}=\boldsymbol{I},\quad \text{and}\quad \boldsymbol{Q}\boldsymbol{Q}^{\rm H}=\boldsymbol{I}, \tag{4-3-2}$$

where $\boldsymbol{I}$ is an $n\times n$ identity matrix.

A MATLAB function $\boldsymbol{Q}=$`orth(`$\boldsymbol{A}$`)` can be used to construct the *orthonormal basis* for the *column space* of matrix $\boldsymbol{A}$. If matrix $\boldsymbol{A}$ is nonsingular, the orthonormal basis matrix $\boldsymbol{Q}$ obtained satisfies the conditions in (4-3-2). If matrix $\boldsymbol{A}$ is singular, however, the columns in matrix $\boldsymbol{Q}$ equals the rank of matrix $\boldsymbol{A}$, and satisfies $\boldsymbol{Q}^{\rm H}\boldsymbol{Q}=\boldsymbol{I}$, other than $\boldsymbol{Q}\boldsymbol{Q}^{\rm H}=\boldsymbol{I}$.

Example 4.31 Compute the orthonormal basis for matrix $\boldsymbol{A}=\begin{bmatrix}5&9&8&3\\0&3&2&4\\2&3&5&9\\3&4&5&8\end{bmatrix}$.

Solution *The orthonormal basis of a given matrix* $\boldsymbol{A}$ *can be established directly with the* `orth()` *function, and the following statements can also be used to verify the properties of the orthogonal matrix obtained.*

```
>> A=[5,9,8,3; 0,3,2,4; 2,3,5,9; 3,4,5,8]; Q=orth(A) % orthonormal basis
   [norm(Q'*Q-eye(4)), norm(Q*Q'-eye(4))]  % verification of the properties
```

The orthonormal basis of $\boldsymbol{A}$ *can then be found as*

$$\boldsymbol{Q}=\begin{bmatrix}-0.61967134&0.77381388&-0.026187275&-0.12858357\\-0.25484758&-0.15505966&0.94903028&0.10173858\\-0.51978107&-0.52982004&-0.15628279&-0.65168555\\-0.52998848&-0.31057898&-0.27245447&0.74055484\end{bmatrix},$$

with calculation errors $||\boldsymbol{Q}^{\rm H}\boldsymbol{Q}-\boldsymbol{I}||=4.6395\times10^{-16}$, $||\boldsymbol{Q}\boldsymbol{Q}^{\rm H}-\boldsymbol{I}||=4.9270\times10^{-16}$.

Example 4.32 Consider the singular matrix $\boldsymbol{A}$ defined in Example 4.10. Compute the orthonormal basis matrix, and then, verify its properties.

Solution *The orthonormal basis of matrix* $\boldsymbol{A}$ *can be obtained easily by the direct use of the function* `orth()`.

```
>> A=[16,2,3,13; 5,11,10,8; 9,7,6,12; 4,14,15,1]; Q=orth(A),
   a=norm(Q'*Q-eye(3)), Q1=orth(sym(A)), norm(Q1'*Q1-eye(3))
```

It can be seen that since $\boldsymbol{A}$ *is a singular matrix with a rank of 3, the orthonormal basis constructed is a rectangular matrix, and in the recent versions of MATLAB, the function* `orth()` *support symbolic computation such that*

$$\boldsymbol{Q}=\begin{bmatrix}-0.5&0.67082039324994&0.5\\-0.5&-0.22360679774998&-0.5\\-0.5&0.22360679774998&-0.5\\-0.5&-0.67082039324994&0.5\end{bmatrix},$$

$$\boldsymbol{Q}_1=\begin{bmatrix}8\sqrt{42}/63&-635\sqrt{21}\sqrt{12178}/767214&109\sqrt{5}\sqrt{6089}/60890\\5\sqrt{42}/126&391\sqrt{21}\sqrt{12178}/383607&-163\sqrt{5}\sqrt{6089}/60890\\\sqrt{42}/14&11\sqrt{21}\sqrt{12178}/42623&-197\sqrt{5}\sqrt{6089}/60890\\2\sqrt{42}/63&1117\sqrt{21}\sqrt{12178}/767214&211\sqrt{5}\sqrt{608960890}\end{bmatrix},$$

and the error $||\boldsymbol{Q}^{\mathrm{H}}\boldsymbol{Q} - \boldsymbol{I}|| = 1.0140\times 10^{-15}$.

4.3.2 Triangular and Cholesky factorizations

I. Row elimination and exchange with matrix multiplications

By left or right multiplying a matrix with deliberately selected matrix, the form of the original matrix is changed. The selection of transformation matrices are demonstrated through examples.

Example 4.33 For a given matrix $\boldsymbol{A}$ given below, please observe the effect of matrix multiplication with a deliberately chosen matrix $\boldsymbol{E}$.

$$\boldsymbol{A} = \begin{bmatrix} 16 & 2 & 3 & 13 \\ 5 & 11 & 10 & 8 \\ 9 & 7 & 6 & 12 \\ 4 & 14 & 15 & 1 \end{bmatrix}, \quad \boldsymbol{E} = \begin{bmatrix} 1 & 0 & 0 & 0 \\ 0 & 1 & 0 & 0 \\ -2 & 0 & 1 & 0 \\ 0 & 0 & 0 & 1 \end{bmatrix}.$$

Solution *The two matrices can be entered first, then, the three matrices,* $\boldsymbol{A}_1 = \boldsymbol{E}\boldsymbol{A}$, $\boldsymbol{A}_2 = \boldsymbol{A}\boldsymbol{E}$ *and* $\boldsymbol{E}_1 = \boldsymbol{E}^{-1}$ *can be computed*

```
>> A=[16,2,3,13; 5,11,10,8; 9,7,6,12; 4,14,15,1]; E=eye(4);
   E1=inv(E), E(3,1)=-2; A1=E*A, A2=A*E
```

The three matrices are

$$\boldsymbol{A}_1 = \begin{bmatrix} 16 & 2 & 3 & 13 \\ 5 & 11 & 10 & 8 \\ -23 & 3 & 0 & -14 \\ 4 & 14 & 15 & 1 \end{bmatrix}, \quad \boldsymbol{A}_2 = \begin{bmatrix} 10 & 2 & 3 & 13 \\ -15 & 11 & 10 & 8 \\ -3 & 7 & 6 & 12 \\ -26 & 14 & 15 & 1 \end{bmatrix}, \quad \boldsymbol{E}_1 = \begin{bmatrix} 1 & 0 & 0 & 0 \\ 0 & 1 & 0 & 0 \\ 2 & 0 & 1 & 0 \\ 0 & 0 & 0 & 1 \end{bmatrix}.$$

Is anything in particular observed? Declare $\boldsymbol{E}$ *as an identity matrix first, the final* $\boldsymbol{E}$ *is generated by setting the element in the third row, first column (denoted by (3,1)th element) to* -2. $\boldsymbol{E}_1$ *is a copy of* $\boldsymbol{E}$*, however, the sign of its (3,1)th element is altered.*

Now let us observe $\boldsymbol{A}_1$*, which is obtained by left multiplying* $\boldsymbol{E}$ *to matrix* $\boldsymbol{A}$*. Comparing* $\boldsymbol{A}$ *and* $\boldsymbol{A}_1$*, it can be seen that only the third row is changed. The new third row is generated by multiplying* -2 *to all the elements in the first row and added to the third row of* $\boldsymbol{A}$*. In* $\boldsymbol{A}_2$*, only the first column is changed into the sum of the first column of* $\boldsymbol{A}$ *and* -2 *times the third column of* $\boldsymbol{A}$*.*

With these rules in mind, matrix $\boldsymbol{E}$ *can be deliberately constructed. For instance, if one wants to eliminate all the elements in the first column, except the first one. The matrix* $\boldsymbol{E}_1$ *can be constructed with the following statements*

```
>> E1=sym(eye(4)); E1(2:4,1)=-A(2:4,1)/A(1,1), A1=E1*A
```

and the constructed matrix and the product are given below, exactly the same as the one we expected.

$$\boldsymbol{E}_1 = \begin{bmatrix} 1 & 0 & 0 & 0 \\ -5/16 & 1 & 0 & 0 \\ -9/16 & 0 & 1 & 0 \\ -1/4 & 0 & 0 & 1 \end{bmatrix}, \quad \boldsymbol{A}_1 = \begin{bmatrix} 16 & 2 & 3 & 13 \\ 0 & 83/8 & 145/16 & 63/16 \\ 0 & 47/8 & 69/16 & 75/16 \\ 0 & 27/2 & 57/4 & -9/4 \end{bmatrix}.$$

Further another matrix $\boldsymbol{E}_2$ *can be selected to eliminate the rest of the elements in column two in* $\boldsymbol{A}_1$*, with a new* $\boldsymbol{E}_2$*, such that the matrices can be obtained with*

```
>> E2=sym(eye(4)); E2([1 3 4],2)=-A1([1 3 4],2)/A1(2,2), A2=E2*A1
   E=E2*E1; A3=E*A % it can be seen that EA is the same as E2E1A
```

and the solutions are

$$E_2 = \begin{bmatrix} 1 & -16/83 & 0 & 0 \\ 0 & 1 & 0 & 0 \\ 0 & -47/83 & 1 & 0 \\ 0 & -108/83 & 0 & 1 \end{bmatrix}, A_2 = \begin{bmatrix} 16 & 0 & 104/83 & 1016/83 \\ 0 & 83/8 & 145/16 & 63/16 \\ 0 & 0 & -68/83 & 204/83 \\ 0 & 0 & 204/83 & -612/83 \end{bmatrix}.$$

The overall transform matrix obtained is $\boldsymbol{E} = \boldsymbol{E}_2\boldsymbol{E}_1$. *This is the foundation of the echelon and triangular factorization approaches.*

Example 4.34 Still use the $\boldsymbol{A}$ matrix in Example 4.33. Now select the transform matrix as follows, and observe what happens in matrix multiplications.

$$\boldsymbol{A} = \begin{bmatrix} 16 & 2 & 3 & 13 \\ 5 & 11 & 10 & 8 \\ 9 & 7 & 6 & 12 \\ 4 & 14 & 15 & 1 \end{bmatrix}, \quad \boldsymbol{E} = \begin{bmatrix} 0 & 0 & 0 & 1 \\ 0 & 1 & 0 & 0 \\ 0 & 0 & 1 & 0 \\ 1 & 0 & 0 & 0 \end{bmatrix}.$$

Solution *The matrix* $\boldsymbol{E}$ *is in fact constructed from an identity matrix, with its first and fourth rows exchanged. The matrices* $\boldsymbol{A}_1 = \boldsymbol{E}\boldsymbol{A}$ *and* $\boldsymbol{A}_2 = \boldsymbol{A}\boldsymbol{E}$ *can be computed with the following statements*

```
>> A=sym([16,2,3,13; 5,11,10,8; 9,7,6,12; 4,14,15,1]);
   E=sym(eye(4)); E([1,4],:)=E([4,1],:); A1=E*A, A2=A*E
```

and the matrices can be obtained

$$\boldsymbol{A}_1 = \begin{bmatrix} 4 & 14 & 15 & 1 \\ 5 & 11 & 10 & 8 \\ 9 & 7 & 6 & 12 \\ 16 & 2 & 3 & 13 \end{bmatrix}, \quad \boldsymbol{A}_2 = \begin{bmatrix} 13 & 2 & 3 & 16 \\ 8 & 11 & 10 & 5 \\ 12 & 7 & 6 & 9 \\ 1 & 14 & 15 & 4 \end{bmatrix}.$$

It can be observed that by left multiplying $\boldsymbol{E}$ *to matrix* $\boldsymbol{A}$*, the first and fourth rows in matrix* $\boldsymbol{A}$ *are swopped, to form the new matrix* $\boldsymbol{A}_1$. *If right multiplication is performed, column swap is made.*

II. Triangular factorizations

The triangular factorization of a matrix is also known as the *LU factorization*, where the original matrix can be factorized into the product of a lower-triangular matrix $\boldsymbol{L}$ and an upper-triangular matrix $\boldsymbol{U}$, such that $\boldsymbol{A} = \boldsymbol{L}\boldsymbol{U}$, where $\boldsymbol{L}$ and $\boldsymbol{U}$ can respectively be written as

$$\boldsymbol{L} = \begin{bmatrix} 1 & & & \\ l_{21} & 1 & & \\ \vdots & \vdots & \ddots & \\ l_{n1} & l_{n2} & \cdots & 1 \end{bmatrix}, \quad \boldsymbol{U} = \begin{bmatrix} u_{11} & u_{12} & \cdots & u_{1n} \\ & u_{22} & \cdots & u_{2n} \\ & & \ddots & \vdots \\ & & & u_{nn} \end{bmatrix}, \tag{4-3-3}$$

where the entities l_{ij} and u_{ij} can be calculated recursively that

$$l_{ij} = \frac{a_{ij} - \sum_{k=1}^{j-1} l_{ik}u_{kj}}{u_{jj}}, \quad (j < i), \quad \text{and } u_{ij} = a_{ij} - \sum_{k=1}^{i-1} l_{ik}u_{kj}, \quad (j \geqslant i), \tag{4-3-4}$$

with initial values defined as $u_{1i} = a_{1i}$, $i = 1, 2, \cdots, n$.

It should be noted that since the pivot element was not selected in the above formula, the direct use of such an algorithm may not be numerically stable, since small values or even 0 might be used as denominators. In MATLAB, a pivot-based LU factorization function `lu()` is provided such that

```
[L,U]=lu(A)      % LU factorization A = LU
[L,U,P]=lu(A)    % P is the permutation matrix, A = P^{-1}LU
```

where $\boldsymbol{L}$ and $\boldsymbol{U}$ are transformed lower- and upper-triangular matrices. In MATLAB, the `lu()` function considers the selection of pivoting element, thus, reliable results will be ensured. The actual matrix $\boldsymbol{L}$ is not necessarily lower-triangular. In recent versions of MATLAB, symbolic version of `lu()` is also supported, while in the first syntax, pivot is not considered.

Example 4.35 Consider the LU factorization problem to Example 4.10. Try to use the two calling syntaxes of `lu()` to compute the triangular factorization, and then, compare the results.

Solution *For matrix* $\boldsymbol{A}$*, the triangular factorization is performed such that*

```
>> A=[16 2 3 13; 5 11 10 8; 9 7 6 12; 4 14 15 1]; [L1,U1]=lu(A)
```

where

$$\boldsymbol{L}_1 = \begin{bmatrix} 1 & 0 & 0 & 0 \\ 0.3125 & 0.76852 & 1 & 0 \\ 0.5625 & 0.43519 & 1 & 1 \\ 0.25 & 1 & 0 & 0 \end{bmatrix}, \quad \boldsymbol{U}_1 = \begin{bmatrix} 16 & 2 & 3 & 13 \\ 0 & 13.5 & 14.25 & -2.25 \\ 0 & 0 & -1.8889 & 5.6667 \\ 0 & 0 & 0 & 3.5527\times 10^{-15} \end{bmatrix}.$$

It can be seen that $\boldsymbol{L}_1$ *matrix is not a lower-triangular matrix. It is the permutated triangular matrix. Now, let us consider the other syntax to the* `lu()` *function*

```
>> [L,U,P]=lu(A) % pivot is considered, and L is not really triangular
```

where

$$\boldsymbol{L}=\begin{bmatrix} 1 & 0 & 0 & 0 \\ 0.25 & 1 & 0 & 0 \\ 0.3125 & 0.7685 & 1 & 0 \\ 0.5625 & 0.4352 & 1 & 1 \end{bmatrix}, \quad \boldsymbol{U}=\begin{bmatrix} 16 & 2 & 3 & 13 \\ 0 & 13.5 & 14.25 & -2.25 \\ 0 & 0 & -1.8889 & 5.6667 \\ 0 & 0 & 0 & 3.55\times 10^{-15} \end{bmatrix}, \quad \boldsymbol{P}=\begin{bmatrix} 1 & 0 & 0 & 0 \\ 0 & 0 & 0 & 1 \\ 0 & 1 & 0 & 0 \\ 0 & 0 & 1 & 0 \end{bmatrix}.$$

It should be noted that the matrix $\boldsymbol{P}$ *is not an identity matrix, and it is just a permutation matrix of such a matrix. The matrix* $\boldsymbol{A}$ *can be transformed back if the statement* `inv(P)*L*U` *is used. If the symbolic* `lu()` *function is used*

```
>> A=sym(A); [L2,U2]=lu(A) % symbolic computation, with L triangular
```

the exact triangular matrices can be found as

$$\boldsymbol{L}_2 = \begin{bmatrix} 1 & 0 & 0 & 0 \\ 5/16 & 1 & 0 & 0 \\ 9/16 & 47/83 & 1 & 0 \\ 1/4 & 108/83 & -3 & 1 \end{bmatrix}, \quad \boldsymbol{U}_2 = \begin{bmatrix} 16 & 2 & 3 & 13 \\ 0 & 83/8 & 145/16 & 63/16 \\ 0 & 0 & -68/83 & 204/83 \\ 0 & 0 & 0 & 0 \end{bmatrix}.$$

Example 4.36 For an arbitrary 3×3 matrix, please find its LU factorization.

Solution *The following commands can be used directly*

```
>> A=sym('a%d%d',3); [L U]=lu(A) % LU factorization for arbitrary matrix
```

and the results are

$$L=\begin{bmatrix}1 & 0 & 0\\ a_{21}/a_{11} & 1 & 0\\ a_{31}/a_{11} & (a_{32}-a_{12}a_{31}/a_{11})(a_{22}-a_{12}a_{21}/a_{11}) & 1\end{bmatrix},$$

$$U=\begin{bmatrix}a_{11} & a_{12} & a_{13}\\ 0 & a_{22}-a_{12}a_{21}/a_{11} & a_{23}-a_{13}a_{21}/a_{11}\\ 0 & 0 & a_{33}-\dfrac{(a_{23}-a_{13}a_{21}/a_{11})\,(a_{32}-a_{12}a_{31}/a_{11})}{a_{22}-a_{12}a_{21}/a_{11}}-\dfrac{a_{13}a_{31}}{a_{11}}\end{bmatrix}.$$

III. Cholesky factorization of symmetrical matrices

If matrix $\boldsymbol{A}$ is a symmetrical matrix, it can be factorized using LU factorization algorithm such that

$$\boldsymbol{A}=\boldsymbol{L}\boldsymbol{L}^{\mathrm{T}}=\begin{bmatrix}l_{11} & & & \\ l_{21} & l_{22} & & \\ \vdots & \vdots & \ddots & \\ l_{n1} & l_{n2} & \cdots & l_{nn}\end{bmatrix}\begin{bmatrix}l_{11} & l_{21} & \cdots & l_{n1}\\ & l_{22} & \cdots & l_{n2}\\ & & \ddots & \vdots\\ & & & l_{nn}\end{bmatrix}, \tag{4-3-5}$$

and the LU factorization algorithm can be simplified such that

$$l_{ii}=\sqrt{a_{ii}-\sum_{k=1}^{i-1}l_{ik}^2},\quad l_{ji}=\frac{1}{l_{jj}}\left(a_{ij}-\sum_{k=1}^{j-1}l_{ik}l_{jk}\right),\quad j<i, \tag{4-3-6}$$

and such an algorithm is referred to as the *Cholesky factorization algorithm*. To start with the algorithm, one should have initially $l_{11}=\sqrt{a_{11}}, l_{j1}=a_{j1}/l_{11}$.

A MATLAB function `chol()` is provided to perform Cholesky factorization such that an upper-triangular matrix $\boldsymbol{D}$ is returned $\boldsymbol{D}$=`chol(`$\boldsymbol{A}$`)`, where $\boldsymbol{D}=\boldsymbol{L}^{\mathrm{T}}$. Again in recent versions of MATLAB, `chol()` function can also be used to deal with symbolic matrices.

Example 4.37 Consider a symmetrical matrix $\boldsymbol{A}$, perform numerical and analytical Cholesky factorizations.

$$\boldsymbol{A}=\begin{bmatrix}9 & 3 & 4 & 2\\ 3 & 6 & 0 & 7\\ 4 & 0 & 6 & 0\\ 2 & 7 & 0 & 9\end{bmatrix}.$$

Solution *The Cholesky factorizations can be obtained directly with*

```
>> A=[9,3,4,2; 3,6,0,7; 4,0,6,0; 2,7,0,9]; D=chol(A), L=chol(sym(A))
```

and the solutions are respectively

$$\boldsymbol{D}=\begin{bmatrix}3 & 1 & 1.3333 & 0.66667\\ 0 & 2.2361 & -0.59628 & 2.8324\\ 0 & 0 & 1.9664 & 0.40684\\ 0 & 0 & 0 & 0.60648\end{bmatrix},\ \boldsymbol{L}=\begin{bmatrix}3 & 0 & 0 & 0\\ 1 & \sqrt{5} & 0 & 0\\ \dfrac{4}{3} & -\dfrac{4\sqrt{5}}{15} & \dfrac{\sqrt{870}}{15} & 0\\ \dfrac{2}{3} & \dfrac{19\sqrt{5}}{15} & \dfrac{2\sqrt{870}}{145} & \dfrac{4\sqrt{174}}{87}\end{bmatrix}.$$

IV. Positive-definite and regular matrices: definitions and tests

The concept of positive-definiteness of matrices is established upon symmetrical matrices. Before introducing the concept, the *leading principal minors* of a given matrix are defined. Assume that a symmetrical matrix $\boldsymbol{A}$ is

$$\boldsymbol{A}=\begin{bmatrix} a_{11} & a_{12} & a_{13} & \cdots & a_{1n} \\ a_{12} & a_{22} & a_{23} & \cdots & a_{2n} \\ a_{13} & a_{23} & a_{33} & \cdots & a_{3n} \\ \vdots & \vdots & \vdots & \ddots & \vdots \\ a_{1n} & a_{2n} & a_{3n} & \cdots & a_{nn} \end{bmatrix}, \tag{4-3-7}$$

with the upper-left cornered sub-matrices defined as the *leading principal sub-matrices*. The determinants of the sub-matrices are referred to as *minors* and can be calculated directly. If all the leading principal minors of the matrix are positive, the matrix is referred to as a *positive-definite matrix*. If they have alternative signs, the matrix is referred to as a *negative-definite matrix*. If all the minors are non-negative, the matrix is referred to as a *positive semi-definite matrix*.

The MATLAB function `[D,p]=chol(A)` can also be used to check whether a matrix is positive-definite or not, where for positive-definite matrix $\boldsymbol{A}$, $p=0$ will be returned. Thus, such a function can be used to check whether a symmetrical matrix is positive-definite or not. For matrices which are not positive-definite, a variable p will be returned, where $p-1$ is the size of the sub-matrix in $\boldsymbol{A}$ which is positive-definite, i.e., the size of matrix $\boldsymbol{D}$.

If a complex matrix $\boldsymbol{A}$ satisfies

$$\boldsymbol{A}^{\rm H}\boldsymbol{A}=\boldsymbol{A}\boldsymbol{A}^{\rm H} \tag{4-3-8}$$

where $\boldsymbol{A}^{\rm H}$ is the Hermitian transpose of matrix $\boldsymbol{A}$, then, the matrix is referred to as a *regular matrix*. The judgement can be made using the following MATLAB statements `norm(A'*A-A*A')<`ϵ, if 1 is returned, then, it can be concluded that $\boldsymbol{A}$ is a regular matrix.

Example 4.38 Judge whether matrix $\boldsymbol{A}$ given below is a positive-definite matrix or not. Then, perform Cholesky factorization to the matrix.

$$\boldsymbol{A}=\begin{bmatrix} 7 & 5 & 5 & 8 \\ 5 & 6 & 9 & 7 \\ 5 & 9 & 9 & 0 \\ 8 & 7 & 0 & 1 \end{bmatrix}.$$

Solution *Cholesky factorization to matrix* $\boldsymbol{A}$ *is performed*

```
>> A=[7,5,5,8; 5,6,9,7; 5,9,9,0; 8,7,0,1]; [D,p]=chol(A)
```

The positive-definite part $\boldsymbol{D}$ *of the matrix can be obtained. It is seen that matrix* $\boldsymbol{A}$ *is not a positive-definite matrix, therefore,* $p\neq 0$. *It is found that* $\begin{bmatrix} 2.6457513 & 1.8898224 \\ 0 & 1.5583874 \end{bmatrix}$, *and* $p=3$.

If one calls the `chol()` function to an asymmetrical matrix $\boldsymbol{A}$, the results are also obtained. However, the results are useless, since it erroneously forced the original matrix into the symmetrical one. Strictly speaking, Cholesky factorization cannot be performed upon asymmetrical matrices.

4.3.3 Companion, diagonal and Jordan transformations

I. Transform an ordinary matrix into companion form

For a given matrix $\boldsymbol{A}$, if there exists a column vector $\boldsymbol{x}$, such that matrix $\boldsymbol{T}$ created by $\boldsymbol{T}=[\boldsymbol{x},\boldsymbol{A}\boldsymbol{x},\cdots,\boldsymbol{A}^{n-1}\boldsymbol{x}]$ is nonsingular, the matrix $\boldsymbol{A}$ can be transformed into a companion-like matrix. The conversion matrix $\boldsymbol{T}$ is not unique.

Example 4.39 Transform the matrix in Example 4.31 into a companion matrix.

Solution *A column vector $\boldsymbol{x}$ can be generated randomly, and then, rounded such that only 0's and 1's appear in the vector. Loop structure can be used to find such an $\boldsymbol{x}$ vector so that the matrix $\boldsymbol{T}$ generated is a nonsingular matrix.*

```
>> A=[5,7,6,5; 7,10,8,7; 6,8,10,9; 5,7,9,10];
   while(1), x=floor(2*rand(4,1)); T=sym([x A*x A^2*x A^3*x]);
   if rank(T)==4, break; end, end
   T, A1=inv(T)*A*T
```

which yields

$$\boldsymbol{T}=\begin{bmatrix}1&11&326&9853\\0&15&453&13696\\1&16&472&14296\\0&14&444&13489\end{bmatrix},\quad \boldsymbol{A}_1=\begin{bmatrix}0&0&0&-1\\1&0&0&100\\0&1&0&-146\\0&0&1&35\end{bmatrix}.$$

It should be noted that the matrix $\boldsymbol{T}$ is not unique. The matrix $\boldsymbol{A}_1$ is quite similar to the companion matrix defined in (4-1-7). If one does need the standard companion form, the following statements can further be given

```
>> T1=inv(T*fliplr(eye(4)))', A2=inv(T1)*A*T1
```

the transformation matrix and companion form are as follows:

$$\boldsymbol{T}=\frac{1}{14053}\begin{bmatrix}-318&10591&-29493&19064\\-176&5243&3298&-11368\\318&-10591&29493&-5011\\75&-1835&-13063&2928\end{bmatrix},\quad \boldsymbol{A}_2=\begin{bmatrix}35&-146&100&-1\\1&0&0&0\\0&1&0&0\\0&0&1&0\end{bmatrix}.$$

II. Diagonal matrix transformation

If there is no repeated eigenvalues in matrix $\boldsymbol{A}$, the eigenvector matrix $\boldsymbol{V}$ obtained with `eig()` function can be used as the transformation matrix, and through similarity transformation, the diagonal matrix can be obtained, and the diagonal elements are in fact the eigenvalues of the matrix.

Example 4.40 Please perform diagonalization to the following matrix

$$\boldsymbol{A}=\begin{bmatrix}3&2&2&2\\1&2&-2&-2\\-1&-2&0&-2\\0&1&3&5\end{bmatrix}.$$

Solution *It can be seen through the following commands that the eigenvalues of $\boldsymbol{A}$ are $1,2,3,4$, and are distinct. Thus, the transformation matrix can be selected as the eigenvector matrix*

```
>> A=[3,2,2,2; 1,2,-2,-2; -1,-2,0,-2; 0,1,3,5];
   [v,d]=eig(sym(A)); A1=inv(v)*A*v % A1 is the same as d
```

and the transformation and diagonalized matrices can be found as

$$\boldsymbol{v}=\begin{bmatrix}1&0&-1&0\\-1&0&1&-1\\-1&-1&1&0\\1&1&-2&1\end{bmatrix},\quad \boldsymbol{A}_1=\begin{bmatrix}1&0&0&0\\0&2&0&0\\0&0&3&0\\0&0&0&4\end{bmatrix}.$$

Example 4.41 Consider the following matrix with complex eigenvalues. Please convert it into a diagonal matrix.

$$\boldsymbol{A}=\begin{bmatrix}1&0&4&0\\0&-3&0&0\\-2&2&-3&0\\0&0&0&-2\end{bmatrix}.$$

Solution *The eigenvalues and eigenvector matrix can both be obtained with* `eig()` *function.*

```
>> A=[1,0,4,0; 0,-3,0,0; -2,2,-3,0; 0,0,0,-2]; [V,D]=eig(sym(A))
```

The transformation matrix and diagonal matrix can be obtained as

$$\boldsymbol{V}=\begin{bmatrix}-1&0&-1+\mathrm{j}&-1-\mathrm{j}\\-1&0&0&0\\1&0&1&1\\0&1&0&0\end{bmatrix},\quad \boldsymbol{J}=\begin{bmatrix}-3&0&0&0\\0&-2&0&0\\0&0&-1-2\mathrm{j}&0\\0&0&0&-1+2\mathrm{j}\end{bmatrix}.$$

It can be seen that the diagonal matrix contains complex values. In the next part, we shall find a way to convert it to some kind of special "diagonal" matrix.

III. Jordan transformation

The matrices containing repeated eigenvalues cannot be decomposed into diagonal matrices. Instead, Jordan decomposition should be used.

Example 4.42 For a given matrix

$$\boldsymbol{A}=\begin{bmatrix}-71&-65&-81&-46\\75&89&117&50\\0&4&8&4\\-67&-121&-173&-58\end{bmatrix},$$

compute its eigenvalues and eigenvector matrix using both numerical and analytical methods.

Solution *With the use of MATLAB, the numerical solutions to the eigenvalues can be found from the following statements*

```
>> A=[-71,-65,-81,-46; 75,89,117,50; 0,4,8,4; -67,-121,-173,-58];
   D=eig(A), [V1,D1]=eig(sym(A)) % repeated eigenvalues
```

where the eigenvalues of matrix $\boldsymbol{A}$ *are* $\lambda(\boldsymbol{A}) = -8.0045, -8 \pm 8 + \mathrm{j}0.004, -7.9955$, *and they seem to be distinct eigenvalues. In fact, the numerical results are misleading, and with symbolic computation, it is found that* -8 *is the quadruple eigenvalue of the matrix, and the eigenvector matrix is a column vector of* $\boldsymbol{v} = [17/8, -13/8, 1, -19/8]^{\mathrm{T}}$. *This means that the eigenvector matrix is not invertible and cannot transform the matrix into a diagonal matrix.*

Due to the restrictions of numerical packages and languages, including the numerical solutions provided by MATLAB language, the best data type is the double-precision one. Thus, in computation with such languages, computation errors are unavoidable.

In order to solve this kind of problem, it is suggested that the symbolic data type be used instead. Using the `jordan()` function in the Symbolic Math Toolbox, the Jordan matrix as well as the nonsingular generalized eigenvector matrix can be obtained. The syntaxes of the function are

J = `jordan(`A`)`	% only the Jordan matrix J returned
`[`V`,`J`]` = `jordan(`A`)`	% Jordan J and generalized vector matrix V

When the generalized eigenvector matrix V is found, the Jordan canonical form can be obtained from $J = V^{-1}AV$. It should be noted that the main diagonal elements in the Jordan matrix are the eigenvalues, with the main sub-diagonal elements taking 1's.

Example 4.43 Perform Jordan decomposition for the matrix in Example 4.42.

Solution *The Jordan decomposition for a given symbolic matrix can be obtained directly by the use of* `jordan()` *function such that*

```
>> A=[-71,-65,-81,-46; 75,89,117,50; 0,4,8,4; -67,-121,-173,-58];
   [V,J]=jordan(sym(A)) % find Jordan decomposition
```

and one has

$$V = \begin{bmatrix} -18496 & 2176 & -63 & 1 \\ 14144 & -800 & 75 & 0 \\ -8704 & 32 & 0 & 0 \\ 20672 & -1504 & -67 & 0 \end{bmatrix}, \quad J = \begin{bmatrix} -8 & 1 & 0 & 0 \\ 0 & -8 & 1 & 0 \\ 0 & 0 & -8 & 1 \\ 0 & 0 & 0 & -8 \end{bmatrix}.$$

The matrix V is now a full-rank matrix and is invertible. Therefore, it is possible to implement some special operations which are very hard to implement using numerical methods. The applications of such a function will be demonstrated later.

Example 4.44 Please revisit the matrix in Example 4.41, where there are complex eigenvalues. The diagonal elements contain complex values. If in the transformation matrix, the real and imaginary parts are extracted to form two columns to replace the complex conjugate columns, the following MATLAB function can be written

```
function V1=realjordan(V)
n=length(V); i=0; vr=real(V); vi=imag(V); n1=n;
while(i<n1), i=i+1; V1(:,i)=vr(:,i), vv=vi(:,i);
    if any(abs(vv)>1e-10), i=i+1; V1(:,i)=vv;
end, end
```

Therefore, for the matrix with complex eigenvalues, alternative real Jordan form can be transformed

```
>> A=[1,0,4,0; 0,-3,0,0; -2,2,-3,0; 0,0,0,-2]; [V,D]=eig(sym(A));
   V1=realjordan(V), A1=inv(V1)*A*V1 % find real Jordan form
```

and the transformation matrix and real Jordan form are

$$\boldsymbol{V}_1=\begin{bmatrix}-1&0&-1&1\\-1&0&0&0\\1&0&1&0\\0&1&0&0\end{bmatrix},\quad \boldsymbol{A}_1=\left[\begin{array}{cc:cc}-3&0&0&0\\0&-2&0&0\\\hdashline 0&0&-1&-2\\0&0&2&-1\end{array}\right].$$

It can be seen that $\boldsymbol{A}_1$ is no longer a diagonal matrix. It is a real matrix with a Jordan block in the lower-right corner of matrix $\boldsymbol{A}_1$.

Example 4.45 Perform Jordan transformation to the following matrix.

$$\boldsymbol{A}=\begin{bmatrix}0&-1&0&0&-1&1\\0.5&0&-0.5&0&-1&0.5\\-0.5&0&-0.5&0&0&0.5\\468.5&452&304.5&577&225&360.5\\-468&-450&-303&-576&-223&-361\\-467.5&-451&-303.5&-576&-223&-361.5\end{bmatrix}.$$

Solution *The eigenvalues of the matrix* $\boldsymbol{A}$ *can be obtained*

```
>> A=[0,-1,0,0,-1,1; 0.5,0,-0.5,0,-1,0.5; -0.5,0,-0.5,0,0,0.5;
      468.5,452,304.5,577,225,360.5; -468,-450,-303,-576,-223,-361;
     -467.5,-451,-303.5,-576,-223,-361.5];
   A=sym(A); eig(A), [v,J]=jordan(A)
```

and the eigenvalues are $-2,-2,-1\pm \mathrm{j}2,-1\pm \mathrm{j}2$. *It can be seen that there exist repeated complex eigenvalues* $-1\pm \mathrm{j}2$. *The simplest way to manipulate is to swap the 4th and 5th column in the transformation matrix* $\boldsymbol{v}$, *and call* `realjordan()` *function to extract the real transformation matrix. In this case, the real Jordan block matrix can be obtained*

```
>> v(:,[4 5])=v(:,[5,4]); V=realjordan(v), J=inv(V)*A*V
```

and the new real transformation matrix and transformed real Jordan matrix can finally be found as

$$\boldsymbol{V}=\begin{bmatrix}423/25&-543/125&851/100&757/100&334/125&-9321/1000\\-423/25&7431/250&2459/100&663/100&-7431/500&-509/1000\\423/5&-471/10&-757/40&851/40&471/20&-1887/80\\4371/25&-70677/250&-47327/400&-9191/100&70677/500&247587/4000\\-4653/25&31353/125&16263/200&15991/200&-31353/250&-96843/2000\\-5922/25&76539/250&22507/200&12399/200&-76539/500&-74767/2000\end{bmatrix},$$

$$\boldsymbol{J}=\left[\begin{array}{cc:cc:cc}-2&1&0&0&0&0\\0&-2&0&0&0&0\\\hdashline 0&0&-1&-2&1&0\\0&0&2&-1&0&1\\\hdashline 0&0&0&0&-1&-2\\0&0&0&0&2&-1\end{array}\right].$$

4.3.4 Singular value decompositions

Singular values of a matrix can be regarded as a measure of the matrix. For any given $n\times m$ matrix $\boldsymbol{A}$, one has

$$\boldsymbol{A}^{\mathrm{T}}\boldsymbol{A}\geqslant 0,\quad \boldsymbol{A}\boldsymbol{A}^{\mathrm{T}}\geqslant 0, \tag{4-3-9}$$

and in theory, it follows that

$$\operatorname{rank}(\boldsymbol{A}^{\mathrm{T}}\boldsymbol{A}) = \operatorname{rank}(\boldsymbol{A}\boldsymbol{A}^{\mathrm{T}}) = \operatorname{rank}(\boldsymbol{A}). \tag{4-3-10}$$

It can further be shown that the matrices $\boldsymbol{A}^{\mathrm{T}}\boldsymbol{A}$ and $\boldsymbol{A}\boldsymbol{A}^{\mathrm{T}}$ have the same non-negative eigenvalues λ_i. The square roots of these non-negative eigenvalues are referred to as the *singular values* of matrix $\boldsymbol{A}$, denoted as

$$\sigma_i(\boldsymbol{A}) = \sqrt{\lambda_i(\boldsymbol{A}^{\mathrm{T}}\boldsymbol{A})}. \tag{4-3-11}$$

Example 4.46 For a matrix $\boldsymbol{A} = \begin{bmatrix} 1 & 1 \\ \mu & 0 \\ 0 & \mu \end{bmatrix}$, where $\mu = 5\texttt{eps}$, find the rank of matrix $\boldsymbol{A}$ using (4-3-10).

Solution *It is obvious that the rank of matrix* $\boldsymbol{A}$ *is 2. The same result can be obtained from the following MATLAB statement.*

```
>> A=[1 1; 5*eps,0; 0,5*eps]; rank(A) % directly assess the rank
```

Now consider the method in (4-3-10) *for calculating the rank of matrix* $\boldsymbol{A}$. *If* $\boldsymbol{A}^{\mathrm{T}}\boldsymbol{A}$ *is used for finding the rank of matrix* $\boldsymbol{A}$, *it can be seen that*

$$\boldsymbol{A}^{\mathrm{T}}\boldsymbol{A} = \begin{bmatrix} 1+\mu^2 & 1 \\ 1 & 1+\mu^2 \end{bmatrix},$$

and under double-precision scheme, since μ^2 *is around* 10^{-30}, *it can be completely neglected when added to 1. Thus, matrix* $\boldsymbol{A}^{\mathrm{T}}\boldsymbol{A}$ *is reduced to a matrix of ones. It can be concluded that the rank of matrix* $\boldsymbol{A}$ *is 1, which is obviously wrong. Therefore, the concepts of singular values for matrix* $\boldsymbol{A}$ *should be introduced to provide a better characterization for the matrices.*

If matrix $\boldsymbol{A}$ is an $n \times m$ matrix, it can be decomposed as

$$\boldsymbol{A} = \boldsymbol{L}\boldsymbol{A}_1\boldsymbol{M} \tag{4-3-12}$$

where matrices $\boldsymbol{L}$ and $\boldsymbol{M}$ are orthogonal matrices, and $\boldsymbol{A}_1 = \operatorname{diag}(\sigma_1, \cdots, \sigma_n)$ is a diagonal matrix, whose elements satisfy the inequality $\sigma_1 \geqslant \sigma_2 \geqslant \cdots \geqslant \sigma_n \geqslant 0$. If $\sigma_n = 0$, then, matrix $\boldsymbol{A}$ is singular. The rank of matrix $\boldsymbol{A}$ is in fact the number of non-zero quantities in the diagonal elements of matrix $\boldsymbol{A}_1$. The transformation is referred to as *singular value decomposition (SVD)*.

A singular value decomposition function `svd()` is provided in MATLAB

```
S=svd(A)              % only singular value decomposition required
[L,A1,M]=svd(A)       % singular value decomposition
```

where $\boldsymbol{A}$ is the original matrix and the returned matrix $\boldsymbol{A}_1$ is a diagonal matrix, while the matrices $\boldsymbol{L}$ and $\boldsymbol{M}$ are orthogonal matrices, satisfying $\boldsymbol{A} = \boldsymbol{L}\boldsymbol{A}_1\boldsymbol{M}^{\mathrm{T}}$. In recent versions of MATLAB, symbolic $\boldsymbol{A}$ is supported, however, there are usually no analytical solutions, only high-precision solutions are obtained.

The singular values of a matrix often determine the properties of the matrix. When some of the singular values are large, while others are small, and they differ significantly, then, very small perturbation to certain elements in the matrix may significantly affect the behavior of the matrix. This kind of matrix is often referred to as an *ill-* or *bad-conditioned* matrix. If zeros exist in the singular values, then, the matrix is a singular matrix. The ratio of maximum singular value $\sigma_{\max}$ and the minimum one $\sigma_{\min}$ is defined as the *condition number* of the matrix, denoted by cond($\boldsymbol{A}$), i.e., cond($\boldsymbol{A}$) $= \sigma_{\max}/\sigma_{\min}$. The larger the condition number of the matrix, the more sensitive the matrix is. The maximum and minimum singular values of the matrix may also be denoted by $\bar{\sigma}(\boldsymbol{A})$ and $\underline{\sigma}(\boldsymbol{A})$, respectively. A MATLAB function `cond(A)` is provided to calculate the condition number of the matrix $\boldsymbol{A}$, and for singular matrices, the condition number is infinity.

Example 4.47 Performing SVD to the matrix $\boldsymbol{A}$ in Example 4.10.

Solution *If the MATLAB function* `svd()` *is used, the matrices* $\boldsymbol{L}$, $\boldsymbol{A}_1$ *and* $\boldsymbol{M}$ *can be obtained. The condition number can also be calculated by the singular values.*

```
>> A=[16,2,3,13; 5,11,10,8; 9,7,6,12; 4,14,15,1]; [L,A1,M]=svd(A)
```

It can be found that the decomposed matrices are

$$\boldsymbol{L}=\begin{bmatrix}-0.5 & 0.67082 & 0.5 & -0.22361\\ -0.5 & -0.22361 & -0.5 & -0.67082\\ -0.5 & 0.22361 & -0.5 & 0.67082\\ -0.5 & -0.67082 & 0.5 & 0.22361\end{bmatrix},\ \boldsymbol{A}_1=\begin{bmatrix}34 & 0 & 0 & 0\\ 0 & 17.889 & 0 & 0\\ 0 & 0 & 4.4721 & 0\\ 0 & 0 & 0 & 0\end{bmatrix},$$

$$\boldsymbol{M}=\begin{bmatrix}-0.5 & 0.5 & 0.67082 & -0.22361\\ -0.5 & -0.5 & -0.22361 & -0.67082\\ -0.5 & -0.5 & 0.22361 & 0.67082\\ -0.5 & 0.5 & -0.67082 & 0.22361\end{bmatrix}.$$

It can be seen that since there exists a zero singular value, the original matrix is a singular matrix. The condition number of the matrix should be infinity; however, since numerical calculation is used, there might be minor errors. The commands `cond(A)` *will result the condition number* 3.2592×10^{16}.

One can convert matrix $\boldsymbol{A}$ *into a symbolic variable, then, with the function* `svd()`, *more accurate singular value decomposition can be obtained.*

Example 4.48 For a rectangular matrix $\boldsymbol{A}=\begin{bmatrix}1 & 3 & 5 & 7\\ 2 & 4 & 6 & 8\end{bmatrix}$, perform singular value decomposition to matrix $\boldsymbol{A}$, and then, verify the results.

Solution *The following statements can be given such that*

```
>> A=[1,3,5,7; 2,4,6,8]; [L,A1,M]=svd(A); A2=L*A1*M'; norm(A-A2)
```

The decomposited matrices are

$$\boldsymbol{L}=\begin{bmatrix}-0.64142 & -0.76719\\ -0.76719 & 0.64142\end{bmatrix},\ \boldsymbol{A}_1=\begin{bmatrix}14.269 & 0 & 0 & 0\\ 0 & 0.62683 & 0 & 0\end{bmatrix},$$

$$\boldsymbol{M}=\begin{bmatrix}-0.15248 & 0.82265 & -0.3945 & -0.37996\\ -0.34992 & 0.42138 & 0.2428 & 0.80066\\ -0.54735 & 0.020103 & 0.69791 & -0.46143\\ -0.74479 & -0.38117 & -0.54621 & 0.040738\end{bmatrix},$$

and it can be seen that the error $||\boldsymbol{L}\boldsymbol{A}_1\boldsymbol{V}^{\mathrm{T}}-\boldsymbol{A}||=9.7277\times10^{-15}$ *and, for this example,* $\boldsymbol{L}\boldsymbol{A}_1\boldsymbol{V}^{\mathrm{T}}$ *will restore the original matrix* $\boldsymbol{A}$, *with very small error.*

4.4 Solving Matrix Equations

Solutions of various matrix equations are discussed in the section. The commonly encountered linear algebraic equation in $\boldsymbol{AX} = \boldsymbol{B}$ is discussed first, followed by Lyapunov equations, Sylvester equations and Riccati equations. Polynomial equations such as Diophantine equations are also discussed.

4.4.1 Solutions to linear algebraic equations

Consider the linear algebraic equation

$$\boldsymbol{Ax} = \boldsymbol{B}, \tag{4-4-1}$$

where $\boldsymbol{A}$ are $\boldsymbol{B}$ given matrices

$$\boldsymbol{A} = \begin{bmatrix} a_{11} & a_{12} & \cdots & a_{1n} \\ a_{21} & a_{22} & \cdots & a_{2n} \\ \vdots & \vdots & \ddots & \vdots \\ a_{m1} & a_{m2} & \cdots & a_{mn} \end{bmatrix}, \quad \boldsymbol{B} = \begin{bmatrix} b_{11} & b_{12} & \cdots & b_{1p} \\ b_{21} & b_{22} & \cdots & b_{2p} \\ \vdots & \vdots & \ddots & \vdots \\ b_{m1} & b_{m2} & \cdots & b_{mp} \end{bmatrix}, \tag{4-4-2}$$

and the target is to find matrix $\boldsymbol{x}$ for the equation. According to linear algebra theory, in some cases, there are unique solutions, and in other cases, the equation may have an infinite number of solutions, or have no solution at all. Thus, here the solutions to the linear equation can be considered thoroughly.

(i) **Unique solutions** If $m = n$ and $\text{rank}(\boldsymbol{A}) = n$, then, the equations in (4-4-1) have unique solutions

$$\boldsymbol{x} = \boldsymbol{A}^{-1}\boldsymbol{B}. \tag{4-4-3}$$

The solution $\boldsymbol{x} = \texttt{inv}(\boldsymbol{A})*\boldsymbol{B}$ can be obtained immediately using MATLAB to the original equation. It should be noted that if the numerical method is used, sometimes the `inv()` function may lead to erroneous results. For instance, if $\text{cond}(\boldsymbol{A})$ is very large, the results may be unreliable. If the Symbolic Math Toolbox is used, then, this problem may be avoided.

Example 4.49 Solve the linear algebraic equations $\begin{bmatrix} 1 & 2 & 3 & 4 \\ 4 & 3 & 2 & 1 \\ 1 & 3 & 2 & 4 \\ 4 & 1 & 3 & 2 \end{bmatrix} \boldsymbol{X} = \begin{bmatrix} 5 & 1 \\ 4 & 2 \\ 3 & 3 \\ 2 & 4 \end{bmatrix}$.

Solution *The analytical solutions to the given equations can be obtained using the following MATLAB statements.*

```
>> A=[1 2 3 4; 4 3 2 1; 1 3 2 4; 4 1 3 2]; B=[5 1; 4 2; 3 3; 2 4];
   x=inv(sym(A))*B % find the analytical solution
```

Therefore, the solutions can be written as $\boldsymbol{x} = \begin{bmatrix} -9/5 & 12/5 \\ 28/15 & -19/15 \\ 58/15 & -49/15 \\ -32/15 & 41/15 \end{bmatrix}$.

Substituting the results back to the equations, one may find that there is no error. If in the above statement, the numerical statement is used instead, i.e., $\boldsymbol{x} = \texttt{inv(sym(}\boldsymbol{A}\texttt{))}*\boldsymbol{B}$ *is replaced by* $\boldsymbol{x} = \texttt{inv}(\boldsymbol{A})*\boldsymbol{B}$*, an error level of* 10^{-15} *may be achieved.*

(ii) **Equations with infinite number of solutions** One may construct first the judging matrix $\boldsymbol{C}$ out of $\boldsymbol{A}$, $\boldsymbol{B}$ matrices

$$\boldsymbol{C} = \begin{bmatrix} a_{11} & a_{12} & \cdots & a_{1n} & b_{11} & b_{12} & \cdots & b_{1p} \\ a_{21} & a_{22} & \cdots & a_{2n} & b_{21} & b_{22} & \cdots & b_{2p} \\ \vdots & \vdots & \ddots & \vdots & \vdots & \vdots & \ddots & \vdots \\ a_{m1} & a_{m2} & \cdots & a_{mn} & b_{m1} & b_{m2} & \cdots & b_{mp} \end{bmatrix}. \tag{4-4-4}$$

If $\text{rank}(\boldsymbol{A}) = \text{rank}(\boldsymbol{C}) = r < n$, the original equations (4-4-1) have an infinite number of solutions. One may find the $n-r$ basic set of solutions $\boldsymbol{x}_i$, $i = 1, 2, \cdots, n-r$ to the homogeneous equations $\boldsymbol{A}\boldsymbol{x} = 0$. From which, for any constant $\alpha_i, i = 1, 2, \cdots, n-r$, the general solutions to the homogeneous equations can be written as

$$\hat{\boldsymbol{x}} = \alpha_1 \boldsymbol{x}_1 + \alpha_2 \boldsymbol{x}_2 + \cdots + \alpha_{n-r} \boldsymbol{x}_{n-r}. \tag{4-4-5}$$

The solution basis can be found directly using MATLAB function `null()` such that $\boldsymbol{Z}$=`null(sym(`$\boldsymbol{A}$`))`, and the function `null()` can also be used in numerical cases, where in this case, the syntax $\boldsymbol{Z}$=`null(`$\boldsymbol{A}$`,'r')` should be used instead. The resulted $\boldsymbol{Z}$ matrix should have $n-r$ columns, to which each one is referred to as a *basic set of solutions* for the given matrix $\boldsymbol{A}$.

Solving equations (4-4-1) is not a difficult task. If a special solution $\boldsymbol{x}_0$ to (4-4-1) can be found, the general solutions to the original equations can be constructed from $\boldsymbol{x} = \hat{\boldsymbol{x}} + \boldsymbol{x}_0$. In fact, the special solution can be found from $\boldsymbol{x}_0$=`pinv(`$\boldsymbol{A}$`)*`$\boldsymbol{B}$.

Example 4.50 Find all the solutions of linear algebraic equation [6]

$$\begin{bmatrix} 1 & 4 & 0 & -1 & 0 & 7 & -9 \\ 2 & 8 & -1 & 3 & 9 & -13 & 7 \\ 0 & 0 & 2 & -3 & -4 & 12 & -8 \\ -1 & -4 & 2 & 4 & 8 & -31 & 37 \end{bmatrix} \boldsymbol{X} = \begin{bmatrix} 3 \\ 9 \\ 1 \\ 4 \end{bmatrix}.$$

Solution *The matrices* $\boldsymbol{A}$ *and* $\boldsymbol{B}$ *can be entered first, and matrix* $\boldsymbol{C}$ *can be constructed. Judging from the ranks of* $\boldsymbol{A}$ *and* $\boldsymbol{C}$*,*

```
>> A=[1,4,0,-1,0,7,-9; 2,8,-1,3,9,-13,7;
      0,0,2,-3,-4,12,-8; -1,-4,2,4,8,-31,37];
   B=[3; 9; 1; 4]; C=[A B]; rank(A), rank(C) % the ranks are equal
```

it can be seen that they both equal to 3, which is smaller than the number of unknowns. It can then be concluded that the equations have an infinite number of solutions, with four free variables. The null space $\boldsymbol{Z}$ *and a particular solution* $\boldsymbol{x}_0$ *can be obtained with*

```
>> Z=null(sym(A)), x0=sym(pinv(A)*B) % find the null space
   a=sym('a%d',[4,1]); x=Z*a+x0, E=A*x-B % construct the general solution
```

The null space and a particular solution of the equations can be found as follows, from which the general solutions of the equations can be found for all free symbolic entities a_1*,* a_2*,* a_3 *and* a_4*.*

$$Z=\begin{bmatrix}-4 & -2 & -1 & 3\\ 1 & 0 & 0 & 0\\ 0 & -1 & 3 & -5\\ 0 & -2 & 6 & -6\\ 0 & 1 & 0 & 0\\ 0 & 0 & 1 & 0\\ 0 & 0 & 0 & 1\end{bmatrix},\ x_0=\begin{bmatrix}92/395\\ 368/395\\ 459/790\\ -24/79\\ 347/790\\ 247/790\\ 303/790\end{bmatrix},\ x=\begin{bmatrix}-4a_1-2a_2-a_3+3a_4+92/395\\ a_1+368/395\\ -a_2+3a_3-5a_4+459/790\\ -2a_2+6a_3-6a_4-24/79\\ a_2+347/790\\ a_3+247/790\\ a_4+303/790\end{bmatrix}.$$

Alternatively, with the reduced row echelon form technique, the general solution can also be found

```
>> C=[A B]; D=rref(C) % resolve the equation with reduced row echelon form
```

with the reduced row echelon form of

$$D=\begin{bmatrix}1 & 4 & 0 & 0 & 2 & 1 & -3 & 4\\ 0 & 0 & 1 & 0 & 1 & -3 & 5 & 2\\ 0 & 0 & 0 & 1 & 2 & -6 & 6 & 1\\ 0 & 0 & 0 & 0 & 0 & 0 & 0 & 0\end{bmatrix}.$$

It can be seen that the variables x_2, x_5, x_6 and x_7 are free variables. The general solutions can be interpreted from matrix D, that $x_1=-4x_2-2x_5-x_6+3x_7+4$, $x_3=-x_5+3x_6-5x_7+2$, $x_4=-2x_5+6x_6-6x_7+1$.

Example 4.51 Please solve the following underdetermined linear algebraic equation

$$\begin{bmatrix}4 & 7 & 1 & 4\\ 3 & 7 & 4 & 6\end{bmatrix}x=\begin{bmatrix}3\\ 4\end{bmatrix}.$$

Solution *The ranks of A and C are all 2 and are equal. Thus, the original equation has an infinite number of solutions. With the two following two approaches, the solutions are found*

$$x_1=\begin{bmatrix}a_1\\ a_2+8/21\\ 6a_1/5+7a_2/5+1/3\\ -13a_1/10-21a_2/10\end{bmatrix},\quad x_2=\begin{bmatrix}3b_1+2b_2-1\\ -13b_1/7-12b_2/7+1\\ b_1\\ b_2\end{bmatrix}.$$

```
>> A=[4,7,1,4; 3,7,4,6]; B=[3; 4]; C=[A B]; rank(A), rank(C)
   syms a1 a2 b1 b2; x1=null(sym(A))*[a1; a2]+sym(A\B), A*x1-B
   a=rref(sym([A B])); x2=[a(:,3:5)*[-b1; -b2; 1]; b1; b2], A*x2-B
```

It can be seen that the two sets of solutions all satisfy the original equation.

(iii) **Equations with no solutions** If $\mathrm{rank}(A) < \mathrm{rank}(C)$, then, the equations in (4-4-1) are conflict equations, and there are no solutions to such equations. Applying Moore–Penrose inverse $x=$`pinv(`A`)*`B, one may obtain the least squares solution to the original equations such that the norm of the error $||Ax-B||$ is minimized.

Example 4.52 Please solve the following algebraic linear equation

$$\begin{bmatrix}1 & 2 & 3 & 4\\ 2 & 2 & 1 & 1\\ 2 & 4 & 6 & 8\\ 4 & 4 & 2 & 2\end{bmatrix}X=\begin{bmatrix}1\\ 2\\ 3\\ 4\end{bmatrix}.$$

Solution *The matrices* $\boldsymbol{A}$ *and* $\boldsymbol{B}$ *can be entered first, and matrix* $\boldsymbol{C}$ *can be constructed. It can be seen that the rank of matrix* $\boldsymbol{C}$ *is 3, which is higher than that of matrix* $\boldsymbol{A}$*, indicating that there is no solution to the original equations*

```
>> A=[1 2 3 4; 2 2 1 1; 2 4 6 8; 4 4 2 2];
   B=[1:4]'; C=[A B]; [rank(A), rank(C)] % the ranks are different
```

One may use function `pinv()` *to evaluate the Moore–Penrose inverse to matrix* $\boldsymbol{A}$*, then, the least squares solutions to the conflict equations can be obtained as*

```
>> x=pinv(A)*B, norm(A*x-B) % validate the result, try also x=pinv(sym(A))*B
```

Using the above statements, one may find the solutions $\boldsymbol{x}^{\mathrm{T}} = [0.9542, 0.7328, -0.0763, -0.29771]$. *Substituting the solution back to the original equations, the norm of the solution error is* 0.4472, *which is the smallest possible error level.*

For linear algebraic equations of the form $\boldsymbol{xA} = \boldsymbol{B}$, one may perform transpose to both sides of the equation to transform the original equation to the following form

$$\boldsymbol{A}^{\mathrm{T}}\boldsymbol{z} = \boldsymbol{B}^{\mathrm{T}}, \tag{4-4-6}$$

where $\boldsymbol{z} = \boldsymbol{x}^{\mathrm{T}}$, the new equations can be transformed into the form in (4-4-1), and the above methods can be applied directly to solve the original problems.

4.4.2 Solutions to Lyapunov equations

I. Continuous Lyapunov equations

A continuous Lyapunov equation can be expressed as

$$\boldsymbol{AX} + \boldsymbol{XA}^{\mathrm{T}} = -\boldsymbol{C}, \tag{4-4-7}$$

and it is known that Lyapunov equations are originated from stability theory of differential equations, where one often expects $-\boldsymbol{C}$ to be symmetrical positive-definite $n \times n$ matrix. It follows that the solution $\boldsymbol{X}$ is also an $n \times n$ symmetrical matrix. Direct solutions to such equations were rather difficult; however, with the use of powerful computer mathematics languages, such a function can be solved easily by using the `lyap()` function provided in the Control Systems Toolbox. The syntax of the function is $\boldsymbol{X}$=`lyap(`$\boldsymbol{A}$`,`$\boldsymbol{C}$`)`, and if one specifies matrices $\boldsymbol{A}$ and $\boldsymbol{C}$, the numerical solutions to the Lyapunov equations can be obtained immediately.

Example 4.53 Assume that in (4-4-7), the matrices $\boldsymbol{A}$ and $\boldsymbol{C}$ are given by

$$\boldsymbol{A} = \begin{bmatrix} 1 & 2 & 3 \\ 4 & 5 & 6 \\ 7 & 8 & 0 \end{bmatrix}, \quad \boldsymbol{C} = -\begin{bmatrix} 10 & 5 & 4 \\ 5 & 6 & 7 \\ 4 & 7 & 9 \end{bmatrix},$$

solve the corresponding Lyapunov equation and check the accuracy of the solutions.

Solution *One may enter the matrices into MATLAB, then, solve the equation using the following MATLAB statements*

```
>> A=[1 2 3;4 5 6; 7 8 0]; C=-[10, 5, 4; 5, 6, 7; 4, 7, 9];
   X=lyap(A,C), norm(A*X+X*A'+C) % numerical solution and validation
```

the solution obtained is

$$X = \begin{bmatrix} -3.9444444444444 & 3.8888888888889 & 0.38888888888889 \\ 3.8888888888889 & -2.7777777777778 & 0.22222222222222 \\ 0.38888888888889 & 0.22222222222222 & -0.11111111111111 \end{bmatrix},$$

and the norm of the error matrix is $||AX + XA^{\mathrm{T}} + C|| = 2.64742\times 10^{-14}$. *It can be seen that the accuracy in the solutions obtained is very high.*

II. Analytical solutions to Lyapunov equations

For simplicity, the matrices in the Lyapunov equation can be rearranged such that

$$X = \begin{bmatrix} x_1 & x_2 & \cdots & x_m \\ x_{m+1} & x_{m+2} & \cdots & x_{2m} \\ \vdots & \vdots & \ddots & \vdots \\ x_{(n-1)m+1} & x_{(n-1)m+2} & \cdots & x_{nm} \end{bmatrix}, \quad C = \begin{bmatrix} c_1 & c_2 & \cdots & c_m \\ c_{m+1} & c_{m+2} & \cdots & c_{2m} \\ \vdots & \vdots & \ddots & \vdots \\ c_{(n-1)m+1} & c_{(n-1)m+2} & \cdots & c_{nm} \end{bmatrix}.$$

The Lyapunov equation can be rewritten as a simple linear equation

$$(A \otimes I + I \otimes A)x = -c, \tag{4-4-8}$$

where $A \otimes B$ denotes the Kronecker product of matrices A and B such that

$$A \otimes B = \begin{bmatrix} a_{11}B & \cdots & a_{1m}B \\ \vdots & \ddots & \vdots \\ a_{n1}B & \cdots & a_{nm}B \end{bmatrix}, \tag{4-4-9}$$

and MATLAB evaluation of the product is C=`kron(`A`,`B`)`.

It can now be seen that the conditions when the equation has a unique solution is no longer that $-C$ is a positive-definite symmetrical matrix. It requires that the matrix $(A \otimes I + I \otimes A)$ is a nonsingular square matrix.

Example 4.54 Consider again the Lyapunov equation in Example 4.53, and find the analytical solutions.

Solution *The analytical solution of the Lyapunov equation can be obtained from the following statements, and if one substitutes the solutions back to the original equation, there is no error.*

```
>> I=eye(3); A0=sym(kron(A,I)+kron(I,A)); % construct (A⊗I + I⊗A) matrix
   c=reshape(C',9,1); x0=-inv(A0)*c; x=reshape(x0,3,3)' % analytical
```

The analytical solutions can be obtained as

$$x = \begin{bmatrix} -71/18 & 35/9 & 7/18 \\ 35/9 & -25/9 & 2/9 \\ 7/18 & 2/9 & -1/9 \end{bmatrix}.$$

Example 4.55 Traditionally in Lyapunov equations, one may assume that matrix C is a real and symmetrically positive-definite matrix. Find whether there are solutions to Lyapunov equations if C is not a symmetrical real matrix.

Solution *Since the Lyapunov equation was originated from stability theory, it was usually assumed that matrix* $-C$ *is a real and symmetrically positive-definite matrix. In fact,*

(4-4-12) *will also have unique solutions even when the above conditions are not satisfied. For instance, if matrix* $\boldsymbol{C}$ *is changed into a complex asymmetrical matrix*

$$\boldsymbol{C} = -\begin{bmatrix} 1+\mathrm{j}1 & 3+\mathrm{j}3 & 12+\mathrm{j}10 \\ 2+\mathrm{j}5 & 6 & 11+\mathrm{j}6 \\ 5+\mathrm{j}2 & 11+\mathrm{j}1 & 2+\mathrm{j}12 \end{bmatrix},$$

while matrix $\boldsymbol{A}$ *remains unchanged in Example 4.53, the complex solution matrix* $\boldsymbol{X}$ *will be immediately obtained.*

```
>> A=[1 2 3;4 5 6; 7 8 0]; % specify the matrices
   C=-[1+1i, 3+3i, 12+10i; 2+5i, 6, 11+6i; 5+2i, 11+1i, 2+12i];
   A0=sym(kron(A,eye(3))+kron(eye(3),A));
   c=reshape(C.',9,1);  x0=-inv(A0)*c; x=reshape(x0,3,3).'
   norm(A*x+x*A.'+C) % validation of the results
```

The solution is

$$\boldsymbol{x} = \begin{bmatrix} -5/102+\mathrm{j}1457/918 & 15/17-\mathrm{j}371/459 & -61/306+\mathrm{j}166/459 \\ 4/17-\mathrm{j}626/459 & -10/51+\mathrm{j}160/459 & 115/153+\mathrm{j}607/459 \\ -55/306+\mathrm{j}166/459 & -26/153-\mathrm{j}209/459 & 203/153+\mathrm{j}719/918 \end{bmatrix},$$

and it can be concluded that there is no error at all in the solution. Thus, if one does not consider the energy concept in the physical model, matrix $\boldsymbol{C}$ *can be generalized into any matrix in the Lyapunov matrix.*

III. Solutions of Stein equations

The typical form of Stein equation is

$$\boldsymbol{AXB} - \boldsymbol{X} + \boldsymbol{Q} = \boldsymbol{0}, \tag{4-4-10}$$

where, all the equations are $n \times n$ square matrices. Similar to the previous discussion, matrix $\boldsymbol{X}$ can be expanded as a vector $\boldsymbol{x}$, and matrix $\boldsymbol{Q}$ can be expanded as another vector $\boldsymbol{q}$. Thus, Stein equation can be solved directly with the following linear algebraic equation

$$\left(\boldsymbol{I}_{n^2\times n^2} - \boldsymbol{B}^{\mathrm{T}} \otimes \boldsymbol{A}\right)\boldsymbol{x} = \boldsymbol{q}. \tag{4-4-11}$$

Example 4.56 Please solve the following Stein equation

$$\begin{bmatrix} -2 & 2 & 1 \\ -1 & 0 & -1 \\ 1 & -1 & 2 \end{bmatrix} \boldsymbol{X} \begin{bmatrix} -2 & -1 & 2 \\ 1 & 3 & 0 \\ 3 & -2 & 2 \end{bmatrix} - \boldsymbol{X} + \begin{bmatrix} 0 & -1 & 0 \\ -1 & 1 & 0 \\ 1 & -1 & -1 \end{bmatrix} = \boldsymbol{0}.$$

Solution *The equation can be solved with the following MATLAB statements*

```
>> A=[-2,2,1; -1,0,-1; 1,-1,2]; B=[-2,-1,2; 1,3,0; 3,-2,2];
   Q=[0,-1,0; -1,1,0; 1,-1,-1]; x=inv(sym(eye(9))-kron(B',A))*Q(:);
   X=reshape(x,3,3), norm(A*X*B-X+Q) % solve and validate
```

and the analytical solution is

$$\boldsymbol{X} = \begin{bmatrix} 4147/47149 & 3861/471490 & -40071/235745 \\ -2613/94298 & 2237/235745 & -43319/235745 \\ 20691/94298 & 66191/235745 & -10732/235745 \end{bmatrix}.$$

IV. Discrete Lyapunov equations

A discrete Lyapunov equation can be expressed in the form

$$\boldsymbol{A}\boldsymbol{X}\boldsymbol{A}^{\mathrm{T}} - \boldsymbol{X} + \boldsymbol{Q} = \boldsymbol{0}, \tag{4-4-12}$$

which can be solved easily by using `dlyap()` function provided by Control Systems Toolbox of MATLAB. The syntax of the function is $\boldsymbol{X}$ = `dlyap(`$\boldsymbol{A}$`,`$\boldsymbol{Q}$`)`. The analytical solution method will be discussed later.

Example 4.57 Solve the discrete Lyapunov equation

$$\begin{bmatrix} 8 & 1 & 6 \\ 3 & 5 & 7 \\ 4 & 9 & 2 \end{bmatrix} \boldsymbol{X} \begin{bmatrix} 8 & 1 & 6 \\ 3 & 5 & 7 \\ 4 & 9 & 2 \end{bmatrix}^{\mathrm{T}} - \boldsymbol{X} + \begin{bmatrix} 16 & 4 & 1 \\ 9 & 3 & 1 \\ 4 & 2 & 1 \end{bmatrix} = \boldsymbol{0}.$$

Solution *The numerical solution to the equation can easily be solved by the use of the function* `dlyap()` *such that*

```
>> A=[8,1,6; 3,5,7; 4,9,2]; Q=[16,4,1; 9,3,1; 4,2,1];
   X=dlyap(A,Q), norm(A*X*A.'-X+Q)   % solve and check accuracy
```

with the error of 2.7778×10^{-14}, *and*

$$\boldsymbol{X} = \begin{bmatrix} -0.16474 & 0.06915 & -0.016785 \\ 0.052843 & -0.029785 & -0.0061542 \\ -0.10198 & 0.044959 & -0.030541 \end{bmatrix}.$$

4.4.3 Solutions to Sylvester equations

A Sylvester equation takes the general form

$$\boldsymbol{A}\boldsymbol{X} + \boldsymbol{X}\boldsymbol{B} = -\boldsymbol{C}, \tag{4-4-13}$$

where $\boldsymbol{A}$ is an $n\times n$ matrix, $\boldsymbol{B}$ is an $m\times m$ matrix, and $\boldsymbol{C}$ and $\boldsymbol{X}$ are $n\times m$ matrices. This equation is also known as the *generalized Lyapunov equation.* The function `lyap()` can still be used such that $\boldsymbol{X}$ = `lyap(`$\boldsymbol{A}$`,`$\boldsymbol{B}$`,`$\boldsymbol{C}$`)`. Schur decomposition is an effective way in solving such an equation.

Similar to the above mentioned Lyapunov equation, the analytical solutions can also be found with the help of Kronecker products

$$(\boldsymbol{A}\otimes \boldsymbol{I}_m + \boldsymbol{I}_n \otimes \boldsymbol{B}^{\mathrm{T}})\boldsymbol{x} = \boldsymbol{c}. \tag{4-4-14}$$

If $(\boldsymbol{A}\otimes \boldsymbol{I}_m + \boldsymbol{I}_n \otimes \boldsymbol{B}^{\mathrm{T}})$ is nonsingular, the Sylvester equation has a unique solution.

Combining the above mentioned analytical solution algorithms, a symbolic-based MATLAB function lyapsym.m for Sylvester equations can be written. The listing of the function is

```
function X=lyapsym(A,B,C)
if nargin==2, C=B; B=A.'; end
[nr,nc]=size(C); A0=kron(A,eye(nc))+kron(eye(nr),B.');
try
   C1=C.'; x0=-inv(A0)*C1(:); X=reshape(x0,nc,nr).';
catch, error('singular matrix found.'), end
```

Considering the discrete Lyapunov equation shown in (4-4-12). If one multiplies both sides of the equation by $(\boldsymbol{A}^{\mathrm{T}})^{-1}$, then, the original discrete Lyapunov equation can be rewritten as

$$\boldsymbol{A}\boldsymbol{X}+\boldsymbol{X}[-(\boldsymbol{A}^{\mathrm{T}})^{-1}]=-\boldsymbol{Q}(\boldsymbol{A}^{\mathrm{T}})^{-1}.$$

Let $\boldsymbol{B}=-(\boldsymbol{A}^{\mathrm{T}})^{-1}$, $\boldsymbol{C}=\boldsymbol{Q}(\boldsymbol{A}^{\mathrm{T}})^{-1}$, the equation can be transformed into a Sylvester equation defined in (4-4-13). Thus, the new `lyapsym()` function can be used to solve such equations. The syntaxes of the function now are

```
X=lyapsym(A,C)                  % continuous Lyapunov equation
X=lyapsym(A,-inv(B),Q*inv(B))   % Stein equation
X=lyapsym(A,-inv(A'),Q*inv(A')) % discrete Lyapunov equation
X=lyapsym(A,B,C)                % Sylvester equation
```

In recent versions of MATLAB, `sylvester()` function is provided for finding numerical solutions of Sylvester equations, using the Kronecker product based algorithm. The syntax of the function is also `X=sylvester(A,B,C)`.

Example 4.58 Solve the following Sylvester equation

$$\begin{bmatrix}8&1&6\\3&5&7\\4&9&2\end{bmatrix}\boldsymbol{X}+\boldsymbol{X}\begin{bmatrix}16&4&1\\9&3&1\\4&2&1\end{bmatrix}=\begin{bmatrix}1&2&3\\4&5&6\\7&8&0\end{bmatrix}.$$

Solution *The original equation can easily be solved by calling* `lyap()` *function*

```
>> A=[8,1,6; 3,5,7; 4,9,2]; B=[16,4,1; 9,3,1; 4,2,1];
   C=-[1,2,3; 4,5,6; 7,8,0]; X=lyap(A,B,C), norm(A*X+X*B+C)
```

and the solution with error norm of 1.0436×10^{-14} *can be obtained*

$$\boldsymbol{X}=\begin{bmatrix}0.074872&0.089913&-0.43292\\0.0080716&0.48144&-0.21603\\0.019577&0.18264&1.1579\end{bmatrix}.$$

If one wants to have the analytical solutions, the following statements can be given

```
>> x=lyapsym(sym(A),B,C), norm(A*x+x*B+C) % find analytical solution
```

the solution now is

$$\boldsymbol{x}=\begin{bmatrix}1349214/18020305&648107/7208122&-15602701/36040610\\290907/36040610&3470291/7208122&-3892997/18020305\\70557/3604061&1316519/7208122&8346439/7208122\end{bmatrix},$$

and it can be seen that there is no longer any error in the solution.

Example 4.59 Consider again the discrete Lyapunov equation given in Example 4.57, now find the analytical solution of the equation.

Solution *The analytical solutions to the original equation can be found*

```
>> A=[8,1,6; 3,5,7; 4,9,2]; Q=[16,4,1; 9,3,1; 4,2,1];
   x=lyapsym(sym(A),-inv(A.'),Q*inv(A.')), norm(A*x*A.'-x+Q)
```

the solution can then be found, and it is with zero error.

$$\boldsymbol{x}=\begin{bmatrix}-22912341/139078240 & 48086039/695391200 & -11672009/695391200\\ 36746487/695391200 & -20712201/695391200 & -4279561/695391200\\ -70914857/695391200 & 31264087/695391200 & -4247541/139078240\end{bmatrix}.$$

Example 4.60 Solve the Sylvester equation with

$$\boldsymbol{A}=\begin{bmatrix}8&1&6\\3&5&7\\4&9&2\end{bmatrix},\quad \boldsymbol{B}=\begin{bmatrix}2&3\\4&5\end{bmatrix},\quad \boldsymbol{C}=\begin{bmatrix}1&2\\3&4\\5&6\end{bmatrix}.$$

Solution *In Sylvester equations, the matrix* $\boldsymbol{C}$ *is not necessarily square. The analytical solutions to the equation can be found using the new* `lyapsym()` *function*

```
>> A=[8,1,6; 3,5,7; 4,9,2]; B=[2,3; 4,5]; C=-[1,2; 3,4; 5,6];
   X=lyapsym(sym(A),B,C), norm(A*X+X*B+C)
```

and the solution with zero error is found

$$\boldsymbol{X}=\begin{bmatrix}-2853/14186 & -11441/56744\\ -557/14186 & -8817/56744\\ 9119/14186 & 50879/56744\end{bmatrix}.$$

If the (2,1)th parameter in $\boldsymbol{B}$ *matrix is changed to a free variable* a*, the Sylvester equation can still be solved with*

```
>> syms a real; B=sym(B); B(2,1)=a;
   X=simplify(lyapsym(A,B,C)), norm(A*X+X*B+C) % solve and validate
```

with the validated solution

$$\boldsymbol{X}=\begin{bmatrix}\dfrac{6\left(3a^3+155a^2-2620a+200\right)}{27a^3-3672a^2+69300a+6800} & -\dfrac{513a^2-10716a+80420}{27a^3-3672a^2+69300a+6800}\\ \dfrac{4\left(9a^3-315a^2+314a+980\right)}{27a^3-3672a^2+69300a+6800} & -\dfrac{3\left(201a^2-7060a+36780\right)}{27a^3-3672a^2+69300a+6800}\\ \dfrac{2\left(27a^3-1869a^2+25472a-760\right)}{27a^3-3672a^2+69300a+6800} & \dfrac{-477a^2+4212a+194300}{27a^3-3672a^2+69300a+6800}\end{bmatrix}.$$

4.4.4 Solutions of Diophantine equations

The equations discussed so far are matrix equations. Now consider a polynomial equation given by

$$A(s)X(s)+B(s)Y(s)=C(s), \tag{4-4-15}$$

where, $A(s)$, $B(s)$ and $C(s)$ are known polynomials given by

$$\begin{aligned}A(s)&=a_1s^n+a_2s^{n-1}+a_3s^{n-2}+\cdots+a_ns+a_{n+1},\\ B(s)&=b_1s^m+b_2s^{m-1}+b_3s^{m-2}+\cdots+b_ms+b_{m+1},\\ C(s)&=c_1s^k+c_2s^{k-1}+c_3s^{k-2}+\cdots+c_ks+c_{k+1}.\end{aligned} \tag{4-4-16}$$

Such a polynomial equation is referred to as a *Diophantine equation.* Without losing generality, assume that $m \leqslant n$. The orders of the unknown polynomials $X(s)$ and $Y(s)$ are respectively $m-1$ and $n-1$, such that

$$\begin{aligned} X(s) &= x_1 s^{m-1} + x_2 s^{m-2} + x_3 s^{m-3} + \cdots + x_{m-1}s + x_m, \\ Y(s) &= y_1 s^{n-1} + y_2 s^{n-2} + y_3 s^{n-3} + \cdots + y_{n-1}s + y_n. \end{aligned} \tag{4-4-17}$$

The matrix form of the Diophantine equation can be written as

$$\underbrace{\begin{bmatrix} a_1 & 0 & \cdots & 0 \\ a_2 & a_1 & \ddots & 0 \\ a_3 & a_2 & \ddots & 0 \\ \vdots & \vdots & \ddots & a_1 \\ a_{n+1} & a_n & \ddots & a_2 \\ 0 & a_{n+1} & \ddots & a_3 \\ \vdots & \vdots & \ddots & \vdots \\ 0 & 0 & \cdots & a_{n+1} \end{bmatrix}}_{m \text{ columns}} \underbrace{\begin{matrix} b_1 & 0 & \cdots & 0 \\ b_2 & b_1 & \ddots & 0 \\ b_3 & b_2 & \ddots & 0 \\ \vdots & \vdots & \ddots & b_1 \\ \cdot & \cdot & \ddots & b_2 \\ \cdot & \cdot & \ddots & b_3 \\ \vdots & \vdots & \ddots & \vdots \\ 0 & 0 & \cdots & b_{m+1} \end{matrix}}_{n \text{ columns}} \begin{bmatrix} x_1 \\ x_2 \\ \vdots \\ x_m \\ y_1 \\ y_2 \\ \vdots \\ y_n \end{bmatrix} = \begin{bmatrix} 0 \\ \vdots \\ 0 \\ c_1 \\ c_2 \\ \vdots \\ c_{k+1} \end{bmatrix}. \tag{4-4-18}$$

The coefficient matrix is the transpose of Sylvester matrix. It can be shown that, if two polynomials $A(s)$ and $B(s)$ are coprime, the Sylvester matrix is nonsingular. Therefore, the equation has a unique solution. To check whether two polynomials are coprime or not, the simplest way is to find the greatest common divisor of the two polynomials and see whether it includes s. If no polynomial is found in greatest common divisor, the two polynomials are coprime.

A MATLAB function can be written to construct a Sylvester matrix

```
function S=sylv_mat(A,B)
n=length(B)-1; m=length(A)-1; S=[];
A1=[A(:); zeros(n-1,1)]; B1=[B(:); zeros(m-1,1)];
for i=1:n, S=[S A1]; A1=[0; A1(1:end-1)]; end
for i=1:m, S=[S B1]; B1=[0; B1(1:end-1)]; end; S=S.';
```

Based on such a function, a MATLAB function `diophantine()` for solving the Diophantine equation can be written as follows

```
function [X,Y]=diophantine(A,B,C,x)
A1=polycoef(A,x); B1=polycoef(B,x); C1=polycoef(C,x);
n=length(B1)-1; m=length(A1)-1; S=sylv_mat(A1,B1);
C2=zeros(n+m,1); C2(end-length(C1)+1:end)=C1(:); x0=inv(S.')*C2;
X=poly2sym(x0(1:n),x); Y=poly2sym(x0(n+1:end),x);
```

Example 4.61 Please solve the Diophantine equation with the polynomials

$$A(s) = s^4 - \frac{27s^3}{10} + \frac{11s^2}{4} - \frac{1249s}{1000} + \frac{53}{250},$$
$$B(s) = 3s^2 - \frac{6s}{5} + \frac{51}{25}, \quad C(s) = 2s^2 + \frac{3s}{5} - \frac{9}{25}.$$

Solution *The following statements can be used to solve the equation*

```
>> syms s; A=s^4-27*s^3/10+11*s^2/4-1249*s/1000+53/250;
   B=3*s^2-6*s/5+51/25; C=2*s^2+3*s/5-9/25;
   [X,Y]=diophantine(A,B,C,s), simplify(A*X+B*Y-C)
```

The solutions to the Diophantine equation can be obtained as follows. If the results are substituted back to the original equation, it can be seen that the error is zero, which validated the results.

$$X(s)=\frac{4280\,s}{4453}+\frac{9480}{4453},\quad Y(s)=-\frac{4280s^3}{13359}+\frac{364s^2}{13359}+\frac{16882s}{13359}-\frac{1771}{4453}.$$

4.4.5 Solutions to Riccati equations

A Riccati equation is a quadratic matrix equation which can be written as

$$\boldsymbol{A}^{\mathrm{T}}\boldsymbol{X}+\boldsymbol{X}\boldsymbol{A}-\boldsymbol{X}\boldsymbol{B}\boldsymbol{X}+\boldsymbol{C}=\boldsymbol{0}. \tag{4-4-19}$$

Due to the quadratic term of matrix $\boldsymbol{X}$, solving such an equation is more difficult than Lyapunov type equations. There are no analytical solutions to such equations. A numerical-based function `are()` is provided in the Control Systems Toolbox, such that $\boldsymbol{X}=$ `are(`$\boldsymbol{A}$`,`$\boldsymbol{B}$`,`$\boldsymbol{C}$`)`.

Example 4.62 Consider the Riccati equation in (4-4-19), where

$$\boldsymbol{A}=\begin{bmatrix}-2&1&-3\\-1&0&-2\\0&-1&-2\end{bmatrix},\ \boldsymbol{B}=\begin{bmatrix}2&2&-2\\-1&5&-2\\-1&1&2\end{bmatrix},\ \boldsymbol{C}=\begin{bmatrix}5&-4&4\\1&0&4\\1&-1&5\end{bmatrix}.$$

Find the numerical solution, and then, validate the results.

Solution *The following commands can be used in solving the Riccati equation*

```
>> A=[-2,1,-3; -1,0,-2; 0,-1,-2]; B=[2,2,-2; -1 5 -2; -1 1 2];
   C=[5 -4 4; 1 0 4; 1 -1 5]; X=are(A,B,C), norm(A.'*X+X*A-X*B*X+C)
```

and the numerical solution is found

$$\boldsymbol{X}=\begin{bmatrix}0.98739491&-0.7983277&0.418869\\0.57740565&-0.13079234&0.57754777\\-0.284045&-0.073036978&0.69241149\end{bmatrix}.$$

It can also be found that the norm of the error matrix is 1.8605×10^{-14}*, which is small enough. More solutions can be given in Chapter 6.*

Of course, there are limitations in solving Riccati equations, since `are()` function can only return one solution. Are there any other solutions? If there are, how many are there and how can they be found? Are there any complex solutions? How to solve different variations of Riccati equations such as

$$\boldsymbol{A}\boldsymbol{X}+\boldsymbol{X}\boldsymbol{D}-\boldsymbol{X}\boldsymbol{B}\boldsymbol{X}+\boldsymbol{C}=\boldsymbol{0},\quad \boldsymbol{A}\boldsymbol{X}+\boldsymbol{X}\boldsymbol{D}-\boldsymbol{X}\boldsymbol{B}\boldsymbol{X}^{\mathrm{T}}+\boldsymbol{C}=\boldsymbol{0}. \tag{4-4-20}$$

The current version of `are()` function fail to find solutions to all these problems. In Chapter 6, methods will be presented in finding all the possible solutions to these problems.

4.5 Nonlinear Functions and Matrix Function Evaluations

Two kinds of functions for matrices are provided in MATLAB. One is for element-by-element calculation of matrices, while the other is for matrix functions. In this section, these two types of functions are explored.

4.5.1 Element-by-element computations

A large number of functions have been provided in MATLAB to carry out element-by-element nonlinear function computation. For instance, the `sin(`x`)` function used in Chapter 2 calculates sinusoidal to each element of the matrix x. The element-by-element computation is useful especially in graphics. The "dot operation" is another example of element-by-element operation. The commonly used element-by-element nonlinear functions are summarized in Table 4.2. The syntax for this kind of operation is very simple,

B= *funname*(A), for instance B=`sin(`A`)` .

TABLE 4.2: Commonly used element-by-element nonlinear functions.

Function name	Meaning	Function name	Meaning
`abs()`	absolute value	`asin(),acos(),atan()`	inverse triangular functions
`sqrt()`	square roots	`log(),log10()`	logarithmic function
`exp()`	exponential function	`real(),imag(),conj()`	real, imaginary or conjugates
`sin(),cos(),tan()`	triangular functions	`round(),floor(),ceil()`	integer functions

Example 4.63 Consider the matrix A in Example 4.10. Call the functions `exp()` and `sin()` to complete nonlinear functions computations.

Solution *The following functions can be used, and the element-by-element exponential and sinusoidal functions can be found*

```
>> A=[16,2,3,13; 5,11,10,8; 9,7,6,12; 4,14,15,1]; exp(A), sin(A)
```

The results obtained are

$$\texttt{exp}(\boldsymbol{A})=\begin{bmatrix} 8.8861\times10^6 & 7.3891 & 20.086 & 4.4241\times10^5\\ 148.41 & 59874 & 22026 & 2981\\ 8103.1 & 1096.6 & 403.43 & 1.6275\times10^5\\ 54.598 & 1.2026\times10^6 & 3.269\times10^6 & 2.7183\end{bmatrix},$$

$$\texttt{sin}(\boldsymbol{A})=\begin{bmatrix} -0.2879 & 0.9093 & 0.14112 & 0.42017\\ -0.95892 & -0.99999 & -0.54402 & 0.98936\\ 0.41212 & 0.65699 & -0.27942 & -0.53657\\ -0.7568 & 0.99061 & 0.65029 & 0.84147\end{bmatrix}.$$

4.5.2 Computations of matrix exponentials

Apart from element-by-element nonlinear computation of matrices, sometimes the nonlinear functions on the whole matrices, i.e., matrix functions, are expected. For instance, if one wants to have the exponential function to a matrix, special algorithms should be required.

The exponential function of matrix $\boldsymbol{A}$ is defined as an infinite series

$$\mathrm{e}^{\boldsymbol{A}} = \sum_{i=0}^{\infty} \frac{1}{i!}\boldsymbol{A}^i = \boldsymbol{I} + \boldsymbol{A} + \frac{1}{2}\boldsymbol{A}^2 + \frac{1}{3!}\boldsymbol{A}^3 + \cdots + \frac{1}{m!}\boldsymbol{A}^m + \cdots. \tag{4-5-1}$$

Nineteen different numerical algorithms have been summarized in Reference [7], and each algorithm has its own advantages. The built-in MATLAB function `expm()` can be used with $\boldsymbol{E}$=`expm(`$\boldsymbol{A}$`)`. Taylor series approximation can also be used to evaluate the exponential function from the above definition with a truncation algorithm.

Example 4.64 Consider a matrix

$$\boldsymbol{A} = \begin{bmatrix} -2 & 1 & 0 & & \\ 0 & -2 & 1 & & \\ 0 & 0 & -2 & & \\ & & & -5 & 1 \\ & & & 0 & -5 \end{bmatrix}.$$

Compute the exponential function $\mathrm{e}^{\boldsymbol{A}}$, then, calculate the logarithmic function and see whether the original matrix $\boldsymbol{A}$ can be restored. Also, find analytically $\mathrm{e}^{\boldsymbol{A}t}$.

Solution *The following MATLAB statements can be used to calculate the exponential matrix* $\boldsymbol{B} = \mathrm{e}^{\boldsymbol{A}}$*, then, the use of* `logm()` *function can be used to calculate the logarithmic function to restore matrix* $\boldsymbol{A}$

```
>> A=[[-2 1 0; 0 -2 1; 0 0 -2], zeros(3,2); zeros(2,3) [-5 1; 0 -5]];
   B=expm(A), C=logm(B), norm(C-A) % numerical solution and validation
```

Thus, one may find that

$$\boldsymbol{B} = \begin{bmatrix} 0.13534 & 0.13534 & 0.067668 & 0 & 0 \\ 0 & 0.13534 & 0.13534 & 0 & 0 \\ 0 & 0 & 0.13534 & 0 & 0 \\ 0 & 0 & 0 & 0.0067379 & 0.0067379 \\ 0 & 0 & 0 & 0 & 0.0067379 \end{bmatrix},$$

and $\boldsymbol{C}$ *is almost equal to* $\boldsymbol{A}$*, with the restoration error* $||\boldsymbol{C}-\boldsymbol{A}|| = 3.9014\times10^{-15}$. *It can be seen that the accuracy of such algorithms is very high.*

The `expm()` *function is also applicable to symbolic matrices. For instance, the exponential function* $\mathrm{e}^{\boldsymbol{A}t}$ *can also be obtained by the direct use of such a function. This kind of problem cannot be solved using the numerical methods of course.*

```
>> syms t; expm(A*t) % symbolic computation
```

So the results can be written as

$$\mathrm{e}^{\boldsymbol{A}t} = \begin{bmatrix} \mathrm{e}^{-2t} & t\mathrm{e}^{-2t} & t^2\mathrm{e}^{-2t}/2 & 0 & 0 \\ 0 & \mathrm{e}^{-2t} & t\mathrm{e}^{-2t} & 0 & 0 \\ 0 & 0 & \mathrm{e}^{-2t} & 0 & 0 \\ 0 & 0 & 0 & \mathrm{e}^{-5t} & t\mathrm{e}^{-5t} \\ 0 & 0 & 0 & 0 & \mathrm{e}^{-5t} \end{bmatrix}.$$

In fact, since matrix $\boldsymbol{A}$ *is a block Jordan matrix, the results can easily be written out, even without the help of computers.*

Example 4.65 For a given matrix $\boldsymbol{A}$, calculate $\mathrm{e}^{\boldsymbol{A}t}$.

$$\boldsymbol{A}=\begin{bmatrix}-3 & -1 & -1\\ 0 & -3 & -1\\ 1 & 2 & 0\end{bmatrix}.$$

Solution *The exponential function of* $\boldsymbol{A}$ *can be obtained directly using the* `expm()` *function such that*

```
>> syms t; A=[-3,-1,-1; 0,-3,-1; 1,2,0]; F=simplify(expm(A*t))
```

and the result is

$$\boldsymbol{F}=\begin{bmatrix}-\mathrm{e}^{-2t}(-1+t) & -t\mathrm{e}^{-2t} & -t\mathrm{e}^{-2t}\\ -t^2\mathrm{e}^{-2t}/2 & -\mathrm{e}^{-2t}(-1+t+t^2/2) & -t\mathrm{e}^{-2t}(2+t/2)\\ t\mathrm{e}^{-2t}/2 & t\mathrm{e}^{-2t}(2+t/2) & \mathrm{e}^{-2t}(1+2t+t^2/2)\end{bmatrix}.$$

Now consider a Jordan transformation technique for solving such a problem. First, Jordan transformation should be obtained such that

```
>> [V,J]=jordan(A)   % Jordan transformation
```

the transformation can then be obtained

$$\boldsymbol{V}=\begin{bmatrix}0 & -1 & 1\\ -1 & 0 & 0\\ 1 & 1 & 0\end{bmatrix},\quad \boldsymbol{J}=\begin{bmatrix}-2 & 1 & 0\\ 0 & -2 & 1\\ 0 & 0 & -2\end{bmatrix},$$

where the matrix $\boldsymbol{V}$ *obtained is no longer singular.*

Since the exponential of a Jordan matrix is known, it can be directly entered, and then, the exponential function of the original matrix can be obtained from

```
>> J1=[exp(-2*t), t*exp(-2*t), 1/2*t^2*exp(-2*t);
            0,    exp(-2*t),         t*exp(-2*t);
            0,            0,           exp(-2*t)];
   A1=simplify(V*J1*inv(V))
```

and the results are exactly the same as the one obtained above.

It should be noted that the use of Jordan matrix to solve the exponential matrix function is not the best method, since it can be calculated directly otherwise. The use of the Jordan function here indicates that this method can be extended to other applications. And in later subsections, the use of Jordan matrices will be explored thoroughly.

4.5.3 Trigonometric functions of matrices

There are no MATLAB functions for trigonometric matrix operations. One may use the universal function `funm()` instead to find the numerical solution to such problems. This function is intended to be used for any nonlinear matrix functions, such that $\boldsymbol{A}_1$ = `funm(`$\boldsymbol{A}$`,` *fun*`)`, where the name of the function should be quoted using single quotation marks. For instance, if one wants to evaluate $\sin\boldsymbol{A}$, the statement $\boldsymbol{B}$ = `funm(`$\boldsymbol{A}$`,'sin')` should be provided. Only in the most recent versions such as MATLAB R2014b, the functions like `funm(`$\boldsymbol{A}$`*`t`,'sin')` can be used directly.

It should be noted that in earlier versions of MATLAB, since the computation used the eigenvector-based algorithm, erroneous results may be obtained if $\boldsymbol{A}$ has repeated eigenvalues.

Example 4.66 Consider again the matrix given in Example 4.64. Find the sinusoidal function of the matrix.

Solution *The function* `funm()` *can be used to calculate the sinusoidal matrix.*

```
>> A=[[-2 1 0; 0 -2 1; 0 0 -2], zeros(3,2); zeros(2,3) [-5 1; 0 -5]];
   funm(A,'sin')
```

Thus, the result can be obtained

$$\sin \boldsymbol{A} = \begin{bmatrix} -0.9093 & -0.41615 & 0.45465 & 0 & 0 \\ 0 & -0.9093 & -0.41615 & 0 & 0 \\ 0 & 0 & -0.9093 & 0 & 0 \\ 0 & 0 & 0 & 0.95892 & 0.28366 \\ 0 & 0 & 0 & 0 & 0.95892 \end{bmatrix}.$$

In fact, some certain matrix functions can be evaluated by the use of Taylor series expansion technique. For instance, sinusoidal function can be evaluated from

$$\sin \boldsymbol{A} = \sum_{i=0}^{\infty} (-1)^i \frac{\boldsymbol{A}^{2i+1}}{(2i+1)!} = \boldsymbol{A} - \frac{1}{3!}\boldsymbol{A}^3 + \frac{1}{5!}\boldsymbol{A}^5 + \cdots. \tag{4-5-2}$$

So the MATLAB function can be written to compute the sinusoidal matrix

```
function E=sinm1(A)
E=zeros(size(A)); F=A; k=1;
while norm(E+F-E,1)>0, E=E+F; F=-A^2*F/((k+2)*(k+1)); k=k+2; end
```

Example 4.67 Solve again the above problem with the new `sinm1()` function.

Solution *From the above explanation, it can be seen that the seemingly complicated sinusoidal matrix functions can easily be solved using a few statements in MATLAB. With such a function,* $\sin \boldsymbol{A}$ *can easily be obtained*

```
>> E=sinm1(A)
```

It can be measured that 39 terms have been added in the above computation, and the results are exactly the same as the one obtained from the previous example.

For sinusoidal and cosine functions, if a scalar a is given, the trigonometric functions satisfy the well-known Euler formula, such that $\mathrm{e}^{\mathrm{j}a} = \cos a + \mathrm{j}\sin a$ and $\mathrm{e}^{-\mathrm{j}a} = \cos a - \mathrm{j}\sin a$. From them, one can find out immediately that

$$\sin a = \frac{1}{\mathrm{j}2}(\mathrm{e}^{\mathrm{j}a} - \mathrm{e}^{-\mathrm{j}a}), \quad \cos a = \frac{1}{2}(\mathrm{e}^{\mathrm{j}a} + \mathrm{e}^{-\mathrm{j}a}), \tag{4-5-3}$$

and the new formula holds also if the scalar a is replaced by a matrix $\boldsymbol{A}$. Some examples will be given to show such computations.

Example 4.68 Consider also the matrix in Example 4.64, evaluate $\sin \boldsymbol{A}$.

Solution *The existing* `expm()` *function can be used to evaluate the sinusoidal function*

```
>> A=[[-2 1 0; 0 -2 1; 0 0 -2], zeros(3,2); zeros(2,3) [-5 1; 0 -5]];
   j=sqrt(-1); A1=(expm(A*j)-expm(-A*j))/(2*j)
```

which yields

$$\sin \boldsymbol{A} = \begin{bmatrix} -0.9093 & -0.41615 & 0.45465 & 0 & 0 \\ 0 & -0.9093 & -0.41615 & 0 & 0 \\ 0 & 0 & -0.9093 & 0 & 0 \\ 0 & 0 & 0 & 0.95892 & 0.28366 \\ 0 & 0 & 0 & 0 & 0.95892 \end{bmatrix},$$

which agrees well with the results in Example 4.67, which means that the algorithm is correct.

Example 4.69 Consider a given matrix

$$\boldsymbol{A} = \begin{bmatrix} -7 & 2 & 0 & -1 \\ 1 & -4 & 2 & 1 \\ 2 & -1 & -6 & -1 \\ -1 & -1 & 0 & -4 \end{bmatrix}.$$

It is known that the matrix contains repeated eigenvalues. Compute matrix functions $\sin \boldsymbol{A}t$ and $\cos \boldsymbol{A}t$. Also, please find $\tan\boldsymbol{A}t$.

Solution *From* (4-5-3), *the following statements can be obtained for the sinusoidal and cosine functions.*

```
>> A=sym([-7,2,0,-1; 1,-4,2,1; 2,-1,-6,-1; -1,-1,0,-4]);
   syms t; j=sym(sqrt(-1)); % declare symbolic variables
   A1=simplify((expm(A*j*t)-expm(-A*j*t))/(2*j)) % sinusoidal function
   A2=simplify((expm(A*j*t)+expm(-A*j*t))/2)      % cosine function
```

The results obtained are

$$\sin \boldsymbol{A}t = \begin{bmatrix} -2/9\sin 3t+(t^2-7/9)\sin 6t-5/3t\cos 6t & -1/3\sin 3t+1/3\sin 6t+t\cos 6t & -2/9\sin 3t+(2/9+t^2)\sin 6t-2/3t\cos 6t & 1/9\sin 3t+(-1/9+t^2)\sin 6t-2/3t\cos 6t \\ -2/9\sin 3t+(t^2+2/9)\sin 6t+1/3t\cos 6t & -1/3\sin 3t-2/3\sin 6t+t\cos 6t & -2/9\sin 3t+(2/9+t^2)\sin 6t+4/3t\cos 6t & 1/9\sin 3t+(-1/9+t^2)\sin 6t+4/3t\cos 6t \\ -2/9\sin 3t+(-2t^2+2/9)\sin 6t+4/3t\cos 6t & -1/3\sin 3t+1/3\sin 6t-2t\cos 6t & -2/9\sin 3t-(7/9+2t^2)\sin 6t-2/3t\cos 6t & 1/9\sin 3t-(1/9+2t^2)\sin 6t-2/3t\cos 6t \\ 4/9\sin 3t+(t^2-4/9)\sin 6t+1/3t\cos 6t & 2/3\sin 3t-2/3\sin 6t+t\cos 6t & 4/9\sin 3t+(-4/9+t^2)\sin 6t+4/3t\cos 6t & -2/9\sin 3t+(-7/9+t^2)\sin 6t+4/3t\cos 6t \end{bmatrix}$$

$$\cos \boldsymbol{A}t = \begin{bmatrix} 2/9\cos 3t+(-t^2+7/9)\cos 6t-5/3t\sin 6t & 1/3\cos 3t-1/3\cos 6t+t\sin 6t & 2/9\cos 3t-(2/9+t^2)\cos 6t-2/3t\sin 6t & -1/9\cos 3t+(1/9-t^2)\cos 6t-2/3t\sin 6t \\ 2/9\cos 3t-(t^2+2/9)\cos 6t+1/3t\sin 6t & 1/3\cos 3t+2/3\cos 6t+t\sin 6t & 2/9\cos 3t-(2/9+t^2)\cos 6t+4/3t\sin 6t & -1/9\cos 3t+(1/9-t^2)\cos 6t+4/3t\sin 6t \\ 2/9\cos 3t+(2t^2-2/9)\cos 6t+4/3t\sin 6t & 1/3\cos 3t-1/3\cos 6t-2t\sin 6t & 2/9\cos 3t+(7/9+2t^2)\cos 6t-2/3t\sin 6t & -1/9\cos 3t+(1/9+2t^2)\cos 6t-2/3t\sin 6t \\ -4/9\cos 3t+(-t^2+4/9)\cos 6t+1/3t\sin 6t & -2/3\cos 3t+2/3\cos 6t+t\sin 6t & -4/9\cos 3t+(4/9-t^2)\cos 6t+4/3t\sin 6t & 2/9\cos 3t+(7/9-t^2)\cos 6t+4/3t\sin 6t \end{bmatrix}.$$

The above results were actually obtained in old versions of Symbolic Math Toolbox, and in new versions, far more complicated results are obtained and cannot be simplified. In later examples, an alternative approach is used.

Since $\sin \boldsymbol{A}t$ *and* $\cos \boldsymbol{A}t$ *are obtained, the matrix function* $\tan\boldsymbol{A}t$ *can be evaluated with* $\boldsymbol{F} = \boldsymbol{A}_1$`*inv(`$\boldsymbol{A}_2$`)`, *however, it might too complicated to present here.*

In recent versions of MATLAB, the matrix functions $\sin \boldsymbol{A}t$, $\cos \boldsymbol{A}t$ and $\tan\boldsymbol{A}t$ can be obtained directly with `funm(A*t,'sin')`, `funm(A*t,'cos')` and `funm(A*t,'tan')`, respectively.

4.5.4 General matrix functions

In this section, a Jordan matrix-based algorithm [8] is presented and implicated for the evaluation of an arbitrary function $\varphi(\boldsymbol{A})$. Also, the matrix functions containing the time variable t can be evaluated. Before presenting such an algorithm, let us first observe the behaviors of a *nilpotent matrix*.

Example 4.70 Please observe the behavior of a nilpotent matrix.

Solution *Observe the power of the nilpotent matrix in a loop*

```
>> H=diag([1 1 1],1), for i=2:4, H^i, end % observe the positions of 1's
```

and the following power matrices can be observed, and it can be seen that from $\boldsymbol{H}^4$ *on, all the subsequent powers are zero matrices*

$$\boldsymbol{H}=\begin{bmatrix}0&1&0&0\\0&0&1&0\\0&0&0&1\\0&0&0&0\end{bmatrix},\ \boldsymbol{H}^2=\begin{bmatrix}0&0&1&0\\0&0&0&1\\0&0&0&0\\0&0&0&0\end{bmatrix},\ \boldsymbol{H}^3=\begin{bmatrix}0&0&0&1\\0&0&0&0\\0&0&0&0\\0&0&0&0\end{bmatrix},\ \boldsymbol{H}^4=\begin{bmatrix}0&0&0&0\\0&0&0&0\\0&0&0&0\\0&0&0&0\end{bmatrix}.$$

With Jordan decomposition, matrix $\boldsymbol{A}$ can be transformed such that

$$\boldsymbol{A}=\boldsymbol{V}\begin{bmatrix}\boldsymbol{J}_1&&&\\&\boldsymbol{J}_2&&\\&&\ddots&\\&&&\boldsymbol{J}_m\end{bmatrix}\boldsymbol{V}^{-1}. \tag{4-5-4}$$

Using Jordan matrix decomposition, the arbitrary function $\varphi(\boldsymbol{A})$ can be evaluated from

$$\varphi(\boldsymbol{A})=\boldsymbol{V}\begin{bmatrix}\varphi(\boldsymbol{J}_1)&&&\\&\varphi(\boldsymbol{J}_2)&&\\&&\ddots&\\&&&\varphi(\boldsymbol{J}_m)\end{bmatrix}\boldsymbol{V}^{-1}. \tag{4-5-5}$$

To solve such a problem, one can first write an $m_i\times m_i$ Jordan block $\boldsymbol{J}_i$ as $\boldsymbol{J}_i=\lambda_i\boldsymbol{I}+\boldsymbol{H}_{m_i}$, where λ_i is a repeated eigenvalue of multiplicity m_i, and $\boldsymbol{H}_{m_i}$ is a nilpotent matrix, i.e., when $k\geqslant m_i$, $\boldsymbol{H}_{m_i}^k\equiv\boldsymbol{0}$. It can be shown that the matrix function $\varphi(\boldsymbol{J}_i)$ can be obtained as follows:

$$\varphi(\boldsymbol{J}_i)=\varphi(\lambda_i)\boldsymbol{I}_{m_i}+\varphi'(\lambda_i)\boldsymbol{H}_{m_i}+\cdots+\frac{\varphi^{(m_i-1)}(\lambda_i)}{(m_i-1)!}\boldsymbol{H}_{m_i}^{m_i-1}. \tag{4-5-6}$$

An algorithm is designed to compute arbitrary matrix function $\varphi(\boldsymbol{A})$.

Require: Matrix $\boldsymbol{A}$, prototype expression $\varphi(x)$, independent variable x
For $\boldsymbol{A}$, find Jordan transform matrices $\boldsymbol{V}$, $\boldsymbol{J}$, and mark m Jordan blocks,
for $i=1$ To m **do**

Extract $\boldsymbol{J}_i$ from $\boldsymbol{J}$, compute $\varphi(\boldsymbol{J}_i)$ from (4-5-6), compose $\varphi(\boldsymbol{J})$
end for
Compute the matrix function $\varphi(\boldsymbol{A})$ from (4-5-5).

Based on the above algorithm, the following new MATLAB `funmsym()` function can be written. This function can be used to find the analytical solution of any matrix function. The listing of the function is

```
function F=funmsym(A,fun,x)
[V,T]=jordan(A); vec=diag(T); v1=[0,diag(T,1)',0];
v2=find(v1==0); lam=vec(v2(1:end-1)); m=length(lam);
for i=1:m,
   k=v2(i):v2(i+1)-1; J1=T(k,k); F(k,k)=funJ(J1,fun,x);
end
F=V*F*inv(V);
function fJ=funJ(J,fun,x)  % sub function to compute φ(J_i)
lam=J(1,1); f1=fun; fJ=subs(fun,x,lam)*eye(size(J));
H=diag(diag(J,1),1); H1=H; % generate nilpotent matrix H_1
for i=2:length(J)          % compute φ(J_i) in a loop
   f1=diff(f1,x); a1=subs(f1,x,lam); fJ=fJ+a1*H1; H1=H1*H/i;
end
```

The syntax of the function is $\boldsymbol{A}_1=$ `funmsym(`$\boldsymbol{A}$`,`*funx*`,`x`)`, where x is a symbolic variable, *funx* is the prototype function $\varphi(\cdot)$ of variable x. For instance, if one wants $\mathrm{e}^{\boldsymbol{A}}$, function *funx* should be defined as `exp(x)`. In fact, the variable *funx* can be assigned to an arbitrarily complicated function, i.e., `exp(x*t)` means that one wants to have $\mathrm{e}^{\boldsymbol{A}t}$, and t could also be a symbolic variable. Composite functions such as `exp(`x`*cos(`x`*t))` can also be used here to indicate that one wants to calculate $\varphi(\boldsymbol{A})=\mathrm{e}^{\boldsymbol{A}\cos\boldsymbol{A}t}$.

Example 4.71 Solve the matrix functions in Example 4.69 with the new function.

Solution *It has been indicated that the matrix functions* $\sin\boldsymbol{A}t$, $\cos\boldsymbol{A}t$ *and* $\tan\boldsymbol{A}t$ *obtained with new Symbolic Math Toolbox are complicated, while in old versions, the results are much simpler. The simplified results can be obtained with the following commands. Thus, this method is recommended.*

```
>> A=sym([-7,2,0,-1; 1,-4,2,1; 2,-1,-6,-1; -1,-1,0,-4]);
   syms t x; C=simplify(funmsym(A,cos(x*t),x))
   S=simplify(funmsym(A,sin(x*t),x)), T=simplify(funmsym(A,tan(x*t),x))
```

Example 4.72 For matrix

$$\boldsymbol{A}=\begin{bmatrix}-7 & 2 & 0 & -1\\ 1 & -4 & 2 & 1\\ 2 & -1 & -6 & -1\\ -1 & -1 & 0 & -4\end{bmatrix},$$

compute the matrix functions $F_1=\mathrm{e}^{\boldsymbol{A}\cos\boldsymbol{A}t}$ and $F_2=\mathrm{e}^{\boldsymbol{A}^2t}\boldsymbol{A}^2+\sin(\boldsymbol{A}^3t)\boldsymbol{A}t+\mathrm{e}^{\sin\boldsymbol{A}t}$.

Solution *If the matrix function* $F_1=\mathrm{e}^{\boldsymbol{A}\cos\boldsymbol{A}t}$ *is required, the following MATLAB statements should be entered*

```
>> A=[-7,2,0,-1; 1,-4,2,1; 2,-1,-6,-1; -1,-1,0,-4];
   syms x t; F1=funmsym(sym(A),exp(x*cos(x*t)),x)
```

Since the results obtained are very lengthy, only the first term of the results is displayed as follows:

$$\varphi_{1,1}(\boldsymbol{A}) = 2/9\mathrm{e}^{-3\cos 3t} + (2t\sin 6t + 6t^2\cos 6t)\mathrm{e}^{-6\cos 6t} + (\cos 6t - 6t\sin 6t)^2\mathrm{e}^{-6\cos 6t} \\ - 5/3(\cos 6t - 6t\sin 6t)\mathrm{e}^{-6\cos 6t} + 7/9\mathrm{e}^{-6\cos 6t}.$$

It can be seen from the results that in $\varphi_{1,1}(t)$, the term $\mathrm{e}^{-6\cos 6t}$ is repeatedly used, thus, one may collect that term using the following statement to simplify the result

```
>> collect(F1(1,1),exp(-6*cos(6*t)))
```

which leads to the following simplified result of $\varphi_{1,1}(\boldsymbol{A})$

$$\varphi_{1,1}(\boldsymbol{A}) = \left[12t\sin 6t + 6t^2\cos 6t + (\cos 6t - 6t\sin 6t)^2 - \frac{5}{3}\cos 6t + \frac{7}{9}\right]\mathrm{e}^{-6\cos 6t} + \frac{2}{9}\mathrm{e}^{-3\cos 3t}.$$

Further, if one assumes $t = 1$, the matrix $\mathrm{e}^{\boldsymbol{A}\cos\boldsymbol{A}}$ can be found such that

```
>> subs(A1,t,1), % which is the same as expm(A*funm(A,'cos'))
```

The complicated function is

$$\mathrm{e}^{\boldsymbol{A}\cos\boldsymbol{A}} = \begin{bmatrix} 4.3583 & 6.5044 & 4.3635 & -2.1326 \\ 4.3718 & 6.5076 & 4.3801 & -2.116 \\ 4.2653 & 6.4795 & 4.2518 & -2.2474 \\ -8.6205 & -12.984 & -8.6122 & 4.3832 \end{bmatrix}.$$

The complicated matrix function $\mathrm{e}^{\boldsymbol{A}^2 t}\boldsymbol{A}^2 + \sin(\boldsymbol{A}^3 t)\boldsymbol{A}t + \mathrm{e}^{\sin\boldsymbol{A}t}$ can also be evaluated directly with `funmsym()` function

```
>> F=funmsym(A,exp(x^2*t)*x^2+sin(x^3*t)*x*t+exp(sin(x*t)),x)
```

4.5.5 Power of a matrix

The target of the section is to find the kth power a given square matrix, i.e., $\boldsymbol{A}^k$, where k is a positive integer. If k is not an integer, it is meaningless discussing the matrix $\boldsymbol{A}^k$ since it is a sum of an infinite number of matrices. There is no use to run MATLAB command $\boldsymbol{A}$^k, since no result can be found. We shall begin this section with a simple example, and then, the idea can be generalized to arbitrary matrices.

Suppose $\boldsymbol{A}$ can be transformed into a Jordan matrix, $\boldsymbol{A} = \boldsymbol{V}\boldsymbol{J}\boldsymbol{V}^{-1}$. Then, $\boldsymbol{A}^k = \boldsymbol{V}\boldsymbol{J}^k\boldsymbol{V}^{-1}$, and we shall concentrate on finding $\boldsymbol{J}^k$.

As it was pointed out in the previous subsection that $\boldsymbol{J} = \lambda\boldsymbol{I} + \boldsymbol{H}_m$, where, $\boldsymbol{H}_m$ is $m \times m$ nilpotent matrix, with $\boldsymbol{H}_m^k \equiv 0$, for $k \geqslant m$. With binomial expansion, it is known that

$$\boldsymbol{J}^k = \lambda^k\boldsymbol{I} + k\lambda^{k-1}\boldsymbol{H}_m + \frac{k(k-1)}{2!}\lambda^{k-2}\boldsymbol{H}_m^2 + \cdots. \tag{4-5-7}$$

Since $\boldsymbol{H}_m^m$ and subsequent terms are all zero, the above equation can be reduced to the sum of m terms, and the analytical solution of $\boldsymbol{J}^k$ can easily be obtained.

Example 4.73 Consider a given matrix studied in Example 4.65, rewritten below. Please find $\boldsymbol{A}^k$ for an integer k.

$$A = \begin{bmatrix} -3 & -1 & -1 \\ 0 & -3 & -1 \\ 1 & 2 & 0 \end{bmatrix}.$$

Solution *Jordan transformation can be carried out first*

```
>> A=sym([-3,-1,-1; 0,-3,-1; 1,2,0]); syms k, [V J]=jordan(sym(A))
```

it is found J is a 3×3 Jordan matrix, with eigenvalue of $\lambda = -2$. Thus, $H = J - \lambda I$ can be used to extract the nilpotent matrix, and the power of the matrix can be easily found with

```
>> A0=-2*eye(3); H=J-A0; % find nilpotent matrix and add up 3 terms
   J1=A0^k+k*A0^(k-1)*H+k*(k-1)/2*A0^(k-2)*H^2, F=simplify(V*J1*inv(V))
```

and the final kth power of A can be found as follows, and it is interesting to note that it applies also for negative integers

$$F = \begin{bmatrix} (-2)^k(k+2)/2 & (-2)^k k/2 & (-2)^k k/2 \\ -(-2)^{(k-2)}k(k-1)/2 & (-2)^k(-k^2+5k+8)/8 & -(-2)^k k(k-5)/8 \\ (-2)^k k(k-5)/8 & (-2)^k k(k-9)/8 & (-2)^k(k^2-9k+8)/8 \end{bmatrix}.$$

Bearing in mind the approach presented in `funmsym()`, a similar MATLAB function can be written as follows, with the kernel `funJ()` replaced by the new function `powJ()`

```
function F=mat_power(A,k)
[V,T]=jordan(A); vec=diag(T); v1=[0,diag(T,1)',0]; v2=find(v1==0);
lam=vec(v2(1:end-1)); m=length(lam);
for i=1:m,
   k0=v2(i):v2(i+1)-1; J1=T(k0,k0); F(k0,k0)=powJ(J1,k);
end
F=simplify(V*F*inv(V));
function fJ=powJ(J,k)  % sub function to compute J_i^k
lam=J(1,1); I=eye(size(J)); H=J-lam*I; fJ=lam^k*I; H1=k*H;
for i=2:length(J)
   fJ=fJ+lam^(k+1-i)*I*H1; H1=H1*H*(k+1-i)/i;
end
```

Example 4.74 Consider the matrix in Example 4.72, compute $F = A^k$.

$$A = \begin{bmatrix} -7 & 2 & 0 & -1 \\ 1 & -4 & 2 & 1 \\ 2 & -1 & -6 & -1 \\ -1 & -1 & 0 & -4 \end{bmatrix},$$

Solution *The matrix can be entered first. With the MATLAB statements*

```
>> A=[-7,2,0,-1; 1,-4,2,1; 2,-1,-6,-1; -1,-1,0,-4];
   syms k, A=sym(A); F=mat_power(A,k) % evaluate directly A^k
```

the power of it can easily be found and manually simplified as

$$F = \begin{bmatrix} 2(-3)^k/9 + (-6)^k(k/4 + 7/9 + k^2/36) & (-3)^k/3 - (-6)^k(k/6 + 1/3) \\ 2(-3)^k/9 + (-6)^k(-k/12 - 2/9 + k^2/36) & (-3)^k/3 + (-6)^k(-k/6 + 2/3) \\ 2(-3)^k/9 + (-6)^k(-k/6 - 2/9 - k^2/18) & (-3)^k/3 + (-6)^k(k/3 - 1/3) \\ -4(-3)^k/9 + (-6)^k(4/9 - k/12 + k^2/36) & -2(-3)^k/3 + (-6)^k(2/3 - k/6) \end{bmatrix}$$

$$\left.\begin{matrix} 2(-3)^k/9+(-6)^k(k/12-2/9+k^2/36) & -(-3)^k/9+(-6)^k(k/12-1/9+k^2/36) \\ 2(-3)^k/9-(-6)^k(k/4+2/9+k^2/36) & -(-3)^k/9+(-6)^k(1/9-k/4+k^2/36) \\ 2(-3)^k/9+(-6)^k(k/6+7/9-k^2/18) & (-6)^k(k/6+1/9-k^2/18)-(-3)^k/9 \\ (-6^k)(4/9-k/4+k^2/36)-4(-3)^k/9 & 2(-3)^k/9+(-6)^k(-k/4+7/9+k^2/36) \end{matrix}\right]$$

To validate the results, the following statement can be issued, and it can be seen that the matrix of $\boldsymbol{A}^{12345}$ *using two methods yield the same result*

```
>> simplify(A^12345-subs(F,k,12345))
```

Exercises

Exercise 4.1 *Please generate a diagonal matrix with diagonal elements* a_1, a_2, $\cdots$, a_{12}.

Exercise 4.2 *Jordan matrix is a very practical matrix in matrix analysis courses. The general form of the matrix is described as*

$$\boldsymbol{J}=\begin{bmatrix} -\alpha & 1 & 0 & \cdots & 0 \\ 0 & -\alpha & 1 & \cdots & 0 \\ \vdots & \vdots & \vdots & \ddots & \vdots \\ 0 & 0 & 0 & \cdots & -\alpha \end{bmatrix}, \quad e.g., \ \boldsymbol{J}_1=\begin{bmatrix} -5 & 1 & 0 & 0 & 0 \\ 0 & -5 & 1 & 0 & 0 \\ 0 & 0 & -5 & 1 & 0 \\ 0 & 0 & 0 & -5 & 1 \\ 0 & 0 & 0 & 0 & -5 \end{bmatrix}.$$

Construct matrix $\boldsymbol{J}_1$ *with the MATLAB function* `diag()`.

Exercise 4.3 *Please enter the two* 20×20 *matrices without loops. Find their determinants, traces and characteristic polynomial coefficients.*

$$\boldsymbol{A}=\begin{bmatrix} a & & & & & b \\ & \ddots & & & \iddots & \\ & & a & b & & \\ & & b & a & & \\ & \iddots & & & \ddots & \\ b & & & & & a \end{bmatrix}, \quad \boldsymbol{B}=\begin{bmatrix} x & a & a & \cdots & a \\ a & x & a & \cdots & a \\ a & a & x & \cdots & a \\ \vdots & \vdots & \vdots & \ddots & a \\ a & a & a & \cdots & x \end{bmatrix}.$$

Exercise 4.4 *Nilpotent matrix is a special matrix defined as* $\boldsymbol{H}_n=\begin{bmatrix} 0 & 1 & 0 & \cdots & 0 \\ 0 & 0 & 1 & \cdots & 0 \\ \vdots & \vdots & \vdots & \ddots & \vdots \\ 0 & 0 & 0 & \cdots & 1 \\ 0 & 0 & 0 & \cdots & 0 \end{bmatrix}$.

Verify for any pre-specified n, $\boldsymbol{H}_n^i=0$ *is satisfied for all* $i\geqslant n$.

Exercise 4.5 *Can you recognize from the way of display whether a matrix is a numeric matrix or a symbolic matrix? If* $\boldsymbol{A}$ *is a numeric matrix and* $\boldsymbol{B}$ *is a symbolic matrix, can you predict whether the product* $\boldsymbol{C}$=$\boldsymbol{A}$∗$\boldsymbol{B}$ *is a numeric matrix or a symbolic matrix? Verify this through a simple example.*

Exercise 4.6 *Compute the determinant of a Vandermonde matrix, and find its simplified representation.*

$$A=\begin{bmatrix} a^4 & a^3 & a^2 & a & 1\\ b^4 & b^3 & b^2 & b & 1\\ c^4 & c^3 & c^2 & c & 1\\ d^4 & d^3 & d^2 & d & 1\\ e^4 & e^3 & e^2 & e & 1\end{bmatrix}.$$

Exercise 4.7 *Input matrices* $\boldsymbol{A}$ *and* $\boldsymbol{B}$ *in MATLAB, and convert them into symbolic matrices.*

$$A=\begin{bmatrix} 5 & 7 & 6 & 5 & 1 & 6 & 5\\ 2 & 3 & 1 & 0 & 0 & 1 & 4\\ 6 & 4 & 2 & 0 & 6 & 4 & 4\\ 3 & 9 & 6 & 3 & 6 & 6 & 2\\ 10 & 7 & 6 & 0 & 0 & 7 & 7\\ 7 & 2 & 4 & 4 & 0 & 7 & 7\\ 4 & 8 & 6 & 7 & 2 & 1 & 7\end{bmatrix},\quad B=\begin{bmatrix} 3 & 5 & 5 & 0 & 1 & 2 & 3\\ 3 & 2 & 5 & 4 & 6 & 2 & 5\\ 1 & 2 & 1 & 1 & 3 & 4 & 6\\ 3 & 5 & 1 & 5 & 2 & 1 & 2\\ 4 & 1 & 0 & 1 & 2 & 0 & 1\\ -3 & -4 & -7 & 3 & 7 & 8 & 12\\ 1 & -10 & 7 & -6 & 8 & 1 & 5\end{bmatrix}.$$

Exercise 4.8 *Check whether the matrices given in the above exercise are singular or not. Find the rank, determinant, trace and inverse matrices for them. Check whether the inverse matrices are correct or not.*

Exercise 4.9 *Please generate a* 15×15 *matrix, whose elements are 0's and 1's, and the determinant of the matrix is 1.*

Exercise 4.10 *Please generate a* 20×20 *matrix as follows, and find the determinants, trace and characteristic polynomial. Can you guess what is the possible determinant of an* $n \times n$ *such a matrix? Please validate your guess for a larger* n*.*

$$A=\begin{bmatrix} x-a & a & a & \cdots & a\\ a & x-a & a & \cdots & a\\ a & a & x-a & \vdots & a\\ \vdots & \vdots & \vdots & \ddots & a\\ a & a & a & \cdots & x-a\end{bmatrix}.$$

Exercise 4.11 *Find the characteristic polynomials, eigenvalues and eigenvectors for the matrices* $\boldsymbol{A}$ *and* $\boldsymbol{B}$ *in Exercise 4.7.*

Exercise 4.12 *Perform singular value decompositions, LU factorizations and orthogonal decompositions to the matrices* $\boldsymbol{A}$ *and* $\boldsymbol{B}$ *in Exercise 4.7.*

Exercise 4.13 *Consider the matrices in Exercise 4.7, please check whether they satisfy Cayley–Hamilton Theorem. If there are errors, is there any way in to reduce errors?*

Exercise 4.14 *For arbitrary matrices*

$$A_1=\begin{bmatrix} a_{11} & a_{12} & a_{13}\\ a_{21} & a_{22} & a_{23}\\ a_{31} & a_{32} & a_{33}\end{bmatrix}, A_2=\begin{bmatrix} a_{11} & a_{12} & a_{13} & a_{14}\\ a_{21} & a_{22} & a_{23} & a_{24}\\ a_{31} & a_{32} & a_{33} & a_{34}\\ a_{41} & a_{42} & a_{43} & a_{44}\end{bmatrix}, A_3=\begin{bmatrix} a_{11} & a_{12} & a_{13} & a_{14} & a_{15}\\ a_{21} & a_{22} & a_{23} & a_{24} & a_{25}\\ a_{31} & a_{32} & a_{33} & a_{34} & a_{35}\\ a_{41} & a_{42} & a_{43} & a_{44} & a_{45}\\ a_{51} & a_{52} & a_{53} & a_{54} & a_{55}\end{bmatrix}$$

verify the Cayley–Hamilton Theorem.

Exercise 4.15 *Please write a low-level MATLAB function to find the inverse of a given square matrix* $\boldsymbol{A}$ *using the rules in Example 4.33.*

Exercise 4.16 *Perform LU factorization and SVD decomposition to the following matrices*

$$\boldsymbol{A}=\begin{bmatrix}8&0&1&1&6\\9&2&9&4&0\\1&5&9&9&8\\9&9&4&7&9\\6&9&8&9&6\end{bmatrix},\quad \boldsymbol{B}=\begin{bmatrix}1&2&2&2\\1&1&2&0\\1&1&1&0\\0&0&2&0\end{bmatrix}.$$

Exercise 4.17 *Please check whether the following matrices are positive-definite matrices. If they are, please find their Cholesky factorizations.*

$$\boldsymbol{A}=\begin{bmatrix}1&3&4&8\\3&2&7&2\\4&7&2&8\\8&2&8&6\end{bmatrix},\quad \boldsymbol{B}=\begin{bmatrix}12&13&24&26\\31&12&27&11\\10&9&22&18\\42&22&10&16\end{bmatrix}.$$

Exercise 4.18 *Compute the eigenvalues, eigenvectors and singular values of the following matrices.*

$$\boldsymbol{A}=\begin{bmatrix}2&7&5&7&7\\7&4&9&3&3\\3&9&8&3&8\\5&9&6&3&6\\2&6&8&5&4\end{bmatrix},\quad \boldsymbol{B}=\begin{bmatrix}703&795&980&137&661\\547&957&271&12&284\\445&523&252&894&469\\695&880&876&199&65\\621&173&737&299&988\end{bmatrix}.$$

Exercise 4.19 *Please check whether the following matrices are positive-definite ones, if so, please find the Cholesky factorized matrices.*

$$\boldsymbol{A}=\begin{bmatrix}9&2&1&2&2\\2&4&3&3&3\\1&3&7&3&4\\2&3&3&5&4\\2&3&4&4&5\end{bmatrix},\quad \boldsymbol{B}=\begin{bmatrix}16&17&9&12&12\\17&12&12&2&18\\9&12&18&7&13\\12&2&7&18&12\\12&18&13&12&10\end{bmatrix}.$$

Exercise 4.20 *Perform the Jordan transformation for the following matrices, and also find the corresponding transformation matrices.*

$$\boldsymbol{A}=\begin{bmatrix}-2&0.5&-0.5&0.5\\0&-1.5&0.5&-0.5\\2&0.5&-4.5&0.5\\2&1&-2&-2\end{bmatrix},\quad \boldsymbol{B}=\begin{bmatrix}-2&-1&-2&-2\\-1&-2&2&2\\0&2&0&3\\1&-1&-3&-6\end{bmatrix}.$$

Exercise 4.21 *Please find the eigenvalues and Jordanian canonical form of the following matrix. It is known that the original matrix contains complex eigenvalues, please find out a real transformation matrix to complete Jordanian transformation.*

$$\boldsymbol{A}=\begin{bmatrix}-5&-2&-4&0&-1&0\\1&-2&2&0&-1&-2\\2&2&0&3&2&0\\1&3&1&0&3&1\\-1&-2&-3&-4&-4&1\\3&4&3&1&2&-1\end{bmatrix}.$$

Exercise 4.22 *Please consider the matrices in Exercise 4.20. Try to select suitable transformation matrices to convert the matrices into companion forms.*

Exercise 4.23 *Find the basic set of solutions of the homogenous equations*

$$\text{(i)}\ \begin{cases} 6x_1+x_2+4x_3-7x_4-3x_5=0\\ -2x_1-7x_2-8x_3+6x_4 \quad =0\\ -4x_1+5x_2+x_3-6x_4+8x_5=0\\ -34x_1+36x_2+9x_3-21x_4+49x_5=0\\ -26x_1-12x_2-27x_3+27x_4+17x_5=0,\end{cases} \qquad \text{(ii)}\ \boldsymbol{A}=\begin{bmatrix} -1 & 2 & -2 & 1 & 0\\ 0 & 3 & 2 & 2 & 1\\ 3 & 1 & 3 & 2 & -1\end{bmatrix}.$$

Exercise 4.24 *Please check whether the following two matrices similar or not. If they are similar, what is the transformation matrix, to transform* $\boldsymbol{A}$ *into* $\boldsymbol{B}$*?*

$$\boldsymbol{A}=\begin{bmatrix} -2 & 1 & -1/2 & 0\\ -1/2 & -7/2 & 0 & -1/2\\ 0 & 0 & -2 & 0\\ 1/2 & 1/2 & -1/2 & -7/2\end{bmatrix},\quad \boldsymbol{B}=\begin{bmatrix} -11/4 & -1/4 & -3/4 & 0\\ 3/4 & -15/4 & -1/4 & 0\\ 1/2 & -1/2 & -5/2 & 0\\ 3/2 & -3/2 & 1/2 & -2\end{bmatrix}.$$

Exercise 4.25 *Find the numerical and analytical solutions to the following linear algebraic equations, and then, validate the results.*

$$\begin{bmatrix} 2 & -9 & 3 & -2 & -1\\ 10 & -1 & 10 & 5 & 0\\ 8 & -2 & -4 & -6 & 3\\ -5 & -6 & -6 & -8 & -4\end{bmatrix}\boldsymbol{X}=\begin{bmatrix} -1 & -4 & 0\\ -3 & -8 & -4\\ 0 & 3 & 3\\ 9 & -5 & 3\end{bmatrix}.$$

Exercise 4.26 *Check whether the equation has a solution.*

$$\begin{bmatrix} 16 & 2 & 3 & 13\\ 5 & 11 & 10 & 8\\ 9 & 7 & 6 & 12\\ 4 & 14 & 15 & 1\end{bmatrix}\boldsymbol{X}=\begin{bmatrix} 1\\ 3\\ 4\\ 7\end{bmatrix}.$$

Exercise 4.27 *Find the analytical solutions to the following linear algebraic equations, and then, validate the results.*

$$\begin{bmatrix} 2 & 9 & 4 & 12 & 5 & 8 & 6\\ 12 & 2 & 8 & 7 & 3 & 3 & 7\\ 3 & 0 & 3 & 5 & 7 & 5 & 10\\ 3 & 11 & 6 & 6 & 9 & 9 & 1\\ 11 & 2 & 1 & 4 & 6 & 8 & 7\\ 5 & -18 & 1 & -9 & 11 & -1 & 18\\ 26 & -27 & -1 & 0 & -15 & -13 & 18\end{bmatrix}\boldsymbol{X}=\begin{bmatrix} 1 & 9\\ 5 & 12\\ 4 & 12\\ 10 & 9\\ 0 & 5\\ 10 & 18\\ -20 & 2\end{bmatrix}.$$

Exercise 4.28 *For the matrices* $\boldsymbol{A}$ *and* $\boldsymbol{B}$*, please calculate* $\boldsymbol{A}\otimes\boldsymbol{B}$ *and* $\boldsymbol{B}\otimes\boldsymbol{A}$*. Are they equal to each other?*

$$\boldsymbol{A}=\begin{bmatrix} -1 & 2 & 2 & 1\\ -1 & 2 & 1 & 0\\ 2 & 1 & 1 & 0\\ 1 & 0 & 2 & 0\end{bmatrix},\quad \boldsymbol{B}=\begin{bmatrix} 3 & 0 & 3\\ 3 & 2 & 2\\ 3 & 1 & 1\end{bmatrix}.$$

Exercise 4.29 *Find the analytical and numerical solutions to the following Sylvester equation, and verify the results.*

$$\begin{bmatrix} 3 & -6 & -4 & 0 & 5 \\ 1 & 4 & 2 & -2 & 4 \\ -6 & 3 & -6 & 7 & 3 \\ -13 & 10 & 0 & -11 & 0 \\ 0 & 4 & 0 & 3 & 4 \end{bmatrix} \boldsymbol{X} + \boldsymbol{X} \begin{bmatrix} 3 & -2 & 1 \\ -2 & -9 & 2 \\ -2 & -1 & 9 \end{bmatrix} = \begin{bmatrix} -2 & 1 & -1 \\ 4 & 1 & 2 \\ 5 & -6 & 1 \\ 6 & -4 & -4 \\ -6 & 6 & -3 \end{bmatrix}.$$

Exercise 4.30 *Find the analytical and numerical solutions to the so-called* Stein equation *given below and verify the results.*

$$\begin{bmatrix} -2 & 2 & 1 \\ -1 & 0 & -1 \\ 1 & -1 & 2 \end{bmatrix} \boldsymbol{X} \begin{bmatrix} -2 & -1 & 2 \\ 1 & 3 & 0 \\ 3 & -2 & 2 \end{bmatrix} - \boldsymbol{X} + \begin{bmatrix} 0 & -1 & 0 \\ -1 & 1 & 0 \\ 1 & -1 & -1 \end{bmatrix} = \mathbf{0}.$$

Exercise 4.31 *Please find the numerical and analytical solutions to the discrete Lyapunov equation* $\boldsymbol{A}\boldsymbol{X}\boldsymbol{A}^{\mathrm{T}} - \boldsymbol{X} + \boldsymbol{Q} = \mathbf{0}$, *where*

$$\boldsymbol{A} = \begin{bmatrix} -2 & -1 & 0 & -3 \\ -2 & -2 & -1 & -3 \\ 2 & 2 & -3 & 0 \\ -3 & 1 & 1 & -3 \end{bmatrix}, \quad \boldsymbol{Q} = \begin{bmatrix} -12 & -16 & 14 & -8 \\ -20 & -25 & 11 & -20 \\ 3 & 1 & -16 & 1 \\ -4 & -10 & 21 & 10 \end{bmatrix}.$$

Exercise 4.32 *Assume that a Riccati equation is given by* $\boldsymbol{P}\boldsymbol{A} + \boldsymbol{A}^{\mathrm{T}}\boldsymbol{P} - \boldsymbol{P}\boldsymbol{B}\boldsymbol{R}^{-1}\boldsymbol{B}^{\mathrm{T}}\boldsymbol{P} + \boldsymbol{Q} = 0$, *where*

$$\boldsymbol{A} = \begin{bmatrix} -27 & 6 & -3 & 9 \\ 2 & -6 & -2 & -6 \\ -5 & 0 & -5 & -2 \\ 10 & 3 & 4 & -11 \end{bmatrix}, \boldsymbol{B} = \begin{bmatrix} 0 & 3 \\ 16 & 4 \\ -7 & 4 \\ 9 & 6 \end{bmatrix}, \boldsymbol{Q} = \begin{bmatrix} 6 & 5 & 3 & 4 \\ 5 & 6 & 3 & 4 \\ 3 & 3 & 6 & 2 \\ 4 & 4 & 2 & 6 \end{bmatrix}, \boldsymbol{R} = \begin{bmatrix} 4 & 1 \\ 1 & 5 \end{bmatrix}.$$

Solve the equation and verify the result.

Exercise 4.33 *Solve the following Diophantine equations and validate the solutions.*

(i) $A(x) = 1 - 0.7x,\ \ B(x) = 0.9 - 0.6x,\ \ C(x) = 2x^2 + 1.5x^3,$

(ii) $A(x) = 1 + 0.6x - 0.08x^2 + 0.152x^3 + 0.0591x^4 - 0.0365x^5,$

$B(x) = 5 - 4x - 0.25x^2 + 0.42x^3,\ \ C(x) = 1.$

Exercise 4.34 *Certain functions can be expressed by polynomial functions, i.e., Taylor series expansions. In these functions, if* x *is substituted by matrix* $\boldsymbol{A}$, *the nonlinear function can also be expressed for matrices. Write M-functions for the matrix function evaluation problems and verify the results with* `funmsym()`.

(i) $\cos\boldsymbol{A} = \boldsymbol{I} - \dfrac{1}{2!}\boldsymbol{A}^2 + \dfrac{1}{4!}\boldsymbol{A}^4 - \dfrac{1}{6!}\boldsymbol{A}^6 + \cdots + \dfrac{(-1)^n}{(2n)!}\boldsymbol{A}^{2n} + \cdots$

(ii) $$\arcsin\boldsymbol{A} = \boldsymbol{A} + \frac{1}{2\cdot 3}\boldsymbol{A}^3 + \frac{1\cdot 3}{2\cdot 4\cdot 5}\boldsymbol{A}^5 + \frac{1\cdot 3\cdot 5}{2\cdot 4\cdot 6\cdot 7}\boldsymbol{A}^7 + \frac{1\cdot 3\cdot 5\cdot 7}{2\cdot 4\cdot 6\cdot 8\cdot 9}\boldsymbol{A}^9 + \cdots + \frac{(2n)!}{2^{2n}(n!)^2(2n+1)}\boldsymbol{A}^{2n+1} + \cdots$$

(iii) $\ln\boldsymbol{A} = \boldsymbol{A} - \boldsymbol{I} - \dfrac{1}{2}(\boldsymbol{A}-\boldsymbol{I})^2 + \dfrac{1}{3}(\boldsymbol{A}-\boldsymbol{I})^3 - \dfrac{1}{4}(\boldsymbol{A}-\boldsymbol{I})^4 + \cdots + \dfrac{(-1)^{n+1}}{n}(\boldsymbol{A}-\boldsymbol{I})^n + \cdots.$

Exercise 4.35 *For an autonomous linear differential equation of the form $\boldsymbol{x}'(t) = \boldsymbol{A}\boldsymbol{x}(t)$, the analytical solution can be written as $\boldsymbol{x}(t) = \mathrm{e}^{\boldsymbol{A}t}\boldsymbol{x}(0)$. Find the analytical solution to the equation*

$$\boldsymbol{x}'(t) = \begin{bmatrix} -3 & 0 & 0 & 1 \\ -1 & -1 & 1 & -1 \\ 1 & 0 & -2 & 1 \\ 0 & 0 & 0 & -4 \end{bmatrix} \boldsymbol{x}(t), \quad \boldsymbol{x}(0) = \begin{bmatrix} -1 \\ 0 \\ 3 \\ 1 \end{bmatrix}.$$

Exercise 4.36 *Compute the logarithmic matrices,* $\ln \boldsymbol{A}$ *and* $\ln \boldsymbol{A}t$ *of the following matrix* $\boldsymbol{A}$. *Please validate the results with the reliable* `expm()` *function.*

$$\boldsymbol{A} = \begin{bmatrix} -1 & -1/2 & 1/2 & -1 \\ -2 & -5/2 & -1/2 & 1 \\ 1 & -3/2 & -5/2 & -1 \\ 3 & -1/2 & -1/2 & -4 \end{bmatrix}.$$

Exercise 4.37 *Please compute the trigonometric functions* $\sin \boldsymbol{A}t$, $\cos \boldsymbol{A}t$ *and* $\tan \boldsymbol{A}t$ *and* $\cot \boldsymbol{A}t$ *for the following matrices*

$$\boldsymbol{A}_1 = \begin{bmatrix} -15/4 & 3/4 & -1/4 & 0 \\ 3/4 & -15/4 & 1/4 & 0 \\ -1/2 & 1/2 & -9/2 & 0 \\ 7/2 & -7/2 & 1/2 & -1 \end{bmatrix}, \boldsymbol{A}_2 = \begin{bmatrix} -1 & 0 & 0 & 0 \\ 0 & -1 & 1 & 0 \\ 2 & 0 & -2 & 1 \\ -1 & 0 & 0 & -2 \end{bmatrix}.$$

Exercise 4.38 *If a block Jordan matrix* $\boldsymbol{A}$ *is given by*

$$\boldsymbol{A} = \begin{bmatrix} \boldsymbol{A}_1 & & \\ & \boldsymbol{A}_2 & \\ & & \boldsymbol{A}_3 \end{bmatrix}, \quad \textit{where } \boldsymbol{A}_1 = \begin{bmatrix} -3 & 1 & 0 \\ 0 & -3 & 1 \\ 0 & 0 & -3 \end{bmatrix},$$

$$\boldsymbol{A}_2 = \begin{bmatrix} -5 & 1 \\ 0 & -5 \end{bmatrix}, \quad \boldsymbol{A}_3 = \begin{bmatrix} -1 & 1 & 0 & 0 \\ 0 & -1 & 1 & 0 \\ 0 & 0 & -1 & 1 \\ 0 & 0 & 0 & -1 \end{bmatrix},$$

find the solutions to $\mathrm{e}^{\boldsymbol{A}t}$, $\sin\left(2\boldsymbol{A}t + \dfrac{\pi}{3}\right)$, $\mathrm{e}^{\boldsymbol{A}^2 t}\boldsymbol{A}^2 + \sin(\boldsymbol{A}^3 t)\boldsymbol{A}t + \mathrm{e}^{\sin \boldsymbol{A}t}$.

Exercise 4.39 *For a given matrix* $\boldsymbol{A}$, *find matrix functions* $\mathrm{e}^{\boldsymbol{A}t}, \sin \boldsymbol{A}t, \mathrm{e}^{\boldsymbol{A}t}\sin\left(\boldsymbol{A}^2\mathrm{e}^{\boldsymbol{A}t}t\right)$, *also please find* $\boldsymbol{A}^k$.

$$\boldsymbol{A} = \begin{bmatrix} -9/2 & 0 & 1/2 & -3/2 \\ -1/2 & -4 & 1/2 & -1/2 \\ 3/2 & 1 & -5/2 & 3/2 \\ 0 & -1 & -1 & -3 \end{bmatrix}.$$

Bibliography

[1] Garbow B S, Boyle J M, Dongarra J J, et al. Matrix eigensystem routines — EISPACK guide extension, Lecture notes in computer sciences, volume 51. New York: Springer-

Verlag, 1977

[2] Dongarra J J, Bunsh J R, Molor C B. LINPACK user's guide. Philadelphia: Society of Industrial and Applied Mathematics, 1979

[3] Hefferon J. Linear algebra. Saint Michael's College, USA: Open source textbook at `http://joshua.smcvt.edu/linearalgebra/`, 2006

[4] Meyer C D. Matrix analysis and applied linear algebra. Philadelphia: Society for Industrial and Applied Mathematics. `http://www.matrixanalysis.com/DownloadChapters.html`, 2001

[5] Dawkins P. Linear algebra. `http://tutorial.math.lamar.edu/pdf/LinAlg/LinAlg_Complete.pdf`, 2007

[6] Beezer R A. A first course in linear algebra, version 2.99. Washington: Department of Mathematics and Computer Science University of Puget Sound, 1500 North Warner, Tacoma, Washington, 98416-1043, `http://linear.ups.edu/`, 2012

[7] Moler C B, Van Loan C F. Nineteen dubious ways to compute the exponential of a matrix. SIAM Review, 1979, 20:801–836

[8] Huang L. Linear algebra in systems and control theory. Beijing: Science Press, 1984 (in Chinese)

Chapter 5

Integral Transforms and Complex-valued Functions

Integral transform technique can usually be used to map a problem from one domain into another, such that the new problem may easily be solved in the transformed domain. For instance, the Laplace transform can be used to map a linear time-invariant (LTI) ordinary differential equation (ODE) into an algebraic equation. Therefore, the properties of the original problem, such as stability, can easily be determined, which lays a foundation of the classical control theory. In many real applications, Fourier transforms as well as Mellin transforms and Hankel transforms are all very useful. Therefore, computer-aided solutions to integral transform problems deserve special attention and are the main topics of this chapter. This chapter is especially useful, if the readers have yet not learnt similar courses. The integral transforms can be obtained directly with MATLAB, and it is, for the readers, as simple as solving calculus problems discussed in Chapter 3.

In Section 5.1, definition and properties of Laplace transform and the inverse Laplace transform are summarized. Our focus is on the MATLAB-based solutions to Laplace transform problems. Numerical Laplace transform and its inverse are also presented for problems with no analytical solutions. Section 5.2 presents Fourier transform and its inverse transform, again with a focus on the MATLAB-based solutions. Moreover, sine and cosine Fourier transforms and discrete Fourier transforms are briefly introduced. Fast Fourier transform, a numerical tool for Fourier transform is also presented. In Section 5.3, Mellin and Hankel transforms are introduced, while z transform and inverse z transform are introduced in Section 5.4 with illustrative examples taken from discrete-time signals and systems. In Section 5.5, a brief introduction to complex-value function is given, to show Riemman surface and mapping of complex-value functions. Problems from typical complex-valued function courses such as singularities, poles and residues are demonstrated together with partial fraction expansions for rational functions in Section 5.6, where an evaluation method of closed-path integrals is introduced based on the concepts of residues. Essential singularities and Laurent series for given functions are also discussed in the section. Section 5.7 devoted to the presentation of difference equations, where analytical solutions and recursive numerical solutions are given.

For readers who wish to check the detailed explanations of complex-valued functions, Laplace and Fourier transforms, we recommend the open source textbook [1] (Chapters 6-13, 31, 32).

5.1 Laplace Transforms and Their Inverses

Integral transform introduced by the French mathematician Pierre-Simon Laplace (1749–1827) can be used to map the ordinary differential equations into algebraic equations. Thus, it established the foundation for many research areas. For instance, the Laplace transform established the basis for the modeling, analysis and synthesis of control systems.

In this section, the definition and basic properties of Laplace transform and inverse Laplace transform are summarized first. Then, we focus on the solutions to Laplace transform problems and their applications using MATLAB. For functions whose analytical representations of Laplace or inverse Laplace transforms do not exist, numerical solutions are introduced.

5.1.1 Definitions and properties

The one-sided Laplace transform of a time function $f(t)$ is defined as

$$\mathscr{L}[f(t)] = \int_0^\infty f(t)\mathrm{e}^{-st}\,\mathrm{d}t = F(s), \tag{5-1-1}$$

where $\mathscr{L}[f(t)]$ is the notation of Laplace transform. The properties of Laplace transform are summarized below without proofs.

(i) **Linear property** $\mathscr{L}[af(t)\pm bg(t)]=a\mathscr{L}[f(t)]\pm b\mathscr{L}[g(t)]$ for scalars a and b.

(ii) **Time-domain shift** $\mathscr{L}[f(t-a)] = \mathrm{e}^{-as}F(s)$.

(iii) **s-domain property** $\mathscr{L}[\mathrm{e}^{-at}f(t)] = F(s+a)$.

(iv) **Differentiation property** $\mathscr{L}[\mathrm{d}f(t)/\mathrm{d}t] = sF(s) - f(0^+)$. Generally, the nth order derivative can be obtained from

$$\mathscr{L}\left[\frac{\mathrm{d}^n}{\mathrm{d}t^n}f(t)\right] = s^nF(s) - s^{n-1}f(0^+) - s^{n-2}\frac{\mathrm{d}f(0^+)}{\mathrm{d}t} - \cdots - \frac{\mathrm{d}^{n-1}f(0^+)}{\mathrm{d}t^{n-1}}. \tag{5-1-2}$$

If the initial values of $f(t)$ and the other derivatives are all zero, (5-1-2) is simplified as

$$\mathscr{L}\left[\frac{\mathrm{d}^nf(t)}{\mathrm{d}t^n}\right] = s^nF(s), \tag{5-1-3}$$

and these properties are the crucial formulas to map ordinary differential equations into algebraic equations.

(v) **Integration property** If zero conditions are assumed, $\mathscr{L}\left[\int_0^t f(\tau)\,\mathrm{d}\tau\right] = \dfrac{F(s)}{s}$.

Generally, the Laplace transform of the multiple integral of $f(t)$ can be obtained from

$$\mathscr{L}\left[\int_0^t \cdots \int_0^t f(\tau)\,\mathrm{d}\tau^n\right] = \frac{F(s)}{s^n}. \tag{5-1-4}$$

(vi) **Initial value property** $\lim_{t\to 0} f(t) = \lim_{s\to\infty} sF(s)$.

(vii) **Final value property** If $F(s)$ has no pole with non-negative real part, i.e., $\mathrm{Re}(s) \geqslant 0$, then

$$\lim_{t\to\infty} f(t) = \lim_{s\to 0} sF(s). \tag{5-1-5}$$

(viii) **Convolution property** $\mathscr{L}[f(t)*g(t)]=\mathscr{L}[f(t)]\mathscr{L}[g(t)]$, where the convolution operator $*$ is defined as

$$f(t)*g(t)=\int_0^t f(\tau)g(t-\tau)\,\mathrm{d}\tau=\int_0^t f(t-\tau)g(\tau)\,\mathrm{d}\tau. \tag{5-1-6}$$

(iv) **Other properties**

$$\mathscr{L}[t^n f(t)]=(-1)^n\frac{\mathrm{d}^n F(s)}{\mathrm{d}s^n},\quad \mathscr{L}\left[\frac{f(t)}{t^n}\right]=\int_s^\infty\cdots\int_s^\infty F(s)\,\mathrm{d}s^n. \tag{5-1-7}$$

If the Laplace transform of a signal $f(t)$ is $F(s)$, the inverse Laplace transform of $F(s)$ is defined as

$$f(t)=\mathscr{L}^{-1}[F(s)]=\frac{1}{\mathrm{j}2\pi}\int_{\sigma-\mathrm{j}\infty}^{\sigma+\mathrm{j}\infty}F(s)\mathrm{e}^{st}\,\mathrm{d}s, \tag{5-1-8}$$

where σ is greater than all the real part of the poles of function $F(s)$. The definitions of poles will be given later.

5.1.2 Computer solution to Laplace transform problems

It is hard if not impossible to write programs with numerical computation-based languages such as C to solve Laplace transform problems. Computer algebra systems should be used instead. For instance, the Symbolic Math Toolbox of MATLAB can be used to solve the problems easily and analytically. The procedures for solving such problems are summarized as follows:

(i) The symbolic variables such as t should be declared using the `syms` command. The time-domain function $f(t)$ should be defined in variable *fun*.

(ii) Call the `laplace()` function to solve the problem. Thus, the Laplace transform can be obtained with the following function call

```
F=laplace(fun)         % the time variable is given in t
F=laplace(fun,v,u)     % with domain variables v, u specified
```

and function `simplify()` can be used to simplify the obtained symbolic result.

(iii) For complicated problems, the returned results are difficult to read. The function `pretty()` can be used to better display the results. Also, the `latex()` function can be used to convert the results into LaTeX string.

(iv) If the Laplace transform function $F(s)$ is known, it can also be described in the symbolic expression *fun*. Then, MATLAB function `ilaplace()` can be used to calculate the inverse Laplace transform of the given function. The syntaxes of the function are

```
f=ilaplace(fun)        % default variable is s
f=ilaplace(fun,u,v)    % specify the domain variables v and u
```

Example 5.1 For a given time domain function $f(t)=t^2\mathrm{e}^{-2t}\sin(t+\pi)$, compute its Laplace transform function $F(s)$.

Solution *From the original problem, it can be seen that the time domain variable t should be declared first. With the MATLAB statements, the function $f(t)$ can be specified. Then, the* `laplace()` *function can be used to derive the Laplace transform of the original function*

```
>> syms t; f=t^2*exp(-2*t)*sin(t+pi); F=laplace(f) % the 3-step procedure
```

and the result is as follows

$$F(s)=\frac{2}{\left[(s+2)^2+1\right]^2}-\frac{2\left(2s+4\right)^2}{\left[(s+2)^2+1\right]^3}.$$

Example 5.2 Assume that the original function is given by $f(x)=x^2\mathrm{e}^{-2x}\sin(x+\pi)$, compute the Laplace transform, and then, take inverse Laplace transform and see whether the original function can be recovered.

Solution *Similarly, the* `laplace()` *function can still be used*

```
>> syms x w; f=x^2*exp(-2*x)*sin(x+pi); F=laplace(f,x,w)
```

and the result is the same as the one obtained in the previous example, albeit the name of the variables used are different.

If the command $f_1=$`simplify(ilaplace(`F`))` *is used, the result* $f_1(t)=-t^2\mathrm{e}^{-2t}\sin t$ *can be obtained. Since the equation* $\sin(t+\pi)=-\sin t$ *holds, the original function is actually restored.*

Example 5.3 Compute the inverse Laplace transform for the following complex-valued function

$$G(x)=\frac{-17x^5-7x^4+2x^3+x^2-x+1}{x^6+11x^5+48x^4+106x^3+125x^2+75x+17}.$$

Solution *The following statements can be used*

```
>> syms x t;                               % declare symbolic variables
   G=(-17*x^5-7*x^4+2*x^3+x^2-x+1)... % specify original function
     /(x^6+11*x^5+48*x^4+106*x^3+125*x^2+75*x+17);
   f=ilaplace(G,x,t)                       % evaluate Laplace transform directly
```

However, the result is far too complicated to display. With the use of `vpa(`f`,7)`, *the analytical form of the solution can be obtained.*

$$\begin{aligned}y(t)&=-556.2565\mathrm{e}^{-3.2617t}+1.7589\mathrm{e}^{-1.0778t}\cos 0.6021t\\&\quad+10.9942\mathrm{e}^{-1.0778t}\sin 0.6021t+0.2126\mathrm{e}^{-0.5209t}\\&\quad+537.2850\mathrm{e}^{-2.5309t}\cos 0.3998t-698.2462\mathrm{e}^{-2.5309t}\sin 0.3998t.\end{aligned}$$

Example 5.4 For the function $f(t)$ given in Example 5.1, explore the relationship between $\mathscr{L}[\mathrm{d}^5 f(t)/\,\mathrm{d}t^5]$ and $s^5\mathscr{L}[f(t)]$.

Solution *To solve the problem, the fifth-order derivative to the given function* $f(t)$ *can be obtained by function* `diff()`. *Then, the Laplace transform can be obtained*

```
>> syms t s; f=t^2*exp(-2*t)*sin(t+pi); % declare variable and function
   F=simplify(laplace(diff(f,t,5)))      % evaluate Laplace transform
```

which yields

$$F(s)=-2\,\frac{3000+6825s+6660s^2+960s^4+3471s^3+110s^5}{\left(s^2+4s+5\right)^3}.$$

Taking Laplace transform to function $f(t)$, *multiplying the result by* s^5, *one may then subtract the above result to find the difference*

```
>> F0=laplace(f); simplify(F-s^5*F0) % simplify the difference
```

and the difference obtained is $6s - 48$.

It is obvious that the difference of the two terms is not zero, which seems not to agree with (5-1-3). This is because the initial conditions here are non-zero. It can easily be found that $f(0) = f'(0) = f''(0) = 0$, *while* $f^{(3)}(0) = -6$, *and* $f^{(4)}(0) = 48$. *Hence, the difference equals* $6s - 48$.

Example 5.5 Display the differentiation property of Laplace transform $\mathscr{L}\left[\dfrac{\mathrm{d}^2 f(t)}{\mathrm{d}t^2}\right]$.

Solution *Some of the properties of Laplace transform can be displayed with the use of the Symbolic Math Toolbox of MATLAB. For this problem, the function* $f(t)$ *should be declared first, then, the second-order derivative can be obtained with the use of function* `diff()`. *The Laplace transform of the second-order derivative can then be obtained*

```
>> syms t f(t); laplace(diff(f,t,2)) % display the formula
```

with the display $s^2F(s) - sf(0) - f'(0)$. *Also, the formula for the Laplace transform of the seventh-order derivative can be derived*

```
>> laplace(diff(f,t,7))
```

and the formula is $s^7F(s)-s^6f(0)-s^5f'(0)-s^4f''(0)-s^3f'''(0)-s^2f^{(4)}(0)-sf^{(5)}(0)-f^{(6)}(0)$.

Example 5.6 For the function $f(t) = \mathrm{e}^{-5t}\cos(2t + 1) + 5$, compute $\mathscr{L}[\mathrm{d}^5 f(t)/\,\mathrm{d}t^5]$.

Solution *For the given function* $f(t)$, *the Laplace and then the inverse Laplace transform can be performed with the following MATLAB statements*

```
>> syms t; f=exp(-5*t)*cos(2*t+1)+5; % declare symbolic variable and function
   F=laplace(diff(f,t,5));          % Laplace transform of the 5th order derivative
   F=simplify(F)                    % simplify the result
```

The result obtained is

$$F(s) = \frac{1475\cos 1s - 1189\cos 1 - 24360\sin 1 - 4282\sin 1s}{s^2 + 10s + 29}.$$

In fact, a simplified result can further be obtained if needed. For instance, collecting the terms in the numerator by using the following statements

```
>> syms s; F1=collect(F) % collect the terms in the result
```

The result can be obtained as

$$F_1 = \frac{(1475\cos 1 - 4282\sin 1)\,s - 1189\cos 1 - 24360\sin 1}{s^2 + 10s + 29}.$$

5.1.3 Numerical solutions of Laplace transforms

The analytic solutions of Laplace transforms of some functions can be obtained with the direct call of `laplace()` or `ilaplace()` functions. However, there exist a great many functions where analytical solutions of their Laplace and inverse transforms cannot be obtained. In this case, numerical approaches should be considered.

The `INVLAP()` function, developed by Juraj Valsa [2,3], can be used in finding numerical solutions of inverse Laplace transforms, with the syntax

`[t,y] = INVLAP(f,t0,tf,N,other parameters)`

where, the original function can be expressed by f, which is a string containing s. The arguments (t_0, t_f) are the interested interval of time t, and N is the number of points of evaluation. Different numbers of N's can be selected to validate the results. *Other parameters* can only be assigned by the experienced users, and it is suggested to use the default parameters.

Example 5.7 Please solve the inverse Laplace transform problem numerically to the function given in Example 5.3.

Solution *It can be seen from the earlier example that, although the analytical solution does not exist, a high-precision numerical solution can be found with Symbolic Math Toolbox. For the same function, the variable x can be substituted by s and can be converted to a string with* `char()` *function. Numerical inverse Laplace transform can be obtained. Compared with exact method, the maximum relative error is* 0.005826%.

```
>> syms x t; % declare symbolic variables and the function
   G=(-17*x^5-7*x^4+2*x^3+x^2-x+1)...
       /(x^6+11*x^5+48*x^4+106*x^3+125*x^2+75*x+17);
   f=ilaplace(G,x,t); fun=char(subs(G,x,'s')); % convert to string of s
   [t1,y1]=INVLAP(fun,0.01,5,100); tic, y0=subs(f,t,t1); toc
   y0=double(y0); err=norm((y1-y0)./y0)          % evaluate the error
```

If the number of points is increased from 100 to 5000, the elapsed time of `INVLAP()` *function is about 1.74 seconds, while* `ilaplace()` *and* `subs()` *functions may need more than 15 seconds. It can be seen that the numerical algorithm is more effective.*

In practical applications, the transfer functions $G(s)$ of a certain complicated system are known, and the Laplace transform is $R(s)$; the output signal $Y(s) = G(s)U(s)$ can be obtained. The method can also be used to find numerical solutions of the output signal.

Example 5.8 Consider the known Laplace expression

$$G(s) = \frac{(s^{0.4} + 0.4s^{0.2} + 0.5)}{\sqrt{s}(s^{0.2} + 0.02s^{0.1} + 0.6)^{0.4}(s^{0.3} + 0.5)^{0.6}}.$$

Please draw the inverse Laplace transform of $G(s)$ in the interval $t \in (0.01, 1)$.

Solution *Unlike the previous example, function* `ilaplace()` *cannot be used in finding the analytical solution to the inverse Laplace transform problems, and numerical approach is the only selection. Select* $N = 1000$, *the following statements can be used to find numerically the inverse Laplace transform, and the time domain function is obtained as shown in Figure 5.1. If the number of points N is increased to* $N = 5000$, *the same curve can be obtained, which validated the results.*

```
>> f='(s^0.4+ 0.4*s^0.2+0.5)/(s^0.2+0.02*s^0.1+0.6)^0.4/(s^0.3+0.5)^0.6';
   [t,y]=INVLAP(f,0.01,1,1000); plot(t,y) % no dot operation in string
```

It should be noted that, fractional-order polynomial $p^\gamma(x)$ is in fact corresponding to infinite series, and it is not possible to find the analytical solution in inverse Laplace transforms. Numerical techniques should be used to solve the original problems. The function is very effective and can be used in practical computations.

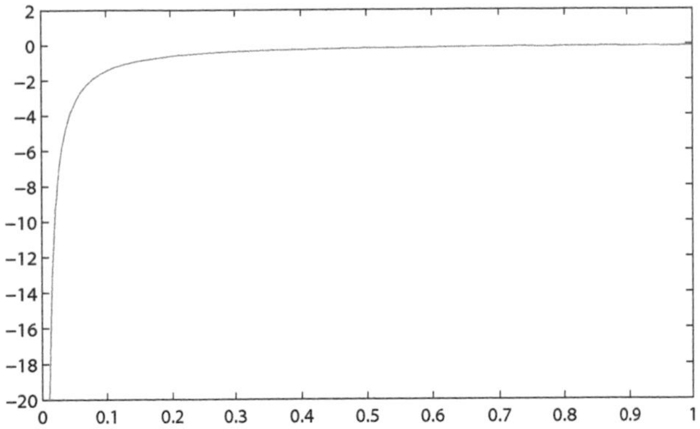

FIGURE 5.1: Time response obtained from numerical Laplace inverse.

If the analytical solution to the Laplace transform of the input $u(t)$ cannot be obtained, numerical approaches should also be used as well. In the source code of `INVLAP()` function, the loop structure is the major structure in the function, and in each step, an $\boldsymbol{s}$ vector is created. Based on the vector, numerical integration is used in Laplace transform

$$\mathscr{L}[u(t)] = \int_0^\infty u(t)\mathrm{e}^{-\boldsymbol{s}t}\,\mathrm{d}t = U(\boldsymbol{s}), \tag{5-1-9}$$

where, $\boldsymbol{s}$ is a vector. Since there is the $\mathrm{e}^{-\boldsymbol{s}t}$ term in the integral, and in practice, if the interval $(0, t_\mathrm{f})$ is large enough, finite time integrals are used to approximate infinite integrals. Thus, numerical Laplace transform can be implemented. If the inputs are described by the sample vectors $\boldsymbol{x}_0$ and $\boldsymbol{u}_0$, the practical input signal $u(t)$ can be approximated with interpolation methods. Details of signal interpolation will be discussed in Chapter 8.

The dot product of the Laplace input signal and the transfer function $G(\boldsymbol{s})$ can be obtained, and it is the Laplace transform of the output signal. In MATLAB 8.0, the MATLAB function `integral()` can be used to take numerical integrals with vector $\boldsymbol{s}$. In earlier versions, `quadv()` function can be used instead. Supposed G is the transfer function of the system, the syntaxes are

$[\boldsymbol{t},\boldsymbol{y}]$ = `num_laplace(`G`,`t_0`,`t_f`,`N`,`f`)`, % input function f
$[\boldsymbol{t},\boldsymbol{y}]$ = `num_laplace(`G`,`t_0`,`t_f`,`N`,`$\boldsymbol{x}_0$`,`$\boldsymbol{u}_0$`)`, % samples of inputs $\boldsymbol{x}_0$, $\boldsymbol{u}_0$

```
function [t,y]=num_laplace(G,t0,tf,nnt,x0,u0)
FF=strrep(strrep(strrep(G,'*','.*'),'/','./'),'^','.^');
a=6; ns=20; nd=19; t=linspace(t0,tf,nnt);
if t0==0, t=t(2:end); nnt=nnt-1; end
n=1:ns+1+nd; alfa=a+(n-1)*pi*1j; beta=-exp(a)*(-1).^n; n=1:nd
bdif=fliplr(cumsum(gamma(nd+1)./gamma(nd+2-n)./gamma(n)))./2^nd;
beta(ns+2:ns+1+nd)=beta(ns+2:ns+1+nd).*bdif; beta(1)=beta(1)/2;
for kt=1:nnt
   tt=t(kt); s=alfa/tt; bt=beta/tt;
   if isnumeric(x0), f=@(x)interp1(x0,u0,x,'spline').*exp(-s.*x);
   else, f=@(x)x0(x).*exp(-s.*x); end
   U=integral(f,t0,tf,'ArrayValued',true);
```

```
    btF=bt.*eval(FF).*U; y(kt)=sum(real(btF));
end
```

Example 5.9 Assume that the transfer function of the fractional-order model $G(s)$ is given in Example 5.8. Please draw the response of the system under the excitation of the input signal $u(t) = \mathrm{e}^{-0.3t}\sin t^2$.

Solution *The transfer function of the generalized fractional-order model can be entered in the same way, and the input function is described by an anonymous function. With the following statements, the output signal can be obtained as shown in Figure 5.2. The embedded numerical integration statements are quite consuming, and the following code takes more than 50 seconds.*

```
>> f=@(t)exp(-0.3*t).*sin(t.^2); % if input function is known
   G='(s^0.4+ 0.4*s^0.2+0.5)/(s^0.2+0.02*s^0.1+0.6)^0.4/(s^0.3+0.5)^0.6';
   tic, [t,y]=num_laplace(G,0,15,400,f); toc, plot(t,y)
```

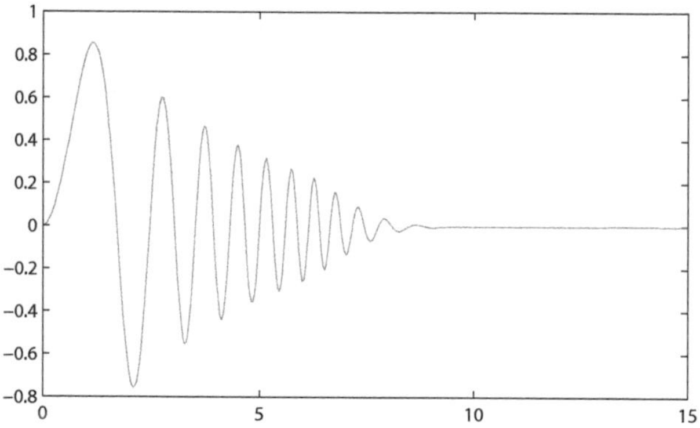

FIGURE 5.2: Output of complex fractional-order system.

Now consider again the input signal. Assume that the mathematical function of the input signal is unknown, only a set of samples in $t \in (0, 15)$ interval is known. Interpolation is used in the input function evaluation, thus, it is quite time consuming, and it may need about 9 minutes to find the numerical solution, even if the number of points in computation is reduced by half. The curve obtained is virtually the same.

```
>> x0=0:0.2:15; u0=exp(-0.3*x0).*sin(x0.^2); % if input samples are known
   tic, [t,y]=num_laplace(G,0,15,200,x0,u0); toc, plot(t,y)
```

5.2 Fourier Transforms and Their Inverses

5.2.1 Definitions and properties

The definition of Fourier transform is

$$\mathscr{F}[f(t)] = \int_{-\infty}^{\infty} f(t)\mathrm{e}^{-\mathrm{j}\omega t}\,\mathrm{d}t = F(\omega). \tag{5-2-1}$$

If the frequency-domain function $F(\omega)$ is known, the inverse Fourier transform can be obtained from

$$f(t) = \mathscr{F}^{-1}[F(\omega)] = \frac{1}{2\pi}\int_{-\infty}^{\infty} F(\omega)\mathrm{e}^{\mathrm{j}\omega t}\mathrm{d}\omega. \tag{5-2-2}$$

The following Fourier transform properties are summarized without proofs.

(i) **Linear property** $\mathscr{F}[af(t)\pm bg(t)]=a\mathscr{F}[f(t)]\pm b\mathscr{F}[g(t)]$ for scalars a and b.

(ii) **Shift property** $\mathscr{F}[f(t\pm a)] = \mathrm{e}^{\pm \mathrm{j}a\omega}F(\omega)$.

(iii) **Complex domain shift** $\mathscr{F}[\mathrm{e}^{\pm \mathrm{j}at}f(t)] = F(\omega \mp a)$.

(iv) **Differentiation property** $\mathscr{F}[\mathrm{d}f(t)/\,\mathrm{d}t] = \mathrm{j}\omega F(\omega)$. Generally, the Fourier transform of the nth derivative can be evaluated from

$$\mathscr{F}\left[\frac{\mathrm{d}^n}{\mathrm{d}t^n}f(t)\right] = (\mathrm{j}\omega)^n\mathscr{F}[f(t)]. \tag{5-2-3}$$

(v) **Integration property** $\mathscr{F}\left[\int_{-\infty}^{t} f(\tau)\,\mathrm{d}\tau\right] = \dfrac{F(\omega)}{\mathrm{j}\omega}$. Generally the Fourier transform to the nth order integral can be obtained from

$$\mathscr{F}\left[\int_{-\infty}^{t}\cdots\int_{-\infty}^{t} f(\tau)\,\mathrm{d}\tau^n\right] = \frac{\mathscr{F}[f(t)]}{(\mathrm{j}\omega)^n}. \tag{5-2-4}$$

(vi) **Scale property** $\mathscr{F}[f(at)] = \dfrac{1}{a}F\left(\dfrac{\omega}{a}\right)$.

(vii) **Convolution property** $\mathscr{F}[f(t)*g(t)] = \mathscr{F}[f(t)]\mathscr{F}[g(t)]$, where the definition of convolution is given in (5-1-6).

5.2.2 Solving Fourier transform problems

Similar to Laplace transform, the variable should be declared first, then, the function to be transformed can be defined in variable *fun*. With the Fourier transform solver `fourier()`, the Fourier transform can be obtained by

```
F=fourier(fun)          % Fourier transform
F=fourier(fun,v,u)      % transform the function of v into a function of u
```

Note that the definition of Fourier transform used in MATLAB is slightly different than the one defined in (5-2-1). In the function `fourier()`, the Fourier transform is defined as

$$\mathscr{F}[f(x)] = \int_{-\infty}^{\infty} f(x)\mathrm{e}^{-\mathrm{j}\omega x}\,\mathrm{d}x = F_1(\omega), \tag{5-2-5}$$

and it can be seen that the difference is the factor $\sqrt{2\pi}$. Throughout this chapter, the definition given in (5-2-5) is used.

Also, the definition of inverse Fourier transform is different from the one defined in (5-2-2). The inverse Fourier transform is defined as

$$f(x) = \mathscr{F}^{-1}[F_1(\omega)] = \frac{1}{2\pi}\int_{-\infty}^{\infty} F_1(\omega)\mathrm{e}^{\mathrm{j}\omega x}\mathrm{d}\omega, \tag{5-2-6}$$

and this definition will be used throughout this chapter.

The inverse Fourier transform can be obtained using `ifourier()` function, with the syntaxes

```
f=ifourier(fun)          % inverse Fourier transform
f=ifourier(fun,u,v)      % transform the function of u into a function of v
```

Example 5.10 Compute the Fourier transform for $f(t) = 1/(t^2 + a^2)$ where $a > 0$.

Solution *The following statements can be used to find the Fourier transform.*

```
>> syms t w; syms a positive; f=1/(t^2+a^2); F=fourier(f,t,w)
```

The result is

$$F = \pi(\mathrm{e}^{-a\omega}\mathrm{Heaviside}(\omega) + \mathrm{e}^{a\omega}\mathrm{Heaviside}(-\omega))/a,$$

where, the function Heaviside(ω) is the step function of ω, which is also referred to as the Heaviside function. *If $\omega \geqslant 0$,* Heaviside(ω) *is 1, otherwise, it is 0. When $\omega \leqslant 0$,* Heaviside($-\omega$) *returns 1, otherwise, it is 0. Assuming that $\omega > 0$, then, F can be simplified as $\pi\mathrm{e}^{-a\omega}/a$. While $\omega < 0$, F can be simplified as $\pi\mathrm{e}^{a\omega}/a$. The result can be summarized as $\mathscr{F}[f(t)] = \pi\mathrm{e}^{-a|\omega|}/a$.*

It should be noted that the simplest result obtained above cannot be derived automatically with MATLAB. The inverse Fourier transform can be obtained using the following statement

```
>> syms t w; syms a positive; f=pi*exp(-a*abs(w))/a; f1=ifourier(f)
```

and the result is $f_1 = \dfrac{1}{a^2 + x^2}$*, showing that the original function can be restored.*

Example 5.11 Compute the Fourier transform for $f(t) = \sin^2(at)/t$ with $a > 0$.

Solution *The Fourier transform can be obtained with*

```
>> syms t w; syms a positive; f=sin(a*t)^2/t; fourier(f,t,w)
```

and the result is

$$F = \mathrm{j}\pi(\mathrm{Heaviside}(\omega - 2a) + \mathrm{Heaviside}(\omega + 2a) - 2\mathrm{Heaviside}(\omega))/2,$$

and again the result is based on the Heaviside() *function. When $\omega > 2a$, the three* Heaviside() *are all 1. Thus, $F(\omega) = 0$. If $\omega \leqslant -2a$, the three functions are all 0, which means $F(\omega) = 0$. If $0 < \omega < 2a$, since the second and third* Heaviside() *functions are 1, $F(\omega) = -\mathrm{j}\pi/2$. When $0 > \omega > -2a$, $F(\omega) = \mathrm{j}\pi/2$. The Fourier transform of the original function can be simplified manually as*

$$\mathscr{F}[f(t)] = \begin{cases} 0, & |\omega| > 2a \\ -\mathrm{j}\pi\ \mathrm{sign}(\omega)/2, & |\omega| < 2a. \end{cases}$$

Example 5.12 Now consider a more complicated problem. Assume that the function is $f(t) = \mathrm{e}^{-a|t|}/\sqrt{|t|}$. Use the MATLAB function `fourier()` and the direct integration method to solve the Fourier transform problem.

Solution *Now considering first the existing function* `fourier()`*, the following statements can be used to solve the Fourier transform problem*

```
>> syms t; syms a positive
   f=exp(-a*abs(t))/sqrt(abs(t)); F=fourier(f) % fourier() function tried
```

which gives the string `fourier(exp(-a*abs(t))/abs(t)^(1/2),t,w)`*, it means nothing has been done, and the function call failed to return a solution. Note that the very latest version of MATLAB yields the analytical result*

$$F(\omega) = \frac{\sqrt{2\pi}}{2\sqrt{|\omega - \mathrm{j}a|}} + \frac{\sqrt{2\pi}}{2\sqrt{|\omega + \mathrm{j}a|}}.$$

Now, let us try the direct integration method based on the definition of Fourier transform (5-2-1)*. The original integration interval can be divided into two such that*

$$\int_{-\infty}^{\infty} f(t)\mathrm{e}^{-\mathrm{j}\omega t}\,\mathrm{d}t = \int_{-\infty}^{0} f(t)\mathrm{e}^{-\mathrm{j}\omega t}\,\mathrm{d}t + \int_{0}^{\infty} f(t)\mathrm{e}^{-\mathrm{j}\omega t}\,\mathrm{d}t.$$

Then, the following statements can be tried to solve the Fourier transform problem

```
>> syms a t positive; syms w real % alternatively try direct integration
   f1=exp(a*t)/sqrt(-t); f2=exp(-a*t)/sqrt(t); j=sym(sqrt(-1));
   F=int(f1*exp(-j*w*t),-inf,0)+int(f2*exp(-j*w*t),0,inf)
```

and it is also claimed that the integral failed. This example shows that not all functions have their corresponding Fourier transforms. For some functions, there exist no Fourier transforms.

5.2.3 Fourier sinusoidal and cosine transforms

The Fourier sinusoidal transform is defined as

$$\mathscr{F}_{\mathscr{S}}[f(t)] = \int_{0}^{\infty} f(t)\sin\omega t\,\mathrm{d}t = F_{\mathrm{s}}(\omega). \tag{5-2-7}$$

Fourier cosine transform is defined as

$$\mathscr{F}_{\mathscr{C}}[f(t)] = \int_{0}^{\infty} f(t)\cos\omega t\,\mathrm{d}t = F_{\mathrm{c}}(\omega). \tag{5-2-8}$$

Similarly, the inverse Fourier sinusoidal/cosine transforms are defined as

$$\mathscr{F}_{\mathscr{S}}^{-1}[F_{\mathrm{s}}(t)] = \frac{2}{\pi}\int_{-\infty}^{\infty} F_{\mathrm{s}}(\omega)\sin\omega t\mathrm{d}\omega \tag{5-2-9}$$

$$\mathscr{F}_{\mathscr{C}}^{-1}[F_{\mathrm{c}}(t)] = \frac{2}{\pi}\int_{-\infty}^{\infty} F_{\mathrm{c}}(\omega)\cos\omega t\mathrm{d}\omega. \tag{5-2-10}$$

There are no functions for Fourier sine and cosine transforms provided in MATLAB. Thus, direct integration can be applied using the Symbolic Math Toolbox. The following examples are given for computing the Fourier sine and cosine transforms.

Example 5.13 Compute the Fourier cosine transforms to the function $f(t) = t^n\mathrm{e}^{-at}, a > 0$ for $n = 1, 2, \cdots, 8$.

Solution *Since different values of n are to be explored, the loop structure of MATLAB can be used to solve the problem. The following statements can be given to evaluate the Fourier cosine transforms, and the results are given in Table 5.1.*

```
>> syms t w; syms a positive % declare symbolic variables
   for n=1:8                 % try Fourier cosine transfor for different n's
      f=t^n*exp(-a*t); F=int(f*cos(w*t),t,0,inf); simplify(F)
   end
```

TABLE 5.1: The Fourier cosine transforms for different values of n.

n	$\mathscr{F}_{\mathscr{C}}[f(t)]$
1~4	$\dfrac{a^2-\omega^2}{(a^2+\omega^2)^2}, -2\dfrac{(-a^2+3\omega^2)a}{(a^2+\omega^2)^3}, 6\dfrac{(-a^2+2a\omega+\omega^2)(-a^2-2a\omega+\omega^2)}{(a^2+\omega^2)^4}, 24\dfrac{a(a^4-10a^2\omega^2+5\omega^4)}{(a^2+\omega^2)^5}$
5,6	$-120\dfrac{(-a+\omega)(a+\omega)(a^2-4\omega a+\omega^2)(a^2+4\omega a+\omega^2)}{(a^2+\omega^2)^6}, 720\dfrac{(a^6-21a^4\omega^2+35\omega^4a^2-7\omega^6)a}{(a^2+\omega^2)^7}$
7	$5040\dfrac{(a^4+4a^3\omega-6a^2\omega^2-4a\omega^3+\omega^4)(a^4-4a^3\omega-6a^2\omega^2+4a\omega^3+\omega^4)}{(a^2+\omega^2)^8}$
8	$40320\dfrac{a(-a^2+3\omega^2)(-a^6+33a^4\omega^2-27a^2\omega^4+3\omega^6)}{(a^2+\omega^2)^9}$

From the mathematics handbook, the result is actually as follows:

$$\mathscr{F}_{\mathscr{C}}\left[t^n \mathrm{e}^{-at}\right] = n!\left(\frac{a}{a^2+\omega^2}\right)^{n+1}\sum_{m=0}^{[n/2]}(-1)^m \mathrm{C}_{n+1}^{2m+1}\left(\frac{\omega}{a}\right)^{2m+1}. \tag{5-2-11}$$

Example 5.14 Compute the Fourier cosine transform for

$$f(t)=\begin{cases}\cos t, & 0<x<a\\ 0, & \text{otherwise.}\end{cases}$$

Solution *It might be difficult to deal with piecewise functions, thus, the direct integration method can be used to solve the problem. From the piecewise function, it can be seen that the value of the integrand outside the interval* $(0,a)$ *is zero, thus, the Fourier cosine transform can be solved with the following statements*

```
>> syms t w; syms a positive; f=cos(t); % original function
   F=simplify(int(f*cos(w*t),t,0,a))    % Fourier cosine transform
```

and the following piecewise function is obtained

$$F(\omega)=\begin{cases}a/2+\sin 2a/4, & \omega\in\{-1,1\}\\ \sin\left[a\left(\omega-1\right)\right]/\left[2\left(\omega-1\right)\right]+\sin\left[a\left(\omega+1\right)\right]/\left[2\left(\omega+1\right)\right], & \omega\notin\{-1,1\}\\ 0, & a=\pi/2\ \&\ \omega=3.\end{cases}$$

5.2.4 Discrete Fourier sine, cosine transforms

Discrete Fourier sine and cosine transforms are also referred to as the *finite Fourier sine, cosine transforms*. The integration interval is changed from $t\in(0,\infty)$ into a finite one $t\in(0,a)$. Thus, the transforms are defined as

$$F_{\mathrm{s}}(k)=\int_0^a f(t)\sin\frac{k\pi t}{a}\,\mathrm{d}t,\quad F_{\mathrm{c}}(k)=\int_0^a f(t)\cos\frac{k\pi t}{a}\,\mathrm{d}t. \tag{5-2-12}$$

Similarly, the inverse transforms are defined as

$$f(t) = \frac{2}{a}\sum_{k=1}^{\infty} F_s(k)\sin\frac{k\pi t}{a}, \tag{5-2-13}$$

$$f(t) = \frac{1}{a}F_c(0) + \frac{2}{a}\sum_{k=1}^{\infty} F_c(k)\cos\frac{k\pi t}{a}. \tag{5-2-14}$$

The inverse transforms are no longer integrals, but infinite series. The finite transforms can also be obtained with the Symbolic Math Toolbox from definitions.

Example 5.15 Compute the transform for the piecewise function ($a > 0$)

$$f(t) = \begin{cases} t, & t \leqslant a/2 \\ a-t, & t > a/2. \end{cases}$$

Solution *The discrete Fourier sine transforms can be obtained directly as the sum of integrals of the two intervals from*

```
>> syms t k; syms a positive; f1=t; f2=a-t;
   Fs=int(f1*sin(k*pi*t/a),t,0,a/2)+int(f2*sin(k*pi*t/a),t,a/2,a);
   F=simplify(Fs) % compute and simplify the result
```

The simplified result is then

$$F = -\frac{2a^2\sin\pi k/2\,(\cos\pi k/2-1)}{\pi^2 k^2}.$$

The simplified result can further be simplified manually. Since it is known that $2\sin a\cos a = \sin 2a$, *the above result can be simplified as*

$$\mathscr{F}_{\mathscr{S}}[f(t)] = \frac{a^2(2\sin k\pi/2 - \sin k\pi)}{k^2\pi^2}.$$

Also, for integer k, $\sin k\pi \equiv 0$. *Therefore, the results can be simplified manually as*

$$\mathscr{F}_{\mathscr{S}}[f(t)] = \frac{2a^2}{k^2\pi^2}\sin\frac{k\pi}{2}.$$

If integer k *is used, and even and odd values of* k *can be tested individually*

```
>> assume(k,'integer'); k1=2*k; k2=2*k-1; % try odd and even seperately
   Fs1=int(f1*sin(k1*pi*t/a),t,0,a/2)+int(f2*sin(k1*pi*t/a),t,a/2,a);
   Fs2=int(f1*sin(k2*pi*t/a),t,0,a/2)+int(f2*sin(k2*pi*t/a),t,a/2,a);
   F1=simplify(Fs1), F2=simplify(Fs2)      % unify the results
```

The result is equivalent to the one obtained above.

5.2.5 Fast Fourier transforms

The Fourier transform for discrete sequences $x_i, i = 1, 2, \cdots, N$ is fundamental to digital signal processing. The discrete Fourier transform is defined as

$$X(k) = \sum_{i=1}^{N} x_i \mathrm{e}^{-2\pi \mathrm{j}(k-1)(i-1)/N}, \quad \text{where } 1 \leqslant k \leqslant N, \tag{5-2-15}$$

and its inverse transform is defined as

$$x(k) = \frac{1}{N}\sum_{i=1}^{N} X(i)\mathrm{e}^{2\pi \mathrm{j}(k-1)(i-1)/N}, \quad \text{where } 1 \leqslant k \leqslant N. \tag{5-2-16}$$

Fast Fourier transform (FFT) technique is the most effective and most practical way in solving Fourier transform. A built-in function `fft()` is provided such that $\boldsymbol{f}$ = `fft(`$\boldsymbol{x}$`)`. A useful feature `fft()` is that the length of the vector does not have to be constrained as 2^n. However, setting length of 2^n indeed makes the computation much faster.

For inverse FFT, the function `ifft()` can be used such that $\hat{\boldsymbol{x}}$ = `ifft(`$\boldsymbol{f}$`)`.

Example 5.16 Given a signal $x(t) = 12\sin(2\pi t + \pi/4) + 5\cos(8\pi t)$ and the step-size selected as h, draw the relationship between the frequency versus FFT magnitude. Observe whether the original signal can be using inverse FFT.

Solution *The time domain response of the function $x(t)$ can be obtained as shown in Figure 5.3 (a). It can be seen that it is hard to find anything about the properties of the time domain function.*

```
>> h=0.01; t=0:h:10; x=12*sin(2*pi*t+pi/4)+5*cos(2*pi*4*t); plot(t,x);
```

For a given sample time h, and the time samples t_i, the fundamental frequency is selected as $f_0 = 1/(ht_\mathrm{f})$, and a frequency vector can be generated as $\boldsymbol{f} = [f_0, 2f_0, 3f_0, \cdots]$. The following statements can be used

```
>> f=t/h/10; X=fft(x); L=1:floor(length(f)/2); stem(f(L),abs(X(L)))
```

and the relationship between frequency and FFT magnitude is shown in Figure 5.3(b). Here only half of the data are used to avoid aliasing phenomenon. It can be seen that there are two peaks in the magnitude plots, and the corresponding frequencies are 1Hz and 4Hz, which are the same as appeared in the original signal. Thus, FFT techniques can be used to examine which frequency components are important in the measured discrete signals.

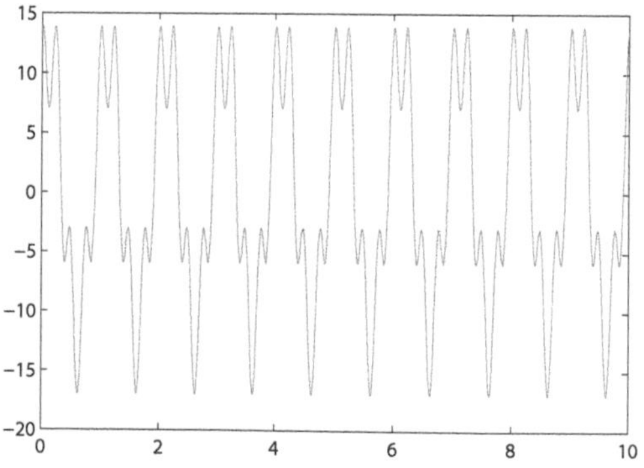

(a) time domain plot of original function

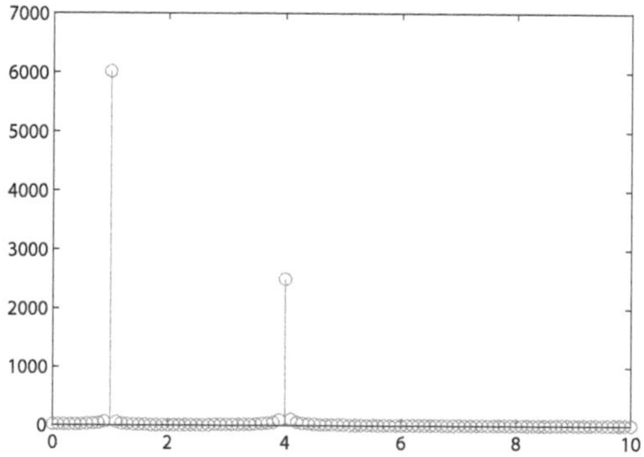

(b) FFT magnitude versus frequency

FIGURE 5.3: Fast Fourier transform analysis.

Inverse FFT can be evaluated directly by

```
>> ix=real(ifft(X)); plot(t,x,t,ix,':'); norm(x-ix) % restore the function
```

and it can be seen that the error in restoring the original function by inverse FFT is as small as 1.0289×10^{-13}. *The restored curve is exactly the same as the one in Figure 5.3 (a).*

Two-dimensional and even high-dimensional FFT and their inverses can be solved with the MATLAB functions `fft2()`, `ifft2()`, `fftn()` and `ifftn()`, respectively.

5.3 Other Integral Transforms

Apart from the well established Laplace and Fourier transforms, there are other integral transforms, such as Mellin transform and Hankel transform. The standard Symbolic Math Toolbox does not provide the functions for these problems. Thus, one can solve the problems by direct integration. It is recommended that the earlier versions of MATLAB, i.e., the versions R2008a or earlier, where Maple was used as its symbolic engine, be used to solve effectively the transformation problems.

5.3.1 Mellin transform

The Mellin transform is defined as

$$\mathscr{M}[f(x)] = \int_0^\infty f(x)x^{z-1}\,\mathrm{d}x = M(z). \tag{5-3-1}$$

Similarly, the inverse Mellin transform is defined as

$$f(x) = \mathscr{M}^{-1}[M(z)] = \frac{1}{\mathrm{j}2\pi}\int_{c-\mathrm{j}\infty}^{c+\mathrm{j}\infty} M(z)x^{-z}\,\mathrm{d}z. \tag{5-3-2}$$

There is no Mellin transform function provided in the Symbolic Math Toolbox. We can solve the problem using the direct integration method. Examples are given below to show the solution process.

Example 5.17 Compute the Mellin transform to $f(t) = \ln t/(t + a)$ with $a > 0$.

Solution *By definition, the following statements can be used to solve the problem*

```
>> syms t z; syms a positive; % try MATLAB R2008a or earlier versions
   f=log(t)/(t+a); M=simplify(int(f*t^(z-1),t,0,inf))
```

and the result can be simplified as

$$\mathscr{M}[f(t)] = a^{z-1}\pi\,(\ln a \sin\pi z - \pi\cos\pi z)\csc^2\pi z.$$

Example 5.18 Compute the Mellin transform to $f(t) = 1/(t + a)^n$, with $a > 0$. For different integer n, find the general form of the transform.

Solution *The following statements can be used to get the Mellin transform problem for* $n = 1, 2, \cdots, 8$,

```
>> syms t z real; syms a positive % try MATLAB R2008a or earlier versions
   for i=1:8, f=1/(t+a)^i; int(f*t^(z-1),t,0,inf), end
```

which gives respectively

$$a^{z-1}\pi\csc\pi z,$$
$$-a^{-2+z}\pi(z-1)\csc\pi z,$$
$$1/2a^{-3+z}\pi(-2+z)(z-1)\csc\pi z,$$
$$-1/6a^{z-4}\pi(z-1)(-2+z)(-3+z)\csc\pi z,$$
$$1/24a^{-5+z}\pi(z-4)(-3+z)(-2+z)(z-1)\csc\pi z,$$
$$-1/120a^{-6+z}\pi(-5+z)(z-4)(-3+z)(-2+z)(z-1)\csc\pi z,$$
$$1/720a^{-7+z}\pi(-6+z)(-5+z)(z-4)(-3+z)(-2+z)(z-1)\csc\pi z,$$
$$-1/5040a^{-8+z}\pi(-7+z)(-6+z)(-5+z)(z-4)(-3+z)(-2+z)(z-1)\csc\pi z.$$

Thus, by inspection, it can be concluded that the Mellin transform is generally given by

$$\mathscr{M}\left[\frac{1}{(t+a)^n}\right]=\frac{(-1)^{k-1}\pi}{(n-1)!}a^{z-n}\prod_{i=1}^{n-1}(z-i)\csc\pi z.$$

As in other transforms, there are a great many functions whose Mellin transforms have no analytical solutions. In this case, numerical Mellin transforms can be introduced. Numerical integrations can be used in solving numerical Mellin transform problems. A MATLAB function `mellin_trans()` is written, with the syntax $\boldsymbol{F}$=`mellin_trans(`*f*`,`*z*`,`*property pairs*`)`, where *property pairs* are the same as the ones used in `integral()` function.

```
function F=mellin_trans(f,z,varargin)
f1=@(x)f(x).*x.^(z-1); % declare integrand for Mellin transform
F=integral(f1,0,Inf,'ArrayValued',true,varargin{:});
```

Example 5.19 For a given function $f(x)=\sin(3x^{0.8})/(x+2)^{1.5}$, please solve numerically to find its Mellin transform.

Solution *It is obvious that there is no analytical solution to the Mellin transform problem for the given function. Numerical Mellin transform is the only choice. The original function can be expressed by an anonymous function first, then, numerical Mellin transform can be taken, and the plot is obtained as shown in Figure 5.4.*

```
>> f=@(x)sin(3*x.^0.8)./(x+2).^1.5; % specify original function
   z=0:0.05:1; F=mellin_trans(f,z); plot(z,F)
```

5.3.2 Hankel transform solutions

Hankel transform is another frequently used integral transform. The νth order Hankel transform is defined as

$$\mathscr{H}[f(t)]=\int_0^\infty tf(t)J_\nu(\omega t)\,\mathrm{d}t=H_\nu(\omega), \tag{5-3-3}$$

where $J_\nu(z)$ is the Bessel function of order ν, which can be evaluated under MATLAB with the statement $\boldsymbol{J}$=`besslej(`ν`,`z`)`. The inverse Hankel transform of order ν is defined as

$$\mathscr{H}^{-1}[H(\omega)]=\int_0^\infty \omega H_\nu(\omega)J_\nu(\omega t)\mathrm{d}\omega. \tag{5-3-4}$$

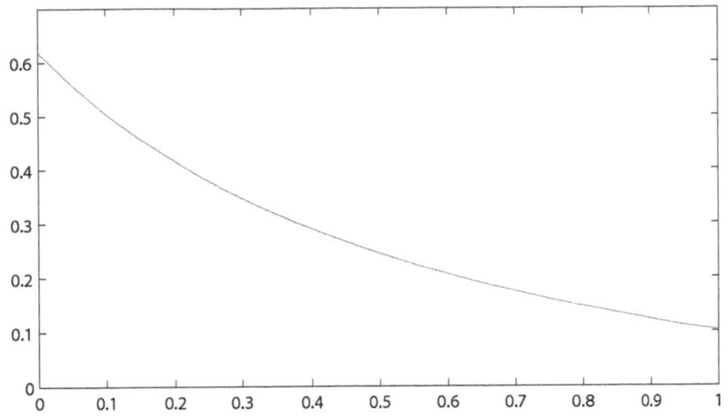

FIGURE 5.4: Numerical Mellin transform.

Example 5.20 Compute the zeroth-order Hankel transform to $f(t) = \mathrm{e}^{-a^2t^2/2}$, with $a > 0$.

Solution *With the existing MATLAB functions, the following statements can be used in finding the zeroth-order Hankel transform*

```
>> syms t w a positive; f=exp(-a^2*t^2/2);
   F=int(f*t*besselj(0,w*t),t,0,inf); F=simplify(F) % Hankel transform
   f1=int(w*F*besselj(0,w*t),w,0,inf) %  inverse Hankel transform
```

and the result obtained is $F = \mathrm{e}^{-w^2/(2a^2)}/a^2$. *The inverse Hankel transform to the result is also performed, and the original function is restored. If the order of Bessel function is changed from 0 to 1, the result contains Bessel function, and the inverse cannot be obtained.*

Currently, the above approach can only be used to a very limited category of functions, and to very low orders of Bessel function, such as $\nu = 0$, which means that most functions have no analytical solutions in Hankel transforms. Thus, numerical Hankel transform should be considered instead.

A MATLAB function can be written to implement Hankel transforms directly, with the syntax of $\boldsymbol{H}$ = `hankel_trans(`f`,`w`,`ν`,`*property pairs*`)`, where *property pairs* are the same as the ones used in `integral()` function.

```
function H=hankel_trans(f,w,nu,varargin)
F=@(t)t.*f(t).*besselj(nu,w*t); % specify integrand of Hankel transform
H=integral(F,0,Inf,'ArrayValued',true); % evaluate numerical integral
```

Example 5.21 Consider the function in the previous example, with $a = 2$. Please draw numerical Hankel transforms for different orders of Bessel functions.

Solution *The theoretical curve obtained earlier can be drawn directly, and superimposed on it, the Hankel transforms with different orders of Bessel functions can be drawn, as shown in Figure 5.5. It can be seen that the 0th order transform are almost identical. Also, the transforms of different orders can be obtained easily and quickly. It can also be seen that the decay of high order Hankel transforms are very slow, and the current curve is not adequate to evaluate the inverse Hankel transform numerically. In order to evaluate inverse transforms, larger ranges of* ω *should be obtained. If the function* $F(\omega)$ *is known, similar MATLAB statements can be used to evaluate numerically the inverse Hankel transforms.*

```
>> F1=subs(F,a,2); ezplot(F1,[0,10]); % theoretic results
```

```
a=2; f=@(t)exp(-a^2*t.^2/2); w=0:0.4:10;
for i=0:4, H=hankel_trans(f,w,i,'ArrayValued',true); line(w,H); end
```

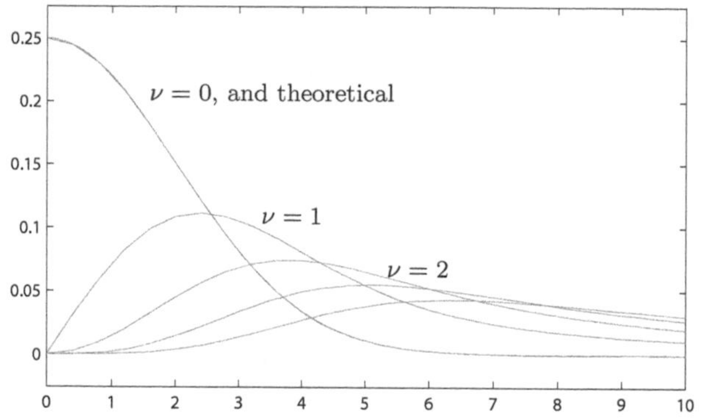

FIGURE 5.5: Numerical Hankel transform with different Bessel orders.

5.4 z Transforms and Their Inverses

5.4.1 Definitions and properties of z transforms and inverses

The z transform of a discrete sequence $f(k)$, $k = 1, 2, \cdots$ is defined as

$$\mathscr{Z}[f(k)] = \sum_{k=0}^{\infty} f(k)z^{-k} = F(z). \tag{5-4-1}$$

Similar to the presentation of Laplace and Fourier transforms, the properties of z transforms are summarized below without proofs.

(i) **Linear property** $\mathscr{Z}[af(k)\pm bg(k)]=a\mathscr{Z}[f(k)]\pm b\mathscr{Z}[g(k)]$ for any scalars a and b.

(ii) **Time domain forward translation property** For nonzero initial value problems, z transform can be calculated with

$$\mathscr{Z}[f(k+n)] = z^n F(z) - \sum_{i=0}^{n-1} z^{n-i} f(i). \tag{5-4-2}$$

Especially, for zero initial condition problems, $\mathscr{Z}[f(k+n)] = z^n F(z)$.

(iii) **Backward translation property** $\mathscr{Z}[f(k-n)] = z^{-n}F(z)$.

(iv) **z-domain proportional property** $\mathscr{Z}[r^{-k}f(k)] = F(rz)$.

(v) **Frequency-domain derivative property** $\mathscr{Z}[kf(k)] = -z\dfrac{\mathrm{d}F(z)}{\mathrm{d}z}$.

(vi) **Frequency-domain integral property** $\mathscr{Z}\left[\dfrac{f(k)}{k}\right] = \displaystyle\int_z^{\infty} \dfrac{F(\omega)}{\omega}\mathrm{d}\omega$.

(vii) **Initial value property** $\lim_{k\to 0} f(k) = \lim_{z\to\infty} F(z)$.

(viii) **Final value property** If $F(z)$ has no poles outside of the unit circle, the final value satisfies

$$\lim_{k\to\infty} f(k) = \lim_{z\to 1}(z-1)F(z). \tag{5-4-3}$$

(ix) **Convolution property** $\mathscr{Z}[f(k)*g(k)] = \mathscr{Z}[f(k)]\mathscr{Z}[g(k)]$ where the operator $*$ for discrete signals is defined as

$$f(k)*g(k) = \sum_{l=0}^{\infty} f(k)g(k-l). \tag{5-4-4}$$

For a z transform function $F(z)$, the inverse z transform is defined as

$$f(k) = \mathscr{Z}^{-1}[f(k)] = \frac{1}{\mathrm{j}2\pi}\oint F(z)z^{k-1}\,\mathrm{d}z. \tag{5-4-5}$$

5.4.2 Computations of z transform

Z transform and its inverse can be obtained directly with the functions `ztrans()` and `iztrans()` provided in the Symbolic Math Toolbox, with the syntaxes

```
F=ztrans(fun), F=ztrans(fun,k,z)     % z transform
F=iztrans(fun), F=iztrans(fun,z,k)   % inverse z transform
```

and if there is only one variable in *fun*, the arguments k and z can be omitted.

Example 5.22 Compute the z transform for $f(kT) = akT - 2 + (akT+2)\mathrm{e}^{-akT}$.

Solution *The z transform can be obtained directly with the statements*

```
>> syms a T k                            % declare the symbolic variables
   f=a*k*T-2+(a*k*T+2)*exp(-a*k*T);  % define the discrete function
   F=ztrans(f)                           % compute the z transform
```

and the result obtained is

$$\mathscr{Z}[f(kT)] = \frac{aTz}{(z-1)^2} - 2\frac{z}{z-1} + \frac{aTz\mathrm{e}^{-aT}}{(z-\mathrm{e}^{-aT})^2} + 2\,z\mathrm{e}^{aT}\left(\frac{z}{\mathrm{e}^{-aT}}-1\right)^{-1}.$$

Example 5.23 Consider the function $F(z) = \dfrac{q}{(z^{-1}-p)^m}, p \neq 0$. Find the inverse z transforms for different values of m, and then, try to summarize the general formula for the transform.

Solution *Let us try to solve the problem for $m = 1, 2, \cdots, 8$. The loop structure should be used, and the inverse z transforms can be obtained by*

```
>> syms p q z; assume(p~=0) % specify p≠0 and try different m's
   for m=1:8, disp(simplify(iztrans(q/(1/z-p)^m))), end
```

It can be seen from the displayed results that in some terms under the new versions, the results contain `nchoosek(n,k)` *function, which is the combination number, meaning that* $\mathrm{C}_n^k = n!/[k!(n-k)!]$.

The results can be simplified manually as

$-q/p(1/p)^n$,

$q/p^2(1+n)(1/p)^n$,

$-1/2q(1/p)^n(1+n)(2+n)/p^3$,

$1/6q(1/p)^n(3+n)(2+n)(1+n)/p^4$,

$-1/24q(1/p)^n(4+n)(3+n)(2+n)(1+n)/p^5$,

$1/120q(1/p)^n(5+n)(4+n)(3+n)(2+n)(1+n)/p^6$,

$-1/720q(1/p)^n(6+n)(5+n)(4+n)(3+n)(2+n)(1+n)/p^7$,

$1/5040q(1/p)^n(7+n)(6+n)(5+n)(4+n)(3+n)(2+n)(1+n)/p^8$.

It is summarized by inspection that the general form of the z transform is

$$\mathscr{Z}^{-1}\left[\frac{q}{(z^{-1}-p)^m}\right] = \frac{(-1)^m q}{(m-1)!\, p^{n+m}} \prod_{i=1}^{m-1}(n+i).$$

5.4.3 Bilateral z transforms

The previously defined z transforms are for the sum $n \geqslant 0$, and also known as *one sided z transform*. If the range of n is extended to the entire set of integers, the bilateral z transforms can be defined

$$\mathscr{Z}[f(k)] = \sum_{k=-\infty}^{\infty} f(k)z^{-k} = F(z). \tag{5-4-6}$$

There are no existing bilateral z transform facilities in MATLAB. Unfortunately, sometimes, the sum over $(-\infty,\infty)$ interval is not supported by the `symsum()` function. The original sum can be divided into the sums of two fields. From the definitions, the expression of bilateral z transform can be evaluated with

$F=$ `symsum(f*z^(-k),k,0,inf)` $+$ `symsum(f*z^(-k),k,-inf,-1)`

Example 5.24 Please find the bilateral z transform of the $f(n)$ [4]

$$f(n) = \begin{cases} 2^n, & n \geqslant 0 \\ -3^n, & n < 0. \end{cases}$$

Solution *From the definition of bilateral z transform, it can be seen that*

$$\mathscr{Z}[f(k)] = \sum_{k=-\infty}^{\infty} f(k)z^{-k} = \sum_{k=0}^{\infty} 2^k z^{-k} + \sum_{-\infty}^{-1} -3^k z^{-k}.$$

The following statements can be issued under old versions of MATLAB

```
>> syms z k; F=symsum(2^k*z^(-k),k,0,inf)+symsum(-3^k*z^(-k),k,-inf,-1)
```

and the bilateral z transform can be written as

$$F = \frac{z}{z-2} + \frac{z}{z-3}.$$

5.4.4 Numerical inverse z transform of rational functions

There are many functions whose inverse z transforms are not analytically obtainable with `iztrans()`. Even though for rational functions, the inverse z transforms cannot be found. Suppose the inverse z transform of a function can be written as

$$F\left(z^{-1}\right) = z^{-d}\frac{b_0 + b_1 z^{-1} + b_2 z^{-2} + \cdots + b_{m-1} z^{-(m-1)} + b_m z^{-m}}{a_0 + a_1 z^{-1} + a_2 z^{-2} + \cdots + a_{n-1} z^{-(n-1)} + a_n z^{-n}}, \tag{5-4-7}$$

and it can be regarded as the power series expansion, of the z^{-k} term, with

$$F\left(z^{-1}\right) = f_0 + f_1 z^{-1} + f_2 z^{-2} + \cdots = \sum_{k=0}^{\infty} f_k z^{-k}, \tag{5-4-8}$$

and it happens to be the definition of z transforms. Long division technique can be used to expand $F\left(z^{-1}\right)$, and a *long division* function is implemented with the syntax y= `inv_z(num,den,`d`,`N`)`, where, d is the pure delay step, and the coefficient vectors `num`= `[`b_0`,`b_1`,`b_2`,`$\cdots$`,`b_m`]`, `den`= `[`a_0`,`a_1`,`a_2`,`$\cdots$`,`a_n`]`, and the number of calculation points is $N = 10$. The inverse z transform sequence can be returned in vector $\boldsymbol{y}$.

```
function y=inv_z(num,den,varargin)
[d,N]=default_pars({0,10},varargin{:}); num=zeros(1,N);
for i=1:N-d, y(d+i)=num(1)/den(1);
   if length(num)>1, ii=2:length(den);
      if length(den)>length(num); num(length(den))=0; end
      num(ii)=num(ii)-y(end)*den(ii); num(1)=[];
end, end
```

Example 5.25 Please find numerically the inverse z transform of

$$G(z) = \frac{z^2 + 0.4}{z^5 - 4.1z^4 + 6.71z^3 - 5.481z^2 + 2.2356z - 0.3645}.$$

Solution *Both the numerator and denominator of $G(z)$ can be multiplied together by z^{-5}, and the following rational expression can be obtained*

$$F\left(z^{-1}\right) = z^{-3}\frac{1 + 0.4z^{-2}}{1 - 4.1z^{-1} + 6.71z^{-2} - 5.481z^{-3} + 2.2356z^{-4} - 0.3645z^{-5}}.$$

The following statements can be used to evaluate numerically the inverse z transform, and the sequence obtained is shown in Figure 5.6.

```
>> num=[1 0 0.4]; den=[1 -4.1 6.71 -5.481 2.2356 -0.3645];
   N=50; y=inv_z(num,den,3,N); t=0:(N-1); stem(t,y)
```

5.5 Essentials of Complex-valued Functions

By name, *complex-valued functions* are those whose independent variables are complex numbers. Since the complex data type is the fundamental data type of MATLAB, the previous MATLAB functions in calculus and linear algebra can be used directly in complex-valued function related problems. In this section, the graphics facilities and mapping display of complex-valued functions are introduced.

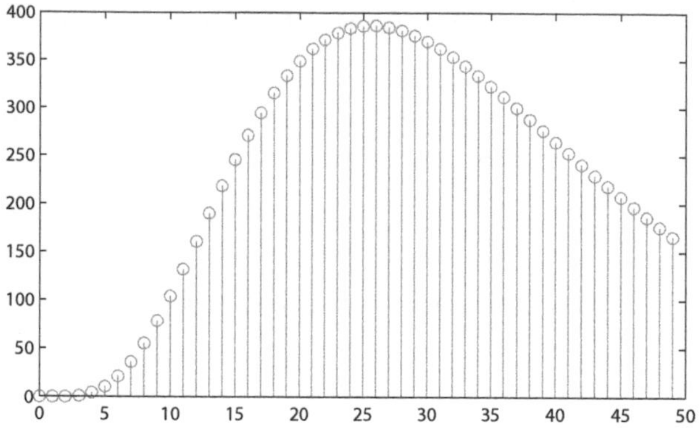

FIGURE 5.6: Numerical solution of inverse z transform.

5.5.1 Complex matrices and their manipulations

It has been pointed out earlier than complex matrices can be used directly in MATLAB in exactly the same way as real matrices. Assume that a complex matrix $\boldsymbol{Z}$ is given, it can be manipulated by the following functions:

(i) Compute complex conjugate with $\boldsymbol{Z}_1$=`conj(`$\boldsymbol{Z}$`)`.

(ii) Extract real and imaginary parts with $\boldsymbol{R}$=`real(`$\boldsymbol{Z}$`)`, $\boldsymbol{I}$=`imag(`$\boldsymbol{Z}$`)`.

(iii) Get magnitude and phase with $\boldsymbol{A}$=`abs(`$\boldsymbol{Z}$`)`, $\boldsymbol{P}$=`angle(`$\boldsymbol{Z}$`)`, while the unit in phase is radians.

Example 5.26 Reconsider the Jordan canonical form in Example 4.41. Since the matrix has complex eigenvalues, the transformation matrix should be modified manually. The following command can also be used to update the transformation matrix, and the same real Jordan matrix can be obtained.

```
>> A=[1,0,4,0; 0,-3,0,0; -2,2,-3,0; 0,0,0,-2]; [V,D]=eig(sym(A));
   V=real(V)+imag(V), D1=inv(V)*A*V % convert to Jordan real matrix
```

5.5.2 Mapping of complex-valued functions

If the independent variable z of function $f(z)$ is complex, the function is referred to as a *complex-valued function*. Since complex matrix is the fundamental data type in MATLAB, most of the algorithms in MATLAB do not specifically distinguish whether they are dealing with real or complex matrices. Therefore, most of the existing functions can be used in the computation of complex matrices directly. For instance, the analytical and numerical computation functions in calculus apply to complex functions directly.

Example 5.27 For complex-valued function $f(z)$, please evaluate $f^{(3)}(-\mathrm{j}\sqrt{5})$.

$$f(z) = \frac{z^2 + 3z + 4}{(z-1)^5}.$$

Solution *The following commands can be used directly, and the result obtained is* $d_3 = 0.8150 - \mathrm{j}0.6646$.

```
>> syms z; f=(z^2+3*z+4)/(z-1)^5; % describe the original symbolic function
   f3=diff(f,z,3); d3=simplify(subs(f3,z,-1i*sqrt(5))) % find derivative
```

Mapping is a very important kind of transformation in complex-valued functions. The so-called *mapping* is to perform variable substitution in complex-valued functions, i.e., by changing z into w, where $z = g(w)$. The commonly used mapping are translation mapping $z = w + \gamma$, inverse mapping $z = 1/w$ and bilinear mapping $z = (aw + b)/(cw + d)$, where γ is a fixed complex number, a, b, c, d are fixed real numbers. The translation mapping translates the origin of the function to point γ; inverse mapping converts a point inside a unit circle to one outside the circle; bilinear transform implements the mapping between lines and circles in different planes. The simplest way in implementing different kinds of mapping is using `subs()` function.

Example 5.28 Consider again the complex function $f(z) = (z^2+3z+4)/(z-1)^5$ in Example 5.27. Perform bilinear transformation $z = (s-1)/(s+1)$.

Solution *The expected bilinear mapping can be obtained directly with function* `subs()`*, and the result should be simplified.*

```
>> syms z s; f=(z^2+3*z+4)/(z-1)^5; F=simplify(subs(f,z,(s-1)/(s+1)))
```

The mapped function is $F(s) = -1/[16(s+1)^3(4s^2+3s+1)]$.

Example 5.29 This example shows how bilinear transformation $s = (z-1)/(z+1)$ can be used to map points inside a unit circle into points on the left-hand-side points in a plane. A series of points inside a unit circle can be generated and stored in complex matrix $\boldsymbol{z}$, as shown in Figure 5.7(a).

```
>> [x,y]=meshgrid(-1:0.1:1); ii=find(x.^2+y.^2<=1); x=x(ii); y=y(ii);
   z=x+sqrt(-1)*y; plot(z,'+'); hold on; ezplot('x^2+y^2=1')
```

With bilinear mapping of $s = (z-1)/(z+1)$, the matrix $\boldsymbol{z}$ can be mapped into complex matrix $\boldsymbol{s}$, as displayed in Figure 5.7(b). It can be seen that the points in the unit circle are converted to the ones in the left-hand-side plane.

```
>> s=(z-1)./(z+1); plot(s,'x') % mapping points in s plane
```

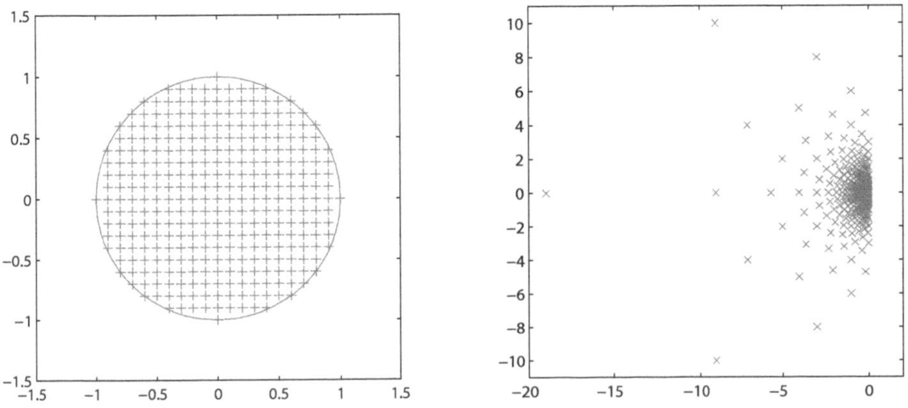

(a) points inside a unit circle (b) mapped points in s domain

FIGURE 5.7: Mapping from z domain into s domain.

5.5.3 Riemann surfaces

The mapping graphics of complex-valued functions are different from the 3D graphics discussed in Chapter 2. One should generate polar grid with the `cplxgrid()` function, and the Riemann surface can be shown with `cplxmap()` function, with the syntaxes

z=`cplxgrid(`n`)`	% generates polar grids
`cplxmap(`z`,`f`)`	% draw the 3D complex mapping Riemann surface

Example 5.30 Draw the Riemann surface of the function $f(z) = z^3 \sin z^2$.

Solution *The Riemann surface of the given function can be displayed with the following statements, as shown in Figure 5.8.*

```
>> z=cplxgrid(50); f=z.^3.*sin(z.^2); cplxmap(z,f) % draw Riemann surface
```

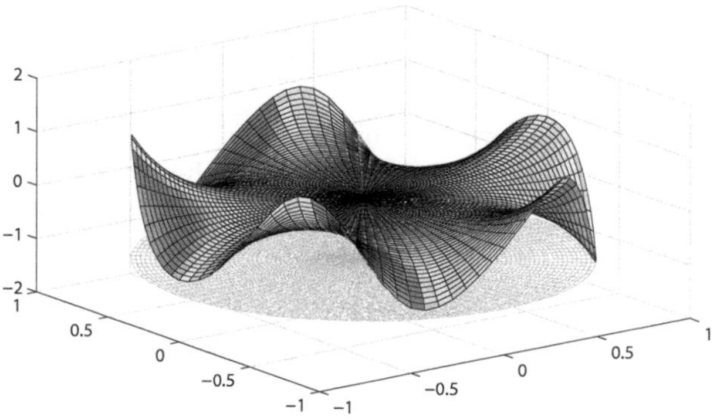

FIGURE 5.8: Riemann surface of a complex-valued function.

For a complex variable z, the Riemann surfaces of a multivalued function may have many branches. For instance, function $f(z) = \sqrt[n]{z}$ has n branches. MATLAB provides a function `cplxroot(`n`)` to draw directly all the branches of Riemann surfaces of the root mapping of $\sqrt[n]{z}$.

Example 5.31 Please draw the Riemann surfaces of $\sqrt[3]{z}$ and $\sqrt[4]{z}$.

Solution *The Riemann surfaces of $\sqrt[3]{z}$ and $\sqrt[4]{z}$ can be directly drawn with function* `cplxroot()`*, and the surfaces are shown respectively in Figures 5.9(a) and (b).*

```
>> subplot(121), cplxroot(3), subplot(122), cplxroot(4) % surfaces of roots
```

The limitations of `cplxroot()` function are that, it can only be used to draw Riemann surfaces of root functions, and cannot be used to draw other multi-valued functions. The existing function `cplxmap()` can be extended and saved as a new function `cplxmap1()`. Then, delete the `mesh()` and `hold` statement. Then, it can be used to draw Riemann surfaces of multi-valued complex functions.

Example 5.32 Use the modified function to draw the Riemann surfaces of $\sqrt[3]{z}$.

Solution *To redraw the Riemann surfaces of function $\sqrt[3]{z}$, it is known that if $f_1(z)$ is a*

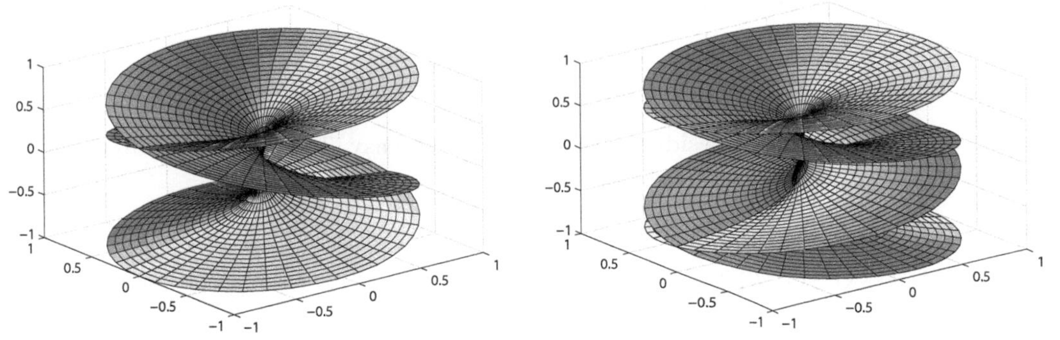

(a) cubic root
(b) the fourth root

FIGURE 5.9: Riemann surfaces of $\sqrt[n]{z}$.

branch of $f(z) = \sqrt[3]{z}$*, then, the other two branches can be obtained by* $f_1(z)\mathrm{e}^{-2\mathrm{j}\pi/3}$ *and* $f_1(z)\mathrm{e}^{-4\mathrm{j}\pi/3}$*. The following statements can be used to draw directly the Riemann surfaces of* $\sqrt[3]{z}$*, and the results are exactly the same as the ones in Figure 5.9(a).*

```
>> z=cplxgrid(30); f1=z.^(1/3); a=exp(-2i*pi/3); cplxmap1(z,f1)
   hold on; cplxmap1(z,a*f1); cplxmap1(z,a^2*f1); zlim([-1 1])
```

5.6 Solving Complex-valued Function Problems

In this section, commonly encountered problems in complex-valued functions will be manipulated. The concept of poles and residues are first studied. Partial fraction expansion and Laurent series expansions are discussed, and finally the closed-path integral of complex-valued function is studied.

5.6.1 Concept and computation of poles and residues

Before introducing the ideas of poles and residues, the concept of analytic complex functions is introduced. The complex function $f(z)$ is said to be analytic, if it is holomorphic, and the derivatives are finite, at all the points within the complex region. The points which make the function $f(z)$ not analytic are referred to as the *singularities*. The *poles* of functions are special singularities which makes the polynomial denominator of $f(z)$ equal zero.

Assume that $z = a$ is a pole of the function $f(z)$, and there exists the smallest positive integer m such that the new function $(z-a)^m f(z)$ is analytic at point $z = a$, then, point $z = a$ is referred to as a *pole of multiplicity* m. The poles can be obtained with `[p,m]=poles(f)`, and if there is more than one pole, they are returned in the column vector $\boldsymbol{p}$, together with their multiplicity vector $\boldsymbol{m}$. The poles within the interested interval (a,b) can be obtained with `[p,m]=poles(f,a,b)`.

If $z = a$ is a single pole, then, the *residue* is defined as

$$\mathrm{Res}\Big[f(z), z=a\Big] = \lim_{z\to a}(z-a)f(z). \tag{5-6-1}$$

If $z = a$ is a pole of multiplicity m, the residue is defined as

$$\operatorname{Res}\Big[f(z), z=a\Big] = \lim_{z\to a}\frac{1}{(m-1)!}\frac{\mathrm{d}^{m-1}}{\mathrm{d}z^{m-1}}\Big[f(z)(z-a)^m\Big]. \tag{5-6-2}$$

Thus, the evaluation of residues can be made very easy. The following statements can be used in evaluating the residues to different kinds of poles.

```
c=limit(F*(z-a),z,a)                              % single pole
c=limit(diff(F*(z-a)^m,z,m-1)/factorial(m-1),z,a)  % multiple poles
```

It can be seen that the first statement is a special case of statement 2, when $m = 1$. Thus, the second statement is used globally in evaluating the residues.

Example 5.33 Compute the residues of the function

$$f(z) = \frac{1}{z^3(z-1)}\sin\left(z+\frac{\pi}{3}\right)\mathrm{e}^{-2z}.$$

Solution *It can be seen from the original function that $z = 0$ is a pole of multiplicity 3, while $z = 1$ is a single pole. Therefore, the following MATLAB statements can be used to evaluate the residues at the two poles*

```
>> syms z; f=sin(z+pi/3)*exp(-2*z)/(z^3*(z-1)); [p,m]=poles(f)
   for i=1:length(p) % find residues for all the poles
      F=limit(diff(f*(z-p(i))^m(i),z,m(i)-1)/factorial(m(i)-1),z,p(i))
   end
```

and the residues are respectively $\dfrac{1}{2}-\dfrac{\sqrt{3}}{4}$ *for* $z=0$, *and* $\mathrm{e}^{-2}\sin\left(1+\dfrac{\pi}{3}\right)$ *for* $z=1$.

An algorithm is designed below to find the poles and residues of a given function

---

Require: Symbolic expression $f(x)$, optional interval (a, b)
  Find all the poles $\boldsymbol{p}$ and their multiplicities $\boldsymbol{m}$
  for $i = 1$ To m **do**
    Compute and store the residues computed from (3-4-6)
  end for.

An automatic pole and residue evaluation function `residuesym()`, with the syntax `[r,p,m]=residuesym(f,a,b)`, can be written based on the above algorithm, where a, and b can be omitted. The listing of function is

```
function [r,p,m]=residuesym(f,a,b), z=symvar(f);
if nargin==1, [p,m]=poles(f); else, [p,m]=poles(f,a,b); end
for k=1:length(p) % compute residues for all the poles
  r(k)=limit(diff(f*(z-p(k))^m(k),z,m(k)-1)/factorial(m(k)-1),z,p(k));
end
```

Example 5.34 Compute the residue of the function $f(z) = \dfrac{\sin z - z}{z^6}$.

Solution *It can be found by the following statements that the point $z = 0$ is a pole of multiplicity 3, and the residue is $r = 1/120$.*

```
>> syms z; f=(sin(z)-z)/z^6; [r,p,m]=residuesym(f)
```

It seems that the multiplicity of the pole is 6 rather than 3. Consider the Taylor series expansion to the $\sin z$ *function. It can be seen that*

$$f(z)=\frac{(z-z^3/6+z^5/120-z^7/5040+\cdots)-z}{z^6}=\frac{-1/6+z^3/120-z^5/5040+\cdots}{z^3}.$$

Therefore, the multiplicity of the pole is indeed 3 rather than 6.

In fact, in practical situations, we can try different values of k *in a loop, from* $k=1$*, to find the smallest* k*, such that*

$$\lim_{z\to a}\frac{\mathrm{d}^{k-1}}{\mathrm{d}z^{k-1}}\Big[(z-a)^k f(z)\Big]<\infty,$$

then, the value of k *can be regarded as the multiplicity of the pole. With function* `poles()`*, the multiplicity can also be measured accurately for this example.*

Also, it is interesting to note that for this example, even if k *is selected as a very large number, the value of the limit is still 1/120. Therefore, in practical applications, even though the multiplicity cannot be assessed correctly, a large number can be selected to find the correct residue.*

```
>> syms z; f=(sin(z)-z)/z^6; R2=limit(diff(f*z^2,z,1)/factorial(1),z,0)
   R3=limit(diff(f*z^3,z,2)/factorial(2),z,0),
   R20=limit(diff(f*z^20,z,19)/factorial(19),z,0)
```

Example 5.35 Compute the residues of the function $f(z)=\dfrac{1}{z\sin z}$.

Solution *Unfortunately the* `poles()` *function failed to find the poles and their multiplicities. It can be seen that there are infinite numbers of poles. At* $z=0$*, the multiplicity is 2, while at* $z=\pm k\pi$*, with integer* $k=1,2,\cdots$*, the multiplicity is 1. For the pole at* $z=0$*, the residue can then be found from*

```
>> syms z; f=1/(z*sin(z)); c0=limit(diff(f*z^2,z,1),z,0) % residue at 0
```

and the residue for $z=0$ *is 0.*

Now for the poles $z=\pm k\pi$*, some integers can be tried, for instance,* $k=1,2,3,4$*, and the residues for these poles can be obtained such that*

```
>> k=[-4 4 -3 3 -2 2 -1 1]; c=[]; % try different integers to find clues
   for kk=k; c=[c,limit(f*(z-kk*pi),z,kk*pi)]; end; c
```

and it can be seen that for the vector $\boldsymbol{k}=[-4,4,-3,3,-2,2,-1,1]$*, the residues evaluated are* $\boldsymbol{c}=\left[-\dfrac{1}{4\pi},\dfrac{1}{4\pi},\dfrac{1}{3\pi},-\dfrac{1}{3\pi},-\dfrac{1}{2\pi},\dfrac{1}{2\pi},\dfrac{1}{\pi},-\dfrac{1}{\pi}\right]$*. It can then be concluded that* $\mathrm{Res}[f(z),z=k\pi]=(-1)^k\dfrac{1}{k\pi}$*. In fact, with recent versions of MATLAB, the same results can be obtained with the following statements*

```
>> syms k; assume(k,'integer'), assumeAlso(k~=0); % k is a nonzero integer
   R=simplify(limit(f*(z-k*pi),z,k*pi))            % evaluate residue directly
```

5.6.2 Partial fraction expansion for rational functions

Consider the rational function

$$G(z)=\frac{B(z)}{A(z)}=\frac{b_1z^m+b_2z^{m-1}+\cdots+b_mz+b_{m+1}}{z^n+a_1z^{n-1}+a_2z^{n-2}+\cdots+a_{n-1}z+a_n}, \tag{5-6-3}$$

where a_i and b_i are all constants. The concept of coprimeness of rational functions is very important. The two polynomials $A(z)$ and $B(z)$ are *coprime* if there does not exist any common divisor containing z. The greatest common divisor of two polynomials can be very difficult to find manually. However, with help of the `gcd()` function provided in the Symbolic Math Toolbox of MATLAB, the greatest common divisor C can easily be found with the syntax C=`gcd(`A,B`)`, where A and B are the two polynomials. If C is a polynomial, then, the two polynomials are not coprime. The two polynomials can then be simplified to A/C and B/C, respectively.

Example 5.36 Check whether the two polynomials $A(z), B(z)$ are coprime or not

$$A(z)=z^4+7z^3+13z^2+19z+20,$$

$$B(z)=z^7+16z^6+103z^5+346z^4+655z^3+700z^2+393z+90.$$

Solution *The greatest common divisor can be obtained with the* `gcd()` *function*

```
>> syms z; A=z^4+7*z^3+13*z^2+19*z+20;
   B=z^7+16*z^6+103*z^5+346*z^4+655*z^3+700*z^2+393*z+90;
   d=gcd(A,B)  % find the greatest common divisor
```

and it can be seen that the greatest common divisor is $d(z)=(z+5)$. *Thus, the two polynomials are not coprime. The two polynomials can easily be reduced with*

```
>> N=simplify(A/d), D=simplify(B/d), F=N/D % find coprime rational function
```

then, $A(z)/d(z)=z^3+2z^2+3z+4$, *and* $B(z)/d(z)=(z+2)(z+3)^2(z+1)^3$. *Thus, the original rational function can be simplified as*

$$F=\frac{z^3+2z^2+3z+4}{(z+2)(z+3)^2(z+1)^3}.$$

If the polynomials $A(z)$ and $B(z)$ are coprime, and the roots $-p_i, i=1,2,\cdots,n$ of the polynomial equation $A(z)=0$ are all distinct, the original rational function $G(z)$ can be expanded into the following form

$$G(z)=\frac{r_1}{z+p_1}+\frac{r_2}{z+p_2}+\cdots+\frac{r_n}{z+p_n}, \tag{5-6-4}$$

and the expansion is referred to as the *partial fraction expansion*. In the expression, r_i are the residues, denoted as $\mathrm{Res}[G(-p_i)]$, which can be obtained from the limit formula such that

$$r_i=\mathrm{Res}[G(z),z=-p_i]=\lim_{z\to -p_i}G(z)(z+p_i). \tag{5-6-5}$$

If the term $(z+p_i)^k$ exists in the denominator, i.e., $-p_i$ is the pole of multiplicity k, the corresponding sub-expansion can be written as

$$\frac{r_i}{z+p_i}+\frac{r_{i+1}}{(z+p_i)^2}+\cdots+\frac{r_{i+k-1}}{(z+p_i)^k}. \tag{5-6-6}$$

Thus, the values r_{i+j-1} can be evaluated from the following formula

$$r_{i+j-1}=\frac{1}{(j-1)!}\lim_{z\to -p_i}\frac{\mathrm{d}^{j-1}}{\mathrm{d}z^{j-1}}\left[G(z)(z+p_i)^k\right],\ j=1,2,\cdots,k. \tag{5-6-7}$$

A numerical function `residue()` is provided in MATLAB, which can be used in the partial fraction expansion of a given rational function $G(z)$. The syntax of the function is `[r,p,K]=residue(b,a)`, where $\boldsymbol{a}=[1,a_1,a_2,\cdots,a_n]$, and $\boldsymbol{b}=[b_1,b_2,\cdots,b_m]$. The returned arguments $\boldsymbol{r}$ and $\boldsymbol{p}$ are vectors containing the r_i coefficients and $-p_i$ poles, in (5-6-4). If repeated poles exist, the r_i terms can be replaced with the coefficients in (5-6-6). The argument k is the direct term, which for the function satisfying $m<n$, K will return an empty matrix. The function can be used to automatically judge whether the pole $-p_i$ is repeated, so as to arrange the values of r_i. It is worth mentioning that, if the function has repeated poles, the numerical approach is unreliable. Symbolic partial fractional expansion should be performed.

Example 5.37 Compute the partial fraction expansion for the following function

$$G(z)=\frac{z^3+2z^2+3z+4}{z^6+11z^5+48z^4+106z^3+125z^2+75z+18}.$$

Solution *The following statements can be used to take partial fraction expansion*

```
>> n=[1,2,3,4]; d=[1,11,48,106,125,75,18]; format long % display format
   [r,p,k]=residue(n,d); [n,d1]=rat(r); [n,d1,p] % numerical solution
```

with $\boldsymbol{n}^{\mathrm{T}}=[-17,-7,2,1,-1,1]$, $\boldsymbol{d}_1^{\mathrm{T}}=[8,4,1,8,2,2]$, $\boldsymbol{p}^{\mathrm{T}}=[-3,-3,-2,-1,-1,-1]$, *where* $\boldsymbol{p}$ *is the vector of poles,* $\boldsymbol{n}$, $\boldsymbol{d}_1$ *are the corresponding numerators and denominators of the coefficients* $\boldsymbol{r}$. *It can be seen that* -3 *is a pole of multiplicity 2,* -2 *is a single pole, while* -1 *is a pole of multiplicity 3. Thus, the partial fraction expansion can be written as*

$$G(z)=-\frac{17}{8(z+3)}-\frac{7}{4(z+3)^2}+\frac{2}{z+2}+\frac{1}{8(z+1)}-\frac{1}{2(z+1)^2}+\frac{1}{2(z+1)^3}.$$

Example 5.38 Compute the partial fraction expansion of the following function

$$G(z)=\frac{2z^7+2z^3+8}{z^8+30z^7+386z^6+2772z^5+12093z^4+32598z^3+52520z^2+45600z+16000}.$$

Solution *With the* `residue()` *function, the numerical solutions can be obtained. And for this example, the partial fraction expansion can be obtained*

```
>> n=[2,0,0,0,2,0,0,8]; d=[1,30,386,2772,12093,32598,52520,45600,16000];
   [r,p]=residue(n,d)
```

and due to the limitations in the numerical results, it might be difficult to find the multiplicity of poles. Thus, the exact partial fraction expansion cannot be obtained. From the results obtained, it can be approximately *assumed that* $p_1=-5$ *and* $p_2=-4$ *are poles of multiplicity 3, while* $p_3=-2$ *and* $p_4=-1$ *are single poles. Thus, the partial fraction expansion can be written as*

$$\frac{49995.9030930686}{(z+5)}+\frac{28488.5832580441}{(z+5)^2}+\frac{13040.9999762507}{(z+5)^3}-\frac{50473.1527861460}{(z+4)}$$
$$\frac{21449.5555022347}{(z+4)^2}-\frac{5481.3333201362}{(z+4)^3}+\frac{1.2222222224}{(z+2)}+\frac{0.0023148148}{(z+1)}.$$

It should be noted that approximations are made in writing the above denominators, the original one is even more imprecise.

Clearly, a better analytical or symbolic function is expected without numerical issues that may cause the pole multiplicity issues. In new versions of MATLAB, the symbolic engine MuPAD provides a low-level function `partfrac()`, and can be called from MATLAB with $F=$ `feval(symengine,'partfrac',`f`)`.

An interface with the same name is written, with the syntaxes

$F=$ `partfrac2(`f`)`, $F=$ `partfrac2(`f,z`)` or $F=$ `partfrac2(`f`,'List')`,

where, the listing of the interface function `partfrac()` is

```
function Y=partfrac2(varargin)
Y=feval(symengine,'partfrac',varargin{:});
```

In the first two synatxes, the partial fraction expansion F is returned, while in the third one, $F(1)$ returns the coefficient vector, while $F(2)$, the denominator of each term.

Example 5.39 Compute the partial fraction expansion to $f(z)$ in Example 5.37.

Solution *The following statements can be used to solve the problem*

```
>> syms z;
   f=(z^3+2*z^2+3*z+4)/(z^6+11*z^5+48*z^4+106*z^3+125*z^2+75*z+18);
   G1=partfrac2(f), % or G2=feval(symengine,'partfrac',f)
```

and the result below is exactly the same as the one obtained earlier

$$G_1(z) = -\frac{17}{8(z+3)} - \frac{7}{4(z+3)^2} + \frac{2}{z+2} + \frac{1}{8(z+1)} - \frac{1}{2(z+1)^2} + \frac{1}{2(z+1)^3}.$$

Example 5.40 Now consider again the rational function $G(z)$ defined in Example 5.38. Write the partial fraction expansion using analytical methods.

Solution *In Example 5.38, the numerical approach was used, and the results may not be accurate. Thus, the problem can be explored with the symbolic function*

```
>> syms z
   G=(2*z^7+2*z^3+8)/(z^8+30*z^7+386*z^6+...
     2772*z^5+12093*z^4+32598*z^3+52520*z^2+45600*z+16000);
   f=partfrac2(G), simplify(f-G)
```

and then, the expected partial fraction expansion is

$$f(z) = \frac{13041}{(z+5)^3} + \frac{341863}{12(z+5)^2} + \frac{7198933}{144(z+5)} - \frac{16444}{3(z+4)^3} + \frac{193046}{9(z+4)^2}$$
$$-\frac{1349779}{27(z+4)} + \frac{11}{9(z+2)} + \frac{1}{432(z+1)}.$$

Compared with the results obtained in Example 5.38, the new result is more convincing. The difference between the expansion and the original function can be found with `simplify(`f-G`)` *which is 0.*

If the following statements are issued

```
>> Y=partfrac2(G,'List'), n=Y(1), d=Y(2)
```

the two vectors are returned,

$$n = \left[\frac{11}{9}, \frac{1}{432}, \frac{7198933}{144}, \frac{341863}{12}, 13041, \frac{-1349779}{27}, \frac{193046}{9}, \frac{-16444}{3}\right],$$

$$d = [z+2, z+1, z+5, (z+5)^2, (z+5)^3, z+4, (z+4)^2, (z+4)^3].$$

Example 5.41 For the non-coprime rational function in Example 5.36, the overloaded `residue()` function can be used to write out the partial fraction expansion to the rational function $G(z) = A(z)/B(z)$.

Solution *The following statements can be specified to solve the problem*

```
>> syms z; A=z^4+7*z^3+13*z^2+19*z+20;
   B=z^7+16*z^6+103*z^5+346*z^4+655*z^3+700*z^2+393*z+90;
   partfrac2(A/B,z)
```

and it can be seen that

$$\frac{A(z)}{B(z)} = -\frac{7}{4(z+3)^2} - \frac{17}{8(z+3)} + \frac{2}{(z+2)} + \frac{1}{2(z+1)^3} - \frac{1}{2(z+1)^2} + \frac{1}{8(z+1)},$$

where in the results, the term regarding $z+5$ does not exist at all, since simplification was performed within the `partfrac2()` *function already.*

Example 5.42 Compute the partial fraction expansion to the following function

$$G(z) = \frac{-17z^5 - 7z^4 + 2z^3 + z^2 - z + 1}{z^6 + 11z^5 + 48z^4 + 106z^3 + 125z^2 + 75z + 17}.$$

Solution *The numerical* `residue()` *can be used first*

```
>> num=[-17 -7 2 1 -1 1]; den=[1 11 48 106 125 75 17];
   [r,p,k]=residue(num,den); [r,p,k] % compute partial fraction numerically
```

and the partial fraction expansion can be written as

$$\frac{-556.256530687201}{z+3.261731010738} + \frac{0.212556796963}{z+0.520859605293}$$
$$\frac{0.879464926195 - \mathrm{j}5.497076257858}{z+2.53094582005 - \mathrm{j}0.39976310545} + \frac{0.879464926195 + \mathrm{j}5.497076257858}{z+2.53094582005 + \mathrm{j}0.39976310545}$$
$$+ \frac{268.64252201892 + \mathrm{j}349.12310949979}{z+2.53094582005 - \mathrm{j}0.39976310545} + \frac{268.64252201892 - \mathrm{j}349.12310949979}{z+2.53094582005 + \mathrm{j}0.39976310545}.$$

However, with the symbolic function `partfrac2()` *used, no analytical solution can be found, since there are no poles found analytically*

```
>> syms z;
   G=(-17*z^5-7*z^4+2*z^3+z^2-z+1)...
      /(z^6+11*z^5+48*z^4+106*z^3+125*z^2+75*z+17);
   G1=partfrac2(G,z)
```

If there exists irrational pole z_0 in the denominator equation $D(z) = 0$, there is no root-finding algorithm to express accurately irrational value of z_0 with finite digits, approximate value $\hat{z}_0$ can be used instead. It should be substituted into (5-6-2) to find the approximate residue

$$\mathrm{Res}[f(z), \hat{z}_0] = \lim_{z \to \hat{z}_0} \frac{1}{(m-1)!} \frac{\mathrm{d}^{m-1}}{\mathrm{d}z^{m-1}} \left[(z-\hat{z}_0)^m f(\hat{z}_0)\right]. \tag{5-6-8}$$

Assume that with vpa() function, all the approximate poles $z_i, i = 1, 2, \cdots, n$ can be obtained, and they can be used to compose the denominator. To replace the original denominator, the approximate function $f_1(z)$ can be used to replace $f(z)$ in (5-6-8). Thus, $f_1(z)$ and $(z - \hat{z}_0)^m$ can be canceled at $\hat{z}_0$ point, and the approximate residue can be found. The listing of the new function is

```
function f=partfrac1(F)
f=sym(0); z=symvar(F);
[num,den]=numden(F); x0=vpasolve(den); [x,ii]=sort(double(x0));
x0=x0(ii); x=[x0; rand(1)]; kvec=find(diff(double(x))~=0);
ee=x(kvec); kvec=[kvec(1); diff(kvec(:,1))];
a0=limit(den/z^length(x0),z,inf); F1=num/(a0*prod(z-x0));
for i=1:length(kvec), for j=1:kvec(i),
   m=kvec(i); z0=ee(i); k=subs(diff(F1*(z-z0)^m,z,j-1),z,z0);
   f=f+k/(z-z0)^(m-j+1)/factorial(j-1);
end, end
```

and the syntax of the function is $f=$partfrac1(F), where, F is the analytic expression of the rational function. The returned f is the approximate partial fractional expansion.

Example 5.43 Solve the partial fraction expansion problem in Example 5.42.

Solution *Since no exact factorization of the denominator exists, the* partfrac2() *function cannot be used. Function* partfrac1() *can be used instead to perform approximate partial fraction expansion*

```
>> syms z;
   G=(-17*z^5-7*z^4+2*z^3+z^2-z+1)...
        /(z^6+11*z^5+48*z^4+106*z^3+125*z^2+75*z+17);
   F=partfrac1(G) % combination of analytical and numerical
```

The approximate partial fraction expansion obtained is

$$F(z) = \frac{0.2125568}{z+0.52086} + \frac{0.8794649+5.49707626\mathrm{j}}{z+1.077759+0.6021066\mathrm{j}} + \frac{268.64252-349.1231095\mathrm{j}}{z+2.530946+0.399763\mathrm{j}}$$
$$+ \frac{556.25653}{z+3.261731} + \frac{0.8794649-5.49707626\mathrm{j}}{z+1.077759-0.6021066\mathrm{j}} + \frac{268.64252+349.1231095\mathrm{j}}{z+2.530946-0.399763\mathrm{j}}.$$

5.6.3 Inverse Laplace transform using PFEs

It has been shown that the Symbolic Math Toolbox function ilaplace() can tackle the inverse Laplace transform to the rational function problems very well. However, for a class of problems where irrational complex roots exist, it has been shown in Example 5.3 that the results obtained directly will have very poor readability.

It is found by observation of the results in Example 5.42 that some terms can be expressed by $(a+\mathrm{j}b)/(s+c+\mathrm{j}d)$, and there is also a complex conjugate term $(a-\mathrm{j}b)/(s+c-\mathrm{j}d)$. Thus, the two terms can be simplified such that

$$(a+\mathrm{j}b)\mathrm{e}^{(c+\mathrm{j}d)t} + (a-\mathrm{j}b)\mathrm{e}^{(c-\mathrm{j}d)t} = \alpha \mathrm{e}^{ct}\sin(dt+\phi), \tag{5-6-9}$$

where $\alpha = -2\sqrt{a^2+b^2}$, and $\phi = -\tan^{-1}(b/a)$.

Based on such an algorithm, a numerical function pfrac() can be written to enhance the facilities provided in the original function residue(). The listings of the function are

```
function [R,P,K]=pfrac(num,den)
[R,P,K]=residue(num,den);
for i=1:length(R),
   if imag(P(i))>eps, a=real(R(i)); b=imag(R(i));
      R(i)=-2*sqrt(a^2+b^2); R(i+1)=-atan2(a,b);
   elseif abs(imag(P(i)))<eps, R(i)=real(R(i));
end, end
```

with the syntax $[\boldsymbol{r},\boldsymbol{p},K]=$`pfrac(num,den)`, where the definitions of $\boldsymbol{p}$ and K are exactly the same as the `residue()` function, and $\boldsymbol{r}$ is defined slightly differently. If p_i is real, then, r_i is the same as the ones described in `residue()` function. However, if p_i is complex, then, r_i and r_{i+1}, return, respectively the values of α and ϕ.

Example 5.44 Reconsider the problem in Example 5.42. Alternatively, we use

```
>> num=[-17,-7,2,1,-1,1]; den=[1,11,48,106,125,75,17];
   [r,p,k]=pfrac(num,den); format long e; [r,p]
```

and the result is shown below

$$\begin{aligned}\mathscr{L}^{-1}[F(s)] &= -556.25653068675\mathrm{e}^{-3.2617310107386t}+2.1255679696\mathrm{e}^{-0.5208596052932t}\\ &-881.03518709\mathrm{e}^{-2.530945820048808t}\sin(0.39976310544995t-0.65585087707)\\ &-11.13396711709\mathrm{e}^{-1.07775887193353t}\sin(0.6021065910608t-2.9829493242804),\end{aligned}$$

which is in a much more readable form.

5.6.4 Laurent series expansions

We have learnt Taylor series in Chapter 3, where the function $f(x)$ can be expanded as an infinite series of $(x-x_0)$ polynomials. Laurent series is an extension of such a series.

If $f(z)$ is analytic within a ring $\mathscr{D}$, $R_1<|z-z_0|<R_2$, where, $0\leqslant R_1<R_2<+\infty$, the Laurent series can be written as

$$f(z)=\sum_{k=-\infty}^{\infty}c_k(z-z_0)^k, \tag{5-6-10}$$

where, the coefficients can be calculated from

$$c_k=\frac{1}{2\pi\mathrm{j}}\int_{|z-z_0|=\rho}\frac{f(\zeta)}{(\zeta-z_0)^{k+1}}\,\mathrm{d}\zeta, \tag{5-6-11}$$

where $|z-z_0|<\rho$ can be any circle, with $R_1<\rho<R_2$. If $f(z)$ is analytic in $\mathscr{D}$, the Laurent series is unique.

It is usually extremely difficult to calculate the coefficients with (5-6-11), an alternative method can be taken, to calculate the series. Suppose the original function $f(z)$ can be partitioned into the product of two subfunctions $f(z)=f_1(z)f_2(z)$, where $f_2(z)$ are suitable for conventional Taylor series expansion, and the other part, $f_1(z)$, is to be expanded as the series of $(z-z_0)^k$, when k is negative. Variable substitution $x=1/(z-z_0)$ can be taken first, from it the original function can be substituted into a function of x, with $z=(1+xz_0)/x$, and Taylor series about x can be obtained and can be denoted by $F_1(x)$. Then, the variable can be substituted back to z with $x=1/(z-z_0)$. The method can be demonstrated with an example.

Example 5.45 Please write out the Laurent series of function $f(z) = z^2 e^{1/z}$.

Solution *There might be odd behavior at $z = 0$, due to the existence of $e^{1/z}$, however, $z = 0$ is not a pole, but an* essential singularity. *From the original function, it can be seen that it is easier to partition the original function into $f_1(z) = e^{1/z}$, and $f_2(z) = z^2$. With variable substitution $z = 1/x$, and the Taylor series expansion of the function $f_1(x)$ can be obtained. Then, the variables in the expression can be converted back to the function of z, by taking $x = 1/z$. The term $f_2(z) = z^2$ can be multiplied back to form the Laurent series*

```
>> syms x z; f1=exp(1/z); f2=z^2; f1a=subs(f1,z,1/x);
   F1a=taylor(f1a,x,'Order',7); F=simplify(f2*subs(F1a,x,1/z))
```

and the Laurent series is

$$F(z) = z^2 + z + \frac{1}{2} + \frac{1}{6z} + \frac{1}{24z^2} + \frac{1}{123z^3} + \frac{1}{720z^4} + \frac{1}{5040z^5} + \cdots.$$

In a relatively large interval of $z \in (-20, 20)$, the curves of the finite term Laurent series $F(z)$ and the original function $f(z)$ can be drawn directly with the following statements, as shown in Figure 5.10, and it can be seen that, apart from the very tiny neighborhood of essential singularity, $z = 0$, the two curves are almost identical.

```
>> ezplot(F,[-20,20]), hold on; ezplot(f1*f2,[-20,20]), plot(0,1/6,'o')
```

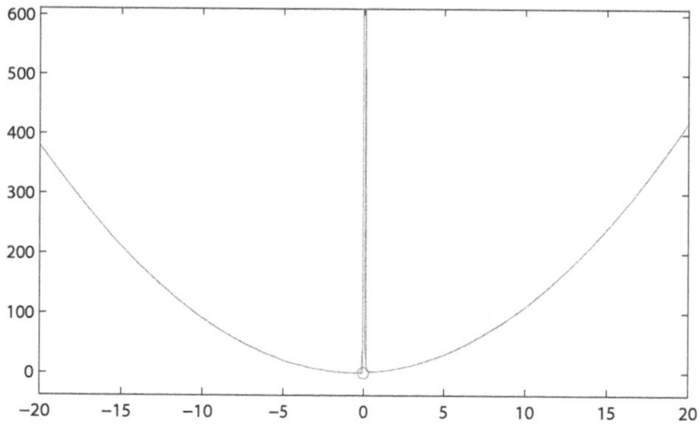

FIGURE 5.10: Output signal of the difference equation.

For a given Laurent series, it can be seen that there are infinite terms in $F(z)$ has z^{-k} as denominators, due to the existence of the essential singularity, and the residue is defined as the value of c_{-1}, the coefficient of z^{-1} term.

Example 5.46 Now consider the complex rational function $f(z) = \dfrac{1}{z-1} + \dfrac{1}{z-2j}$. How to find the Laurent series expansion?

Solution *The function $f(z)$ is clearly not analytic at the poles $z = 1$ and $z = 2j$. The regions can be partitioned as, specifically, (i) the disk $|z| < 1$, (ii) $1 < |z| < 2$, the ring, and (iii) $\infty > |z| > 2$, the ring. In the three cases*

(i) If $|z| < 1$, it is implied that $|z| < 2$, or $|z/2| < 1$ is satisfied. Taylor series is sufficient, and with the well-known expansion formula

$$\frac{1}{1-u} = \sum_{k=0}^{\infty} u^k, \text{ convergent when } |u| < 1,$$

the Taylor series can be written as

$$F_1(z) = \frac{-1}{1-z} + \frac{-1/2\mathrm{j}}{1-z/(2\mathrm{j})} = -\sum_{k=0}^{\infty}\left(1+\frac{1}{(2\mathrm{j})^{k+1}}\right)z^k.$$

(ii) If $1 < |z| < 2$, function $f(z)$ is analytic, and since the convergent condition is not satisfied for the first term, $1/(z-1)$, it can be rewritten as $(1/z)/(1-1/z)$, Taylor series expansion to $1/z$ can be made, while the Taylor expansion to the second term is valid. Thus, the Laurent series can be written as

```
>> syms z x; f1=1/(z-1); f1a=subs(f1,z,1/x);
   F2a=taylor(f1a,'Order',6); F2=subs(F2a,x,1/z)
```

and the Laurent series can be written as

$$F_2(z) = \sum_{k=-\infty}^{-1} z^k - \sum_{k=0}^{\infty}\frac{1}{(2\mathrm{j})^{k+1}}z^k.$$

(iii) If $|z| > 2$, expansions cannot be made on z, and the Taylor series expansions on $1/z$ should be made

```
>> f3=1/(z-1)+1/(z-2i); f3a=subs(f3,z,1/x);
   F3a=taylor(f3a,'Order',6); F3=subs(F3a,x,1/z)
```

and the Laurent series can be written as

$$F_3(z) = \sum_{k=-\infty}^{-1}\left(1+\frac{1}{(2\mathrm{j})^{k+1}}\right)z^k.$$

It can be seen from this example that in the three different analytic regions, three different Laurent series can be obtained. Thus, the Laurent series expansion is a piecewise function.

Following the cases studied earlier, an algorithm for obtaining Laurent series for a given rational function is presented

---

Require: Rational function $f(z)$, the order n, with default 6
  Find all the poles $\boldsymbol{p}$ and sort according to $|\boldsymbol{p}|$
  Find different radius $\boldsymbol{R}$, and make a vector $\boldsymbol{v} = [0, \boldsymbol{R}]$
  for $i=1$ To the length of $\boldsymbol{v}$ **do**
    Partition $f(z) = F_1(z) + F_2(z)$, $F_2(z)$ with all the poles $|z| < R_{i+1}$
    Compute $f_1 \leftarrow$ Taylor series for $F_1(z)$, and $f_2 \leftarrow$ Taylor for $F_2(z)$ about $1/z$
    Compose $F(i) = f_1 + f_2$ as the symbolic expansion for $R_i < |z| < R_{i+1}$
  end for
  Generate the piecewise function F_0 of Laurent series expansion.

A MATLAB function `laurent_series()` is written based on the algorithm, and a piecewise function with different expansions can be returned.

```
function [F0,p,m,F]=laurent_series(f,n), [p,m]=poles(f);
STR=''; if nargin==1, n=6; end
syms z x; assume(z~=0); assume(x~=0); F2=0;
if length(p)==0, error('The poles cannot be found, failed.'); end
v=sort(unique([sym(0); abs(p)])); v0=[v; inf];
```

```
Fx=partfrac2(f,'List'); nv=Fx(1); dv=Fx(2); f=partfrac2(f,z)
for i=1:length(v), F1=f-F2;
   f1=taylor(F1,'Order',n); f2=subs(F2,z,1/x);
   f2=taylor(f2,'Order',n); f2=subs(f2,x,1/z); F(i)=f1+f2;
   v1=[char(v(i)) '<abs(z)']; F2=0;
   if i==length(v), str1=v1;
   else, str1=[v1 ' and abs(z)<' char(v(i+1))]; end
   str2=char(F(i)); STR=[STR, '''' str1 ''',''' str2 ''','];
   for j=1:length(nv), x0=solve(dv(j)); x0=x0(1);
      if abs(x0)<v0(i+1)+eps, F2=F2+nv(j)/dv(j); end
end, end
F0=eval(['piecewise(' STR(1:end-1) ');'])
```

For a given Laurent series, it can be seen that there are infinite terms in $F(z)$ has z^{-k} as denominators, due to the existence of essential singularities, and the residue is defined as the value of c_{-1}, the coefficient of z^{-1}.

Example 5.47 Consider again the rational function in Example 5.38. Please compute the Laurent series about $z = 0$.

Solution *The problem can be solved easily with the new function* `laurent_series()`

```
>> syms z
   G=(2*z^7+2*z^3+8)/(z^8+30*z^7+386*z^6+...
      2772*z^5+12093*z^4+32598*z^3+52520*z^2+45600*z+16000);
   F=laurent_series(G) % express Laurent series with a piecewise function
```

The mathematical interpretation of the piecewise results is

(i) When $|z| < 1$

$$F_1(z) = -\frac{22818679z^5}{6400000000} + \frac{221063z^4}{64000000} - \frac{4981z^3}{1600000} + \frac{121z^2}{50000} - \frac{57z}{40000} + \frac{1}{2000};$$

(ii) When $1 < |z| < 2$

$$F_2(z) = -\frac{216104333z^5}{172800000000} + \frac{1968701z^4}{1728000000} - \frac{34487z^3}{43200000} + \frac{71z^2}{675000} + \frac{961z}{1080000} - \frac{49}{27000} + \frac{1}{432z} - \frac{1}{432z^2} + \frac{1}{432z^3} - \frac{1}{432z^4} + \frac{1}{432z^5};$$

(iii) When $2 < |z| < 4$

$$F_3(z) = \frac{3083895667z^5}{172800000000} - \frac{64031299z^4}{1728000000} + \frac{3265513z^3}{43200000} - \frac{51527z^2}{337500} + \frac{330961z}{1080000} - \frac{16549}{27000} + \frac{529}{432z} - \frac{1057}{432z^2} + \frac{2113}{432z^3} - \frac{4225}{432z^4} + \frac{8449}{432z^5};$$

(iv) When $4 < |z| < 5$

$$F_4(z) = -\frac{342479989z^5}{56250000} + \frac{62140157z^4}{2250000} - \frac{56159897z^3}{450000} + \frac{252776089z^2}{450000} - \frac{226630597z}{90000} + \frac{202363009}{18000} + \frac{31883669}{144z^2} - \frac{7198645}{144z} - \frac{140679637}{144z^3} + \frac{618454229}{144z^4} - \frac{2709385813}{144z^5};$$

(v) When $|z| > 5$

$$F_5(z) = \frac{2}{z} - \frac{60}{z^2} + \frac{1028}{z^3} - \frac{13224}{z^4} + \frac{142048}{z^5} - \frac{1346208}{z^6} + \frac{11631876}{z^7}.$$

5.6.5 Computing closed-path integrals

Now consider the closed-path integral

$$I = \oint_\Gamma f(z)\,\mathrm{d}z, \tag{5-6-12}$$

where Γ is a closed-path in a counterclockwise direction. Suppose that the closed-path encircles m poles, p_i, $(i = 1, 2, \cdots, m)$. The residues $\mathrm{Res}[f(p_i)]$ of the poles can be obtained using the MATLAB statements given earlier. The closed-path integral of the $f(z)$ function can be calculated from

$$I = \oint_\Gamma f(z)\,\mathrm{d}z = \mathrm{j}2\pi \sum_{i=1}^{m} \mathrm{Res}[f(p_i)]. \tag{5-6-13}$$

It should be mentioned that the closed-path Γ can be of any shape, any curvature, the integral is only related to the poles it encircles. In other words, if the closed-path is quite complicated, it might be difficult to evaluate the integral using the curve integral method presented in Chapter 3, however, it can be evaluated directly by taking curve integrals of other simple closed-paths, such as circles, as long as they encircle the same poles.

If Γ is in a clockwise direction, the integral I should be multiplied by -1.

Example 5.48 Compute the closed-path integral on $|z|=6$, and $f(z)$ is given by

$$f(z) = \frac{2z^7 + 2z^3 + 8}{z^8+30z^7+386z^6+2772z^5+12093z^4+32598z^3+52520z^2+45600z+16000}.$$

Solution *It can be seen from Example 5.40 that the partial fraction expansion of the original function is*

$$f(z) = \frac{13041}{(z+5)^3} + \frac{341863}{12(z+5)^2} + \frac{7198933}{144(z+5)} - \frac{16444}{3(z+4)^3} + \frac{193046}{9(z+4)^2} - \frac{1349779}{27(z+4)} + \frac{11}{9(z+2)} + \frac{1}{432(z+1)}.$$

Therefore, the poles $p_1 = -1$ and $p_2 = -2$ are single poles, and $p_3 = -4$, $p_4 = -5$ are poles of multiplicity 3. Thus, the residues of the poles are the coefficients of the first degree terms in the expression. Also, it is known that the poles are all encircled by the $|z| = 6$ path. The closed-path integral solution can be found from

$$\oint_{|z|=6} f(z)\,\mathrm{d}z = \mathrm{j}2\pi \left[\frac{7198933}{144} - \frac{1349779}{27} + \frac{11}{9} + \frac{1}{432}\right] = \mathrm{j}4\pi.$$

The same result can also be obtained with the function `residuesym()`

```
>> syms z
   G=(2*z^7+2*z^3+8)/(z^8+30*z^7+386*z^6+2772*z^5+12093*z^4+...
       32598*z^3+52520*z^2+45600*z+16000);
   [r,p,m]=residuesym(G); I=2i*pi*sum(r) % compute the integral
```

With the path integral method discussed in Chapter 3, the integral can be evaluated directly. The path Γ of the circle $|z| = 6$ can be expressed as $z = 6\cos t + \mathrm{j}6\sin t$, $t \in [0, 2\pi]$. Thus, the following statements can be given to find $I = \mathrm{j}4\pi$.

```
>> syms t; F=subs(G,z,6*cos(t)+6*sin(t)*sqrt(-1));          % integrand
   I=int(F*diff(6*cos(t)+6*sin(t)*sqrt(-1),t),t,0,2*pi) % path integral
```

If the closed-path is described by $|z| = 3$, *in counterclockwise direction, the two poles* p_3 *and* p_4 *should be excluded, since they are outside the circle. The integral can then be obtained with the following statements, and the result is*

$$I = 2\pi\mathrm{j}\left(\frac{11}{9}+\frac{1}{431}\right) = \frac{529\pi}{216}\mathrm{j}.$$

Also, the same result can be obtained with

```
>> [r,p,m]=residuesym(G,-3,3); I=2i*pi*sum(r)
```

The same result can also be obtained with the following integral evaluation

```
>> F=subs(G,z,3*cos(t)+3*sin(t)*sqrt(-1)); % alternative solution
   I=int(F*diff(3*cos(t)+3*sin(t)*sqrt(-1),t),t,0,2*pi)
```

Example 5.49 Compute the following closed-path integral, where the path Γ is the counterclockwise closed-path $|z| = 2$.

$$I = \oint_{\Gamma} \frac{1}{(z+\mathrm{j}1)^{10}(z-1)(z-3)}\,\mathrm{d}z.$$

Solution *It can be seen that the original function has single poles at* $z = 1$ *and* $z = 3$. *Also, the pole at* $z = -\mathrm{j}1$ *is a pole of multiplicity 10. The poles* $z = 1$ *and* $z = -\mathrm{j}1$ *are encircled by* Γ, *and* $z = 3$ *is not. Therefore, the closed-path integral can be evaluated with the following statements*

```
>> syms z t; f=1/((z+1i)^10*(z-1)*(z-3)); [r,p,m]=residuesym(f)
```

with the poles located at $\boldsymbol{p} = [-\mathrm{j}, 3, 1]$, *and it can be seen that since the first and the third poles are encircled by the* $|z| = 2$ *circle, the integral can be evaluated with*

```
>> R=2i*pi*sum(r([1,3])) % evaluate the integral
```

and the integral can be evaluated as $R = (237/312500000 + \mathrm{j}779/78125000)\pi$.

With direct path integral method, the same result can be obtained.

```
>> F=subs(f,z,2*cos(t)+2*sin(t)*sqrt(-1))                 % integrand
   I=int(F*diff(2*cos(t)+2*sin(t)*sqrt(-1),t),t,0,2*pi) % path integral
```

If the path Γ *is changed to* $|z| = 4$, *all three poles are encircled by the path. Therefore, the closed-path integral can be obtained with*

```
>> R=2i*pi*sum(r) % evaluate the integral through residues
```

and the integral is 0, which can also be confirmed by direct integration method.

```
>> F=subs(f,z,4*cos(t)+4*sin(t)*sqrt(-1))
   I=int(F*diff(4*cos(t)+4*sin(t)*sqrt(-1),t),t,0,2*pi)
```

Example 5.50 Find the closed-path integral $I = \displaystyle\int_{|z|=1} z^2\mathrm{e}^{1/z}\,\mathrm{d}z$.

Solution *The Laurent series of the function has been obtained in Example 5.45, and for this function, it is clear that* $z = 0$ *is an essential singularity of the function. However, since it is not a pole, it cannot be obtained with* `poles()` *function. The residue of* $z = 0$ *is the Laurent series coefficient* $c_{-1} = 1/6$, *therefore, the closed-path integral can be obtained as* $I = 2\pi\mathrm{j}c_{-1} = \pi\mathrm{j}/3$. *The result can be validated with path integral commands*

```
>> syms z t; f=z^2*exp(1/z); F=subs(f,z,cos(t)+sin(t)*sqrt(-1));
   I=int(F*diff(cos(t)+sin(t)*sqrt(-1),t),t,0,2*pi) % direct path integral
```

In traditional complex function courses, the univariate integrals can also be converted to the closed-path integrals of complex functions by variable substitutions. For instance, the integral $\int_0^\infty \frac{\sin x}{x}\,dx$ is often solved in this way. With the use of MATLAB function `int()`, the above problem can be solved directly. Therefore, it is not recommended to convert it to complex functions.

```
>> syms x; I=int(sin(x)/x,x,0,inf) % direct integral is recommended
```

5.7 Solutions of Difference Equations

A linear difference equation is given by

$$\begin{aligned} &y[(k+n)T]+a_1y[(k+n-1)T]+a_2y[(k+n-2)T]+\cdots+a_ny(kT)\\ &\quad =b_1u[(k-d)T]+b_2u[(k-d-1)T]+\cdots+b_mu[(k-d-m+1)T], \end{aligned} \tag{5-7-1}$$

where, T is the sample time, also known as *sampling period*. Similar to continuous systems described by differential equations, the coefficients a_i and b_i are all constants, and the system is referred to as a *linear time-invariant discrete system*. Besides, the corresponding input and output signals can be expressed as $u(kT)$ and $y(kT)$, where $u(kT)$ is the input signal in the kth sample time, and $y(kT)$ is the output at the same time. For simplicity, denote $y(t) = y(kT)$, and denote $y[(k+i)T]$ by $y(t+i)$, the difference equation can be simply rewritten as

$$\begin{aligned} &y(t+n)+a_1y(t+n-1)+a_2y(t+n-2)+\cdots+a_ny(t)\\ &\quad =b_1u(t+m-d)+b_2u(t+m-d-1)+\cdots+b_{m+1}u(t-d). \end{aligned} \tag{5-7-2}$$

5.7.1 Analytical solutions of linear difference equations

For the linear difference equation given earlier, if the initial values of $y(0)$, $y(1)$, $\cdots$, $y(n-1)$ contain nonzero elements, z transform can be taken to both sides of (5-7-2), and we have

$$\begin{aligned} &z^nY(z)-\sum_{i=0}^{n-1}z^{n-i}y(i)+a_1z^{n-1}Y(z)-a_1\sum_{i=0}^{n-2}z^{n-i}y(i)+\cdots+a_nY(z)\\ &\quad =z^{-d}\left[b_1z^mU(z)-b_1\sum_{i=0}^{m-1}z^{n-i}u(i)+\cdots+b_{m+1}U(z)\right]. \end{aligned} \tag{5-7-3}$$

It can be found that

$$Y(z)=\frac{(b_1z^m+b_2z^{m-1}+\cdots+b_{m+1})z^{-d}U(z)+E(z)}{z^n+a_1z^{n-1}+a_2z^{n-2}+\cdots+a_n}, \tag{5-7-4}$$

where, $E(z)$ is the expression obtained from the input and output according to (5-4-2)

$$E(z)=\sum_{i=0}^{n-1} z^{n-i}y(i)-a_1\sum_{i=0}^{n-2} z^{n-i}y(i)-a_2\sum_{i=0}^{n-3} z^{n-i}y(i)-\cdots-a_{n-}zy(0)+\hat{u}(n), \quad (5\text{-}7\text{-}5)$$

where

$$\hat{u}(n)=-b_1\sum_{i=0}^{m-1} z^{n-i}u(i)-\cdots-b_mzu(0). \quad (5\text{-}7\text{-}6)$$

Taking inverse z transform to $Y(z)$, the analytical solution $y(t)$ can be obtained. Based on the previous approach, the MATLAB function for solving linear time-invariant difference equation can be written

```
function y=diff_eq(A,B,y0,U,d)
E=0; n=length(A)-1; syms z; if nargin==4, d=0; end
m=length(B)-1; u=iztrans(U); u0=subs(u,0:m-1);
for i=1:n, E=E+A(i)*y0(1:n+1-i)*[z.^(n+1-i:-1:1)].'; end
for i=1:m, E=E-B(i)*u0(1:m+1-i)*[z.^(m+1-i:-1:1)].'; end
Y=(poly2sym(B,z)*U*z^(-d)+E)/poly2sym(A,z); y=iztrans(Y);
```

with the syntax $\boldsymbol{y}$=diff_eq($\boldsymbol{A}$,$\boldsymbol{B}$,$\boldsymbol{y}_0$,U,d), where $\boldsymbol{A}$ and $\boldsymbol{B}$ are coefficient vectors, U is the z transform of the input signal, and $\boldsymbol{y}_0$ is the initial values of the output vector, d is the delay constant, with a default value of 0. The function can be used in finding the analytical solution of the difference equation.

Example 5.51 Please solve the following difference equation

$$48y(n+4)-76y(n+3)+44y(n+2)-11y(n+1)+y(n)=2u(n+2)+3u(n+1)+u(n),$$

where, $y(0)=1$, $y(1)=2$, $y(2)=0$, $y(3)=-1$, and the input is $u(n)=(1/5)^n$.

Solution *The vectors* $\boldsymbol{A}$ *and* $\boldsymbol{B}$ *can be extracted directly from the equation, and the initial vector and input can also be entered into MATLAB environment. Function* `diff_eq()` *can then be used to solve the difference equation*

```
>> syms z n; u=(1/5)^n; U=ztrans(u);        % evaluate z transform of the input
   y=diff_eq([48 -76 44 -11 1],[2 3 1],[1 2 0 -1],U) % analytical solution
   n0=0:20; y0=subs(y,n,n0); stem(n0,y0) % numerical response and plot
```

and it can be seen that the analytical solution is

$$y(n)=\frac{432}{5}\left(\frac{1}{3}\right)^n-\frac{26}{5}\left(\frac{1}{2}\right)^n-\frac{752}{5}\left(\frac{1}{4}\right)^n+\frac{175}{3}\left(\frac{1}{5}\right)^n-\frac{42}{5}\left(\frac{1}{2}\right)^n(n-1).$$

The output signal can also be obtained as shown in Figure 5.11, and it can be seen that the initial points are all obtained in the solution.

5.7.2 Numerical solutions of linear time varying difference equations

Linear time varying state space equation can be written as

$$\begin{cases} \boldsymbol{x}(k+1)=\boldsymbol{F}(k)\boldsymbol{x}(k)+\boldsymbol{G}(k)\boldsymbol{u}(k), \quad \boldsymbol{x}(0)=\boldsymbol{x}_0 \\ \boldsymbol{y}(k)=\boldsymbol{C}(k)\boldsymbol{x}(k)+\boldsymbol{D}(k)\boldsymbol{u}(k), \end{cases} \quad (5\text{-}7\text{-}7)$$

and with recursive approach, we have

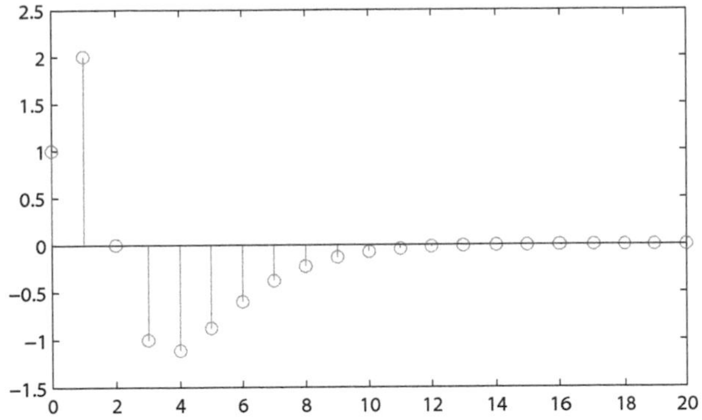

FIGURE 5.11: Output signal of the difference equation.

$$\begin{aligned}\boldsymbol{x}(1) &= \boldsymbol{F}(0)\boldsymbol{x}_0 + \boldsymbol{G}(0)\boldsymbol{u}(0)\\ \boldsymbol{x}(2) &= \boldsymbol{F}(1)\boldsymbol{x}(1) + \boldsymbol{G}(1)\boldsymbol{u}(1) = \boldsymbol{F}(1)\boldsymbol{F}(0)\boldsymbol{x}_0 + \boldsymbol{F}(1)\boldsymbol{G}(0)\boldsymbol{u}(0) + \boldsymbol{G}(1)\boldsymbol{u}(1)\\ &\vdots\end{aligned}$$

and the solutions of equation can be eventually obtained with

$$\begin{aligned}\boldsymbol{x}(k) &= \boldsymbol{F}(k-1)\boldsymbol{F}(k-2)\cdots\boldsymbol{F}(0)\boldsymbol{x}_0+\boldsymbol{G}(k-1)\boldsymbol{u}(k-1)\\ &\quad+\boldsymbol{F}(k-1)\boldsymbol{G}(k-2)\boldsymbol{u}(k-2)+\cdots+\boldsymbol{F}(k-1)\cdots\boldsymbol{F}(0)\boldsymbol{G}(0)\boldsymbol{u}(0)\\ &=\prod_{j=0}^{k-1}\boldsymbol{F}(j)\boldsymbol{x}_0+\sum_{i=0}^{k-1}\left[\prod_{j=i+1}^{k-1}\boldsymbol{F}(j)\right]\boldsymbol{G}(i)\boldsymbol{u}(i).\end{aligned} \tag{5-7-8}$$

If $\boldsymbol{F}(i), \boldsymbol{G}(i)$ are known, the above recursive algorithm can be used directly in the solution of time varying difference equation. Also, the iterative process can be used to solve the equation, from the known $\boldsymbol{x}(0)$, and find $\boldsymbol{x}(1)$ from (5-7-7), and then find $\boldsymbol{x}(2), \cdots$, so that the solutions at all the time instances can be iteratively obtained.

Example 5.52 Please solve the following time varying difference equation [5]

$$\begin{bmatrix}x_1(k+1)\\ x_2(k+1)\end{bmatrix}=\begin{bmatrix}0 & 1\\ 1 & \cos(k\pi)\end{bmatrix}\begin{bmatrix}x_1(k)\\ x_2(k)\end{bmatrix}+\begin{bmatrix}\sin(k\pi/2)\\ 1\end{bmatrix}u(k),$$

where, $\begin{bmatrix}x_1(0)\\ x_2(0)\end{bmatrix}=\begin{bmatrix}1\\ 1\end{bmatrix}$, and $u(k)=\begin{cases}1, & k=0,2,4,\cdots\\ -1, & k=1,3,5,\cdots\end{cases}$.

Solution *With iteration method, the* `for` *loop structure can be used, and the state variables at all the time instances can be obtained directly, as shown in Figure 5.12.*

```
>> x0=[1; 1]; x=x0;   % initial states
   for k=0:100, if rem(k,2)==0, u=1; else, u=-1; end % compute input
     F=[0 1; 1 cos(k*pi)]; G=[sin(k*pi/2); 1];      % compute F and G
     x1=F*x0+G*u; x0=x1; x=[x x1];                   % iterate for one step
   end
   subplot(211), stairs(x(1,:)), subplot(212), stairs(x(2,:))
```

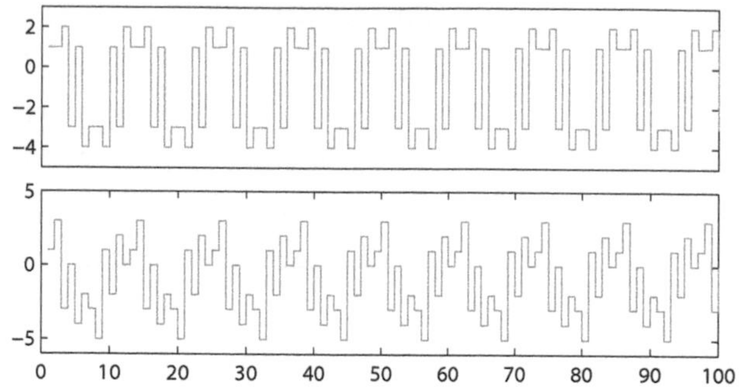

FIGURE 5.12: Response of a discrete time varying system.

5.7.3 Solutions of linear time-invariant difference equations

A linear time-invariant difference equation is $\boldsymbol{F}(k) = \cdots = \boldsymbol{F}(0) = \boldsymbol{F}$, $\boldsymbol{G}(k) = \cdots = \boldsymbol{G}(0) = \boldsymbol{G}$, and it can be found from (5-7-8) that

$$\boldsymbol{x}(k) = \boldsymbol{F}^k \boldsymbol{x}_0 + \sum_{i=0}^{k-1} \boldsymbol{F}^{k-i-1}\boldsymbol{G}\boldsymbol{u}(i). \tag{5-7-9}$$

However, it is not an easy thing to compute the sum in the above equation, an alternative method needs to be considered.

Now, let us reconsider the equation in (5-7-7), its time-invariant form can be rewritten as

$$\begin{cases} \boldsymbol{x}(k+1) = \boldsymbol{F}\boldsymbol{x}(k) + \boldsymbol{G}\boldsymbol{u}(k), \quad \boldsymbol{x}(0) = \boldsymbol{x}_0 \\ \boldsymbol{y}(k) = \boldsymbol{C}\boldsymbol{x}(k) + \boldsymbol{D}\boldsymbol{u}(k). \end{cases} \tag{5-7-10}$$

Take z transform to both side of the equation, with the property in (5-4-2), it can be found that

$$\boldsymbol{X}(z) = (z\boldsymbol{I} - \boldsymbol{F})^{-1}[z\boldsymbol{x}_0 + \boldsymbol{G}\boldsymbol{U}(z) - \boldsymbol{G}z\boldsymbol{u}_0], \tag{5-7-11}$$

and the analytical solution of the discrete equation can be derived as

$$\boldsymbol{x}(k) = \mathscr{Z}^{-1}\left[z(z\boldsymbol{I}-\boldsymbol{F})^{-1}\right]\boldsymbol{x}_0 + \mathscr{Z}^{-1}\left\{(z\boldsymbol{I}-\boldsymbol{F})^{-1}[\boldsymbol{G}\boldsymbol{U}(z)-\boldsymbol{G}z\boldsymbol{u}_0]\right\}. \tag{5-7-12}$$

The kth power of a constant matrix $\boldsymbol{F}$ can also be computed with

$$\boldsymbol{F}^k = \mathscr{Z}^{-1}\left[z(z\boldsymbol{I} - \boldsymbol{F})^{-1}\right]. \tag{5-7-13}$$

Example 5.53 For a state space model of a given discrete equation given below, find the analytical solution of the step responses of the states.

$$\boldsymbol{x}(k+1) = \begin{bmatrix} 11/6 & -5/4 & 3/4 & -1/3 \\ 1 & 0 & 0 & 0 \\ 0 & 1/2 & 0 & 0 \\ 0 & 0 & 1/4 & 0 \end{bmatrix} \boldsymbol{x}(k) + \begin{bmatrix} 4 \\ 0 \\ 0 \\ 0 \end{bmatrix} u(k), \quad \boldsymbol{x}_0 = 0.$$

Solution *With the following statements, the analytical solutions can be found*

```
>> F=sym([11/6 -5/4 3/4 -1/3; 1 0 0 0; 0 1/2 0 0; 0 0 1/4 0]);
   G=sym([4; 0; 0; 0]); syms z k; U=ztrans(sym(1));
   x=iztrans(inv(z*eye(4)-F)*G*U,z,k)
```

and the analytical solutions can be written as

$$\boldsymbol{x}(k)=\begin{bmatrix} 48(1/3)^k-48(1/2)^k k-72(1/2)^k-24(1/2)^k\mathrm{C}_{k-1}^2+48 \\ 144(1/3)^k-48(1/2)^k k-144(1/2)^k-48(1/2)^k\mathrm{C}_{k-1}^2+48 \\ 216(1/3)^k-192(1/2)^k-48(1/2)^k\mathrm{C}_{k-1}^2+24 \\ 24(1/2)^k k-24(1/2)^k\mathrm{C}_{k-1}^2-144(1/2)^k+162(1/3)^k+6. \end{bmatrix}$$

In fact, in the returned results, combination `nchoosek(n,k)` *is expressed, mathematically written as* C_n^k*, and it can be evaluated from* $\mathrm{C}_n^k=n!/(n-k)!/k!$*. For instance,* C_{k-1}^2 *can be simplified as a polynomial* $(k-1)(k-2)/2=(k^2-3k+2)/2$*. Therefore, the solution can be manually simplified as*

$$\boldsymbol{x}(k)=\begin{bmatrix} -12(8+k+k^2)(1/2)^k+48(1/3)^k+48 \\ 24(-8+k+2k^2)(1/2)^k+144(1/3)^k+48 \\ 24(-10+3k-k^2)(1/2)^k+216(1/3)^k \\ 12(-14+5k-k^2)(1/2)^k+162(1/3)^k+6 \end{bmatrix}.$$

Alternatively, since there is only the C_{k-1}^2 *term in the result, the following statement can be used to simplify the results, and the simplified version is equivalent to the one obtained manually.*

```
>> x1=simplify(subs(x,nchoosek(k-1,2),(k-1)*(k-2)/2))
```

Example 5.54 Consider the $\boldsymbol{A}^k$ problem in Example 4.74, please find $\boldsymbol{A}^k$ using inverse z transform.

Solution *It can be seen from* (5-7-13) *that, the matrix* $\boldsymbol{A}^k$ *can be found with the following statements, and the result is the same as the one obtained in Example 4.74.*

```
>> A=[-7,2,0,-1; 1,-4,2,1; 2,-1,-6,-1; -1,-1,0,-4]; % enter original matrix
   syms z k; F1=iztrans(z*inv(z*eye(4)-A),z,k);      % inverse z transform
   F2=simplify(subs(F1,nchoosek(k-1,2),(k-1)*(k-2)/2)) % simplification
```

5.7.4 Numerical solutions of nonlinear difference equations

Assume that the difference equation is given in explicit form

$$y(t)=f(t,y(t-1),\cdots,y(t-n),u(t),\cdots,u(t-m)), \tag{5-7-14}$$

recursive method can be used to find the solutions numerically.

Example 5.55 Assume a nonlinear difference equation is given by

$$y(t)=\frac{y(t-1)^2+1.1y(t-2)}{1+y(t-1)^2+0.2y(t-2)+0.4y(t-3)}+0.1u(t),$$

and assume the input signal is a sinusoidal function $u(t)=\sin t$, and the sample time is $T=0.05\,\mathrm{s}$, please find the numerical solutions of the difference equation.

Solution *A vector* $\boldsymbol{y}_0$ *for storing the past information is introduced, and its three components* $\boldsymbol{y}_0(1)$*,* $\boldsymbol{y}_0(2)$ *and* $\boldsymbol{y}_0(3)$ *store respectively* $y(t-3)$*,* $y(t-2)$ *and* $y(t-1)$*. Within the loop structure, the vector* $\boldsymbol{y}_0$ *is updated. Thus, with the following loop structure, the difference equation can be solved recursively, and the input and output signals are shown in Figure 5.13.*

```
>> y0=zeros(1,3); T=0.05; t=0:T:8*pi; u=sin(t); % initial setting
   for i=1:length(t)
      y(i)=(y0(3)^2+1.1*y0(2))/(1+y0(3)^2+0.2*y0(2)+0.4*y0(1))+...
            0.1*u(i); y0=[y0(2:3), y(i)]; % recursively update the vector
   end
   plot(t,y,t,u)
```

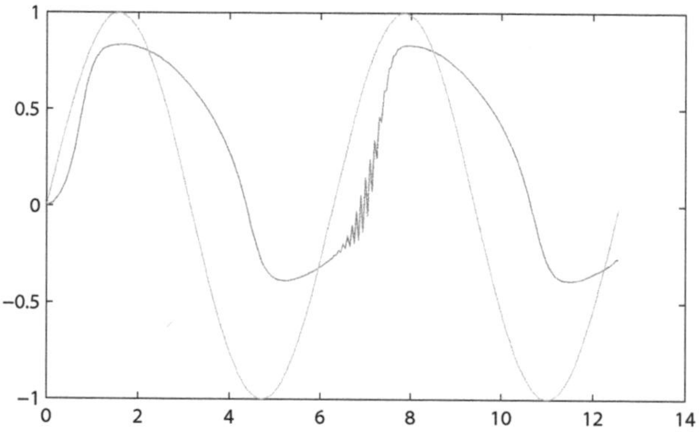

FIGURE 5.13: Numerical solutions of nonlinear difference equation.

It can be seen that under the excitation of sinusoidal signal, distortions are observed in the output signal of nonlinear systems.

Exercises

Exercise 5.1 *Perform Laplace transforms for the following functions*

$$(i)\ \ f(t)=\frac{\sin\alpha t}{t},\quad (ii)\ \ f(t)=t^5\sin\alpha t,\quad (iii)\ \ f(t)=t^8\cos\alpha t,$$

$$(iv)\ \ f(t)=t^6\mathrm{e}^{\alpha t},\quad (v)\ \ f(t)=5\mathrm{e}^{-at}+t^4\mathrm{e}^{-at}+8\mathrm{e}^{-2t},$$

$$(vi)\ \ f(t)=\mathrm{e}^{\beta t}\sin(\alpha t+\theta),\quad (vii)\ \ f(t)=\mathrm{e}^{-12t}+6\mathrm{e}^{9t}.$$

Exercise 5.2 *Take inverse transforms for the problems solved above and see whether the corresponding original function can be restored.*

Exercise 5.3 *The following properties are also given for Laplace transforms. Verify for different values of n, that the following formula are satisfied.*

$$(i)\ \ \mathscr{L}[t^n f(t)]=(-1)^n\frac{\mathrm{d}^n\mathscr{L}[f(t)]}{\mathrm{d}s^n},\quad (ii)\ \ \mathscr{L}[t^{n-1/2}]=\frac{\sqrt{\pi}(2n-1)!}{2^n}s^{-n-1/2}.$$

Exercise 5.4 *Perform inverse Laplace transforms to the following F(s).*

$$(i)\ \ F(s)=\frac{1}{\sqrt{s^2(s^2-a^2)(s+b)}},\quad (ii)\ \ F(s)=\sqrt{s-a}-\sqrt{s-b},$$

(iii) $F(s) = \ln\dfrac{s-a}{s-b}$, *(iv)* $F(s) = \dfrac{1}{\sqrt{s}(s+a)}$, *(v)* $F(s) = \dfrac{3a^2}{s^3+a^3}$,

(vi) $F(s) = \dfrac{(s-1)^8}{s^7}$, *(vii)* $F(s) = \ln\dfrac{s^2+a^2}{s^2+b^2}$,

(viii) $F(s) = \dfrac{s^2+3s+8}{\prod_{i=1}^{8}(s+i)}$, *(ix)* $F(s) = \dfrac{1}{2}\dfrac{s+\alpha}{s-\alpha}$.

Exercise 5.5 *Show the Laplace transforms where the non-integer power of s is introduced, which is the fundamental of fractional-order calculus.*

(i) $\mathscr{L}[t^\gamma] = \dfrac{\Gamma(\gamma+1)}{s^{\gamma+1}}$, *one should check different values of* γ

(ii) $\mathscr{L}\left[\dfrac{1}{\sqrt{t}\,(1+at)}\right] = \dfrac{\pi}{a}\,\mathrm{e}^{s/a}\mathrm{erfc}\left(\sqrt{s/a}\right)$ *for* $a > 0$.

Exercise 5.6 *One of the applications of Laplace transform is that it can be used in solving linear constant differential equations with zero initial conditions, using the property* $\mathscr{L}[\mathrm{d}^n f(t)/\,\mathrm{d}t^n] = s^n\mathscr{L}[f(t)]$. *Solve the differential equations*

(i) $y''(t)+3y'(t)+2y(t) = \mathrm{e}^{-t}$, $y(0) = y'(0) = 0$;

(ii) $y''-y = 4\sin t+5\cos 2t$, $y(0) = -1$, $y'(0) = -2$;

(iii) $\begin{cases} x''-x+y+z=0 \\ x+y''-y+z=0 \\ x+y+z''-z=0, \end{cases}$ $x(0)=1, x'(0)=y(0)=y'(0)=z(0)=z'(0)=0.$

Exercise 5.7 *Assume that a fractional-order system is constructed of two sub models* $G_1(s)$ *and* $G_2(s)$, *in parallel connection, and the overall model can be obtained with* $G(s) = G_1(s)+G_2(s)$. *Please find the step response of the overall system, where*

$$G_1(s) = \frac{(s^{0.4}+2)^{0.8}}{\sqrt{s}(s^2+3s^{0.9}+4)^{0.3}},\quad G_2(s) = \frac{s^{0.4}+0.6s+3}{(s^{0.5}+3s^{0.4}+5)^{0.7}}.$$

Exercise 5.8 *If the two sub models* $G_1(s)$ *and* $G_2(s)$ *in the previous problem are in series connection, and the overall model can be expressed as* $G(s) = G_2(s)G_1(s)$, *please draw the step response of the overall system.*

Exercise 5.9 *Perform Fourier transforms to the following functions, and then, perform inverse Fourier transforms to see whether the original functions can be restored.*

(i) $f(x) = x^2(3\pi-2|x|)$, *(ii)* $f(t) = t^2(t-2\pi)^2$,

(iii) $f(t) = \mathrm{e}^{-t^2}$, *(iv)* $f(t) = t\mathrm{e}^{-|t|}$.

Exercise 5.10 *Perform Fourier sine and cosine transforms for the following functions and then, perform inverse transformation and see whether the original functions can be restored.*

(i) $f(t) = \mathrm{e}^{-t}\ln t$, *(ii)* $f(x) = \dfrac{\cos x^2}{x}$, *(iii)* $f(x) = \ln\dfrac{1}{\sqrt{1+x^2}}$,

(iv) $f(x) = x(a^2-x^2)$, $a > 0$, *(v)* $f(x) = \cos kx$.

Exercise 5.11 *Compute the discrete Fourier sine and cosine transforms for the functions (i)* $f(x) = \mathrm{e}^{kx}$, *and (ii)* $f(x) = x^3$.

Exercise 5.12 *Write the Mellin transform for the function*

$$f(x)=\begin{cases}\sin(a\ln x), & x\leqslant 1\\ 0, & \textit{otherwise}.\end{cases}$$

Exercise 5.13 *Perform z transforms to the time sequences $f(kT)$, and verify the results.*

(i) $f(kT)=\cos(kaT)$, *(ii)* $f(kT)=(kT)^2\mathrm{e}^{-akT}$, *(iii)* $f(kT)=\dfrac{1}{a}(akT-1+\mathrm{e}^{-akT})$,

(iv) $f(kT)=\mathrm{e}^{-akT}-\mathrm{e}^{-bkT}$, *(v)* $f(kT)=1-\mathrm{e}^{-akT}(1+akT)$.

Exercise 5.14 *Perform inverse z transforms to the following functions.*

(i) $F(z)=\dfrac{10z}{(z-1)(z-2)}$, *(ii)* $F(z)=\dfrac{z^{-1}(1-\mathrm{e}^{-aT})}{(1-z^{-1})(1-z^{-1}\mathrm{e}^{-aT})}$,

(iii) $F(z)=\dfrac{z}{(z-a)(z-1)^2}$, *(iv)* $F(z)=\dfrac{Az[z\cos\beta-\cos(\alpha T-\beta)]}{z^2-2z\cos(\alpha T)+1}$.

Exercise 5.15 *Take inverse Laplace transform to the following functions, then, take z transform and verify the results.*

(i) $G(s)=\dfrac{b}{s^2(s+a)}$, *(ii)* $G(s)=\dfrac{b}{s^2(s+a)^2}\dfrac{1-\mathrm{e}^{-Ts}}{s}$.

Exercise 5.16 *For $G(s)=1/(s+1)^3$, if one substitutes $s=2(z-1)/[T(z+1)]$ into $G(s)$, the function $H(z)$ can be obtained. This kind of transform is referred to as* bilinear transform. *For $T=1/2$, find $H(z)$. One may also assume that $z=(1+Ts/2)/(1-Ts/2)$, inverse bilinear transform can be performed. Check whether the original function can be restored.*

Exercise 5.17 *Assume that $z=x+\mathrm{j}y$, where x and y satisfy $x^2+(y-1)^2=1$. It is obvious that z is a circle. Please use $w=1/z$ to map the circle into other kind of curve. What is the mapped curve?*

Exercise 5.18 *Show that*

$$\mathscr{Z}\left\{1-\mathrm{e}^{-akT}\left[\cos bkT+\frac{a}{b}\sin bkT\right]\right\}=\frac{z(Az+B)}{(z-1)(z^2-2z\mathrm{e}^{-aT}\cos bT+\mathrm{e}^{-2aT})},$$

where

$$A=1-\mathrm{e}^{-aT}\cos bT-\frac{a}{b}\mathrm{e}^{-aT}\sin bT,$$

$$B=\mathrm{e}^{-2aT}+\frac{a}{b}\mathrm{e}^{-aT}\sin bT-\mathrm{e}^{-aT}\cos bT.$$

Exercise 5.19 *For the function*

$$f(z)=\frac{z^2+4z+3}{z^5+4z^4+3z^3+2z^2+5z+2}\mathrm{e}^{-5z},$$

find the poles and their multiplicities and compute the residues for each pole.

Exercise 5.20 *Judge whether the following pairs of polynomials are coprime or not. If not, find the terms which can simplify $B(z)/A(z)$.*

(i) $B(z) - 3z^4 + z^5 - 11z^3 + 51z^2 - 62z + 24,$

$A(z) = z^7 - 12z^6 + 26z^5 + 140z^4 - 471z^3 - 248z^2 + 1284z - 720,$

(ii) $B(z) = 3z^6 - 36z^5 + 120z^4 + 90z^3 - 1203z^2 + 2106z - 1080,$

$A(z) = z^9+15z^8+79z^7+127z^6-359z^5-1955z^4-3699z^3-3587z^2-1782z-360.$

Exercise 5.21 *Please draw Riemann surfaces for the following functions*

(i) $f(z) = z\cos z^2$, *(ii)* $f(z) = z\mathrm{e}^{-z^2}(\cos z - \sin z)$.

Exercise 5.22 *Please write out the Laurent series of the functions, and find the residues*

(i) $f(z) = z\mathrm{e}^{-1/z^2}[\sin(1/z) - \cos(1/z)]$, *(ii)* $f(z) = z^5\cos(1/z^2)$.

Exercise 5.23 *Please write the Laurent series for the following function*

$$f(z) = \frac{3}{z-1} + \frac{1}{(z-1)^2} + \frac{1}{z-2} + \frac{1}{(z-2)^2} + \frac{5}{z+\mathrm{i}} + \frac{5}{z-\mathrm{i}}.$$

Exercise 5.24 *Perform partial fraction expansions for the following functions, and find the poles, the multiplicities and the residures of the functions*

(i) $$f(z) = \frac{3z^4 - 21z^3 + 45z^2 - 39z + 12}{z^7 + 15z^6 + 96z^5 + 340z^4 + 720z^3 + 912z^2 + 640z + 192},$$

(ii) $$f(z) = \frac{z+5}{z^8+21z^7+181z^6+839z^5+2330z^4+4108z^3+4620z^2+3100z+1000},$$

(iii) $$f(z) = \frac{3z^6 - 36z^5 + 120z^4 + 90z^3 - 1203z^2 + 2106z - 1080}{z^7 + 13z^6 + 52z^5 + 10z^4 - 431z^3 - 1103z^2 - 1062z - 360},$$

(iv) $$f(z) = \frac{(z^2 + 4z + 3)\mathrm{e}^{-5z}}{z^5 + 7z^4 - 2z^3 - 100z^2 - 232z - 160}.$$

Exercise 5.25 *Find the poles, multiplicities and residues of the following functions*

(i) $$f(z) = \frac{1 - \sin z\mathrm{e}^{-2z}}{z^7\sin(z - \pi/3)}(z^4 + 10z^3 + 35z^2 + 50z + 24),$$

(ii) $$f(z) = \frac{(z-3)^4}{z^4 + 5z^3 + 9z^2 + 7z + 2}(\sin z - \mathrm{e}^{-3z}),$$

(iii) $$f(z) = \frac{(1 - \cos 2z)(1 - \mathrm{e}^{-z^2})}{z^3\sin z}.$$

Exercise 5.26 *Evaluate the closed-path integrals*

(i) $\displaystyle\oint_\Gamma \frac{z^{15}}{(z^2+1)^2(z^4+2)^3}\,\mathrm{d}z$, *where* Γ *is the positive circle* $|\,z\,| = 3$;

(ii) $\displaystyle\oint_\Gamma \frac{z^3}{1+z}\mathrm{e}^{1/z}\,\mathrm{d}z$, *where* Γ *is the positive circle* $|\,z\,| = 2$;

(iii) $\displaystyle\oint_\Gamma \frac{\cos z(1 - \mathrm{e}^{-z^2})\sin(3z+2)}{z\sin z}\,\mathrm{d}z$, *where* Γ *is the positive circle* $|\,z\,| = 1$,

(iv) $\displaystyle\int_{|\,z\,|=2} \frac{z-2}{z^3(z-1)(z-3)}\,\mathrm{d}z.$

Exercise 5.27 *Fibonacci sequence,* $a(1)=a(2)=1,\ a(t+2)=a(t)+a(t+1),\ t=1,2,\cdots,$ *is a linear difference equation. Please find the analytical solution of its general term* $a(t)$.

Exercise 5.28 *Solve the following linear difference equations.*

(i) $72y(t)+102y(t-1)+53y(t-2)+12y(t-3)+y(t-4)=12u(t)+7u(t-1)$, *where* $u(t)$ *is step input signal, and* $y(-3)=1, y(-2)=-1,\ y(-1)=y(0)=0$;

(ii) $y(t)-0.6y(t-1)+0.12y(t-2)+0.008y(t-3)=u(t),\ u(t)=\mathrm{e}^{-0.1t}$, *and the initial value of* $y(t)$ *is 0.*

Exercise 5.29 *Please solve the following nonlinear difference equation*

$$y(t)=u(t)+y(t-2)+3y^2(t-1)+\frac{y(t-2)+4y(t-1)+2u(t)}{1+y^2(t-2)+y^2(t-1)},$$ *and when* $t\leqslant 0$, *the initial values are* $y(t)=0$, *and* $u(t)=\mathrm{e}^{-0.2t}$.

Exercise 5.30 *Assume the state space models of the discrete systems are given below, please find the analytical solutions of the step responses of the systems, and compare with numerical solutions.*

(i) $$\boldsymbol{x}(t+1)=\begin{bmatrix}0 & 1\\ -0.16 & -1\end{bmatrix}\boldsymbol{x}(t)+\begin{bmatrix}1\\1\end{bmatrix}u(t),\ \boldsymbol{x}^{\mathrm{T}}(0)=[1,-1],$$

(ii) $$\boldsymbol{x}(t+1)=\begin{bmatrix}11/6 & -1/4 & 25/24 & -2\\ 1 & 1 & -1 & -1\\ 0 & 1 & -1 & 0\\ 0 & 1 & -3/4 & 0\end{bmatrix}\boldsymbol{x}(t)+\begin{bmatrix}2\\1/2\\-3/8\\1/4\end{bmatrix}u(t),\boldsymbol{x}^{\mathrm{T}}(0)=[0,0,1,1].$$

Bibliography

[1] Mauch S. Advanced mathematical methods for scientists and engineers. Open source textbook at `http://www.its.caltech.edu/~sean/ applied_math.pdf`, 2004

[2] Valsa J, Brančik L. Approximate formulae for numerical inversion of Laplace transforms. International Journal of Numerical Modelling: Electronic Networks, Devices and Fields, 1998, 11(3):153–166

[3] Valsa J. Numerical inversion of Laplace transforms in MATLAB, MATLAB Central File ID: #32824, 2011

[4] Song S N, Sun T, Zhang G W. Complex-valued function and integral transforms. Beijing: Science Press, 2006. (in Chinese)

[5] Zheng D Z. Linear system theory (2nd edition). Beijing: Tsinghua University Press, 2002. (in Chinese)

Chapter 6

Nonlinear Equations and Numerical Optimization Problems

The solutions to linear algebraic equations have been discussed extensively in Chapter 4. However, most frequently encountered are nonlinear equations in science and engineering problems. Solving nonlinear equations could be computationally expensive; therefore, the solving of approximate linear equations was indispensable especially in the early times when the computers were not powerful enough. Today, with the rapid development of computing technology, directly solving nonlinear equations is becoming increasingly important. In this chapter, we will focus on MATLAB solutions to nonlinear equations and optimization problems. In Sections 6.1 and 6.2, solutions to nonlinear algebraic equations will be presented. The graphical method for nonlinear equations with one and two unknown variables will be given first, and quasi-analytical solutions will be studied for polynomial equations and equations convertible to polynomial equations. Numerical solutions to nonlinear equations and nonlinear matrix equations will also be discussed in this section. Efforts are made for the user to find all possible solutions to the equations with simple methods.

The so-called *optimization* is to find the values of certain variables such that the preselected objective function takes maximum or minimum. Optimization technique is very useful in scientific research and engineering practice. Equipped with the ideas and methods, the user's research capabilities can be significantly promoted, since by that time, he may no longer be satisfied with finding a solution. His target may become "how to find the best solution."

Optimization problems can be classified as unconstrained optimization problems and constrained optimization problems. In Section 6.3, MATLAB-based solutions to unconstrained optimization problems will be given. Graphical and numerical methods will be presented and the concept of global optimum solutions and local optimum solutions will be illustrated. A new solver is proposed aiming at finding global minimum of unconstrained problems. In Section 6.4, constrained optimization problems will be studied and MATLAB-based solution methods will be presented. The concept of feasible regions will be introduced. In this section, the linear programming, quadratic programming and general nonlinear programming will be studied and MATLAB-based solutions will be illustrated through examples. In Section 6.5, the idea of programming problems will be further extended to integer programming and mixed integer programming problems. An algorithm is proposed and implemented in MATLAB, aiming at finding global optimal solutions of constrained nonlinear programming problems. Mixed linear programming solutions will be studied and also a MATLAB solver based on the *branch-and-bound* algorithm is used for solving nonlinear mixed integer programming problems. Binary programming problems are also studied. In Section 6.6, a special type of optimization problem — the linear matrix inequality problem, is fully discussed and the solution methods are presented. An introductory presentation to multiple objective programming and dynamic programming will be given in Sections 6.7 and 6.8. Note that in Chapter 10, the global optimization methods will be presented based on evolutionary computing methods.

For readers who wish to check the detailed explanations of various solution techniques for nonlinear equations, we recommend the free textbook [1] (Chapter 9). For optimization theory and numerical optimization methods, the free textbook [2] is highly recommended. For LMIs, we suggest the free textbook [3]. We also found the online resource for deciding the right software for optimization problems [4] useful and interesting.

6.1 Nonlinear Algebraic Equations

6.1.1 Graphical method for solving nonlinear equations

In has been shown in Chapter 2 that implicit functions with one and two variables can be drawn easily using the MATLAB function `ezplot()`. With `ezplot()`, the nonlinear equations can be shown graphically and its real solutions can be obtained by extracting the coordinates of the intersections of the curves. The graphical methods are restricted only to nonlinear equations with one or two variables. Nonlinear equations with more than two variables have to be solved numerically or for some special cases, symbolically.

I. Graphical solution of univariate nonlinear equations

The function `ezplot()` can be used to draw the curve from the implicit function $f(x) = 0$. The real solutions can be identified from the intersections of the curves with the line $y = 0$.

Example 6.1 Solve the equation $\mathrm{e}^{-3t}\sin(4t+2)+4\mathrm{e}^{-0.5t}\cos 2t = 0.5$ using graphical method and examine the accuracy of the solutions.

Solution *The function* `ezplot()` *can be used to draw the curve of the function as shown in Figure 6.1 (a). The intersections with the horizontal axis are the solutions to the original nonlinear equation.*

```
>> syms t; f=exp(-3*t)*sin(4*t+2)+4*exp(-0.5*t)*cos(2*t)-0.5;
   ezplot(f,[0 5])    % draw directly the implicit function
   line([0,5],[0,0]) % draw the horizontal axis as well
```

From the curve it can be observed that there are three real solutions, p_1, p_2, p_3, over the interval $t \in (0,5)$. One may zoom the area around a particular solution until the horizontal axis reads the same scale. The horizontal scale is then regarded as a solution. An example of the zoomed curve is shown in Figure 6.1 (b) where a solution $t = 0.6738$ can be obtained. Substituting this reading back to the equation

```
>> t0=0.6738; e=double(subs(f,t,t0)) % find the error
```

the error can be found as $e = -2.9852\times10^{-4}$. So, for this example, the achieved accuracy of the solution is not quite as high. Similar methods can be used to find and validate other solutions.

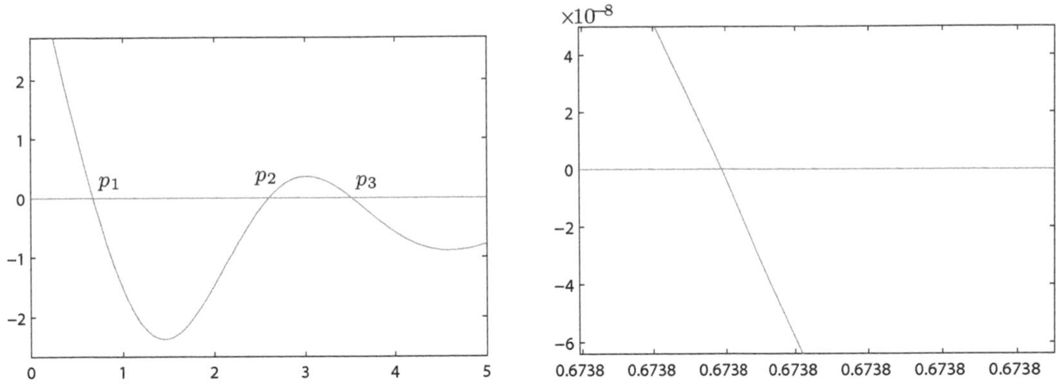

(a) the curve for implicit function over $t \in (0, 5)$ (b) zoomed curve and the solution

FIGURE 6.1: Graphical solutions to a univariate equation.

II. Graphical solution of nonlinear equations with two variables

Nonlinear equations with two variables can also be solved easily using the graphical method. Use `ezplot()` function to draw the solutions to the first equation. Then, use `hold on` command to hold the graphics window such that the plot from `ezplot()` for the second nonlinear equation is superimposed to the first one. The intersections of the two sets of curves are then the solutions to the original nonlinear equations. The solutions can be read out graphically using the zooming method illustrated earlier.

Example 6.2 Solve graphically the following simultaneous equations

$$\begin{cases} x^2 e^{-xy^2/2} + e^{-x/2}\sin(xy) = 0 \\ y^2\cos(y + x^2) + x^2 e^{x+y} = 0. \end{cases}$$

Solution *The graphical method can be used to solve the above nonlinear simultaneous equations. The first equation can be displayed with the direct use of the implicit function drawing command* `ezplot()`*, as shown in Figure 6.2 (a). Unfortunately, the color of the plots generated by* `ezplot()` *function may become very light, and it can be set manually with specific MATLAB commands as follows.*

```
>> h=ezplot('x^2*exp(-x*y^2/2)+exp(-x/2)*sin(x*y)') % the first equation
   set(h,'Color','b'); hold on;                      % hold the coordinate
```

Use the command `hold on` *to ensure the curves will not be removed. Then,* `ezplot()` *draws the solutions to the second equation. The curves will then be superimposed on the curves obtained earlier, as shown in Figure 6.2 (b).*

```
>> h=ezplot('y^2*cos(y+x^2)+x^2*exp(x+y)'), set(h,'Color','r')
```

The intersections are then the solutions to the nonlinear equation sets. In this way, all the real solutions to the given simultaneous equations in the interested ranges can be displayed. To get the coordinates of a certain point, for instance point B in Figure 6.2 (b), one may zoom the area around the point again and again until all the scales on the x- and y-axes read the same, as shown in Figure 6.3 (a). Thus, the solution at point B is $x = -0.7327, y = 1.5619$.

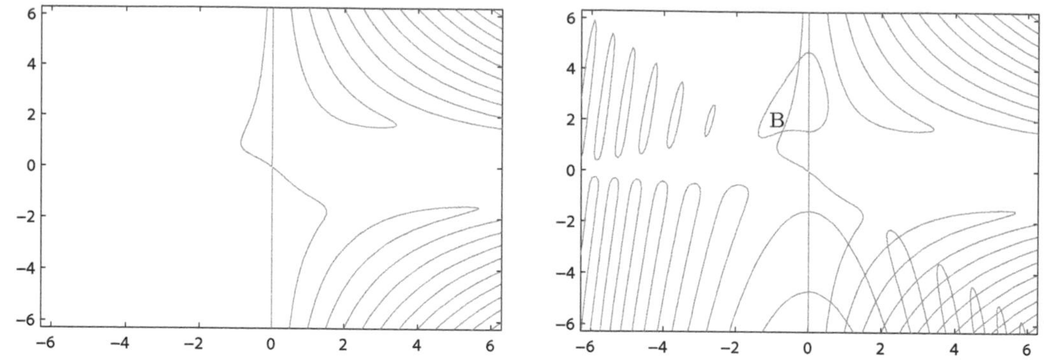

(a) solutions to the first equation (b) superimposed curves

FIGURE 6.2: Graphical solutions to the nonlinear equations.

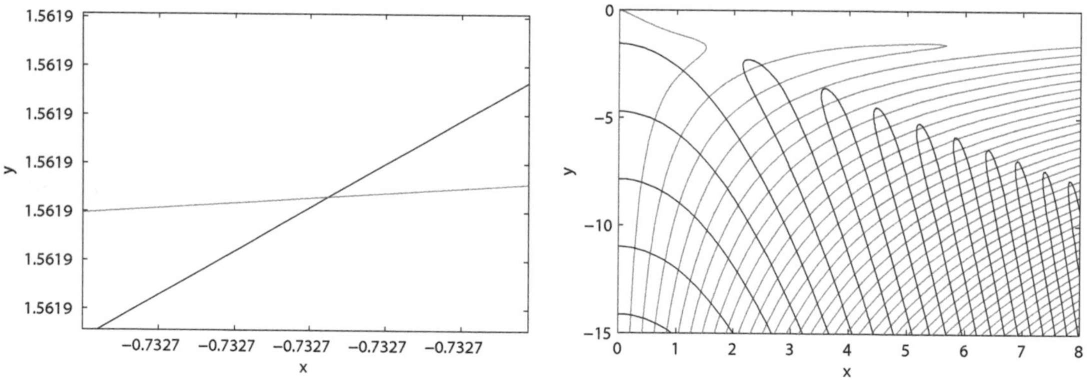

(a) zooming for a solution (b) a larger region

FIGURE 6.3: More on the graphical solutions.

From Figure 6.2 (b) it is also found that most of the solutions are located in the fourth quadrant. Thus, a large area can be chosen. For instance, the rectangular region $(0,0),(8,-10)$ *can be selected, and the solutions in the new area can be displayed as shown in Figure 6.3 (b).*

```
>> h1=ezplot('x^2*exp(-x*y^2/2)+exp(-x/2)*sin(x*y)',[0,8,-15,0])
   hold on, h2=ezplot('y^2*cos(y+x^2)+x^2*exp(x+y)',[0,8,-15,0])
   set(h1,'Color','b'); set(h2,'Color','r') % set the colors of the curves
```

6.1.2 Quasi-analytic solutions to polynomial-type equations

Before illustrating solutions to polynomial equations, let us consider two simple-to-solve equations.

Example 6.3 Consider a 1500-year-old ancient Chinese math legend: In a cage, there are chicks and rabbits, with a total of 35 heads and 94 feet. How many chicks and rabbits in the cage?

Solution *With modern mathematics, simultaneous equations can be established,* $x+y=35$

and $2x + 4y = 94$, where x and y are the numbers of chicks and rabbits, respectively. Of course, this can be solved with simple methods such as linear algebra. Is there any better way to solve this kind of problem with MATLAB?

Example 6.4 Solve the following equations using the graphical method.

$$\begin{cases} x^2 + y^2 - 1 = 0 \\ 0.75x^3 - y + 0.9 = 0. \end{cases}$$

Solution *Using the graphical method, the two curves for the two equations can be displayed easily using the following statements, as shown in Figure 6.4. The intersections are the solutions to the original equations.*

```
>> h1=ezplot('x^2+y^2-1'); hold on % solutions to the first equation
   h2=ezplot('0.75*x^3-y+0.9')      % the second equation
   set(h1,'Color','b'); set(h2,'Color','r') %  set the colors of the curves
```

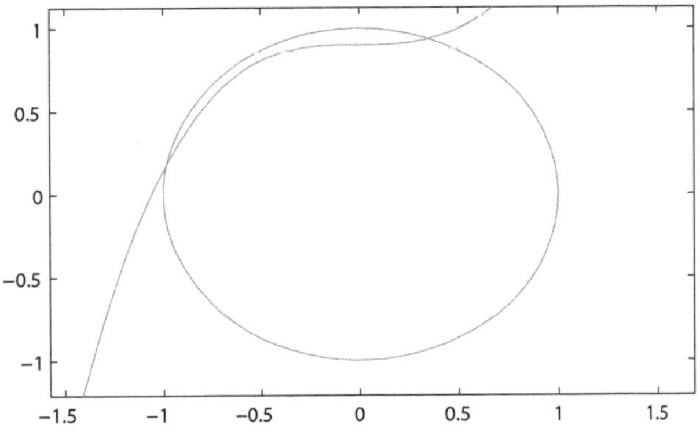

FIGURE 6.4: Solutions using graphical method.

It can be seen from the curves in Figure 6.4 that there are two intersections. However, it cannot be simply concluded that the original equations have only two solutions. One may solve y from the second equation and find that y is a function of x^3. Substituting the equation into the first one, it can be concluded that the equation can be converted into a polynomial equation of x, with the highest degree of 6. Thus, the polynomial equation must have 6 roots. Where are the other 4 roots?

The new function `vpasolve()` provided in the Symbolic Math Toolbox of MATLAB is quite effective in finding all the solutions, real and complex, to polynomial-type equations, while the `solve()` function can only be used in finding analytical solutions, if they exist. The function can be used in finding solutions to the simultaneous equations which can be converted to polynomial equations. The syntaxes of the function are as follows. Please note that the syntaxes of the function are different from the ones in old versions.

```
S=vpasolve([eqn1,eqn2,···,eqnn])
[x,y,···]=vpasolve([eqn1,eqn2,···,eqnn])
[x,y,···]=vpasolve([eqn1,eqn2,···,eqnn],[x,y,···])
[x,y,···]=vpasolve([eqn1,eqn2,···,eqnn],[x,y,···],[x0,y0,···])
```

where *eqn*$_i$ is the symbolic expression, with == representing the equal signs, of the ith equation to be solved. In old versions, strings can be used to express the equations, and it seems that string descriptions are not recommended in future versions. Also, in the function call, initial search points in the independent variables can be assigned with the vector `[`x_0`,`y_0`,`$\cdots$`]`.

In this way, simultaneous equations can easily be represented. In the first statement, a structured variable S is returned, and the solutions are fields of S. For instance, S`.`$\boldsymbol{x}$ and S`.`$\boldsymbol{y}$.

Example 6.5 Solve again the two equations in Examples 6.3 and 6.4.

Solution *For the rabbit–chick cage problem, the MATLAB statements can be used, and the solution is* $x = 23$, $y = 12$, *meaning there are 23 chicks and 12 rabbits in the cage.*

```
>> syms x y; [x0 y0]=vpasolve(x+y==35,2*x+4*y==94) % or with solve()
```

The `vpasolve()` *function can be used in solving the equations in Example 6.4 with*

```
>> syms x y; [x0,y0]=vpasolve(x^2+y^2-1==0,0.75*x^3-y+0.9==0)
```

and the solutions are found as

$$\boldsymbol{x} = \begin{bmatrix} .35696997189122287798839037801365 \\ .86631809883611811016789809418 65 + \mathrm{j}1.2153712664671427801318378544 4 \\ -.55395176056834560077984413882 7 + \mathrm{j}.35471976465080793456863789934 9 \\ -.98170264842676789676449828873194 \\ -.553951760568345600779844138827 - \mathrm{j}.354719764650807934568637899349 \\ .866318098836118110167898094186 5 - \mathrm{j}1.215371266467142780131837854 44 \end{bmatrix}$$

$$\boldsymbol{y} = \begin{bmatrix} .93411585960628007548796029415446 \\ -1.4916064075658223174787216959 + \mathrm{j}.70588200721402267753918827138 8 \\ .929338302266743628529852766772 + \mathrm{j}.21143822185895923615623381762 21 \\ .19042035099187730240977756415289 \\ .929338302266743628529852766772 - \mathrm{j}.2114382218589592361562338176221 \\ -1.4916064075658223174787216959 - \mathrm{j}.705882007214022677539188271388 \end{bmatrix}.$$

For this high-degree polynomial-type equation, according to the well-known Abel–Ruffini Theorem, there exist no analytical solutions. The Symbolic Math Toolbox can be used to obtain high-precision solutions. These types of solutions are referred to as the quasi-analytical solutions. *It can be seen that apart from the two sets of real solutions, there are yet other sets of complex conjugate solutions to the original nonlinear equations. These solutions cannot be obtained using graphical methods.*

If the solutions are to be validated, it is better to express the equations with symbolic expressions, rather than with equations. For this example, the equations can be solved and validated with the following statements, and the norm of the error matrix is about 3.67×10^{-38}, *far more accurate than the ones obtained under double-precision scheme.*

```
>> syms x y; F=[x^2+y^2-1, 0.75*x^3-y+0.9]; % describe the equations
   [x0,y0]=vpasolve(F,[x,y]), norm(subs(F,{x,y},{x0,y0})) % solve & verify
```

It can be seen in the example that, although the solution process in the two problems are completely different, the interface for the user is exactly the same. This is the beauty of MATLAB — the equations can be described in a unified form, and the problems can be solved with the same solver.

Example 6.6 The polynomial-type equations with more variables can also be obtained using the vpasolve() function. Find the solutions to the following equations

$$\begin{cases} x+3y^3+2z^2=1/2 \\ x^2+3y+z^3=2 \\ x^3+2z+2y^2=2/4. \end{cases}$$

Solution *The equations given are with three variables x, y, z. It can be seen that there are only polynomial terms, thus, it can theoretically be converted into a univariate polynomial equation. Even for a complicated set of equations, it is as easy as the simple cage equation to the user, since, the quasi-analytical solutions of the equations can be solved with*

```
>> syms x y z;
   F=[x+3*y^3+2*z^2-1/2, x^2+3*y+z^3-2, x^3+2*z+2*y^2-2/4];
   [x0,y0,z0]=vpasolve(F,[x,y,z]), size(x0) % find and count solutions
```

It can be found that the obtained $\boldsymbol{x}_0$ is a 27×1 column vector, which means that the original equations can be converted into a univariate polynomial equation with a degree of 27. Thus, the quasi-analytical solutions can be obtained using the above statements. The solutions can be validated, and it can be seen that, the norm of the error matrix is as low as 6.16×10^{-34}.

```
>> norm(subs(F,{x,y,z},{x0,y0,z0})) % substitute back to find the error
```

In fact, the terms given in the equations can also be written as a product of polynomials. For instance, if the last equation in the third equation is given by $x^3+2zy^2=2/4$, with the product of polynomials such as zy^2, the solutions to the original equations can still be found by the direct use of the solve() *function. The statements used are*

```
>> F(3)=x^3+2*z*y^2-2/4; [x0,y0,z0]=vpasolve(F,[x,y,z])
   norm(subs(F,{x,y,z},{x0,y0,z0})) % substitute back to find the error
```

and quasi-analytical solutions can be found, and the norm of the error for the new equations can be as small as 7.3485×10^{-34}.

Example 6.7 Solve the following equations where the reciprocals to the variables are involved

$$\begin{cases} \dfrac{1}{2}x^2+x+\dfrac{3}{2}+2\dfrac{1}{y}+\dfrac{5}{2y^2}+3\dfrac{1}{x^3}=0 \\ \dfrac{y}{2}+\dfrac{3}{2x}+\dfrac{1}{x^4}+5y^4=0. \end{cases}$$

Solution *It is not likely possible to solve this kind of complicated equation without the help of powerful computer mathematics languages. However, with the following statements the quasi-analytical solutions can be obtained easily.*

```
>> syms x y; clear F; F(1)=x^2/2+x+3/2+2/y+5/(2*y^2)+3/x^3;
   F(2)=y/2+3/(2*x)+1/x^4+5*y^4; [x0,y0]=vpasolve(F), size(x0)
   e=norm(subs(F,{x,y},{x0,y0})) % substitute back to find the error
```

and it can be seen that there are 26 pairs of solutions. Substituting all the solutions back to the original equations, one can immediately find that the norm of the error is 1.7374×10^{-33}, *which means that the solutions are very accurate.*

Function `solve()` was the major algebraic equation solver in old versions of Symbolic Math Toolbox. In recent versions, this function can only be used in solving equations with analytical solutions. If there is no analytical solution, `vdpsolve()` function is recommended.

Example 6.8 Solve the equations with constants

$$\begin{cases} x^2 + ax^2 + 6b + 3y^2 = 0 \\ y = a + x + 3. \end{cases}$$

Solution *The* `solve()` *function can be used directly to solve the equations, even if it contains extra variables. The solutions to the problem can be obtained by the direct use of the function calls such that*

```
>> syms a b x y; % find analytical solution
   F=[x^2+a*x^2+6*b+3*y^2, y-a-(x+3)]; [x0,y0]=solve(F,[x,y])
```

and the solutions can be written as

$$x = \frac{-6a - 18 \pm 2\sqrt{-21a^2 - 45a - 27 - 24b - 6ab - 3a^3}}{2(4+a)}, \quad y = a + (x+3).$$

In fact, the method may apply to third- or fourth-degree equations as well. However, the solutions are usually too complicated to display.

It should be noted that the analytical or quasi-analytical solution methods introduced in the previous subsections are not general-purpose. They can only be used in dealing with problems convertible to high-degree univariate polynomial equations. Furthermore, for most nonlinear equations, we cannot expect to find all the possible solutions.

6.1.3 Numerical solutions to general nonlinear equations

A numerical solution function `fsolve()` provided in MATLAB can be used to search for a real solution to given nonlinear equations. The syntaxes of the function are

```
x = fsolve(fun,x0)                                 % simplest syntax
[x,f,flag,out] = fsolve(fun,x0,opts,p1,p2,···)   % formal full syntax
```

where *fun* can either be an M-function, an anonymous function or an inline function describing the equations to be solved. The variable $\boldsymbol{x}_0$ is the initial search point for the solution. A numerical solution to the equations can be obtained by searching method from the initial point $\boldsymbol{x}_0$ using numerical algorithms. If a solution is successfully found, the returned `flag` is greater than 0, otherwise, the search is not successful.

For more complicated problems, the solution control option *opts* can be used to select methods and control accuracies in searching the solution. The *opts* variable is defined as a structured variable, with the commonly used fields explained in Table 6.1. The following syntaxes can be used in modifying the contents in the control options

```
opts = optimset;                                   % get default controls
opts.TolX = 1e-10; or set(opts,'TolX',1e-10) % set control parameters
```

where, some of the fields such as `MaxFunEvals` are problem dependent, which is usually set to 100 to 200 times the number of variables. The user may change the options using the above mentioned function calls.

TABLE 6.1: Control options for equation solutions and optimizations.

field name	explanation to the options
`Display`	To control whether the intermediate results are displayed, with the values `'off'` for no display, `'iter'` for display in each iteration, `'notify'` for alert at none convergence, and `'final'` for final results display only
`GradObj`	To indicate whether the gradient information is used in optimization. The options are `'off'` and `'on'`, with `'off'` the default
`LargeScale`	To indicate whether large-scale algorithms are used, with options `'on'` and `'off'`. For problems with only a few variables, it should be set to `'off'`
`MaxIter`	The maximum allowed iterations for equation solution and optimization. This value can be increased for problems failed to converge within the current control options
`MaxFunEvals`	The maximum allowed times of objective function calls
`TolFun`	The error tolerance of objective functions
`TolX`	The error tolerance of the solutions

Additional parameters $p_1, p_2, \cdots, p_m$ are also allowed in the function calls.

Example 6.9 Consider again the equation $e^{-3t}\sin(4t+2)+4e^{-0.5t}\cos 2t = 0.5$ defined in Example 6.1. Find the solutions using numerical methods for a better accuracy.

Solution *Using the* `vpasolve()` *function*

```
>> syms t x; f=exp(-3*t)*sin(4*t+2)+4*exp(-0.5*t)*cos(2*t)-0.5;
   t0=vpasolve(f), subs(f,t,t0) % find a solution with vpasolve()
```

it can be found that the solution is $t_0 = 0.67374570500134756702960220427474$, *with an error of* 6.5×10^{-35}. *It is obvious that the nonlinear equation has no analytical solutions. Graphical methods shown in Example 6.1 can be used to find the numerical solutions. However, the accuracy achieved by graphical method may not be very high. From the approximate solution by graphical approach* $t = 3.5203$, *better results can be obtained by directly using the* `fsolve()` *function.*

By combining the graphical and numerical methods, it can be seen that a better solution can be found

```
>> y=@(t)exp(-3*t).*sin(4*t+2)+4*exp(-0.5*t).*cos(2*t)-0.5;
   [t,f]=fsolve(y,3.5203)
```

such that $t = 3.52026389294877$ *and* $f = -6.06378\times10^{-10}$. *The solution found is much more accurate than the graphical method. To get even better approximations, one can further modify the control options with the following statements*

```
>> ff=optimset; ff.TolX=1e-16; ff.TolFun=1e-30; % higher precision
   [t,f]=fsolve(y,3.5203,ff)                     % solve again
```

and the new solution is $t = 3.52026389244155$ *with* $f = 0$.

Example 6.10 Solve the equations in Example 6.4 using numerical algorithms.

Solution *Before solving such equations, the variables should be selected such that the unknowns to be solved are assigned as a vector. Selecting the variables* $p_1 = x$, $p_2 = y$, *the original simultaneous equations can be rewritten in matrix form as*

$$F(p) = [p_1^2 + p_2^2 - 1,\ 0.75p_1^3 - p_2 + 0.9]^{\mathrm{T}} = 0.$$

An anonymous function can be written to describe the two equations as a column vector, and then, one may select the initial values at $p_0 = [1, 2]^{\mathrm{T}}$*. The function* `fsolve()` *can be used directly to solve the original equations and find a solution.*

```
>> f=@(p)[p(1)^2+p(2)^2-1; 0.75*p(1)^3-p(2)+0.9]; % describe equation
   [x,Y,c,d]=fsolve(f,[1; 2])                      % numerical solution
```

The solution found is $x = [0.35696997, 0.93411586]^{\mathrm{T}}$*, with the error* $Y = [0.1215\times10^{-9}, 0.0964\times10^{-9}]$*. It can also be found by examining the* `d` *argument that 21 function calls are made. Thus, the algorithm is quite effective.*

Similarly, the original equations can also be described by the inline function or by M-file. With the anonymous functions, there is no need to create a separate M-file for each problem, which makes the file management more tidy and convenient.

If the initial values are changed to $p_0 = [-1, 0]^{\mathrm{T}}$*, then, by using*

```
>> [x,Y,c,d]=fsolve(f,[-1,0]') % search solution from another initial point
```

another solution is found at $x = [-0.981703, 0.1904204]^{\mathrm{T}}$*, and this time 15 function calls are made and the norm of the error vector is* 0.5618×10^{-10}*. In this example, it can be seen that the selection of initial values may lead to other solutions.*

If a complex initial value is selected, for instance, $x0 = [-1-\mathrm{j}1, 1-\mathrm{j}1]$*, complex root can be found with* `fsolve()` *function*

```
>> x0=[-1-1i; 1-1i]; [x,Y,c,d]=fsolve(f,x0) % search from complex point
```

with $x = [-0.5540 - \mathrm{j}0.3547, 0.9293 - \mathrm{j}0.2114]^{\mathrm{T}}$*. With different complex initial value selections, all the other 3 complex solutions can be found.*

Example 6.11 Consider the well-known Lambert W equation $W(x)\mathrm{e}^{W(x)} = x$, where x is a fixed value, and $W(x)$ is the unknown variable to be solved. Please draw the curve $W(x)$ for $x \in (0, 5)$.

Solution *If the parameter* x *changes, the equation itself also changes. If an M-function or other form is used to describe the equation, the file should be modified, which makes the description of equation complicated. In this case, if* x *is represented as an additional parameter, and there is no more need to modify the M-function. The following statements can be used to draw Lambert W function. In fact, the MATLAB function* `lambertw()` *can be used to evaluate Lambert W function directly, with* $y_1 =$ `lambertw(`x_0`)`.

```
>> f=@(W,x)W.*exp(W)-x; x0=0:0.1:5; y=[]; opts=optimset;
   for x=x0, y0=fsolve(f,x,opts,x); y=[y y0]; end, plot(x0,y)
```

6.2 Nonlinear Equations with Multiple Solutions

In Section 4.4.5, a special form of nonlinear matrix equation, algebraic Riccati equation, is discussed. However, the solution is based on a very specialized algorithm, which cannot be extended to other forms of nonlinear matrix equations. For instance, if the equation is changed to

$$AX + XD - XBX + C = 0, \qquad (6\text{-}2\text{-}1)$$

or even a tricky form

$$AX + XD - XBX^{\mathrm{T}} + C = 0, \tag{6-2-2}$$

the `are()` function is no longer applicable. So here, a nonlinear matrix equation solution method is given for solving general nonlinear matrix equations.

Also, think about the plots given in Figure 6.2. How can we find all the solutions to the given simultaneous equations in the interested area?

In this section, we shall explore generalized ways in finding solutions to nonlinear equations, using numerical, or even high-precision approaches.

6.2.1 Numerical solutions

Assume that a matrix equation is given by $F(X) = 0$, where both X and $F(\cdot)$ are both $n \times m$ matrices. Anonymous functions or M-functions can be used to describe the equations. The function `fsolve()` can be used in solving directly the equations.

In order to solve matrix or other nonlinear equations with multiple solutions, the following algorithm can be formulated

---

Require: Anonymous function $Y = F(X)$, initial solution X, range of interested region A, error tolerance ϵ
  Initialization, construct initial stored solution set X, in a 3D array
  while true **do**
    Randomly generate an x_0, and find a solution x with `fsolve()`
    if this solution does not exist in X **then**
      Store it in X
    end if
    if no new solution found in t_{lim} seconds **then**
      Terminate the while loop with `break` command
    end if
  end while

A MATLAB function `more_sols()` is written to implement the algorithm. The user may terminate the function at any time by pressing Ctrl-C keys. The listing of the function is as follows

```
function more_sols(f,X0,varargin)
[A,tol,tlim]=default_vals({1000,eps,30},varargin{:});
if length(A)==1, a=-0.5*A; b=0.5*A; else, a=A(1); b=A(2); end
ar=real(a); br=real(b); ai=imag(a); bi=imag(b);
ff=optimset; ff.Display='off'; [n,m,i]=size(X0);
ff1=ff; ff.TolX=tol; ff.TolFun=tol; X=X0;
try, err=evalin('base','err');
catch, err=0; end, if i<=1; err=0; end, tic
while (1),
   x0=ar+(br-ar)*rand(n,m);
   if abs(imag(A))>1e-5, x0=x0+(ai+(bi-ai)*rand(n,m))*1i; end
   [x,aa,key]=fsolve(f,x0,ff1);
   t=toc; if t>tlim, break; end
   if key>0, N=size(X,3);
```

```
    for j=1:N, if norm(X(:,:,j)-x)<1e-5; key=0; break; end, end
    if key>0, [x1,aa,key]=fsolve(f,x,ff);
      if norm(x-x1)<1e-5 & key>0; X(:,:,i+1)=x1;
        assignin('base','X',X); err=max([norm(aa),err]);
        assignin('base','err',err); i=i+1, tic
end, end, end, end
```

The syntax of the function is `more_sols(`f`,`$\boldsymbol{X}_0$`,`A`,`ϵ`,`$t_{\rm lim}$`)`, and the low-level function `default_vals()` was defined in Chapter 3, which can be used to assign default variables. The argument A can be a number or an interval `[`a`,`b`]`, specifying the interested solution region. If it is a number, the interested region is $(-A/2, A/2)$, and the default value of $A = 1000$, usually meaning to search solutions in large region. If the argument A contains imaginary part, the variables of A or a, b pair specify the range of solutions, and complex solutions are also expected. The default value of the argument ϵ is `eps`. The default value of the argument $t_{\rm lim}$ is 30, meaning if within 30 seconds there is no new solution found, the function is terminated normally. The user may also terminate the function at any time, by pressing the Ctrl-C keys.

Compared with other functions, this function is very special, since infinite loop is used. It may be terminated at any time by the user by pressing the Ctrl-C keys, it is not suitable to assign returned arguments. Function `assignin()` is used in the loop, so that each time a new solution is found, the variables $\boldsymbol{X}$ and `err` in MATLAB workspace are updated. The variable $\boldsymbol{X}$ is a three-dimensional array, with $\boldsymbol{X}$`(:,:,`i`)` storing the ith solution found, and variable `err` stores the maximum norm of errors so far found.

If the function is terminated, and the user wants to resume the function, the argument $\boldsymbol{X}$ in MATLAB workspace can be used as the initial $\boldsymbol{X}_0$ in the function call. For instance, `more_sols(`f`,`$\boldsymbol{X}$`)` can be used to resume.

Example 6.12 Solve again the Riccati equation in Example 4.62, where the matrices are given below

$$\boldsymbol{A}=\begin{bmatrix}-2&1&-3\\-1&0&-2\\0&-1&-2\end{bmatrix},\ \boldsymbol{B}=\begin{bmatrix}2&2&-2\\-1&5&-2\\-1&1&2\end{bmatrix},\ \boldsymbol{C}=\begin{bmatrix}5&-4&4\\1&0&4\\1&-1&5\end{bmatrix}.$$

Solution *To find all the real solutions, the Riccati equation should be described as an anonymous function, and* `more_sols()` *function can be used, in the statement, the initial matrix* $\boldsymbol{X}_0$ *is set to an empty* $3\times 3\times 0$ *array, with the first two 3's indicating the size of the solution matrix.*

```
>> A=[-2,1,-3; -1,0,-2; 0,-1,-2]; B=[2,2,-2; -1 5 -2; -1 1 2];
   C=[5 -4 4; 1 0 4; 1 -1 5]; f=@(X)A'*X+X*A-X*B*X+C;
   more_sols(f,zeros(3,3,0)); X, err % find all the real solutions
```

If this function is left running, after some time, it can be terminated normally, and the maximum error norm is 1.4904×10^{-12}*, and all the 8 real matrix solutions can be found*

$$\boldsymbol{X}_1=\begin{bmatrix}0.9874&-0.7983&0.4189\\0.5774&-0.1308&0.5775\\-0.2840&-0.0730&0.6924\end{bmatrix},\ \boldsymbol{X}_2=\begin{bmatrix}1.2213&-0.4165&1.9775\\0.3578&-0.4894&-0.8863\\-0.7414&-0.8197&-2.3560\end{bmatrix},$$

$$\boldsymbol{X}_3=\begin{bmatrix}0.6665&-1.3223&-1.7200\\0.3120&-0.5640&-1.1910\\-1.2273&-1.6129&-5.5939\end{bmatrix},\ \boldsymbol{X}_4=\begin{bmatrix}-2.1032&1.2978&-1.9697\\-0.2467&-0.3563&-1.4899\\-2.1494&0.7190&-4.5465\end{bmatrix},$$

$$X_5 = \begin{bmatrix} -0.1538 & 0.1087 & 0.4623 \\ 2.0277 & -1.7437 & 1.3475 \\ 1.9003 & -1.7513 & 0.5057 \end{bmatrix}, \quad X_6 = \begin{bmatrix} 0.8878 & -0.9609 & -0.2446 \\ 0.1072 & -0.8984 & -2.5563 \\ -0.0185 & 0.3604 & 2.4620 \end{bmatrix},$$

$$X_7 = \begin{bmatrix} 23.9467 & -20.6673 & 2.4529 \\ 30.1460 & -25.9830 & 3.6699 \\ 51.9666 & -44.9108 & 4.6410 \end{bmatrix}, \quad X_8 = \begin{bmatrix} -0.7619 & 1.3312 & -0.8400 \\ 1.3183 & -0.3173 & -0.1719 \\ 0.6371 & 0.7885 & -2.1996 \end{bmatrix}.$$

If complex solutions are also expected, the following commands should be specified, and altogether 20 real and complex solutions can be found

```
>> more_sols(f,X,1000+1000i); X, err % find all the complex solutions
```

Example 6.13 Now consider the new Riccati-like equation given in (6-2-2), where

$$A = \begin{bmatrix} 2 & 1 & 9 \\ 9 & 7 & 9 \\ 6 & 5 & 3 \end{bmatrix}, \quad B = \begin{bmatrix} 0 & 3 & 6 \\ 8 & 2 & 0 \\ 8 & 2 & 8 \end{bmatrix}, \quad C = \begin{bmatrix} 7 & 0 & 3 \\ 5 & 6 & 4 \\ 1 & 4 & 4 \end{bmatrix}, \quad D = \begin{bmatrix} 3 & 9 & 5 \\ 1 & 2 & 9 \\ 3 & 3 & 0 \end{bmatrix}.$$

Find and verify all the possible solutions.

Solution *To date, there are no other existing algorithms for solving such types of equations. With the function* `more_sols()`*, and selecting the default search region, all the 16 solutions can be found, and the function call may terminate normally, with maximum error norm of* 1.5504×10^{-10}.

```
>> A=[2 1 9; 9 7 9; 6 5 3]; B=[0 3 6; 8 2 0; 8 2 8];
   C=[7 0 3; 5 6 4; 1 4 4]; D=[3 9 5; 1 2 9; 3 3 0];
   f=@(X)A*X+X*D-X*B*X.'+C; more_sols(f,zeros(3,3,0)); X, err
```

If complex solutions are also allowed, the following commands can be issued, and all the 38 solutions can eventually be found.

```
>> more_sols(f,X,1000+1000i); X, err % find all the complex solutions
```

Example 6.14 Consider again the nonlinear simultaneous equations in Example 6.2. Please find all the numerical solutions within the range of $-2\pi \leqslant x, y \leqslant 2\pi$.

$$\begin{cases} x^2 e^{-xy^2/2} + e^{-x/2}\sin(xy) = 0 \\ y^2\cos(y + x^2) + x^2 e^{x+y} = 0. \end{cases}$$

Solution *In Example 6.2, the graphical method was used, and many intersections were witnessed in the interested area. If the graphical method is to be used to find all the interested solutions, it might be a time-consuming task, since each time, only one solution can be found by the zooming facilities. Furthermore, the accuracy of the solutions found by the graphical method is extremely low. The new* `more_sols()` *function can be used instead to find numerically all the solutions in the interested area. Since the interested area is* $(-2\pi, 2\pi)$*, we can set* $A=$ `[-2*pi,2*pi]`*. Also, since* $[0, 0]$ *is one solution of the equation, it can be used as the starting point in the solution process. The following statements can be used to describe the equations with anonymous function, then, the solution function can be called. It can be seen that the maximum error norm is* 1.8829×10^{-13}.

```
>> f=@(x)[x(1)^2*exp(-x(1)*x(2)^2/2)+exp(-x(1)/2)*sin(x(1)*x(2));
          x(2)^2*cos(x(2)+x(1)^2)+x(1)^2*exp(x(1)+x(2))];
   more_sols(f,[0; 0],[-2*pi,2*pi]); err
```

All the solutions obtained are marked on top of the graphical solutions, as shown in Figure 6.5. It can be seen that all the real solutions in this area are found, and the accuracy is much higher than the graphical solutions.

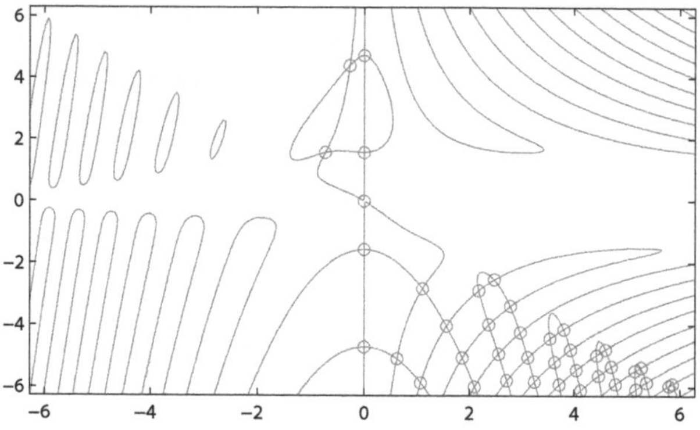

FIGURE 6.5: All the solutions of the nonlinear simultaneous equations.

```
>> h1=ezplot('x^2*exp(-x*y^2/2)+exp(-x/2)*sin(x*y)=0');
   hold on; h2=ezplot('y^2*cos(y+x^2)+x^2*exp(x+y)=0');
   x=X(1,1,:); x=x(:); y=X(2,1,:); y=y(:); plot(x,y,'o')
   set(h1,'Color','b'); set(h2,'Color','r')
```

It is worth mentioning that, some of the solutions thus searched are not in the interested area. We can exclude those solutions with the following statements, and finally it can be found that there are 41 solutions located in the interested area.

```
>> ii=find(abs(X(1,1,:))<=2*pi & abs(X(2,1,:))<=2*pi);
   X_sol=X(:,:,ii); size(ii) % find and count the roots in the area
```

Example 6.15 Consider the simultaneous equations $\sin(x-y)=0$ and $\cos(x+y)=0$. Please find all the real solutions for $0 \leqslant x, y \leqslant 4\pi$.

Solution *Graphical approach can be used first, and the distribution of the solutions in the interested area, as depicted as shown in Figure 6.6(a).*

```
>> h1=ezplot('sin(x-y)',[0,4*pi]); set(h1,'Color','b'); hold on
   h2=ezplot('cos(x+y)',[0,4*pi]); set(h2,'Color','r');
```

The original equations can be described by anonymous function, and select the interested area $A=$ `[0,4*pi]`. *With the* `more_sols()` *function, all the solutions in the area can be obtained, as shown in Figure 6.6(b). It can be seen that all the real solutions in the interested area are found.*

```
>> f=@(x)[sin(x(1)-x(2)); cos(x(1)+x(2))];        % describe the equations
   more_sols(f,zeros(2,1,0),[0,4*pi]);    % find the solutions in (0,4π) area
   x=X(1,1,:); y=X(2,1,:); plot(x(:),y(:),'o') % draw the solutions
```

Example 6.16 Solve the fractional-order polynomial equation

$$x^{2.3}+5x^{1.6}+6x^{1.3}-5x^{0.4}+7=0.$$

Solution *An "obvious" way is to introduce* $z=x^{0.1}$, *and the original equation is mapped*

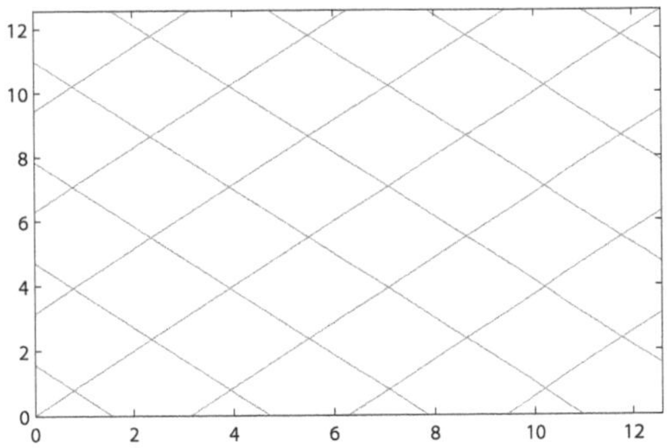

(a) graphical method

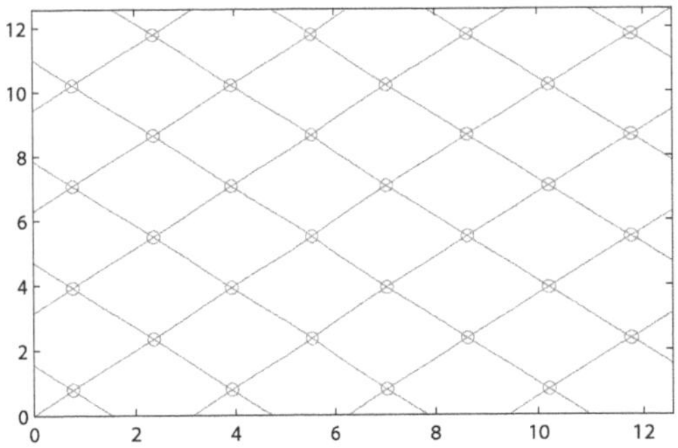

(b) all the solutions

FIGURE 6.6: All the solutions in the interested area.

into an ordinary polynomial equation of z, with 23 solutions. It seems that $x = z^{10}$ *are the solutions of the original equation, and the idea can be implemented in MATLAB*

```
>> syms x z; f1=z^23+5*z^16+6*z^13-5*z^4+7; p=sym2poly(f1); r=roots(p);
   f=x^2.3+5*x^1.6+6*x^1.3-5*x^0.4+7; r1=r.^10, double(subs(f,x,r1))
```

Unfortunately, most of the "solutions" thus obtained do not satisfy the original equation. How many solutions are there in the original equation? This question can be answered with the following statements, and believe it or not, there are only two solutions in the equation $x = -0.1076 \pm \mathrm{j}0.5562$. *The remaining 21 solutions are* extraneous roots.

```
>> f=@(x)x.^2.3+5*x.^1.6+6*x.^1.3-5*x.^0.4+7;
   more_sols(f,zeros(1,1,0),100+100i), x0=X(:)
```

6.2.2 Finding high-precision solutions

In fact, `vpasolve()` function can be used to solve more accurately nonlinear solutions as well, with the user-selected initial search points.

$[x,y,\cdots]=$ `vpasolve(`$[eqn_1,eqn_2,\cdots,eqn_n]$`,`$[x,y,\cdots]$`,`$[x_0,y_0,\cdots]$`)`

Note that, the equations must be specified in symbolic expressions, or alternatively, == can be used for equal sign in equations, and when omitted, it means = 0. Variable list $[x,y,\cdots]$ can be omitted, and retrieved with `symvar()` automatically. Initial search point is allowed to be assigned in $[x_0,y_0,\cdots]$. More than that, for polynomial types of equations, all the high-precision solutions can also be found.

Example 6.17 Consider again the Riccati equation in Example 4.62. Try to find all the high-precision solutions with `vpasolve()`.

Solution *The function* `vpasolve()` *can be tried, and a total of 20 solutions can be found, with 8 of them real solutions. The solutions are the same as the ones obtained in Example 6.12, but with much higher precision.*

```
>> A=[-2,1,-3; -1,0,-2; 0,-1,-2]; B=[2,2,-2; -1 5 -2; -1 1 2];
   C=[5 -4 4; 1 0 4; 1 -1 5]; X=sym('x%d%d',3);
   F=A'*X+X*A-X*B*X+C; Y=vpasolve(F) % find all the solutions of Riccati equation
```

The returned variable Y is a structured variable, and the solutions can be converted to cells, and then, can be rearranged in a matrix with the following statements

```
>> Z=struct2cell(Y); [n,m]=size(Z); V=[]; for i=1:n, V=[V Z{i}]; end
```

In this case, each row in matrix $\boldsymbol{V}$ is a solution. Now let us validate the 5th solution. It can be seen that when the 5th solution is substituted back into the original equation, the norm of the error matrix is 6.2×10^{-38}, which is far beyond the capabilities of any numerical algorithms under double-precision scheme.

```
>> x=V(5,:); X0=reshape(x,3,3).'; norm(subs(F,X,X0)) % the 5th solution
```

For the problem in Example 6.13, since $\boldsymbol{X}^{\mathrm{T}}$ is involved, a tremendous amount of time is needed, and all the 38 solutions in high-precision scheme can be found.

Example 6.18 Revisit the nonlinear equations in Example 6.2. Now, consider solving the equations with `vpasolve()` function. If initial search point is not specified, only one solution can be found, at $x = y = 0$.

```
>> syms x y;
   F=[x^2*exp(-x*y^2/2)+exp(-x/2)*sin(x*y),y^2*cos(y+x^2)+x^2*exp(x+y)];
   [x0,y0]=vpasolve(F) % try to solve the equation
```

In the function `more_sols()`, MATLAB function `fsolve()` is used as the kernel. Similarly, `vpasolve()` can also be used as the kernel, for finding high-precision solutions of nonlinear equations. A new solver can be written

```
function more_vpasols(f,X0,varargin)
[A,tlim]=default_vals({1000,60},varargin{:}); X=X0;
if length(A)==1, a=-0.5*A; b=0.5*A; else, a=A(1); b=A(2); end
ar=real(a); br=real(b); ai=imag(a); bi=imag(b); [i,n]=size(X0); tic
while (1),
   x0=ar+(br-ar)*rand(1,n);
   if abs(imag(A))>1e-5, x0=x0+(ai+(bi-ai)*rand(1,n))*1i; end
   V=vpasolve(f,x0); N=size(X,1); key=1;
   if length(V)==0, continue, else, x=sol2vec(V); end
   t=toc; if t>tlim, break; end
   for j=1:N, if norm(X(j,:)-x)<1e-5; key=0; break; end, end
   if key>0, i=i+1;
      X=[X; x]; disp(['i=',int2str(i)]); assignin('base','X',X); tic
end, end, end
function v=sol2vec(A)
v=[]; A=struct2cell(A); for i=1:length(A), v=[v, A{i}]; end
```

The syntax of the function is `more_vpasols(`f`,`$\boldsymbol{X}_0$`,`A`,`t_{lim}`)`, with a low-level supporting function `sol2vec()` embedded in the same file.

The argument f should be specified as a symbolic row vector in describing the simultaneous equations, and the initial $\boldsymbol{X}_0$ should be assigned to `zeros(0,`n`)`, with n the number of unknowns. The other arguments are the same as the ones on `more_sols()`, discussed earlier. The returned argument $\boldsymbol{X}(i,:)$ storing the ith solution found. It is worth mentioning that the speed of `more_vdpsols()` is much slower than that of `more_sols()`.

Example 6.19 Consider the simultaneous equations in Example 6.14. Find all the solutions in the interval $-2\pi < x, y < 2\pi$ under high-precision scheme.

Solution *The following commands can be used directly and all the solutions in the interested area can be found with the following statements*

```
>> syms x y; % finding all the high-precision solutions
   F=[x^2*exp(-x*y^2/2)+exp(-x/2)*sin(x*y),y^2*cos(y+x^2)+x^2*exp(x+y)];
   more_vpasols(F,zeros(0,2),4*pi)
```

To check the accuracy of one of the solutions, the solutions of $\boldsymbol{x}_0$ *and* $\boldsymbol{y}_0$ *in the interested area can be extracted and sorted first. The following statements can be used, and the norm of the errors is* 7.79×10^{-32}*, the accuracy is much higher than the ones obtained in Example 6.14. It took about nearly half an hour, much longer than with* `more_sols()` *function, to find all the 41 solutions.*

```
>> x0=X(:,1); y0=X(:,2); ii=find(abs(x0)<2*pi & abs(y0)<2*pi);
   x0=x0(ii); y0=y0(ii); [x0 ii]=sort(x0); y0=y0(ii);
   double(norm(subs(F,{x,y},{x0,y0}))), size(x0)
   ezplot(F(1)), hold on, ezplot(F(2)), plot(x0,y0,'o')
```

6.2.3 Solutions of underdetermined equations

Algebraic equations are referred to as *underdetermined*, if there are fewer equations than the number of unknowns. The implicit equation $f(x, y) = 0$ discussed earlier is an example of underdetermined equation, and it has been shown that all the points on the curves drawn with `ezplot()` are the solutions of the implicit equation. In this section, a further example is given to show the possible solutions of typical underdetermined equations.

Example 6.20 Please solve the following underdetermined equations

$$\begin{cases} x^2 z\mathrm{e}^{-xy^2z^2/2} + \mathrm{e}^{-x/2}z^2\sin(xy) = 0 \\ y^2\cos(y+x^2) + x^2\mathrm{e}^{x+y}z = 0. \end{cases}$$

Solution *Since there are two equations, with three unknowns* x, y *and* z*, the original equations are underdetermined. One may first fix the value of* z*, and find all the solutions of* x *and* y*, with* `more_sols()` *function. Then, fix another value of* z *and perform the same process again. In this way, the possible solutions of the original equations can be found. This process is quite time-consuming, however, it may be valuable in solving such kind of equations. The solutions for the* ith *value of* z *can also be retrieved with* $\boldsymbol{A}\{i\}$ *and* $\boldsymbol{B}\{i\}$*. The points of the solutions are shown in Figure 6.7 (a).*

```
>> z0=0.1:0.1:1;
   for i=1:length(z0), z=z0(i);
      f=@(x)[x(1)^2*z*exp(-x(1)*x(2)^2*z^2/2)+exp(-x(1)/2)*z^2*sin(x(1)*x(2));
          x(2)^2*cos(x(2)+x(1)^2)+x(1)^2*exp(x(1)+x(2))*z];
      more_sols(f,zeros(2,1,0),[-2*pi,2*pi]); x=X(1,1,:); y=X(2,1,:);
      A{i}=x(:); B{i}=y(:); plot3(x(:),y(:),z*ones(size(x(:))),'o'), hold on
   end
   axis([-2*pi, 2*pi, -2*pi, 2*pi, 0,1])
```

```
>> for i=1:length(z0), z=z0(i);
      [x,ii]=sort(A{i}); y=B{i}(ii); zz=z*ones(size(x));
      plot3(x,y,zz,x,y,zz,'o'), hold on
   end
```

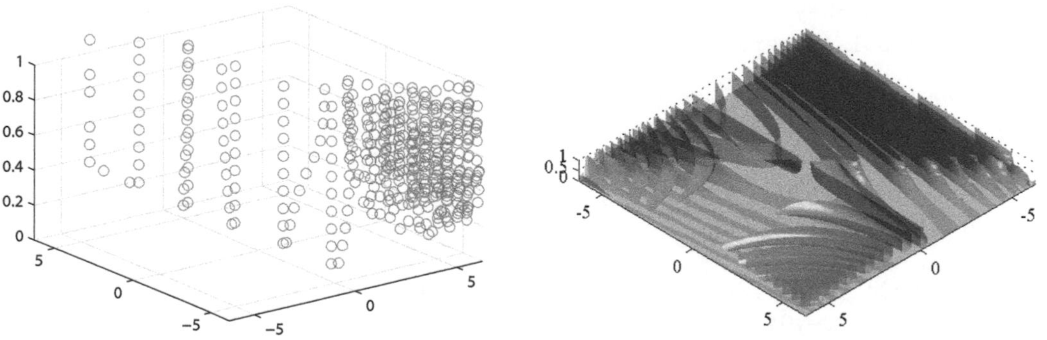

(a) samples of the solutions (b) intersections of surfaces

FIGURE 6.7: Solutions of underdetermined equations.

Alternatively, the original equations can be solved graphically with `ezimplot3()` *function, and the results are shown in Figure 6.7 (b), the intersection curves of the two sets of surfaces are the solutions of the equations.*

```
>> f1=@(x,y,z)x^2*z*exp(-x*y^2*z^2/2)+exp(-x/2)*z^2*sin(x*y);
   f2=@(x,y,z)x^2*cos(y+x^2)+x^2*exp(x+y)*z;
   ezimplot3(f1,[-2*pi,2*pi,-2*pi,2*pi,0,1])
   ezimplot3(f2,[-2*pi,2*pi,-2*pi,2*pi,0,1])
```

6.3 Unconstrained Optimization Problems

Unconstrained optimization problems are considered as simpler than constrained optimization problems. The mathematical description of unconstrained problems is that

$$\min_{\boldsymbol{x}} f(\boldsymbol{x}), \tag{6-3-1}$$

where $\boldsymbol{x} = [x_1, x_2, \cdots, x_n]^{\mathrm{T}}$ are referred to as *decision variables*, or *optimization variables*, and scalar function $f(\cdot)$ is referred to as the *objective function*. The task is to find a vector $\boldsymbol{x}$ such that the value of the objective function $f(\boldsymbol{x})$ is minimized. Thus, the optimization problem is also referred to as the *minimization problem*. In fact, the definition of minimization problem may not lose the generality. For instance, the maximization problem can be converted to minimization problem by multiplying -1 to its objective function. Therefore, the optimization problems studied in this book are for minimization problems.

6.3.1 Analytical solutions and graphical solution methods

It is known from advanced mathematics courses that the necessary conditions for an unconstrained optimization problem are that at the optimum point $\boldsymbol{x}^*$, the first-order derivatives of the objective function are all 0's. Thus, the following simultaneous equations can be established

$$\left.\frac{\partial f}{\partial x_1}\right|_{\boldsymbol{x}=\boldsymbol{x}^*}=0,\quad \left.\frac{\partial f}{\partial x_2}\right|_{\boldsymbol{x}=\boldsymbol{x}^*}=0,\quad \cdots,\quad \left.\frac{\partial f}{\partial x_n}\right|_{\boldsymbol{x}=\boldsymbol{x}^*}=0. \tag{6-3-2}$$

Solving the above equations, the extremum points can be obtained. In fact, the extremum points obtained may not be all minimum points, some of the points may actually be maximum points. Minimum points can be judged by taking the second-order derivatives, where positive second-order derivative means that minimum points are obtained. For univariate functions, the analytical method can be considered. However, for multivariate problems, solving the equations derived may be even more difficult than solving the optimization problem itself.

The graphical solutions to optimization problems with one variable are quite straightforward. The derivative function can be drawn first and the minimum points can be read from the curves. Functions with two variables may also be solved using graphical methods. However, for problems with three or more variables, graphical methods may not be applicable.

Example 6.21 From the equation $f(t)=\mathrm{e}^{-3t}\sin(4t+2)+4\mathrm{e}^{-0.5t}\cos(2t)-0.5$ studied in Example 6.1, use graphical and analytical methods to study the optimality of the function.

Solution *The first-order derivative of the objective function can be derived first and with the function* `ezplot()`*, the first-order derivative can be drawn over the interval* $t\in[0,4]$ *as shown in Figure 6.8 (a).*

```
>> syms t; y=exp(-3*t)*sin(4*t+2)+4*exp(-0.5*t)*cos(2*t)-0.5;
   y1=diff(y,t); ezplot(y1,[0,4]) % find intersection of derivative with x-axis
```

In fact, finding the solution points of $y_1(t)=0$ *is not easier than the direct finding of optimum points. With the graphical method, two points* A_1 *and* A_2 *can be found. From the first-order derivative, it can be seen that the point* A_1 *has a positive second-order derivative, thus, it corresponds to the minimum point, while for point* A_2*, a negative second-order derivative corresponds to the maximum point. The value of* A_1 *can be obtained using the following statements*

```
>> t0=vpasolve(y1), ezplot(y,[0,4]) % draw the first-order derivative
   y2=diff(y1); y0=subs(y2,t,t0)    % verify positive 2nd-order derivative
```

where $t_0=1.4528424981725411893375778$*, and* $y_0=7.8553420253336013794644$*, which means that the point has a positive second-order derivative. Therefore, the minimum point has been obtained. It can further be confirmed from Figure 6.8 (b) that* A_1 *is the minimum point, and* A_2 *is the maximum point.*

Since the original problem is a nonlinear function, analytical solutions may not be obtained. Note that the solution to the first-order derivative equation is not easier than the direct solution to the unconstrained optimization problem. For practical applications rather than demonstrations, such a graphical method is not recommended. Direct solutions to the optimization problems are recommended instead.

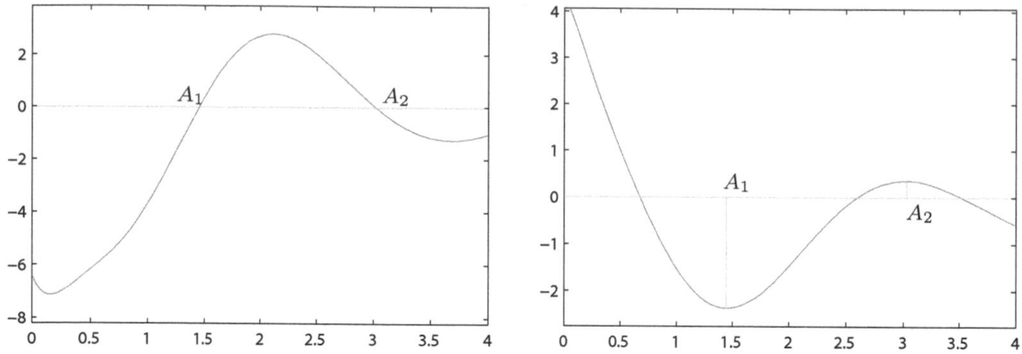

(a) first-order derivative (b) objective function

FIGURE 6.8: Minimum solution using graphical method.

6.3.2 Solution of unconstrained optimization using MATLAB

An unconstrained optimization problem solver, `fminsearch()`, is provided in MATLAB. Moreover, a similar function `fminunc()` is also provided in the Optimization Toolbox. Both functions have the same syntaxes. For instance,

```
x=fminunc(fun,x0)                              % simplest call
[x,f,flag,out]=fminunc(fun,x0,opt,p1,p2,...)   % more general form
```

where the input and output arguments are very similar to the `fsolve()` function described earlier. The control options are also the same. Furthermore, the whole optimization problem can be described with a structured variable *problem*, and the problem can be solved with `x=fminunc(problem)`.

The improved simplex algorithm in Reference [5] is used to solve the optimization problem. This method is an effective one in solving unconstrained optimization problems. The following examples will be given for illustrations.

Example 6.22 For a function with two variables given by $z = (x^2 - 2x)e^{-x^2-y^2-xy}$, find the minimum with MATLAB functions, and interpret the solutions graphically.

Solution *The surface of the function was studied in Chapter 2, and the target of minimization is to find the values of x and y where the surface is at its valley.*

The variables in the objective function are x, y, not the same as in the standard unconstrained optimization definition. A vector $\boldsymbol{x}$ should be defined by the variable substitutions such that $x_1 = x$, and $x_2 = y$. Thus, the objective function can be rewritten as $f(\boldsymbol{x}) = (x_1^2 - 2x_1)e^{-x_1^2-x_2^2-x_1x_2}$. Describing it with the anonymous function, the following statements can be used to find the optimal solution, and the solution is $\boldsymbol{x} = [0.6110, -0.3056]^{\mathrm{T}}$.

```
>> f=@(x)(x(1)^2-2*x(1))*exp(-x(1)^2-x(2)^2-x(1)*x(2));
   x0=[2; 1]; x=fminsearch(f,x0) % minimization from an initial point
```

Similarly, the same problem can be solved with the `fminunc()` *function, where*

```
>> x=fminunc(f,x0) % a more effective function with the same syntax
```

and the solution obtained is $\boldsymbol{x} = [0.6110, -0.3055]^{\mathrm{T}}$.

Normally, the number of the objective function calls in the `fminunc()` *function is much less than the* `fminsearch()` *function, since a more effective algorithm is used in* `fminunc()`*. Therefore, if one installs the Optimization Toolbox, it is suggested that the function* `fminunc()` *be used for unconstrained optimization problems.*

To illustrate and monitor the solution process graphically, an output function can be written in the following format

```
function stop=myout(x,optimValues,state), stop=false;
switch state % monitoring intermediate results
   case 'init', hold on  % preparation
   case 'iter', plot(x(1),x(2),'o'), % display intermediate results
      text(x(1)+0.1,x(2),int2str(optimValues.iteration));
   case 'done', hold off % finishing up display process
end
```

In the function, at each iteration, the intermediate point found is marked and numbered. To monitor the optimization process, the `OutputFcn` *option must be set to* `@myout`.

To demonstrate the optimization process, the contour plot of the objective function can be drawn first. If $\boldsymbol{x}_0 = [2, 1]^{\mathrm{T}}$ *is used as the initial search point, the following statements can be used to monitor the searching process, the search trajectory can be superimposed on the contours of the given function using the following statements, as shown in Figure 6.9, where the overlapped numbers mean that at the points, the searching step-size is very small, and the points found are close to each other.*

```
>> [x,y]=meshgrid(-3:.1:3, -2:.1:2); % draw contour and search trajectory
   z=(x.^2-2*x).*exp(-x.^2-y.^2-x.*y); contour(x,y,z,30);
   ff=optimset; ff.OutputFcn=@myout; x0=[2 1]; x=fminunc(f,x0,ff)
```

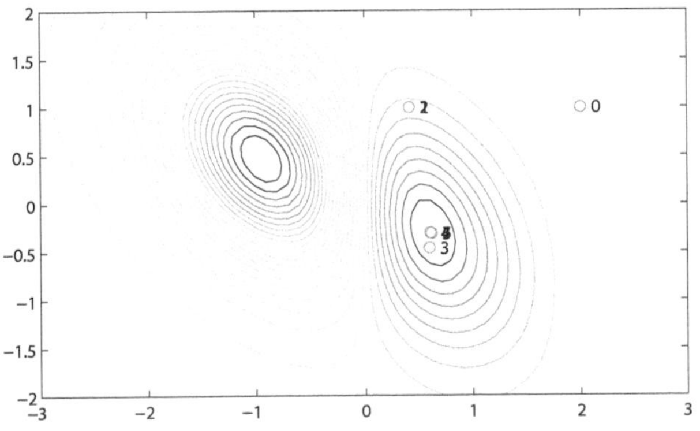

FIGURE 6.9: Search trajectory.

With Optimization Toolbox of MATLAB, structured variable can be used to describe optimization problems, which makes the solution process more standardized. All the fields, `objective`, `x0`, `solver` and `options`, are essential fields in the function. A structured variable P can be created, with its field P.`objective` to describe objective function, and its field $P.\boldsymbol{x}_0$ to describe initial search point. The other two fields can usually be set to

P.options = optimset, P.solver = 'fminunc'.

Then, the variable P can be used to describe the optimization problem. Function [x,$f_{\rm m}$,flag] = fminunc(P) can be used to solve directly the unconstrained optimization problems.

Example 6.23 Please solve again the unconstrained optimization problem in Example 6.22, with the use of structured variables.

Solution *The structured variable P describing the whole optimization problem can be established with the following statements. Then, the function* fminunc() *can be used directly in solving the optimization problem, and the result is exactly the same as the one obtained in Example 6.22. Please note that in the first sentence, make sure the variable P is cleared from MATLAB workspace before use. Also, the* options *field in P must be assigned. The four fields are essential for* fminunc().

```
>> clear P; P.solver='fminunc'; P.options=optimset;
   P.objective=@(x)(x(1)^2-2*x(1))*exp(-x(1)^2-x(2)^2-x(1)*x(2));
   P.x0=[2; 1]; [x,b,c,d]=fminunc(P) % the 4 fields are essential
```

6.3.3 Global minimum and local minima

According to multivariable calculus, the necessary condition for a minimum point to exist is that $df(x)/dx = 0$. However, the points satisfying such a condition may not be unique, and in many real applications, there might be many such points. If a search method is used, only one such point at a time can be found from a given initial point, and this point may not be the global minimum point. The concepts of global minimum and local minima are illustrated through the following example.

Example 6.24 Consider the minimization problem of Rastrigin function [6]

$$f(x_1,x_2) = 20 + x_1^2 + x_2^2 - 10(\cos\pi x_1 + \cos\pi x_2).$$

Draw the surface of the objective function, and try to find the optimal solution with simple searching algorithm, and see what may happen.

Solution *The surface of the function can be immediately obtained with the following statements, as shown in Figure 6.10 (a). It can be seen that the surface changes irregularly, and there are a great amount of peaks and valleys.*

```
>> ezsurf('20+x1^2+x2^2-10*(cos(pi*x1)+cos(pi*x2))')
```

The planform view of the surface can be obtained as shown in Figure 6.10 (b). It can be distinguished by the colors that the valley in the center is the global minimum, and the rest of the valleys are local minima. Furthermore, the four neighbors of the global minimum point can be regarded as suboptimal *local minima.*

```
>> view(0,90), shading flat
```

A few initial points are selected, and optimizations can be performed from these initial points

```
>> f=@(x)20+x(1)^2+x(2)^2-10*(cos(pi*x(1))+cos(pi*x(2)));
   x1=fminunc(f,[2,3]), f(x1), x2=fminunc(f,[-1,2]), f(x2)
   x3=fminunc(f,[8 2]), f(x3), x4=fminunc(f,[-4,6]), f(x4)
```

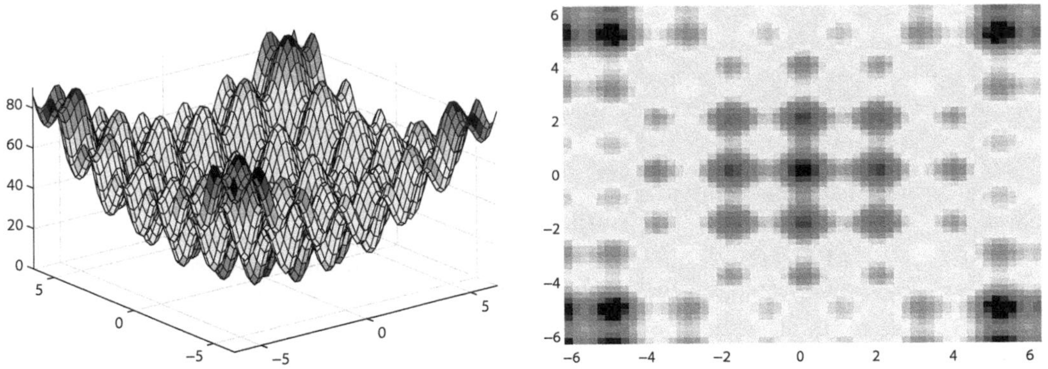

(a) 3D view (b) planform view

FIGURE 6.10: Surface of Rastrigin function.

and the solutions are

$$\boldsymbol{x}_1=[1.9602,1.9602],\ f(\boldsymbol{x}_1)=7.8409,\quad \boldsymbol{x}_2=[-0.0000,1.9602], f(\boldsymbol{x}_2)=3.9205,$$

$$\boldsymbol{x}_3=[7.8338,1.9602],\ f(\boldsymbol{x}_3)=66.6213,\ \boldsymbol{x}_4=[-3.9197,5.8779],\ f(\boldsymbol{x}_4)=50.9570.$$

It can be seen from the optimization results that the "optimal" objective functions have significant differences. Most of them, if not all, are local minima. If the initial values are not set properly, it is very likely to find a local minima.

It can be seen from the above example that sometimes the solution obtained directly from the function calls may be a local minimum, rather than global minimum. Thus, different initial conditions should be tested to possibly improve the solutions. Genetic algorithm-based optimization techniques can significantly improve the globalness of the minima, since many initial conditions are tested simultaneously. However, even genetic algorithm-based algorithms cannot guarantee the global solution. In Section 10.4, introductions and application illustrations will be given on genetic algorithms and other evolutionary optimization approaches.

An alternative method is proposed here, aiming at finding the global optimal solution of such an objective function.

Require: Objective function $f(\boldsymbol{x})$, interval (a,b), number of decision variables n, number of runs N

Initialization: Assign $f_0=\infty$

for $i=1$ To N **do**

  Randomly generate an $\boldsymbol{x}_0$, and find a solution $\boldsymbol{x}$ and $f_1=f(\boldsymbol{x})$

  if $f_1<f_0$ **then**

    Store $\boldsymbol{x}$, and let $f_0=f_1$

  end if

end for

A MATLAB function can be written based on the algorithm.

```
function [x,f0]=fminunc_global(f,a,b,n,N,varargin)
k0=0; f0=Inf; if strcmp(class(f),'struct'), k0=1; end
```

```
for i=1:N, x0=a+(b-a)*rand(n,1);
   if k0==1, f.x0=x0; [x1 f1 key]=fminunc(f);
   else, [x1 f1 key]=fminunc(f,x0,varargin{:}); end
   if key>0 & f1<f0, x=x1; f0=f1; end
end
```

The syntax of the function is [$\boldsymbol{x}$,$f_{\min}$] = fminunc_global(*fun*,a,b,n,N), where *fun* is the MATLAB description of the objective function, implemented either as an anonymous function, or an M-function, and it can also be the structured variable for the entire optimization problem. When the number of N is properly selected, the returned arguments $\boldsymbol{x}$ and $f_{\min}$ are quite likely the global solution of the problem.

Example 6.25 Consider again the optimization problem in Example 6.24. The number of loops N can be set to 50, it can be seen that global minimum at $x_1 = x_2 = 0$ can be found.

```
>> f=@(x)20+x(1)^2+x(2)^2-10*(cos(pi*x(1))+cos(pi*x(2))); F=[];
   [x,f0]=fminunc_global(f,-2*pi,2*pi,2,50); % try to find global solution
```

To further validate the new global optimization problem solver, we can call the function 100 times. It is found that each time, the function is successful in finding the global optimum.

```
>> F=[];
   for i=1:100, % try the function 100 times and assess successful rate
      [x,f0]=fminunc_global(f,-2*pi,2*pi,2,50); F=[F,f0];
   end
```

Of course, the global optimal point is $x_1 = x_2 = 0$, and uniformly distributed random numbers in each run is quite likely to have an initial point near $(0,0)$, so that global optimal solution can be found each time. Therefore, it is unfair to compare this method with other ones through such an example.

Example 6.26 Now suppose the original Rastrigin function is modified as

$$f(x_1,x_2) = 20 + (x_1/30-1)^2 + (x_2/20-1)^2 - 10[\cos(x_1/30-1)\pi + \cos(x_2/20-1)\pi],$$

run the global search function 100 times, and see the successful rate in finding the global optimal solutions.

Solution *If the searching area is extended to* ± 100, *the following statements can be tested*

```
>> f=@(x)20+(x(1)/30-1)^2+(x(2)/20-1)^2-10*(cos(pi*(x(1)/30-1))+...
        cos(pi*(x(2)/20-1))); F=[]; tic
   for i=1:100 % try to solve the new problem 100 times
      [x,f0]=fminunc_global(f,-100,100,2,50); F=[F,f0];
   end, toc
```

It can be seen that in the 100 runs of the solver, only 3 times failed to find the global optimal point at $(30,20)$, *the rest are successful. The unsuccessful 3 times all stopped at the neighbors of the global optimum, and, in fact, they can be regarded as the suboptimal solutions. Therefore, this method can be considered reliable and trustworthy. If the* fminunc_global() *function is executed alone, it is quite likely to find the global optimum.*

```
>> [x,f0]=fminunc_global(f,-100,100,2,50) % try to find global solution
```

6.3.4 Solving optimization problems with gradient information

Sometimes, the convergence speed for solving optimization problems may be very low, and the exact optimum may not even be obtained using the information provided in the objective function alone. Thus, the gradient information can be used to improve the optimization process.

In some functions of the Optimization Toolbox, the gradient information can also be provided in the MATLAB function describing the objective function. In this case, two arguments are returned with the first one still describing the objective function, and the second one for the gradients. Furthermore, in this case, the `GradObj`, a field in the control options, should be set to `'on'`. The optimization solver can then be used to solve the optimization problems with the gradient information. Anonymous functions are no longer usable, since only one returned variable is allowed.

Example 6.27 Consider the Rosenbrock function

$$f(x_1, x_2) = 100(x_2 - x_1^2)^2 + (1 - x_1)^2.$$

Solve the unconstrained minimization problem for the function.

Solution *It can be seen that the objective function is made of two squared terms. Thus, when* $x_2 = x_1 = 1$, *the objective function takes its global minimum. Three-dimensional contours of the function can be drawn as shown in Figure 6.11.*

```
>> [x,y]=meshgrid(0.5:0.01:1.5); z=100*(y.^2-x).^2+(1-x).^2;
   contour3(x,y,z,100), zlim([0,310]) % draw 3D contour of objective func
```

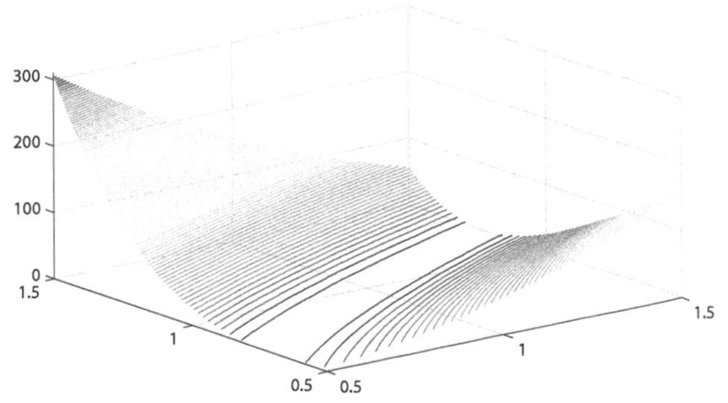

FIGURE 6.11: Three-dimensional contours for the Rosenbrock function.

From the contours, the minimum is located in a very flat and narrow valley. Thus, Rosenbrock function is also known as the banana function. *In the valley, the change of the objective function is extremely slow, therefore, it is a very challenging problem to optimization algorithms. This function is often used as a* benchmark problem *to test whether an optimization algorithm is good or not. The following statements can be used to solve the optimization problem without the gradient information.*

```
>> f=@(x)100*(x(2)-x(1)^2)^2+(1-x(1))^2;
   ff=optimset; ff.TolX=1e-10; ff.TolFun=1e-20; x=fminunc(f,[0;0],ff)
```

The best solution obtained is $x_1 = 0.9999956$, *and* $x_2 = 0.9999912$. *It can be seen that although very tough error tolerances are used, the exact minimum still cannot be obtained. A warning is also given, expecting the gradient information.*

For the given Rosenbrock function, the gradient vector can easily be obtained by

```
>> syms x1 x2; f=100*(x2-x1^2)^2+(1-x1)^2;
   J=jacobian(f,[x1,x2]) % find analytically the gradients
```

and the Jacobian matrix is $\boldsymbol{J} = [-400(x_2 - x_1^2)x_1 - 2 + 2x_1, 200x_2 - 200x_1^2]$. *Therefore, the gradient matrix* $\boldsymbol{J}$ *can be described in the objective function, and it is rewritten as*

```
function [y,Gy]=c6fun3(x)
y=100*(x(2)-x(1)^2)^2+(1-x(1))^2;
Gy=[-400*(x(2)-x(1)^2)*x(1)-2+2*x(1); 200*x(2)-200*x(1)^2];
```

It should be noted that since two returned arguments are expected in the function, neither inline function nor anonymous function can be used to describe the new objective function. M-function is the only choice in describing the gradients. The following statements can be used to solve the optimization problem:

```
>> ff.GradObj='on'; x=fminunc(@c6fun3,[0;0],ff) % use gradient information
```

and the solution found is $\boldsymbol{x} = [1.000000000000018, 1.000000000000036]^{\mathrm{T}}$. *It can be seen that with the gradient information, the optimization is significantly speeded up, and the accuracy is also significantly improved, which approaches the analytical values. Such an accuracy cannot be obtained without specifying the gradient information. However, in some other applications, the derivation and programming of gradient information could be very difficult, if not impossible, and one has to directly solve the optimization problem without the gradient information. The global solver* `fminunc_global()` *can also be used to replace* `fminunc()` *function, and a much more accurate solution can normally be found.*

```
>> x=fminunc_global(@c6fun3,-10,10,2,50,ff) % global solver with gradients
```

Structured variable can also be used in describing the optimization problem, and the following statements yield the same result.

```
>> clear P; P.solver='fminunc'; ff=optimset; P.objective=@c6fun3;
   ff.GradObj='on'; ff.TolX=1e-20; ff.TolFun=1e-20; P.options=ff;
   P.x0=[2; 1]; [x,b,c,d]=fminunc(P) % structured variable description
```

If `fminunc_global()` *is used, a more accurate solution can be found, and the cost is, the elapsed time may be increased by 100 times, around 45 seconds.*

```
>> tic, [x1,b]=fminunc_global(P,-10,10,2,100), toc
```

It is worth mentioning that, Rosenbrock function was deliberately designed to examine the effectiveness of the optimization algorithms, and it is an artificial function. In practical applications, gradient information is not needed in a great amount of optimization algorithms, and optimization problems can equally be solved successfully.

6.4 Constrained Optimization Problems

A general description to constrained optimization problems is

$$\min_{\boldsymbol{x} \text{ s.t. } \boldsymbol{G}(\boldsymbol{x})\leqslant\mathbf{0}} f(\boldsymbol{x}), \tag{6-4-1}$$

where $\boldsymbol{x} = [x_1, x_2, \cdots, x_n]^{\mathrm{T}}$. The interpretation of such a description is that, under the constraints $\boldsymbol{G}(\boldsymbol{x}) \leqslant 0$, a decision vector $\boldsymbol{x}$ is expected which minimizes the objective function $f(\boldsymbol{x})$. In practical optimization problems, the constraints could be very complicated. For instance, it can be equalities or inequalities, and it can also be linear or nonlinear. Sometimes, these functions may not easily be described by mathematical functions.

6.4.1 Constraints and feasibility regions

The solution area $\boldsymbol{x}$ satisfying all the constraints $\boldsymbol{G}(\boldsymbol{x}) \leqslant \mathbf{0}$ is referred to as the *feasible region*. A function of two variables will be given below, and the feasible regions will be illustrated graphically.

Example 6.28 Consider the optimization problem with two variables given below. Study the optimization problem using the graphical method.

$$\max_{\boldsymbol{x} \text{ s.t. } \begin{cases} 9\geqslant x_1^2+x_2^2 \\ x_1+x_2\leqslant 1 \end{cases}} -x_1^2 - x_2.$$

Solution *From the given constraints, the initial square region* $[-3, 3]$ *can be selected and mesh grids can be made. Then, the objective function for unconstraint problems can be obtained using the following statements.*

```
>> [x1,x2]=meshgrid(-3:.1:3);  % grid data generation
   z=-x1.^2-x2;                % calculate objective function over the grids
```

When the constraints are introduced, one may remove the points outside the feasible region from the three-dimensional surface by setting the values to `NaN`*, and the solutions can then be found graphically using the following statements.*

```
>> i=find(x1.^2+x2.^2>9); z(i)=NaN; % find x1^2+x2^2>9 points and assign NaN
   i=find(x1+x2>1); z(i)=NaN;   % find x1+x2>1 points, assign values to NaN
   surf(x1,x2,z); shading interp; % draw the allowed surface
```

The three-dimensional surface plot can be drawn as shown in Figure 6.12 (a). If one wants to observe the three-dimensional surface from the top, then, the command `view(0,90)` *can be used, and the two-dimensional projection can be obtained as shown in Figure 6.12 (b). The region shown on the graph is the* feasible region. *The maximum value in this feasible region is the solution to the constrained optimization problem. Using the graphical method, it can be found that the solution is* $x_1 = 0$, $x_2 = -3$, *with the maximum value 3.*

For problems with one or two variables, the graphical solution is of course the most effective and straightforward method. However, for problems with more variables, the graphical solution method cannot be used. Numerical searching methods should be used instead and, again, there is to date no method to judge whether the solutions obtained is global or not.

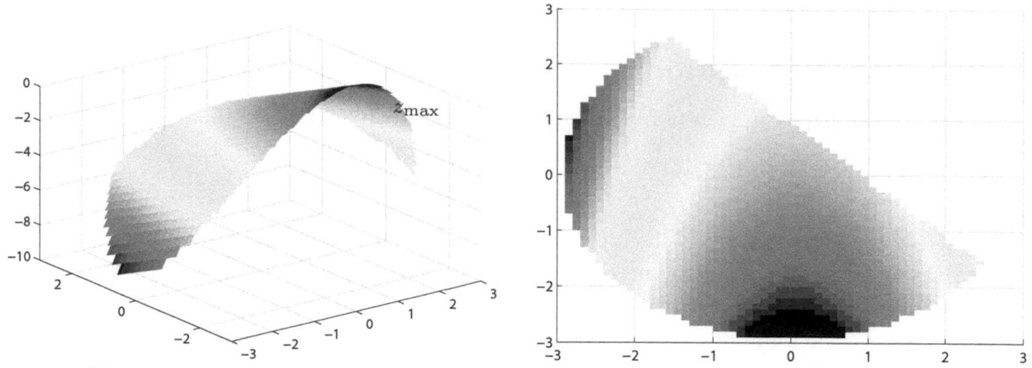

(a) three-dimensional surface plot (b) feasible region

FIGURE 6.12: Graphical solution to a 2D optimization problem.

6.4.2 Solving linear programming problems

I. Solutions of standard linear programming problems

Linear programming problems are special constrained programming problems whose objective function and constraints are all linear with respect to $\boldsymbol{x}$. The general mathematical description is

$$\min_{\boldsymbol{x}\ \text{s.t.}\ \begin{cases}\boldsymbol{A}\boldsymbol{x}\leqslant\boldsymbol{B}\\ \boldsymbol{A}_{\mathrm{eq}}\boldsymbol{x}=\boldsymbol{B}_{\mathrm{eq}}\\ \boldsymbol{x}_{\mathrm{m}}\leqslant\boldsymbol{x}\leqslant\boldsymbol{x}_{\mathrm{M}}\end{cases}} \boldsymbol{f}^{\mathrm{T}}\boldsymbol{x}. \tag{6-4-2}$$

In order to solve the problem effectively, the constraints can further be classified to linear equality constraints $\boldsymbol{A}_{\mathrm{eq}}\boldsymbol{x}=\boldsymbol{B}_{\mathrm{eq}}$ and linear inequality constraints $\boldsymbol{A}\boldsymbol{x}\leqslant\boldsymbol{B}$. Moreover, the upper- and lower-bounds $\boldsymbol{x}_{\mathrm{M}}$ and $\boldsymbol{x}_{\mathrm{m}}$ can be introduced, such that $\boldsymbol{x}_{\mathrm{m}}\leqslant\boldsymbol{x}\leqslant\boldsymbol{x}_{\mathrm{M}}$.

For inequality constraints, the standard description under MATLAB is the "$\leqslant$" relationship. If certain constraints are given by "$\geqslant$" relationship, then, -1 can be multiplied to both sides of the inequality such that the inequalities can be converted to the "$\leqslant$" relationship.

Linear programming problems are the simplest and most widely used constrained optimization problems. There are many algorithms suggested for this kind of optimization problem. The simplex algorithm is proven to be the most effective algorithm and with the well established function `linprog()`, the linear programming problems can easily be solved, with the syntaxes

```
[x,f_opt,flag,c] = linprog(problem)
[x,f_opt,flag,c] = linprog(f,A,B,A_eq,B_eq,x_m,x_M,x_0,opts)
```

where $\boldsymbol{f}$, $\boldsymbol{A}$, $\boldsymbol{B}$, $\boldsymbol{A}_{\mathrm{eq}}$, $\boldsymbol{B}_{\mathrm{eq}}$, $\boldsymbol{x}_{\mathrm{m}}$, $\boldsymbol{x}_{\mathrm{M}}$ are the same as in the above general formulation. The argument $\boldsymbol{x}_0$ is the user selected initial search point, which can be omitted. If certain constraints do not exist, replace the matrices with empty ones. After the function call, the optimized results are returned in variable $\boldsymbol{x}$, and the optimum objective function is returned in f_{opt}. A positive `flag` indicates the successful function call. Structured variable description to linear programming problems can also be used.

It should be noted that a linear programming problem is a *convex problem*, and in simple terms, if a minimum is found, it must be the global minimum. Thus, the initial choice of $\boldsymbol{x}_0$ is not important and can usually be omitted.

Example 6.29 Solve the following linear programming problem:

$$\min_{\boldsymbol{x} \text{ s.t. } \begin{cases} 2x_2+x_3+4x_4+2x_5 \leqslant 54 \\ 3x_1+4x_2+5x_3-x_4-x_5 \leqslant 62 \\ x_1,x_2 \geqslant 0,\ x_3 \geqslant 3.32,\ x_4 \geqslant 0.678,\ x_5 \geqslant 2.57 \end{cases}} \quad -2x_1 - x_2 - 4x_3 - 3x_4 - x_5.$$

Solution *For this linear programming problem, the objective function can be defined by the vector* $\boldsymbol{f}=[-2,-1,-4,-3,-1]^{\mathrm{T}}$. *There are two inequality constraints*

$$\boldsymbol{A} = \begin{bmatrix} 0 & 2 & 1 & 4 & 2 \\ 3 & 4 & 5 & -1 & -1 \end{bmatrix}, \quad \textit{and } \boldsymbol{B} = \begin{bmatrix} 54 \\ 62 \end{bmatrix}.$$

Since there are no equality constraints, the matrices $\boldsymbol{A}_{\mathrm{eq}}$ *and* $\boldsymbol{B}_{\mathrm{eq}}$ *can be declared as empty matrices. From the original problem, it is also seen that the lower-bound of* $\boldsymbol{x}$ *can be defined as* $\boldsymbol{x}_{\mathrm{m}} = [0,0,3.32,0.678,2.57]^{\mathrm{T}}$, *and since there is no upper-bound,* $\boldsymbol{x}_{\mathrm{M}}$ *can be assigned to an empty vector. Thus, the problem can be solved directly using the following MATLAB statements, and the result can be obtained immediately*

```
>> f=-[2 1 4 3 1]'; A=[0 2 1 4 2; 3 4 5 -1 -1];
   B=[54; 62]; Ae=[]; Be=[]; xm=[0,0,3.32,0.678,2.57];
   ff=optimset; ff.TolX=1e-15; ff.TolFun=1e-20; ff.TolCon=1e-20;
   [x,f_opt,key,c]=linprog(f,A,B,Ae,Be,xm,[],[],ff)
```

where $\boldsymbol{x} = [19.785,0,3.32,11.385,2.57]^{\mathrm{T}}$, $f_{\mathrm{opt}} = -89.5750$, *and* key=*1, meaning that the solution is successful. In fact, only 5 steps are used in finding the optimal solution for this problem.*

Linear programming problems can also be described by a structured variable P, and its fields `lb` and `ub` are used to describe the bounds $\boldsymbol{x}_{\mathrm{m}}$ and $\boldsymbol{x}_{\mathrm{M}}$, while the fields `Aineq`, `bineq`, `Aeq`, `beq` are used respectively to describe matrices $\boldsymbol{A}$, $\boldsymbol{B}$, $\boldsymbol{A}_{\mathrm{eq}}$, $\boldsymbol{B}_{\mathrm{eq}}$. The `solver` field can be set to `'linprog'`, and objective function vector $\boldsymbol{f}$ can be described by the `f` field. Function `linprog(`P`)` can be used to solve directly the linear programming problem P. In this function, `f`, `solver` and `options` are essential fields, while others are optional ones.

Example 6.30 Structured variables can be used to model the linear programming problems. Since the fields `Aeq` and `Beq` are empty ones, there is no need to describe them in P variable. The following statements can be used, and the same results can be obtained.

```
>> clear P; P.f=-[2 1 4 3 1]'; P.Aineq=[0 2 1 4 2; 3 4 5 -1 -1];
   P.bineq=[54; 62]; P.lb=[0,0,3.32,0.678,2.57]; P.solver='linprog';
   ff=optimset; ff.TolX=1e-15; ff.TolFun=1e-20; ff.TolCon=1e-20;
   P.options=ff; [x,f_opt,key,c]=linprog(P)
```

Example 6.31 Consider the linear programming problem with four variables defined as follows. Solve it using the Optimization Toolbox of MATLAB.

$$\max_{\boldsymbol{x} \text{ s.t. } \begin{cases} x_1/4-60x_2-x_3/50+9x_4 \leqslant 0 \\ -x_1/2+90x_2+x_5/50-3x_4 \geqslant 0 \\ x_3 \leqslant 1,\ x_1 \geqslant -5,\ x_2 \geqslant -5,\ x_3 \geqslant -5,\ x_4 \geqslant -5 \end{cases}} \quad \frac{3}{4}x_1 - 150x_2 + \frac{1}{50}x_3 - 6x_4.$$

Solution *In the original problem, the maximization is required. It should be converted first to the minimization problem by multiplying −1 to the objective function. The new objective function can be rewritten as* $-3x_1/4+150x_2-x_3/50+6x_4$. *From the linear programming problem formulation, it can be seen that* $\boldsymbol{f}=[-3/4,150,-1/50,6]^{\mathrm{T}}$.

For the given constraints, $x_i \geqslant -5$ *will lead to a lower-bound vector* $\boldsymbol{x}_{\mathrm{m}}=$ `[-5;-5;-5;-5]`. *Similarly, the upper-bound vector* $\boldsymbol{x}_{\mathrm{M}}=$ `[Inf;Inf;1;Inf]`, *with* `Inf` *for* $+\infty$. *Also, in the two inequality constraints, the second one is given by the* $\geqslant$ *relationship. We should multiply both sides by −1 to convert it into the* $\leqslant$ *inequality. Thus, the matrices for the inequalities can be written as*

$$\boldsymbol{A}=\begin{bmatrix}1/4 & -60 & -1/50 & 9\\ 1/2 & -90 & -1/50 & 3\end{bmatrix},\ \boldsymbol{B}=\begin{bmatrix}0\\0\end{bmatrix}.$$

Since there are no equality constraints, one should assume that $\boldsymbol{A}_{\mathrm{eq}}=$ `[]`, $\boldsymbol{B}_{\mathrm{eq}}=$ `[]`. *Therefore, the original linear programming problem can be solved using the following statements*

```
>> clear P; P.f=[-3/4,150,-1/50,6];
   P.Aineq=[1/4,-60,-1/50,9; 1/2,-90,-1/50,3]; P.bineq=[0;0];
   P.lb=[-5;-5;-5;-5]; P.ub=[Inf;Inf;1;Inf]; ff=optimset;
   ff.TolX=1e-15; ff.TolFun=1e-20; TolCon=1e-20; P.options=ff;
   P.solver='linprog'; [x,a,b,c]=linprog(P)
```

and within 19 iterations, the solution $\boldsymbol{x}=[-5,-0.194667,1,-5]^{\mathrm{T}}$ *can be found. The objective function reaches its minimum at* $f_{\mathrm{opt}}=-55.47$.

II. Linear programming with double subscripts

In real applications, one may find some linear programming problems are presented, where the decision variable is given by a matrix, rather than a vector. Conversions must be made first to convert such kind of problems into standard linear programming problems, then, `linorog()` can be used to solve the original problem. After solution, the vectorized decision variable should be converted back to the decision matrix.

Example 6.32 Consider a tricky double subscripted linear programming problem given below. Find the optimal solution to the problem.

$$\min\quad 2800(x_{11}+x_{21}+x_{31}+x_{41})+4500(x_{12}+x_{22}+x_{32})+6000(x_{13}+x_{23})+7300x_{14}.$$

$$\boldsymbol{x}\ \text{s.t.}\begin{cases}x_{11}+x_{12}+x_{13}+x_{14}\geqslant 15\\ x_{12}+x_{13}+x_{14}+x_{21}+x_{22}+x_{23}\geqslant 10\\ x_{13}+x_{14}+x_{22}+x_{23}+x_{31}+x_{32}\geqslant 20\\ x_{14}+x_{23}+x_{32}+x_{41}\geqslant 12\\ x_{ij}\geqslant 0,\ (i=1,2,3,4, j=1,2,3,4)\end{cases}$$

Solution *Since the* `linprog()` *function can only be used to solve single subscripted problems, the original problem should be converted first by rearranging the variables such that* $x_1=x_{11}, x_2=x_{12}, x_3=x_{13}, x_4=x_{14}, x_5=x_{21}, x_6=x_{22}, x_7=x_{23}, x_8=x_{31}, x_9=x_{32}, x_{10}=x_{41}$. *The original problem can then be rewritten as*

$$\min \quad 2800(x_1+x_5+x_8+x_{10})+4500(x_2+x_6+x_9)+6000(x_3+x_7)+7300x_4.$$

$$\boldsymbol{x} \text{ s.t.} \begin{cases} -(x_1+x_2+x_3+x_4)\leqslant -15 \\ -(x_2+x_3+x_4+x_5+x_6+x_7)\leqslant -10 \\ -(x_3+x_4+x_6+x_7+x_8+x_9)\leqslant -20 \\ -(x_4+x_7+x_9+x_{10})\leqslant -12 \\ x_i\geqslant 0,\ i=1,2,\cdots,10 \end{cases}$$

The following MATLAB statements can then be used in finding the solutions to the optimization problem

```
>> clear P; P.solver='linprog'; P.options=optimset;
   P.f=2800*[1 0 0 0 1 0 0 1 0 1]+4500*[0 1 0 0 0 1 0 0 1 0]+...
     6000*[0 0 1 0 0 0 1 0 0 0]+7300*[0 0 0 1 0 0 0 0 0 0];
   P.Aineq=-[1 1 1 1 0 0 0 0 0 0; 0 1 1 1 1 1 1 0 0 0;
      0 0 1 1 0 1 1 1 1 0; 0 0 0 1 0 0 1 0 1 1];
   P.bineq=-[15; 10; 20; 12]; P.lb=[0 0 0 0 0 0 0 0 0 0];
   x=linprog(P)
```

and it is found that $\boldsymbol{x} = [4.2069, 0, 0, 10.7931, 0, 0, 0, 8, 1.2069, 0.0000]$. *Converting the solutions back to the double subscripted format, one has* $x_{11} = 4.2069$, $x_{14} = 10.7931$, $x_{31} = 8$, $x_{32} = 1.2069$, *and the rest of the variables are all zeros.*

III. Modeling and solution of transportation problems

Transportation problems are essential problems both in "Operations Research" courses and real applications. A typical transportation problem is illustrated in Table 6.2 [7]. Suppose it is required to transport the n kinds of products from the m suppliers. The transportation cost of each unit of product is specified as c_{ij}, and the total requirement for the ith product is d_i, while the total of products in the ith supplier is s_i. The target of the transportation problem is to decide how many units of each product is required from each supplier, with the total transportation cost minimized.

TABLE 6.2: Typical table for a transportation problem.

		cost per unit distributed different kinds of products				supply
		1	2	$\cdots$	n	s_i
	1	c_{11}	c_{12}	$\cdots$	c_{1n}	s_1
sources	2	c_{21}	c_{22}	$\cdots$	c_{2n}	s_2
		$\cdots$	$\cdots$	$\cdots$	$\cdots$	$\cdots$
	m	c_{m1}	c_{m2}	$\cdots$	c_{mn}	s_m
	demond	d_1	d_2	$\cdots$	d_n	

To model such a problem mathematically, the decision variables x_{ij} should be selected, according to each cost c_{ij}. Therefore, the transportation problem can be modeled by linear programming problems with double subscripts as

$$\min \sum_{i=1}^{m}\sum_{j=1}^{n} c_{ij}x_{ij}. \tag{6-4-3}$$

$$\boldsymbol{X} \text{ s.t. } \begin{cases} \sum\limits_{j=1}^{n} x_{ij}=s_i, i=1,2,\cdots,m \\ \sum\limits_{i=1}^{m} x_{ij}=d_j, j=1,2,\cdots,n \\ x_{ij}\geqslant 0, i=1,2,\cdots,m, j=1,2,\cdots,n \end{cases}$$

To solve the linear programming problem with double subscripts, it is rather a tedious and error-prone task to convert it first into single subscript problems. An automatic modeling and solution function is written, to allow direct use of double subscript matrix $\boldsymbol{C}$, and vectors $\boldsymbol{s}$, $\boldsymbol{d}$. The syntax of the function is $\boldsymbol{X}$ = `transport_linprog(`$\boldsymbol{C}$`,`$\boldsymbol{s}$`,`$\boldsymbol{d}$`)`. With the function, the transportation problem can be solved directly, and the optimal solution is returned in $\boldsymbol{X}$.

```
function [x,f,key]=transport_linprog(F,s,d,intkey)
[m,n]=size(F); X=zeros(n*m,m); Y=zeros(n*m,n);
for i=0:m-1, X(i*(n+n*m)+1:i*(n+n*m)+n)=1; end
for k=1:n*m+1:n*m*n, i=0:m-1; Y(k+n*i)=1; end
Aeq=[X Y]'; xm=zeros(1,n*m); F1=F.'; f=F1(:).'; Beq=[s(:); d(:)];
if nargin==3, [x,f,key]=linprog(f,[],[],Aeq,Beq,xm);
else, [x,f,key]=intlinprog(f,1:n*m,[],[],Aeq,Beq,xm); x=round(x); end
x=reshape(x,n,m).';
```

Example 6.33 Assume that a department store is planning to purchase clothes from three different cities I, II and III. Four kinds of clothes, A, B, C and D, are to be purchased, with the amounts required for A, B, C and D are respectively 1500, 2000, 3000 and 3500. It is known that the maximum supplies of the three cities are respectively 2500, 2500 and 5000. It is also known that the expected profit of the clothes are given in Table 6.3. Please design a purchase plan, which maximizes the total profit.

TABLE 6.3: Expected profit for each type of clothes.

cities	different types of clothes				total supply
	A	B	C	D	s_i
I	10	5	6	7	2500
II	8	2	7	6	2500
III	9	3	4	8	5000
total requirement d_i	1500	2000	3000	3500	

Solution *The original problem is a maximum profit problem, therefore, the objective function should be multiplied by* -1 *to convert it to a minimization problem. The lower bounds of the decision variables are zeros. The following statements can be used to solve the problem directly.*

```
>> C=[10 5 6 7; 8 2 7 6; 9 3 4 8]; s=[2500 2500 5000];
   d=[1500 2000 3000 3500]; X=transport_linprog(-C,s,d)
   f=sum(C(:).*X(:)) % compute the maximized profit
```

and the obtained matrix $\boldsymbol{X}$ *is*

$$\boldsymbol{X}=\begin{bmatrix}0 & 2000 & 500 & 0\\ 0 & 0 & 2500 & 0\\ 1500 & 0 & 0 & 3500\end{bmatrix},\quad f=72000,$$

meaning to purchase 2000 B and 500 C clothes from city I; purchase 2500 C from city II; 1500 A and 3500 D from city III, with a total of expected profit $f = 72000$.

If the demands are changed to $\boldsymbol{d} = [1500, 2500, 3000, 3500]$, *the decision variables returned may become fractional numbers. In that case, integer programming should be introduced. This will be discussed later.*

6.4.3 Solving quadratic programming problems

Quadratic programming problems are another category of simple constrained optimization problems. In quadratic programming problems, the objective function contains the quadratic form of vector $\boldsymbol{x}$. The constraints are still linear. The general form of quadratic programming is

$$\min_{\boldsymbol{x}\ \text{s.t.}\ \begin{cases}\boldsymbol{A}\boldsymbol{x}\leqslant\boldsymbol{B}\\ \boldsymbol{A}_{\rm eq}\boldsymbol{x}=\boldsymbol{B}_{\rm eq}\\ \boldsymbol{x}_{\rm m}\leqslant\boldsymbol{x}\leqslant\boldsymbol{x}_{\rm M}\end{cases}} \frac{1}{2}\boldsymbol{x}^{\rm T}\boldsymbol{H}\boldsymbol{x}+\boldsymbol{f}^{\rm T}\boldsymbol{x}. \tag{6-4-4}$$

Comparing the problem in linear programming, it can be seen that a quadratic form $\boldsymbol{x}^{\rm T}\boldsymbol{H}\boldsymbol{x}$ is introduced to describe x_i^2 and x_ix_j terms. The `quadprog()` function can be used to solve the quadratic programming problems, with the following syntaxes

```
[x,fopt,flag,c] = quadprog(problem)
[x,fopt,flag,c] = quadprog(H,f,A,B,Aeq,Beq,xm,xM,x0,opts)
```

where in the function call, $\boldsymbol{H}$ matrix should be declared, while the other arguments are exactly the same as those in linear programming problems. Quadratic programming problems can also be modeled with structured variables, in a similar manner, as in the linear programming problems. $\boldsymbol{H}$ matrix should be assigned to its `H` field, and its `solver` field can be set to `'quadprog'`.

Example 6.34 Solve the following quadratic programming problems with four variables:

$$\min_{\boldsymbol{x}\ \text{s.t.}\ \begin{cases}x_1+x_2+x_3+x_4\leqslant 5\\ 3x_1+3x_2+2x_3+x_4\leqslant 10\\ x_1,x_2,x_3,x_4\geqslant 0\end{cases}} (x_1-1)^2+(x_2-2)^2+(x_3-3)^2+(x_4-4)^2.$$

Solution *In order to solve the original problem, the objective function should be rewritten such that*

$$\begin{aligned}f(x)&=x_1^2-2x_1+1+x_2^2-4x_2+4+x_3^2-6x_3+9+x_4^2-8x_4+16\\ &=(x_1^2+x_2^2+x_3^2+x_4^2)+(-2x_1-4x_2-6x_3-8x_4)+30.\end{aligned}$$

Since the constant term does not affect the optimization problem solving, it can be omitted safely. Thus, the matrices for quadratic programming can then be defined as

$\boldsymbol{H} = \text{diag}([2,2,2,2])$, *and* $\boldsymbol{f}^{\rm T} = [-2,-4,-6,-8]$. *Therefore, the following statements can be used to solve this quadratic programming problem*

```
>> clear P; P.f=[-2,-4,-6,-8]; P.H=diag([2,2,2,2]);
   OPT=optimset; OPT.LargeScale='off'; % turn off the large-scale algorithm
   P.options=OPT; P.solver='quadprog'; P.lb=zeros(4,1);
   P.Aineq=[1,1,1,1; 3,3,2,1]; P.Bineq=[5;10]; [x,f_opt]=quadprog(P)
```

with the solutions $\boldsymbol{x} = [0, 0.6667, 1.6667, 2.6667]^{\rm T}$, $f_{\rm opt} = -23.6667$.

It should be noted that since there is a factor $1/2$ *in the quadratic term, the generation of matrix* $\boldsymbol{H}$ *should be prepared with care. Thus, the diagonal terms in* $\boldsymbol{H}$ *should be 2's instead of 1's in this example. It should also be noted that the constant term was removed in performing the optimization computation. The constant should be added back to the optimum value such that the objective function is 6.3333.*

6.4.4 Solving general nonlinear programming problems

The general nonlinear programming problems are formulated as follows:

$$\min_{\boldsymbol{x} \text{ s.t. } \boldsymbol{G}(\boldsymbol{x})\leqslant \boldsymbol{0}} f(\boldsymbol{x}), \tag{6-4-5}$$

where $\boldsymbol{x} = [x_1, x_2, \cdots, x_n]^{\rm T}$. For simplicity, the constraints can further be classified into linear inequalities and equalities, upper- and lower-bounds, and nonlinear equalities and inequalities. The original constraints can thus be rewritten as

$$\min_{\boldsymbol{x} \text{ s.t. } \begin{cases} \boldsymbol{A}\boldsymbol{x}\leqslant \boldsymbol{B} \\ \boldsymbol{A}_{\rm eq}\boldsymbol{x}=\boldsymbol{B}_{\rm eq} \\ \boldsymbol{x}_{\rm m}\leqslant \boldsymbol{x}\leqslant \boldsymbol{x}_{\rm M} \\ \boldsymbol{C}(\boldsymbol{x})\leqslant \boldsymbol{0} \\ \boldsymbol{C}_{\rm eq}(\boldsymbol{x})=\boldsymbol{0} \end{cases}} f(\boldsymbol{x}). \tag{6-4-6}$$

A MATLAB function `fmincon()` can be used to solve general nonlinear programming problems. The syntaxes of the function are

$[\boldsymbol{x}$,$f_{\rm opt}$,`flag`,`c`$]$ = `fmincon(`*problem*`)`
$[\boldsymbol{x}$,$f_{\rm opt}$,`flag`,`c`$]$ = `fmincon(`*fun*,$\boldsymbol{x}_0$,$\boldsymbol{A}$,$\boldsymbol{B}$,$\boldsymbol{A}_{\rm eq}$,$\boldsymbol{B}_{\rm eq}$,$\boldsymbol{x}_{\rm m}$,$\boldsymbol{x}_{\rm M}$,*CFun*,*opts*,p_1,p_2,$\cdots$`)`

where *fun* is the M-function or anonymous function to describe the objective function. The argument $\boldsymbol{x}_0$ is the initial search point. The definitions of $\boldsymbol{A}$, $\boldsymbol{B}$, $\boldsymbol{A}_{\rm eq}$, $\boldsymbol{B}_{\rm eq}$, $\boldsymbol{x}_{\rm m}$, $\boldsymbol{x}_{\rm M}$ are the same as in (6-4-6). The argument *CFun* is the M-function to describe the nonlinear constraints, with two returned arguments, indicating the inequality and equality constraints, respectively. The argument *opts* is the control options. The returned arguments are exactly the same as in other optimization functions in MATLAB.

Nonlinear programming problems can also be modeled with structured variable *problem*, whose `solver` field must be set to `'fmincon'`, and nonlinear constraint functions can be assigned to the `nonlcon` field.

Example 6.35 Solve the following nonlinear programming problem:

$$\min_{\boldsymbol{x} \text{ s.t. } \begin{cases} x_1^2+x_2^2+x_3^2-25=0 \\ 8x_1+14x_2+7x_3-56=0 \\ x_1,x_2,x_3 \geqslant 0 \end{cases}} 1000 - x_1^2 - 2x_2^2 - x_3^2 - x_1x_2 - x_1x_3.$$

Solution *In this problem, there are nonlinear constraints, thus, the original problem is not a quadratic programming problem. Nonlinear programming solvers should be used to solve this problem. The objective function can be expressed by an anonymous function. Moreover, since the two constraints are all equalities, thus, the constraint functions can be expressed as*

```
function [c,ceq]=opt_con1(x), c=[]; % specify the two constraints
ceq=[x(1)*x(1)+x(2)*x(2)+x(3)*x(3)-25; 8*x(1)+14*x(2)+7*x(3)-56];
```

where the nonlinear inequalities and nonlinear equalities are returned respectively in arguments `c` *and* `ceq`. *Since there is no inequality constraint, the argument* `c` *is then assigned to an empty matrix.*

Having declared the constraints, the matrices $\boldsymbol{A}$, $\boldsymbol{B}$, $\boldsymbol{A}_{\rm eq}$, $\boldsymbol{B}_{\rm eq}$ *are now all empty matrices. The lower-bound vector can be written as* $\boldsymbol{x}_{\rm m} = [0, 0, 0]^{\rm T}$. *If the initial point is selected as* $\boldsymbol{x}_0 = [1, 1, 1]^{\rm T}$, *the problem can then be solved using the following statements*

```
>> clear P; ff=optimset; ff.LargeScale='off'; ff.TolFun=1e-30;
   ff.TolX=1e-15; ff.TolCon=1e-20; P.x0=[1;1;1]; P.lb=[0;0;0];
   P.objective=@(x)1000-x(1)*x(1)-2*x(2)*x(2)-x(3)*x(3)-x(1)*x(2)-x(1)*x(3);
   P.solver='fmincon'; P.nonlcon=@opt_con1; P.options=ff;
   [x,f_opt,c,d]=fmincon(P)
```

with $\boldsymbol{x} = [3.5121, 0.2170, 3.5522]^{\rm T}$, *and* $f_{\rm opt} = 961.7151$. *Totally 111 calls to the objective functions are made during the solution process.*

Since the second constraint is in fact linear, it can be removed from the nonlinear constraint function. Thus, the constraint function can be simplified as

```
function [c,ceq]=opt_con2(x)
ceq=x(1)*x(1)+x(2)*x(2)+x(3)*x(3)-25; c=[];
```

and the linear equality constraint can be declared in the following statements, and the solution obtained using the following is exactly the same as the one obtained previously.

```
>> P.Aeq=[8,14,7]; P.beq=56; P.nonlcon=@opt_con2; x=fmincon(P)
```

Example 6.36 Please solve the following nonlinear programming problem [8]

$$\min_{\boldsymbol{q},w,k \text{ s.t. } \begin{cases} q_3+9.625q_1w+16q_2w+16w^2+12-4q_1-q_2-78w=0 \\ 16q_1w+44-19q_1-8q_2-q_3-24w=0 \\ 2.25-0.25k \leqslant q_1 \leqslant 2.25+0.25k \\ 1.5-0.5k \leqslant q_2 \leqslant 1.5+0.5k \\ 1.5-1.5k \leqslant q_3 \leqslant 1.5+1.5k \end{cases}} k.$$

Solution *It can be seen from the model that the decision variables are* $\boldsymbol{q}$, w *and* k, *while the standard nonlinear programming problem solver can only solve vector decision variable problems. Variable substitutions should be made first to convert the problem into a standard* *problem. The following variables are assigned $x_1 = q_1, x_2 = q_2, x_3 = q_3, x_4 = w, x_5 = k$. Also, the inequality constraints should be rewritten. The original problem can be manually written as*

$$\min_{\boldsymbol{x} \text{ s.t.} \begin{cases} x_3+9.625x_1x_4+16x_2x_4+16x_4^2+12-4x_1-x_2-78x_4=0 \\ 16x_1x_4+44-19x_1-8x_2-x_3-24x_4=0 \\ -0.25x_5-x_1\leqslant-2.25 \\ x_1-0.25x_5\leqslant2.25 \\ -0.5x_5-x_2\leqslant-1.5 \\ x_2-0.5x_5\leqslant1.5 \\ -1.5x_5-x_3\leqslant-1.5 \\ x_3-1.5x_5\leqslant1.5 \end{cases}} x_5.$$

The following statements can be used to describe the nonlinear constraints

```
function [c,ceq]=c6exnls(x), c=[];
ceq=[x(3)+9.625*x(1)*x(4)+16*x(2)*x(4)+16*x(4)^2+12-4*x(1)-x(2)-78*x(4);
     16*x(1)*x(4)+44-19*x(1)-8*x(2)-x(3)-24*x(4)];
```

Structured variable P is used to describe the original problem, and random numbers are used as the initial searching vector, and the results of optimization process is

$$\boldsymbol{x} = [1.9638, 0.9276, -0.2172, 0.0695, 1.1448],$$

optimal objective function is 1.1448, and the value of `flag` *is 1, meaning the solution process is successful.*

```
>> clear P; P.objective=@(x)x(5); P.nonlcon=@c6exnls;
   P.Aineq=[-1 0 0 0 -0.25; 1 0 0 0 -0.25; 0 -1 0 0 -0.5;
            0 1 0 0 -0.5; 0 0 -1 0 -1.5; 0 0 1 0 -1.5];
   P.bineq=[-2.25; 2.25; -1.5; 1.5; -1.5; 1.5]; P.solver='fmincon';
   P.options=optimset; P.x0=rand(5,1); [x,fm,flag]=fmincon(P)
```

It should be noted that the above solution may still be a local minimum of the problem. The method in Section 6.3.3 can be extended to constrained optimization problems, and the new function is

```
function [x,f0]=fmincon_global(f,a,b,n,N,varargin)
x0=rand(n,1); k0=0; if strcmp(class(f),'struct'), k0=1; end
if k0==1, f.x0=x0; [x f0]=fmincon(f);
else, [x f0]=fmincon(f,x0,varargin{:}); end
for i=1:N, x0=a+(b-a)*rand(n,1);
   if k0==1, f.x0=x0; [x1 f1 key]=fmincon(f);
   else, [x1 f1 key]=fmincon(f,x0,varargin{:}); end
   if key>0 & f1<f0, x=x1; f0=f1; end
end
```

The syntax of the function is

$[\boldsymbol{x}, f_{\min}] =$ `fmincon_global`$(\textit{fun}, a, b, n, N, \textit{others})$

where *fun* can be a structured variable, or the handle of the objective function. In latter case, *others* arguments can be used to specify constraints, as in the case of `fmincon()` function.

Example 6.37 With the above function, the problem can be solved with the following statements, and global optimum can be obtained, as $\boldsymbol{x} = [2.4544, 1.9088, 2.7263, 1.3510, 0.8175]^{\mathrm{T}}$, whose fifth value x_5 is the value of the objective function.

```
>> [x,f0]=fmincon_global(P,0,5,5,50) % find global optimum solution
```

If the function call is executed 100 times, it is quite likely all the runs find the global optimal solutions, which means that the new solver is successful in finding global optimal solutions. The total test time is about 15 minutes.

```
>> tic, X=[];
   for i=1:100, % run 100 times the function and assess the successful rate
      [x,f0]=fmincon_global(P,0,5,5,50); X=[X; x'];
   end, toc
```

Example 6.38 Consider again the optimization problem in Example 6.35. Solve the problem using gradients and compare the results with the original example.

Solution *For the given objective function $f(x)$, the gradient vector, or the Jacobian matrix, can be derived*

```
>> syms x1 x2 x3; f=1000-x1*x1-2*x2*x2-x3*x3-x1*x2-x1*x3;
   J=jacobian(f,[x1,x2,x3]) % find the gradients analytically
```

and the gradient matrix can be written as

$$\boldsymbol{J} = \left[\frac{\partial f}{\partial x_1}, \frac{\partial f}{\partial x_2}, \frac{\partial f}{\partial x_3}\right]^{\mathrm{T}} = \begin{bmatrix} -2x_1 - x_2 - x_3 \\ -4x_2 - x_1 \\ -2x_3 - x_1 \end{bmatrix}.$$

With the gradients, the objective function can then be rewritten as

```
function [y,Gy]=opt_fun2(x)
y=1000-x(1)*x(1)-2*x(2)*x(2)-x(3)*x(3)-x(1)*x(2)-x(1)*x(3);
Gy=[-2*x(1)-x(2)-x(3); -4*x(2)-x(1); -2*x(3)-x(1)];
```

where, `Gy` *returns the gradient vector of the objective function. The following statements can be used*

```
>> clear P; ff=optimset; ff.LargeScale='off'; ff.GradObj='on';
   ff.TolFun=1e-30; ff.TolX=1e-15; ff.TolCon=1e-20;
   P.x0=[1;1;1]; P.lb=[0;0;0];
   P.objective=@opt_fun2; P.solver='fmincon'; P.nonlcon=@opt_con1;
   P.options=ff; [x,f_opt,c,d]=fmincon(P)
```

find the solution $\boldsymbol{x} = [3.5121, 0.2170, 3.5522]^{\mathrm{T}}$, and $f_{\mathrm{opt}} = 961.7151$. In total, 79 calls to the objective functions are made during the solution process.

For this example, if the gradients are used, 13 iterations are needed which are fewer than the one without gradients, where 16 iterations are needed. However, taking into account the time needed in deriving and coding the gradients, the actual time required may be much greater. Therefore, sometimes it may not be worthwhile to use gradient information for certain applications.

6.5 Mixed Integer Programming Problems

In many applications, all or part of the decision variables may be required to be integers for the optimization problems, which are referred to as the *integer programming problems.* If only part of the decision variables are required to be integers, they are referred to as *mixed integer programming problems.* In some applications, the allowable values for the decision variables are either 0's or 1's, and are referred to as *binary programming problems*, also known as the *0-1 programming problems.*

6.5.1 Enumerate method in integer programming problems

Enumerate method, also known as *brutal force method*, means that all the possible feasible combinations of the decision variables are enumerated, from which the global optimal solution can be found. This method is only applicable to integer programming problems.

If the intervals of the decision variables are known, theoretically all the combinations can be enumerated, from which the feasible combinations can be found. The objective function values can then be sorted, and the optimal solution can be found from the sorting results. This method seems to be very simple and straightforward, however, for large-scale problems, i.e., with too many decision variables, and/or the range of the variables are too large to enumerate, computation or even storage burden are too much to be solved even by the most powerful computers in the world. The related mathematical problems are referred to as *non-polynomial hard* (or NP hard) problems. Enumerate methods are only suitable for solving small-scale problems.

Example 6.39 Consider the linear programming problem studied in Example 6.29.

$$\min \quad -2x_1 - x_2 - 4x_3 - 3x_4 - x_5.$$
$$\boldsymbol{x} \text{ s.t. } \begin{cases} 2x_2+x_3+4x_4+2x_5 \leqslant 54 \\ 3x_1+4x_2+5x_3-x_4-x_5 \leqslant 62 \\ x_1,\ x_2 \geqslant 0,\ x_3 \geqslant 3.32,\ x_4 \geqslant 0.678,\ x_5 \geqslant 2.57 \end{cases}$$

If all the decision variables x_i must be integers, the original problem is an integer linear programming problem. Please solve the integer programming problem with enumerate method.

Solution *For this kind of small-scale problem, enumerate method should be used. We can assume that all the components in* $\boldsymbol{x}_{\rm M}$ *are 25, the following statements can be used, and it can be found that the global optimum is* $\boldsymbol{x} = [19, 0, 4, 10, 5]^{\rm T}$, *with* $f_{\min} = -89$.

```
>> N=25; [x1,x2,x3,x4,x5]=ndgrid(1:N,0:N,4:N,1:N,3:N);
   i=find((2*x2+x3+4*x4+2*x5<=54) & (3*x1+4*x2+5*x3-x4-x5<=62));
   x1=x1(i); x2=x2(i); x3=x3(i); x4=x4(i); x5=x5(i);
   f=-2*x1-x2-4*x3-3*x4-x5; [fmin,ii]=sort(f);
   index=ii(1); x=[x1(index),x2(index),x3(index),x4(index),x5(index)]
```

There are two problems to be noted. One is that the search region is set to $x_1 \in [0, 25]$. *If the region is extended from 25 to 30, there will be significant increase in both computational effort and memory allocations. For instance, the memory required for the five variables* x_i *will be* $31^5 \times 5 \times 8/2^{20} = 1092.1$*MB. Thus, enumerate methods are not suitable for this kind of problem.*

The other problem is that there are other combinations of $\boldsymbol{x}$ which may have slightly larger objective functions than the global optimum one. These combinations can be referred to as the suboptimal solutions. *When enumerate method is used, suboptimal can also be found as shown in Table 6.4.*

```
>> L=15; fx=fmin(1:L)' % global and sub-optimal solutions
   in=ii(1:L); x=[x1(in),x2(in),x3(in),x4(in),x5(in),fmin(1:15)]
```

TABLE 6.4: Optimum and sub-optimum solutions.

x_1	x_2	x_3	x_4	x_5	f	x_1	x_2	x_3	x_4	x_5	f	x_1	x_2	x_3	x_4	x_5	f
19	0	4	10	5	−89	19	0	4	9	7	−88	11	0	8	10	3	−87
18	0	4	11	3	−88	16	0	6	8	8	−88	10	0	9	9	4	−87
17	0	5	10	4	−88	20	0	4	7	11	−88	8	0	10	9	4	−87
15	0	6	10	4	−88	15	0	6	10	3	−87	5	0	12	8	5	−87
12	0	8	9	5	−88	13	0	7	10	3	−87	18	0	4	10	5	−87

Example 6.40 Please solve the following nonlinear integer programming problem with enumerate method

$$\min \quad x_1^2+x_2^2+2x_3^2+x_4^2-5x_1-5x_2-21x_3+7x_4$$

$$\boldsymbol{x} \text{ s.t. } \begin{cases} -x_1^2-x_2^2-x_3^2-x_4^2-x_1+x_2-x_3+x_4+8\geqslant 0 \\ -x_1^2-2x_2^2-x_3^2-2x_4^2+x_1+x_4+10\geqslant 0 \\ -2x_1^2-x_2^2-x_3^2-2x_4^2+x_2+x_4+5\geqslant 0 \end{cases}$$

Solution *Select the interested regions for the decision variables as $-N \sim N$, and let $N=30$, all the possible x_i's can be generated, and all the feasible solutions can be found directly. The feasible solutions can then be sorted, and the global optimal solution can be found as $\boldsymbol{x}=[0,1,2,0]^{\mathrm{T}}$, with objective function -38. Apart from the global solution, some suboptimal solutions can also be found $[0,0,2,0]$, $[0,1,2,1]$, $[0,1,1,-1]$ and $[1,2,1,0]$ can also be found.*

```
>> N=30; [x1 x2 x3 x4]=ndgrid(-N:N); % generate all possible combinations
   ii=find(-x1.^2-x2.^2-x3.^2-x4.^2-x1+x2-x3+x4+8>=0 & ...
            -x1.^2-2*x2.^2-x3.^2-2*x4.^2+x1+x4+10>=0 & ...
            -2*x1.^2-x2.^2-x3.^2-2*x4.^2+x2+x4+5>=0); % all the constraints
   x1=x1(ii); x2=x2(ii); x3=x3(ii); x4=x4(ii); % all the feasible solutions
   ff=x1.^2+x2.^2+2*x3.^2+x4.^2-5*x1-5*x2-21*x3+7*x4; % first 5 solutions
   [fm,ii]=sort(ff); k=ii(1:5); X=[x1(k),x2(k),x3(k),x4(k)], fm(1:5)
```

It should be pointed out that, enumerate method cannot be used in solving problems with mixed integer programming problems, since some of the decision variables are continuous and cannot be enumerated. In this case, searching methods should be used instead.

6.5.2 Solutions of linear integer programming problems

In the recent versions of Optimization Toolbox, a function `intlinprog()` is provided in solving linear mixed integer programming problems, with the syntaxes

$[\boldsymbol{x}, f_{\rm m}, \texttt{key}, \texttt{c}] = \texttt{intlinprog}(\textit{problem})$

$[\boldsymbol{x}, f_{\rm m}, \texttt{key}, \texttt{c}] = \texttt{intlinprog}(f, \texttt{intcon}, \boldsymbol{A}, \boldsymbol{b}, \boldsymbol{A}_{\rm eq}, \boldsymbol{b}_{\rm eq}, \boldsymbol{x}_{\rm m}, \boldsymbol{x}_{\rm M}, \textit{options})$

The syntaxes are very similar to the ones in `linprog()` function. The new input argument `intcon` actually indicates the indexes of the decision variables to be integers. In the description of structured variable *problem*, the essential fields are `f`, `intcon`, `solver` and `options`. Compared with the one in `linprog()`, the differences are that, `intcon` field must be specified, and

problem.`solver='intlinprog'`,

problem.`options=optimoptions('intlinprog')`

Due to the limitations in the new function `intlinprog()`, the requested integer decision variables are not really integers. There might be slight errors. The best way to correct the results is to use $\boldsymbol{x}$`(intcon)=round(`$\boldsymbol{x}$`(intcon))` command after the function call, when it is claimed successful.

Example 6.41 Consider again the linear integer programming problem in Example 6.29. Please solve also the mixed linear programming problem, when the 1st, 4th and 5th decision variables are requested as integers.

Solution *The ordinary integer linear programming problem can be solved directly with the following statements*

```
>> clear P; P.solver='intlinprog'; P.options=optimoptions('intlinprog');
   P.lb=[0; 0; 3.32; 0.678; 2.57]; P.f=[-2 -1 -4 -3 -1];
   P.Aineq=[0 2 1 4 2; 3 4 5 -1 -1]; P.Bineq=[54; 62];
   P.intcon=[1 2 3 4 5]; [x,f,a,b]=intlinprog(P), x=round(x)
```

It can be seen that the result is exactly the same as the one obtained in Example 6.29, however, this result is more reliable, since the one with enumerate method was obtained with the assumption that $N \leqslant 25$.

Now consider the mixed integer linear programming program. According to the request, with `intcon` *should be set to* $[1,4,5]$, *and the problem can be solved with the following statements, with the result of* $\boldsymbol{X} = [19, 0, 3.8, 11, 3]^{\rm T}$. *It should be noted also that the problem cannot be solved with enumerate method.*

```
>> P.intcon=[1 4 5]; [x,f,a,b]=intlinprog(P) % integer LP solution
   x(P.intcon)=round(x(P.intcon)) % fine tuning the solutions
```

Example 6.42 Consider the transportation problem in Example 6.33. If the decision variables are required to be integers, please solve the problem again.

Solution *In the function* `transport_linprog()` *discussed in Example 6.33, we left a tuning knob, i.e., to use the fourth input argument in the function call. In this case, the linear integer programming problem can be solved automatically, and the results obtained this way are integers*

```
>> C=[10,5,6,7; 8,2,7,6; 9,3,4,8]; b=[1500 2500 3000 3500];
   a=[2500 2500 5000]; x=transport_linprog(-C,a,b,1) % maximization
   f=sum(C(:).*X(:)) % evaluate the expected profit
```

6.5.3 Solutions of nonlinear integer programming problems

One of the most frequently used algorithms for mixed integer programming problems is the *branch-and-bound algorithm*. Details of the algorithm are omitted and an existing MATLAB function `bnb20()`, developed by Koert Kuipers of Groningen University in The Netherlands, can be used directly. The function can be downloaded freely from MathWorks file-exchange website. The function is updated and some bugs are fixed in the book, and now the function allows the use of anonymous functions and models described by structured variables. The syntax is

`[err,`f`,`$\boldsymbol{x}$`]` $=$ `BNB20_new(`*fun*`,`$\boldsymbol{x}_0$`,intcon,`$\boldsymbol{x}_{\rm m}$`,`$\boldsymbol{x}_{\rm M}$`,`$\boldsymbol{A}$`,`$\boldsymbol{B}$`,`$\boldsymbol{A}_{\rm eq}$`,`$\boldsymbol{B}_{\rm eq}$`,`*CFun*`)`

where, most of the arguments in the function call are the same as in other optimization functions. The function `fmincon()` is called by the `BNB20_new()` function. The variables $\boldsymbol{x}$ and f are returned respectively and if `err` is empty, the results obtained are correct. The argument `intcon` is the same as the one defined in `intlinprog()` function.

If the original optimization problem is described by a structured variable P, its `intcon` field should be set to the `intcon` vector discussed earlier. The problem can be solved with `[err,`f`,`$\boldsymbol{x}$`]` $=$ `BNB20_new(`P`)`.

Example 6.43 Solve again the integer and mixed integer linear programming problems in Example 6.29, with the nonlinear mixed integer programming solver.

Solution *In the new* `BNB20_new()` *function, anonymous functions are supported. Since all the* x_i*'s are expected to be integers, the* `intcon` *vector should be set to* $[1,2,3,4,5]$. *Also, the upper-bound can no longer be set to* `Inf`*, and finite values should be assigned instead. For instance, the upper-bounds can be assigned to 20000. The following statements can then be used*

```
>> f=@(x)-[2 1 4 3 1]*x; xm=[0,0,3.32,0.678,2.57]'; x0=ceil(xm);
   A=[0 2 1 4 2; 3 4 5 -1 -1]; intcon=1:5; Aeq=[]; Beq=[];
   B=[54; 62]; xM=20000*ones(5,1);
   [errmsg,fm,X]=BNB20_new(f,x0,intcon,xm,xM,A,B,Aeq,Beq)
```

and the solution obtained is $\boldsymbol{x} = [19,0,4,10,5]^{\rm T}$.

When the problem is described by a structured variable P*, the problem can be solved with the following statements, and the same results can be obtained*

```
>> clear P; P.objective=f; P.lb=xm; P.x0=x0; P.ub=xM; % structured variable
   P.Aineq=A; P.Bineq=B; P.intcon=intcon; [errmsg,fm,X]=BNB20_new(P)
```

If x_1, x_4, x_5 *are constrained to be integers while the other two variables can be arbitrarily chosen, then, the original problem is a mixed integer programming problem. The argument* `intcon` *should be modified accordingly to* `intcon`$=$ `[1,4,5]`*, then, the problem can be solved with the following statements, and the result is* $\boldsymbol{X} = [19,0,3.8,11,3]^{\rm T}$. *Again the result is the same as the one obtained with the* `intlinprog()` *function.*

```
>> P.intcon=[1,4,5]; [errmsg,fm,X]=BNB20_new(P) % solve again
```

Example 6.44 Please solve the following integer programming problem [9]

$$\min \quad x_1^3+x_2^2-4x_1+4+x_3^4.$$
$$\boldsymbol{x} \text{ s.t.} \begin{cases} x_1-2x_2+12+x_3\geqslant 0 \\ -x_1^2+3x_2-8-x_3\geqslant 0 \\ x_1\geqslant 0, x_2\geqslant 0, x_3\geqslant 0 \end{cases}$$

Solution *Since the original problem contains nonlinear constraints, the following MATLAB function can be written*

```
function [c,ce]=c6exinl(x)
ce=[]; c=[-x(1)+2*x(2)-12-x(3); x(1)^2-3*x(2)+8+x(3)];
```

With the following statements, the problem can be solved directly, and the solution obtained is $\boldsymbol{x}=[1,3,0]^{\mathrm{T}}$.

```
>> clear P; P.objective=@(x)x(1)^3+x(2)^2-4*x(1)+4+x(3)^4;
   P.intcon=1:3; P.nonlcon=@c6exinl; P.lb=[0;0;0];
   P.ub=100*[1;1;1]; P.x0=P.ub; [err,fm x]=BNB20_new(P)
```

Since the original problem is a small-scale problem, enumerate method can be used to ensure global optimal solution can be found. The result obtained is exactly the same as the searching method. Besides, some suboptimal solutions can also be found, which cannot be obtained with searching methods.

```
>> N=200; [x1 x2 x3]=meshgrid(0:N); % generate all the possible combinations
   ii=find(x1-2*x2+12+x3>=0 & -x1.^2+3*x2-8-x3>=0); % evaluate constraints
   x1=x1(ii); x2=x2(ii); x3=x3(ii); % find all the feasible solutions
   ff=x1.^3+x2.^2-4*x1+4+x3.^4; [fm,ij]=sort(ff); % sort the solutions
   k=ij(1:5); [x1(k) x2(k) x3(k)], fm(1:5)  % extract the first 5 solutions
```

Example 6.45 Solve the discrete optimization problem [10], where x_1 are multiples of 0.25, and x_2 are multiples of 0.1

$$\min \quad 2x_1^2+x_2^2-16x_1-10x_2.$$
$$\boldsymbol{x} \text{ s.t.} \begin{cases} x_1^2-6x_1+x_2-11\leqslant 0 \\ -x_1x_2+3x_2+\mathrm{e}^{x_1-3}-1\leqslant 0 \\ x_2\geqslant 3 \end{cases}$$

Solution *MATLAB cannot be used to solve directly discrete optimization problems. In fact, two new variables* $y_1=4x_1$, *and* $y_2=10x_2$ *can be introduced, and with variable substitution, we have* $x_1=y_1/4$, $x_2=y_2/10$. *The original problem can be rewritten as the integer programming problem of new decision variables* y_i

$$\min \quad 2y_1^2/16+y_2^2/100-4y_1-y_2.$$
$$\boldsymbol{y} \text{ s.t.} \begin{cases} y_1^2/16-6y_1/4+y_2/10-11\leqslant 0 \\ -y_1y_2/40+3y_2/10+\mathrm{e}^{y_1/4-3}-1\leqslant 0 \\ y_2\geqslant 30 \end{cases}$$

The nonlinear constraints can be expressed in MATLAB

```
function [c,ceq]=c6mdisp(y), ceq=[];
c=[y(1)^2/10-6*y(1)/4+y(2)/10-11;
   -y(1)*y(2)/40+3*y(2)/10+exp(y(1)/4-3)-1];
```

Assume that the lower and upper bounds of the variable y_1 are ± 200 (i.e., the bounds of x_1 are ± 50), and the upper bound of y_2 is 200, and lower bound is 30 (i.e., $3 \leqslant x_2 \leqslant 20$). With `BNB20_new()` *function, the following statements can be used to solve the problem directly, and the solution is $\boldsymbol{x} = [4,5]^{\mathrm{T}}$, and it is slightly different from the one, $(4, 4.75)$, given in [9]. Since both the solutions are feasible solutions, and the one obtained here has slightly smaller objective function.*

```
>> clear P; P.objective=@(y)2*y(1)^2/16+y(2)^2/100-4*y(1)-y(2);
   P.nonlcon=@c6mdisp; P.lb=[-200;30]; P.ub=[200;200]; P.intcon=[1,2];
   P.x0=[12;30]; [errmsg,ym,y]=BNB20_new(P); x=[y(1)/4,y(2)/10]
```

The enumerate method can also be used in solving discrete optimization problems. Assume that the range of decision variables are $(-20, 20)$, the following statements can be used to solve the problem, and the global optimal solution is $(4, 5)$. Meanwhile some suboptimal points, $(4, 5.1), (4, 4.9), (4, 4.8), (4, 5.2)$ can also be found.

```
>> [x1 x2]=meshgrid(-20:0.25:20,3:0.1:20);
   ii=find(x1.^2-6*x1+x2-11<=0 & -x1.*x2+3*x2+exp(x1-3)-1<=0);
   x1=x1(ii); x2=x2(ii); ff=2*x1.^2+x2.^2-16*x1-10*x2; [fm,ij]=sort(ff);
   k=ij(1:5); X=[x1(k) x2(k)], fm(1:5)
```

6.5.4 Solving binary programming problems

The so-called binary programming is the optimization problem, where the decision variables x_i to be optimized are 0's and 1's. It seems quite easy to solve the binary programming problems, since one may try to substitute all the possible combinations of each variable, and the optimum point can be found by comparing the calculated objective functions. In fact, this brutal force method is only applicable to small-scale problems. For large-scale problems, for instance, if there are n variables to be optimized, the size of the possible combinations will be $2^n n$ which might be prohibitive to run on an average computer. Thus, specially design search methods should be used instead.

A function `bintprog()` could be used to solve binary linear programming problems, with $\boldsymbol{x} =$ `bintprog(`$\boldsymbol{f}$`,`$\boldsymbol{A}$`,`$\boldsymbol{B}$`,`$\boldsymbol{A}_{\mathrm{eq}}$`,`$\boldsymbol{B}_{\mathrm{eq}}$`)`. However, this function has been removed in R2014b, and `intlinprog()` is recommended for binary linear programming problems. Binary nonlinear programming problems can be solved with `BNB20_new()`.

Example 6.46 Solve the following binary linear programming problem.

$$\min \quad -3x_1 + 2x_2 + 5x_3.$$

$$\boldsymbol{x} \text{ s.t. } \begin{cases} x_1+2x_2-x_3 \leqslant 2 \\ x_1+4x_2+x_3 \leqslant 4 \\ x_1+x_2 \leqslant 3 \\ 4x_2+x_3 \leqslant 6 \end{cases}$$

Solution *From the linear programming model, the required vectors and matrices $\boldsymbol{f}$, $\boldsymbol{A}$ and $\boldsymbol{B}$ can easily be constructed and the binary linear programming problem can be solved from*

```
>> f=[-3,2,-5]; A=[1 2 -1; 1 4 1; 1 1 0; 0 4 1]; B=[2;4;5;6];
   xm=[0,0,0]; xM=[1,1,1]; intcon=[1,2,3]; % setting for binary programming
   x=intlinprog(f,intcon,A,B,[],[],xm,xM)  % binary programming
```

and the solution is $\boldsymbol{x} = [1, 0, 1]^{\mathrm{T}}$.

For such a small-scale problem, the enumerate method can be used to test for all possible combinations of the variables whether the constraints are satisfied. Then, by simple sorting to the feasible combinations, according to the values of the objective function, the optimum solution can be found. In this case, the global optimal solution can be obtained

```
>> [x1,x2,x3]=meshgrid([0,1]); % generate all the possible combinations
   i=find((x1+2*x2-x3<=2)&(x1+4*x2+x3<=4)&(x1+x2<=3)&(4*x1+x3<=6));
   f=-3*x1(i)+2*x2(i)-5*x3(i); [fmin,ii]=sort(f); % sort feasible solutions
   index=i(ii(1)); x=[x1(index),x2(index),x3(index)] % global optimal solution
```

with $\boldsymbol{x} = [1\ 0\ 1]^{\mathrm{T}}$. *Moreover, all feasible solutions to the problem can be found using the following statements, which were not possible to be found using other methods*

```
>> x=[x1(i(ii)),x2(i(ii)),x3(i(ii))]; [x fmin]
```

where $\boldsymbol{x}_1 = [1, 0, 1]^{\mathrm{T}}, f(\boldsymbol{x}_1) = -8$, $\boldsymbol{x}_2 = [0, 0, 1]^{\mathrm{T}}, f(\boldsymbol{x}_2) = -5, \boldsymbol{x}_3 = [1, 0, 0]^{\mathrm{T}}, f(\boldsymbol{x}_3) = -3, \boldsymbol{x}_4 = [0, 0, 0]^{\mathrm{T}}, f(\boldsymbol{x}_4) = 0, \boldsymbol{x}_5 = [0, 1, 0]^{\mathrm{T}}, f(\boldsymbol{x}_5) = 2$.

Direct calls to `BNB20_new()` functions, with the lower- and upper-bounds $\boldsymbol{x}_{\mathrm{m}}$, $\boldsymbol{x}_{\mathrm{M}}$ set to zeros and ones vectors, respectively, can also solve the binary programming problems.

Example 6.47 Solve the binary linear programming problem in Example 6.46 with the `BNB20_new()` function.

Solution *An anonymous function can be used to express the objective function. When the lower- and upper-bounds* $\boldsymbol{x}_{\mathrm{m}}$, $\boldsymbol{x}_{\mathrm{M}}$ *set to zeros and ones vectors, respectively, the binary programming problem can also be solved for the example using the following statements*

```
>> f=@(x)[-3 2 -5]*x; xm=[0;0;0]; xM=[1;1;1]; x0=xm;
   intcon=[1,2,3]; A=[1 2 -1; 1 4 1; 1 1 0; 0 4 1];
   B=[2;4;5;6]; [err,f,x]=BNB20_new(f,x0,intcon,xm,xM,A,B)
```

where $\boldsymbol{x} = [1, 0, 1]^{\mathrm{T}}$ *and* $f = -8$. *The result is exactly the same as the one in Example 6.46. In fact, the last two constraints are redundant and can be removed.*

Example 6.48 Please solve the mixed binary programming problem [9].

$$\min_{\boldsymbol{x},\boldsymbol{y}\ \text{s.t.}\ \begin{cases} 0.8\ln(x_2+1)+0.96\ln(x_1-x_2+1)-0.8x_3 \leqslant 0 \\ \ln(x_2+1)+1.2\ln(x_1-x_2+1)-x_3-2y_3 \geqslant -2 \\ x_2-x_1 \leqslant 0 \\ x_2-2y_1 \leqslant 0 \\ x_1-x_2-2y_2 \leqslant 0 \\ y_1+y_2 \leqslant 1 \\ 0 \leqslant \boldsymbol{x} \leqslant [2,2,1]^{\mathrm{T}}, \boldsymbol{y} \in \{0,1\} \end{cases}} 5y_1+6y_2+8y_3+10x_1-7x_3-18\ln(x_2+1)-19.2\ln(x_1-x_2+1)+10.$$

Solution *Since there exists nonlinear constraints and objective functions, function* `intlinprog()` *cannot be used. Nonlinear mixed integer programming solvers should be used instead. Similar to the previous discussed problems, two decision vectors* $\boldsymbol{x}$, $\boldsymbol{y}$ *are involved, and cannot be handled with MATLAB functions directly. The two decision vectors should be converted to the one with only one decision vector. A new decision vector* $\boldsymbol{x}$ *is introduced, whose first three components are the original* $\boldsymbol{x}$ *vector, while the other three components* *should be* $x_4 = y_1$, $x_5 = y_2$ *and* $x_6 = y_3$. *Therefore, the original problem can be manually rewritten as*

$$\min \quad 5x_4 + 6x_5 + 8x_6 + 10x_1 - 7x_3 - 18\ln(x_2+1) - 19.2\ln(x_1 - x_2 + 1) + 10.$$

$$\boldsymbol{x} \ \text{s.t.} \begin{cases} 0.8\ln(x_2+1)+0.96\ln(x_1-x_2+1)-0.8x_3\geqslant 0 \\ \ln(x_2+1)+1.2\ln(x_1-x_2+1)-x_3-2x_6\geqslant -2 \\ x_2-x_1\leqslant 0 \\ x_2-2x_4\leqslant 0 \\ x_1-x_2-2x_5\leqslant 0 \\ x_4+x_5\leqslant 1 \\ 0\leqslant \boldsymbol{x}\leqslant [2,2,1,1,1,1]^{\mathrm{T}} \end{cases}$$

The nonlinear constraints can be expressed in the following MATLAB function

```
function [c,ceq]=c6mmibp(x), ceq=[]; % no equality constraints
c=[-0.8*log(x(2)+1)-0.96*log(x(1)-x(2)+1)+0.8*x(3); % first 2 constraints
   -log(x(2)+1)-1.2*log(x(1)-x(2)+1)+x(3)+2*x(6)-2];
```

Structured variable can be used to express the binary programming problem, and the following statements can be used, and the result is $\boldsymbol{x} = [1.301, 0, 1, 0, 1, 0]^{\mathrm{T}}$, *with the optimal objective function 6.098. The solution* $\boldsymbol{x} = [1.301, 0, 1, 1, 0, 1]^{\mathrm{T}}$ *recommended in [9] is incorrect, since its objective function is 13.0098.*

```
>> clear P; P.intcon=[4,5,6]; P.x0=[0,0,0,0,0,0]';
   P.objective=@(x)5*x(4)+6*x(5)+8*x(6)+10*x(1)-7*x(3) ...
             -18*log(x(2)+1)-19.2*log(x(1)-x(2)+1)+10; % objective function
   P.ub=[2 2 1 1 1 1]'; P.lb=[0 0 0 0 0 0]'; P.bineq=[0;0;0;1];
   P.Aineq=[-1 1 0 0 0 0; 0 1 0 -2 0 0; 1 -1 0 0 -2 0; 0 0 0 1 1 0];
   P.nonlcon=@c6mmibp; [errmsg,fm,x]=BNB20_new(P) % solve the problem
```

6.5.5 Assignment problems

The *assignment problem* is a special type of binary linear programming problem where assignees are being assigned to perform tasks, under the following assumptions [7]:

(i) The number of assignees and the number of tasks are the same;

(ii) Each assignee is to be assigned to exactly one task;

(iii) Each task is to be performed by exactly one assignee.

Assume that the costs c_{ij} associated with assignee i performing task j are all known, and the objective is to determine how all n assignments should be made to minimize the total cost. The mathematical model of the typical assignment problem is expressed as

$$\min \quad \sum_{i=1}^{n}\sum_{j=1}^{n} c_{ij}x_{ij}, \tag{6-5-1}$$

$$\boldsymbol{X} \ \text{s.t.} \begin{cases} \sum\limits_{j=1}^{n} x_{ij}=1, i=1,2,\cdots,n \\ \sum\limits_{i=1}^{n} x_{ij}=1, j=1,2,\cdots,n \\ x_{ij} \text{ are binary numbers}, i=1,2,\cdots,n, j=1,2,\cdots,n \end{cases}$$

and it can be seen that it is a special case of the transportation problem, with $m = n$, $s_i = 1$ and $d_i = 1$. Besides, the decision variables are binary numbers.

A MATLAB function `assignment_prog()` is written to solve the assignment problems. The syntax of the function is $\boldsymbol{X}=$`assignment_prog(`$\boldsymbol{C}$`)`, where $\boldsymbol{C}$ is the cost matrix.

```
function [x,fv,key]=assignment_prog(C)
[n,m]=size(C); c=C(:); A=[];b=[]; Aeq=zeros(2*n,n^2);
for i=1:n, Aeq(i,(i-1)*n+1:n*i)=1; Aeq(n+i,i:n:n^2)=1; end
beq=ones(2*n,1); xm=zeros(n^2,1); xM=ones(n^2,1);
[x,fv,key]=intlinprog(c,1:n^2,A,b,Aeq,beq,xm,xM); x=reshape(x,n,m).';
```

Example 6.49 Please solve the assignment problem with a cost matrix

$$C = \begin{bmatrix} 12 & 7 & 9 & 7 & 9 \\ 8 & 9 & 6 & 6 & 6 \\ 7 & 17 & 12 & 14 & 9 \\ 15 & 14 & 6 & 6 & 10 \\ 4 & 10 & 7 & 10 & 9 \end{bmatrix}.$$

Solution *The cost matrix can be entered first, and the problem can be solved directly*

```
>> C=[12,7,9,7,9; 8,9,6,6,6; 7,17,12,14,9; 15,14,6,6,10; 4,10,7,10,9];
   [X fv]=assignment_prog(C) % solve directly the assignment problem
```

The solution of the assignment problem can be found as

$$X = \begin{bmatrix} 0 & 0 & 0 & 0 & 1 \\ 1 & 0 & 0 & 0 & 0 \\ 0 & 1 & 0 & 0 & 0 \\ 0 & 0 & 0 & 1 & 0 \\ 0 & 0 & 1 & 0 & 0 \end{bmatrix}, \text{ with } f_{\rm v} = 32,$$

meaning the first task is assigned to assignee 5, the second task is assigned to assignee 1, and so on. In this case, the total cost is minimized to 32.

6.6 Linear Matrix Inequalities

The theory of linear matrix inequalities (LMI) has been attracting the attention of research communities for a decade especially from researchers in the control systems community [3]. The concept of LMI and its applications are based on the fact that LMIs can be reduced to linear programming problems which can easily be solved by computers [11].

In this section, the concepts of LMI will be presented first, followed by the MATLAB solution examples using the Robust Control Toolbox, as well as a free YALMIP Toolbox.

6.6.1 A general introduction to LMIs

Linear matrix inequalities can be generally described as

$$\boldsymbol{F}(\boldsymbol{x}) = \boldsymbol{F}_0 + x_1\boldsymbol{F}_1 + \cdots + x_m\boldsymbol{F}_m < \boldsymbol{0}, \tag{6-6-1}$$

where $\boldsymbol{x} = [x_1, \cdots, x_m]^{\rm T}$ is the coefficient vector of a polynomial. It is also referred to as a *decision vector*. The matrices $\boldsymbol{F}_i$ are Hermitian matrices. If the LMI matrix $\boldsymbol{F}(\boldsymbol{x})$ is a negative-definite matrix, the solution set is convex, i.e.,

$$\boldsymbol{F}[\alpha\boldsymbol{x}_1 + (1-\alpha)\boldsymbol{x}_2] = \alpha\boldsymbol{F}(\boldsymbol{x}_1) + (1-\alpha)\boldsymbol{F}(\boldsymbol{x}_2) < \boldsymbol{0}, \tag{6-6-2}$$

where $\alpha > 0, 1-\alpha > 0$. The solution is also referred to as the *feasible solution*. From two such LMIs $\boldsymbol{F}_1(\boldsymbol{x}) < 0$ and $\boldsymbol{F}_2(\boldsymbol{x}) < 0$, a single LMI can be constructed such that

$$\begin{bmatrix} \boldsymbol{F}_1(\boldsymbol{x}) & 0 \\ 0 & \boldsymbol{F}_2(\boldsymbol{x}) \end{bmatrix} < \boldsymbol{0}. \tag{6-6-3}$$

It can be seen that several LMIs $\boldsymbol{F}_i(\boldsymbol{x}) < \boldsymbol{0}, i = 1, 2, \cdots, k$ can be combined into a single LMI such that $\boldsymbol{F}(\boldsymbol{x}) < \boldsymbol{0}$, where

$$\boldsymbol{F}(\boldsymbol{x}) = \begin{bmatrix} \boldsymbol{F}_1(\boldsymbol{x}) & & & \\ & \boldsymbol{F}_2(\boldsymbol{x}) & & \\ & & \ddots & \\ & & & \boldsymbol{F}_k(\boldsymbol{x}) \end{bmatrix} < \boldsymbol{0}. \tag{6-6-4}$$

6.6.2 Lyapunov inequalities

Consider first the Lyapunov stability problem. Lyapunov theory states that for a given positive-definite matrix $\boldsymbol{Q}$, if the Lyapunov equation

$$\boldsymbol{A}^{\mathrm{T}}\boldsymbol{X} + \boldsymbol{X}\boldsymbol{A} = -\boldsymbol{Q} \tag{6-6-5}$$

has positive-definite solution $\boldsymbol{X}$, the matrix $\boldsymbol{A}$ is stable, i.e., all the eigenvalues of the matrix are located in the left-hand-side of the complex plane. The previous equation can also be converted into a Lyapunov inequality

$$\boldsymbol{A}^{\mathrm{T}}\boldsymbol{X} + \boldsymbol{X}\boldsymbol{A} < \boldsymbol{0}. \tag{6-6-6}$$

Since $\boldsymbol{X}$ is a symmetrical matrix, a vector $\boldsymbol{x}$ containing the $n(n+1)/2$ elements can be used to describe the original matrix such that

$$x_i = X_{i,1}, \quad i = 1, \cdots, n, \; x_{n+i} = X_{i,2}, \; i = 2, \cdots, n, \cdots, \tag{6-6-7}$$

and it follows that

$$x_{(2n-j+2)(j-1)/2+i} = X_{i,j}, \quad j = 1, 2, \cdots, n, \quad i = j, j+1, \cdots, n. \tag{6-6-8}$$

A MATLAB function can be created for the above conversion as follows:

```
function F=lyap2lmi(A0)
if prod(size(A0))==1, n=A0; A=sym('a%d%d',n); % symbolic matrix
else, n=size(A0,1); A=A0; end
vec=0; for i=1:n, vec(i+1)=vec(i)+n-i+1; end
for k=1:n*(n+1)/2,
   X=zeros(n); i=find(vec>=k); i=i(1)-1; j=i+k-vec(i)-1;
   X(i,j)=1; X(j,i)=1; F(:,:,k)=A.'*X+X*A;
end
```

The function can be called by $\boldsymbol{F}$=`lyap2lmi(`$\boldsymbol{A}$`)`, where $\boldsymbol{A}$ can be a double-precision matrix. Note that if $\boldsymbol{A}$ is simply an integer, it indicates a square symbolic matrix to be established. The returned argument $\boldsymbol{F}$ is a three-dimensional array, and $\boldsymbol{F}(:,:,i)$ is the $\boldsymbol{F}_i$ matrix.

Example 6.50 If in the Lyapunov inequality $A = \begin{bmatrix} 1 & 2 & 3 \\ 4 & 5 & 6 \\ 7 & 8 & 0 \end{bmatrix}$, find its LMI representation. For a 3×3 matrix A, display its LMI form.

Solution *The matrix A is entered by the statements*

```
>> A=[1,2,3; 4,5,6; 7,8,0]; F=lyap2lmi(A)
```

and it can be found that the F_i matrices are

$$x_1 \begin{bmatrix} 2 & 2 & 3 \\ 2 & 0 & 0 \\ 3 & 0 & 0 \end{bmatrix} + x_2 \begin{bmatrix} 8 & 6 & 6 \\ 6 & 4 & 3 \\ 6 & 3 & 0 \end{bmatrix} + x_3 \begin{bmatrix} 14 & 8 & 1 \\ 8 & 0 & 2 \\ 1 & 2 & 6 \end{bmatrix}$$
$$+ x_4 \begin{bmatrix} 0 & 4 & 0 \\ 4 & 10 & 6 \\ 0 & 6 & 0 \end{bmatrix} + x_5 \begin{bmatrix} 0 & 7 & 4 \\ 7 & 16 & 5 \\ 4 & 5 & 12 \end{bmatrix} + x_6 \begin{bmatrix} 0 & 0 & 7 \\ 0 & 0 & 8 \\ 7 & 8 & 0 \end{bmatrix} < \mathbf{0}.$$

For a symbolic 3×3 matrix, the following statement can be given instead

```
>> F=lyap2lmi(3) % generate symbolic expression
```

such that the LMI can be written as

$$x_1 \begin{bmatrix} 2a_{11} & a_{12} & a_{13} \\ a_{12} & 0 & 0 \\ a_{13} & 0 & 0 \end{bmatrix} + x_2 \begin{bmatrix} 2a_{21} & a_{22}+a_{11} & a_{23} \\ a_{22}+a_{11} & 2a_{12} & a_{13} \\ a_{23} & a_{13} & 0 \end{bmatrix} + x_3 \begin{bmatrix} 2a_{31} & a_{32} & a_{33}+a_{11} \\ a_{32} & 0 & a_{12} \\ a_{33}+a_{11} & a_{12} & 2a_{13} \end{bmatrix}$$
$$+ x_4 \begin{bmatrix} 0 & a_{21} & 0 \\ a_{21} & 2a_{22} & a_{23} \\ 0 & a_{23} & 0 \end{bmatrix} + x_5 \begin{bmatrix} 0 & a_{31} & a_{21} \\ a_{31} & 2a_{32} & a_{33}+a_{22} \\ a_{21} & a_{33}+a_{22} & 2a_{23} \end{bmatrix} + x_6 \begin{bmatrix} 0 & 0 & a_{31} \\ 0 & 0 & a_{32} \\ a_{31} & a_{32} & 2a_{33} \end{bmatrix} < \mathbf{0}.$$

Some nonlinear inequalities can be converted into LMIs, too. For instance, for a partitioned matrix $F(x) = \begin{bmatrix} F_{11}(x) & F_{12}(x) \\ F_{21}(x) & F_{22}(x) \end{bmatrix}$, if $F_{11}(x)$ is a square matrix, the following three cases are equivalent:

$$F(x) < \mathbf{0} \tag{6-6-9}$$
$$F_{11}(x) < \mathbf{0}, \quad F_{22}(x) - F_{21}(x)F_{11}^{-1}(x)F_{12}(x) < \mathbf{0} \tag{6-6-10}$$
$$F_{22}(x) < \mathbf{0}, \quad F_{11}(x) - F_{12}(x)F_{22}^{-1}(x)F_{21}(x) < \mathbf{0}. \tag{6-6-11}$$

The above property is known as *Schur complement.*

Consider an algebraic Riccati inequality

$$A^{\mathrm{T}}X + XA + (XB - C)R^{-1}(XB - C^{\mathrm{T}})^{\mathrm{T}} < \mathbf{0}, \tag{6-6-12}$$

where $R = R^{\mathrm{T}} > 0$. Due to its quadratic term, it is not an LMI. However, with Schur complement, the original nonlinear inequality can be converted equivalently into the following LMIs

$$X > \mathbf{0}, \quad \begin{bmatrix} A^{\mathrm{T}}X + XA & XB - C^{\mathrm{T}} \\ B^{\mathrm{T}}X - C & -R \end{bmatrix} < \mathbf{0}. \tag{6-6-13}$$

6.6.3 Classification of LMI problems

LMI problems can be classified into three typical problems, i.e., the feasible solution problems, linear objective function minimization problems and the generalized eigenvalue problems.

(i) **Feasible solution problem** The so-called *feasible solution problem* is in fact the feasible region problem in optimization, i.e., for the inequality

$$\boldsymbol{F}(\boldsymbol{x}) < \boldsymbol{0}, \tag{6-6-14}$$

find a feasible solution. The feasible solution problem is to find the solution $\boldsymbol{F}(\boldsymbol{x}) < t_{\rm min}\boldsymbol{I}$, where the minimum $t_{\rm min}$ is to be found. If $t_{\rm min} < 0$ can be found, there exist solutions to the original problem, otherwise, there is no feasible solution.

(ii) **Linear objective function minimization problems** Consider the problem

$$\min_{\boldsymbol{x} \text{ s.t. } \boldsymbol{F}(\boldsymbol{x})<\boldsymbol{0}} \boldsymbol{c}^{\rm T}\boldsymbol{x}. \tag{6-6-15}$$

Since the constraints are given as LMIs, and the objective function is also linear, this problem can also be solved using ordinary linear programming methods.

(iii) **The generalized eigenvalue problems** The generalized eigenvalue problem is most commonly seen in LMI optimizations. Recall the generalized eigenvalue problem in Chapter 4, which is expressed as $\boldsymbol{A}\boldsymbol{x} = \lambda\boldsymbol{B}\boldsymbol{x}$. Such a problem can be expressed by general matrix functions as $\boldsymbol{A}(\boldsymbol{x}) < \lambda\boldsymbol{B}(\boldsymbol{x})$, and λ can be regarded as the generalized eigenvalue. Therefore, the optimization problem becomes

$$\min_{\lambda,\boldsymbol{x} \text{ s.t. } \begin{cases} \boldsymbol{A}(\boldsymbol{x})<\lambda\boldsymbol{B}(\boldsymbol{x}) \\ \boldsymbol{B}(\boldsymbol{x})>\boldsymbol{0} \\ \boldsymbol{C}(\boldsymbol{x})<\boldsymbol{0} \end{cases}} \lambda. \tag{6-6-16}$$

Other constraints can be written as $\boldsymbol{C}(\boldsymbol{x}) < \boldsymbol{0}$. The generalized eigenvalue problem can be expressed as a special LMI problem.

6.6.4 LMI problem solutions with MATLAB

The LMI solver in MATLAB is currently provided in the Robust Control Toolbox. However, the way of describing the LMIs are quite complicated. An example is presented to show in detail the uses of the LMI solver.

The following procedures are used to describe LMIs in MATLAB:

(i) **Create an LMI model** An LMI framework can be established with `setlmis([])` function. Therefore, a framework can be established in MATLAB workspace.

(ii) **Define the decision variables** The decision variables can be declared by `lmivar()` function, with $\boldsymbol{P}$=`lmivar(key,[`$\boldsymbol{n}_1$`,`$\boldsymbol{n}_2$`])`, where `key` specifies the type of the decision matrix, with `key`$=2$ for an ordinary $n_1 \times n_2$ matrix $\boldsymbol{P}$, while `key`$=1$ for an $n_1 \times n_1$ symmetrical matrix. If `key`$=1$, $\boldsymbol{n}_1$ and $\boldsymbol{n}_2$ are both vectors, $\boldsymbol{P}$ is a block diagonal symmetrical matrix. If `key`$=3$, $\boldsymbol{P}$ is a special matrix, which is not discussed in this book. The interested readers may refer to the Robust Control Toolbox manual [12].

(iii) **Describe LMIs in partitioned form** The LMIs can be described by the `lmiterm()` function and its syntax is quite complicated

```
lmiterm([k,i,j,P],A,B,flag)
```

where k is the number of the LMIs. Since an LMI problem may be described by several

LMIs, one should number each of them. If an LMI is given $G_k(\boldsymbol{x}) > 0$, then, k should be described by $-k$. A term in a block in the partitioned matrix can be described by `lmiterm()` function, with i, j representing respectively the row and column numbers of the block. $\boldsymbol{P}$ is the declared decision variables, and the matrices $\boldsymbol{A}, \boldsymbol{B}$ indicate the matrices in the term $\boldsymbol{APB}$. If `flag` is assigned as `'s'`, the symmetrical term $\boldsymbol{APB} + (\boldsymbol{APB})^{\mathrm{T}}$ is specified. If the whole term is a constant matrix, $\boldsymbol{P}$ is set to 0, and matrix $\boldsymbol{B}$ is omitted.

(iv) **Confirm the LMI model** After all the LMIs are declared by the `lmiterm()` function, G = `getlmis` can be used to confirm the model G.

(v) **Solve the LMI problem** For the declared G model, the LMI optimization problems can be solved in one of the following three forms

```
[t_min, x] = feasp(G, options, target)              % feasible solution
[c_opt, x] = mincx(G, c, options, x0, target)       % linear objective function
[λ, x] = gevp(G, nlfc, options, λ0, x0, target)     % generalized eigenvalues
```

The solution $\boldsymbol{x}$ thus obtained is a vector, and the `dec2mat()` function can be used to extract the matrix. The control variable `options` is defined as a 5-element vector, whose first element declares the precision requirement, with its default value of 10^{-5}.

Example 6.51 For the Riccati inequality $\boldsymbol{A}^{\mathrm{T}}\boldsymbol{X} + \boldsymbol{X}\boldsymbol{A} + \boldsymbol{X}\boldsymbol{B}\boldsymbol{R}^{-1}\boldsymbol{B}^{\mathrm{T}}\boldsymbol{X} + \boldsymbol{Q} < \boldsymbol{0}$, where

$$\boldsymbol{A} = \begin{bmatrix} -2 & -2 & -1 \\ -3 & -1 & -1 \\ 1 & 0 & -4 \end{bmatrix}, \quad \boldsymbol{B} = \begin{bmatrix} -1 & 0 \\ 0 & -1 \\ -1 & -1 \end{bmatrix}, \quad \boldsymbol{Q} = \begin{bmatrix} -2 & 1 & -2 \\ 1 & -2 & -4 \\ -2 & -4 & -2 \end{bmatrix}, \quad \boldsymbol{R} = \boldsymbol{I}_2,$$

find a feasible positive-definite solution $\boldsymbol{X}$.

Solution *The original nonlinear matrix inequality is obvious not an LMI. Using the Schur complement, this Riccati inequality can be expressed by a partitioned LMI as follows. Also, since a positive-definite solution is expected, the second LMI can be established*

$$\left[\begin{array}{c:c} \boldsymbol{A}^{\mathrm{T}}\boldsymbol{X}+\boldsymbol{X}\boldsymbol{A}+\boldsymbol{Q} & \boldsymbol{X}\boldsymbol{B} \\ \hdashline \boldsymbol{B}^{\mathrm{T}}\boldsymbol{X} & -\boldsymbol{R} \end{array}\right] < \boldsymbol{0}, \quad \textit{and} \quad \boldsymbol{X} > \boldsymbol{0}.$$

One may number the Riccati inequality as no. 1, and the positive-definite inequality as no. 2. Thus, one can set k to 1 and 2 respectively in using the `lmiterm()` *function. It should also be noted that the matrix* $\boldsymbol{X}$ *is a* 3×3 *symmetrical matrix. Therefore, the feasible solution to the original problem can be obtained with the following statements. It should also be noted that since the second inequality is* $\boldsymbol{X} > 0$, *its number should be* -2 *instead of 2.*

```
>> A=[-2,-2,-1; -3,-1,-1; 1,0,-4]; B=[-1,0; 0,-1; -1,-1];
   Q=[-2,1,-2; 1,-2,-4; -2,-4,-2]; R=eye(2); % enter the matrices
   setlmis([]);                      % create a blank LTI framework
   X=lmivar(1,[3 1]);                % declare X as a 3 × 3 symmetrical matrix
   lmiterm([1 1 1 X],A',1,'s')       % (1,1)th block, 's' means A^T X + XA
   lmiterm([1 1 1 0],Q)              % (1,1)th, appended by constant matrix Q
   lmiterm([1 1 2 X],1,B)            % (1,2)th, meaning XB
   lmiterm([1 2 2 0],-1)             % (2,2)th, meaning -R
   lmiterm([-2,1,1,X],1,1)           % the second inequality meaning X > 0
   G=getlmis;                        % complete the LTI framwork setting
   [tmin b]=feasp(G);                % solve the feasible problem
   X=dec2mat(G,b,X)                  % extract the solution matrix X
```

It is found that $t_{\min} = -0.2427$, *and the feasible solution to the original problem is*

$$\boldsymbol{X} = \begin{bmatrix} 1.0329 & 0.4647 & -0.23583 \\ 0.4647 & 0.77896 & -0.050684 \\ -0.23583 & -0.050684 & 1.4336 \end{bmatrix}.$$

It is worth mentioning that, due to possible problems in the new Robust Control Toolbox, if the command `lmiterm([1 2 1 X],B',1)` *is used to describe the symmetrical term in the first inequality, wrong results were found. Thus, the symmetrical terms should not be described again in solving LMI problems.*

6.6.5 Optimization of LMI problems by YALMIP Toolbox

The YALMIP Toolbox released by Dr. Johan Löfberg is a more flexible general purpose optimization language in MATLAB, with support also for LMI problems [13]. The description of LMI problems is much simpler and more straightforward than the ones in the Robust Control Toolbox. The YALMIP Toolbox can be downloaded for free from The MathWorks Inc.'s file-exchange site. The toolbox is also provided with the companion CD.

Decision variables can be declared in YALMIP Toolbox with `sdpvar()` function, which can be called in the following ways

```
X=sdpvar(n)               % symmetrical matrix description
X=sdpvar(n,m)             % rectangular matrix declaration
X=sdpvar(n,n,'full')      % declaration of a square asymmetrical matrix
```

The decision variables declared previously can further be treated, for instance, `hankel()` function can be applied on a decision vector to form the decision matrix in Hankel form. Similarly, the functions `intvar()` and `binvar()` can be used to declare integer and binary variables, respectively; thus, integer programming and binary programming problems can be handled.

For `sdpvar` decision variables, the symbols `[` and `]` can be used to describe LMIs. If there are many LMIs, they can be joined together with the `,` sign, to form a single LMI representation.

An objective function, when necessary, can also be described, and the LMI optimization programs can be solved with the following syntaxes

```
s=solvesdp(F)             % find a feasible solution
s=solvesdp(F,f)           % optimization with objective function f
s=solvesdp(F,f,options)   % options allowed such as algorithm selection
```

where F is the collection of constraints. After the solution, the `X=double(X)` command can be used to extract the solution matrix $\boldsymbol{X}$.

Example 6.52 With the use of the YALMIP Toolbox, the problem in Example 6.51 can be solved using simpler commands such that

```
>> A=[-2,-2,-1; -3,-1,-1; 1,0,-4]; B=[-1,0; 0,-1; -1,-1];
   Q=[-2,1,-2; 1,-2,-4; -2,-4,-2]; R=eye(2); X=sdpvar(3);
   F=[[A'*X+X*A+Q, X*B; B'*X, -R]<0, X>0];
   sol=solvesdp(F); X=double(X) % solve the LMI problem
```

and it can be seen that the results are exactly the same.

Example 6.53 Solve the linear programming problem in Example 6.29 again.

Solution *For simplicity, the original problem can be rewritten as*

$$\min_{\boldsymbol{x} \text{ s.t. } \begin{cases} 2x_2+x_3+4x_4+2x_5\leqslant 54 \\ 3x_1+4x_2+5x_3-x_4-x_5\leqslant 62 \\ x_1,x_2\geqslant 0,x_3\geqslant 3.32,x_4\geqslant 0.678,x_5\geqslant 2.57 \end{cases}} -2x_1 - x_2 - 4x_3 - 3x_4 - x_5.$$

It is obvious that $\boldsymbol{x}$ is a 5×1 column vector, and the original problem can be solved with the following statements

```
>> x=sdpvar(5,1); % declare decision variable is a 5 × 1 double vector
   F=[2*x(2)+x(3)+4*x(4)+2*x(5)<=54,          % specify the two constraints
      3*x(1)+4*x(2)+5*x(3)-x(4)-x(5)<=62,
      x>=[0;0;3.32;0.678;2.57]];               % specify the lower bounds
   sol=solvesdp(F,-[2 1 4 3 1]*x); x=double(x) % solve directly the problem
```

and the solution is $\boldsymbol{x} = [19.785, 0, 3.32, 11.385, 2.57]^{\rm T}$, which is exactly the same as the one obtained in the original example. Now assuming that the decision variables are integers with the `intvar()` *function, the integer linear programming problem can be solved by the following scripts:*

```
>> x=intvar(5,1); % declare decision variable is a 5 × 1 integer vector
   F=[2*x(2)+x(3)+4*x(4)+2*x(5)<=54,       % specify the two constraints
      3*x(1)+4*x(2)+5*x(3)-x(4)-x(5)<=62,
      x>=[0;0;3.32;0.678;2.57]];            % specify the lower bounds
   sol=solvesdp(F,-[2 1 4 3 1]*x); x=double(x) % solve LMI problem
```

which gives the solution $\boldsymbol{x} = [19, 0, 4, 10, 5]^{\rm T}$, the same as the one in Example 6.39.

Example 6.54 For a linear system $(\boldsymbol{A}, \boldsymbol{B}, \boldsymbol{C}, \boldsymbol{D})$, its $\mathcal{H}_\infty$ norm can directly be evaluated by `norm()` function. The norm evaluation problem can also be posed into the following LMI framework

$$\min_{\gamma,\, \boldsymbol{P} \text{ s.t. } \begin{cases} \begin{bmatrix} \boldsymbol{A}^{\rm T}\boldsymbol{P}+\boldsymbol{P}\boldsymbol{A} & \boldsymbol{P}\boldsymbol{B} & \boldsymbol{C}^{\rm T} \\ \boldsymbol{B}^{\rm T}\boldsymbol{P} & -\gamma\boldsymbol{I} & \boldsymbol{D}^{\rm T} \\ \boldsymbol{C} & \boldsymbol{D} & -\gamma\boldsymbol{I} \end{bmatrix} < 0 \\ \boldsymbol{P} > 0 \end{cases}} \gamma. \tag{6-6-17}$$

Find the $\mathcal{H}_\infty$ norm for the system

$$\boldsymbol{A} = \begin{bmatrix} -4 & -3 & 0 & -1 \\ -3 & -7 & 0 & -3 \\ 0 & 0 & -13 & -1 \\ -1 & -3 & -1 & -10 \end{bmatrix},\ \boldsymbol{B} = \begin{bmatrix} 0 \\ -4 \\ 2 \\ 5 \end{bmatrix},\ \boldsymbol{C} = [\ 0,\ 0,\ 4,\ 0\ ],\ D = 0.$$

Solution *With the YALMIP Toolbox, the $\mathcal{H}_\infty$ norm is computed to 0.4640, which is quite close to the value obtained by the* `norm()` *function.*

```
>> A=[-4,-3,0,-1; -3,-7,0,-3; 0,0,-13,-1; -1,-3,-1,-10];
   B=[0; -4; 2; 5]; C=[0,0,4,0]; D=0; gam=sdpvar(1); P=sdpvar(4);
   F=[[A*P+P*A',P*B,C'; B'*P,-gam,D'; C,D,-gam]<0, P>0];
   sol=solvesdp(F,gam); double(gam), norm(ss(A,B,C,D),'inf')
```

6.7 Solutions of Multi-objective Programming Problems

The optimization problems discussed so far are assumed scalar objective function $f(\boldsymbol{x})$, and those problems are single-objective function problems. If the objective function is a vector $\boldsymbol{F}(\boldsymbol{x})$, the problem is referred to as *multi-objective programming* problems. In this section, an introduction is made on multi-objective programming problems.

6.7.1 Multi-objective optimization model

The general form of multi-objective programming problem is

$$J = \min_{\boldsymbol{x} \text{ s.t. } \boldsymbol{G}(\boldsymbol{x}) \leqslant \boldsymbol{0}} \boldsymbol{F}(\boldsymbol{x}), \tag{6-7-1}$$

where, $\boldsymbol{F}(\boldsymbol{x}) = [f_1(\boldsymbol{x}), f_2(\boldsymbol{x}), \cdots, f_p(\boldsymbol{x})]^{\mathrm{T}}$. An example of multi-objective modeling will be given, and physical interpretation will be illustrated.

Example 6.55 Assume that three kinds of candies, A_1, A_2, A_3, are supplied in the shop, and the prices are respectively \$4, \$2.8 and \$2.4 per kilogram. If one wants to buy some candies, total cost cannot exceed \$20, total weight cannot be less than 6kg, and the total weight of candies A_1 and A_2 cannot be less than 3kg. How should he design the best purchasing plan? (source of data from [14])

Solution *The first thing in solving such a problem is how to define the "best" purchasing plan. In this example, the best plan is to have the least cost, while buy the heaviest candies. Thus, the two objective functions should be introduced, and it is obvious that the these two objective functions are in conflict. The other conditions can be regarded as the constraints.*

The modeling procedure is needed in this example. Assume that the purchased weights of the three kinds of candies A_1, A_2, A_3 are respectively x_1, x_2 and x_3 kg's, respectively, the two objective functions can be written as

$$\textit{cost:}\ f_1(\boldsymbol{x}) = 4x_1 + 2.8x_2 + 2.4x_3 \ \rightarrow \min,$$
$$\textit{total weight:}\ f_2(\boldsymbol{x}) = x_1 + x_2 + x_3 \ \rightarrow \max.$$

If they are unified as a minimization problem, and considering the constraints, the multi-objective programming model can be established as

$$\min_{\boldsymbol{x} \text{ s.t. } \begin{cases} 4x_1+2.8x_2+2.4x_3 \leqslant 20 \\ x_1+x_2+x_3 \geqslant 6 \\ x_1+x_2 \geqslant 3 \\ x_1,x_2,x_3 \geqslant 0 \end{cases}} \begin{bmatrix} 4x_1 + 2.8x_2 + 2.4x_3 \\ -(x_1 + x_2 + x_3) \end{bmatrix}.$$

6.7.2 Least squares solutions of unconstrained multi-objective programming problems

Assume that multi-objective function is given by

$$\boldsymbol{F}(\boldsymbol{x}) = [f_1(\boldsymbol{x}), f_2(\boldsymbol{x}), \cdots, f_p(\boldsymbol{x})]^{\mathrm{T}},$$

the unconstrained optimization problem can be converted into the following single-objective programming problem

$$\min_{\boldsymbol{x} \text{ s.t. } \boldsymbol{x}_m \leqslant \boldsymbol{x} \leqslant \boldsymbol{x}_M} f_1^2(\boldsymbol{x}) + f_2^2(\boldsymbol{x}) + \cdots + f_p^2(\boldsymbol{x}). \tag{6-7-2}$$

The converted single-objective programming problem can be solved directly with the approaches discussed earlier. Also, the `lsqnonlin()` in MATLAB can be used to solve the problem, with the syntax

$[\boldsymbol{x}, n_f, \boldsymbol{f}_{opt}, \texttt{flag}, \texttt{c}] = \texttt{lsqnonlin}(fun, \boldsymbol{x}_0, \boldsymbol{x}_m, \boldsymbol{x}_M)$

where, *fun* is the MATLAB description of the objective function vector, it can either be M-functions or anonymous functions. The argument $\boldsymbol{x}_0$ is the initial search point vector. The optimal solution and objective function are returned in vectors $\boldsymbol{x}$ and $\boldsymbol{f}_{opt}$, and the norm is returned in variable n_f.

Example 6.56 Please find the least squares solution for the following unconstrained multi-objective nonlinear programming problem.

$$\min_{\boldsymbol{x} \text{ s.t. } \begin{bmatrix}0\\0\\0\end{bmatrix} \leqslant \boldsymbol{x} \leqslant \begin{bmatrix}3\\\pi\\5\end{bmatrix}} \begin{bmatrix} (x_1+2x_2+3x_3)\sin(x_1+x_2)\mathrm{e}^{-x_1^2-x_3^2}+5x_3 \\ \mathrm{e}^{-x_2^2-4x_2^3}\cos(4x_1+x_2) \end{bmatrix}.$$

Solution *The vector form of the objective function can be specified first, and the problem can be solved with the following statements, and the result obtained is* $\boldsymbol{x} = [2.9998, 3.1415, 0]$.

```
>> f=@(x)[(x(1)+2*x(2)+3*x(3))*sin(x(1)+x(2))*...
                 exp(-x(1)^2-x(3)^2)+5*x(3); % the 2 objective functions
           exp(-x(2)^2-4*x(2)^3)*cos(4*x(1)+x(2))];
   xm=[0; 0; 0]; xM=[3; pi; 5]; x0=xM; x=lsqnonlin(f,x0,xm,xM)
```

In fact, the objective function can be redefined, and with the `fmincon()` *function, the problem can be solved and the solution obtained is* $\boldsymbol{x} = [3, 3.1416, 0]$. *It is worth mentioning that, the latter method can also be used in finding least squares solutions to problems with constraints.*

```
>> G=@(x)f(x)'*f(x); x=fmincon(G,x0,[],[],[],[],xm,xM)
```

6.7.3 Converting multi-objective problems into single-objective ones

There are a great amount of numerical algorithms used in solving single-objective optimization problems, to use them in solving multi-objective programming problems, the latter must be converted to single-objective programming problems. In this section, the conversion methods such as weighting method, least squares method are presented.

I. Linear weighting transforms

In order to convert a multi-objective programming problem into a single-objective one, the simplest way is to introduce weighting on each objective function and sum them up.

The new objective function in scalar form can be written as

$$f(\boldsymbol{x}) = w_1 f_1(\boldsymbol{x}) + w_2 f_2(\boldsymbol{x}) + \cdots + w_p f_p(\boldsymbol{x}), \tag{6-7-3}$$

where, $w_1 + w_2 + \cdots + w_p = 1$, and $0 \leqslant w_1, w_2, \cdots, w_p \leqslant 1$.

Example 6.57 Try different weights, and solve the problem in Example 6.55.

Solution *The original problem can be converted to the following single-objective linear programming problem*

$$\min \quad (w_1[4, 2.8, 2.4] - w_2[1,1,1])\boldsymbol{x}.$$
$$\boldsymbol{x} \text{ s.t. } \begin{cases} 4x_1+2.8x_2+2.4x_3 \leqslant 20 \\ -x_1-x_2-x_3 \leqslant -6 \\ -x_1-x_2 \leqslant -3 \\ x_1,x_2,x_3 \geqslant 0 \end{cases}$$

The best purchasing plans can be computed under different weights, using the loop structure, and the results are shown in Table 6.5. It can be seen that under different selections of the weighting coefficients, the best purchasing plans are also different. Besides, $x_1 \equiv 0$, this is because there are no specific constraints on x_1, the value of it should be made as small as possible.

```
>> f1=[4,2.8,2.4]; f2=[-1,-1,-1]; Aeq=[]; Beq=[]; xm=[0;0;0]; C=[];
   A=[4 2.8 2.4; -1 -1 -1; -1 -1 0]; B=[20;-6;-3]; ww1=[0:0.1:1];
   for w1=ww1, w2=1-w1; % try different weighting
      x=linprog(w1*f1+w2*f2,A,B,Aeq,Beq,xm); C=[C; w1 w2 x' f1*x -f2*x]
   end
```

TABLE 6.5: Optimal plans under different weighting coefficients.

w_1	w_2	x_1	x_2	x_3	cost	weight	w_1	w_2	x_1	x_2	x_3	cost	weight
0	1	0	3	4.8333	20	7.8333	0.6	0.4	0	3	3	15.6	6
0.1	0.9	0	3	4.8333	20	7.8333	0.7	0.3	0	3	3	15.6	6
0.2	0.8	0	3	4.8333	20	7.8333	0.8	0.2	0	3	3	15.6	6
0.3	0.7	0	3	3	15.6	6	0.9	0.1	0	3	3	15.6	6
0.4	0.6	0	3	3	15.6	6	1	0	0	3	3	15.6	6
0.5	0.5	0	3	3	15.6	6							

II. Best compromise solutions to linear programming problems

Consider the following multi-objective linear programming problem

$$J = \max \quad \boldsymbol{C}\boldsymbol{x}, \tag{6-7-4}$$
$$\boldsymbol{x} \text{ s.t. } \begin{cases} \boldsymbol{A}\boldsymbol{x} \leqslant \boldsymbol{B} \\ \boldsymbol{A}_{\rm eq}\boldsymbol{x} = \boldsymbol{B}_{\rm eq} \\ \boldsymbol{x}_{\rm m} \leqslant \boldsymbol{x} \leqslant \boldsymbol{x}_{\rm M} \end{cases}$$

where, the objective function coefficients are a matrix, rather than a vector. Each objective function $f_i(\boldsymbol{x}) = \boldsymbol{c}_i\boldsymbol{x},\ i = 1, 2, \cdots, p$ can be considered as the benefit of the ith party, and the best compromise solution is the compromise of the decision, on the benefit of all the parties. Of course, under the constraints and the interactions among the parties, it is not possible to let every party gain its maximum benefit. Some compromise must be made by all parties. The best compromise solution can be found uniquely. The procedures in finding the best compromise solution is as follows:

(i) Solve individually the linear programming problems for each objective function, and find the values of the objective functions f_k, $k = 1, 2, \cdots, p$.

(ii) The single-objective function can be constructed

$$f(\boldsymbol{x}) = -\frac{1}{f_1}\boldsymbol{c}_1\boldsymbol{x} - \frac{1}{f_2}\boldsymbol{c}_2\boldsymbol{x} - \cdots - \frac{1}{f_p}\boldsymbol{c}_p\boldsymbol{x}. \tag{6-7-5}$$

(iii) The best compromise solution can be converted into the following single-objective linear programming problem

$$J = \min_{\boldsymbol{x} \text{ s.t. } \begin{cases} \boldsymbol{A}\boldsymbol{x}\leqslant\boldsymbol{B} \\ \boldsymbol{A}_{\rm eq}\boldsymbol{x}=\boldsymbol{B}_{\rm eq} \\ \boldsymbol{x}_{\rm m}\leqslant\boldsymbol{x}\leqslant\boldsymbol{x}_{\rm M} \end{cases}} f(\boldsymbol{x}). \tag{6-7-6}$$

Based on the above algorithm, a MATLAB function can be written to find the best compromise solution of multi-objective linear programming problems, where the maximum value is found

```
function [x,f,flag,cc]=linprog_c(C,A,B,Aeq,Beq,xm,xM)
[p,m]=size(C); c=0;
for i=1:p, [x,f]=linprog(C(i,:),A,B,Aeq,Beq,xm,xM); c=c-C(i,:)/f; end
[x,f,flag,cc]=linprog(c,A,B,Aeq,Beq,xm,xM);
```

Example 6.58 Find the best compromise solution of the problem in Example 6.55.

Solution *It can be seen that the best compromise solution can be obtained with the following statements, and the results are* $\boldsymbol{x} = [0, 3, 4.8333]^{\rm T}$, *with total cost of \$20, and total weight of candies 7.8333 kg.*

```
>> C=[-4 -2.8 -2.4; 1 1 1]; A=[4 2.8 2.4; -1 -1 -1; -1 -1 0];
   B=[20; -6; -3]; Aeq=[]; Beq=[]; xm=[0;0;0]; xM=[];
   x=linprog_c(C,A,B,Aeq,Beq,xm,xM), C*x
```

Example 6.59 Find the best compromise solution of the following problem

$$\min_{\boldsymbol{x} \text{ s.t. } \begin{cases} 2x_1+4x_2+x_4\leqslant 110 \\ 5x_3+3x_4\geqslant 180 \\ x_1+2x_2+6x_3+5x_4\leqslant 250 \\ x_1,x_2,x_3,x_4\geqslant 0 \end{cases}} \begin{bmatrix} 3x_1+x_2+6x_4 \\ 10x_2+7x_4 \\ 2x_1+x_2+8x_3 \\ x_1+x_2+3x_3+2x_4 \end{bmatrix}.$$

Solution *From the multi-objective linear programming problem, matrix* $\boldsymbol{C}$ *can be constructed, together with the other constraints, the best compromise solution can be found with* `linprog_c()` *function, and the result is* $\boldsymbol{x} = [0, 26.087, 32.6087, 5.6522]^{\rm T}$, *and the compromised objective functions* $[60, 300.4348, 286.9565, 135.2174]^{\rm T}$.

```
>> C=-[3,1,0,6; 0,10,0,7; 2,1,8,0; 1,1,3,2];
   A=[2,4,0,1; 0,0,-5,-3; 1,1,6,5]; B=[110; -180; 250];
   Aeq=[]; Beq=[]; xm=[0;0;0;0]; xM=[];
   x=linprog_c(C,A,B,Aeq,Beq,xm,xM), -C*x
```

III. Least squares solutions of linear programming problems

Consider the least squares representation of the multi-objective programming problem given by

$$\min_{\boldsymbol{x} \text{ s.t.} \begin{cases} \boldsymbol{A}\boldsymbol{x}\leqslant\boldsymbol{B} \\ \boldsymbol{A}_{\rm eq}\boldsymbol{x}=\boldsymbol{B}_{\rm eq} \\ \boldsymbol{x}_{\rm m}\leqslant\boldsymbol{x}\leqslant\boldsymbol{x}_{\rm M} \end{cases}} \frac{1}{2}||\boldsymbol{C}\boldsymbol{x}-\boldsymbol{d}||^2, \tag{6-7-7}$$

the least squares solution can be obtained with

$\boldsymbol{x}$=`lsqlin(`$\boldsymbol{C}$,$\boldsymbol{d}$,$\boldsymbol{A}$,$\boldsymbol{B}$,$\boldsymbol{A}_{\rm eq}$,$\boldsymbol{B}_{\rm eq}$,$\boldsymbol{x}_{\rm m}$,$\boldsymbol{x}_{\rm M}$,$\boldsymbol{x}_0$,`options)`

and since the problem is a convex problem, the selection of $\boldsymbol{x}_0$ is not important.

Example 6.60 Consider again the multi-objective linear programming problem in Example 6.59, please find its least squares solution.

Solution *From the given problem, the matrix $\boldsymbol{C}$ is established, and other constraints can be specified. The* `lsqlin()` *function can be used to solve the problem, and the result is $\boldsymbol{x} = [0, 0, 28.4456, 12.5907]^{\rm T}$, and the optimal objective functions are* $[75.544, 88.1347, 227.5648, 110.5181]^{\rm T}$.

```
>> C=[3,1,0,6; 0,10,0,7; 2,1,8,0; 1,1,3,2]; d=zeros(4,1);
   A=[2,4,0,1; 0,0,-5,-3; 1,1,6,5]; B=[110; -180; 250]; Aeq=[]; Beq=[];
   xm=[0;0;0;0]; xM=[]; x=lsqlin(C,d,A,B,Aeq,Beq,xm,xM), C*x
```

Please note that, since the solution algorithm and target are different from the best compromise method, the solutions are also different.

6.7.4 Pareto front of multi-objective programming problems

It can be seen from the previous discussion that the solutions to multi-objective programming problems are not unique. Now, let us consider the multi-objective programming problem as an alternative. Assume that one of the objective functions is expressed by some scatters, then, the number of objective functions is reduced by one. New results may be found, if multi-objective programming problems are treated in this way.

Example 6.61 Reexamine the multi-objective optimization problem studied in Example 6.55 using scatters.

Solution *Assume that the total cost in the original problem is represented by a series of scatters $m_i \in (15, 20)$, then, the original problem can be converted to a single-objective linear programming problem.*

$$\min \quad -[\,1,\ 1,\ 1\,]\boldsymbol{x}.$$
$$\boldsymbol{x}\ \text{s.t.}\ \begin{cases} 4x_1+2.8x_2+2.4x_3=m_i \\ -x_1-x_2-x_3\leqslant-6 \\ -x_1-x_2\leqslant-3 \\ x_1,x_2,x_3\geqslant 0 \end{cases}$$

The above formula means that, if the total cost is m_i, what is the maximum weight of candies that can be purchased under the constraints. In this way, the largest weight n_i can be obtained. For different values of m_i, different n_i can be obtained. The relationship between cost and total weight are shown in Figure 6.13. Please note that not all the scatters in the plane the solutions of multi-objective optimization problems, since the optimization of m_i was not yet considered.

```
>> f2=[-1,-1,-1]; Aeq=[4 2.8 2.4]; xm=[0;0;0];
   A=[-1 -1 -1; -1 -1 0]; B=[-6;-3]; mi=15:0.1:20; ni=[];
   for m=mi, Beq=m; x=linprog(f2,A,B,Aeq,Beq,xm); ni=[ni,-f2*x]; end
   plot(mi,ni)
```

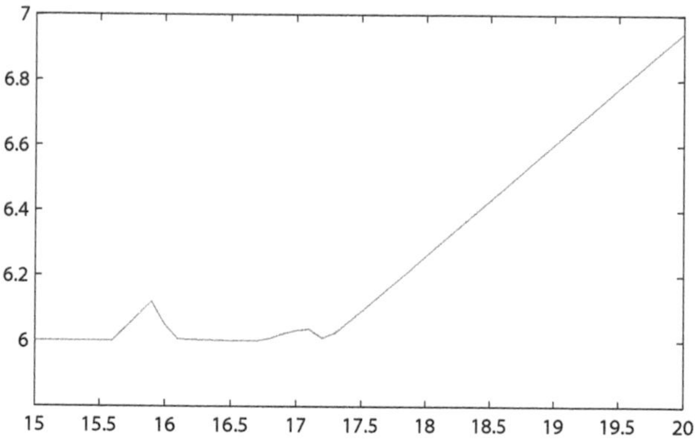

FIGURE 6.13: The relationship of the two objective functions.

Consider double-objective programming problem. The scatters of feasible solutions can be obtained first, and shown in x-y plane, as shown in Figure 6.14. Since the targets of the original problem is to let the two axes f_1 and f_2 take smallest values, a curve can be extracted from the scatters. All the points on the curve are the solutions of the original problem, and the curve is referred to as *Pareto set* or *Pareto front*. Based on the Pareto front extracting function contributed by Gianluca Dorini, Yi Cao developed an improved fast extracting package [15], the syntax of the main function is $\boldsymbol{K}$ = `paretofront([`$\boldsymbol{f}_1$`,`$\boldsymbol{f}_2$`,`$\cdots$`,`$\boldsymbol{f}_p$`])`, where, $\boldsymbol{f}_1$, $\boldsymbol{f}_2$, $\cdots$, $\boldsymbol{f}_p$ are column vectors composed of scatters of the feasible solutions.

Example 6.62 Please extract the Pareto front from Example 6.55.

Solution *Similar to the enumerate method, mesh grids of the variables x_1, x_2, x_3 are generated, and those points outside the feasible regions are removed. The Pareto front can be extracted with* `paretofront()`, *as shown in Figure 6.15.*

```
>> [x1,x2,x3]=meshgrid(0:0.1:4);
   ii=find(4*x1+2.8*x2+2.4*x3<=20&x1+x2+x3>=6&x1+x2>=3);
```

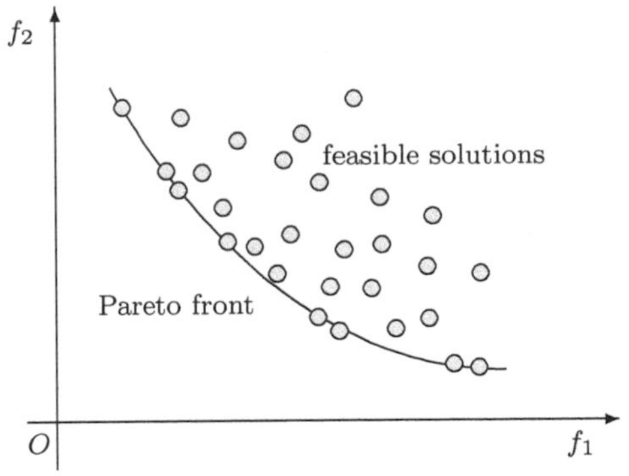

FIGURE 6.14: Illustration of Pareto front.

```
xx1=x1(ii); xx2=x2(ii); xx3=x3(ii); f1=4*xx1+2.8*xx2+2.4*xx3;
f2=-(xx1+xx2+xx3); k=paretofront([f1 f2]);
plot(f1,f2,'x'), hold on; plot(f1(k),f2(k),'o')
```

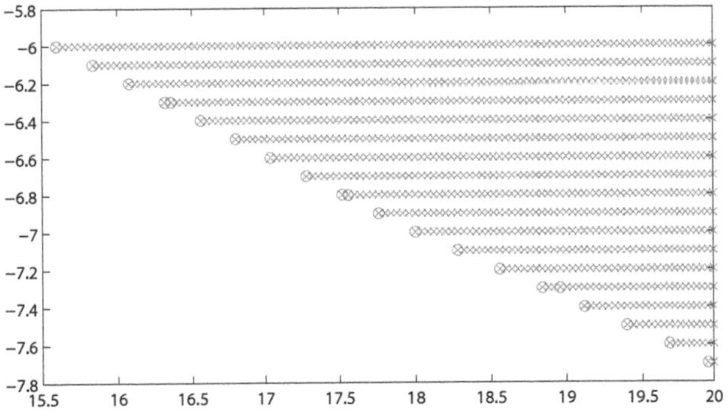

FIGURE 6.15: Pareto front in Example 6.55.

6.7.5 Solutions of minimax problems

A class of important problems in multi-objective programming problems is the minimax problems. Assume that there are a group of p objective functions $f_i(\boldsymbol{x}), i = 1, 2, \cdots, p$, a maximum value can be extracted from each objective function, i.e., $\max\limits_{\boldsymbol{x} \text{ s.t. } \boldsymbol{G}(\boldsymbol{x}) \leqslant 0} f_i(\boldsymbol{x})$, and the set of maximum values are still a function of $\boldsymbol{x}$. The minimum value from the set of maximum values is expected, i.e.,

$$J = \min \left[\max_{\boldsymbol{x} \text{ s.t. } \boldsymbol{G}(\boldsymbol{x}) \leqslant 0} f_i(\boldsymbol{x}) \right], \tag{6-7-8}$$

and this kind of problem is referred to as a *minimax problem*. If the maximum value is interpreted as a worst case, or maximum loss, minimax problem is to find a way to minimize the losses. Constrained minimax problems are generally written as

$$J = \min_{\boldsymbol{x} \text{ s.t. } \begin{cases} \boldsymbol{Ax} \leqslant \boldsymbol{B} \\ \boldsymbol{A}_{\text{eq}}\boldsymbol{x} = \boldsymbol{B}_{\text{eq}} \\ \boldsymbol{x}_{\text{m}} \leqslant \boldsymbol{x} \leqslant \boldsymbol{x}_{\text{M}} \\ \boldsymbol{C}(\boldsymbol{x}) \leqslant \boldsymbol{0} \\ \boldsymbol{C}_{\text{eq}}(\boldsymbol{x}) = \boldsymbol{0} \end{cases}} \max \quad f_i(\boldsymbol{x}). \tag{6-7-9}$$

Function `fminimax()` provided in MATLAB Optimization Toolbox can be used in solving minimax problems. The syntax of the function is

$[\boldsymbol{x}, f_{\text{opt}}, \texttt{flag}, \texttt{c}] = \texttt{fminimax}(\mathit{fun}, \boldsymbol{x}_0, \boldsymbol{A}, \boldsymbol{B}, \boldsymbol{A}_{\text{eq}}, \boldsymbol{B}_{\text{eq}}, \boldsymbol{x}_{\text{m}}, \boldsymbol{x}_{\text{M}}, \mathit{CFun}, \mathit{options}, p_1, p_2, \cdots)$

and it can be seen that the syntax is very close to the `fmincon()` function. The multi-objective function can also be expressed as M-functions or anonymous functions.

Example 6.63 Please solve the following minimax problem

$$\min_{\boldsymbol{x} \text{ s.t. } \begin{cases} 4.3x_1 + 3.8x_2 \leqslant 4.9 \\ x_1 + x_2 \leqslant 3 \end{cases}} \max \begin{bmatrix} x_1^2 \sin x_2 + x_2 - 3x_1x_2 \cos x_1 \\ -x_1^2 \mathrm{e}^{-x_2} - x_2^2 \mathrm{e}^{-x_1} + x_1x_2 \cos x_1x_2 \\ x_1^2 + x_2^2 - 2x_1x_2 + x_1 - x_2 \\ -x_1^2 - x_2^2 \cos x_1x_2 \end{bmatrix}.$$

Solution *The above minimax problem can be solved with the following statements directly, with randomly selected initial search point, the solution of the original problem is* $\boldsymbol{x} = [0.5319, 0.6876]$.

```
>> f=@(x)[x(1)^2*sin(x(2))+x(2)-3*x(1)*x(2)*cos(x(1));
        -x(1)^2*exp(-x(2))-x(2)^2*exp(-x(1))+x(1)*x(2)*cos(x(1)*x(2));
        x(1)^2+x(2)^2-2*x(1)*x(2)+x(1)-x(2);
        -x(1)^2-x(2)^2*cos(x(1)*x(2))];
   A=[4.3 3.8; 1 1]; B=[4.9; 3]; x=fminimax(f,rand(2,1),A,B)
```

In fact, with function `fminimax()`, other variations of minimax problems can also be solved, such as minimin problem

$$J = \min \left[\min_{\boldsymbol{x} \text{ s.t. } \boldsymbol{G}(\boldsymbol{x}) \leqslant 0} f_i(\boldsymbol{x}) \right]. \tag{6-7-10}$$

The problem can be converted to the following minimax problem

$$J = \min \left[\max_{\boldsymbol{x} \text{ s.t. } \boldsymbol{G}(\boldsymbol{x}) \leqslant 0} -f_i(\boldsymbol{x}) \right]. \tag{6-7-11}$$

6.7.6 Solutions of multi-objective goal attainment problems

In practical optimization problems, sometimes the feasible solutions cannot be found, and the constraints should be relaxed to some extent. for instance, the inequality constraints $\boldsymbol{Ax} \leqslant \boldsymbol{B}$ can be rewritten as $\boldsymbol{Ax} \leqslant \boldsymbol{B} + \boldsymbol{d}^- - \boldsymbol{d}^+$. In the practical solution process, the biases ($\boldsymbol{d}^-, \boldsymbol{d}^+$) are introduced into the objective function and to minimize the biases. This type of optimization problem is referred to as a *multi-objective goal attainment problem.*

Function `fgoalattain()` provided in MATLAB Optimization Toolbox can be used in solving multi-objective goal attainment problems, whose mathematical representation is

$$\min_{\boldsymbol{x},\gamma \text{ s.t. } \begin{cases} \boldsymbol{F}(\boldsymbol{x})-\boldsymbol{w}\gamma\leqslant \boldsymbol{g} \\ \boldsymbol{A}\boldsymbol{x}\leqslant \boldsymbol{B} \\ \boldsymbol{A}_{\rm eq}\boldsymbol{x}=\boldsymbol{B}_{\rm eq} \\ \boldsymbol{c}(\boldsymbol{x})\leqslant 0 \\ \boldsymbol{c}_{\rm eq}(\boldsymbol{x})=0 \\ \boldsymbol{x}_{\rm m}\leqslant \boldsymbol{x}\leqslant \boldsymbol{x}_{\rm M} \end{cases}} \gamma, \tag{6-7-12}$$

where, $\boldsymbol{F}(\boldsymbol{x})$ is the objective function vector, and $\boldsymbol{w}$ are the weights on the objective functions. The vector $\boldsymbol{g}$ contains the values that the objectives attempt to attain. The syntax of the function is

$\boldsymbol{x}=$ `fgoalattain`$(fun, \boldsymbol{x}_0, \boldsymbol{g}, \boldsymbol{w}, \boldsymbol{A}, \boldsymbol{B}, \boldsymbol{A}_{\rm eq}, \boldsymbol{B}_{\rm eq}, \boldsymbol{x}_{\rm m}, \boldsymbol{x}_{\rm M}, CFun, options, p_1, p_2, \cdots)$

Example 6.64 Solve the problem in Example 6.55 using goal attainment method.

Solution *It is obvious that the accepted goals for the two objective functions are 20 and −6. Apart from that, weights on the two objective functions should be introduced. If "less cost" is concentrated, the weight of it can be assigned to 80%, while the weight on the other is 20%. The following statements can be used in solving the original problem, and the result obtained is* $\boldsymbol{x} = [0, 3, 3.6875]^{\rm T}$, *with* $\boldsymbol{f}(\boldsymbol{x}) = [17.25, -6.6875]^{\rm T}$.

```
>> f=@(x)[[4,2.8,2.4]*x; [-1 -1 -1]*x]; Aeq=[]; Beq=[]; xm=[0;0;0];
   x0=xm; w=[0.8,0.2]; goal=[20; -6]; A=[-1 -1 0]; B=[-3];
   x=fgoalattain(f,x0,goal,w,A,B,[],[],xm), f(x)
```

6.8 Dynamic Programming and Shortest Path Planning

The optimization problems discussed so far can be classified as static optimization problems, since the objective functions and constraints are fixed in the whole solution process. In scientific research and engineering, we often meet another category of optimization problem, where the programming problems consist of several overlapped subproblems. The subproblems should be solved and combined to get the overall solution of the whole problem. In ordinary problem solving strategies, since the subproblems are related to each other, the subproblems are usually to be solved several times. It is the job of dynamic programming to solve each subproblem only once, so as to minimize the time used in the whole problem solution process.

Dynamic programming, which was proposed by Richard Bellman in the 1940s and get matured in the late 1950s [16], is another area of optimization. The theory has many applications, and in particular, it is useful in computer programming. In this section, the application and solution of dynamic programming problems in shortest path planning of oriented and other graphs are presented.

6.8.1 Matrix representation of graphs

Before introducing the representation of graphs, some ideas about graphs are presented. An example of a graph is shown in Figure 6.16 [17].

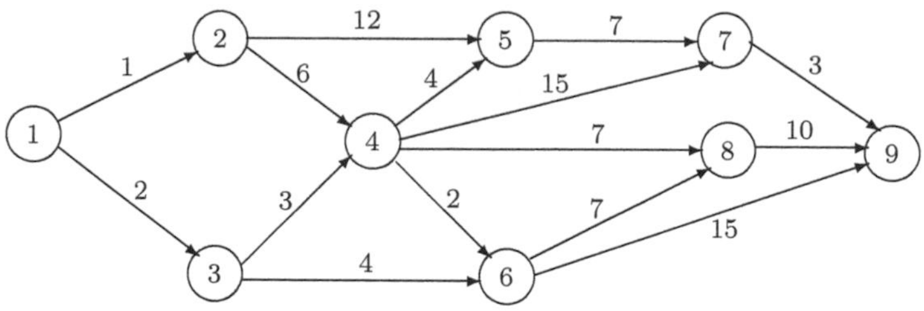

FIGURE 6.16: A typical oriented graph.

In graph theory, the graphs are constructed by *nodes* and *edges*. In the graph shown in Figure 6.16, the nodes are the encircled numbers. The edge is the path that connects directly to the nodes, and in the graph, the edges are the arrows between the nodes. If the edge is one-directional, then, the graph is referred to as *oriented* or *directed* (also known as a *digraph*), otherwise, it is referred to as an *undigraph*.

Apart from nodes and edges, the entities *weights* are also essential in graphs. The weights are the numbers above the edges, which usually stand for the distance, or traveling time in the edges.

There are different ways to represent a graph, and the representation most suitable for computer modeling is the incidence matrix method. If there are n nodes in a graph, it can be described by an $n \times n$ matrix $\boldsymbol{R}$. Assuming that the edge from node i to node j has a weight of k, the matrix element can be expressed by $\boldsymbol{R}(i,j) = k$. Such a matrix is referred to as an incidence matrix. If there is no edge from node i to j, we can assign $\boldsymbol{R}(i,j) = 0$. However, certain algorithms may assign the element value to $\boldsymbol{R}(i,j) = \infty$.

The sparse form representation of incidence matrices can also be used in MATLAB to describe graphs. Suppose that a graph is composed of n nodes and m edges. It is known that the ith edge is from node a_i to node b_i, and the edge has a weight of w_i, $i = 1, 2, \cdots, m$. Three vectors can be established, and an incidence matrix can be declared in MATLAB with the following statements

$\boldsymbol{a}=$ `[`a_1`,`a_2`,`$\cdots$`,`a_m`,`n`];` $\boldsymbol{b}=$ `[`b_1`,`b_2`,`$\cdots$`,`b_m`,`n`];` % start and end nodes

$\boldsymbol{w}=$ `[`w_1`,`w_2`,`$\cdots$`,`w_m`,0];` $\boldsymbol{R}=$`sparse(`$\boldsymbol{a}$`,`$\boldsymbol{b}$`,`$\boldsymbol{w}$`);` % edge, incidence matrix

Note that the last element in each vector ensures a square incidence matrix $\boldsymbol{R}$. Sparse matrices and ordinary ones can be converted with functions `full()` and `sparse()`, respectively.

6.8.2 Optimal path planning of oriented graphs

The *oriented graph* representation of optimal path searching can also be encountered in many application areas. With dynamic programming theory, the method of backward derivation is usually adopted from the destination to the starting node. An example will be given to demonstrate the backward derivation methods. Then, a computer solution to the same problem will be used with relevant functions in the Bioinformatics Toolbox, which is the MATLAB implementation of the Dijkstra algorithms.

I. Solutions by dynamic programming technique

An example of an oriented graph is used to show dynamic programming with applications to the shortest path problem. The problem is to be demonstrated with a manual solution method first.

Example 6.65 Please find the shortest path in the oriented graph shown in Figure 6.16, from nodes ① to ⑨, using dynamic programming method.

Solution *The number on top of each edge indicates the shortest distance required to travel from its starting node to the ending node. The shortest path from node ① to node ⑨ has to be calculated.*

Consider the destination node, node ⑨. Mark the distance at the node as (0). Connected to this node, there are three nodes, respectively, nodes ⑥, ⑦ and ⑧. Since there is only one edge to travel to the destination node, the shortest distances of these nodes are, respectively, 15, 3 and 10, that is, the weights of the edges. From node ⑤ to node ⑦, there is only one edge, therefore, the shortest distance from node ⑤ is 10, that is, the sum of the distance labeled on node ⑦ plus the weight of the edge. Now let us find the label on node ④. From node ④, there are edges to nodes ⑤, ⑥, ⑦ and ⑧, respectively. Summing the weights separately to the corresponding labels, the sums are, respectively, 14, 18, 17 and 17, being 14 the smallest. Therefore, the label on node ④ should be (14). In a similar way, the labels on node ② and ③ can be marked to (20) and (17), and the label on node ① can be marked as (19), the shortest distance expected. From the above, it can be seen that the shortest path is nodes ① → ③ → ④ → ⑤ → ⑦ → ⑨, as shown in Figure 6.17.

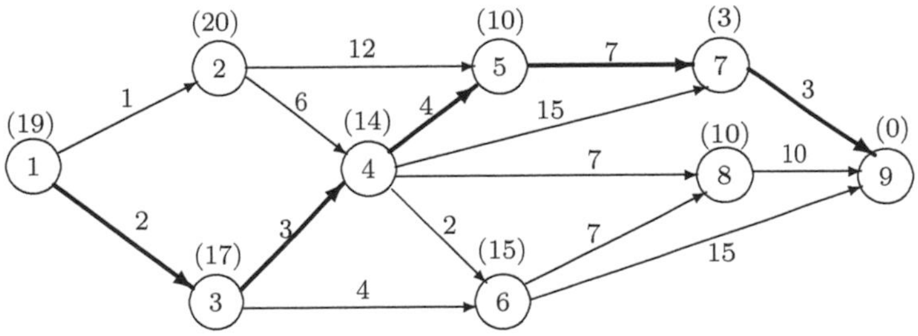

FIGURE 6.17: Manual solution to the oriented graph problem.

From the above, it can be seen that the method is quite straightforward and easy to understand. However, for large-scale problems, manual derivation is extremely complicated and error-prone. Computer solutions are necessary and will be explained later.

II. Search and illustration of oriented graphs

Some relevant functions are provided in the Bioinformatics Toolbox to solve graphs and shortest path searching problems. For instance, to establish an object for an oriented graph, the function `view()` is used to display the object, and the function `graphshortestpath()` is used to solve directly the shortest path problems. The syntaxes of these functions are

```
P=biograph(R)                        % create an object P
[d,p]=graphshortestpath(P,n1,n2)     % find the shortest path
```

where $\boldsymbol{R}$ is the incidence matrix of an oriented graph, expressed as either an ordinary or a sparse matrix. The matrix can be processed by the `biograph()` function to establish an object for the oriented graph. For the oriented graph shown in Figure 6.16, the matrix element $\boldsymbol{R}(i,j)$ refers to the weight of the edge from node i to node j. Having established the object P, function `graphshortestpath()` can be used to solve the shortest path problem directly. The arguments n_1 and n_2 are, respectively, the starting and terminal node numbers. The returned variable d is the shortest distance, while vector $\boldsymbol{p}$ returns the node numbers on the shortest path. Other functions can be used to further show the search results graphically.

Example 6.66 Considering again the problem in Example 6.65, the MATLAB functions in the Bioinformatics Toolbox can be used to reexamine the original problem.

Solution *From Figure 6.16, it can be seen that the information of each edge in the graph is summarized as shown in Table 6.6, where the starting node, ending node and the weight of each edge are obtained. The following commands can be used to enter the incidence matrix* $\boldsymbol{R}$*, from which the graph is generated automatically with the* `view()` *function, and the handle is assigned to* `h`*. The automatically drawn graph is shown in Figure 6.18(a). Note that when constructing the incidence matrix* $\boldsymbol{R}$*, you have to make sure that it is a square matrix.*

```
>> a=[1 1 2 2 3 3 4 4 4 4 5 6 6 7 8]; b=[2 3 5 4 4 6 5 7 8 6 7 8 9 9 9];
   w=[1 2 12 6 3 4 4 15 7 2 7 7 15 3 10]; R=sparse(a,b,w); R(9,9)=0;
   obj=biograph(R); h=view(obj); h.ShowWeights='on'; % show the graph
```

TABLE 6.6: Edge data.

start	end	weight	start	end	weight	start	end	weight	start	end	weight
1	2	1	1	3	2	2	5	12	2	4	6
3	4	3	3	6	4	4	5	7	4	7	15
4	8	7	4	6	2	5	7	7	6	8	7
6	9	15	7	9	3	8	9	10			

For a given oriented graph object defined in $\boldsymbol{R}$*, the function* `graphshortestpath()` *can be used to find the shortest path, and with the help of the* `view()` *function, the shortest path is shown in red; the result is shown in Figure 6.18(b). It can be seen that the result obtained is exactly the same as the one obtained manually.*

```
>> [d,p]=graphshortestpath(R,1,9) % shortest path from node ① to node ⑨
   set(h.Nodes(p),'Color',[1 0.4 0.4])
   edges=getedgesbynodeid(h,get(h.Nodes(p),'ID'));
   set(edges,'LineColor',[1 0 0])  % the path in red is the shortest
```

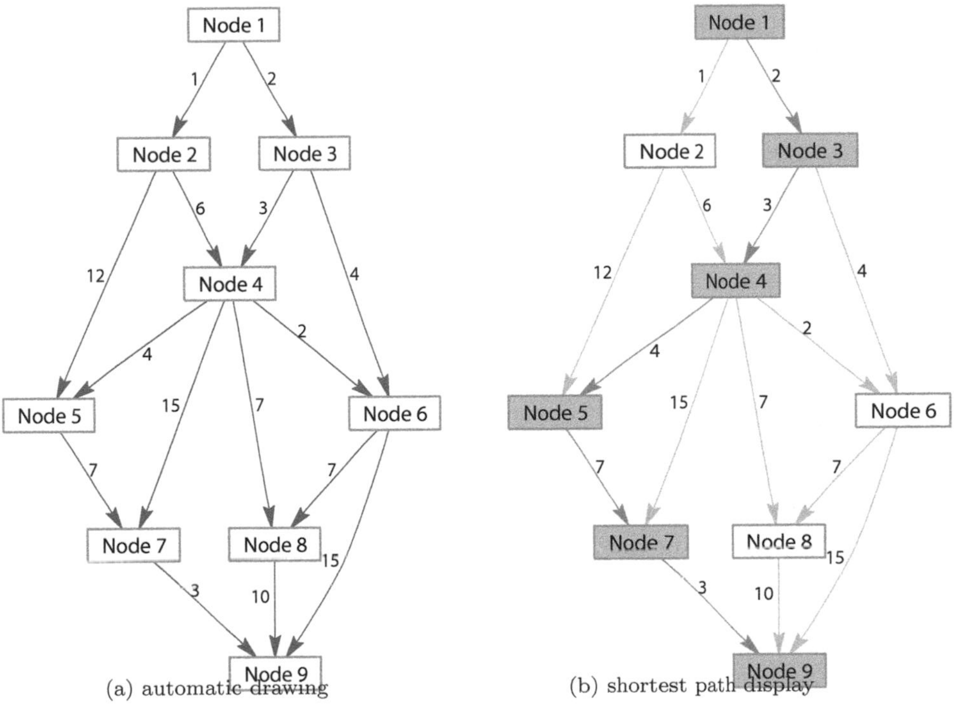

(a) automatic drawing (b) shortest path display

FIGURE 6.18: Oriented graph and solution of shortest path program.

III. Dijkstra shortest path algorithm and its implementation

The shortest path between two nodes can be obtained directly with Dijkstra algorithm [18]. In fact, if the starting node is assigned, the shortest paths to all the other nodes can be obtained together, without sacrificing searching speed. In shortest path algorithms, Dijkstra algorithm is the most effective one. Based on Dijkstra algorithm, the following function is written

```
function [d,path]=dijkstra(W,s,t)
[n,m]=size(W); ix=(W==0); W(ix)=Inf;
if n~=m, error('Square W required'); end
visited(1:n)=0; dist(1:n)=Inf; parent(1:n)=0; dist(s)=0; d=Inf;
for i=1:(n-1),  % find the relation between each node and the origin
   ix=(visited==0); vec(1:n)=Inf; vec(ix)=dist(ix);
   [a,u]=min(vec); visited(u)=1;
   for v=1:n, if (W(u,v)+dist(u)<dist(v)),
      dist(v)=dist(u)+W(u,v); parent(v)=u;
end; end; end
if parent(t)~=0, path=t; d=dist(t); % trace back to find the shortest path
   while t~=s, p=parent(t); path=[p path]; t=p; end
end
```

The function can be called with $[d,\boldsymbol{p}]=$ `dijkstra(`$\boldsymbol{W},s,t$`)`, where $\boldsymbol{W}$ is the square incidence matrix, and s, t are, respectively, the serial number of the starting node, and ending node. The returned argument d is the shortest path, while $\boldsymbol{p}$ is a vector containing the serial numbers of nodes in the shortest path. Please note that in the function, the entities

0 in the $\boldsymbol{W}$ matrix are reassigned automatically to ∞, so that the Dijkstra algorithm can work normally.

Example 6.67 Solve again the problem in Example 6.65 with Dijkstra algorithm.

Solution *The following statements can be used, and the results are exactly the same as the one obtained in Example 6.65.*

```
>> a=[1 1 2 2 3 3 4 4 4 4 5 6 6 7 8];b=[2 3 5 4 4 6 5 7 8 6 7 8 9 9 9];
   w=[1 2 12 6 3 4 4 15 7 2 7 7 15 3 10]; R=sparse(a,b,w); R(9,9)=0;
   W=ones(9); [d,p]=dijkstra(R.*W,1,9) % find the shortest path
```

6.8.3 Optimal path planning of undigraphs

In practical applications, for instance in route finding in cities, the relevant graphs can also be described by undigraphs, since for nodes A and B, the distance from node A to node B is exactly the same as that from node B to node A. Manipulating undigraphs is also simple. Assume that the edges are assigned first to be one-directional, and the incidence matrix $\boldsymbol{R}_1$ can then be entered. The incidence matrix for the undigraph can be obtained directly from $\boldsymbol{R} = \boldsymbol{R}_1 + \boldsymbol{R}_1^{\mathrm{T}}$. If there are one-way streets in the city, manual modification can be made. For instance, if the edge from node i to j is one way, the element of $\boldsymbol{R}(i,j)$ is retained, and set $\boldsymbol{R}(j,i) = 0$.

For some special undigraphs, the weights from node i to node j may be different from the weight from node j to node i. The elements in matrix $\boldsymbol{R}$ can be modified manually, too, so that even more general problems can be solved with the functions.

6.8.4 Optimal path planning for graphs described by coordinates

If the nodes are specified by its absolute coordinates (x_i, y_i), and the connections of these notes are also specified, the weights of the sides can be calculated directly as the Euclidian distances of the nodes. Thus, the incidence matrix can be established, and the shortest path problems can be solved.

Example 6.68 Assume that there are 11 cities, distributed at the following coordinates $(4,49)$, $(9,30)$, $(21,56)$, $(26,26)$, $(47,19)$, $(57,38)$, $(62,11)$, $(70,30)$, $(76,59)$, $(76,4)$, $(96,4)$, the roads between the cities are shown in Figure 6.19. Please find the shortest distance from city A to city B. If the road between cities ⑥ and ⑧ are in construction, please find a new shortest path.

Solution *The incidence matrix should be entered first. The sparse matrix can be created based on the oriented graph, and the weighting matrix can be set to a matrix of ones. The incidence matrix of the undigraph can be obtained from the oriented graph, and the actual weights of the sides can be calculated directly from the Euclidian distance, i.e.,* $d_{ij} = \sqrt{(x_i - x_j)^2 + (y_i - y_j)^2}$. *Thus, the effective weighting matrix can be obtained by the product of the two matrices. With the following statements, the shortest path can be found as from city A → ② → ④ → ⑥ → ⑧ → city B, and the shortest distance is 111.6938.*

```
>> x=[4,9,21,26,47,57,62,70,76,76,96];
   y=[49,30,56,26,19,38,11,30,59,4,32];
   for i=1:11, for j=1:11,  % calculate Euclidian distance matrix
```

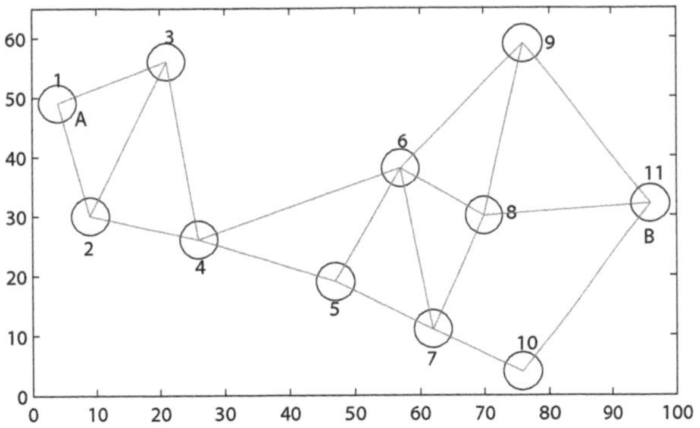

FIGURE 6.19: Map of city locations and traffic.

```
    D(i,j)=sqrt((x(i)-x(j))^2+(y(i)-y(j))^2);
end, end
n1=[1 1 2 2 3 4 4 5 5 6 6 6 7 8 7 10 8 9];
n2=[2 3 3 4 4 5 6 6 7 7 8 9 8 9 10 11 11 11];
R=sparse(n1,n2,1); R(11,11)=0; R=R+R'; [d,p]=dijkstra(R.*D,1,11)
```

If the route from node ⑥ to node ⑧ is not usable, we can set $\boldsymbol{R}(8,6)=\boldsymbol{R}(6,8)=\infty$. *In this case, the following statements can be used to find the optimal path again, and the new result is city A → ② → ④ → ⑤ → ⑦ → ⑧ → city B, with the shortest distance of 122.9394.*

```
>> R(6,8)=Inf; R(8,6)=Inf; [d,p]=dijkstra(R.*D,1,11)
```

Exercises

Exercise 6.1 *Find the solutions to the following equations, and verify the accuracy of the solutions.*

$$\text{(i)}\ \begin{cases} 24xy - x^2 - y^2 - x^2y^2 = 13 \\ 24xz - x^2 - z^2 - x^2z^2 = 13 \\ 24yz - y^2 - z^2 - y^2z^2 = 13, \end{cases} \qquad \text{(ii)}\ \begin{cases} x^2y^2 - zxy - 4x^2yz^2 = xz^2 \\ xy^3 - 2yz^2 = 3x^3z^2 + 4xzy^2 \\ y^2x - 7xy^2 + 3xz^2 = x^4zy. \end{cases}$$

Exercise 6.2 *Solve and validate the following equations with the given parameter t* [19]

$$\begin{cases} t^{31} + t^{23}y + t^{17}x + t^{11}y^2 + t^5xy + t^2x^2 = 0 \\ t^{37} + t^{29}y + t^{19}x + t^{13}y^2 + t^7xy + t^3x^2 = 0. \end{cases}$$

Exercise 6.3 *Solve graphically the following equations, and verify the results.*

$$\text{(i)}\ \mathrm{e}^{-(x+1)^2+\pi/2}\sin(5x+2) = 0, \quad \text{(ii)}\ \begin{cases} (x^2+y^2+10xy)\mathrm{e}^{-x^2-y^2-xy} = 0 \\ x^3 + 2y = 4x + 5. \end{cases}$$

Exercise 6.4 *Find by numerical methods the solutions to the above problems, and verify the results.*

Exercise 6.5 *Find and validate all the solutions to the following modified Riccati equation, and verify the results. Is it possible to find all solutions with high-precision?*

$$\boldsymbol{AX}+\boldsymbol{XD}-\boldsymbol{XBX}+\boldsymbol{C}=\boldsymbol{0},$$

where

$$\boldsymbol{A}=\begin{bmatrix}2&1&9\\9&7&9\\6&5&3\end{bmatrix},\ \boldsymbol{B}=\begin{bmatrix}0&3&6\\8&2&0\\8&2&8\end{bmatrix},\ \boldsymbol{C}=\begin{bmatrix}7&0&3\\5&6&4\\1&4&4\end{bmatrix},\ \boldsymbol{D}=\begin{bmatrix}3&9&5\\1&2&9\\3&3&0\end{bmatrix}.$$

Exercise 6.6 *Find c such that the integral* $\displaystyle\int_0^1 (\mathrm{e}^x-cx)^2\,\mathrm{d}x$ *is minimized.*

Exercise 6.7 *Solve the unconstrained optimization problems*

$$\min_{\boldsymbol{x}}\ \begin{array}{l}100(x_2-x_1^2)^2+(1-x_1)^2+90(x_4-x_3^2)+(1-x_3^2)^2+\\ \quad 10.1\left[(x_2-1)^2+(x_4-1)^2\right]+19.8(x_2-1)(x_4-1).\end{array}$$

Exercise 6.8 *Try to find the global for the following objective function*

$$f(x_1,x_2)=-\frac{\sin\left(0.1+\sqrt{(x_1-4)^2+(x_2-9)^2}\right)}{1+(x_1-4)^2+(x_2-9)^2}.$$

Exercise 6.9 *A set of challenging benchmark problems for evaluating optimization algorithms can be solved using MATLAB. Solve the following unconstrained optimization problems with MATLAB:*

(i) De Jong's problems [20]

$$J=\min_{\boldsymbol{x}}\boldsymbol{x}^{\mathrm{T}}\boldsymbol{x}=\min_{\boldsymbol{x}}(x_1^2+x_2^2+\cdots+x_p^2),\quad \text{where } x_i\in[-512,512],\ i=1,\cdots,p,$$

with theoretic solution $x_1=\cdots=x_p=0$.

(ii) Griewangk's benchmark problem

$$J=\min_{\boldsymbol{x}}\left(1+\sum_{i=1}^{p}\frac{x_i^2}{4000}-\prod_{i=1}^{p}\cos\frac{x_i}{\sqrt{i}}\right),\quad \text{where } x_i\in[-600,600].$$

(iii) Ackley's benchmark problem [21]

$$J=\min_{\boldsymbol{x}}\left[20+10^{-20}\exp\left(-0.2\sqrt{\frac{1}{p}\sum_{i=1}^{p}x_i^2}\right)-\exp\left(\frac{1}{p}\sum_{i=1}^{p}\cos 2\pi x_i\right)\right].$$

(iv) Kursawe's benchmark problem

$$J=\min_{\boldsymbol{x}}\sum_{i=1}^{p}|x_i|^{0.8}+5\sin^3 x_i+3.5828,\ p=2,\ \text{and } p=20.$$

Exercise 6.10 *Solve the nonlinear programming problem with graphical methods, and verify the results using numerical methods.*

$$\min_{\boldsymbol{x}\ \text{s.t.}\begin{cases}x_1-x_2+2\geqslant 0\\ -x_1^2+x_2-1\geqslant 0\\ x_1\geqslant 0,x_2\geqslant 0\end{cases}} x_1^3+x_2^2-4x_1+4.$$

Exercise 6.11 *Try to solve the following linear programming problems:*

(i) $$\min_{\boldsymbol{x} \text{ s.t.} \begin{cases} 4x_1-x_2+2x_3-x_4=-2 \\ x_1+x_2-x_3+2x_4\leqslant 14 \\ 2x_1-3x_2-x_3-x_4\geqslant -2 \\ x_{1,2,3}\geqslant -1 \end{cases}} -3x_1+4x_2-2x_3+5x_4,$$

(ii) $$\min_{\boldsymbol{x} \text{ s.t.} \begin{cases} x_1+x_2+x_3+x_4=4 \\ -2x_1+x_2-x_3-x_6+x_7=1 \\ 3x_2+x_3+x_5+x_7=9 \\ x_{1,2,\cdots,7}\geqslant 0 \end{cases}} x_6+x_7.$$

Exercise 6.12 *Please solve the following optimization problem:*

$$\min_{\boldsymbol{x} \text{ s.t.} \begin{cases} -\sum_{i=1}^{5}(9+i)x_i+50000\geqslant 0 \\ x_i\geqslant 0, i=1,2,3,4,5 \end{cases}} -(x_1+x_2+x_3+x_4+x_5).$$

Exercise 6.13 *Please solve the following transportation problems.*

(i)

	destination				*supply*
S1	*3*	*7*	*6*	*4*	*5*
S2	*2*	*4*	*3*	*2*	*2*
S3	*4*	*3*	*8*	*5*	*3*
D	*3*	*3*	*2*	*2*	

(ii)

	shipping cost				*output*
S1	*464*	*513*	*654*	*867*	*75*
S2	*352*	*416*	*690*	*791*	*125*
S3	*995*	*682*	*388*	*685*	*100*
D	*80*	*65*	*70*	*85*	

Exercise 6.14 *Solve the following quadratic programming problems and also illustrate the solutions using graphical methods:*

(i) $$\min_{\boldsymbol{x} \text{ s.t.} \begin{cases} x_1+x_2\leqslant 3 \\ 4x_1+x_2\leqslant 9 \\ x_{1,2}\geqslant 0 \end{cases}} 2x_1^2-4x_1x_2+4x_2^2-6x_1-3x_2,$$

(ii) $$\min_{\boldsymbol{x} \text{ s.t.} \begin{cases} -x_1+x_2=1 \\ x_1+x_2\leqslant 2 \\ x_{1,2}\geqslant 0 \end{cases}} (x_1-1)^2+(x_2-2)^2.$$

Exercise 6.15 *Try to solve the following optimization problem* [19]*. Is there any chance to find high-precision solutions to the problem?*

$$\max_{x,y,z \text{ s.t.} \begin{cases} 8+5z^3x-4z^8y+3x^2y-xy^2=0 \\ 1-z^9-z^3x+y+3z^5xy+7x^2y+2xy^2=0 \\ -1-5z-5z^9x-5z^8y-2z^9xy+x^2y+4xy^2=0 \end{cases}} z.$$

Exercise 6.16 *Solve the constrained optimization problem* $\boldsymbol{q}$ *and* k [8]*.*

$$\min_{\boldsymbol{q},k \text{ s.t.} \begin{cases} g(\boldsymbol{q})\leqslant 0 \\ 800-800k\leqslant q_1\leqslant 800+800k \\ 4-2k\leqslant q_2\leqslant 4+2k \\ 6-3k\leqslant q_3\leqslant 6+3k \end{cases}} k,$$

where

$$\begin{aligned} g(\boldsymbol{q}) &= 10q_2^2q_3^3+10q_2^3q_3^2+200q_2^2q_3^2+100q_2^3q_3+q_1q_2q_3^2+q_1q_2^2q_3+1000q_2q_3^3 \\ &\quad +8q_1q_3^2+1000q_2^2q_3+8q_1q_2^2+6q_1q_2q_3-q_1^2+60q_1q_3+60q_1q_2-200q_1. \end{aligned}$$

Exercise 6.17 *Solve numerically the following nonlinear programming problems:*

$$\min_{\boldsymbol{x} \text{ s.t.} \begin{cases} x_1+x_2\leqslant 0 \\ -x_1x_2+x_1+x_2\geqslant 1.5 \\ x_1x_2\geqslant -10 \\ -10\leqslant x_1,x_2\leqslant 10 \end{cases}} \quad e^{x_1}(4x_1^2+2x_2^2+4x_1x_2+2x_2+1).$$

Exercise 6.18 *Solve the following nonlinear programming problems:*

(i) $$\min_{\boldsymbol{x} \text{ s.t.} \begin{cases} 41.67x_3(1-x_2^2/x_1^2)/x_1^3\leqslant 1 \\ 2.5(1-x_2^2/x_1^2)/x_1^3\leqslant 1 \\ 10-x_3\leqslant 0 \end{cases}} \quad 0.00613(x_1^2-x_2^2)x_3,$$

(ii) $$\min_{\boldsymbol{x} \text{ s.t.} \begin{cases} x_1+x_2+x_3\leqslant 40 \\ 5\leqslant x_1\leqslant 20, 3\leqslant x_2\leqslant 11 \\ 10\leqslant x_3\leqslant 40 \end{cases}} \quad x_1^2+3x_2^2+x_3^3.$$

Exercise 6.19 *Please try to find global optimal solution*

$$\min_{\boldsymbol{x} \text{ s.t.} \begin{cases} 0.003079x_1^3x_2^3x_5-\cos^3 x_6\geqslant 0 \\ 0.1017x_3^3x_4^3-x_5^2\cos^3 x_6\geqslant 0 \\ 0.09939(1+x_5)x_1^3x_2^2-\cos^2 x_6\geqslant 0 \\ 0.1076(31.5+x_5)x_3^3x_4^2-x_5^2\cos^2 x_6\geqslant 0 \\ x_3x_4(x_5+31.5)-x_5[2(x_1+5)\cos x_6+x_1x_2x_5]\geqslant 0 \\ 0.2\leqslant x_1\leqslant 0.5, 14\leqslant x_2\leqslant 22, 0.35\leqslant x_3\leqslant 0.6 \\ 16\leqslant x_4\leqslant 22, 5.8\leqslant x_5\leqslant 6.5, 0.14\leqslant x_6\leqslant 0.2618 \end{cases}} \quad \frac{1}{2\cos x_6}\left[x_1x_2(1+x_5)+x_3x_4\left(1+\frac{31.5}{x_5}\right)\right].$$

Exercise 6.20 *Solve the following integer linear programming problems:*

(i) $$\max_{\boldsymbol{x} \text{ s.t.} \begin{cases} \boldsymbol{x}\geqslant 0 \\ 3534x_1+2356x_2+1767x_3+589x_4+528x_5+451x_6+304x_7\leqslant 119567 \end{cases}} \quad 592x_1+381x_2+273x_3+55x_4+48x_5+37x_6+23x_7,$$

(ii) $$\max_{\boldsymbol{x} \text{ s.t.} \begin{cases} x_1+x_2+x_3=30 \\ x_4+x_5+x_6=18 \\ x_1+x_4=10 \\ x_2+x_5\leqslant 18 \\ x_3+x_6\geqslant 30 \\ x_{1,\cdots,6}\geqslant 0 \end{cases}} \quad 120x_1+66x_2+72x_3+58x_4+132x_5+104x_6.$$

Exercise 6.21 *Solve the following nonlinear integer programming problems* [9], *and validate the results with enumerate method.*

(i) $$\min_{\boldsymbol{x} \text{ s.t. } 12\leqslant x_i\leqslant 32} \quad \left(\frac{1}{6.931}-\frac{x_2x_3}{x_1x_4}\right)^2,$$

(ii) $$\min_{\boldsymbol{x} \text{ s.t.} \begin{cases} -2x_1^2-3x_2^4-x_3-4x_4^2-5x_5+127\geqslant 0 \\ 7x_1-3x_2-10x_3^2-x_4+x_5+282\geqslant 0 \\ 23x_1-x_2^2-6x_6^2+8x_7+196\geqslant 0 \\ -4x_1^2-x_2^2+3x_1x_2-2x_3^2-5x_6+11x_7\geqslant 0 \end{cases}} \quad (x_1-10)^2+5(x_2-12)^2+x_3^4+3(x_4-11)^2+10x_5^6+7x_6^2+x_7^4-10x_6-8x_7.$$

Exercise 6.22 *Solve the following binary linear programming problems and verify the results in problems (i) and (ii) using the enumerate methods.*

(i)
$$\min_{\boldsymbol{x} \text{ s.t. } \begin{cases} x_1-x_2+5x_3+x_4-4x_5\geqslant 2 \\ -2x_1+6x_2-3x_3-2x_4+2x_5\geqslant 0 \\ -2x_2+2x_3-x_4-x_5\leqslant 1 \end{cases}} \quad 5x_1 + 7x_2 + 10x_3 + 3x_4 + x_5,$$

(ii)
$$\min_{\boldsymbol{x} \text{ s.t. } \begin{cases} x_1-x_6\leqslant 0 \\ x_1-x_5\leqslant 0 \\ x_2-x_4\leqslant 0 \\ x_2-x_5\leqslant 0 \\ x_3-x_4\leqslant 0 \\ x_1+x_2+x_3\leqslant 2 \end{cases}} \quad -3x_1 - 4x_2 - 5x_3 + 4x_4 + 4x_5 + 2x_6.$$

Exercise 6.23 *Solve the following binary linear programming problem:*

$$\max_{\boldsymbol{x} \text{ s.t. } \boldsymbol{A}\boldsymbol{x}\leqslant \begin{bmatrix} 600 \\ 600 \end{bmatrix}} \quad -\boldsymbol{f}\boldsymbol{x},$$

where

$$\boldsymbol{A} = \left[\begin{matrix} 45 & 0 & 85 & 150 & 65 & 95 & 30 & 0 & 170 & 0 & 40 & 25 & 20 & 0 \\ 30 & 20 & 125 & 5 & 80 & 25 & 35 & 73 & 12 & 15 & 15 & 40 & 5 & 10 \end{matrix}\right.$$
$$\left.\begin{matrix} 0 & 25 & 0 & 0 & 25 & 0 & 165 & 0 & 85 & 0 & 0 & 0 & 0 & 100 \\ 10 & 12 & 10 & 9 & 0 & 20 & 60 & 40 & 50 & 36 & 49 & 40 & 19 & 150 \end{matrix}\right],$$

$$\boldsymbol{f} = [1898, 440, 22507, 270, 14148, 3100, 4650, 30800, 615, 4975, 1160, 4225,$$
$$510, 11880, 479, 440, 490, 330, 110, 560, 24355, 2885, 11748, 4550,$$
$$750, 3720, 1950, 10500].$$

Exercise 6.24 *Solve the following optimization problem using Robust Control Toolbox and YALIMP:*

$$\min_{\boldsymbol{X} \text{ s.t. } \begin{cases} \begin{bmatrix} \boldsymbol{A}^{\mathrm{T}}\boldsymbol{X}+\boldsymbol{X}\boldsymbol{A}+\boldsymbol{Q} & \boldsymbol{X}\boldsymbol{B} \\ \boldsymbol{B}^{\mathrm{T}}\boldsymbol{X} & -\boldsymbol{I} \end{bmatrix} < 0 \\ \boldsymbol{X}<0 \end{cases}} \quad \mathrm{tr}(\boldsymbol{X}),$$

where $\boldsymbol{A} = \begin{bmatrix} -1 & -2 & 1 \\ 3 & 2 & 1 \\ 1 & -2 & -1 \end{bmatrix}$, $\boldsymbol{B} = \begin{bmatrix} 1 \\ 0 \\ 1 \end{bmatrix}$, $\boldsymbol{Q} = \begin{bmatrix} 1 & -1 & 0 \\ -1 & -3 & -12 \\ 0 & -12 & -36 \end{bmatrix}$.

Exercise 6.25 *Solve the following linear matrix inequalities*

$$\begin{cases} \boldsymbol{P}^{-1} > \boldsymbol{0}, \quad \textit{or equavelently} \ \ \boldsymbol{P} > \boldsymbol{0} \\ \boldsymbol{A}_1\boldsymbol{P} + \boldsymbol{P}\boldsymbol{A}_1^{\mathrm{T}} + \boldsymbol{B}_1\boldsymbol{Y} + \boldsymbol{Y}^{\mathrm{T}}\boldsymbol{B}_1^{\mathrm{T}} < \boldsymbol{0} \\ \boldsymbol{A}_2\boldsymbol{P} + \boldsymbol{P}\boldsymbol{A}_2^{\mathrm{T}} + \boldsymbol{B}_2\boldsymbol{Y} + \boldsymbol{Y}^{\mathrm{T}}\boldsymbol{B}_2^{\mathrm{T}} < \boldsymbol{0}, \end{cases}$$

where

$$\boldsymbol{A}_1 = \begin{bmatrix} -1 & 2 & -2 \\ -1 & -2 & 1 \\ -1 & -1 & 0 \end{bmatrix}, \ \boldsymbol{B}_1 = \begin{bmatrix} -2 \\ 1 \\ -1 \end{bmatrix}, \ \boldsymbol{A}_2 = \begin{bmatrix} 0 & 2 & 2 \\ 2 & 0 & 2 \\ 2 & 0 & 1 \end{bmatrix}, \ \boldsymbol{B}_2 = \begin{bmatrix} -1 \\ -2 \\ -1 \end{bmatrix}.$$

Exercise 6.26 *Solve the following unconstrained multi-objective minimization problem, with*

$$f_1(\boldsymbol{x}) = x_1,\ f_2(\boldsymbol{x}, \boldsymbol{z}) = g(\boldsymbol{z})h(f_1(\boldsymbol{x}), g(\boldsymbol{z})),\ p = 20,$$

where

$$g(\boldsymbol{z}) = 1 + \sum_{i=1}^{p} \frac{z_i}{p},\ \textit{and}\ h(f_1(\boldsymbol{x}), g(\boldsymbol{z})) = 1 - \sqrt{f_1(\boldsymbol{x})/g(\boldsymbol{z})}.$$

Exercise 6.27 *Find the best compromise solutions for the following multi-objective linear programming problems:*

(i)

$$\begin{aligned} \max\quad & z_1 = 100x_1 + 90x_2 + 80x_3 + 70x_4 \\ \min\quad & z_2 = 3x_2 + 2x_4, \\ \boldsymbol{x}\ \text{s.t.} & \begin{cases} x_1+x_2 \geqslant 30 \\ x_3+x_4 \geqslant 30 \\ 3x_1+2x_2 \leqslant 120 \\ 3x_2+2x_4 \leqslant 48 \\ x_1,x_2,x_3,x_4 \geqslant 0 \end{cases} \end{aligned}$$

(ii)

$$\max \begin{bmatrix} 50x_1 + 20x_2 + 100x_3 + 60x_4 \\ 20x_1 + 70x_2 + 5x_3 \\ 3x_2 + 5x_4 \\ 2x_1 + 20x_3 + 2x_4 \end{bmatrix}.$$

$$\boldsymbol{x}\ \text{s.t.} \begin{cases} 2x_1+5x_2+10x_3 \leqslant 100 \\ x_1+6x_2+8x_4 \leqslant 250 \\ 5x_1+8x_2+7x_3+10x_4 \leqslant 350 \\ x_1,x_2,x_3,x_4 \geqslant 0 \end{cases}$$

Exercise 6.28 *Find the shortest path from node A to node B, in the graphs given in Figure 6.20 (a) and (b).*

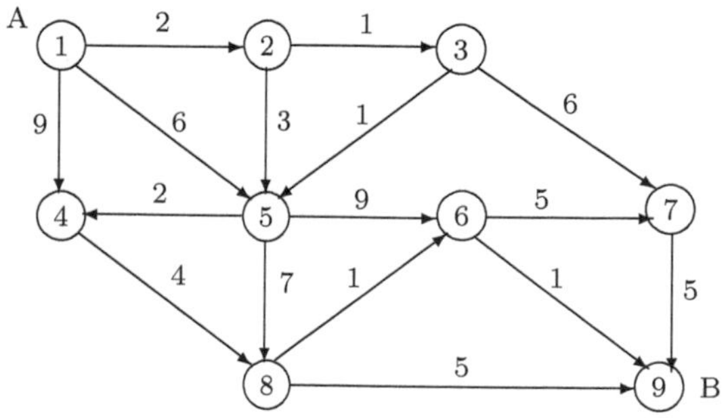

(a) oriented graph

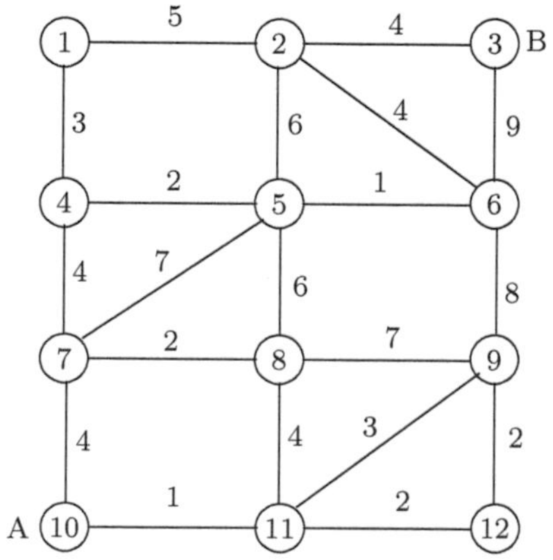

(b) undigram

FIGURE 6.20: Graphs of the shortest path problems.

Exercise 6.29 *Assume that a factory needs to import a machine from an overseas manufacturer. There are three ports of exit to select from the manufacturer, and three ports of entry to select. The machine can then be transported to the factory via one of the two cities. The transportation costs are shown in Figure 6.21. Please find the route with minimized total transportation cost.*

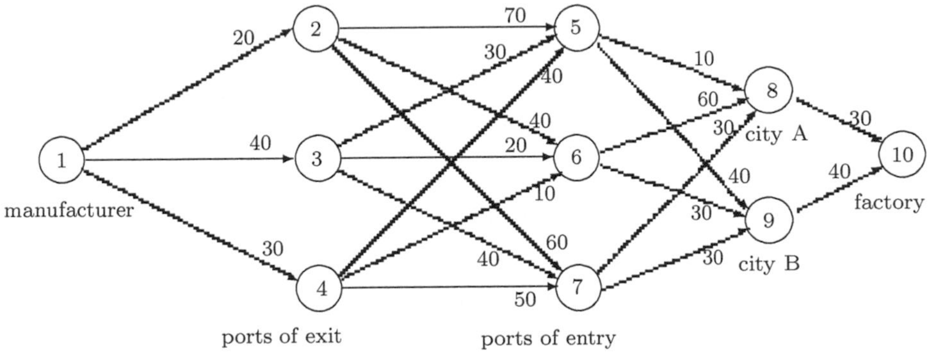

FIGURE 6.21: Roadmap and transportation costs.

Exercise 6.30 *Assume that a person usually lives in city C_1. He often needs to visit other cities $C_2, \cdots, C_8$. The travel costs from city C_i to city C_j are given in the incidence matrix $R_{i,j}$. Please design the cheapest road map for him to travel from city C_1 to other cities.*

$$\boldsymbol{R} = \begin{bmatrix} 0 & 364 & 314 & 334 & 330 & \infty & 253 & 287 \\ 364 & 0 & 396 & 366 & 351 & 267 & 454 & 581 \\ 314 & 396 & 0 & 232 & 332 & 247 & 159 & 250 \\ 334 & 300 & 232 & 0 & 470 & 50 & 57 & \infty \\ 330 & 351 & 332 & 470 & 0 & 252 & 273 & 156 \\ \infty & 267 & 247 & 50 & 252 & 0 & \infty & 198 \\ 253 & 454 & 159 & 57 & 273 & \infty & 0 & 48 \\ 260 & 581 & 220 & \infty & 156 & 198 & 48 & 0 \end{bmatrix}.$$

Bibliography

[1] Press W H, Flannery B P, Teukolsky S A, et al. Numerical recipes, the art of scientific computing. Cambridge: Cambridge University Press, 1986. Free textbook at `http://www.nrbook.com/a/bookcpdf.php`

[2] Boyd S, Vandenberghe L. Convex optimization. Cambridge University Press, 2004. Free textbook at `http://www.stanford.edu/~boyd/cvxbook/bv_cvxbook.pdf`

[3] Boyd S, Ghaoui L El, Feron E, et al. Linear matrix inequalities in systems and control theory. Philadelphia: SIAM books, Volume 15 of Studies in Applied Mathematics. Free textbook at `http://www.stanford.edu/~boyd/lmibook/lmibook.pdf`, 1994

[4] Mittelmann H D. Decision tree for optimization software. `http://plato.asu.edu/guide.html`, 2007

[5] Nelder J A, Mead R. A simplex method for function minimization. Computer Journal, 1965, 7:308–313

[6] Goldberg D E. Genetic algorithms in search, optimization and machine learning. Reading, MA: Addison-Wesley, 1989

[7] Hillier F S, Lieberman G J. Introduction to operations research, 10th edition. New York: McGraw-Hill Education, 2015

[8] Henrion D. A review of the global optimization toolbox for Maple, 2006. `http://www.laas.fr/~henrion/Papers/mapleglobopt.pdf`

[9] Leyffer S. Deterministic methods for mixed integer nonlinear programming. Ph.D. thesis, Department of Mathematics & Computer Science, University of Dundee, U.K., 1993

[10] Cha J Z, Mayne R W. Optimization with discrete variables via recursive quadratic programming: Part 2 algorithms and results. Transactions of the ASME, Journal of Mechanisms, Transmissions, and Automation in Design, 1994, 111:130–136

[11] Willems J C. Least squares stationary optimal control and the algebraic Riccati equation. IEEE Transactions on Automatic Control, 1971, 16(6):621~634

[12] The MathWorks Inc. Robust control toolbox user's manual, 2007

[13] Löfberg J. YALMIP: a toolbox for modeling and optimization in MATLAB. Proceedings of IEEE International Symposium on Computer Aided Control Systems Design. Taipei, 2004, 284~289

[14] Gao L F. Optimization theory and methods. Shenyang: Northeastern University Press, 2005. (in Chinese)

[15] Cao Y. Pareto set. MATLAB Central File ID: # 15181, 2007

[16] Bellman R. Dynamic programming. Princeton, NJ: Princeton University Press, 1957

[17] Lin Y X. Dynamic programming and sequential optimization. Zhengzhou: Henan University Press, 1997. (in Chinese)

[18] Dijkstra E W. A note on two problems in connexion with graphs. Numerische Mathematik, 1959, 1:269–271

[19] Sturmfels B. Solving systems of polynomial equations. CBMS Conference on Solving Polynomial Equations, Held at Texas A & M University, American Mathematical Society, 2002

[20] Chipperfield A, Fleming P. Genetic algorithm toolbox user's guide. Department of Automatic Control and Systems Engineering, University of Sheffield, 1994

[21] Ackley D H. A connectionist machine for genetic hillclimbing. Boston, USA: Kluwer Academic Publishers, 1987

Chapter 7

Differential Equation Problems

Differential equations include ordinary differential equations (ODEs) and partial differential equations (PDEs). It is interesting to note that the phrase "extraordinary differential equations" is sometimes used to indicate differential equations of non-integer orders based on fractional calculus, which is covered in Section 10.6 of the last chapter. In this chapter, we focus on integer order differential equations.

Before Isaac Newton and Gottfried Wilhelm Leibniz invented calculus, people characterized the nature simply by geometry and algebra. With calculus, a dynamic system view of the nature around us becomes possible. Today, differential equations are the most widely used mathematical tools for describing dynamic systems evolving either along time or along spatial variables. Differential equations also provide a solid foundation for mathematical modeling in many scientific and engineering disciplines. Linear differential equations and a few special low-order nonlinear differential equations may have analytical solutions. However, generally speaking, most nonlinear equations have no analytical or close form solutions. Thus, numerical techniques should be adopted for solving these equations. In Section 7.1, analytical solutions to a special class of ordinary differential equations (ODEs) will be explored. It will be explained how linear time-invariant (LTI) differential equations can be solved in MATLAB. Moreover, analytical solutions to a very special first-order nonlinear differential equation is presented. For linear state space equations the analytical solutions can be obtained by vector integration methods. In Section 7.2, numerical algorithms for first-order explicit differential equations will be presented, with an illustrative MATLAB implementation. The conversion of various different types of differential equations, including matrix differential equations, into first-order explicit ones will be discussed. Most interestingly, numerical solutions to several different types of special differential equations will be discussed in Section 7.4, which include stiff differential equations, differential algebraic equations (DAEs), implicit differential equations as well as switching differential equations, stochastic differential equations. In Sections 7.5 to 7.7, delay differential equations (DDE), boundary value problems (BVPs) of ordinary differential equations and partial differential equations (PDEs) will be discussed respectively with extensive demonstrative examples. Section 7.8 briefly discusses the modeling and solution of differential equations using Simulink environment with different kinds of examples. Solutions to the Simulink models are obtained by automatic simulation methods based on Simulink's extensive libraries of building blocks. We demonstrate that, theoretically speaking, ordinary differential equations of almost any complexity can be solved numerically in Simulink.

For readers who wish to check the detailed explanations of various aspects of differential equations, we recommend the free textbooks [1] (Chapters 17, 18 and 20) and [2]. The open source free textbook [3] is also an excellent resource for learning differential equations (Chapters 14-34 for ordinary differential equations and Chapters 35-47 for partial differential equations.)

7.1 Analytical Solution Methods for Some Ordinary Differential Equations

7.1.1 Linear time-invariant ordinary differential equations

The general description of the linear time-invariant (LTI) ordinary differential equations is given by

$$\begin{aligned}\frac{\mathrm{d}^n y(t)}{\mathrm{d}t^n}+a_1\frac{\mathrm{d}^{n-1}y(t)}{\mathrm{d}t^{n-1}}+a_2\frac{\mathrm{d}^{n-2}y(t)}{\mathrm{d}t^{n-2}}+\cdots+a_{n-1}\frac{\mathrm{d}y(t)}{\mathrm{d}t}+a_n y(t)=\\ b_n\frac{\mathrm{d}^n u(t)}{\mathrm{d}t^n}+b_{n-1}\frac{\mathrm{d}^{n-1}u(t)}{\mathrm{d}t^{n-1}}+\cdots+b_1\frac{\mathrm{d}u(t)}{\mathrm{d}t}+b_0u(t),\end{aligned} \tag{7-1-1}$$

where a_i, b_i are constants. For practical reasons, the derivative order in the right-hand-side should be less than n for the system to be strictly proper. In this case, b_n, b_{n-1} and so on can be set to zeros.

If the right-hand-side of the equation is 0, the equation is referred to as a *homogeneous ordinary differential equation.*

From the properties of Laplace transform introduced in Section 5.1, for zero initial condition problems, one has

$$\mathscr{L}\left[\frac{\mathrm{d}^m y(t)}{\mathrm{d}t^m}\right]=s^m\mathscr{L}[y(t)]=s^mY(s),\ \mathscr{L}\left[\frac{\mathrm{d}^m u(t)}{\mathrm{d}t^m}\right]=s^m\mathscr{L}[u(t)]=s^mU(s).$$

Therefore, the above linear time-invariant differential equation can be mapped into the following algebraic polynomial equation:

$$Y(s)Q(s)=U(s)P(s), \tag{7-1-2}$$

where $Q(s)=s^n+a_1s^{n-1}+a_2s^{n-2}+\cdots+a_{n-1}s+a_n$ and $P(s)=b_ns^n+b_{n-1}s^{n-1}+\cdots+b_1s+b_0$. Note that, if initial conditions are not all zeros, extra attention should be paid during the conversion. The neat form of (7-1-2) with the same coefficients as (7-1-1) is still possible. However, by rearranging the terms, we can get

$$Y(s)Q(s)=Y_0(s)+U_0(s)+U(s)P(s), \tag{7-1-3}$$

where $Y_0(s)$ and $U_0(s)$ are due to nonzero initial conditions in signals $y(t)$ and $u(t)$, respectively.

The following polynomial equation

$$Q(s)=s^n+a_1s^{n-1}+a_2s^{n-2}+\cdots+a_{n-1}s+a_n=0 \tag{7-1-4}$$

is referred to as the *characteristic equation.* If all the roots r_i in the characteristic equation can be obtained and they are all distinct, the general form of the analytical solution to the corresponding ordinary differential equation can be written as

$$y(t)=y_1(t)+y_0(t), \tag{7-1-5}$$

where $y_0(t)$ is a particular solution of the original equation, and $y_1(t)$ is the general solution of the homogeneous equation, which can be expressed as

$$y_1(t)=C_1\mathrm{e}^{r_1t}+C_2\mathrm{e}^{r_2t}+\cdots+C_n\mathrm{e}^{r_nt}, \tag{7-1-6}$$

where C_i's are the undetermined constants related to initial conditions. When r_j is a repeated root with multiplicity m_j, the term $C_j e^{r_j t}$ should be replaced by

$$\sum_{k=0}^{m_j-1} C_{jk} t^k e^{r_j t}. \tag{7-1-7}$$

In general, assume that there are k distinct roots r_j, $j = 1, 2, \cdots, k$, and r_j has its multiplicity m_j with $m_j \geqslant 1$ and $\sum_{j=1}^{k} m_j = n$. Then, the general solution of the homogeneous differential equation can be written as

$$y_1(t) = \sum_{j=1}^{k} \sum_{k=0}^{m_j-1} C_{jk} t^k e^{r_j t}. \tag{7-1-8}$$

Let us investigate (7-1-2). From the well-known Abel–Ruffini Theorem, it is known that the polynomial equation with degree less than or equal to 4 has analytical solutions. Thus, it can be concluded that low-order linear time-invariant ordinary differential equations have analytical solutions. For higher-order equations, one may combine the numerical and analytical approaches to find quasi-analytical solutions with high accuracy. In this section, symbolic math-based analytical solution approaches will be presented first.

7.1.2 Analytical solution with MATLAB

The function `dsolve()` provided in the Symbolic Math Toolbox in MATLAB can be used to symbolically solve a class of ordinary differential equations with mixed initial and boundary conditions. The syntaxes of the function are

```
y=dsolve(fun1,fun2,···,funm)          % default independent variable t
y=dsolve(fun1,fun2,···,funm,'x')      % independent variable x assigned
```

where the string variables *fun*$_i$ can be used to describe not only differential equations, but also initial and boundary conditions. When describing differential equations in `dsolve()`, the symbol `D4y` is used to denote $y^{(4)}(t)$. One can also use `D2y(2)=3` to denote given condition as $y''(2) = 3$. With `dsolve()`, the analytical solutions to a class of ordinary differential equations can easily be found. If the independent variable in the differential equations is x rather than t, this should be declared in the function call statement.

Alternatively, the differential equations *eqn*$_i$ can also be described by symbolic expressions. This kind of descriptions are suitable for validations. Unfortunately, this kind of description are not quite suitable for describing equations with high-order boundary conditions.

Example 7.1 Let the input signal be defined by $u(t) = e^{-5t}\cos(2t + 1) + 5$. Find the general solution to the following ordinary differential equation

$$y^{(4)}(t) + 10y^{(3)}(t) + 35y''(t) + 50y'(t) + 24y(t) = 5u''(t) + 4u'(t) + 2u(t).$$

Solution *To solve the original differential equation using* `dsolve()`*, the right-hand-side of the equation should be evaluated first given* $u(t)$*. Then, the original equation can be solved using the following statements*

```
>> syms t; u=exp(-5*t)*cos(2*t+1)+5; % declare symbolic variable and input
   uu=5*diff(u,t,2)+4*diff(u,t)+2*u; % evaluate the right-hand side of ODE
   y=dsolve(['D4y+10*D3y+35*D2y+50*Dy+24*y=',char(uu)]) % solve directly
```

where in the above statements, the variable uu *is a symbolic expression. Since the differential equation should be expressed in a string,* char() *function is used to convert it, and square brackets are used to join the left- and right-hand sides of the equation together to form a complete string description. The final solution to this differential equation is found as*

$$y(t)=\frac{5}{12}-\frac{343}{520}\mathrm{e}^{-5t}\cos(2t+1)-\frac{547}{520}\mathrm{e}^{-5t}\sin(2t+1)+\mathrm{C}_1\mathrm{e}^{-4t}+\mathrm{C}_2\mathrm{e}^{-3t}+\mathrm{C}_3\mathrm{e}^{-2t}+\mathrm{C}_4\mathrm{e}^{-t},$$

where C_i*'s are undetermined constants. Given 4 independent initial or boundary conditions, the constants* C_i *can then be solved uniquely.*

If the equations are described by symbolic expressions, the following MATLAB commands can be used instead, and the same results can be found.

```
>> syms y(t); % describe the ODE in an symbolic expression
   y0=dsolve(diff(y,4)+10*diff(y,3)+35*diff(y,2)+50*diff(y)+24*y==uu)
```

Now let us assume that the following conditions are given $y(0)=3, y'(0)=2,\ y''(0)=y^{(3)}(0)=0$*. Use the following statements to find the solution to the original ordinary differential equation.*

```
>> y=dsolve(['D4y+10*D3y+35*D2y+50*Dy+24*y=',char(uu)],'y(0)=3',...
      'Dy(0)=2','D2y(0)=0','D3y(0)=0') % specify initial conditions
```

which is

$$\begin{aligned}y(t) &= \frac{5}{12}-\frac{343}{520}\mathrm{e}^{-5t}\cos\left(2t+1\right)-\frac{547}{520}\mathrm{e}^{-5t}\sin(2t+1)\\ &\quad+\left(-\frac{445}{26}\cos 1-\frac{51}{13}\sin 1-\frac{69}{2}\right)\mathrm{e}^{-2t}+\left(-\frac{271}{30}\cos 1+\frac{41}{15}\sin 1-\frac{25}{4}\right)\mathrm{e}^{-4t}\\ &\quad+\left(\frac{179}{8}\cos 1+\frac{5}{8}\sin 1+\frac{73}{3}\right)\mathrm{e}^{-3t}+\left(\frac{133}{30}\cos 1+\frac{97}{60}\sin 1+19\right)\mathrm{e}^{-t}.\end{aligned}$$

The solution can be validated with the following statements

```
>> simplify(diff(y,4)+10*diff(y,3)+35*diff(y,2)+50*diff(y)+24*y-uu)
   for i=0:3, subs(diff(y,i),t,0), end % validations of the solution
```

To describe the differential equations with symbolic expressions, extra signals should be introduced, and the following statements can be used, and the same results can be obtained. Please note that, the definitions of the extra signals Dy, D2y *and* D3y *cannot be omitted.*

```
>> syms t y(t); Dy=diff(y); D2y=diff(y,2); D3y=diff(y,3);
   u=exp(-5*t)*cos(2*t+1)+5; uu=5*diff(u,t,2)+4*diff(u,t)+2*u;
   y1=dsolve(diff(y,4)+10*diff(y,3)+35*diff(y,2)+50*diff(y)+24*y==uu,...
             y(0)==3,Dy(0)==2,D2y(0)==0,D3y(0)==0)
```

With the powerful Symbolic Math Toolbox, solutions to some seemingly impossible-to-solve ordinary differential equations can be obtained easily. For instance, let $y(0)=1/2, y'(\pi)=1,\ y''(2\pi)=0, y'(2\pi)=1/5$*, the analytical solution to this differential equation can be obtained*

```
>> y=dsolve(['D4y+10*D3y+35*D2y+50*Dy+24*y=',char(uu)],'y(0)=1/2',...
     'Dy(pi)=1','D2y(2*pi)=0','Dy(2*pi)=1/5')
```

It is possible to display the analytical form of the undetermined constants C_i *but each coefficient will take about 10 lines at least. In this situation, a fairly accurate approximate representation to these terms can be used such that the quasi-analytical solution to the equation can be found with* `vpa(ans)`*, where the solution obtained above can be expressed by*

$$y(t) = \frac{5}{12} - \frac{343}{520}\mathrm{e}^{-5t}\cos(2t+1) - \frac{547}{520}\mathrm{e}^{-5t}\sin(2t+1) - 219.1291604\mathrm{e}^{-t} + 442590.9052\mathrm{e}^{-4t} + 31319.63786\mathrm{e}^{-2t} - 473690.0889\mathrm{e}^{-3t}.$$

Example 7.2 The above differential equation contains only real poles. In fact, in the `dsolve()` function, differential equations containing complex poles can also be solved. Now consider the following equation

$$y^{(5)}(t) + 5y^{(4)}(t) + 12y^{(3)}(t) + 16y''(t) + 12y'(t) + 4y(t) = 3u'(t) + 3u(t),$$

where the input signal $u(t) = \sin t$, and $y(0) = y'(0) = y''(0) = y^{(3)}(0) = y^{(4)}(0) = 0$. Try to use the analytical solution approach for this equation.

Solution *One can solve the differential equation using the following statements*

```
>> syms t; u=sin(t); uu=3*diff(u)+3*u; % compute right-hand side
   y=dsolve(['D5y+5*D4y+12*D3y+16*D2y+12*Dy+4*y=' char(uu)],...
     'y(0)=0','Dy(0)=0','D2y(0)=0','D3y(0)=0','D4y(0)=0')
   simplify(y) % solve linear time-invariant ODE with repeated complex poles
```

The analytical solution can then be found

$$y(t) = -\frac{2}{5}\sin t - \frac{1}{5}\cos t + \mathrm{e}^{-t} + \frac{11}{10}\mathrm{e}^{-t}\sin t - \frac{4}{5}\mathrm{e}^{-t}\cos t - \frac{1}{2}\mathrm{e}^{-t}t\cos t.$$

Example 7.3 Find the analytical solution to the following simultaneous ordinary differential equations:

$$\begin{cases} x''(t) + 2x'(t) = x(t) + 2y(t) - \mathrm{e}^{-t} \\ y'(t) = 4x(t) + 3y(t) + 4\mathrm{e}^{-t}. \end{cases}$$

Solution *The linear differential equation set can also be solved directly using* `dsolve()` *function. For instance, the above equation can be solved with*

```
>> [x,y]=dsolve('D2x+2*Dx=x+2*y-exp(-t)','Dy=4*x+3*y+4*exp(-t)')
```

and it can be found that the solutions are

$$\begin{cases} x(t) = -6t\mathrm{e}^{-t} + C_1\mathrm{e}^{-t} + C_2\mathrm{e}^{(1+\sqrt{6})t} + C_3\mathrm{e}^{-(-1+\sqrt{6})t} \\ y(t) = 6t\mathrm{e}^{-t} - C_1\mathrm{e}^{-t} + 2\left(2+\sqrt{6}\right)C_2\mathrm{e}^{(1+\sqrt{6})t} + 2\left(2-\sqrt{6}\right)C_3\mathrm{e}^{-(-1+\sqrt{6})t} + \frac{1}{2}\mathrm{e}^{-t}. \end{cases}$$

Example 7.4 Solve the following high-order linear differential equations

$$\begin{cases} x'' - x + y + z = 0, \\ x + y'' - y + z = 0, \\ x + y + z'' - z = 0, \end{cases} \quad x(0) = 1, y(0) = z(0) = x'(0) = y'(0) = z'(0) = 0.$$

Solution *With the following statements, the differential equations can be solved directly with*

```
>> [x,y,z]=dsolve('D2x-x+y+z=0','x+D2y-y+z=0','x+y+D2z-z=0',...
        'x(0)=1, y(0)=0, z(0)=0','Dx(0)=0, Dy(0)=0, Dz(0)=0')
```

and the solutions of the equations obtained are

$$x(t)=\frac{\mathrm{e}^{\sqrt{2}t}}{3}+\frac{\mathrm{e}^{-\sqrt{2}t}}{3}+\frac{\cos t}{3},\ y(t)=\frac{\cos t}{3}-\frac{\mathrm{e}^{-\sqrt{2}t}}{6}-\frac{\mathrm{e}^{\sqrt{2}t}}{6},\ z(t)=\frac{\cos t}{3}-\frac{\mathrm{e}^{-\sqrt{2}t}}{6}-\frac{\mathrm{e}^{\sqrt{2}t}}{6}.$$

Example 7.5 Consider the time-varying differential equation given by

$$x^2(2x-1)\frac{\mathrm{d}^3y}{\mathrm{d}t^3}-(4x-3)x\frac{\mathrm{d}^2y}{\mathrm{d}x^2}-2x\frac{\mathrm{d}y}{\mathrm{d}x}+2y=0.$$

Solution *It seems that the equation can be solved easily with*

```
>> syms x; y=dsolve('x^2*(2*x-1)*D3y-(4*x-3)*x*D2y-2*x*Dy+2*y=0')
```

It should be noted that in the original equation, the independent variable is x rather than the default t*, thus, the equation cannot be solved correctly with the above statement. To solve the equation, the notation* `'x'` *should be used in the function call, otherwise, the solution obtained is wrong. The equation can be solved with*

```
>> y=simplify(dsolve('x^2*(2*x-1)*D3y+(4*x-3)*x*D2y-2*x*Dy+2*y=0','x'))
   simplify(x^2*(2*x-1)*diff(y,3)+(4*x-3)*x*diff(y,2)-2*x*diff(y)+2*y)
```

The solution of the differential equation is as follows. Substituting the results back to the equation, it can be seen that the error is 0, which means that the result y is the solution of the original equation.

$$y(x)=-\frac{1}{16x}(2\mathrm{C}_1-\mathrm{C}_3+8\mathrm{C}_3x+32\mathrm{C}_2x^2+8\mathrm{C}_3x^2\ln x).$$

If recent versions of MATLAB are used, the differential equation can alternatively be specified as a symbolic expression. In this case, there is no need to declare the variable x again in the function call. The same results can be obtained.

```
>> syms x y(x); % an alternative way in describing and solution of ODE
   y0=dsolve(x^2*(2*x-1)*diff(y,3)+(4*x-3)*x*diff(y,2)-2*x*diff(y)+2*y)
   simplify(y0) % simplify the results
```

7.1.3 Analytical solutions of linear state space equations

Assume that the linear time-invariant state space model is given by

$$\begin{cases} \boldsymbol{x}'(t)=\boldsymbol{A}\boldsymbol{x}(t)+\boldsymbol{B}\boldsymbol{u}(t) \\ \boldsymbol{y}(t)=\boldsymbol{C}\boldsymbol{x}(t)+\boldsymbol{D}\boldsymbol{u}(t), \end{cases} \tag{7-1-9}$$

where, $\boldsymbol{A},\boldsymbol{B},\boldsymbol{C}$ and $\boldsymbol{D}$ are constant matrices, and the initial state vector $\boldsymbol{x}_0$ is given. The analytical solution can be written mathematically as

$$\boldsymbol{x}(t)=\mathrm{e}^{\boldsymbol{A}(t-t_0)}\boldsymbol{x}(t_0)+\int_{t_0}^{t}\mathrm{e}^{\boldsymbol{A}(t-\tau)}\boldsymbol{B}\boldsymbol{u}(\tau)\,\mathrm{d}\tau. \tag{7-1-10}$$

It can be seen that the input signal $u(t)$ is a function of t, while in the above equation, the function of τ is expected, the MATLAB function subs() can be used in variable substitution. Also, the exponential of matrices and definite integrals are involved; these can be processed with expm() and int() functions. Thus, the solution can be obtained with

x = expm(A*t)*x_0 + int(expm(A*(t-τ))*B*subs(u,t,τ),τ,0,t)

An example is given to show how to get the analytical solutions of the equations, with the direct use of appropriate MATLAB functions.

Example 7.6 Assume that the input signal is $u(t) = 2 + 2e^{-3t}\sin 2t$, and the matrices in the state space equation are given below. Find the analytical solution.

$$A = \begin{bmatrix} -19 & -16 & -16 & -19 \\ 21 & 16 & 17 & 19 \\ 20 & 17 & 16 & 20 \\ -20 & -16 & -16 & -19 \end{bmatrix},\ B = \begin{bmatrix} 1 \\ 0 \\ 1 \\ 2 \end{bmatrix},\ C^{\mathrm{T}} = \begin{bmatrix} 2 \\ 1 \\ 0 \\ 0 \end{bmatrix},\ D = 0,\ x_0 = \begin{bmatrix} 0 \\ 1 \\ 1 \\ 2 \end{bmatrix}.$$

Solution *With the direct used of* (7-1-10), *the analytical solution of the equation can be obtained with the following statements*

```
>> syms t tau; u=2+2*exp(-3*t)*sin(2*t); % declare symbolic variables/input
   A=[-19,-16,-16,-19; 21,16,17,19; 20,17,16,20; -20,-16,-16,-19];
   B=[1; 0; 1; 2]; C=[2 1 0 0]; x0=[0; 1; 1; 2]; % enter matrices
   x=expm(A*t)*x0+int(expm(A*(t-tau))*B*subs(u,t,tau),tau,0,t);
   y=simplify(C*x) % compute the output signal
```

The analytical solution obtained is

$$y(t) = \frac{119}{8}e^{-t} + 57e^{-3t} + \frac{127t}{4}e^{-t} + 4t^2e^{-t} - \frac{135}{8}e^{-3t}\cos 2t + \frac{77}{4}e^{-3t}\sin 2t - 54.$$

With recent versions of MATLAB, the matrix differential equations can alternatively be solved with the following statements, and the same results can be obtained

```
>> syms x1(t) x2(t) x3(t) x4(t); X=[x1; x2; x3; x4]; % alternatively
   R=dsolve(diff(X)==A*X+B*u,X(0)==x0); y=C*[R.x1; R.x2; R.x3; R.x4]
```

7.1.4 Analytical solutions to special nonlinear differential equations

Very few nonlinear differential equations have analytical solutions. If there are, the solutions can be obtained with the dsolve() function. An example for the analytical solution to a special nonlinear differential equation is demonstrated. Another example is also presented where the analytical solution is not possible.

Example 7.7 Find the analytical solution to the first-order nonlinear differential equation $x'(t) = x(t)(1 - x^2(t))$.

Solution *Such a simple nonlinear differential equation can be solved analytically using the function* dsolve().

```
>> syms x; x=dsolve('Dx=x*(1-x^2)') % solve directly
```

Therefore, the analytical solutions are $x(t) = \sqrt{\dfrac{1}{1 - e^{C-2t}}}$. *Apart from this solution,*

$x(t) = \pm 1$, and $x(t) = 0$ are all solutions of the equation.

This is a lucky case. Now, let us slightly change the original ordinary differential equation, for instance, by adding 1 to the right-hand side of the original equation. The following statements can be used to solve the modified differential equation. With no surprise, no solution can be found by using `dsolve()`.

```
>> syms x; x=dsolve('Dx=x*(1-x^2)+1') % no analytical solution exists
```

Example 7.8 Find the analytical solution to the well-known *Van der Pol equation*

$$\frac{\mathrm{d}^2 y(t)}{\mathrm{d}t^2} + \mu(y^2(t) - 1)\frac{\mathrm{d}y(t)}{\mathrm{d}t} + y(t) = 0.$$

Solution *From the previous examples, it seems that the function* `dsolve()` *is quite powerful in finding analytical solutions to many differential equations. Here we are trying to solve the Van der Pol nonlinear equation. The following statements*

```
>> syms mu; y=dsolve('D2y+mu*(y^2-1)*Dy+y=0') % no analytical solution exists
```

can be executed but from the message "Explicit solution could not be found," which means that there is no analytical solution at all to the given Van der Pol equation.

It can be seen that the function `dsolve()` cannot be used to solve general nonlinear ordinary differential equations. Thus, to solve nonlinear equations, numerical methods have to be used. In the rest of this chapter, we shall concentrate on numerical solutions of various differential equations.

7.2 Numerical Solutions to Ordinary Differential Equations

In the previous section, analytical solutions for a limited class of ordinary differential equations were discussed and demonstrated. It is also noted that there is no analytical solutions to most nonlinear differential equations. Therefore, numerical algorithms should be used.

7.2.1 Overview of numerical solution algorithms

A large category of numerical solution algorithms for ordinary differential equations is for the so-called *initial value problems* (IVPs). The standard vector form of *first-order explicit differential equations* is given by

$$\boldsymbol{x}'(t) = \boldsymbol{f}(t, \boldsymbol{x}(t)), \quad \boldsymbol{x}(t_0) = \boldsymbol{x}_0, \tag{7-2-1}$$

where $\boldsymbol{x}^{\mathrm{T}}(t) = [x_1(t), x_2(t), \cdots, x_n(t)]$ is referred to as the *state vector*, $\boldsymbol{x}(t_0) = [x_1(t_0), \cdots, x_n(t_0)]^{\mathrm{T}}$ is the given initial value vector, and $\boldsymbol{f}^{\mathrm{T}}(\cdot) = [f_1(\cdot), f_2(\cdot), \cdots, f_n(\cdot)]$ is the vector of any nonlinear functions.

In this section, we shall explore numerical algorithms for solving $\boldsymbol{x}(t)$, $t \in [t_0, t_{\mathrm{f}}]$, where t_{f} is also referred to as the *terminal time*.

For the above initial value problems, the Euler's algorithm is obviously the most straightforward algorithm. Although the algorithm is very simple, understanding such an algorithm will help us to better understand other complicated algorithms.

Assume that at time t_0, the state vector is written as $\boldsymbol{x}(t_0)$. Given a small calculation step-size h, the left-hand-side of the differential equation can be approximately written as $\hat{\boldsymbol{x}}(t_0+h) = [\boldsymbol{x}(t_0+h) - \boldsymbol{x}(t_0)]/h$, at least it is true when $h \to 0$. Substituting it back to the original equation, the approximate solution at time $t_0 + h$ can be written as

$$\hat{\boldsymbol{x}}(t_0 + h) \approx \boldsymbol{x}(t_0) + h\boldsymbol{f}(t_0, \boldsymbol{x}(t_0)). \tag{7-2-2}$$

The above approximate solution certainly contains errors. Thus, the state vector at time $t_0 + h$ should be written as

$$\boldsymbol{x}(t_0 + h) = \hat{\boldsymbol{x}}(t_0 + h) + \boldsymbol{R}_0 = \boldsymbol{x}_0 + h\boldsymbol{f}(t, \boldsymbol{x}_0) + \boldsymbol{R}_0, \tag{7-2-3}$$

where the vector $\boldsymbol{R}_0$ is the approximation error. Denoting $\boldsymbol{x}_1 = \boldsymbol{x}(t_0 + h)$, $\hat{\boldsymbol{x}}_1 = \hat{\boldsymbol{x}}(t_0 + h)$ approximates the state vector at time $t_0 + h$. In this numerical procedure, one can simply denote the numerical solution by $\boldsymbol{x}_1$.

The state vector at time t_k is denoted by $\boldsymbol{x}_k$. At time $t_k + h$, the Euler's algorithm generates a new state vector

$$\boldsymbol{x}_{k+1} = \boldsymbol{x}_k + h\boldsymbol{f}(t, \boldsymbol{x}_k). \tag{7-2-4}$$

Thus, a recursive algorithm can be used to evaluate the solutions in each time instant over the given time interval $t \in [0, t_f]$. In this way, the numerical solutions at time instances $t_0 + h, t_0 + 2h, \cdots$ can be obtained.

To increase the accuracy of the solutions, one may reduce the step-size h. However, one cannot expect to reduce the step-size unlimitedly. The following two reasons should be considered:

(i) **Slowing down in computation** If an extremely small step-size is selected, the number of computation points over the solution interval will significantly increase.

(ii) **Increasing the cumulative errors** No matter how small the step-size is, the roundoff error in numerical solutions is unavoidable. As pointed out earlier, decreasing the step-size means the increase in computation points, which also means that the cumulative error and error propagation may increase. The relationship among the step-size, roundoff error, cumulative error and the overall error are sketched in Figure 7.1 (a) for illustration.

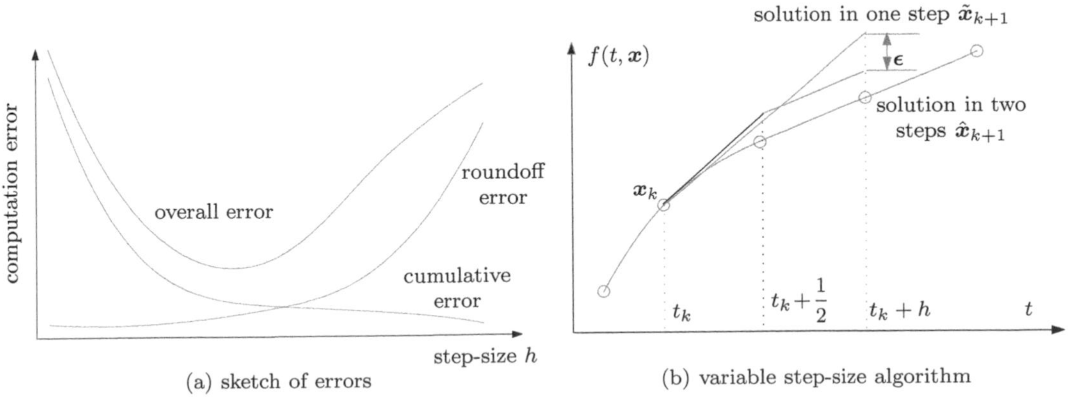

FIGURE 7.1: Errors and step-sizes.

Thus, in order to effectively solve the ordinary differential equations, the following considerations should be in place:

(i) **Suitable step-size selection** If simple algorithms such as the Euler's algorithm are used, one should choose suitable step-sizes, neither too large, nor too small.

(ii) **Improved algorithms** Since the above simple Euler's algorithm is a trapezoidal approximation of the original integration problem, the accuracy is rather low when the step-size is not small enough. It usually cannot effectively solve the original ordinary differential equation problems. Other advanced algorithms should be adopted to replace the Euler's algorithm. Successful algorithms include the Runge–Kutta's algorithm and the Adams' algorithm, etc.

(iii) **Variable step-size algorithms** It has been suggested that one should "suitably" choose the step-size. The concept of "suitable step-size" is very vague. The selection of step-size sometimes depends on the experience and the differential equation at hand. In fact, many numerical algorithms allow the use of variable-step-size computation. When the error detected is small, a relatively large step-size can be chosen. However, when the detected error is large, a smaller step-size will be used instead. Using this variable step-size mechanism, fast and accurate algorithms can be devised for ordinary differential equations.

The basic idea of variable step-size algorithm is illustrated in Figure 7.1 (b). If at the current time instant t_k, the state vector is $\boldsymbol{x}_k$, then, the state vector $\tilde{\boldsymbol{x}}_{k+1}$ at time instant t_k+h can be obtained. On the other hand, if one divides the step-size by two halves, the state vector at time $t_k + h$ obtained by integrating two steps using half step-size $h/2$ is denoted by $\hat{\boldsymbol{x}}_{k+1}$. The error of the state vectors using the two methods is $\epsilon = ||\hat{\boldsymbol{x}}_{k+1} - \tilde{\boldsymbol{x}}_{k+1}||$, and if it is smaller than the preassigned error tolerance, then, the original step-size can be used, and accordingly, the step-size can be increased. If the error is too large, the step-size should be decreased to ensure the computational accuracy. Thus, the above method considers both the computation speed and the computation accuracy. Such algorithms are referred to as *adaptive* or *variable step-size algorithm.*

7.2.2 Fixed-step Runge–Kutta algorithm and its MATLAB implementation

Fourth-order fixed-step Runge–Kutta algorithm is a classical differential equation solving algorithm, often taught in numerical analysis courses. It had been considered as an effective and easy-to-implement numerical algorithm.

The above fourth-order Runge–Kutta algorithm calculates the state vector at the next step using

$$\boldsymbol{x}_{k+1} = \boldsymbol{x}_k + \frac{1}{6}(\boldsymbol{k}_1 + 2\boldsymbol{k}_2 + 2\boldsymbol{k}_3 + \boldsymbol{k}_4), \tag{7-2-5}$$

where the four additional intermediate variables are introduced such that

$$\begin{cases} \boldsymbol{k}_1 = h\boldsymbol{f}(t_k, \boldsymbol{x}_k) \\ \boldsymbol{k}_2 = h\boldsymbol{f}\left(t_k + \dfrac{h}{2}, \boldsymbol{x}_k + \dfrac{\boldsymbol{k}_1}{2}\right) \\ \boldsymbol{k}_3 = h\boldsymbol{f}\left(t_k + \dfrac{h}{2}, \boldsymbol{x}_k + \dfrac{\boldsymbol{k}_2}{2}\right) \\ \boldsymbol{k}_4 = h\boldsymbol{f}(t_k + h, \boldsymbol{x}_k + \boldsymbol{k}_3), \end{cases} \tag{7-2-6}$$

where h is the fixed step-size.

Therefore, with the use of the recursive algorithm, the initial value problem at time instants $t_0 + h, t_0 + 2h, \cdots$, within the time interval $t \in [t_0, t_f]$ can be solved numerically step by step.

Based on the above algorithm, its MATLAB implementation is written as:

```
function [tout,yout]=rk_4(fun,tspan,y0)
ts=tspan; t0=ts(1); tf=ts(2); yout=[]; tout=[]; y0=y0(:);
if length(ts)==3, h=ts(3); else, h=(ts(2)-ts(1))/100; tf=ts(2); end
for t=[t0:h:tf]
   k1=h*feval(fun,t,y0); k2=h*feval(fun,t+h/2,y0+0.5*k1);
   k3=h*feval(fun,t+h/2,y0+0.5*k2); k4=h*feval(fun,t+h,y0+k3);
   y0=y0+(k1+2*k2+2*k3+k4)/6; yout=[yout; y0']; tout=[tout; t];
end
```

where `tspan` can be given in two ways. One way is to specify an evenly distributed time vector and the other is to use `tspan`$=$ [t_0,t_f,h], where t_0 and t_f are the initial and final time, h the step-size. The argument *fun* is the function handle for describing the function $\boldsymbol{f}(t,\boldsymbol{x})$. The argument $\boldsymbol{y}_0$ provides the initial state vector. The time vector and the matrix containing states at each time instant are returned in arguments `tout` and `yout`, respectively.

The above algorithm seems to be rather simple. However, from a numerical accuracy point of view, it may not be a good algorithm, since numerical accuracy during the computation is not monitored. Thus, the overall numerical accuracy of the algorithm cannot be guaranteed. In later examples, this algorithm will be compared with variable-step algorithm.

7.2.3 Numcrical solution to first-order vector ODEs

I. Runge–Kutta–Felhberg algorithm

German mathematician Erwin Felhberg proposed an improved version of the algorithm based on the traditional Runge–Kutta algorithm [4]. Within each computation step, six evaluations of the $f_i(\cdot)$ function are performed to ensure high-precision and numerical stability. The algorithm is also known as the *4/5 Runge–Kutta–Felhberg algorithm.* Assuming that the current step-size is h_k, the six intermediate variables $\boldsymbol{k}_i$ are evaluated by the following formula:

$$\boldsymbol{k}_i = h_k \boldsymbol{f}\left(t_k + \alpha_i h_k, \boldsymbol{x}_k + \sum_{j=1}^{i-1} \beta_{ij}\boldsymbol{k}_j\right), \quad i = 1, 2, \cdots, 6, \tag{7-2-7}$$

where t_k is the current time instant, and intermediate parameters α_i, β_{ij} and other parameters are given in Table 7.1. The parameter pairs α_i, β_{ij} are also referred to as the *Dormand–Prince pairs.* The state vector at the next time instant is obtained from

$$\boldsymbol{x}_{k+1} = \boldsymbol{x}_k + \sum_{i=1}^{6} \gamma_i \boldsymbol{k}_i. \tag{7-2-8}$$

Of course, this algorithm seems to be a fixed-step type like the 4th order Runge–Kutta algorithm in the previous subsection. However, in practical applications, an error vector $\boldsymbol{\epsilon}_k = \sum_{i=1}^{6}(\gamma_i - \gamma_i^*)\boldsymbol{k}_i$ can be obtained, and the step-size can be adjusted according to this error vector. This algorithm is often referred to as the *adaptive step-size algorithm.* It cannot only ensure the accuracy of numerical solutions but also provide faster speed.

In contrast with the concepts in feedback control, the fixed-step algorithm is somewhat like an open-loop control structure. It does not care whether the error of solution is

TABLE 7.1: Coefficients in 4/5 Runge–Kutta-Felhberg algorithm.

α_i	β_{ij}					γ_i	γ_i^*
0						16/135	25/216
1/4	1/4					0	0
3/8	3/32	9/32				6656/12825	1408/2565
12/13	1932/2197	−7200/2197	7296/2197			28561/56430	2197/4104
1	439/216	−8	3680/513	−845/4104		−9/50	−1/5
1/2	−8/27	2	−3544/2565	1859/4104	−11/40	2/55	0

acceptable or not using the identical step-size throughout the computation period. However, the variable-step algorithm is similar to the closed-loop control concept. It monitors the errors in the solution process and whenever necessary, the step-size can be adjusted according to the estimated computation error.

II. MATLAB functions for solving ordinary differential equations

To solve the ODEs numerically, the following procedures are needed

(i) **Standard form** Express the equations in the form of first-order explicit equations, $\boldsymbol{x}'(t) = \boldsymbol{f}(t, \boldsymbol{x})$, with known $\boldsymbol{x}_0$.

(ii) **Describe the equations** The standard form, i.e., $\boldsymbol{x}'(t) = \boldsymbol{f}(t, \boldsymbol{x})$, should be described correctly in MATLAB. MATLAB functions with the leading statements should be written

```
function xd = fun(t,x)              % without additional variables
function xd = fun(t,x,p1,p2,···)    % with additional variables
```

where t is the time variable. Note that even the original equation is time-independent, and the argument t should still be used, to avoid argument mismatch problems. The variable $\boldsymbol{x}_\mathrm{d}$ is the derivative of the state vector. Alternatively, anonymous functions can also be used to describe the equations.

(iii) **Solve the equations** The MATLAB function `ode45()` is the most widely used initial value problem solver. In the algorithm, the variable-step 4/5 Runge–Kutta–Felhberg algorithm is implemented. The syntaxes of the function are

```
[t,x] = ode45(fun,[t0,tf],x0)                    % direct solutions
[t,x] = ode45(fun,[t0,tf],x0,opts)               % solver with control options
[t,x] = ode45(fun,[t0,tf],x0,opts,p1,p2,···)     % solver with additional variables
```

where the differential equations should be expressed using an M-function *fun*, an anonymous function. The structures of the functions will be explained through examples later. The time span $[t_0, t_\mathrm{f}]$ describes the time interval for numerical solutions. If only one value is provided, it means that the final value t_f with the default setting of initial time at $t_0 = 0$. For initial value problems, one should also specify the initial state vector $\boldsymbol{x}_0$.

Note that this function allows the selection of $t_\mathrm{f} < t_0$, where t_0 can be regarded as the final time, with t_f the initial time. Also, $\boldsymbol{x}_0$ can be regarded as the final value of the equations. Thus, this function can also be used for final value problems.

(iv) **Validate the solutions** This procedure is crucial in real applications and will be demonstrated with examples later.

In actual differential equation solving processes, sometimes one may further assign some control options. This can be done with the use of the variable *opts*. The initial *opts* template can be obtained with the function `odeset()`. This template is a structured variable with many fields. Some of the frequently used fields are given in Table 7.2.

TABLE 7.2: Control parameters in differential equation solutions.

Parameters	Descriptions to the parameters
`RelTol`	Upper-bound of the relative error tolerance. The default value is 0.001, i.e., 0.1% relative error. In most applications, this value should be reduced to ensure that the results are accurate
`AbsTol`	A vector controlling the permissible absolute error in states. The default value is 10^{-6}. Of course, this value can be changed to improve the accuracy of solution
`MaxStep`	Maximum allowed step-size
`Mass`	Mass matrix, which is used in describing differential algebraic equations
`Jacobian`	The function describing the Jacobian matrix $\partial \boldsymbol{f}/\partial \boldsymbol{x}$. If the Jacobian matrix is known, the simulation process can speed up

Two methods can be used in modifying the *opts* template. One is by the use of the function `odeset()`, and the other is by the modification of the fields directly. For instance, if one wants to assign the relative error tolerance to 10^{-7}, either of the following statements can be used

```
opts=odeset('RelTol',1e-7);          % assign it by odeset() function
opts=odeset; opts.RelTol=1e-7;   % by direct assignment
```

In many differential equation solution applications, sometimes additional variables may be used for flexible testing of various parameter combinations. These additional variables, denoted by $p_1, p_2, \cdots, p_m$, should be properly declared and passed to the MATLAB description for the $\boldsymbol{f}(t, \boldsymbol{x})$ function. When calling the differential equation solver, the parameters should also be matched in the same order.

Apart from the solver `ode45()`, other solvers can also be used, and among them, there are solvers `ode15s()`, `ode23()`, `ode113()`, `ode23t()`, `ode23tb()`, `ode23s()`.

Example 7.9 Consider the well-known *Lorenz equation* given by

$$\begin{cases} x_1'(t) = -\beta x_1(t) + x_2(t)x_3(t) \\ x_2'(t) = -\rho x_2(t) + \rho x_3(t) \\ x_3'(t) = -x_1(t)x_2(t) + \sigma x_2(t) - x_3(t), \end{cases}$$

where $\beta = 8/3, \rho = 10, \sigma = 28$. The initial values are given by $x_1(0) = x_2(0) = 0,\ x_3(0) = \epsilon$, and ϵ is a very small positive number, i.e., $\epsilon = 10^{-10}$. Find the numerical solutions to the initial value problem of the given differential equations.

Solution *It is obvious that these ordinary differential equations are nonlinear time-invariant with no analytical solutions. Numerical solutions should be pursued. For the equations, an anonymous function can be prepared as follows for use with the function* `ode45()` *to directly solve the equations.*

```
>> f=@(t,x)[-8/3*x(1)+x(2)*x(3); -10*x(2)+10*x(3); ...
        -x(1)*x(2)+28*x(2)-x(3)]; % 3 rows for 3 equations in vector form
   t_final=100; x0=[0;0;1e-10];        % terminate tine and initial states
   [t,x]=ode45(f,[0,t_final],x0); subplot(121), plot(t,x)  % solve & plot
   subplot(122), plot3(x(:,1),x(:,2),x(:,3)); grid % phase space trajectory
```

In the above MATLAB statements, `t_final` *denotes the terminating time,* $\boldsymbol{x}0$ *is the initial state vector. The first drawing command shows the dynamical curves of the states versus time, as in Figure 7.2 (a). In the second drawing command, the three-dimensional phase space trajectory is visualized as shown in Figure 7.2 (b). So, the seemingly difficult nonlinear differential equations with known initial values can be solved using only a few MATLAB statements.*

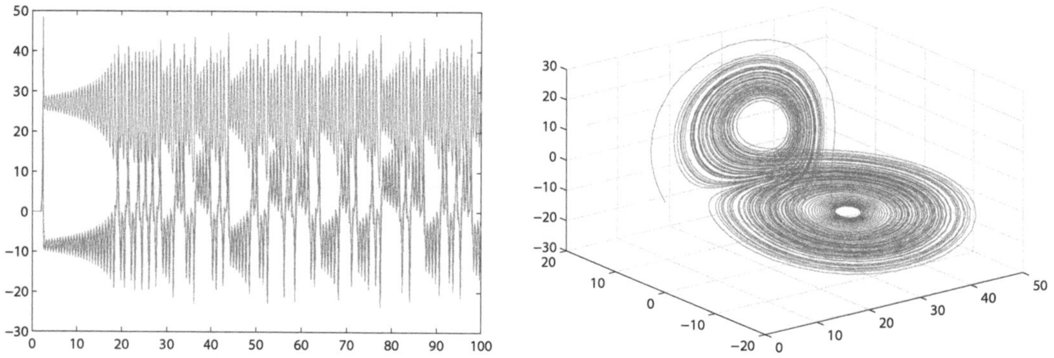

(a) time responses of the state variables (b) 3D phase space trajectory

FIGURE 7.2: Numerical solutions of the Lorenz equations.

Furthermore, the best command to visualize the three-dimensional trajectory is by the use of the function `comet3(`$\boldsymbol{x}$`(:,1),`$\boldsymbol{x}$`(:,2),`$\boldsymbol{x}$`(:,3))`*, where the animated display shows the trace of the trajectory.*

From the above example, it can be observed that if the differential equations to be solved can be expressed by first-order explicit ones, using the numerical solutions can immediately be found with function `ode45()`. Therefore, preparing a MATLAB function describing the equations is a crucial step in solving the initial value problems.

III. Solving ODEs with additional variables in MATLAB

In the ordinary differential equation solving process, frequently one may introduce additional variables so that when the parameters in the equations are changed, they can be modified through the additional variables rather than having to modify the MATLAB function itself. For instance, the Lorenz equations in Example 7.9 contain parameters such as β, ρ and σ and they can all be considered as additional variables. Thus, when their values change, one does not have to modify the MATLAB function describing the Lorenz equations. The following example illustrates the method and benefits of using additional variables in differential equation solvers.

Example 7.10 Write a MATLAB function to describe the Lorenz equations in Example

7.9 with additional variables. Then, use the new function to find the numerical solutions for another set of parameters $\beta = 2, \rho = 5$ and $\sigma = 20$.

Solution *Select the variables β, ρ and σ as the additional variables. The new anonymous function for the differential equations can then be written as*

```
>> f=@(t,x,beta,rho,sigma)[-beta*x(1)+x(2)*x(3);
             -rho*x(2)+rho*x(3); -x(1)*x(2)+sigma*x(2)-x(3)];
```

From the new anonymous function, the following statements can be used instead for the numerical solutions

```
>> t_final=100; x0=[0;0;1e-10]; % terminate tine and initial states
   b1=8/3; r1=10; s1=28;   % one may also use other variable names
   [t,x]=ode45(f,[0,t_final],x0,[],b1,r1,s1); subplot(121), plot(t,x)
   subplot(122); plot3(x(:,1),x(:,2),x(:,3)); % phase space trajectory
```

With the above statements, it can be seen that the additional variables can easily be passed to function $\boldsymbol{f}(t, \boldsymbol{x})$. Moreover, the opts *variable is replaced by an empty matrix, which means that the control parameters need not be changed.*

Using MATLAB functions with additional variables, the Lorenz equation with other values of β, ρ and σ can be solved directly, without the need of modifying the existing anonymous function, and with $\beta = 1, \rho = 5$ and $\sigma = 20$, the following statements can be used to find the numerical solutions. The time responses and phase space trajectory obtained are shown in Figures 7.3 (a) and (b), respectively.

```
>> tf=100; x0=[0;0;1e-10]; b2=2; r2=5; s2=20;     % additional parameters
   [t2,x2]=ode45(f,[0,tf],x0,[],b2,r2,s2); subplot(121), plot(t2,x2)
   subplot(122); plot3(x2(:,1),x2(:,2),x2(:,3)); % phase space trajectory again
```

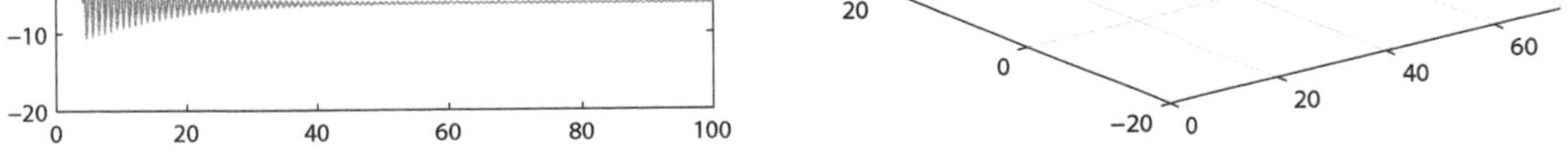

(a) time responses of state variables (b) three-dimensional phase trajectory

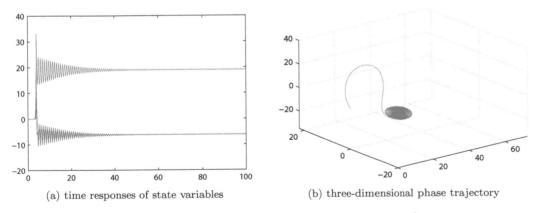

FIGURE 7.3: Results of Lorenz equations under a new set of parameters.

If the differential equation can be described by anonymous function, it is not necessary to use additional variables, since the variables in MATLAB workspace can be used directly. For instance, the equation can be solved with

```
>> b=8/3; r=10; s=28; % anonymous function uses workspace variables
   f=@(t,x)[-b*x(1)+x(2)*x(3); -r*x(2)+r*x(3); -x(1)*x(2)+s*x(2)-x(3)];
   [t,x]=ode45(f,[0,t_final],x0); % solve ODE for another set of data
```

7.3 Transforms to Standard Differential Equations

From the above description and introduction, it can be seen that only the first-order vector form explicit ordinary differential equations $\boldsymbol{x}'(t) = \boldsymbol{f}(t, \boldsymbol{x})$ can be solved using the relevant differential equation solvers in MATLAB. If the equations are described by high-order ones, one has to convert the equations first into the first-order vector form explicit equations. In this subsection, two different cases are explored.

7.3.1 Manipulating a single high-order ODE

Assume that a high-order differential equation can be expressed as

$$y^{(n)} = f(t, y, y', \cdots, y^{(n-1)}), \tag{7-3-1}$$

and the initial conditions of the output signal and its derivatives are $y(0), y'(0), \cdots, y^{(n-1)}(0)$. One may select a set of state variables

$$x_1 = y, x_2 = y', \cdots, x_n = y^{(n-1)}. \tag{7-3-2}$$

Taking first order derivatives for the above variables, it is immediately found that $x_1' = y' = x_2$, $x_2' = y'' = x_3$, $\cdots$, $x_{n-1}' = y^{(n-1)} = x_n$. Finally, $x_n' = y^{(n)}$. Taking into account of the original differential equation, it is found that $x_n' = f(t, x_1, x_2, \cdots, x_n)$. Summarizing the above results, the high-order differential equation can be converted into the following first-order vector form explicit equations

$$\begin{cases} x_1' = x_2 \\ x_2' = x_3 \\ \quad\vdots \\ x_n' = f(t, x_1, x_2, \cdots, x_n), \end{cases} \tag{7-3-3}$$

with initial states $x_1(0) = y(0), x_2(0) = y'(0), \cdots, x_n(0) = y^{(n-1)}(0)$. Therefore, the converted differential equations can directly be solved using the numerical methods introduced earlier.

Example 7.11 Consider again the Van der Pol equation $y'' + \mu(y^2 - 1)y' + y = 0$. If the initial conditions $y(0) = -0.2, y'(0) = -0.7$ are given, solve numerically the Van der Pol equation for different values of μ's.

Solution *Since the MATLAB functions illustrated earlier can only deal with first-order explicit differential equations, conversion should be made before the problem can be solved numerically using the MATLAB ODE solvers. For this example, by choosing the state variables $x_1 = y, x_2 = y'$, the original differential equation can be converted to*

$$\begin{cases} x_1' = x_2 \\ x_2' = -\mu(x_1^2 - 1)x_2 - x_1. \end{cases}$$

It is not wise to write one function for each interested value of μ. The concept of additional variables should be introduced so as to pass the value of μ to the function $\boldsymbol{f}(t, \boldsymbol{x})$. Thus, an anonymous function can be written. The calling syntax of `ode45()` *is by additional* *variables. The initial state vector is given by* $\boldsymbol{x} = [-0.2, -0.7]^{\mathrm{T}}$, *and the final solution can be obtained from*

```
>> f=@(t,x,mu)[x(2); -mu*(x(1)^2-1)*x(2)-x(1)]; x0=[-0.2;-0.7];
   t_final=20; mu=1; [t1,y1]=ode45(f,[0,t_final],x0,[],mu);
   mu=2; [t2,y2]=ode45(f,[0,t_final],x0,[],mu); plot(t1,y1,t2,y2,':')
   figure; plot(y1(:,1),y1(:,2),y2(:,1),y2(:,2),':')
```

and for $\mu = 1, 2$, *the time responses and the phase plane trajectories are shown respectively in Figures 7.4 (a) and (b). It can be seen that both the phase plane trajectories settle down on corresponding closed-paths. The closed-path is referred to as the* limit cycle.

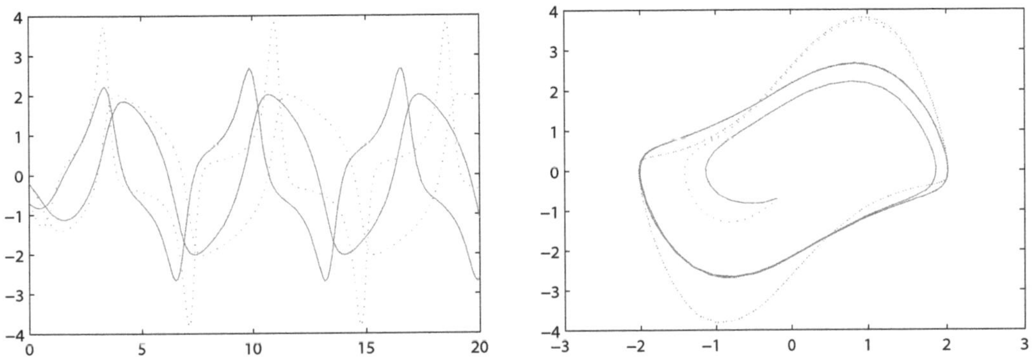

(a) time responses for different values of μ (b) phase plane trajectories

FIGURE 7.4: Van der Pol equation solutions for different values of μ's.

If one changes the value of μ, *say,* $\mu = 1000$, *and sets the terminate time* $t_{\mathrm{f}} = 3000$, *the following statements can be used to find the numerical solution to the corresponding Van der Pol equation.*

```
>> x0=[2;0]; t_final=3000; % a counter example, do not run the following
   mu=1000; [t,y]=ode45(f,[0,t_final],x0,[],mu);
```

However, even after a long wait, the solutions cannot be found, since the step-size used might be too small and too many computation points may be involved. Thus, this kind of ordinary differential equation with known initial values may not be suitably solvable by `ode45()` *functions. In the next section, stiff equations-based algorithms will be presented to solve the problem.*

7.3.2 Manipulating multiple high-order ODEs

Now consider the differential equation sets composed of several explicit high-order differential equations. For example,

$$\begin{cases} x^{(m)} = f(t, x, x', \cdots, x^{(m-1)}, y, \cdots, y^{(n-1)}) \\ y^{(n)} = g(t, x, x', \cdots, x^{(m-1)}, y, \cdots, y^{(n-1)}). \end{cases} \tag{7-3-4}$$

Let us still select the state variables $x_1 = x, x_2 = x', \cdots, x_m = x^{(m-1)}, x_{m+1} =$

$y, x_{m+2} = y', \cdots, x_{m+n} = y^{(n-1)}$. Therefore, the original high-order differential equations can be converted to

$$\begin{cases} x_1' = x_2 \\ \quad\vdots \\ x_m' = f(t, x_1, x_2, \cdots, x_{m+n}) \\ x_{m+1}' = x_{m+2} \\ \quad\vdots \\ x_{m+n}' = g(t, x_1, x_2, \cdots, x_{m+n}). \end{cases} \tag{7-3-5}$$

Therefore, the desirable first-order vector form explicit ordinary differential equations can be obtained. The following example illustrates the conversion and solution process.

Example 7.12 The trajectory (x, y) of the Apollo satellite satisfies the following differential equation sets [5]

$$x'' = 2y' + x - \frac{\mu^*(x+\mu)}{r_1^3} - \frac{\mu(x-\mu^*)}{r_2^3}, \quad y'' = -2x' + y - \frac{\mu^* y}{r_1^3} - \frac{\mu y}{r_2^3},$$

where $\mu = 1/82.45$, $\mu^* = 1 - \mu$, $r_1 = \sqrt{(x+\mu)^2 + y^2}$, $r_2 = \sqrt{(x-\mu^*)^2 + y^2}$. It is known that the initial values are given by $x(0) = 1.2$, $x'(0) = 0$, $y(0) = 0$, $y'(0) = -1.04935751$. Solve the differential equations and draw the trajectory of (x, y).

Solution *The state variables are chosen as* $x_1 = x, x_2 = x', x_3 = y, x_4 = y'$. *Thus, the first-order explicit differential equations can be found as follows*

$$\begin{cases} x_1' = x_2 \\ x_2' = 2x_4 + x_1 - \mu^*(x_1+\mu)/r_1^3 - \mu(x_1-\mu^*)/r_2^3 \\ x_3' = x_4 \\ x_4' = -2x_2 + x_3 - \mu^* x_3/r_1^3 - \mu x_3/r_2^3, \end{cases}$$

where $r_1 = \sqrt{(x_1+\mu)^2 + x_3^2}$, $r_2 = \sqrt{(x_1-\mu^*)^2 + x_3^2}$, *and* $\mu = 1/82.45, \mu^* = 1 - \mu$.

From the above mathematical equations obtained, the MATLAB function describing the original differential equations is prepared as follows:

```
function dx=apolloeq(t,x) % not suitable for anonymous function
mu=1/82.45; mu1=1-mu; r1=sqrt((x(1)+mu)^2+x(3)^2);
r2=sqrt((x(1)-mu1)^2+x(3)^2);  % intermediate variables computation
dx=[x(2); 2*x(4)+x(1)-mu1*(x(1)+mu)/r1^3-mu*(x(1)-mu1)/r2^3;
     x(4); -2*x(2)+x(3)-mu1*x(3)/r1^3-mu*x(3)/r2^3];
```

Since there are some intermediate computation steps involved, the anonymous or inline functions are not suitable. Using `ode45()` *function, the numerical solutions of the equations are as follows, with the trajectory shown in Figure 7.5 (a).*

```
>> x0=[1.2; 0; 0; -1.04935751];    % initial states
   tic, [t,y]=ode45(@apolloeq,[0,20],x0); toc % measure time elapsed
   length(t), plot(y(:,1),y(:,3)) % count number of points computed
```

For this example, the elapsed time is 0.33 seconds. Also, the number of points calculated is returned by the use of `length()` *function. For this example, the number of points is 689.*

In fact, the obtained trajectory is not correct, since in the simulation control parameters, the relative error tolerance `RelTol` *is too large. To have more accurate results, one usually should reduce that value, for instance, to* 10^{-6}. *Let us use the following statements to solve the differential equations again:*

```
>> options=odeset; options.RelTol=1e-6; % set controls and try again
   tic, [t1,y1]=ode45(@apolloeq,[0,20],x0,options); toc % measure time
   length(t1), plot(y1(:,1),y1(:,3))    % count number of points
```

The elapsed time is 0.701 seconds, with 1873 computation points. The new trajectory obtained is shown in Figure 7.5 (b). It can be seen that the different results are obtained and further reducing the error tolerance will yield the same numerical results. It can be concluded that it is always necessary to validate the simulation results after simulation by reducing the error tolerance `RelTol`.

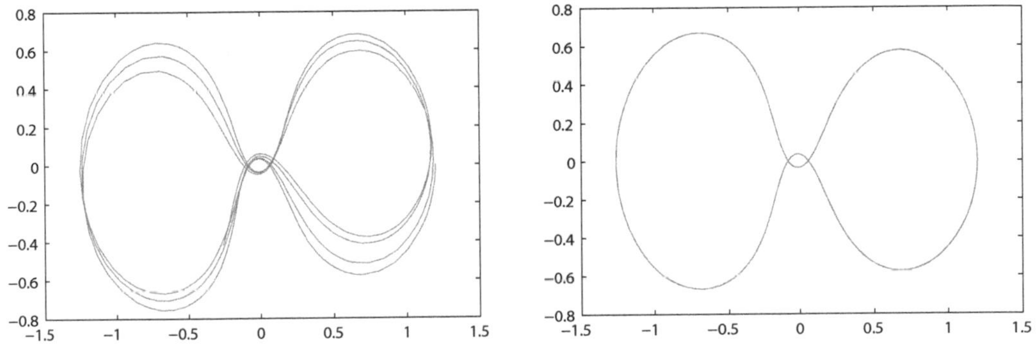

(a) results with default parameters (erroneous) (b) improved results

FIGURE 7.5: The trajectories of Apollo with different numerical accuracy specifications.

The following statements can be used to find the step-sizes used in the simulation process, as shown in Figure 7.6. The minimum step-size is 1.8927×10^{-4}.

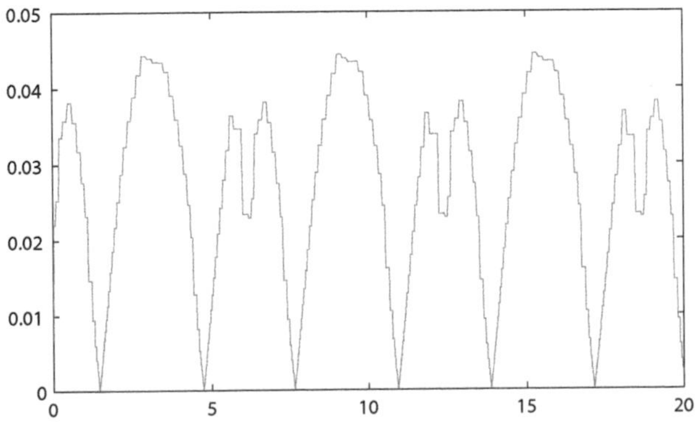

FIGURE 7.6: The step-sizes over time in the simulation process.

```
>> plot(t1(1:end-1),diff(t1)), min(diff(t1)) % plot step-size
```

From the above step-size curve, it can be seen that when variable step-size algorithms are used, the step-size can be adapted according to the error accuracy requirements. In part

of the simulation period, step-sizes larger than 0.03 are used. However, in order to keep precision under control, at some points, very small step-size of 2×10^{-4} *is used. If fixed-step algorithms are used, in order to make sure that the simulation results reliable, a small step-size of* 2×10^{-4} *should be used in the whole simulation process. The computation required becomes a total of* 10^5 *steps, which is 56 times more than the number of points used in the variable-step algorithm. Therefore, variable step-size approaches are more effective.*

Example 7.13 Solve the Apollo problem using fixed-step Runge–Kutta algorithm.

Solution *To use fixed-step algorithms, one should consider two problems first: (i) how to select the step-size, (ii) how to ensure the accuracy in computation. The first problem can only be solved by trial-and-error approach. When selecting step-sizes, small step-size can be tested. However, the computational cost could be very high. For instance, a step-size of 0.01 will produce the trajectory shown in Figure 7.7 (a). The computation time is 2.654 seconds.*

```
>> x0=[1.2; 0; 0; -1.04935751]; % initial states and solve the ODE
   tic, [t,y]=rk_4(@apolloeq,[0,20,0.01],x0); toc, plot(y(:,1),y(:,3))
```

It is obvious that the results thus obtained are wrong, thus, a smaller step-size should be used instead. If one selects a smaller step-size 0.001, more accurate results can be obtained and the trajectory is shown in Figure 7.7 (b). However, the time required is 97.079 seconds, which is about 138 times the time required for the variable-step computation.

```
>> tic, [t2,y2]=rk_4(@apolloeq,[0,20,0.001],x0); toc % solve again
   plot(y2(:,1),y2(:,3))                              % draw trajectory
```

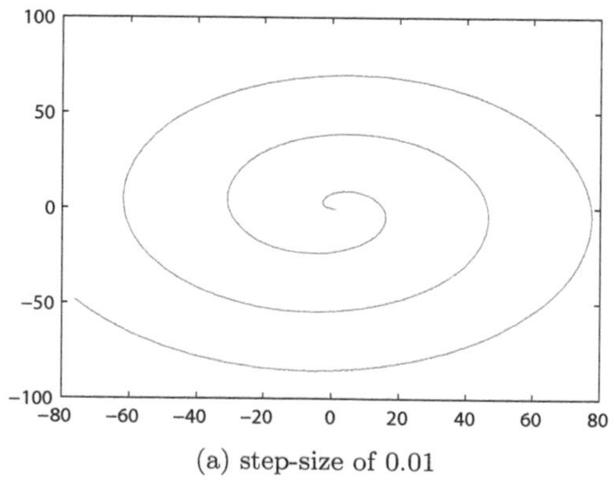

(a) step-size of 0.01

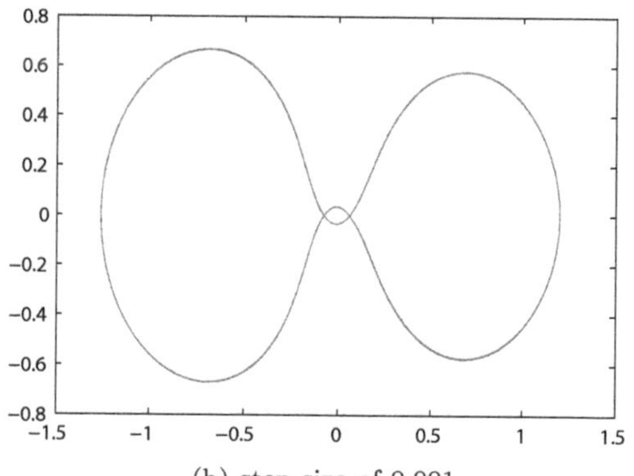

(b) step-size of 0.001

FIGURE 7.7: The Apollo trajectories under different step-sizes.

In fact, strictly speaking, the results obtained using the fixed-step algorithm cannot satisfy the requirement of 10^{-6} *relative error tolerance, although the shape of the responses is similar to the one obtained from variable-step ODE solvers.*

The following example illustrates how to transform two simultaneous high-order implicit differential equations into the standard first-order vector form ones involving symbolic computation.

Example 7.14 Consider the following high-order implicit differential equations

$$\begin{cases} x''+2y'x=2y'' \\ x''y'+3x'y''+xy'-y=5. \end{cases}$$

Convert it into the first-order explicit differential equations.

Solution *In the above two equations, both x'' and y'' are cross-related. Let us still select the state variables as $x_1=x, x_2=x', x_3=y, x_4=y'$. Our purpose is to eliminate one of the high-order terms and solve for the other. Thus, it can be found that the analytical expression of y'' is*

$$y''=y'x+\frac{x''}{2}.$$

Substituting it into the section equation, x'' can be found that

$$x''=\frac{2y+10-2xy'-6xx'y'}{2y'+3x'}.$$

Thus, the state equation can be written as

$$x_2'=\frac{2x_3+10-2x_1x_4-6x_1x_2x_4}{2x_4+3x_2},\quad x_4'=\frac{x_3+5-x_1x_4+2x_1x_4^2}{2x_4+3x_2}.$$

The converted first-order explicit differential equation can then be written as

$$\begin{cases} x_1'=x_2 \\ x_2'=\dfrac{2x_3+10-2x_1x_4-6x_1x_2x_4}{2x_4+3x_2} \\ x_3'=x_4 \\ x_4'=\dfrac{x_3+5-x_1x_4+2x_1x_4^2}{2x_4+3x_2}. \end{cases}$$

In fact, some of the above equations may not be easily solved manually. Sometimes one may solve the conversion problem using Symbolic Math Toolbox. For simplicity, denote `dx`$=x''$ *and* `dy`$=y''$*, thus,* `dx` *and* `dy` *are x_2' and x_4'. The following statements can be used to solve the equations*

```
>> syms x1 x2 x3 x4 dx dy % solve analytically for dx and dy
   [y1,y2]=solve(dx+2*x4*x1==2*dy,dx*x4+3*x2*dy+x1*x4-x3==5,dx,dy)
```

The derivatives of the variables can be written as

$$x_2'=-2\frac{3x_4x_1x_2-5+x_4x_1-x_3}{3x_2+2x_4},\quad x_4'=\frac{2x_4^2x_1+5-x_4x_1+x_3}{3x_2+2x_4}.$$

It can be seen that the results are exactly the same as the previous results.

For more complicated problems, solving the corresponding algebraic equations manually might be extremely difficult, if not impossible. Therefore, algebraic equations may be embedded in the MATLAB function describing the first-order explicit equations. These related numerical solution methods will be discussed in the following section.

7.3.3 Validation of numerical solutions to ODEs

It has been shown through examples that if the control parameters are not properly chosen, the solutions may not even be correct. Thus, the numerical solutions should be validated. However, since the equations have no analytic solutions, an alternative method to validate the solution is by setting the control parameters to different values and checking whether they yield the same results. The most effective control parameter is the `RelTol` property. The default value for it is 10^{-3}, which is usually too large for many applications. It can be set to 10^{-6} or even 10^{-8}. This approach may usually not add too much computation effort. Alternatively, selecting different ODE solvers may also cross-validate the results.

7.3.4 Transformation of differential matrix equations

In some real applications, matrix-type differential equations are preferred. For instance, the Lagrange equation can be expressed as

$$\boldsymbol{M}\boldsymbol{X}'' + \boldsymbol{C}\boldsymbol{X}' + \boldsymbol{K}\boldsymbol{X} = \boldsymbol{F}u(t), \tag{7-3-6}$$

where $\boldsymbol{M}$, $\boldsymbol{C}$, $\boldsymbol{K}$ are $n \times n$ matrices, and $\boldsymbol{X}, \boldsymbol{F}$ are $n \times 1$ column vectors. Introducing the vectors $\boldsymbol{x}_1 = \boldsymbol{X}$ and $\boldsymbol{x}_2 = \boldsymbol{X}'$, it is found that $\boldsymbol{x}_1' = \boldsymbol{x}_2$, and $\boldsymbol{x}_2' = \boldsymbol{X}''$. From (7-3-6), it is found that $\boldsymbol{X}'' = \boldsymbol{M}^{-1}\Big[\boldsymbol{F}u(t) - \boldsymbol{C}\boldsymbol{X}' - \boldsymbol{K}\boldsymbol{X}\Big]$. The new state vector can be expressed as $\boldsymbol{x} = [\boldsymbol{x}_1^{\mathrm{T}}, \boldsymbol{x}_2^{\mathrm{T}}]^{\mathrm{T}}$, the state space equation can be rewritten as

$$\boldsymbol{x}'(t) = \begin{bmatrix} \boldsymbol{x}_2(t) \\ \boldsymbol{M}^{-1}\Big[\boldsymbol{F}u(t) - \boldsymbol{C}\boldsymbol{x}_2(t) - \boldsymbol{K}\boldsymbol{x}_1(t)\Big] \end{bmatrix}, \tag{7-3-7}$$

which is exactly an explicit vector form first-order differential equations directly solvable with the corresponding MATLAB ODE solver functions.

Example 7.15 Consider the inverted pendulum model expressed as [6]

$$\boldsymbol{M}(\boldsymbol{\theta})\boldsymbol{\theta}'' + \boldsymbol{C}(\boldsymbol{\theta}, \boldsymbol{\theta}')\boldsymbol{\theta}' = \boldsymbol{F}(\boldsymbol{\theta}),$$

where $\boldsymbol{\theta} = [a, \theta_1, \theta_2]^{\mathrm{T}}$, and a is the position of the cart, and θ_1, θ_2 are the angles of the two bars. The matrices are given by

$$\boldsymbol{M}(\boldsymbol{\theta}) = \begin{bmatrix} m_{\mathrm{c}} + m_1 + m_2 & (0.5m_1 + m_2)L_1\cos\theta_1 & 0.5m_2L_2\cos\theta_2 \\ (0.5m_1 + m_2)L_1\cos\theta_1 & (m_1/3 + m_2)L_1^2 & 0.5m_2L_1L_2\cos\theta_1 \\ 0.5m_2L_2\cos\theta_2 & 0.5m_2L_1L_2\cos\theta_1 & m_2L_2^2/3 \end{bmatrix},$$

$$\boldsymbol{C}(\boldsymbol{\theta}, \boldsymbol{\theta}') = \begin{bmatrix} 0 & -(0.5m_1 + m_2)L_1\theta_1'\sin\theta_1 & -0.5m_2L_2\theta_2'\sin\theta_2 \\ 0 & 0 & 0.5m_2L_1L_2\theta_2'\sin(\theta_1 - \theta_2) \\ 0 & -0.5m_2L_1L_2\theta_1'\sin(\theta_1 - \theta_2) & 0 \end{bmatrix},$$

$$\boldsymbol{F}(\boldsymbol{\theta}) = \begin{bmatrix} u(t) \\ (0.5m_1 + m_2)L_1 g\sin\theta_1 \\ 0.5m_2L_2 g\sin\theta_2 \end{bmatrix}.$$

The parameters for a particular experimental system are $m_{\mathrm{c}} = 0.85$kg, $m_1 = 0.04$kg, $m_2 = 0.14$kg, $L_1 = 0.1524$m, and $L_2 = 0.4318$m. Find the step response of the system.

Solution *The coefficient matrices $\boldsymbol{M}(\theta_1, \theta_2), \boldsymbol{C}(\theta_1, \theta_2)$ and $\boldsymbol{F}(\theta_1, \theta_2)$ contain the nonlinear terms of the state vector $\boldsymbol{x}$, for instance the cosine terms of θ_1. Introducing additional variables $\boldsymbol{x}_1 = \boldsymbol{\theta}$ and $\boldsymbol{x}_2 = \boldsymbol{\theta}'$, the new state vector $\boldsymbol{x} = [\boldsymbol{x}_1^{\mathrm{T}}, \boldsymbol{x}_2^{\mathrm{T}}]^{\mathrm{T}}$ can be constructed and the explicit first-order differential equation can be established*

```
function dx=inv_pendulum(t,x,u,mc,m1,m2,L1,L2,g)
```

```
M=[mc+m1+m2, (0.5*m1+m2)*L1*cos(x(2)), 0.5*m2*L2*cos(x(3))
   (0.5*m1+m2)*L1*cos(x(2)),(m1/3+m2)*L1^2,0.5*m2*L1*L2*cos(x(2))
   0.5*m2*L2*cos(x(3)),0.5*m2*L1*L2*cos(x(2)),m2*L2^2/3];
C=[0,-(0.5*m1+m2)*L1*cos(x(5))*sin(x(2)),-0.5*m2*L2*x(6)*sin(x(3))
   0, 0, 0.5*m2*L1*L2*x(6)*sin(x(2)-x(3))
   0, -0.5*m2*L1*L2*x(5)*sin(x(2)-x(3)), 0];
F=[u; (0.5*m1+m2)*L1*g*sin(x(2)); 0.5*m2*L2*g*sin(x(3))];
dx=[x(4:6); inv(M)*(F-C*x(4:6))];
```

The input signal $u(t)$ is a step signal, and the following statements can be used to solve numerically the differential equations. The results are shown in Figures 7.8 (a) and (b).

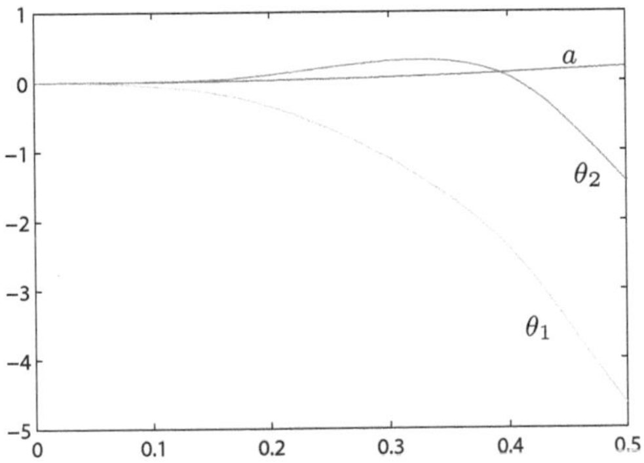

(a) $\boldsymbol{\theta}$ curves

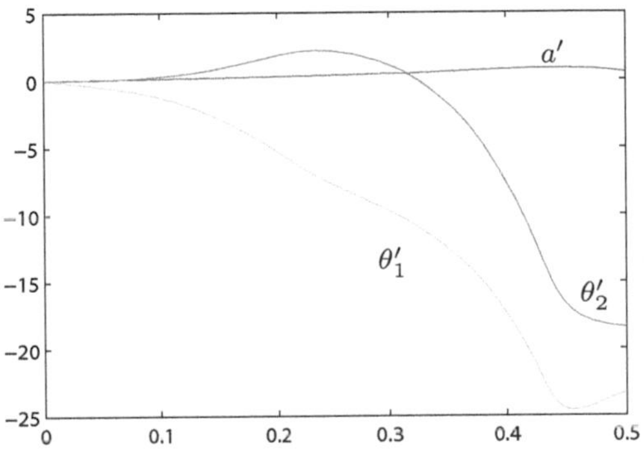

(b) $\boldsymbol{\theta}'$ curves

FIGURE 7.8: Step responses of the inverted pendulum.

```
>> opt=odeset; opt.RelTol=1e-8; u=1; mc=0.85;
   m1=0.04; m2=0.14; L1=0.1524; L2=0.4318; g=9.81; x0=zeros(6,1);
   [t,x]=ode45(@inv_pendulum,[0,0.5],x0,opt,u,mc,m1,m2,L1,L2,g);
   subplot(121), plot(t,x(:,1:3)), subplot(122), plot(t,x(:,4:6))
```

It should be noted that the inverted pendulum system is naturally unstable. A properly designed input signal constructed from the state vector information should be applied to stabilize the system.

Furthermore, if the matrices $\boldsymbol{M}$, $\boldsymbol{C}$, $\boldsymbol{K}$ and $\boldsymbol{F}$ are independent of $\boldsymbol{X}$, the original matrix type differential equations become linear time-invariant differential equations, given by

$$\begin{bmatrix} \boldsymbol{x}_1'(t) \\ \boldsymbol{x}_2'(t) \end{bmatrix} = \begin{bmatrix} \boldsymbol{0} & \boldsymbol{I} \\ -\boldsymbol{M}^{-1}\boldsymbol{K} & -\boldsymbol{M}^{-1}\boldsymbol{C} \end{bmatrix} \begin{bmatrix} \boldsymbol{x}_1(t) \\ \boldsymbol{x}_2(t) \end{bmatrix} + \begin{bmatrix} \boldsymbol{0} \\ \boldsymbol{M}^{-1}\boldsymbol{F} \end{bmatrix} u(t). \tag{7-3-8}$$

Riccati differential equations are another commonly encountered matrix differential equations. The general form of the equation is

$$\boldsymbol{P}'(t) = \boldsymbol{A}^{\mathrm{T}}\boldsymbol{P}(t) + \boldsymbol{P}(t)\boldsymbol{A} + \boldsymbol{P}(t)\boldsymbol{B}\boldsymbol{P}(t) + \boldsymbol{C}, \tag{7-3-9}$$

where, $\boldsymbol{B}$, $\boldsymbol{C}$ are symmetrical matrices. Assume that the condition $\boldsymbol{P}(t_{\mathrm{f}})$ at terminal time t_{f} is known, and the numerical solution over the time interval (t_0, t_{f}) is expected. Since the function $\boldsymbol{P}(t)$ is in matrix form and should be converted to vector form, the function `reshape()` can be used to convert a vector to a matrix, or use `P(:)` to convert a matrix into a vector. The following command can be used to describe the Riccati differential equation,

```
function dy=ric_de(t,x,A,B,C)
P=reshape(x,size(A)); Y=A'*P+P*A+P*B*P+C; dy=Y(:);
```

Luckily, the function `ode45()` allows the actual "starting time" larger than the "terminal time," thus, in solving the ODE, the time span `tspan`= [t_f,0], with the following syntax [t,p] = `ode45(@ric_de,[`t_f`,0],`P_f`(:),`$opts$`,`A,B,C`)`.

Example 7.16 Suppose the matrices and terminal values of a Riccati differential equation are given below. Please solve numerically the Riccati differential equation.

$$A = \begin{bmatrix} 6 & 6 & 17 \\ 1 & 0 & -1 \\ -1 & 0 & 0 \end{bmatrix}, \; B = \begin{bmatrix} 0 & 0 & 0 \\ 0 & 4 & 2 \\ 0 & 2 & 1 \end{bmatrix}, \; C = \begin{bmatrix} 1 & 2 & 0 \\ 2 & 8 & 0 \\ 0 & 0 & 4 \end{bmatrix}, \; P_1(0.5) = \begin{bmatrix} 1 & 0 & 0 \\ 0 & 3 & 0 \\ 0 & 0 & 5 \end{bmatrix}.$$

Solution *With the relevant statements, the Riccati differential equations can be solved, and the results are shown in Figure 7.9.*

```
>> A=[6,6,17; 1,0,-1; -1,0,0]; B=[0,0,0; 0,4,2; 0,2,1];
   C=[1,2,0; 2,8,0; 0,0,4]; P1=[1,0,0; 0,3,0; 0,0,5];
   [t,p]=ode45(@ric_de,[0.5,0],P1(:),[],A,B,C); plot(t,p)
```

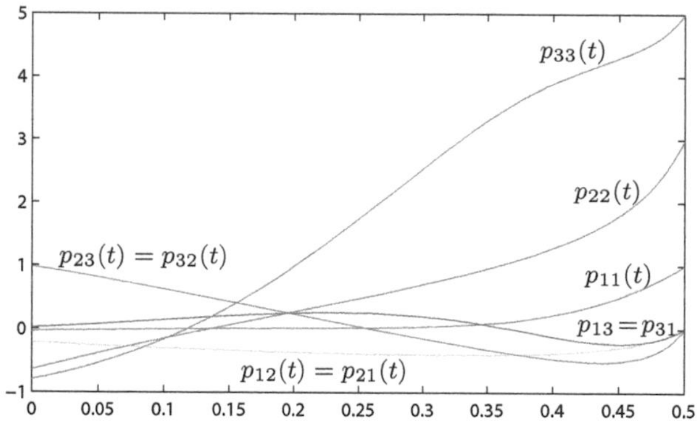

FIGURE 7.9: Numerical solutions of Riccati differential equation.

Now, with the value obtained at $t = 0$ *in the previous solution used as the initial vector, and Riccati equation can be solved again, and the results are also shown in Figure 7.9, and it can be seen that exactly the same results can be obtained.*

```
>> P=p(end,:); P0=reshape(P,size(A)); % the final value matrix
   [t1,p1]=ode45(@ric_de,[0,0.5],P0(:),[],A,B,C); plot(t1,p1) % solution
```

7.4 Solutions to Special Ordinary Differential Equations

From the introduction and examples in Section 7.2, one can easily convert a given ordinary differential equation into first-order vector form explicit one. The function `ode45()` can then be used to solve the equations. However, it is also shown from some examples, for instance the Van der Pol equation with $\mu = 1000$ cannot be solved by using the `ode45()`

function. Thus, other types of differential equations should be introduced, for instance, the stiff equations. Special MATLAB functions can be used to solve the stiff equation problems. Moreover, some special types of differential equations such as differential algebraic equations, implicit differential equations and delay differential equations will be discussed in this section.

7.4.1 Solutions of stiff ordinary differential equations

In many differential equations, some states change very rapidly while others may change very slowly. This type of differential equation is usually referred to as a *stiff equation*. The function `ode45()` is not suitable for stiff equations. An alternative function, `ode15s()`, can be used instead and this function has exactly the same syntax as that of the `ode45()` function.

Example 7.17 Find the numerical solutions to the Van der Pol equation, when $\mu = 1000$, which was used as a counter example in Example 7.11.

Solution *Similar to the statements given earlier, if the function* `ode15s()` *is used, the states can be obtained directly in 3.15 seconds, and the time responses of the states are shown in Figure 7.10.*

```
>> h_opt=odeset; h_opt.RelTol=1e-6; x0=[2;0]; t_final=3000;
   f=@(t,x,mu)[x(2); -mu*(x(1)^2-1)*x(2)-x(1)];
   tic, mu=1000; [t,y]=ode15s(f,[0,t_final],x0,h_opt,mu); toc
   subplot(121), plot(t,y(:,1)); subplot(122); plot(t,y(:,2))
```

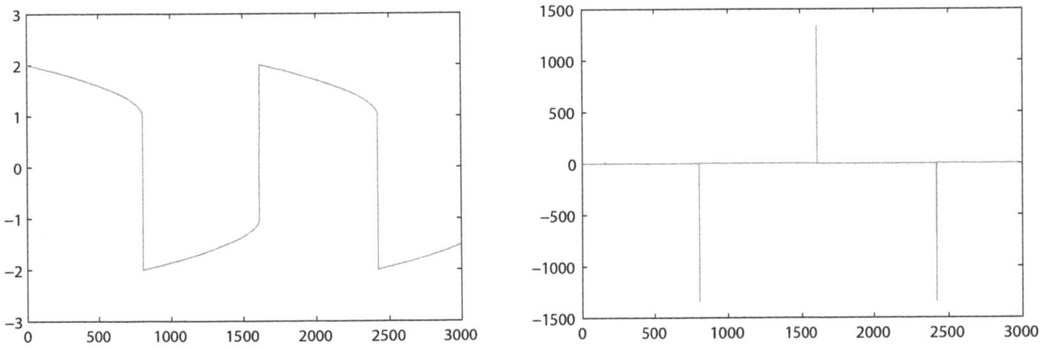

(a) state variable $x_1(t)$ (b) state variable $x_2(t)$

FIGURE 7.10: Solutions of Van der Pol equation with $\mu = 1000$.

It can be seen that, stiff ODE solver can be used to solve this type of equation rapidly. It can be seen that that the two equations at some particular points have almost vertical time responses, which makes the selection of step-size difficult.

Example 7.18 In classical textbooks regarding numerical solutions to ordinary differential equations, the following equation is regarded as stiff.

$$\boldsymbol{y}' = \begin{bmatrix} -21 & 19 & -20 \\ 19 & -21 & 20 \\ 40 & -40 & -40 \end{bmatrix} \boldsymbol{y}, \quad \boldsymbol{y}_0 = \begin{bmatrix} 1 \\ 0 \\ -1 \end{bmatrix}.$$

Find the numerical solutions with MATLAB.

Solution *The analytical solutions of the equations can be found symbolically by the following statements*

```
>> syms t; A=sym([-21,19,-20; 19,-21,20; 40,-40,-40]);
   y0=[1; 0; -1]; y=expm(A*t)*y0 % analytical solution of linear equation
```

which yields

$$y(t) = \begin{bmatrix} 0.5e^{-2t} + 0.5e^{-40t}(\cos 40t + \sin 40t) \\ 0.5e^{-2t} - 0.5e^{-40t}(\cos 40t + \sin 40t) \\ e^{-40t}(\sin 40t - \cos 40t) \end{bmatrix}.$$

Now let us consider the numerical solutions to the same problem. Prepare an anonymous function, and find the numerical solutions by the following statements

```
>> opt=odeset; opt.RelTol=1e-6;
   f=@(t,x)[-21,19,-20; 19,-21,20; 40,-40,-40]*x;
   tic,[t,y]=ode45(f,[0,1],[1;0;-1],opt); toc % find numerical solution
   x1=exp(-2*t); x2=exp(-40*t).*cos(40*t); x3=exp(-40*t).*sin(40*t);
   y1=[0.5*x1+0.5*x2+0.5*x3, 0.5*x1-0.5*x2-0.5*x3, -x2+x3];
   plot(t,y,t,y1,':') % compare exact and numerical solution
```

and it can be seen that only 0.16 seconds are used to find the solutions using the `ode45()` *function. The analytical and numerical solutions can be shown in Figure 7.11 (a). It can be seen that the accuracy of the results are rather high, and computation speed is also very high. The stiffness of the problem seems to be not very obvious. This is because the variable-step algorithm is used, and the step-size can adaptively be modified so the stiffness of the problem may not cause serious issues. However, if a fixed-step algorithm is used, for instance, the fourth-order Runge–Kutta algorithm is applied, the following statements*

```
>> tic, [t2,y2]=rk_4(f,[0,1,0.01],[1;0;-1]); toc, plot(t,y1,t2,y2,':')
```

produce numerical results shown in Figure 7.11 (b) together with the analytical curves. The elapsed time is 0.21 seconds, which is slightly longer than the direct use of `ode45()` *function. From the results, it can be seen that the fixed-step approach gives erroneous results if the step-size if large. Further reducing the step-size until* $h = 0.0001$ *seconds, one can still see the difference between the analytical and numerical results. However, this time, the time required is about 26 seconds, which is 162 times longer than the time required for the* `ode45()` *function. Thus, in practical applications, the variable-step algorithm should be used whenever possible.*

It can further be tested that in the fixed-step algorithm, if the step-size is 0.00001, the time required may as long as 8124 seconds, more than two hours.

It can be concluded that for many conventional stiff equations, one may still try to use the ordinary ODE solvers such as `ode45()` rather than the stiff equation solver. If for some examples, the time consumed using `ode45()` is too much, the stiff equation solver should be used instead. This will be illustrated in the following Example 7.19.

Example 7.19 Consider the following ordinary differential equations

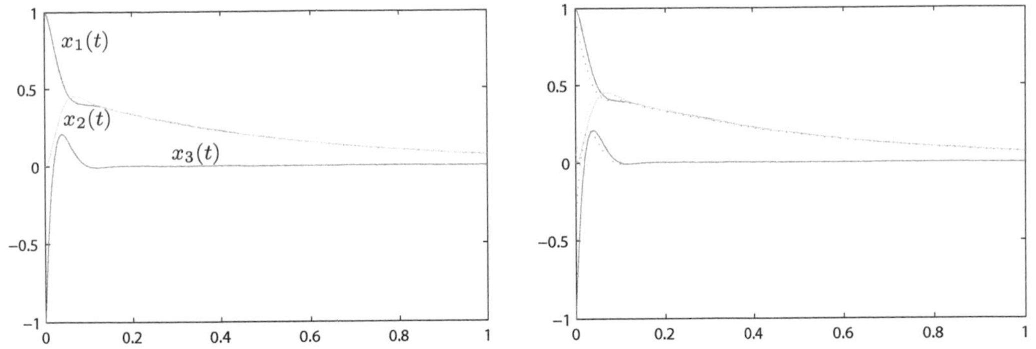

(a) variable-step solutions (b) fixed-step solutions

FIGURE 7.11: Comparisons of solutions of a traditional stiff equation.

$$\begin{cases} y_1'(t) = 0.04(1-y_1(t)) - (1-y_2(t))y_1(t) + 0.0001(1-y_2(t))^2 \\ y_2'(t) = -10^4 y_1(t) + 3000(1-y_2(t))^2, \end{cases}$$

where the initial values are $y_1(0)=0, y_2(0)=1$. Within the time interval $t \in (0,100)$, find a suitable algorithm for the initial value problem.

Solution *For the given differential equations, an anonymous function can be written to describe the right-hand-side functions, then, the following statements can be issued in the MATLAB command window to find the numerical solutions.*

```
>> f=@(t,y)[0.04*(1-y(1))-(1-y(2))*y(1)+0.0001*(1-y(2))^2;
            -10^4*y(1)+3000*(1-y(2))^2];
   tic, [t2,y2]=ode45(f,[0,100],[0;1]); toc % measure time elapsed
   length(t2), plot(t2,y2)                   % count number of points
```

After a long wait for about 43 seconds, the numerical solutions can be obtained. The number of points computed is 356,941. The numerical solutions are shown in Figure 7.12 (a). It is found that the function `ode45()` *takes too much time. Also, the step-sizes can be obtained from the following statements*

```
>> plot(t2(1:end-1),diff(t2)) % draw step-size
```

and the step-sizes are shown in Figure 7.12 (b). It can be seen that during the whole simulation period the step-sizes are very small. In most of the time interval, the step-sizes are about 0.004, which increases significantly the computation time. The adaptation in the step-size also consumes a tremendous amount of time.

Now consider the use of `ode15s()` *instead, the following statements can be used*

```
>> opt=odeset; opt.RelTol=1e-6; % use another solver and try again
   tic,[t,y]=ode15s(f,[0,100],[0;1],opt); toc, length(t), plot(t,y)
```

and it can be seen that the time required is reduced significantly to 0.26 seconds, which is about 1/160 of the time by `ode45()`. *The number of points computed is 169. The curves obtained are almost identical to the previous results.*

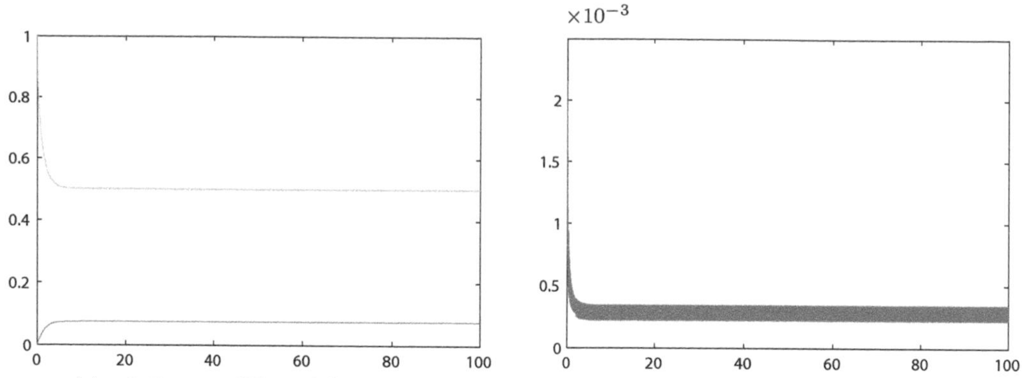

(a) solutions to differential equations (b) step-sizes in simulation

FIGURE 7.12: Solution of using the 4/5 RKF algorithm.

7.4.2 Solutions of implicit differential equations

The so-called *implicit differential equations* are those not convertible into the first-order vector form explicit differential equations as in (7-2-1). In earlier versions of MATLAB, solvers for these implicit equations were not provided. The following two examples demonstrate the implicit differential equation solution procedures.

Example 7.20 Consider the following implicit differential equation

$$\begin{cases} x_1'(t)\sin x_1(t) + x_2'(t)\cos x_2(t) + x_1(t) = 1 \\ -x_1'(t)\cos x_2(t) + x_2'(t)\sin x_1(t) + x_2(t) = 0, \end{cases}$$

where $x_1(0) = x_2(0) = 0$. Find the numerical solutions to the initial value problem.

Solution *Let* $\boldsymbol{x} = [x_1, x_2]^{\mathrm{T}}$. *The matrix form of the equation can be written as*

$$\boldsymbol{A}(\boldsymbol{x})\boldsymbol{x}' = \boldsymbol{B}(\boldsymbol{x}),\ \textit{where}\ \boldsymbol{A}(\boldsymbol{x}) = \begin{bmatrix} \sin x_1 & \cos x_2 \\ -\cos x_2 & \sin x_1 \end{bmatrix},\quad \boldsymbol{B}(\boldsymbol{x}) = \begin{bmatrix} 1 - x_1 \\ -x_2 \end{bmatrix}.$$

If $\boldsymbol{A}(\boldsymbol{x})$ *is a nonsingular matrix for all* $\boldsymbol{x}$*, the first-order vector form explicit equation can be expressed by* $\boldsymbol{x}' = \boldsymbol{A}^{-1}(\boldsymbol{x})\boldsymbol{B}(\boldsymbol{x})$. *Using the methods stated earlier, the numerical solutions can be obtained. However, it is not possible to prove theoretically that the matrix* $\boldsymbol{A}(\boldsymbol{x})$ *is a nonsingular matrix. Thus, one may initially assume that matrix* $\boldsymbol{A}(\boldsymbol{x})$ *is nonsingular. During the differential equation solving process, if there is no error message displayed, it will indicate that for the solutions the matrix* $\boldsymbol{A}(\boldsymbol{x})$ *is nonsingular. Thus, the solutions obtained are valid. If, however, error messages appear, then, the solutions obtained are useless.*

For the equations to be solved, an anonymous function can be written, and the following statements can be used

```
>> f=@(t,x)inv([sin(x(1)) cos(x(2)); -cos(x(2)) sin(x(1))])...
        *[1-x(1); -x(2)]; opt=odeset; opt.RelTol=1e-6;
   [t,x]=ode45(f,[0,10],[0; 0],opt); plot(t,x) % try to solve
```

The obtained state variables are shown in Figure 7.13. Since in the solution process, no error messages are given, this indicates that for the solution points, the matrix $\boldsymbol{A}(t)$ *is not singular. Thus, the solutions obtained may be valid.*

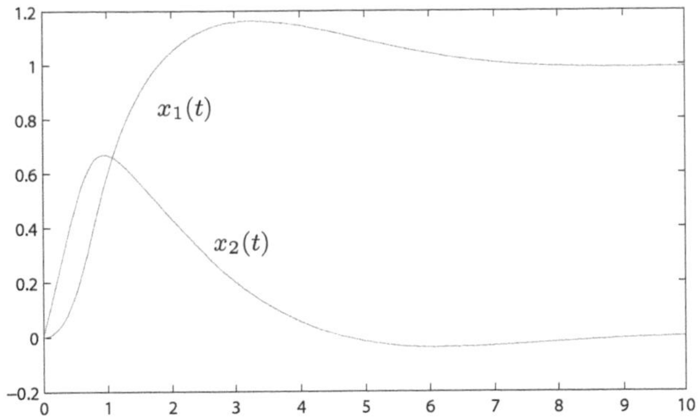

FIGURE 7.13: Time responses of the state variables.

Example 7.21 The previous example is very simple and can be converted into explicit differential equations immediately. Now consider a set of more complicated implicit equations given by

$$\begin{cases} x''(t)\sin y'(t) + (y''(t))^2 = -2x(t)y(t)\mathrm{e}^{-x'(t)} + x(t)x''(t)y'(t) \\ x(t)x''(t)y''(t) + \cos y''(t) = 3y(t)x'(t)\mathrm{e}^{-x(t)}. \end{cases}$$

The state variables can be selected as $x_1 = x, x_2 = x', x_3 = y, x_4 = y'$. The initial states of the equation are $\boldsymbol{x} = [1, 0, 0, 1]^{\mathrm{T}}$. Find the numerical solutions of the initial value problem.

Solution *Obviously the analytical solutions for state derivatives x_2' and x_4' cannot be written out as in Example 7.14. Hence, a numerical-based converting algorithm is introduced such that the numerical solutions for each state variable $\boldsymbol{x}'$ can be obtained.*

From the original equations, assume that $p_1 = x''$, $p_2 = y''$, then, the following equation can be written as

$$\begin{cases} p_1 \sin x_4 + p_2^2 + 2x_1x_3\mathrm{e}^{-x_2} - x_1p_1x_4 = 0 \\ x_1p_1p_2 + \cos p_2 - 3x_3x_2\mathrm{e}^{-x_1} = 0, \end{cases}$$

and from the following MATLAB statements, the algebraic equation solution statements can be embedded into the MATLAB function describing the differential equations. The MATLAB function can then, be written as

```
function dy=c7impode(t,x)
dx=@(p,x)[p(1)*sin(x(4))+p(2)^2+2*x(1)*x(3)*exp(-x(2))-x(1)*p(1)*x(4);
           x(1)*p(1)*p(2)+cos(p(2))-3*x(3)*x(2)*exp(-x(1))];
ff=optimset; ff.Display='off'; dx1=fsolve(dx,x([1,3]),ff,x);
dy=[x(2); dx1(1); x(4); dx1(2)]; % embed algebraic equation solver
```

Once calling the function, from the input arguments $\boldsymbol{x}$, the newly defined variables p_1, p_2 can be used in the anonymous function, and with the use of the `fsolve()` *function, the variables p_1 and p_2 can be numerically solved. In the above MATLAB statements, $\boldsymbol{x}$ is used as the additional variable. Thus, from this function, the variables p_i can be obtained. Since the obtained p_1, p_2 are in fact the derivatives of the state variables such that $p_1 = x_2'$, $p_2 = x_4'$, the explicit differential equations can then be obtained and the corresponding MATLAB scripts are given in the function as well.*

Once the explicit differential equations are obtained, the solutions to the implicit differential equations can be solved numerically using the following statements and the time responses of the states are shown in Figure 7.14.

```
>> [t,x]=ode15s(@c7impode,[0,2],[1,0,0,1]); plot(t,x) % solve ODE
```

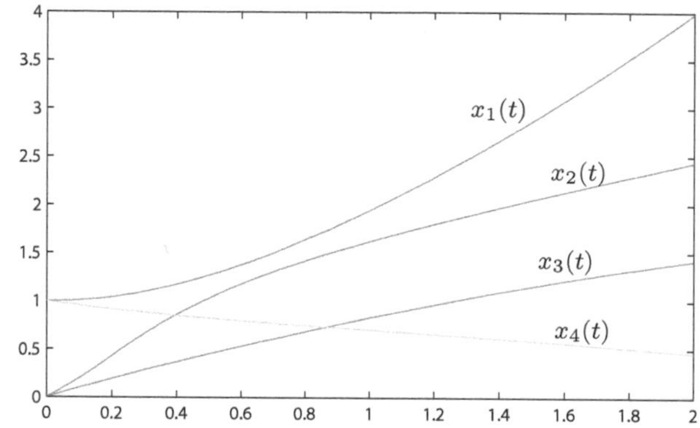

FIGURE 7.14: Time responses of the implicit differential equations.

Function `ode15i()` is provided for the solutions of implicit differential equations. If the mathematical descriptions to the implicit differential equations is written as

$$\boldsymbol{F}[t,\boldsymbol{x}(t),\boldsymbol{x}'(t)]=0, \text{ and } \boldsymbol{x}(t_0)=\boldsymbol{x}_0,\ \boldsymbol{x}'(t_0)=\boldsymbol{x}_0', \tag{7-4-1}$$

then, the function *fun* can be used to describe the implicit equations. The new MATLAB function `decic()` can be used to solve the compatible undefined initial conditions. Then, the solver function `ode15i()` can be used to solve the implicit differential equations.

$[\boldsymbol{x}_0^*,\boldsymbol{x}_0'^*]=$ `decic(`$fun, t_0, \boldsymbol{x}_0, \boldsymbol{x}_0^{\rm F}, \boldsymbol{x}_0', \boldsymbol{x}_0'^{\rm F}$`)` % find consistent initial values

`res=ode15i(`fun`,tspan,`$\boldsymbol{x}_0^*,\boldsymbol{x}_0'^*$`)` % solve implicit equations

The solution process of the implicit differential equations is different from the explicit equations. In numerically solving implicit equations, the initial state variables and their derivatives should both be declared, and they cannot be assigned arbitrarily, otherwise, there might be conflicting initial conditions. Before the solution, the $2n$ initial values $(\boldsymbol{x}_0, \boldsymbol{x}_0')$ can only have n independent ones. The rest of the values should be solved from the corresponding implicit algebraic equations. Thus, in the actual solution process, if one cannot determine the values $\boldsymbol{x}_0'^*$, the function `decic()` can be used to solve for compatible initial values. In the function call, $(\boldsymbol{x}_0, \boldsymbol{x}_0')$ can be any initial values, while $\boldsymbol{x}_0^{\rm F}$ and $\boldsymbol{x}_0'^{\rm F}$ are both n-dimensional column vectors and when the vector element is 1, it means that the corresponding initial value is to be maintained, otherwise, it indicates that the initial value should be resolved. From the corresponding algebraic equation solver, the compatible initial values $\boldsymbol{x}_0^*$ and $\boldsymbol{x}_0'^*$ can be obtained. The implicit differential equations can then be solved with the function `ode15i()`. The returned variables `res.x` and `res.y` are respectively $\boldsymbol{t}$ and $\boldsymbol{x}$ as in other ODE solvers. The following example demonstrates the numerical solution process of implicit differential equations.

Example 7.22 Solve the implicit differential equations given in Example 7.21 using the implicit ODE solver `ode15i()`.

Solution *Still select the state variables* $x_1 = x, x_2 = x', x_3 = y, x_4 = y'$ *and the original equations can be converted into the following:*

$$\begin{cases} x_1' - x_2 = 0 \\ x_2' \sin x_4 + x_4'^2 + 2\mathrm{e}^{-x_2} x_1 x_3 - x_1 x_2' x_4 = 0 \\ x_3' - x_4 = 0 \\ x_1 x_2' x_4' + \cos x_4' - 3\mathrm{e}^{-x_1} x_3 x_2 = 0. \end{cases}$$

Thus, the implicit differential equations can be entered in MATLAB using anonymous function. The definitions of the initial values $\boldsymbol{x}_0$ *are exactly the same as before. The function* `decic()` *can be used to determine the initial derivatives. Thus,* $\boldsymbol{x}_0^{\mathrm{F}}$ *should be assigned to a vector of ones. Since a consistent* $\boldsymbol{x}_0'$ *is expected, the indicator* $\boldsymbol{x}_0'^{\mathrm{F}}$ *should be set to an all zero vector. The following statements can be used to solve the implicit equations:*

```
>> f=@(t,x,xd)[xd(1)-x(2);
        xd(2)*sin(x(4))+xd(4)^2+2*exp(-x(2))*x(1)*x(3)-x(1)*xd(2)*x(4);
        xd(3)-x(4);
        x(1)*xd(2)*xd(4)+cos(xd(4))-3*exp(-x(1))*x(3)*x(2)];
   x0=[1,0,0,1]; xd0=[0;1;1;-1]; x0F=[1 1 1 1]; xd0F=[];  % retain x0
   [x0,xd0]=decic(f,0,x0,x0F,xd0,xd0F) % compute compatible conditions
   r=ode15i(f,[0,2],x0,xd0); plot(r.x,r.y) % draw the time responses
```

With the `decic()` *function, the consistent initial derivatives* $\boldsymbol{x}_0' = [0, 1.6833, 1, -0.5166]^{\mathrm{T}}$ *can be obtained. Then, the implicit equations can be solved, and the time responses of the states can be drawn. It can be seen that the results are exactly the same as the ones shown in Figure 7.14.*

7.4.3 Solutions to differential algebraic equations

The so-called *differential algebraic equation* (DAE) means that some of the differential equations are degenerated to algebraic equations. Thus, these algebraic equations appear as the constraints in the differential equations. Differential algebraic equations cannot be solved directly using the methods presented earlier.

The general form of the differential equations is given by

$$\boldsymbol{M}(t, \boldsymbol{x})\boldsymbol{x}' = \boldsymbol{f}(t, \boldsymbol{x}),\ \boldsymbol{x}(t_0) = \boldsymbol{x}_0, \tag{7-4-2}$$

where the $\boldsymbol{f}(t, \boldsymbol{x})$ function description is exactly the same as in the previous sections. For differential algebraic equations, the matrix $\boldsymbol{M}(t, \boldsymbol{x})$ is singular. Thus, in the solutions options for MATLAB functions, the `Mass` property can be used to describe the matrix $\boldsymbol{M}(t, \boldsymbol{x})$ in MATLAB. Then, the differential algebraic equations can be solved.

Example 7.23 Find the solutions to the following differential algebraic equation:

$$\begin{cases} x_1' = -0.2x_1 + x_2 x_3 + 0.3x_1 x_2 \\ x_2' = 2x_1 x_2 - 5x_2 x_3 - 2x_2^2 \\ 0 = x_1 + x_2 + x_3 - 1, \end{cases}$$

with initial conditions $x_1(0) = 0.8$, $x_2(0) = x_3(0) = 0.1$.

Solution *The last equation is an algebraic equation. It can also be regarded as a constraint among the three state variables. The matrix form of the differential algebraic equations can be written as*

$$\begin{bmatrix} 1 & 0 & 0 \\ 0 & 1 & 0 \\ 0 & 0 & 0 \end{bmatrix} \begin{bmatrix} x_1' \\ x_2' \\ x_3' \end{bmatrix} = \begin{bmatrix} -0.2x_1 + x_2x_3 + 0.3x_1x_2 \\ 2x_1x_2 - 5x_2x_3 - 2x_2^2 \\ x_1 + x_2 + x_3 - 1 \end{bmatrix}.$$

Clearly in MATLAB, function $\boldsymbol{f}(t, \boldsymbol{x})$ can be expressed by an anonymous function. The matrix $\boldsymbol{M}$ can also be entered into MATLAB workspace, and the following statements can be entered as well into MATLAB command window. In this problem, the solver `ode45()` *generates wrong results. Thus, the stiff equation based algorithms should be used instead.*

```
>> f=@(t,x)[-0.2*x(1)+x(2)*x(3)+0.3*x(1)*x(2);
       2*x(1)*x(2)-5*x(2)*x(3)-2*x(2)*x(2); x(1)+x(2)+x(3)-1];
   M=[1,0,0; 0,1,0; 0,0,0]; options=odeset; options.Mass=M;
   x0=[0.8; 0.1; 0.1]; [t,x]=ode15s(f,[0,20],x0,options); plot(t,x)
```

From the above statements, the differential algebraic equations can be directly solved and the time responses of the states are shown in Figure 7.15.

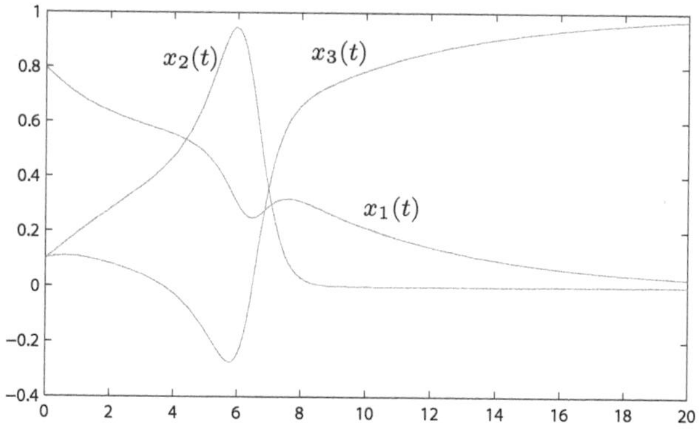

FIGURE 7.15: Numerical solutions to differential algebraic equations.

In fact, some of the differential algebraic equations can be converted into lower-order explicit differential equations. For instance, in the example, from the constraint, one may immediately find that $x_3(t) = 1 - x_1(t) - x_2(t)$. Substituting it into the other two equations yields

$$\begin{cases} x_1' = -0.2x_1 + x_2(1 - x_1 - x_2) + 0.3x_1x_2 \\ x_2' = 2x_1x_2 - 5x_2(1 - x_1 - x_2) - 2x_2^2, \end{cases}$$

and the original three-state differential algebraic equations can be converted into a second-order differential equation. A new anonymous function can be written to describe the equations. The results using the two methods are exactly the same.

```
>> x0=[0.8; 0.1]; % input initial conditions
   fDae=@(t,x)[-0.2*x(1)+x(2)*(1-x(1)-x(2))+0.3*x(1)*x(2);
                  2*x(1)*x(2)-5*x(2)*(1-x(1)-x(2))-2*x(2)*x(2)];
   [t1,x1]=ode45(fDae,[0,20],x0); plot(t1,x1,t1,1-sum(x1'))
```

Note that, in this converted case, even with the use of function `ode45()`, *the results are still valid.*

The implicit differential equation solver `ode15i()` *can also be used to solve this problem. An anonymous function can be written to describe the implicit equation. Let* $\boldsymbol{x}_0 =$ `[0.8,0.1,*]'` *and* $\boldsymbol{x}_0^{\rm F} =$ `[1,1,0]'`, *where* `*` *is used to denote free values. The following statements can be used to find compatible initial conditions. The original differential algebraic equation can be solved in this way, and the same results can be obtained. It can be seen that with the use of the method, a more straightforward solution process can be enjoyed.*

```
>> f=@(t,x,xd)[xd(1)+0.2*x(1)-x(2)*x(3)-0.3*x(1)*x(2);
     xd(2)-2*x(1)*x(2)+5*x(2)*x(3)+2*x(2)^2; x(1)+x(2)+x(3)-1];
   x0=[0.8;0.1;2]; x0F=[1;1;0]; xd0=[1;1;1]; xd0F=[];
   [x0,xd0]=decic(f,0,x0,x0F,xd0,xd0F) % compatible initial conditions
   res=ode15i(f,[0,20],x0,xd0); plot(res.x,res.y) % solve equations
```

With the above solution commands, the compatible initial values are $\boldsymbol{x}(0)=[0.8, 0.1, 0.1]^{\rm T}$ *and* $\boldsymbol{x}'(0) = [-0.126, 0.09, 1]^{\rm T}$.

Example 7.24 Solve the implicit differential equations given in Example 7.20 using the differential algebraic equation solver.

Solution *In Example 7.20, one converts the original equations to first-order explicit differential equations by inverting matrix* $\boldsymbol{A}(\boldsymbol{x})$. *In fact, an assumption has already been made that the matrix* $\boldsymbol{A}(t)$ *is nonsingular. Although it happens that the assumption is correct for this example, the numerical algorithm used is not quite reliable and convincing, strictly speaking. For this kind of problem, the differential algebraic equation solvers can also be used.*

For the original equations, an anonymous function can be written for the differential equation, and another anonymous function for the mass matrix. The differential algebraic equation can then be solved using the following statements:

```
>> f=@(t,x)[1-x(1); -x(2)]; % describe ODE with an anonymous function
   fM=@(t,x)[sin(x(1)),cos(x(2)); -cos(x(2)),sin(x(1))]; % mass matrix
   options=odeset; options.Mass=fM; options.RelTol=1e-6;
   [t,x]=ode45(f,[0,10],[0;0],options); plot(t,x) % solve and plot
```

and the results obtained are exactly the same as the ones shown in Figure 7.13.

7.4.4 Solutions of switching differential equations

The research on switching systems is an active research area in control system theory [7]. The so-called *switching system* is a system composed of several subsystems, and the system is switched among the subsystems under certain conditions, known as *switching laws*. The general form of the subsystems is expressed as

$$\boldsymbol{x}'(t) = \boldsymbol{f}_i(t, \boldsymbol{x}), \quad i = 1, \cdots, m. \tag{7-4-3}$$

The overall system is switched among various subsystems under the given switching laws. With the proper design of switching laws, the whole system $\boldsymbol{f}_i(\cdot)$ may be stabilized.

Example 7.25 Assume that the subsystems are described by $\boldsymbol{x}' = \boldsymbol{A}_i\boldsymbol{x}$, where

$$\boldsymbol{A}_1 = \begin{bmatrix} 0.1 & -1 \\ 2 & 0.1 \end{bmatrix}, \quad \boldsymbol{A}_2 = \begin{bmatrix} 0.1 & -2 \\ 1 & 0.1 \end{bmatrix}.$$

It can be seen that the two subsystems are unstable. The switching laws are:
(i) When $x_1x_2 < 0$, i.e., the states are in the II and IV quadrants, switch to $\boldsymbol{A}_1$
(ii) When $x_1x_2 \geqslant 0$, i.e., the states are in the I and III quadrants, switch to $\boldsymbol{A}_2$.
Under the initial states $x_1(0) = x_2(0) = 5$, please solve the switching differential equation with MATLAB.

Solution *Under the given switching laws, the switching system can be expressed by the following MATLAB function*

```
function dx=switch_sys(t,x)
if x(1)*x(2)<0, A=[0.1 -1; 2 0.1]; else, A=[0.1 -2; 1 0.1]; end
dx=A*x; % describe the switching ODE
```

The following commands can be used to solve the switching system, the time response and phase plane plot are obtained as shown in Figure 7.16. It can be seen that, the overall system composed of two unstable subsystems is stabilized under suitable switching laws.

```
>> [t,x]=ode45(@switch_sys,[0,30],[5;5]); % solve equations
   plot(t,x), figure; plot(x(:,1),x(:,2))
```

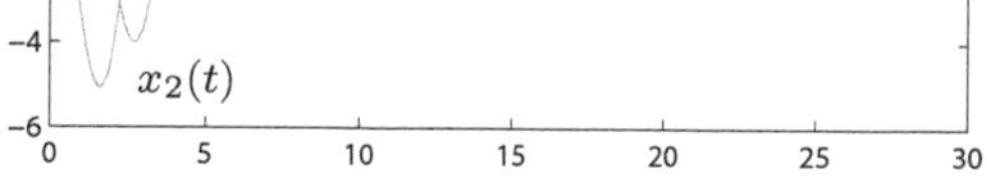

(a) time response

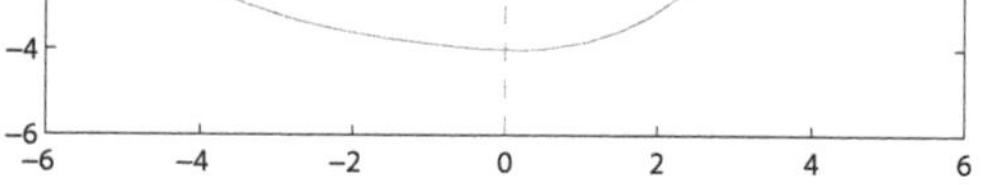

(b) phase plane trajectory

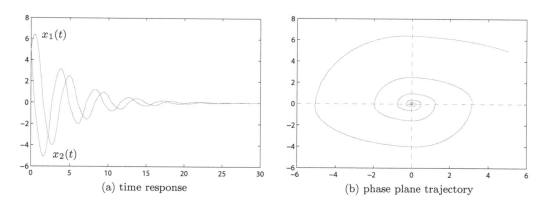

FIGURE 7.16: Time response and switching effect of the system.

In fact, for simplicity, the switching system can alternatively be described with the following anonymous function

```
>> f=@(t,x)(x(1)*x(2)<0)*[0.1 -1; 2 0.1]*x+...
          (x(1)*x(2)>=0)*[0.1 -2; 1 0.1]*x;
```

7.4.5 Solutions to linear stochastic differential equations

Consider a first-order linear differential equation described by

$$y'(t) + ay(t) = \gamma(t), \tag{7-4-4}$$

where a is a given constant. Assume that $\gamma(t)$ is Gaussian white noise with zero mean and a variance of σ^2. It is known that the output signal $y(t)$ is also Gaussian, with zero mean, and a variance of $\sigma_y^2 = \sigma^2/(2a)$. Assume that the input signal is kept a constant e_k within a computation step-size, the original system can be discretized as

$$y_{k+1} = \mathrm{e}^{-\Delta t/a} y_k + (1 - \mathrm{e}^{-\Delta t/a})\sigma e_k, \tag{7-4-5}$$

where Δt is a computation step-size, and e_k is a pseudorandom number which satisfies standard normal distribution $N(0,1)$. It can be seen that

$$E[y_{k+1}^2] = \mathrm{e}^{-2\Delta t/a} E[y_k^2] + 2\sigma \mathrm{e}^{-\Delta t/a} E[e_k y_k] + \sigma^2 (1 - \mathrm{e}^{-\Delta t/a})^2 E[e_k^2]. \tag{7-4-6}$$

If the input and output signals are stationary processes, then, $E[y_{k+1}^2]=E[y_k^2]=\sigma_y^2$, and since y_k and e_k are independent, then, $E[y_k e_k] = 0$. Also, $E[e_k^2] = 1$. It can be shown that

$$\sigma_y^2 = \frac{\sigma^2(1-\mathrm{e}^{-\Delta t/a})^2}{(1-\mathrm{e}^{-2\Delta t/a})} = \frac{\sigma^2(1-\mathrm{e}^{-\Delta t/a})}{1+\mathrm{e}^{-\Delta t/a}}. \tag{7-4-7}$$

If $\Delta t/a \to 0$, and the numerator and denominator in (7-4-7) are approximated by power series, it can be seen that

$$\sigma_y^2 = \lim_{\Delta t/a \to 0} \frac{\Delta t/a + o[(\Delta t/a)^2]}{2 + o(\Delta t/a)} \sigma^2 = \frac{\Delta t}{2a}\sigma^2. \tag{7-4-8}$$

It can be seen that the variance of the output signal depends on the computation step-size Δt. Of course, the results are not correct. This means that, when the input signal is random, conventional methods cannot be used in simulation.

Because of this, in some simulation software, other kinds of processes are used to replace Gaussian white noise. For instance, in the ACSL language, the Ornstein–Uhlenbeck process was used to approximate Gaussian white noise so that the input signal could maintain a constant within a certain frequency range. However, the simulation result thus obtained is not satisfactory either.

Assume that the linear state space representation

$$\boldsymbol{x}'(t) = \boldsymbol{A}\boldsymbol{x}(t) + \boldsymbol{B}[d(t) + \gamma(t)],\ y(t) = \boldsymbol{C}\boldsymbol{x}(t), \tag{7-4-9}$$

where $\boldsymbol{A}$ is an $n \times n$ matrix, $\boldsymbol{B}$ is an $n \times m$ matrix, $\boldsymbol{C}$ is an $r \times n$ matrix and $\boldsymbol{d}(t)$ is an $m \times 1$ deterministic input signal, $\gamma(t)$ is an $m \times 1$ Gaussian white noise vector, satisfying

$$E[\gamma(t)] = 0,\ E[\gamma(t)\gamma^{\mathrm{T}}(t)] = \boldsymbol{V}_\sigma \delta(t-\tau). \tag{7-4-10}$$

Introducing a vector $\gamma_{\mathrm{c}}(t) = \boldsymbol{B}\gamma(t)$, it can be shown that $\gamma_{\mathrm{c}}(t)$ is also a Gaussian white noise, satisfying

$$E[\gamma_{\mathrm{c}}(t)] = 0,\ E[\gamma_{\mathrm{c}}(t)\gamma_{\mathrm{c}}^{\mathrm{T}}(t)] = \boldsymbol{V}_{\mathrm{c}}\delta(t-\tau), \tag{7-4-11}$$

where $\boldsymbol{v}_{\mathrm{c}} = \sigma \boldsymbol{B}\boldsymbol{V}\boldsymbol{B}^{\mathrm{T}}$ is an $m \times m$ covariance matrix, then, (7-4-9) can be rewritten as

$$\boldsymbol{x}'(t) = \boldsymbol{A}\boldsymbol{x}(t) + \boldsymbol{B}\boldsymbol{d}(t) + \gamma_{\mathrm{c}}(t),\ y(t) = \boldsymbol{C}\boldsymbol{x}(t). \tag{7-4-12}$$

The analytical solutions of the states can be written as

$$\boldsymbol{x}(t) = \mathrm{e}^{-\boldsymbol{A}t}\boldsymbol{x}(t_0) + \int_{t_0}^{t} \mathrm{e}^{\boldsymbol{A}(t-\tau)}\boldsymbol{d}(\tau)\boldsymbol{B}\mathrm{d}\tau + \int_{t_0}^{t} \gamma_{\mathrm{c}}(t)\mathrm{d}\tau. \tag{7-4-13}$$

Assume that $t_0 = k\Delta t$, $t = (k+1)\Delta t$, where Δt is the computation step-size, and assume that within a computation step-size, the deterministic input $d(t)$ is constant, that is, if $\Delta t \leqslant t \leqslant (k+1)\Delta t$, $d(t) = d(k\Delta t)$. The discrete form of (7-4-13) can be written as

$$\boldsymbol{x}[(k+1)\Delta t] = \boldsymbol{F}\boldsymbol{x}(k\Delta t)+\boldsymbol{G}d(k\Delta t)+\gamma_{\rm d}(k\Delta t),\ \ y(k\Delta t) = \boldsymbol{C}\boldsymbol{x}(k\Delta t), \tag{7-4-14}$$

where $\boldsymbol{F} = \mathrm{e}^{\boldsymbol{A}\Delta t}$, $\boldsymbol{G} = \displaystyle\int_0^{\Delta t} \mathrm{e}^{\boldsymbol{A}(\Delta t-\tau)}\boldsymbol{B}\,\mathrm{d}\tau$, and

$$\gamma_{\rm d}(k\Delta t) = \int_{k\Delta t}^{(k+1)\Delta t} \mathrm{e}^{\boldsymbol{A}[(k+1)\Delta t-\tau]}\gamma_{\rm c}(t)\mathrm{d}\tau = \int_0^{\Delta t} \mathrm{e}^{\boldsymbol{A}t}\gamma_{\rm c}[(k+1)\Delta t-\tau]\,\mathrm{d}\tau. \tag{7-4-15}$$

It can be seen that $\boldsymbol{F}$ and $\boldsymbol{G}$ matrices are exactly the same as those in deterministic systems. When there exists a stochastic input, discretization of the system is slightly different. It can be shown that $\gamma_{\rm d}(t)$ is also a Gaussian white noise vector, satisfying

$$E[\gamma_{\rm d}(k\Delta t)] = 0,\ \ E[\gamma_{\rm d}(k\Delta t)\gamma_{\rm d}^{\rm T}(j\Delta t)] = \boldsymbol{V}\delta_{kj}, \tag{7-4-16}$$

where $\boldsymbol{V} = \displaystyle\int_0^{\Delta t} \mathrm{e}^{\boldsymbol{A}t}\boldsymbol{V}_{\rm c}\mathrm{e}^{\boldsymbol{A}^{\rm T}t}\mathrm{d}t$. With Taylor series techniques

$$\boldsymbol{V} = \int_0^{\Delta t} \sum_{k=0}^{\infty} \frac{\boldsymbol{R}_k(0)}{k!}t^k\,\mathrm{d}t = \sum_{k=0}^{\infty} \boldsymbol{V}_k, \tag{7-4-17}$$

where $\boldsymbol{R}_k(0)$ and $\boldsymbol{V}_k$ can recursively be obtained as

$$\begin{cases} \boldsymbol{R}_k(0) = \boldsymbol{A}\boldsymbol{R}_{k-1}(0) + \boldsymbol{R}_{k-1}(0)\boldsymbol{A}^{\rm T} \\ \boldsymbol{V}_k = \dfrac{\Delta t}{k+1}(\boldsymbol{A}\boldsymbol{V}_{k-1} + \boldsymbol{V}_{k-1}\boldsymbol{A}^{\rm T}), \end{cases} \tag{7-4-18}$$

with initial values $\boldsymbol{R}_0(0) = \boldsymbol{R}(0) = \boldsymbol{V}_{\rm c}$, $\boldsymbol{V}_0 = \boldsymbol{V}_{\rm c}\Delta t$. With the singular value decomposition technique, matrix $\boldsymbol{V}$ can be written as $\boldsymbol{V} = \boldsymbol{U}\boldsymbol{\Gamma}\boldsymbol{U}^{\rm T}$, where $\boldsymbol{U}$ is an orthogonal matrix, and $\boldsymbol{\Gamma}$ is a diagonal matrix containing nonzero elements. Cholesky factorization can be performed such that $\boldsymbol{V} = \boldsymbol{D}\boldsymbol{D}^{\rm T}$, and $\gamma_{\rm d}(k\Delta t) = \boldsymbol{D}\boldsymbol{e}(k\Delta t)$, where $\boldsymbol{e}(k\Delta t)$ is an $n \times 1$ vector, and $\boldsymbol{e}(k\Delta t) = [e_k, e_{k+1}, \cdots, e_{k+n-1}]^{\rm T}$, such that the components e_k satisfy a standard normal distribution, that is, $e_k \sim \mathrm{N}(0,1)$. A recursive solution can be obtained

$$\boldsymbol{x}[(k+1)\Delta t] = \boldsymbol{F}\boldsymbol{x}(k\Delta t) + \boldsymbol{G}d(k\Delta t) + \boldsymbol{D}\boldsymbol{e}(k\Delta t),\ \ y(k\Delta t) = \boldsymbol{C}\boldsymbol{x}(k\Delta t). \tag{7-4-19}$$

Based on the above algorithm, the discretization algorithm for continuous linear stochastic differential equation can be written as

```
function [F,G,D,C]=sc2d(G,V,T)
G=ss(G); G=balreal(G); A=G.a; B=G.b; C=G.c; [F,G]=c2d(A,B,T);
V0=B*V*B'*T; Vd=V0; vmax=sum(sum(abs(Vd))); vv=vmax; v0=1; i=1;
while (v0<1e-10*vmax) % terminate condition
   V1=T/(i+1)*(A*V0+V0*A'); v0=sum(abs(V1(:)));
   Vd=Vd+V1; V0=V1; vv=[vv v0]; i=i+1;
end
[U,S,V0]=svd(Vd); V0=sqrt(diag(S)); Vd=diag(V0); D=U*Vd;
```

In simulation, a set of pseudorandom numbers can be generated, and the vector $\boldsymbol{e}(k\Delta t)$

can be constructed. Then, the state vector $\boldsymbol{x}[(k+1)\Delta t]$ at the next time instance can be evaluated, and the current output signal $y(k\Delta t)$ can be obtained

$$y'(t) = -\frac{1}{a}y(t) + \frac{1}{a}\gamma_0(t). \tag{7-4-20}$$

The discrete form of the output signal can be written as

$$y_{k+1} = \mathrm{e}^{\Delta t/a}y_k + \sigma\sqrt{\frac{1}{2a}\left(1-\mathrm{e}^{-2\Delta t/a}\right)}e_k. \tag{7-4-21}$$

Example 7.26 Consider a transfer function model defined as

$$G(s) = \frac{s^3+7s^2+24s+24}{s^4+10s^3+35s^2+50s+24}.$$

If a white noise signal is used to excite the system, solve the differential equation.

Solution *We can select a sample time* $T = 0.001$, *the discretized model can be obtained with the following MATLAB statements*

```
>> G=tf([1,7,24,24],[1,10,35,50,24]); T=0.02; [F,G0,D,C]=sc2d(G,1,T)
```

The matrices in the discretized model can be found as

$$\boldsymbol{F} = \begin{bmatrix} 0.9838 & -0.0067 & 0.0132 & 0.0013 \\ 0.0067 & 0.9883 & 0.0702 & 0.0036 \\ 0.0132 & -0.0702 & 0.8653 & -0.0257 \\ 0.0013 & -0.0036 & -0.0257 & 0.9684 \end{bmatrix},$$

$$\boldsymbol{G}_0 = \begin{bmatrix} 0.0182 \\ -0.0036 \\ -0.0076 \\ -0.0007 \end{bmatrix}, \quad \boldsymbol{D} = \begin{bmatrix} -0.1303 & 0 & 0 & 0 \\ 0.0235 & 0 & 0 & 0 \\ 0.0594 & 0 & 0 & 0 \\ 0.0061 & 0 & 0 & 0 \end{bmatrix},$$

and $\boldsymbol{C} = [0.9216, 0.1663, -0.4201, -0.0431]$. *From the discrete model, 30,000 simulation points can be calculated with the following statements, and the time response of the system is shown in Figure 7.17(a).*

```
>> n_point=30000; r=randn(n_point+4,1); r=r-mean(r);
   y=zeros(n_point,1); x=zeros(4,1); d0=0;
   for i=1:n_point, x=F*x+G0*d0+D*r(i:i+3); y(i)=C*x; end
   t=0:.02:(n_point-1)*0.02; subplot(121), plot(t,y)
   v=covar(G,1); xx=linspace(-2.5,2.5,30); yy=hist(y,xx);
   yy=yy/(30000*(xx(2)-xx(1))); yp=exp(-xx.^2/(2*v))/sqrt(2*pi*v);
   subplot(122), bar(xx,yy), hold on; plot(xx,yp)
```

It can be seen that the output signal behaves in a disorderly way. For systems with random signal, statistical analysis may be more informative. Histograms can be used to approximate probability density functions, as shown in Figure 7.17(b). The result obtained from simulation data agrees well with the theoretical results, and this means that the simulation results are valid for stochastic differential equation.

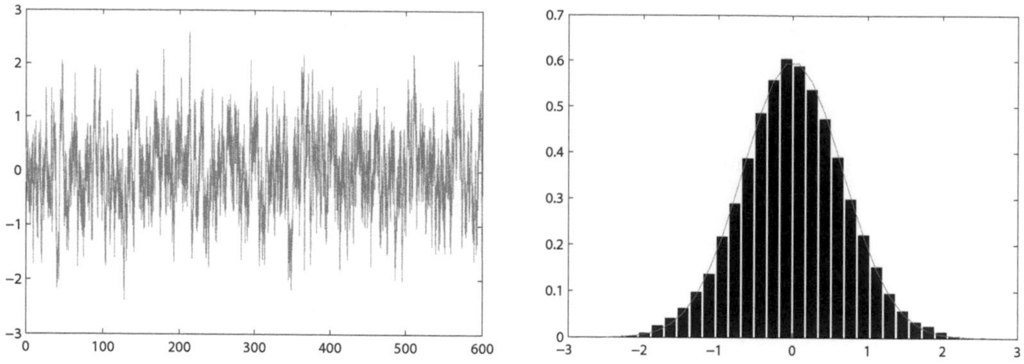

(a) time response (b) probability density function

FIGURE 7.17: Response to stochastic inputs.

7.5 Solutions to Delay Differential Equations

The differential equations presented so far have the standard form of $\boldsymbol{x}'(t) = \boldsymbol{f}(t, \boldsymbol{x}(t))$, where all the signals in the equations happen at the same time t. If some of the signal contains not only the values at current time t, but also have values in the past, the differential equations are referred to as *delay differential equations*. In this section, numerical solutions of various delay differential equations are presented, including general delay differential equations, neutral-type delay differential equations and variable delay-time differential equations.

7.5.1 Solutions of typical delay differential equations

The general form of the delay differential equations is given by

$$\boldsymbol{x}'(t) = \boldsymbol{f}(t, \boldsymbol{x}(t), \boldsymbol{x}(t-\tau_1), \boldsymbol{x}(t-\tau_2), \cdots, \boldsymbol{x}(t-\tau_n)), \tag{7-5-1}$$

where $\tau_i \geqslant 0$ are the delay constants for state variables $\boldsymbol{x}(t)$.

A MATLAB function `dde23()` [8] is provided to solve numerically delay differential equations using implicit Runge–Kutta algorithms, with the syntax

```
sol=dde23(fun1,τ,fun2,[t0,tf],options)
```

where $\boldsymbol{\tau} = [\tau_1, \tau_2, \cdots, \tau_n]$, *fun*$_1$ is the function describing the delay differential equations and *fun*$_2$ is used to describe the history of the state vector for $t \leqslant t_0$, which can either be a MATLAB function or a constant. The returned variable `sol` is a structure, with its `sol.x` and `sol.y` fields describing the time vector $\boldsymbol{t}$ and states matrix $\boldsymbol{x}$, respectively. The returned variable $\boldsymbol{x}$ is different from the $\boldsymbol{x}$ matrix from the `ode45()` function. It is arranged on a row basis instead of a column.

The entrance of function *fun*$_1$ is *fun*$_1$=`@(t,x,Z)`, where $\boldsymbol{Z}$ is used to describe delayed states, with the kth column, $\boldsymbol{Z}$`(:,k)`, storing the state vector $\boldsymbol{x}(t-\tau_k)$.

Example 7.27 The delay differential equations are given by

$$\begin{cases} x'(t) = 1 - 3x(t) - y(t-1) - 0.2x^3(t-0.5) - x(t-0.5) \\ y''(t) + 3y'(t) + 2y(t) = 4x(t), \end{cases}$$

where when $t \leqslant 0$, $x(t) = y(t) = y'(t) = 0$. Solve numerically the delay differential equations.

Solution *The values of $x(t)$, $y(t)$ at time instants t, $t-1$ and $t-0.5$ are involved in the delay differential equations, thus, special functions are required for the equations. One straightforward way is to introduce a set of state variables, such that $x_1(t) = x(t), x_2(t) = y(t), x_3(t) = y'(t)$, then, the original equation can be transformed into the following first-order vector form explicit delay differential equations such that*

$$\begin{cases} x_1'(t) = 1 - 3x_1(t) - x_2(t-1) - 0.2x_1^3(t-0.5) - x_1(t-0.5) \\ x_2'(t) = x_3(t) \\ x_3'(t) = 4x_1(t) - 2x_2(t) - 3x_3(t). \end{cases}$$

Two delay constants $\tau_1 = 1$ and $\tau_2 = 0.5$ can be defined. Thus, from the first state equation, the delay constant to the first state $x_1(t)$, both the delay constants τ_1 and τ_2 are involved. However, for the second state $x_2(t)$, only the delay constant τ_2 is used. The delay differential equation can be expressed by

```
function dx=c7exdde(t,x,Z)
z1=Z(:,1); z2=Z(:,2);
dx=[1-3*x(1)-z1(2)-0.2*z2(1)^3-z2(1); x(3); 4*x(1)-2*x(2)-3*x(3)];
```

where $\boldsymbol{z}_k$ is the state vector of $\boldsymbol{x}(t-\tau_k)$, and can be extracted from the input argument `Z(:,k)`*. Alternatively, the delay equation can be rewritten as*

$$\begin{cases} x_1'(t) = 1 - 3x_1(t) - z_{2,1}(t) - 0.2z_{1,2}^3(t) - z_{1,2}(t) \\ x_2'(t) = x_3(t) \\ x_3'(t) = 4x_1(t) - 2x_2(t) - 3x_3(t), \end{cases}$$

where, $z_{i,k} =$ `Z(i,k)`*. Thus, the differential equation can be implemented with anonymous function as*

```
>> f=@(t,x,Z)[1-3*x(1)-Z(2,1)-0.2*Z(1,2)^3-Z(1,2);
                x(3); 4*x(1)-2*x(2)-3*x(3)];
```

Then, the following statements can be used to find the numerical solutions to the delay differential equations

```
>> lags=[1 0.5]; tx=dde23(@c7exdde,lags,zeros(3,1),[0,10]);
   plot(tx.x,tx.y(2,:)) % please note the fields x and y in solution tx
```

and the time response of $y(t)$ can be obtained as shown in Figure 7.18 (a).

It should be noted that, if constant vector $\boldsymbol{x}_0$ is used to describe the history function fun_2, it means that for $t \leqslant t_0$, the history value of the states remain at $\boldsymbol{x}_0$. In the next example, the delay equations with time-varying history functions will be demonstrated.

Example 7.28 Reconsider the delay differential equation in Example 7.27. If the history functions of the state are given by $x_1(t) = \mathrm{e}^{2.1t}, x_2(t) = \sin t, x_3(t) = \cos t$, for $t \leqslant 0$. Please solve the delay equation solution again.

Solution *Anonymous functions for describing the history of states at $t \leqslant 0$ are given below.*

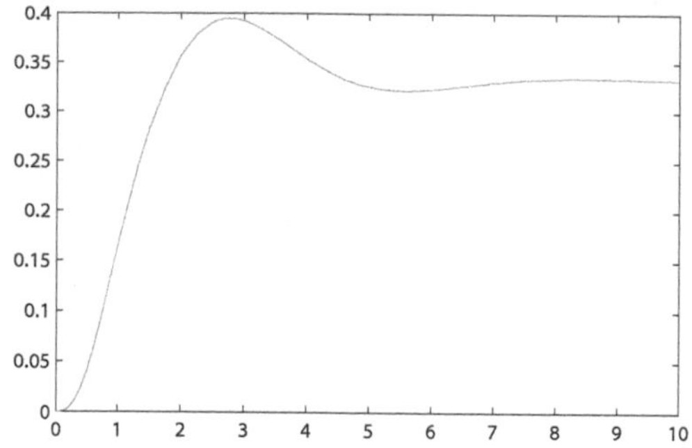

(a) zero history states

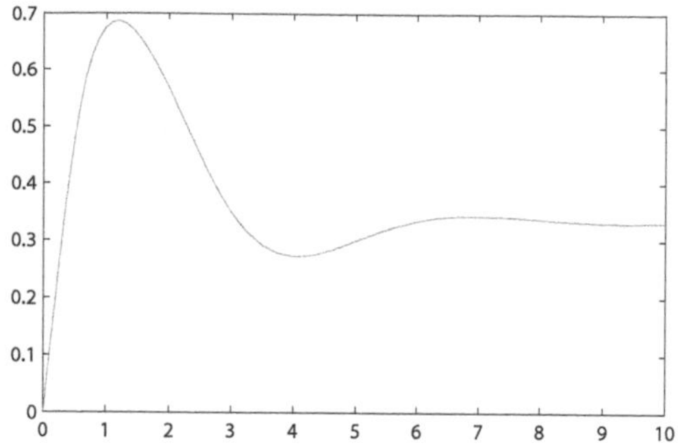

(b) history state function

FIGURE 7.18: Numerical solutions to the delay differential equations.

The new solution of the delay differential equation can also be found, as shown in Figure 7.18 (b).

```
>> f=@(t,x,Z)[1-3*x(1)-Z(2,1)-0.2*Z(1,2)^3-Z(1,2);
              x(3); 4*x(1)-2*x(2)-3*x(3)]; % describe the delay ODE
   f2=@(t,x)[exp(2.1*t); sin(t); cos(t)];  % describe the history functions
   lags=[1 0.5]; tx=dde23(f,lags,f2,[0,10]); plot(tx.x,tx.y(2,:))
```

7.5.2 Solutions of differential equations with variable delays

Function `ddesd()` provided in MATLAB can be used to solve differential equations with variable delays. The time delay terms can be described with MATLAB functions, so that the differential equations with variable delays can be described. Of course, the method can be extended, such that function `ddensd()` can be used to solve neutral-type differential equations. The syntax of function `ddesd()` is

`sol = ddesd(`fun_1`,`f_τ`,`fun_2`,[`t_0`,`t_f`],`*options*`)`

where, f_τ is the function handle of the delay function, and it can be described with M-functions or anonymous functions.

Example 7.29 If the history of the states are zero, please solve the following delay differential equations

$$\begin{cases} x_1'(t) = -2x_2(t) - 3x_1(t-0.2|\sin t|) \\ x_2'(t) = -0.05x_1(t)x_3(t) - 2x_2(t-0.8) + 2 \\ x_3'(t) = 0.3x_1(t)x_2(t)x_3(t) + \cos(x_1(t)x_2(t)) + 2\sin 0.1t^2. \end{cases}$$

Solution *It is obvious that there exists variable time delays, i.e., the state signal at time instance* $t-0.2|\sin t|$. *Thus, function* `dde23()` *is not suitable in the solutions of these equations. It can be seen that the first delay term is at time* $t-0.2|\sin t|$, *the second is at constant* $t-0.8$, *and history states for* $t \leqslant 0$ *are zeros. The following statements can be used to solve the equation with variable delays, and the result is shown in Figure 7.19.*

```
>> tau=@(t,x)[t-0.2*abs(sin(t)); t-0.8];
   f=@(t,x,Z)[-2*x(2)-3*Z(1,1); -0.05*x(1)*x(3)-2*Z(2,2)+2;
```

```
            0.3*x(1)*x(2)*x(3)+cos(x(1)*x(2))+2*sin(0.1*t^2)];
   sol=ddesd(f,tau,zeros(3,1),[0,10]); plot(sol.x,sol.y)
```

Different control parameters or algorithms can be used to validate the results. For instance relative error tolerance is used, and it can be seen that exactly the same result is obtained.

```
>> ff=odeset; ff.RelTol=1e-12; sol=ddesd(f,tau,zeros(3,1),[0,10],ff);
   hold on; plot(sol.x,sol.y)
```

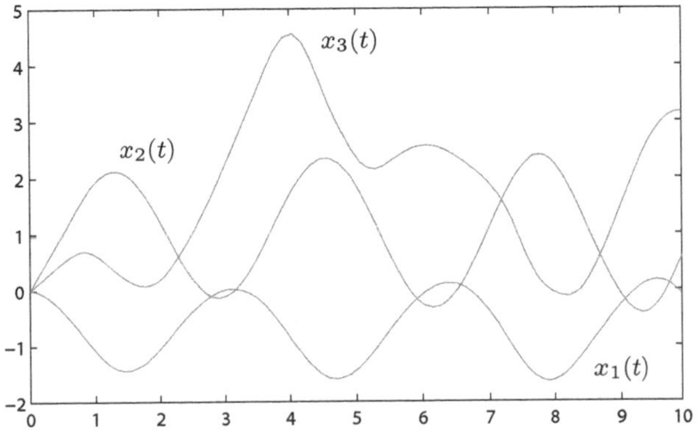

FIGURE 7.19: Numerical solutions of equations with variable delays.

It should be noted that, although the second delay time constant is 0.8, it should be expressed as $t-0.8$, *rather than 0.8, otherwise, wrong results will be returned. In fact, if it is written as 0.8, then,* $x_2(0.8)$ *is used, rather than the correct* $x_2(t-0.8)$.

Example 7.30 Consider the problem in the previous equations again. If the history of states are given by $x_1(t)=\sin(t+1), x_2(t)=\cos t, x_3(t)=\mathrm{e}^{3t}$, for $t\leqslant 0$, please solve the equations. Also, assume that the history of the states at $t<0$ are zeros, while at $t=0$, the initial values of the states satisfy the history formula, please solve again the equations.

Solution *Is it was demonstrated earlier, the history of states can be expressed by an anonymous function. Thus, the differential equations with nonzero history of states can be solved with the following statements, and the result is shown in Figure 7.20 (a).*

```
>> tau=@(t,x)[t-0.2*abs(sin(t)); t-0.8];   % the delay functions
   f=@(t,x,Z)[-2*x(2)-3*Z(1,1); -0.05*x(1)*x(3)-2*Z(2,2)+2;
             0.3*x(1)*x(2)*x(3)+cos(x(1)*x(2))+2*sin(0.1*t^2)];
   f2=@(t,x)[sin(t+1); cos(t); exp(3*t)]; % the history functions
   sol=ddesd(f,tau,f2,[0,10]); plot(sol.x,sol.y) % solve the DDE
```

If the history of states are zeros, and only at $t=0$, *the initial values are nonzero, the history can be expressed again, and the results are obtained as shown in 7.20(b). It can be seen that the two cases have large differences.*

```
>> f2=@(t,x)[sin(t+1); cos(t); exp(3*t)]*(t==0);
   sol=ddesd(f,tau,f2,[0,10]); plot(sol.x,sol.y)
```

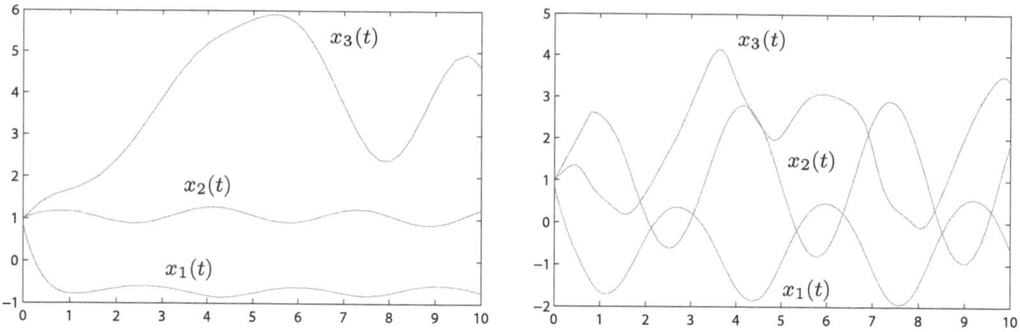

(a) nonzero history (b) zero history, nonzero initial states

FIGURE 7.20: Solutions of differential equations with nonzero initial values.

Example 7.31 Solve the slightly modified differential equations

$$\begin{cases} x_1'(t) = -2x_2(t) - 3x_1(t - 0.2|\sin t|) \\ x_2'(t) = -0.05x_1(t)x_3(t) - 2x_2(\alpha t) + 2, \quad \text{where } \alpha = 0.77 \\ x_3'(t) = 0.3x_1(t)x_2(t)x_3(t) + \cos(x_1(t)x_2(t)) + 2\sin 0.1t^2. \end{cases}$$

Solution *It can be seen that in the second equation, there is a term $x_2(0.77t)$, it means that the equation used the value at $0.77t$ of signal x_2. It could be an extremely complicated problem to solve, however, the user can just fill $0.77{*}t$ in the time delay function, and the difficult tasks can be handled internally by MATLAB. The following statements can be used, and the result can be obtained as shown in Figure 7.21.*

```
>> tau=@(t,x)[t-0.2*abs(sin(t)); 0.77*t]; % describe the delay time
   f=@(t,x,Z)[-2*x(2)-3*Z(1,1); -0.05*x(1)*x(3)-2*Z(2,2)+2;
              0.3*x(1)*x(2)*x(3)+cos(x(1)*x(2))+2*sin(0.1*t^2)];
   sol=ddesd(f,tau,zeros(3,1),[0,10]); plot(sol.x,sol.y) % solve DDE
```

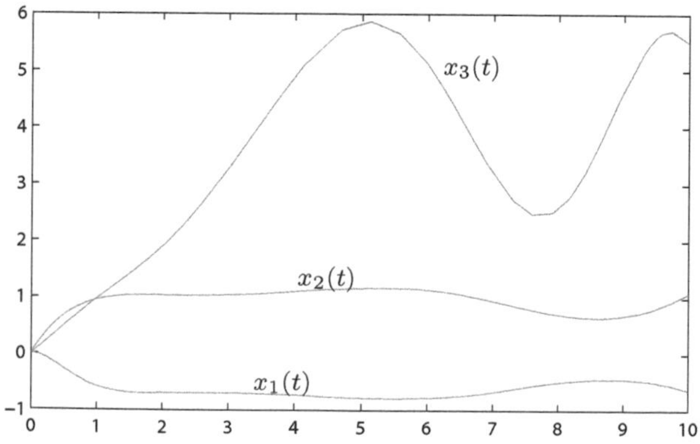

FIGURE 7.21: Numerical solutions of equations with variable delays.

Please note that in the examples, the delay signals $x_i(\alpha t)$ and $x_i(t - \beta(t))$ can be used, however, the delay time should be less than t, i.e., $\alpha \leqslant 1$ and $\beta(t) \geqslant 0$, meaning the signal happened in the past, otherwise, the signal value $x_i(t)$ is used automatically in the DDE solver, without any warning.

7.5.3 Solutions of neutral-type delay differential equations

The general form of neutral-type delay differential equation is

$$\boldsymbol{x}'(t) = \boldsymbol{f}(t, \boldsymbol{x}(t), \boldsymbol{x}(t-\boldsymbol{\tau}_1), \boldsymbol{x}'(t-\boldsymbol{\tau}_2)), \tag{7-5-2}$$

where, the current and past values of the derivatives of the state variables appear simultaneously in the equation. Delay vectors $\boldsymbol{\tau}_1 = [\tau_{p1}, \tau_{p2}, \cdots, \tau_{pm}]$ and $\boldsymbol{\tau}_2 = [\tau_{q1}, \tau_{q2}, \cdots, \tau_{qk}]$ are used to describe the delays in the states and derivatives of the states, respectively. Neutral-type delay differential equations cannot be solved with function `dde23()`. It can only be solved with `ddensd()` function (MATLAB 8.0 onwards), with the syntax

`sol = ddensd(`*fun*$_1$`,`$\boldsymbol{\tau}_1$`,`$\boldsymbol{\tau}_2$`,`*fun*$_2$`,[`t_0`,`t_f`],`*options*`)`

If the delay time are not constants, references to `ddesd()` function call, and change $\boldsymbol{\tau}_1$ and $\boldsymbol{\tau}_2$ to function handles. They can be described by M-functions or anonymous functions.

Example 7.32 Please solve the following neutral-type delay differential equation

$$\boldsymbol{x}'(t) = \boldsymbol{A}_1\boldsymbol{x}(t-0.15) + \boldsymbol{A}_2\boldsymbol{x}'(t-0.5) + \boldsymbol{B}u(t),$$

where, the input signal is $u(t) \equiv 1$, and the matrices are given by

$$\boldsymbol{A}_1 = \begin{bmatrix} -13 & 3 & -3 \\ 106 & -116 & 62 \\ 207 & -207 & 113 \end{bmatrix}, \quad \boldsymbol{A}_2 = \begin{bmatrix} 0.02 & 0 & 0 \\ 0 & 0.03 & 0 \\ 0 & 0 & 0.04 \end{bmatrix}, \quad \boldsymbol{B} = \begin{bmatrix} 0 \\ 1 \\ 2 \end{bmatrix}.$$

Solution *Since the derivative terms at different time, $\boldsymbol{x}'(t)$ and $\boldsymbol{x}'(t-0.5)$, appear in the equation simultaneously, the function* `dde23()` *cannot be used. Function* `ddensd()` *can be used directly in solving the problem. The delay constants for the states and its derivatives are $\tau_1 = 0.15$ and $\tau_2 = 0.5$. The following anonymous can be used to describe neutral-type delay differential equations, and then, the equations can be solved directly. The states are obtained as shown in Figure 7.22.*

```
>> A1=[-13,3,-3; 106,-116,62; 207,-207,113]; u=1;
   A2=diag([0.02,0.03,0.04]); B=[0; 1; 2]; % input matrices
   f=@(t,x,z1,z2)A1*z1+A2*z2+B*u; x0=zeros(3,1); % neutral-type DDE
   sol=ddensd(f,0.15,0.5,x0,[0,15]); plot(sol.x,sol.y) % solve and plot
```

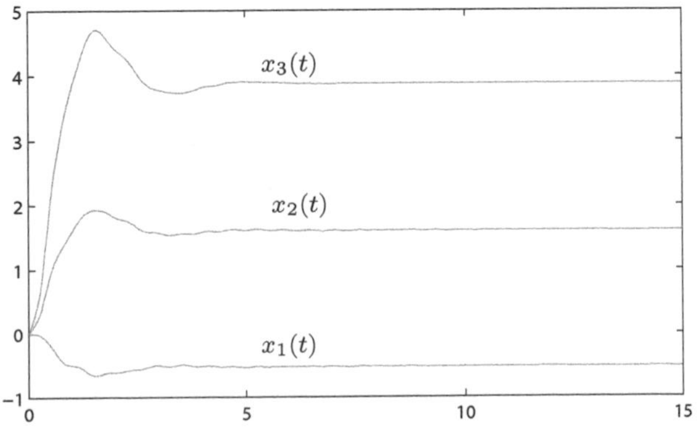

FIGURE 7.22: Numerical solutions of neutral-type delay differential equations.

Example 7.33 Consider again the non-neutral-type delay differential equation with variable delays, studied in Example 7.31. Please solve it again with neutral-type DDE solver.

Solution *The delay of the derivative signal can be expressed with an empty matrix* `[]`. *Thus, the following statements can be used to solve the original problem, and the results obtained are exactly the same as the one shown in Figure 7.19.*

```
>> f=@(t,x,Z,z)[-2*x(2)-3*Z(1,1); -0.05*x(1)*x(3)-2*Z(2,2)+2;
                0.3*x(1)*x(2)*x(3)+cos(x(1)*x(2))+2*sin(0.1*t^2)];
   sol=ddensd(f,tau,[],zeros(3,1),[0,10]); plot(sol.x,sol.y)
```

7.6 Solving Boundary Value Problems

In the previous sections, only initial value problems were considered, i.e., from the given $\boldsymbol{x}_0$, the state vector $\boldsymbol{x}$ at other time instants is to be evaluated. In practical situations, some states at time $t = 0$ are given while other states at $t = t_{\rm f}$ are also given. This kind of differential equation problem involving mixed initial and final conditions is referred to as the *boundary value problems* (BVPs). Boundary value problems cannot be solved directly from `ode45()` type functions. In this section, we focus on how to solve boundary value problems of ordinary differential equations.

The two-point boundary value problem (TPBVP) of a given differential equation involves mixed initial and final conditions. In this section, we illustrate the two-point boundary value problem solution procedures using a second order differential equation as examples. General boundary value problems are considered in the next subsection.

7.6.1 Shooting algorithm for linear equations

Consider a simple case. A linear differential equation is given by

$$y''(t) + p(t)y'(t) + q(t)y(t) = f(t), \tag{7-6-1}$$

where, $p(t)$, $q(t)$ and $f(t)$ are all known functions, the boundary conditions in (7-6-11) are also in their simplest form

$$y(a) = \gamma_{\rm a}, \quad y(b) = \gamma_{\rm b}. \tag{7-6-2}$$

The basic idea of a shooting algorithm is to select an initial condition for $y'(0)$, and solve the initial value problem, then, use the error between the obtained final value and the given $y(b)$, to adjust the initial condition $y'(0)$ recursively, so that the error approaches zero eventually.

For the boundary value problem of the linear differential equation, the main procedures of the shooting algorithm are

(i) Solve the following initial value problem and find $y_1(b)$

$$y_1''(t) + p(t)y_1'(t) + q(t)y_1(t) = 0, \quad y_1(a) = 1, \; y_1'(a) = 0. \tag{7-6-3}$$

(ii) Solve the following initial value problem and find $y_2(b)$

$$y_2''(t) + p(t)y_2'(t) + q(t)y_2(t) = 0, \quad y_2(a) = 0, \; y_2'(a) = 1. \tag{7-6-4}$$

(iii) Solve the following equation to find the numerical solution $y_{\rm p}(b)$

$$y_{\rm p}''(t)+p(t)y_{\rm p}'(t)+q(t)y_{\rm p}(t)=f(t),\quad y_{\rm p}(a)=0,\ y_{\rm p}'(a)=1. \tag{7-6-5}$$

(iv) If $y_2(b)\neq 0$, compute m, the appropriate initial value

$$m=\frac{\gamma_{\rm b}-\gamma_{\rm a}y_1(b)-y_{\rm p}(b)}{y_2(b)}. \tag{7-6-6}$$

(v) Solve the following initial value problem, and the solution of $y(x)$ is the solution of the boundary value problem

$$y''(t)+p(t)y'(t)+q(t)y(t)=f(t),\quad y(a)=\gamma_{\rm a},\ y'(a)=m. \tag{7-6-7}$$

The corresponding explicit form of (7-6-3) and (7-6-4) are obtained

$$\begin{cases} x_1'=x_2\\ x_2'=-q(t)x_1-p(t)x_2, \end{cases} \tag{7-6-8}$$

with initial vectors $[1,0]^{\rm T}$ and $[0,1]^{\rm T}$, respectively.

The explicit form of (7-6-5) and (7-6-7) is

$$\begin{cases} x_1'=x_2\\ x_2'=-q(t)x_1-p(t)x_2+f(t), \end{cases} \tag{7-6-9}$$

with initial vectors $[0,0]^{\rm T}$ and $[\gamma_{\rm a},m]^{\rm T}$, respectively. The following MATLAB function can be written as

```
function [t,y]=ln_shooting(p,q,f,tspan,x0f,varargin)
if isnumeric(p), p=@(t)p; end, if isnumeric(q), q=@(t)q; end
if isnumeric(f), f=@(t)f; end, % for constants, establish function handles
t0=tspan(1); tfinal=tspan(2); ga=x0f(1);gb=x0f(2);
f1=@(t,x)[x(2); -q(t)*x(1)-p(t)*x(2)]; f2=@(t,x)f1(t,x)+[0; f(t)];
[t,y1]=ode45(f1,tspan,[1; 0],varargin{:});
[t,y2]=ode45(f1,tspan,[0; 1],varargin{:});
[t,yp]=ode45(f2,tspan,[0; 0],varargin{:});
m=(gb-ga*y1(end,1)-yp(end,1))/y2(end,1);
[t,y]=ode45(f2,tspan,[ga;m],varargin{:});
```

and the syntax of the function is

`[t,y] = ln_shooting(p,q,f,tspan,x0f,others)`

where p, q and f are the function handles of the given functions of the equation. The argument `tspan` $=[a,b]$ is the interval of computation. The vector $\boldsymbol{x}_{\rm 0f}=[\gamma_{\rm a},\gamma_{\rm b}]$ is the boundary value vector. Other arguments such as `options` in the ODE solver can also be used in the function call.

Example 7.34 Solve the boundary value problem of the following equation

$$y''(x)-\left(2-\frac{1}{x}\right)y'(x)+\left(1-\frac{1}{x}\right)y(x)=x^2{\rm e}^{-5x},\quad \text{with } y(1)=\pi,\ y(\pi)=1.$$

Solution *The analytical solution of the equation can be found directly, however, the result is very complicated, and contains special function. The curve of signal $y(t)$ is shown in Figure 7.23.*

```
>> y=dsolve('D2y-(2-1/x)*Dy+(1-1/x)*y=x^2*exp(-5*x)',...
      'y(1)=pi','y(pi)=1','x') % analytical solution
   ezplot(y,[1,pi])                % plot of the solution
```

Compare with the descriptions of the original equation, the functions p, q and f can be found, and described by anonymous functions. The numerical solution of the boundary value problem can be obtained, also as shown in Figure 7.23, almost identical to the one expressed by the analytical solution.

```
>> p=@(x)-(2-1./x); q=@(x)1-1./x; f=@(x)x.^2.*exp(-5*x);
   [t,y]=ln_shooting(p,q,f,[1,pi],[pi; 1]); line(t,y(:,1))
```

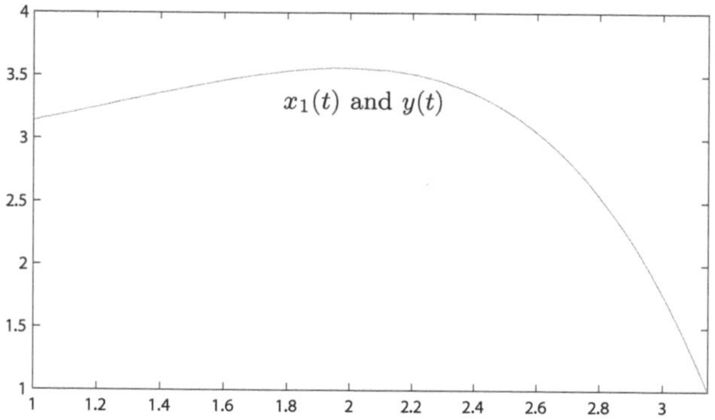

FIGURE 7.23: Numerical solutions of linear boundary value problem.

7.6.2 Boundary value problems of nonlinear equations

The two-point boundary value problem for second-order differential equation can be mathematically described as follows:

$$y''(x) = F(x, y, y'). \tag{7-6-10}$$

In the interested interval $[a, b]$, the following two boundary conditions are known:

$$\alpha_a y(a) + \beta_a y'(a) = \eta_a, \quad \alpha_b y(b) + \beta_b y'(b) = \eta_b. \tag{7-6-11}$$

In some simple cases, the boundary conditions are specified as

$$y(a) = \gamma_a, \ y(b) = \gamma_b. \tag{7-6-12}$$

Assume that the original two-point boundary value problem can be converted to an initial value problem as follows:

$$y'' = F(x, y, y'), \ \ y(a) = \gamma_a, \ y'(a) = m. \tag{7-6-13}$$

That is, the two-point boundary value problem is converted to solving $y(b|m) = \gamma_b$, which means to evaluate the value of $y(b)$ based on the information of m. Using the following Newton's iterative algorithm the value of m can be obtained:

$$m_{i+1} = m_i - \frac{y(b|m_i) - \gamma_b}{(\partial y/\partial m)(b|m_i)} = m_i - \frac{v_1(b) - \gamma_b}{v_3(b)}, \tag{7-6-14}$$

where $v_1 = y(x|m_i), v_2 = y'(x|m_i), v_3 = (\partial y/\partial m)(x|m_i), v_4 = (\partial y'/\partial m)(x|m_i)$, and clearly, the original boundary value problem can be converted to a series of initial value problems that can be solved using the numerical algorithms given in the previous sections.

$$\begin{cases} v_1' = v_2, & v_1(a) = \gamma_a \\ v_2' = F(x, v_1, v_2), & v_2(a) = m \\ v_3' = v_4, & v_3(a) = 0 \\ v_4' = \dfrac{\partial F}{\partial y}(x, v_1, v_2)v_3 + \dfrac{\partial F}{\partial y'}(x, v_1, v_2)v_4, & v_4(a) = 1, \end{cases} \tag{7-6-15}$$

where in order to solve explicitly $\partial F/\partial y$, $\partial F/\partial y'$, an auxiliary value of m can be introduced such that the initial value problems in (7-6-15) can be solved. The results can be substituted into (7-6-14) for one step. Then, in turn the results can be substituted into (7-6-15). When the values of m computed in the two algorithms meet the pre-specified error tolerance, the iteration stops and the required m is found. In this way, from (7-6-13), the original boundary value problem can be considered as solved. The above algorithm can be implemented in the following MATLAB function

```
function [t,y]=nlbound(funcn,funcv,tspan,x0f,tol,varargin)
t0=tspan(1);tfinal=tspan(2); ya=x0f(1); yb=x0f(2); m=1; m0=0;
while (norm(m-m0)>tol), m0=m;
   [t,v]=ode45(funcv,tspan,[ya;m;0;1],varargin{:});
   m=m0-(v(end,1)-yb)/(v(end,3));
end
[t,y]=ode45(funcn,tspan,[ya;m],varargin{:});
```

where a MATLAB user defined function `funcv()` must be prepared to describe the initial value problems defined in (7-6-15). An example is given below to illustrate the coded algorithm.

Example 7.35 Solve the following boundary value problem for the nonlinear differential equation

$$y'' = F(x, y, y') = 2yy', \ y(0) = -1, \ y(\pi/2) = 1.$$

Solution *The partial derivatives can easily be found as* $\partial F/\partial y = 2y'$, $\partial F/\partial y' = 2y$. *Substituting them back to the fourth equation in* (7-6-15), *it can immediately be found that* $v_4' = 2v_2v_3 + 2v_1v_4$. *Thus, the related functions can be expressed by anonymous functions. The following statements can be used to solve this two-point boundary value problem. The time histories of the states are shown in Figure 7.24. It can be seen that the solution* $x_1(t)$ *satisfies the two given boundary conditions.*

```
>> f1=@(t,v)[v(2); 2*v(1)*v(2); v(4); 2*v(2)*v(3)+2*v(1)*v(4)];
   f2=@(t,x)[x(2); 2*x(1)*x(2)]; % describe the two equations
   [t,y]=nlbound(f2,f1,[0,pi/2],[-1,1],1e-8); plot(t,y); xlim([0,pi/2]);
```

It is known that the analytical solution of the two-point boundary value problem is $y(x) = \tan(x - \pi/4)$. *The following statements can be used to check the accuracy of the above numerical results.*

```
>> y0=tan(t-pi/4); norm(y(:,1)-y0) % compare with analytical solution
```

The norm of the error vector is 1.6629×10^{-5}, *which is satisfactory.*

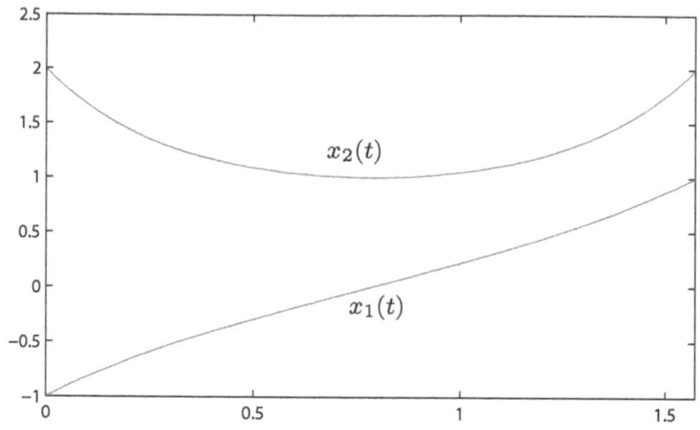

FIGURE 7.24: Solutions to a simple two-point boundary value problem.

7.6.3 Solutions to general boundary value problems

The two-point boundary value problems are quite restricted, since they can only be used to deal with second-order differential equations with known parameters. Assume the differential equation to be analyzed is given by

$$\boldsymbol{y}' = \boldsymbol{f}(t, \boldsymbol{y}, \boldsymbol{\theta}), \tag{7-6-16}$$

where, the time $t \in [a, b]$, $\boldsymbol{y}$ is the state vector and $\boldsymbol{\theta}$ is the vector of unknown parameters. The general boundary conditions are given by

$$\boldsymbol{\phi}[\boldsymbol{y}(a), \boldsymbol{y}(b), \boldsymbol{\theta}] = 0. \tag{7-6-17}$$

To convert the original problems into initial value problems, several algebraic equations should be solved. If the numbers of unknowns and equations are the same, numerical algebraic equation solution algorithms can be used. The boundary value problems to be solved here are more general, since the equations and undetermined constants can be handled at the same time.

The `bvp5c()` function provided in MATLAB can be used to solve the above general boundary value problems [9]. The procedures of problem solution are summarized below:

(i) **Parameter initialization** The `bvpinit()` function can be used to initialize the BVP. The equation and the undetermined constants can be described together in this function such that `sinit=bvpinit(`$\boldsymbol{v}$`,`$\boldsymbol{x}_0$`,`$\boldsymbol{\theta}_0$`)`, where $\boldsymbol{v}$ contains the sample times generated by $\boldsymbol{v}$`=linspace(`a`,`b`,`M`)`. M should be set to small integers for computation speed, e.g., $M = 5$. Apart from the vector $\boldsymbol{v}$, the initial search points of the state vector $\boldsymbol{x}_0$ and undetermined constants $\boldsymbol{\theta}_0$ should also be provided.

(ii) **MATLAB descriptions to ODEs and BVPs** The description of differential equations is exactly the same as the one in the initial value problems illustrated in the previous sections. The description of boundary values in (7-6-17) will be demonstrated through examples in the following.

(iii) **Solving the BVPs** The `bvp5c()` function can be used in solving boundary value problems

`sol=bvp5c(`fun_1`,`fun_2`,sinit,options,`p_1`,`p_2`,`$\cdots$`)`

where fun_1 and fun_2 are respectively the differential equations and the boundary values. The returned argument `sol` is a structured variable, whose fields `sol.`$\boldsymbol{x}$ and `sol.`$\boldsymbol{y}$ store respectively the t vector and the state matrix. The field `sol.parameters` stores the undetermined constant vector $\boldsymbol{\theta}$.

Example 7.36 Solve again the boundary value problem in Example 7.35, rewritten below, with the `bvp5c()` solver

$$y'' = F(x, y, y') = 2yy', \ y(0) = -1, \ y(\pi/2) = 1.$$

Solution *Let $x_1 = y$, $x_2 = y'$. The first-order vector form explicit ODE can be written as $x'_1 = x_2$, $x'_2 = 2x_1x_2$. Anonymous functions can be used to describe the differential equations and the boundary values.*

For the boundary conditions, suppose the left and right bounds can be denoted by a and b, $y(0) = -1$ condition can be denoted as $x_1(a) + 1 = 0$, or denoted as $\boldsymbol{x}_a(1) + 1 = 0$. The condition $y(\pi/2) = 1$ can be expressed by $x_1(b) - 1 = 0$, or $\boldsymbol{x}_b(1) - 1 = 0$. Therefore, the boundary conditions can be written as follows

```
>> f2=@(xa,xb)[xa(1)+1; xb(1)-1]; % describe the boundary conditions
```

The `bvp5c()` function can be used to solve directly the boundary value problem, and the result is shown in Figure 7.25 (a). It can be seen that the curve obtained is not smooth, this is because the number of points selected in `bvpinit()` function is too small. If the number of points is changed from 5 to 20, the curve will be smooth, and the result is exactly the same as the one in the previous example.

```
>> f1=@(t,x)[x(2); 2*x(1)*x(2)]; f2=@(xa,xb)[xa(1)+1; xb(1)-1];
   sinit=bvpinit(linspace(0,pi/2,5),rand(2,1)); % parameter initialization
   sol=bvp5c(f1,f2,sinit); plot(sol.x,sol.y)    % solve and plot
```

(a) $y(0) = -1, y(\pi/2) = 1$ (b) $y'(\pi/2) = 1, y(\pi/2) - y(0) = 1$

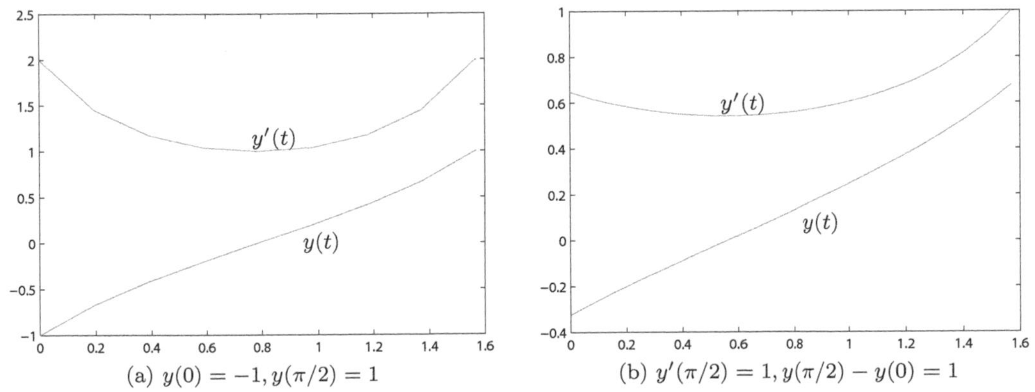

FIGURE 7.25: Solutions of boundary value problem.

The `bvp5c()` function can also be used to solve more complicated boundary value problems. For instance, if the boundary value problem is changed to $y'(\pi/2) = 1$, $y(\pi/2) - y(0) = 1$, it can be denoted as $x_2(b) - 1 = 0$, $x_1(b) - x_1(a) - 1 = 0$, and the function fun$_2$ should be changed as follows

```
>> f2=@(xa,xb)[xb(2)-1; xb(1)-xa(1)-1];        % boundary conditions
   sol=bvp5c(f1,f2,sinit); plot(sol.x,sol.y) % solve and plot
```

and the results are shown in Figure 7.25 (b).

Example 7.37 Consider the problem in the previous example. A random initial value $y'(0)$ can be selected, from which the differential equation can be solved to find $\hat{y}(t_f)$. An objective function can be designed to minimize $|y(t_f) - \hat{y}(t_f)|$, and find a suitable initial value $y'(0)$. In this case, the boundary value problem can be converted into an optimization problem. Solve the problem again using this method.

Solution *The objective function can be written as follows*

```
function y=c7meqopt(x,f,tspan,x0,ypf)
x0vec=[x0; x]; [t,y]=ode45(f,tspan,x0vec); y=abs(ypf-y(end,2));
```

With such a function, optimization can be performed to find the exact initial value $x_2(0)$*, as the decision variable. In this case, the exact* $x_2(0)$ *can be found* $x_2(0) = 1.7585$ *and the differential problem can be solved again under such an initial condition. The obtained solution is exactly the same as the one obtained in the previous example.*

```
>> x0=-1; ypf=1; tspan=[0,pi/2]; f1=@(t,x)[x(2); 2*x(1)*x(2)];
   x=fminunc(@c7meqopt,rand(1),'',f1,tspan,x0,ypf) % perform optimization
   x0a=[x0; x]; [t,y]=ode45(f1,tspan,x0a); plot(t,y) % solve and plot
```

Example 7.38 Given the ordinary differential equation

$$\begin{cases} x'(t) = 4x(t) - \alpha x(t)y(t) \\ y'(t) = -2y(t) + \beta x(t)y(t), \end{cases}$$

with initial and final conditions $x(0) = 2, y(0) = 1, x(3) = 4, y(3) = 2$, find the parameters α and β and solve the boundary value problem.

Solution *Choosing the state variables* $x_1 = x, x_2 = y$*, the original problem can be converted into an explicit differential equation with respect to* $\boldsymbol{x}$*. Let* $v_1 = \alpha$ *and* $v_2 = \beta$*. The ODE and boundary value problem can then be expressed by the following anonymous functions*

```
>> f=@(t,x,v)[4*x(1)+v(1)*x(1)*x(2); -2*x(2)+v(2)*x(1)*x(2)];
   g=@(ya,yb,v)[ya(1)-2; ya(2)-1; yb(1)-4; yb(2)-2]; % boundaries
```

and it can be seen that the description to boundary value problem is quite straightforward in MATLAB. The `bvpinit()` *function should be called to initialize the time array. The initial states and parameters* α *and* β *should also be specified. Since there are two states and two undetermined constants, they both can be set to* `rand(2,1)`*. With these initial parameters, the function* `bvp5c()` *can be called to solve the boundary value problem as well as the undetermined constants* α *and* β*.*

```
>> x1=[1;1]; x2=[-1;1]; sinit=bvpinit(linspace(0,3,20),x1,x2);
   options=bvpset; options.RelTol=1e-8; S=bvp5c(f,g,sinit,options);
   S.parameters % show undermined parameters
   subplot(121), plot(S.x,S.y); subplot(122), plot(S.y(1,:),S.y(2,:));
```

The results are shown in Figure 7.26. Meanwhile, it is found that $\alpha = -2.3721, \beta = 0.8934$*. From the simulation results, it is found that the boundary conditions are satisfied, which verifies the obtained results. It should be noted that if the initial vectors* $\boldsymbol{x}_1$ *and* $\boldsymbol{x}_2$ *are not properly chosen, the Jacobian matrix generated might be singular. In this case, the initial vectors should be chosen again differently so that convergent results can be obtained.*

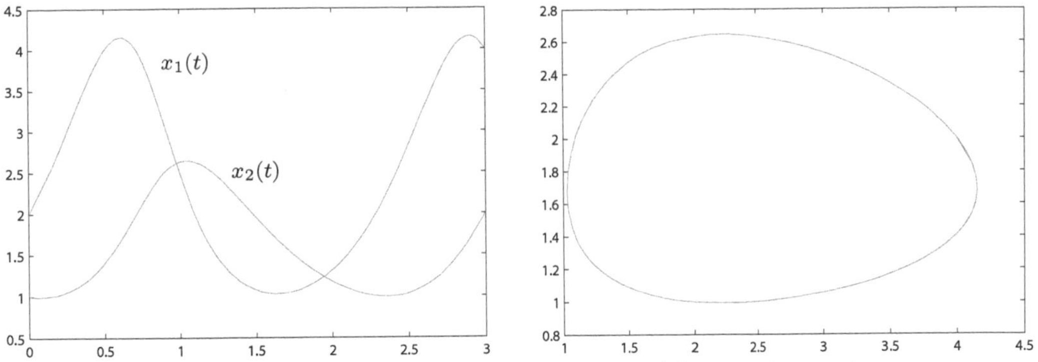

(a) time responses of the states

(b) phase plane trajectory

FIGURE 7.26: Solution to the boundary value problem.

7.7 Introduction to Partial Differential Equations

Apart from ordinary differential equations, partial differential equations (PDEs) are also very useful in science and engineering. In MATLAB many partial differential equations can be solved by the Partial Differential Equation Toolbox. In this section, a general introduction to partial differential equations is given and some particular types of the two-dimensional partial differential equations solvable in the easy-to-use GUI of the PDE Toolbox are illustrated.

A PDE contains 4 elements: the PDE, the initial conditions (ICs), the *boundary conditions* (BCs) and the domain of interest. In the following, we will introduce the solution procedures for 1D PDEs first and then 2D PDEs.

7.7.1 Solving a set of one-dimensional partial differential equations

A partial differential equation solver `pdepe()` provided in MATLAB PDE Toolbox can be used to solve numerically the general 1D partial differential equations of the following form

$$\boldsymbol{c}\left(x,t,\boldsymbol{u},\frac{\partial \boldsymbol{u}}{\partial x}\right)\frac{\partial \boldsymbol{u}}{\partial t}=x^{-m}\frac{\partial}{\partial x}\left[x^m\boldsymbol{f}\left(x,t,\boldsymbol{u},\frac{\partial \boldsymbol{u}}{\partial x}\right)\right]+\boldsymbol{s}\left(x,t,\boldsymbol{u},\frac{\partial \boldsymbol{u}}{\partial x}\right), \tag{7-7-1}$$

where the partial differential equation to be solved should be modeled in the following function $[\boldsymbol{c},\boldsymbol{f},\boldsymbol{s}]=\texttt{pdefun}(x,t,\boldsymbol{u},\boldsymbol{u}_x)$, where `pdefun` is the function name, the functions $\boldsymbol{c}$, $\boldsymbol{f}$ and $\boldsymbol{s}$ can be calculated from the given arguments.

Boundary conditions can be described by the following function

$$\boldsymbol{p}(x,t,\boldsymbol{u})+\boldsymbol{q}(x,t,\boldsymbol{u}).*\boldsymbol{f}\left(x,t,\boldsymbol{u},\frac{\partial \boldsymbol{u}}{\partial x}\right)=0, \tag{7-7-2}$$

where `.*` operation is the MATLAB element-by-element dot product. These boundary conditions can be described by the following MATLAB function

$[\boldsymbol{p}_a,\boldsymbol{q}_a,\boldsymbol{p}_b,\boldsymbol{q}_b]=\texttt{pdebc}(x,t,\boldsymbol{u},\boldsymbol{u}_x)$

Apart from the above two MATLAB functions, the initial conditions should also be described. The initial conditions are usually defined as $\boldsymbol{u}(x,t_0) = \boldsymbol{u}_0$. Thus, a simple MATLAB function can be prepared by using u_0=`pdeic(`x`)`.

One can also select the variable vectors $\boldsymbol{x}$ and $\boldsymbol{t}$. With the use of the above functions, the `pdepe()` function can be used to solve the PDE. The syntax of the function is `sol=pdepe(`m`,@pdefun,@pdeic,@pdebc,`$\boldsymbol{x}$`,`$\boldsymbol{t}$`)`.

Example 7.39 Solve the following partial differential equations:

$$\begin{cases} \dfrac{\partial u_1}{\partial t} = 0.024\dfrac{\partial^2 u_1}{\partial x^2} - F(u_1 - u_2) \\ \dfrac{\partial u_2}{\partial t} = 0.17\dfrac{\partial^2 u_2}{\partial x^2} + F(u_1 - u_2), \end{cases}$$

where $F(x) = \mathrm{e}^{5.73x} - \mathrm{e}^{-11.46x}$, and the initial conditions are given by $u_1(x,0) = 1$, $u_2(x,1) = 0$ and the boundary conditions are

$$\frac{\partial u_1}{\partial x}(0,t) = 0,\ u_2(0,t) = 0,\ u_1(1,t) = 1,\ \frac{\partial u_2}{\partial x}(1,t) = 0.$$

Solution *Comparing the given differential equation with the standard form described in (7-7-1), we can rewrite the equation as*

$$\begin{bmatrix}1\\1\end{bmatrix} .* \frac{\partial}{\partial t}\begin{bmatrix}u_1\\u_2\end{bmatrix} = \frac{\partial}{\partial x}\begin{bmatrix}0.024\partial u_1/\partial x\\0.17\partial u_2/\partial x\end{bmatrix} + \begin{bmatrix}-F(u_1-u_2)\\F(u_1-u_2)\end{bmatrix},$$

with $m = 0$, and

$$\boldsymbol{c} = \begin{bmatrix}1\\1\end{bmatrix},\quad \boldsymbol{f} = \begin{bmatrix}0.024\partial u_1/\partial x\\0.17\partial u_2/\partial x\end{bmatrix},\quad \boldsymbol{s} = \begin{bmatrix}-F(u_1-u_2)\\F(u_1-u_2)\end{bmatrix}.$$

Thus, the M-function describing the equation is written as

```
function [c,f,s]=c7mpde(x,t,u,du)
c=[1; 1]; y=u(1)-u(2); F=exp(5.73*y)-exp(-11.46*y); s=[-F; F];
f=[0.024*du(1); 0.17*du(2)]; % describe the three functions in PDE
```

Referring to the boundary conditions in (7-7-2), the following boundary conditions can be constructed such that

$$\text{left bounds}\quad \begin{bmatrix}0\\u_2\end{bmatrix} + \begin{bmatrix}1\\0\end{bmatrix} .* \boldsymbol{f} = \begin{bmatrix}0\\0\end{bmatrix},\qquad \text{right bounds}\quad \begin{bmatrix}u_1-1\\0\end{bmatrix} + \begin{bmatrix}0\\1\end{bmatrix} .* \boldsymbol{f} = \begin{bmatrix}0\\0\end{bmatrix},$$

and the MATLAB function for the boundary conditions can be prepared as

```
function [pa,qa,pb,qb]=c7mpbc(xa,ua,xb,ub,t)
pa=[0; ua(2)]; qa=[1;0]; pb=[ub(1)-1; 0]; qb=[0;1]; % the boundaries
```

The initial conditions for this partial differential equation can be given by an anonymous function.

With these three functions, if the vectors $\boldsymbol{x}$ and $\boldsymbol{t}$ are used, the following statements solve the given differential equation where the solutions u_1 and u_2 are visualized in Figures 7.27 (a) and (b), respectively.

```
>> x=0:.05:1; t=0:0.05:2; m=0; u0=@(x)[1; 0];
   S=pdepe(m,@c7mpde,u0,@c7mpbc,x,t); subplot(121), surf(x,t,S(:,:,1))
   subplot(122), surf(x,t,S(:,:,2))
```

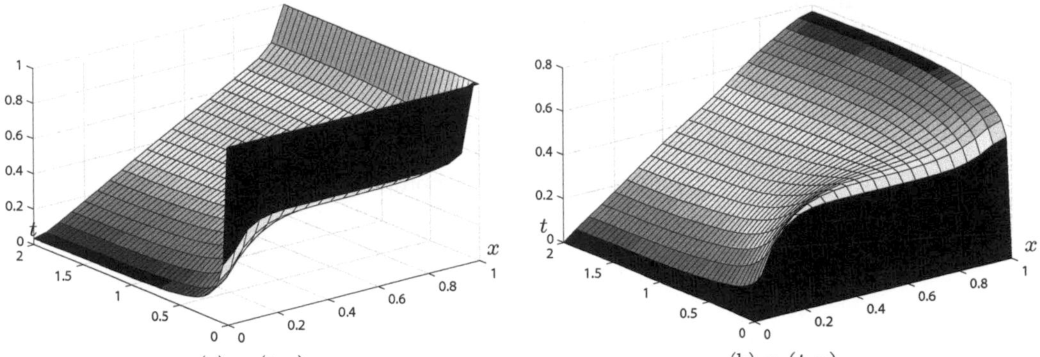

(a) $u_1(t,x)$ (b) $u_2(t,x)$

FIGURE 7.27: Solution surfaces in partial differential equations.

7.7.2 Mathematical description to two-dimensional PDEs

The PDE Toolbox can be used to solve 2D partial differential equations via a very handy graphical user interface `pdetool`. Some of the 2D partial differential equations solvable via the PDE Toolbox are introduced and examples are given to demonstrate the use of the PDE GUI.

I. Elliptic partial differential equations

The general form of an elliptic partial differential equation is given by

$$-\text{div}(c\nabla u) + au = f(\boldsymbol{x}, t), \tag{7-7-3}$$

where $u=u(x_1, x_2, \cdots, x_n, t)=u(\boldsymbol{x}, t)$, and ∇u is the gradient of u defined as

$$\nabla u = \left[\frac{\partial}{\partial x_1}, \frac{\partial}{\partial x_2}, \cdots, \frac{\partial}{\partial x_n}\right] u. \tag{7-7-4}$$

The divergence $\text{div}(v)$ is defined as

$$\text{div}(v) = \left(\frac{\partial}{\partial x_1} + \frac{\partial}{\partial x_2} + \cdots + \frac{\partial}{\partial x_n}\right) v. \tag{7-7-5}$$

Thus, $\text{div}(c\nabla u)$ can further be defined as

$$\text{div}(c\nabla u) = \left[\frac{\partial}{\partial x_1}\left(c\frac{\partial u}{\partial x_1}\right) + \frac{\partial}{\partial x_2}\left(c\frac{\partial u}{\partial x_2}\right) + \cdots + \frac{\partial}{\partial x_n}\left(c\frac{\partial u}{\partial x_n}\right)\right]. \tag{7-7-6}$$

If c is a constant, the above equation can be simplified as

$$\text{div}(c\nabla u) = c\left(\frac{\partial^2}{\partial x_1^2} + \frac{\partial^2}{\partial x_2^2} + \cdots + \frac{\partial^2}{\partial x_n^2}\right) u = c\Delta u, \tag{7-7-7}$$

where Δ is also referred to as the *Laplacian operator*. Thus, the elliptic PDE can further be simplified as

$$-c\left(\frac{\partial^2}{\partial x_1^2} + \frac{\partial^2}{\partial x_2^2} + \cdots + \frac{\partial^2}{\partial x_n^2}\right) u + au = f(\boldsymbol{x}, t). \tag{7-7-8}$$

II. Parabolic partial differential equations

The general form of a parabolic partial differential equation is given by

$$d\frac{\partial u}{\partial t} - \mathrm{div}(c\nabla u) + au = f(\boldsymbol{x}, t). \tag{7-7-9}$$

If c is a constant, the above equation can be simplified as

$$d\frac{\partial u}{\partial t} - c\left(\frac{\partial^2 u}{\partial x_1^2} + \frac{\partial^2 u}{\partial x_2^2} + \cdots + \frac{\partial^2 u}{\partial x_n^2}\right) + au = f(\boldsymbol{x}, t). \tag{7-7-10}$$

III. Hyperbolic partial differential equations

The general form of a hyperbolic partial differential equation is given by

$$d\frac{\partial^2 u}{\partial t^2} - \mathrm{div}(c\nabla u) + au = f(\boldsymbol{x}, t). \tag{7-7-11}$$

If c is a constant, the above equation can be simplified as

$$d\frac{\partial^2 u}{\partial t^2} - c\left(\frac{\partial^2 u}{\partial x_1^2} + \frac{\partial^2 u}{\partial x_2^2} + \cdots + \frac{\partial^2 u}{\partial x_n^2}\right) + au = f(\boldsymbol{x}, t). \tag{7-7-12}$$

It can be seen from the above three types of partial differential equations that the most significant difference is the order of the derivative of the function u with respect to t. If the derivative term is zero, the equation is elliptic. The first- and second-order derivative terms correspond to parabolic and hyperbolic equations, respectively.

Finite element-based algorithms are implemented in the PDE Toolbox. In the elliptic equations, the variables c, a, d and f can all be defined as any given functions, while in the other two types of PDEs, they must be constants.

IV. Eigenvalue problem

An eigenvalue partial differential equation problem is defined as

$$-\mathrm{div}(c\nabla u) + au = \lambda du. \tag{7-7-13}$$

If c is a constant, the above equation can be simplified as

$$-c\left(\frac{\partial^2 u}{\partial x_1^2} + \frac{\partial^2 u}{\partial x_2^2} + \cdots + \frac{\partial^2 u}{\partial x_n^2}\right) + au = \lambda du. \tag{7-7-14}$$

7.7.3 The GUI for the PDE Toolbox — an introduction

I. PDE Toolbox — an overview

A GUI for solving partial differential equations is provided in the PDE Toolbox. The interface can be used in solving 2D partial differential equations. The solution regions of interest can be drawn freely by the GUI tools using combinations of circles, ellipsis, rectangles and polygons. Moreover, the regions of interest can be organized by set operations such as union, difference and intersect, etc. Two-dimensional partial differential equations can easily be solved using the GUI.

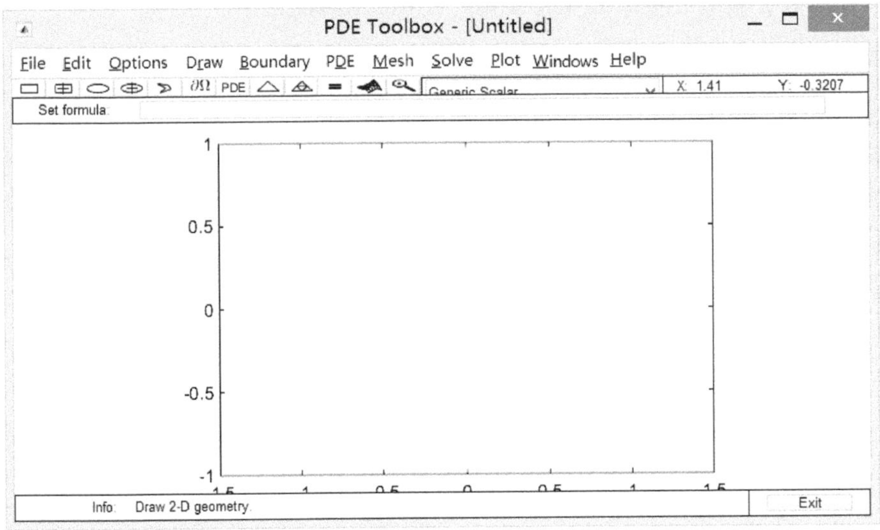

FIGURE 7.28: Graphical user interface of the PDE solver.

Type **pdetool** in the MATLAB prompt, and a user interface shown in Figure 7.28 will be displayed. This interface can be used in solving the given partial differential equation.

The PDE user interface has the following functions:

(i) **Menu system** A complete and comprehensive menu system is provided in the interface, which makes most of the functions callable directly from the menu items and toolbar buttons.

(ii) **Toolbar** The detailed explanation to the buttons on the toolbar is shown in Figure 7.29. The toolbar can be used to define the solution regions of interest, to set parameters in the partial differential equation, to solve the equation and to visualize the results. The right-hand side of the interface provides a list box containing different types of solvable partial differential equations.

(iii) **Set formula** Set formula edit box can be used to define set operations, such as union, intersect and difference operations.

(iv) **Solution regions** The user can draw solution regions, and then, solve the two-dimensional equations within the solution region. 3D display can also be obtained.

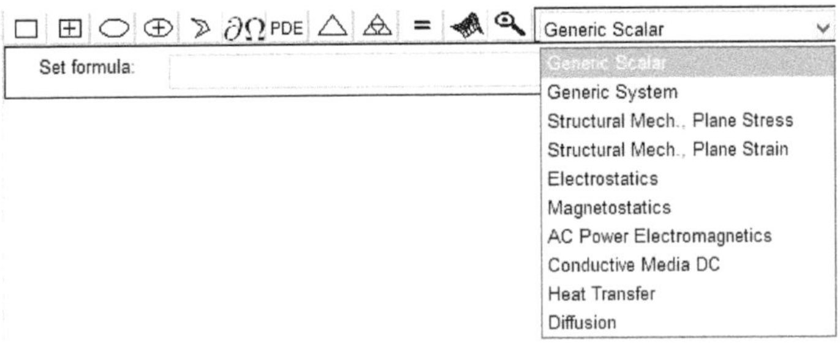

FIGURE 7.29: Toolbar in the PDE solver GUI.

II. Drawing and defining the PDE solution region

An illustrative example is introduced to define the solution region in the GUI. One can first select the ellipses and rectangle buttons and draw the regions, defined as *sets*, as shown in Figure 7.30 (a). Then, the solution region can be defined using the set operation edit box such that the contents are changed to `(R1+E1+E3)-E2`, which means removal of the set `E2` from the union of the rectangle `R1`, the ellipses `E1`, and `E3`. Therefore, the button labeled as $\partial\Omega$ can be used to define the solution region. The menu item Boundary → Remove All Subdomain Borders can be used to remove the curves within the adjacent regions. Thus, the solution region can then be obtained as shown in Figure 7.30 (b).

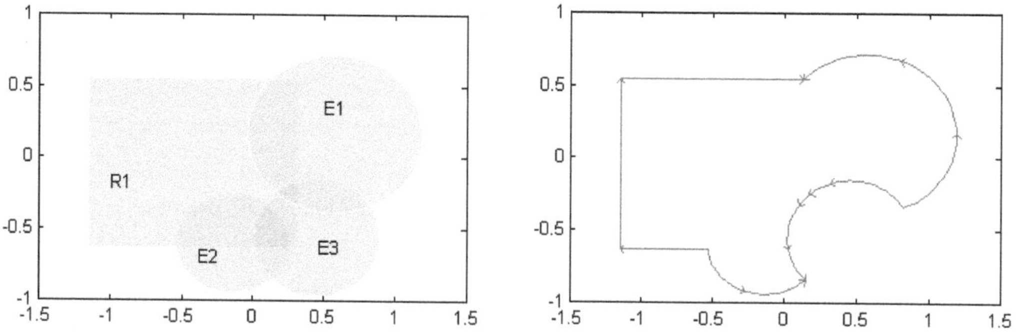

(a) region definition (b) actual solution region

FIGURE 7.30: Solution region for partial differential equations.

From the given solution region, if one clicks the Δ button, triangular mesh can be generated within the solution region, as shown in Figure 7.31 (a). If one is not satisfied with the mesh, the refine mesh button can be used to add more grids to the region and the final grids can be shown in Figure 7.31 (b). It is worth mentioning that the finer the mesh, the more accurate solutions can be obtained. However, the cost is that the longer computation time is required.

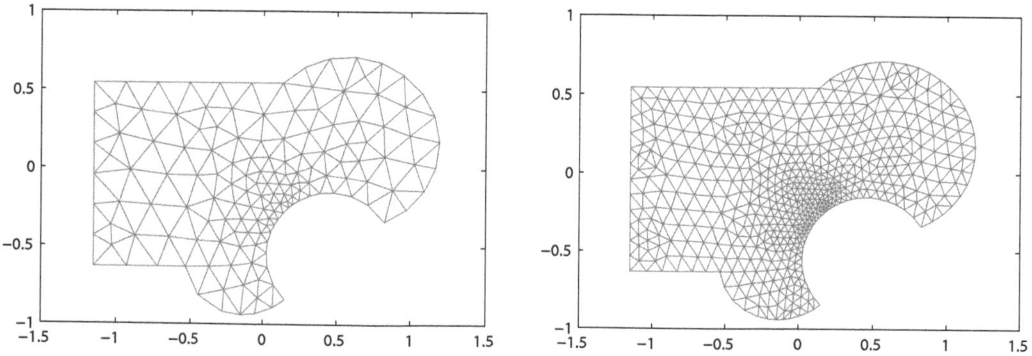

(a) default mesh specifications (b) refined mesh

FIGURE 7.31: Generation of the grids within the solution region.

III. Boundary conditions for 2D PDEs

In the interface, the boundary conditions can be represented by the $\partial\Omega$ button. Generally speaking, the Dirichlet and Neumann types of boundary conditions are supported. These two types of boundary conditions are discussed below:

(i) **Dirichlet conditions** Dirichlet boundary conditions are described as follows:

$$h\left(\boldsymbol{x},t,u,\frac{\partial u}{\partial \boldsymbol{x}}\right)u\bigg|_{\partial\Omega} = r\left(\boldsymbol{x},t,u,\frac{\partial u}{\partial \boldsymbol{x}}\right), \tag{7-7-15}$$

where $\partial\Omega$ denotes the boundary of the solution region. Assume that certain conditions must be satisfied on the solution boundary, one can simply specify the functions r and h which can be either constants or functions of $\boldsymbol{x}$ and u, $\partial u/\partial \boldsymbol{x}$. For convenience, one may assume that $h=1$. In the following example, how to specify the Dirichlet boundary conditions in MATLAB will be illustrated.

(ii) **Neumann conditions** The extended form is given by

$$\left[\frac{\partial}{\partial \boldsymbol{n}}(c\nabla u)+qu\right]\bigg|_{\partial\Omega} = g, \tag{7-7-16}$$

where $\partial u/\partial \boldsymbol{n}$ is the partial derivative of vector $\boldsymbol{x}$ in the normal direction.

If Boundary → Specify Boundary Conditions menu item is selected, a dialog box as in Figure 7.32 will show up. The boundary conditions can be specified through the dialog box. If ones wants to assign zero values on all the boundaries, fill 0 in the r edit box.

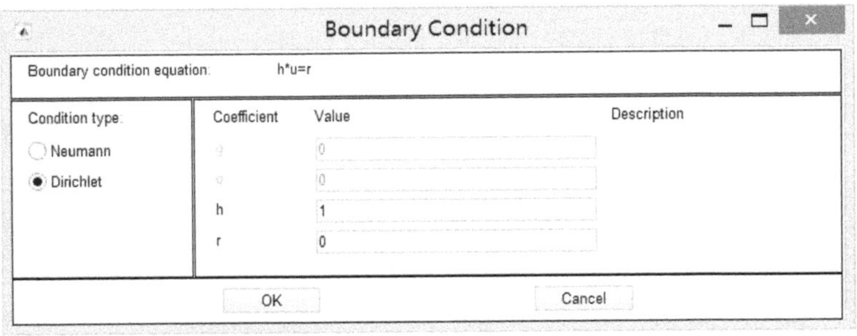

FIGURE 7.32: Boundary condition setting dialog box.

IV. PDE solution examples via PDE Toolbox

The solution region, boundary conditions can be specified using the methods discussed earlier. If the partial differential equations can be specified, the = button in the toolbar can be used to solve the partial differential equations. An example will be given below to show the solution procedures.

Example 7.40 Solve the following hyperbolic partial differential equation

$$\frac{\partial^2 u}{\partial t^2} - \frac{\partial^2 u}{\partial x^2} - \frac{\partial^2 u}{\partial y^2} + 2u = 10.$$

Solution *From the given partial differential equation, one can immediately find that* $c=1, a=2, f=10$ *and* $d=1$. *Click the* PDE *button in the toolbar. A dialog box as in*

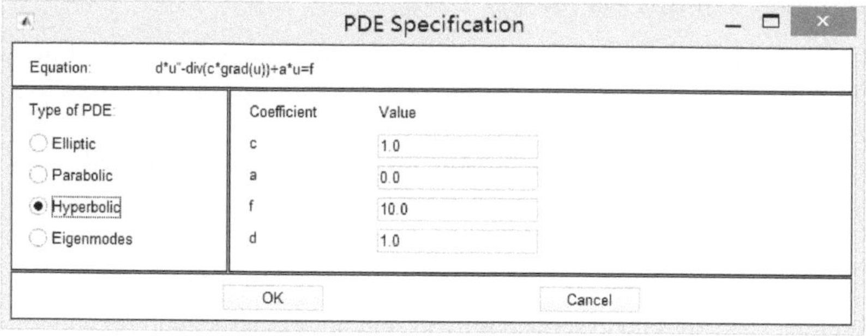

FIGURE 7.33: PDE parameters setting dialog box.

Figure 7.33 shows. From the radio button on the left, select the Hyperbolic *option, and the parameters can be entered in the dialog box.*

If the numerical solutions are required, click the = button in the toolbar and the solutions can immediately be obtained as shown in Figure 7.34 (a), where pseudocolors are used to denote the values of $u(x, y)$. It should be noted that only the solution $u(x, y)$ at $t = 0$ is displayed. Later on, it will be shown how the solutions at other t values can be visualized.

The boundary value conditions can be modified. For instance, the Dirichlet conditions can still be used, and assume $r = 5$ on the boundaries. Use the dialog box to fill 5 into the r *edit box. Solve the partial differential equations again, and the obtained results are shown in Figure 7.34 (b).*

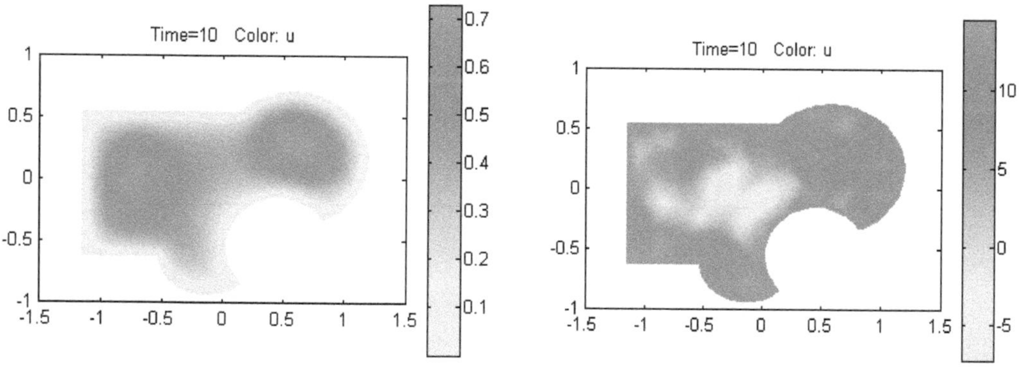

(a) solutions (b) solutions after boundary changes

FIGURE 7.34: Solutions to the partial differential equations.

Numerical results can be visualized in many different ways. If the 3D button is clicked, a dialog box shown in Figure 7.35 appears. One can select the Contour icon to show the contours of the solutions. If the Arrows option is selected, attraction curves will be displayed. When the two options are selected simultaneously, the results are visualized in Figure 7.36 (a).

It should be noted that, in the dialog box shown in Figure 7.35, list boxes are provided for each item of the Property column. For instance, the default for the first item is u, indicating all the analyses are made for the function $u(\cdot)$. The results displayed are the function of

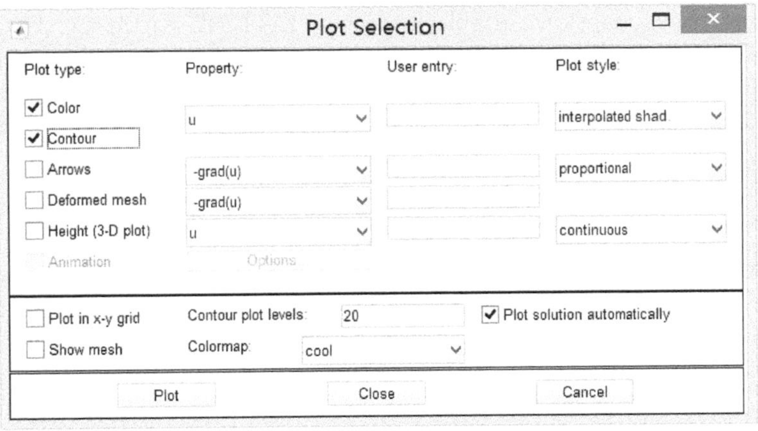

FIGURE 7.35: Dialog box for results display format settings.

$u(x, y)$. If other functions are to be visualized, the button ▼ to the right should be clicked. Another list box is then shown and other functions can be selected for visualization.

In the menu item Height (3d-plot), a graphics window is opened and a three-dimensional mesh grid plot is generated, as shown in Figure 7.36 (b).

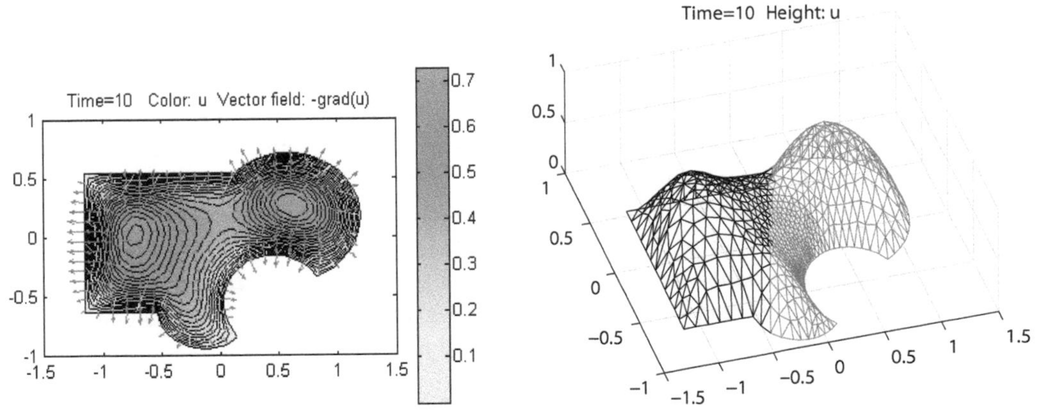

(a) contour representation (b) three-dimensional representations

FIGURE 7.36: Different representations of the solutions.

V. Animation of the time varying solutions

The default time vector is defined as t=0:10. The results shown in Figure 7.34 are only at the final time $t = 10$. From the hyperbolic equations, it can be seen that the solutions should also be a function of time t. Thus, animation should be used to dynamically display the solutions. Let us still use the PDE in Example 7.40 to illustrate the animation process.

Use the menu item Solve → Parameters and a dialog box is displayed so as to specify the time vector. For instance, if in the edit box, one specifies the vector 0:0.1:4, the solutions will then be made over the new time vector instead. In the dialog box shown in Figure 7.35, the Animation check box can be selected. The Options button can be used to adjust the speed in animations (the default speed is 6 frames per second). The animation can then be obtained directly. The Plot → Export Movie menu item can be used to export the animation variable into the MATLAB workspace. For instance, the animation results can be saved to MATLAB variable `M`, and with the use of the `movie(M)` function, animation can be played in a MATLAB graphics window. Furthermore, using the `movie2avi(M,'myavi.avi')` command, the animation can be saved in the `myavi.avi` file for later play.

VI. Solving PDEs when parameters are not constants

In the partial differential equations discussed earlier, c, a, d, f coefficients are all assumed constant. In practical applications, however, c, a, d, f may be functions. For elliptic PDEs, the solvers currently allow the use of function to describe the above-mentioned coefficients. The variables `x` and `y` are used to represent x_1, x_2 or x, y, while the variables `ux` and `uy` are for $\partial u/\partial x$ and $\partial u/\partial y$, respectively. They can be described by any nonlinear functions. The following example illustrates the solution process when c, a, d, f are not constants.

Example 7.41 Assume that the partial differential equations are described as

$$-\mathrm{div}\left(\frac{1}{1+|\nabla u|^2}\nabla u\right)+(x^2+y^2)u=\mathrm{e}^{-x^2-y^2},$$

with boundaries at 0. Solve numerically the partial differential equations.

Solution *It can be found that the original partial differential equation is elliptic, with*

$$c=\frac{1}{\sqrt{(1+\left(\dfrac{\partial u}{\partial x}\right)^2+\left(\dfrac{\partial u}{\partial y}\right)^2}},\quad a=x^2+y^2,\quad f=\mathrm{e}^{-x^2-y^2},$$

and the boundary conditions are 0. One may still use the `pdetool` *interface shown in Figure 7.33. In the partial differential equation type dialog box, select the* Elliptic *item, and in the* c *edit box, specify* `1./sqrt(1+ux.^2+uy.^2)`*. In edit boxes of* a *and* f*, specify respectively* `x.^2+y.^2` *and* `exp(-x.^2-y.^2)`*. Then, open the* Solve → Parameters *dialog box and select the* Use nonlinear solver *property (it should be noted that this option is only applicable to elliptic equations), and the equal sign icon can be clicked to solve the equation. The results obtained are shown in Figures 7.37 (a) and (b).*

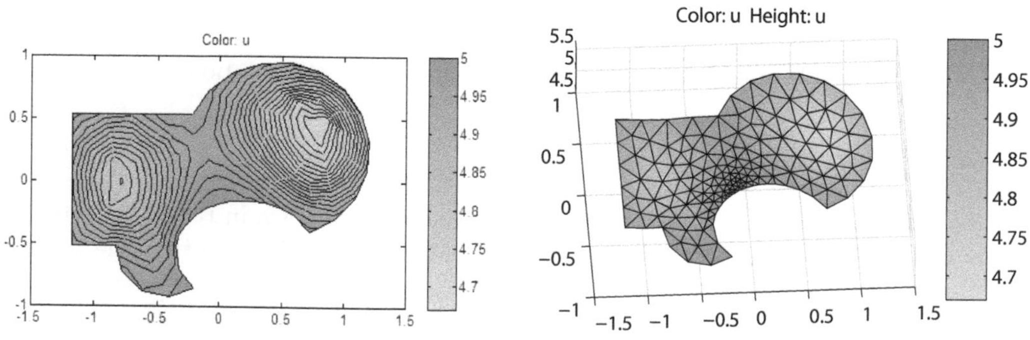

(a) contour representation (b) three-dimensional grid representation

FIGURE 7.37: Different presentations of the PDE solutions.

7.8 Solving ODEs with Block Diagrams in Simulink

7.8.1 A brief introduction to Simulink

The Simulink environment was first proposed by The MathWorks Inc., in around 1990. The original name was SimuLAB [10], and it took the current name in 1992. From the name Simulink, we can see two meanings: "simu" and "link," which means to connect the blocks together, then, to perform simulation for the system thus constructed. Simulink is an effective tool useful in defining different types of ODEs and other algebraic equations.

Of course, the functions provided in Simulink are not limited to ODE solvers. It can also be used to construct control systems with the existing and extended blocks. Moreover, modeling and simulation to engineering systems, such as motor and drive systems, mechanical systems and communication systems, can be carried out very easily, with the help of the Simulink blocksets and MATLAB toolboxes. Simulink, a very powerful environment, can be used to model and simulate dynamic systems of almost any complexity using Simulink block libraries and user-defined blocks [11,12]. Only materials related to ordinary differentia equation solutions are discussed in this section.

The most commonly used Simulink blocks will be introduced, and then, examples will be given to show the modeling and simulation procedures.

7.8.2 Simulink — relevant blocks

One can issue the command `open_system('simulink')` in the MATLAB window, so that the block library window of Simulink can be opened as shown in Figure 7.38. It can be seen that block groups are also provided. For instance, the groups such as Sources, Continuous are provided. Each of the groups further contains sub-groups or blocks. Theoretically speaking, the systems of almost any complexity can be modeled and simulated using the facilities provided in Simulink.

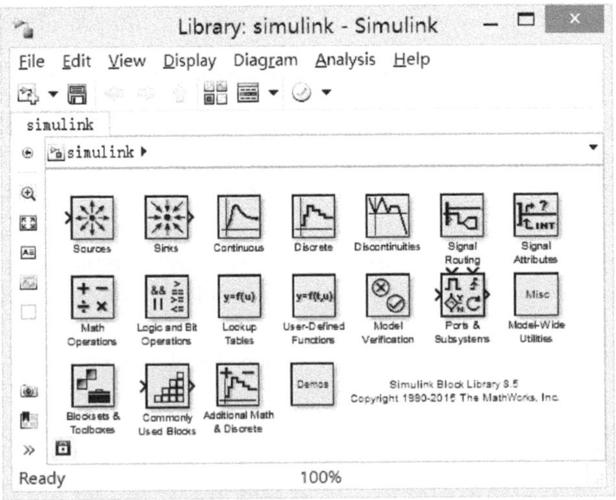

FIGURE 7.38: Block library window in Simulink environment.

Hundreds of Simulink blocks are provided, and it is not possible to describe all of them.

Here only the blocks related to differential equation modeling are summarized and some of the commonly used blocks are made into a user-group, named as `odegroup`. The command `odegroup` can be used to open the user defined blockset, as shown in Figure 7.39.

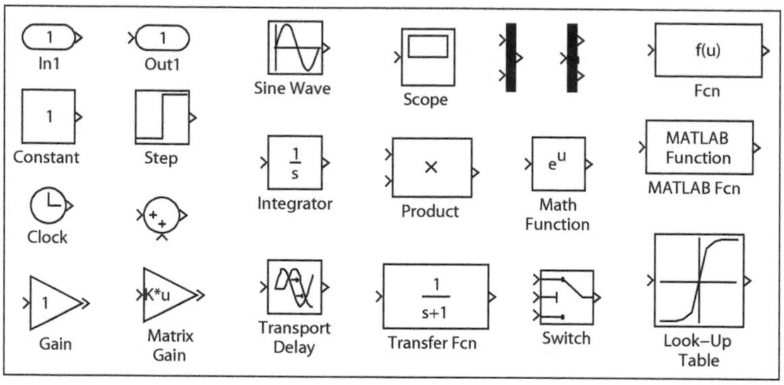

FIGURE 7.39: User-defined most commonly used blocks.

The frequently used blocks are summarized below

(i) **Input and output port blocks** (In1, Out1) These blocks can generate a variable `yout` in MATLAB workspace. Any signal in the simulation model can be connected to the Scope blocks.

(ii) **Clock block** generates time t, and it can be used in the modeling of time varying differential equations.

(iii) **Commonly used input blocks** The Sine block can be used in generating sinusoidal signals. The block Step can be used in generating the step signal, and the Constant block can be used to define the constant signals.

(iv) **Integrator block** (Int) The block is used to evaluate integral to the input signal. For instance, assume that the input to the ith integrator is $x_i'(t)$, the output of the integrator is then $x_i(t)$. The use of integrator is a crucial step in the modeling of differential equations. For high-order linear differential equations, it can also be modeled by Transfer Function blocks.

(v) **Transport delay blocks** (Transport Delay) The output signal is the value of the input signal at time $t-\tau$. It can be used in modeling of delay differential equations.

(vi) **Gain blocks** (Gain, Sliding Gain **and** Matrix Gain) These gain blocks are very useful in Simulink modeling. The definitions of them are different. The Gain block is used to amplify the input signal. If the input signal is u, the output of the block is Ku. The Matrix Gain block is used for vector input signal $\boldsymbol{u}$, whose output is $\boldsymbol{Ku}$. The Sliding Gain block implements a scroll bar so that the gain of the block can be changed arbitrarily using mouse dragging.

(vii) **Mathematical operation blocks** These blocks can be used to perform algebraic operations such as plus, minus and times as well as logical operations.

(viii) **Mathematical function blocks** These blocks can be used to perform nonlinear functions such as trigonometry functions or exponential functions.

(ix) **Signal vectorization blocks** The Mux block can be used as the vector signal composed of individual input signals. The Demux block can be used to extract scalar signals from the vector signal.

7.8.3 Using Simulink for modeling and simulation of ODEs

Using suitable blocks, the differential equations can be constructed using Simulink. The solution of the ODEs can be obtained with the function `sim()`, with the following syntax

`[t,x,y] = sim(model_name,tspan,options)`

which is quite similar to the function `ode45()`. Examples will be given in the following to demonstrate the modeling and simulation of different types of differential equations. The Lorenz equation will be studied first, followed by the delay differential equations.

Example 7.42 Consider again the Lorenz equation studied in Example 7.9

$$\begin{cases} x_1'(t) = -\beta x_1(t) + x_2(t)x_3(t) \\ x_2'(t) = -\rho x_2(t) + \rho x_3(t) \\ x_3'(t) = -x_1(t)x_2(t) + \sigma x_2(t) - x_3(t), \end{cases}$$

where $\beta = 8/3, \rho = 10$ and $\sigma = 28$, and the initial states are $x_1(0) = x_2(0) = 0$, and $x_3(0) = 10^{-10}$. Model the differential equations with Simulink, and then, find the solutions using simulation technique.

Solution *Since there are three first-order derivative terms, three integrators are used to describe respectively $x_1'(t)$, $x_2'(t)$ and $x_3'(t)$. The outputs of these blocks are $x_1(t)$, $x_2(t)$ and $x_3(t)$. The framework of the system can be established as shown in Figure 7.40 (a). Double click each integrator block and the initial state variables can be specified for each integrator.*

With the framework, the states and their derivatives are defined, and with the use of Mux block, the state vector can be defined such that $\boldsymbol{x}(t) = [x_1(t), x_2(t), x_3(t)]^{\mathrm{T}}$. Then, the Lorenz equation can be established when the derivative terminals are assigned to suitable signals. For instance, the first equation, $x_1'(t) = -\beta x_1(t) + x_2(t)x_3(t)$, can be specified with the use of Fcn block. One can fill the edit box of Fcn block with the string `-beta*u[1]+u[2]*u[3]`*. The other two equations can also be modeled in a similar way, such that the original ODEs can be constructed in Simulink as shown in Figure 7.40 (b). The numerical solutions of the equations can be obtained using simulation methods.*

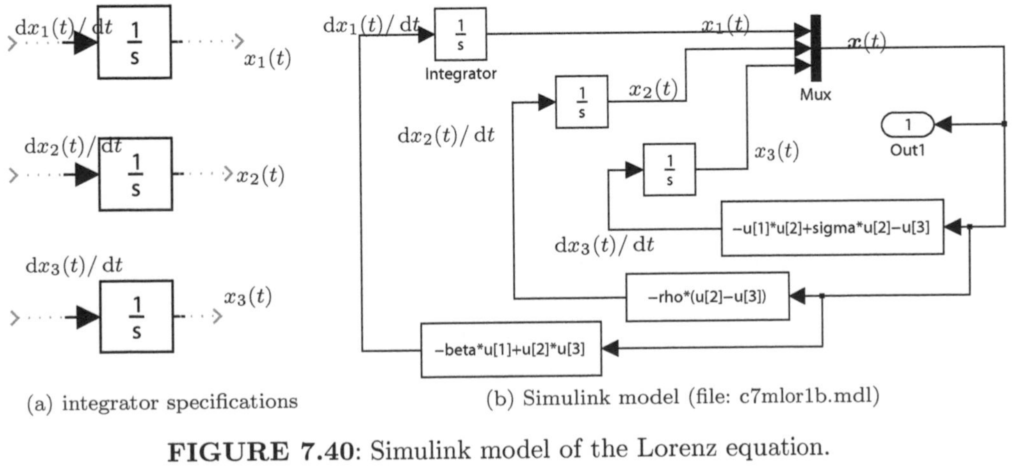

(a) integrator specifications (b) Simulink model (file: c7mlor1b.mdl)

FIGURE 7.40: Simulink model of the Lorenz equation.

The following statements can be used to solve the Simulink model. The results obtained are exactly the same as the ones given in Figures 7.2 (a) and (b).

```
>> beta=8/3; rho=10; sigma=28; % assign another set of parameters
   [t,x]=sim('c7mlor1b',[0,100]); subplot(121), plot(t,x) % simulation
   subplot(122), plot3(x(:,1),x(:,2),x(:,3)) % draw 3D phase space trajectory
```

It should be noted that the variables beta, rho *and* sigma *can be entered directly into MATLAB workspace, rather than specifying them as additional variables. The initial states can be specified in relevant dialog boxes.*

It is noted that for problems of small scales, the modeling of the differential equations in Simulink is more complicated than the direct use of ode45() *function. However, for complicated ordinary differential equations, using Simulink may make the equation solution process straightforward.*

In fact, Interpreted MATLAB Function block in User-Defined Functions group can be used in modeling the differential equations easily. A MATLAB function can be written to describe the right-hand-side of the explicit differential equation, as a static function of $\boldsymbol{y} = \boldsymbol{f}(\boldsymbol{x})$, and the file can be embedded in the block.

Example 7.43 For the right-hand-side of the equation in the previous example can be expressed in MATLAB as the following static function

```
function y=c7model(x), b=8/3; r=10; s=28;
y=[-b*x(1)+x(2)*x(3); -r*x(2)+r*x(3); -x(1)*x(2)+s*x(2)-x(3)];
```

and the file can be saved in file c7model.m. The model in Figure 7.41 (a) can be constructed, where a vectorized integrator is used to accept the vector signal $\boldsymbol{x}(t)$. The initial values of the equation can be set as a vector [0;0;1e−10] in the integrator. The model yields exactly the same result, as the one obtained in the previous example.

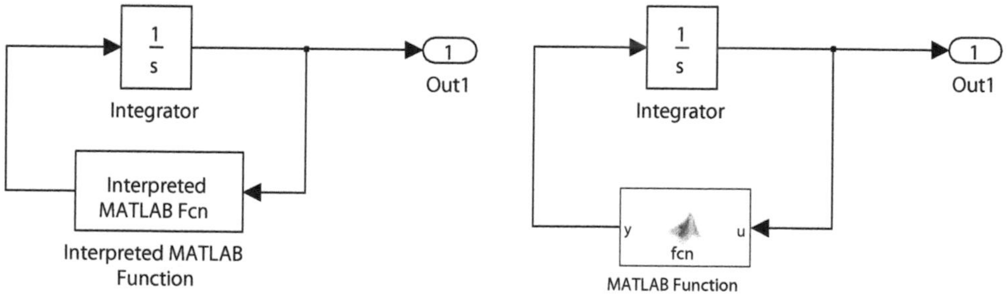

(a) file c7modela.mdl (b) another file: c7modelb.mdl

FIGURE 7.41: The new Simulink models for the Lorenz equation.

Alternatively, if the MATLAB Function block in the same group is used, the MATLAB function can be embedded in the block, and no extra MATLAB function files are necessary. Unfortunately in this case, the embedded function will be compiled with Stateflow and a lot of other files are generated. The first method is recommended. Besides, the additional variables are not supported, and S-function or extra input ports will be used, if the extra parameters are used.

Example 7.44 Consider the delay differential equations in Example 7.27, where

$$\begin{cases} x'(t) = 1 - 3x(t) - y(t-1) - 0.2x^3(t-0.5) - x(t-0.5) \\ y''(t) + 3y'(t) + 2y(t) = 4x(t). \end{cases}$$

Solve numerically the delay differential equations using Simulink.

Solution *The equation has already been solved in the previous example by the use of function* `dde23()`*, where some MATLAB functions are used. However, this method is not quite straightforward.*

Now consider the first equation. One may move the $-3x(t)$ *term to the left-hand side, and the original equation can be converted to*

$$x'(t) + 3x(t) = 1 - y(t-1) - 0.2x^3(t-0.5) - x(t-0.5).$$

Thus, the state $x(t)$ *can be regarded as the output signal from the transfer function* $1/(s+3)$*, while the input signal to the transfer function is* $1-y(t-1)-0.2x^3(t-0.5)-x(t-0.5)$*. In the second equation, the* $y(t)$ *signal can be regarded as the output of block* $4/(s^2+3s+2)$*, while the input to the block is* $x(t)$*. The transport delay blocks can be connected to the signals* $x(t)$ *and* $y(t)$ *to generate the delayed signals from them. From the above analysis, the Simulink model shown in Figure 7.42 can be constructed.*

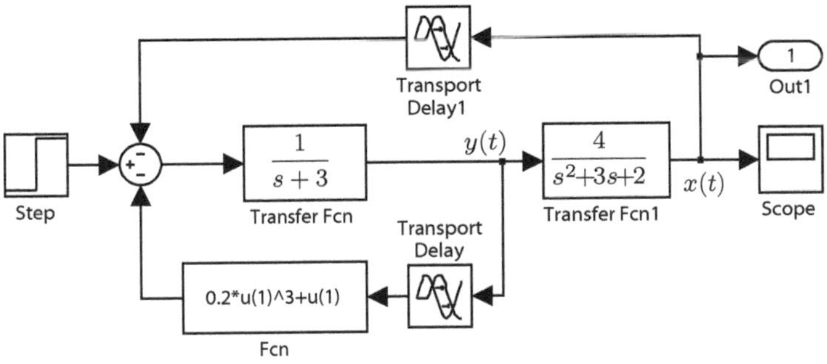

FIGURE 7.42: Simulink model (file: c7mdde2.mdl).

When the model is established, the following statements can be used to solve the differential equations. The output signal $y(t)$ *obtained is exactly the same as the one obtained in Example 7.27. Also, the solutions can be displayed on the* Scope *in the simulation model.*

```
>> [t,x]=sim('c7mdde2',[0,10]); plot(t,x)
```

Of course, if one is not used to the transfer function representation, one can still assume that $x_1 = x, x_2 = y, x_3 = y'$*; thus, the original ordinary differential equations are converted to the first-order explicit differential equations*

$$\begin{cases} x_1'(t) = 1 - x_1(t) - x_2(t-1) + 0.2x_1^3(t-0.5) - x_1(t-0.5) \\ x_2'(t) = x_3(t) \\ x_3'(t) = -4x_1(t) - 3x_3(t) - 2x_2(t). \end{cases}$$

Since there are three equations, three integrators should be used, and the Simulink model can then be established using the descriptions given earlier. The same numerical results are obtained.

Example 7.45 Now let us consider the neutral-type delay differential equations defined in Example 7.32, where

$$A_1 = \begin{bmatrix} -13 & 3 & -3 \\ 106 & -116 & 62 \\ 207 & -207 & 113 \end{bmatrix}, \quad A_2 = \begin{bmatrix} 0.02 & 0 & 0 \\ 0 & 0.03 & 0 \\ 0 & 0 & 0.04 \end{bmatrix}, \quad B = \begin{bmatrix} 0 \\ 1 \\ 2 \end{bmatrix}.$$

Model and solve it with Simulink.

Solution *The equations cannot be directly solved using the* `dde23()` *function. Simulink-based modeling can be used to describe the equations, and it can also be simulated via Simulink. Before building the Simulink model, the following statements can be entered to define the given matrices*

```
>> A1=[-13,3,-3; 106,-116,62; 207,-207,113];
   A2=diag([0.02,0.03,0.04]); B=[0; 1; 2];
```

Now consider the original equations. Define an integrator, and then, define the output signal to the integrator as the state vector, $\boldsymbol{x}(t)$ *and the input signal is automatically the derivative of the states* $\boldsymbol{x}'(t)$. *Transport delay blocks can be appended to the two signals to describe the signals* $\boldsymbol{x}(t-\tau_1)$ *and* $\boldsymbol{x}'(t-\tau_2)$. *Thus, the Simulink model can be constructed as shown in Figure 7.43.*

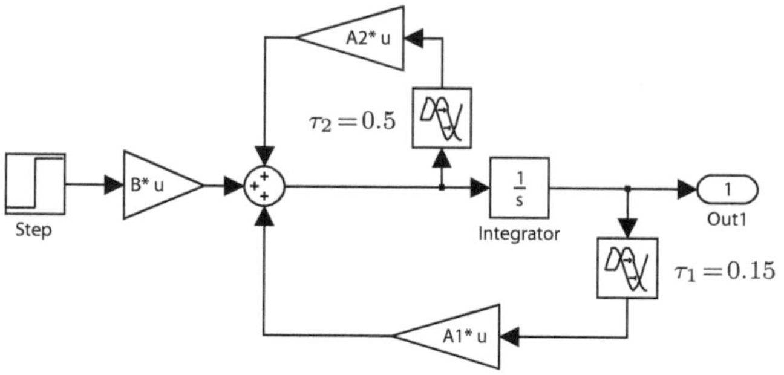

FIGURE 7.43: Neutral-type DDE model (file: c7mdde3.mdl).

This neutral type delay differential equation can easily be solved and the time responses are shown in Figure 7.44.

```
>> [t,x]=sim('c7mdde3',[0,8]); plot(t,x) % direct simulation model
```

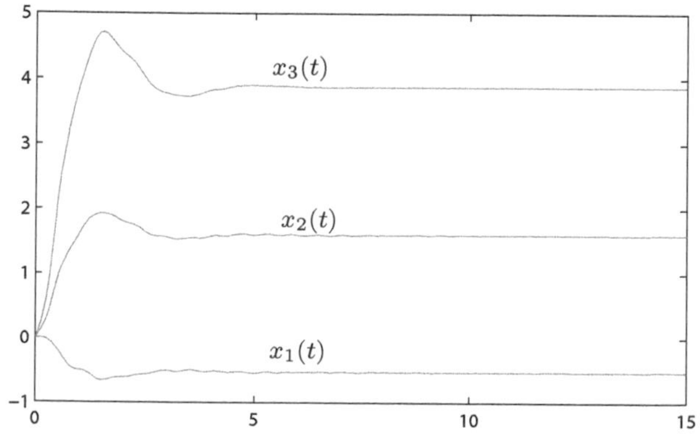

FIGURE 7.44: Solutions to delay differential equations.

Example 7.46 Assume that the initial states are zeros, please solve the following variable delay differential equations

$$\begin{cases} x_1'(t) = -2x_2(t) - 3x_1(t - 0.2|\sin t|) \\ x_2'(t) = -0.05x_1(t)x_3(t) - 2x_2(t - 0.8) \\ x_3'(t) = 0.3x_1(t)x_2(t)x_3(t) + \cos(x_1(t)x_2(t)) + 2\sin 0.1t^2. \end{cases}$$

Solution *Similar to the modeling process of other differential equations, a vectorized integrator is needed to describe the state vector $\boldsymbol{x}(t)$. The six signals composed of $\boldsymbol{x}(t)$, $x_1(t - 0.2|\sin t|)$, $x_2(t - 0.8)$ and t can be grouped together with* Mux *block to form the output signal vector. The following MATLAB static function can be written and saved in the* Interpreted MATLAB Function *block.*

```
function y=c7fdde6(x)
y=[-2*x(2)-3*x(4); -0.05*x(1)*x(3)-2*x(5)+2;
    0.3*x(1)*x(2)*x(3)+cos(x(1)*x(2))+2*sin(0.1*x(6)^2)];
```

Thus, thc Simulink model shown in Figure 7.45 can be established, and in the system, the variable delay term can be expressed by the Variable Time Delay *block, with its second input port driven by the signal $0.2|\sin t|$. It can be seen that the simulation results are exactly the same as the one obtained in Figure 7.19.*

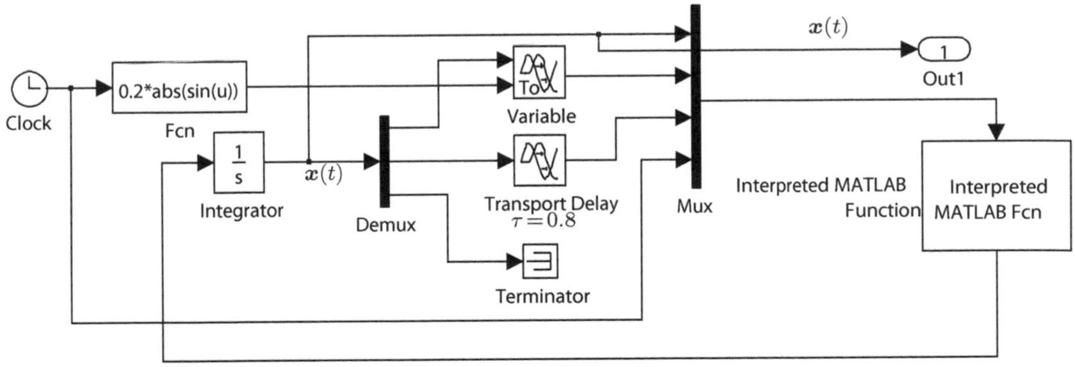

FIGURE 7.45: Simulink model of variable delay differential equations (file: c7mdde6.mdl).

If there exist nonzero initial values in the state vectors, and the history data for $t < 0$ are zeros, the initial vector of the integrator can be set to a nonzero vector, for instance, $\boldsymbol{x}_0 = [\sin 1, \cos 1, 1]^{\mathrm{T}}$. The simulation results obtained are exactly the same as the one obtained in Figure 7.20 (b).

In Section 7.4.5, linear systems driven by stochastic inputs are studied with time domain approach, which cannot be extended in use for nonlinear systems. Simulink modeling and simulation can be adopted for nonlinear systems, and the stochastic input can be generated by the Band-limited White Noise block, in the Sources group, and cannot be generated by other random number generator blocks. Fixed-step simulation is recommended and the step-size in simulation can be assigned to the block. Besides, since the system is driven by stochastic signals, and the number of points in simulation should be set to a large number, for instance, 30,000.

Example 7.47 Consider again the linear stochastic differential equation in Example 7.26. Please solve the problem with Simulink.

Solution *The Simulink model is shown in Figure 7.46. Assume that the computation step-size is $T=0.02$ s, the Simulink model can be executed, and the following statements can be used to draw the histogram, and the results are exactly the same as the one shown in Figure 7.17 (b), and the simulation results are validated.*

```
>> w=0.2; x=-2.5:w:2.5; y1=hist(yout,x); bar(x,y1/length(yout)/w);
   x1=-2.5:0.05:2.5; v=0.6655; y2=1/sqrt(2*pi)/v*exp(-x1.^2/2/v^2);
   line(x1,y2)  % superimposed theoretical probability density function
```

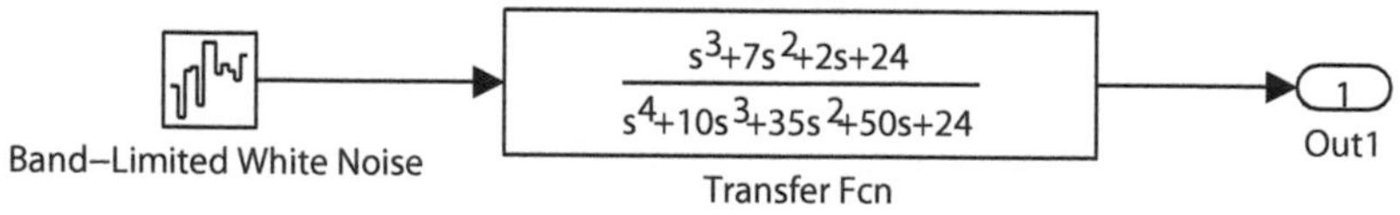

FIGURE 7.46: Simulink model of linear stochastic differential equation (c7mrand.mdl).

Example 7.48 Consider the nonlinear system structure as shown in Figure 7.47, where the transfer function and saturation can be described as

$$G(s)=\frac{s^3+7s^2+24s+24}{s^4+10s^3+35s^2+50s+24},\ \text{nonlinearity}\ \mathcal{N}(e)=\begin{cases}2\,\text{sign}(e), & |e|>1\\ 2e, & |e|\leqslant 1.\end{cases}$$

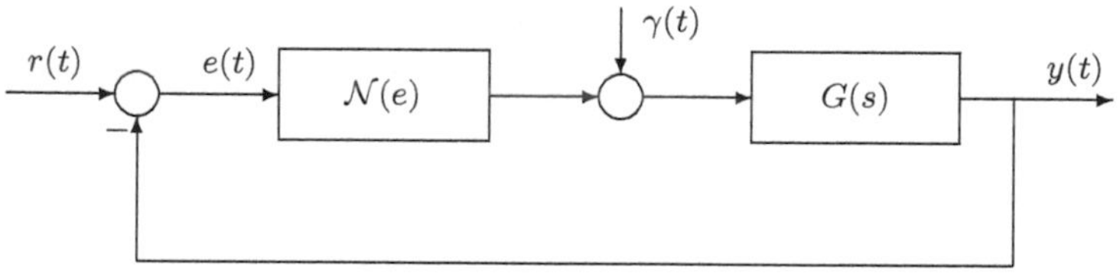

FIGURE 7.47: Block diagram of a stochastic nonlinear system model.

Assume that the mean and variance of the Gaussian white noise disturbance $\gamma(t)$ are 0 and 3, respectively, and the deterministic signal is $r(t)=0$. Perform simulation to the system, and find the distribution of the error signal $e(t)$.

Solution *The Simulink model can be established as shown in Figure 7.48.*

After simulation, the two variables `tout` *and* `yout` *are returned back into MATLAB workspace. The time response of the last 500 points in simulation are shown in Figure 7.49 (a), and it can be seen that the results are meaningless. The histogram obtained from the whole simulation results can also be obtained with the following statements, as shown in Figure 7.49 (b), and it is the probabilistic distribution of the error signal $e(t)$.*

```
>> plot(tout(end-500:end),yout(end-500:end)); c=linspace(-2,2,20);
   y1=hist(yout,c); figure; bar(c,y1/(length(tout)*(c(2)-c(1))))
```

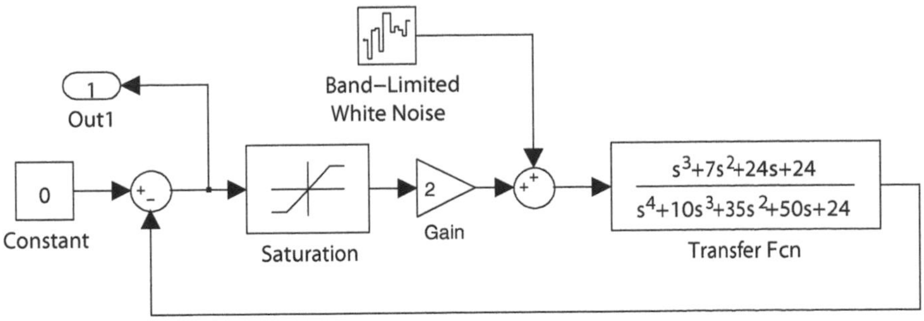

FIGURE 7.48: Simulink model of the nonlinear stochastic differential equation (file: c7mnlrsys.mdl).

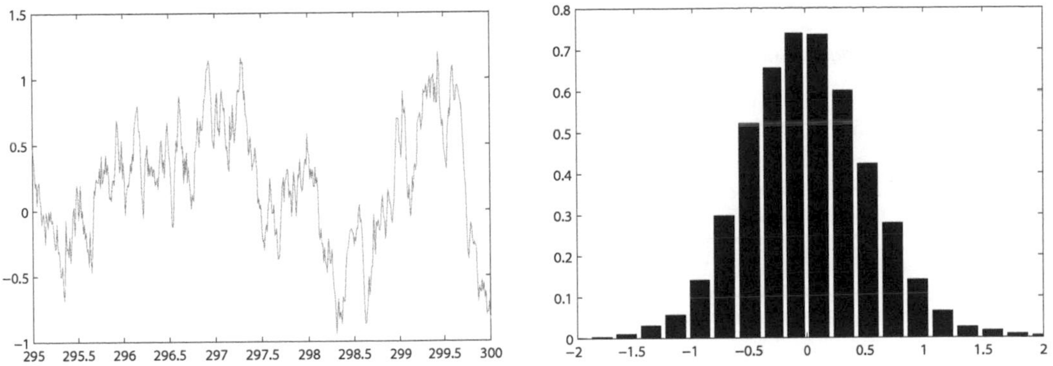

(a) time responses (b) probability density function

FIGURE 7.49: Simulation result of signal $e(t)$.

Exercises

Exercise 7.1 *Find the general solution to the following linear time-invariant ordinary differential equation:*

$$\frac{\mathrm{d}^5y(t)}{\mathrm{d}t^5}+13\frac{\mathrm{d}^4y(t)}{\mathrm{d}t^4}+64\frac{\mathrm{d}^3y(t)}{\mathrm{d}t^3}+152\frac{\mathrm{d}^2y(t)}{\mathrm{d}t^2}+176\frac{\mathrm{d}y(t)}{\mathrm{d}t}+80y(t)$$
$$=\mathrm{e}^{-2t}\left[\sin\left(2t+\frac{\pi}{3}\right)+\cos 3t\right].$$

With the initial conditions $y(0)=1$, $y(1)=3$, $y(\pi)=2, y'(0)=1, y'(1)=2$, *find the analytical solution. Verify the obtained results by back-substitution.*

Exercise 7.2 *Please find the general solutions of the following equations*

$$(i)\ \begin{cases} x''(t)-2y''(t)+y'(t)+x(t)-3y(t)=0\\ 4y''(t)-2x''(t)-x'(t)-2x(t)+5y(t)=0,\end{cases}$$

$$(ii)\quad \begin{cases} 2x''(t)+2x'(t)-x(t)+3y''(t)+y'(t)+y(t)=0 \\ x''(t)+4x'(t)-x(t)+3y''(t)+2y'(t)-y(t)=0. \end{cases}$$

Exercise 7.3 *Find the general analytical solution to the following linear time-invariant simultaneous ordinary differential equations:*

$$\begin{cases} x''(t)+5x'(t)+4x(t)+3y(t)=\mathrm{e}^{-6t}\sin 4t \\ 2y'(t)+y(t)+4x'(t)+6x(t)=\mathrm{e}^{-6t}\cos 4t. \end{cases}$$

If initial and boundary conditions are $x(0)=1, x(\pi)=2, y(0)=0$, *find the corresponding specific solution. Verify the results by back-substitution.*

Exercise 7.4 *Find the analytical solutions to the following linear time-varying ordinary differential equations:*

(i) Legendre equation: $(1-t^2)\dfrac{\mathrm{d}^2x(t)}{\mathrm{d}t^2}-2t\dfrac{\mathrm{d}x(t)}{\mathrm{d}t}+n(n+1)x(t)=0,$

(ii) Bessel equation: $t^2\dfrac{\mathrm{d}^2x(t)}{\mathrm{d}t^2}+t\dfrac{\mathrm{d}x(t)}{\mathrm{d}t}+(t^2-n^2)x(t)=0.$

Exercise 7.5 *Find the analytical solution to the following nonlinear differential equation:*

$$y''(x)-\left(2-\frac{1}{x}\right)y'(x)+\left(1-\frac{1}{x}\right)y(x)=x^2\mathrm{e}^{-5x}.$$

Furthermore, if the boundaries are specified by $y(1)=\pi,\ y(\pi)=1$, *find the analytical solution.*

Exercise 7.6 *Solve the following linear time-invariant differential equations using Laplace transform method:*

$$\begin{cases} x''(t)+y''(t)+x(t)+y(t)=0,\ x(0)=2, y(0)=1 \\ 2x''(t)-y''(t)-x(t)+y(t)=\sin t,\ x'(0)=y'(0)=-1. \end{cases}$$

Compare the obtained results with those by numerical methods.

Exercise 7.7 *Find the general solutions to the following ordinary differential equations.*

(i) $x''(t)+2tx'(t)+t^2x(t)=t+1,$ *(ii)* $y'(x)+2xy(x)=x\mathrm{e}^{-x^2},$

(iii) $y^{(3)}(t)+3y''(t)+3y'(t)+y(t)=\mathrm{e}^{-t}\sin t.$

Exercise 7.8 *Please find analytical solutions for the following nonlinear differential equations*

(i) $y'(x)=y^4(x)\cos x+y(x)\tan x,$

(ii) $xy^2(x)y'(x)=x^2+y^2(x),\ xy'(x)+2y(x)+x^5y^3(x)\mathrm{e}^x=0.$

Exercise 7.9 *Please find the solutions of the following differential equations, and plot the trajectory of the variables* (x,y).

$$\begin{cases} (2x''(t)-x'(t)+9x(t))-(y''(t)+y'(t)+3y(t))=0, \\ (2x''(t)+x'(t)+7x(t))-(y''(t)-y'(t)+5y(t))=0, \end{cases},$$

where, $x(0)=x'(0)=1$ *and* $y(0)=y'(0)=0$.

Exercise 7.10 *Please solve the following time-varying differential equations*

(i) $(x^2-2x+3)y'''(x)-(x^2+1)y''(x)+2xy'(x)-2y(x)=0$,

(ii) $x^2\ln x y''(x)-xy'(x)+y(x)=0$, (iii) $(\mathrm{e}^t+1)y''(t)-2y'(t)-\mathrm{e}^t y(t)=0$.

Exercise 7.11 *Please solve the following differential equations*

$$\text{(i)}\ \frac{\mathrm{d}x(t)}{x^3(t)+3x(t)y^2(t)}=\frac{\mathrm{d}y(t)}{2y^3(t)}=\frac{\mathrm{d}z(t)}{2y^2(t)z(t)},\quad \text{(ii)}\ \begin{cases}\mathrm{d}^2y(x)/\,\mathrm{d}x^2+2y(x)+4z(x)=\mathrm{e}^x\\ \mathrm{d}^2z(x)/\,\mathrm{d}x^2-y(x)-3z(x)=-x.\end{cases}$$

Exercise 7.12 *Limit cycle is a common phenomenon in nonlinear differential equations. For dynamic systems governed by nonlinear differential equations, no matter what initial values are selected, the phase trajectory will settle down on the same closed path, which is referred to as the* limit cycle. *Solve the following differential equations and draw the limit cycle in the x-y plane:*

$$\begin{cases}x'(t)=y(t)+x(t)(1-x^2(t)-y^2(t))\\ y'(t)=-x(t)+y(t)(1-x^2(t)-y^2(t)).\end{cases}$$

Try different initial values, and check whether the phase plane plot converges to the same limit cycle.

Exercise 7.13 *Consider the following nonlinear differential equations. It was pointed out in [13] that there are multiple limit cycles,* $r=1/(n\pi), n=1,2,3,\cdots$. *Please solve the equations numerically and observe the status of the limit cycles.*

$$\begin{cases}x'(t)=-y(t)+x(t)f\left(\sqrt{x^2(t)+y^2(t)}\right)\\ y'(t)=x(t)+y(t)f\left(\sqrt{x^2(t)+y^2(t)}\right),\end{cases}\quad \textit{where,}\quad f(r)=r^2\sin(1/r).$$

Exercise 7.14 *Consider the well-known Rössler equation described as*

$$\begin{cases}x'(t)=-y(t)-z(t)\\ y'(t)=x(t)+ay(t)\\ z'(t)=b+(x(t)-c)z(t),\end{cases}$$

where $a=b=0.2$, $c=5.7$ *and* $x_1(0)=x_2(0)=x_3(0)$. *Draw the 3D phase trajectory and also its projection on the x-y plane. In preparing the MATLAB scripts, the parameters* a,b,c *are suggested to be used as additional variables. If the parameters are changed to* $a=0.2, b=0.5, c=10$, *solve the problem again and observe the change of phase behavior.*

Exercise 7.15 *Consider the following Chua's circuit equation well-known in chaos studies:*

$$\begin{cases}x'(t)=\alpha[y(t)-x(t)-f(\xi)]\\ y'(t)=x(t)-y(t)+z(t)\\ z'(t)=-\beta y(t)-\gamma z(t),\end{cases}$$

where the nonlinear function $f(\xi)$ *is described by*

$$f(\xi)=b\xi+\frac{1}{2}(a-b)(|\xi+1|-|\xi-1|),\ \textit{and}\ a<b<0.$$

Prepare an M-function to describe the above equations, and draw the phase trajectory for the parameters $\alpha = 9$, $\beta = 100/7$, $\gamma = 0$, $a = -8/7$, $b = -5/7$, *and initial conditions* $x(0) = -2.121304, y(0) = -0.066170, z(0) = 2.881090$.

Exercise 7.16 *For the Lotka–Volterra's predator–prey equations*

$$\begin{cases} x'(t) = 4x(t) - 2x(t)y(t) \\ y'(t) = x(t)y(t) - 3y(t), \end{cases}$$

with initial conditions $x(0) = 2, y(0) = 3$, *solve the time responses of* $x(t)$ *and* $y(t)$, *and also plot the phase plane trajectory.*

Exercise 7.17 *Consider Duffing equation*

$$x''(t) + \mu_1 x'(t) - x(t) + 2x^3(t) = \mu_2 \cos t, \ \ \textit{with } x_1(0) = \gamma, \ x_2(0) = 0.$$

(i) Assume that $\mu_1 = \mu_2 = 0$, *please solve the equation numerically and draw the phase plots in the same coordinates for different initial values, for instance,* $\gamma = [0.1 : 0.1 : 2]$;

(ii) If $\mu_1 = 0.01, \mu_2 = 0.001$, *please draw the phase plots for different initial conditions, for instance,* $\gamma = 0.99, 1.01$;

(iii) If $x_2(0) = 0.2$, *please draw the phase plots again for different* γ.

Exercise 7.18 *Select appropriate state variables to convert the following high-order ordinary differential equations into first-order vector-form explicit ones. Solve the equations and plot the phase trajectories.*

$$(i) \begin{cases} x''(t) = -x(t) - y(t) - (3x'(t))^2 + (y'(t))^3 + 6y''(t) + 2t \\ y^{(3)} = -y'' - x' - \mathrm{e}^{-x} - t, \end{cases}$$

with $x(1) = 2, x'(1) = -4, y(1) = -2, y'(1) = 7, \ddot{y}(1) = 6$,

$$(ii) \begin{cases} x''(t) - 2x(t)z(t)x'(t) = 3x^2(t)y(t)t^2 \\ y''(t) - \mathrm{e}^{y(t)}y'(t) = 4x(t)t^2 z(t) \\ z''(t) - 2tz'(t) = 2t\mathrm{e}^{-x(t)y(t)}, \end{cases} \quad \textit{with} \begin{cases} z'(1) = x'(1) = y'(1) = 2 \\ z'(1) = x(1) = y(1) = 3. \end{cases}$$

$$(iii) \begin{cases} x^{(4)}(t) - 8\sin t y(t) = 3t - \mathrm{e}^{-2t} \\ y^{(4)}(t) + 3t\mathrm{e}^{-5t}x(t) = 12\cos t, \end{cases} \quad \textit{with} \begin{cases} x(0) = y(0) = 0, \ x'(0) = y'(0) = 0.3 \\ x''(0) = y''(0) = 1, x'''(0) = y'''(0) = 0.1. \end{cases}$$

Exercise 7.19 *Solve the following nonlinear time-varying differential equation*

$$y^{(3)}(t) + ty(t)y''(t) + t^2 y'(t)y^2(t) = \mathrm{e}^{-ty(t)}, \ y(0) = 2, \ y'(0) = y''(0) = 0,$$

and plot $y(t)$. *Select the fixed-step Runge–Kutta algorithm for solving the same problem as the baseline and compare in speed and accuracy with other MATLAB ODE solvers for this problem.*

Exercise 7.20 *Find the analytical and numerical solutions of the high-order simultaneous linear time-invariant ordinary differential equations and compare the results:*

$$\begin{cases} x''(t) = -2x(t) - 3x'(t) + \mathrm{e}^{-5t}, & x(0) = 1, x'(0) = 2 \\ y''(t) = 2x(t) - 3y(t) - 4x'(t) - 4y'(t) - \sin t, & y(0) = 3, y'(0) = 4. \end{cases}$$

Exercise 7.21 *Consider the following nonlinear differential equations:*

$$\begin{cases} u''(t) = -u(t)/r^3(t) \\ v''(t) = -v(t)/r^3(t), \end{cases}$$

where $r(t) = \sqrt{u^2(t)+v^2(t)}$, *and* $u(0)=1, u'(0)=2, v'(0)=2, v(0)=1$. *Select a set of state variables and convert the original ordinary differential equations to the form solvable by MATLAB. Plot the curves of* $u(t)$, $v(t)$, *and the phase plane trajectory.*

Exercise 7.22 *Consider the following implicit ODEs* [14]

$$\begin{cases} u_1'(t) = u_3(t) \\ u_2'(t) = u_4(t) \\ 2u_3'(t) + \cos(u_1(t)-u_2(t))u_4'(t) = -\mathrm{g}\sin u_1(t) - \sin(u_1(t)-u_2(t))u_4^2(t) \\ \cos(u_1(t)-u_2(t))u_3'(t) + u_4'(t) = -\mathrm{g}\sin u_2(t) + \sin(u_1(t)-u_2(t))u_3^2(t), \end{cases}$$

where $u_1(0)=45, u_2(0)=30, u_3(0)=u_4(0)=0$ *and* g$=9.81$. *Solve the above initial value problem and plot the time responses to the given initial states.*

Exercise 7.23 *Numerically solve the following implicit differential equations*

$$\begin{cases} x_1'(t)x_2''(t)\sin(x_1(t)x_2(t)) + 5x_1''(t)x_2'(t)\cos(x_1^2(t)) + t^2x_1(t)x_2^2(t) = \mathrm{e}^{-x_2^2(t)} \\ x_1''(t)x_2(t) + x_2''(t)x_1'(t)\sin(x_1^2(t)) + \cos(x_2''(t)x_2(t)) = \sin t, \end{cases}$$

where $x_1(0)=1, x_1'(0)=1, x_2(0)=2, x_2'(0)=2$. *Plot the phase trajectory.*

Exercise 7.24 *The following linear time-invariant differential equations are considered as stiff. Solve the initial value problems using ordinary and stiff ODE solver provided in MATLAB. Meanwhile, try to solve these equations using analytical methods and verify the accuracy of the numerical results.*

(i)
$$\begin{cases} y_1'(t) = 9y_1(t) + 24y_2(t) + 5\cos t - \dfrac{1}{3}\sin t,\ y_1(0) = \dfrac{1}{3} \\ y_2'(t) = -24y_1(t) - 51y_2(t) - 9\cos t + \dfrac{1}{3}\sin t,\ y_2(0) = \dfrac{2}{3}, \end{cases}$$

(ii)
$$\begin{cases} y_1'(t) = -0.1y_1(t) - 49.9y_2(t),\ y_1(0) = 1 \\ y_2'(t) = -50y_2(t),\ y_2(0) = 2 \\ y_3'(t) = 70y_2(t) - 120y_3(t),\ y_3(0) = 1. \end{cases}$$

Exercise 7.25 *Consider the following chemical reaction differential equations*

$$\begin{cases} y_1'(t) = -0.04y_1(t) + 10^4y_2(t)y_3(t) \\ y_2'(t) = 0.04y_1(t) - 10^4y_2(t)y_3(t) - 3\times10^7y_2^2(t) \\ y_3'(t) = 3\times10^7y_2^2(t), \end{cases}$$

where the initial values are $y_1(0)=1, y_2(0)=y_3(0)=0$. *Note that the above equations might be regarded as stiff. Solve the initial value problem with* `ode45()` *and the stiff solver* `ode15s()` *and check whether the obtained results are comparable or not.*

Exercise 7.26 *Solve the boundary value problem in Exercise 7.5 using numerical methods and draw the solution $y(t)$. Compare the accuracy of the results with the analytical results obtained earlier.*

Exercise 7.27 *Solve numerically the following differential equations with zero initial conditions*

$$\text{(i)}\ \begin{cases} x'(t) = \sqrt{x^2(t) - y(t) + 3} - 3 \\ y'(t) = \arctan(x^2(t) + 2x(t)y(t)), \end{cases} \qquad \text{(ii)}\ \begin{cases} x'(t) = \ln(2 - y(t) + 2y^2(t)) \\ y'(t) = 4 - \sqrt{x(t) + 4x^2(t)}. \end{cases}$$

Exercise 7.28 *Please solve the following differential equation with zero initial conditions*

$$\begin{cases} \cos x''(t)y'''(t) - \cos x''(t) - y''(t) - x(t)y'(t) + \mathrm{e}^{-x(t)}y(t) = 2 \\ \sin x''(t)\cos y'''(t) - x(t)y'(t) + x''(t)y(t) - y^2(t)y'(t) = 5. \end{cases}$$

Exercise 7.29 *Please solve the following delay differential equation, where for $t \leqslant 0$, it is known that $x(t) = t$, and $y(t) = \mathrm{e}^t$.*

$$\begin{cases} x'(t) = x^2(t - 0.2) + y^2(t - 0.2) - 6x(t - 0.5) - 8y(t - 0.1) \\ y'(t) = x(t)[2y(t - 0.2) - x(t) + 5 - 2x^2(t - 0.1)]. \end{cases}$$

If the last term in the equation is changed from $x^2(t - 0.1)$ to $x'(t - 0.1)$, solve the equation again.

Exercise 7.30 *Solve the boundary value problems where*

$$\text{(i)}\ x'' + \frac{1}{t}x' + \left(1 - \frac{1}{4t^2}\right)x = \sqrt{t}\cos t,\ \text{with } x(1) = 1, x(6) = -0.5,$$

$$\text{(ii)}\ -u''(x) + 6u(x) = \mathrm{e}^{10x}\cos 12x,\ u(0) = u(1) = 1.$$

Exercise 7.31 *For the Van der Pol equation $y'' + \mu(y^2 - 1)y' + y = 0$, if $\mu = 1$, find the numerical solutions for boundary conditions $y(0) = 1, y(5) = 3$. If μ is an undetermined constant, an extra condition $y'(5) = -2$ can be used. Solve the parameter μ as well as the equation. Plot the obtained solution and verify the results.*

Exercise 7.32 *Solve the boundary value problem with undetermined parameter c, where*

$$\begin{cases} x'(t) = x^2(t) - y(t) \\ y'(t) = [x(t) - y(t)][x(t) - y(t) - c], \end{cases} \quad \text{with} \begin{cases} x(0) = y(0) = 0 \\ y(5) = 1. \end{cases}$$

Exercise 7.33 *Please solve the boundary value problem given by*

$$y''(x) = \lambda^2(y^2(x) + \cos^2 \pi x) + 2\pi^2 \cos 2\pi x,\ \text{where } y(0) = y(1) = 0,\ \text{and } y'(0) = 1.$$

Exercise 7.34 *An infectious disease with periodic outbreak can be described by Kermack–McKendrick model [8]*

$$\begin{cases} y_1'(t) = -y_1(t)y_2(t - 1) + y_2(t - 10) \\ y_2'(t) = y_1(t)y_2(t - 1) - y_2(t) \\ y_3'(t) = y_2(t) - y_2(t - 10), \end{cases}$$

where, for $t \leqslant 0$, the history is described by $y_1(t) = 5$, $y_2(t) = 0.1$, $y_3(t) = 1$, please solve numerically the equation for $t \in [0, 40]$.

Exercise 7.35 *Consult the method described in Example 7.37, extend the method to handle high-order differential equations with undetermined constants. Please validate the method with the equation in Example 7.38.*

Exercise 7.36 *Solve numerically the partial differential equations below and draw the surface plot of the solution u.*

$$\begin{cases} \dfrac{\partial^2 u}{\partial x^2} + \dfrac{\partial^2 u}{\partial y^2} = 0 \\ u\big|_{x=0,y>0} = 1, \quad u\big|_{y=0,\ x\geqslant 0} = 0 \\ x > 0, \quad y > 0. \end{cases}$$

Exercise 7.37 *Consider the following simple linear time-invariant differential equation*

$$y^{(4)}(t) + 4y^{(3)}(t) + 6y''(t) + 4y'(t) + y(t) = e^{-3t} + e^{-5t}\sin(4t + \pi/3),$$

with the initial conditions given by $y(0) = 1, y'(0) = y''(0) = 1/2, y^{(3)}(0) = 0.2$. Construct the simulation model with Simulink, and find the simulation results.

Exercise 7.38 *Consider the following time varying differential equation*

$$y^{(4)}(t) + 4ty^{(3)}(t) + 6t^2y''(t) + 4y'(t) + y(t) = e^{-3t} + e^{-5t}\sin(4t + \pi/3),$$

with the initial conditions $y(0) = 1$, $y'(0) = y''(0) = 1/2, y^{(3)}(0) = 0.2$. Construct a Simulink model, solve the differential equation and plot the solution trajectories.

Exercise 7.39 *Consider the following delay differential equation*

$$y^{(4)}(t) + 4y^{(3)}(t-0.2) + 6y''(t-0.1) + 6y''(t) + 4y'(t-0.2) + y(t-0.5) = e^{-t^2}.$$

It is assumed that for $t \leqslant 0$, the above delay differential equation has zero initial conditions. Construct a Simulink model and solve the delay differential equations. Moreover, use the function `dde23()` *to solve the same problem and compare the results from these two methods.*

Bibliography

[1] Press W H, Flannery B P, Teukolsky S A, et al. Numerical recipes, the art of scientific computing. Cambridge: Cambridge University Press, 1986

[2] Dawkins P. Differential equations. `http://tutorial.math.lamar.edu/pdf/DE/DE_Complete.pdf`, 2007

[3] Mauch S. Advanced mathematical methods for scientists and engineers. Open source textbook at `http://www.its.caltech.edu/~sean/applied_math.pdf`, 2004

[4] Fehlberg E. Low-order classical Runge–Kutta formulas with step size control and their application to some heat transfer problems. Technical Report 315, NASA, 1969

[5] Forsythe G E, Malcolm M A, Moler C B. Computer methods for mathematical computations. Englewood Cliffs: Prentice-Hall, 1977

[6] Bogdanov A. Optimal control of a double inverted pendulum on a cart. Technical Report CSE-04-006, Department of Computer Science & Electrical Engineering, OGI School of Science & Engineering, OHSU, 2004

[7] Liberzon D, Morse A S. Basic problems in stability and design of switched systems. IEEE Control Systems Magazine, 1999, 19(5):59–70

[8] Shampine L F, Thompson S. Solving DDEs in MATLAB. Applied Numerical Mathematics, 2001, 37(4):441–458

[9] Shampine L F, Kierzenka J, Reichelt M W. Solving boundary value problems for ordinary differential equation problems in MATLAB with bvp4c, 2000

[10] The MathWorks Inc. SimuLAB, a program for simulating dynamic systems, user's guide, 1990

[11] The MathWorks Inc. Simulink user's manual, 2007

[12] Xue D, Chen Y Q. System simulation techniques with MATLAB/Simulink. London: Wiley, 2013

[13] Enns R H, McGuire G C. Nonlinear physics with MAPLE for scientists and engineers. Boston: Birkhäuser, second edition, 2000

[14] Molor C B. Numerical computing with MATLAB. MathWorks Inc, 2004

Chapter 8

Data Interpolation and Functional Approximation Problems

In scientific research and engineering practice, a lot of experimental data are generated. Based on the experimental data, the problems of data interpolation and function fitting may always be encountered. The existing data can be regarded as the samples, the so-called *data interpolation* is to numerically generate new data points from a discrete set of known samples. In Section 8.1, one-dimensional, two-dimensional or even high-dimensional interpolation problems are solved in MATLAB. An interpolation-based numerical integration method is also introduced. In Section 8.2, two of the most widely used splines for interpolation are introduced, the cubic spline and the B-spline. Spline function-based numerical differentiations and integrations are also introduced. The integration results are even more accurate than those presented in Chapter 3 and Section 8.1. Data interpolation problems can easily be solved by following the examples in these two sections.

The so-called *function approximation problem* is to extract the function representation from the measured sample points. Polynomial approximations and the least squares method for nonlinear function approximation will be studied. Furthermore, Padé approximations and continued fraction approximations for given functions are explored in Section 8.3. In Section 8.6, the correlation analysis of signals and experimental data are introduced. Fast Fourier transforms and filter-based de-noising and other signal processing problems are introduced.

At this point, let us note the difference between "interpolation" and "fitting." The key distinguishing feature for "interpolation" is that the obtained interpolation function passes through all given data points, while for "fitting" the obtained fitting function may not pass all the given data points.

Some problems in the chapter are also related to those topics presented in other chapters, with Chapter 10 in particular. For instance, the data fitting problems can be solved with neural networks to be discussed in Section 10.3. Data filtering and de-noising problems can also be solved with wavelet transform techniques to be introduced in Section 10.5.

For readers who wish to check the detailed explanations of various data interpolation techniques and function fitting methods, we recommend the free textbook [1] (Chapters 3, 5 and 15). We also found the online resource [2] useful and interesting (Chapter 5).

8.1 Interpolation and Data Fitting

8.1.1 One-dimensional data interpolation

I. Solving one-dimensional data interpolation problems

Assume that the function $f(x)$ is a one-dimensional function and the mathematical formula is not known. If at a set of N distinct points $x_1, x_2, \cdots, x_N$, the values of the function are measured as $y_1, y_2, \cdots, y_N$. The points (x_1, y_1), (x_2, y_2), $\cdots$, (x_N, y_N) are often referred to as *sample points*. The idea of using sample points for finding other unknown points is called *interpolation*.

Many interpolation functions have been provided in MATLAB, such as the one-dimensional interpolation function `interp1()`, two-dimensional `interp2()` and the polynomial fitting function `polyfit()`.

One-dimensional interpolation function `interp1()` can be called, with the syntax $\boldsymbol{y}_1$ = `interp1(`$\boldsymbol{x}$`,`$\boldsymbol{y}$`,`$\boldsymbol{x}_1$`,method)`, where $\boldsymbol{x} = [x_1, x_2, \cdots, x_N]^{\mathrm{T}}$ and $\boldsymbol{y} = [y_1, y_2, \cdots, y_N]^{\mathrm{T}}$. The two vectors provide the information of the sample points. The argument $\boldsymbol{x}_1$ is the points of independent variable to be interpolated. It can be a scalar, a vector or a matrix. The interpolated results are returned in argument $\boldsymbol{y}_1$. The default of `method` argument is `'linear'`. Alternative options for the `method` argument can be `'nearest'`, `'pchip'` (in old versions, `'cubic'`) and `'spline'`. Usually it is recommended that the `'spline'` option be used. Extrapolations, i.e., the interpolation point is outside of the interval $[x_1, x_N]$ can also be performed with $\boldsymbol{y}_1$ = `interp1(`$\boldsymbol{x}$`,`$\boldsymbol{y}$`,`$\boldsymbol{x}_1$`,method,'extrap')`.

Example 8.1 Assume the sample points are generated from the function $f(x) = (x^2 - 3x + 5)\mathrm{e}^{-5x} \sin x$. Interpolate the sample points and see whether a smoother curve can be reconstructed.

Solution *The sample data can be generated with the following statements, where the spacing can be selected as a large value, such as 0.12. The sample points are then obtained as shown in Figure 8.1 (a).*

```
>> x=0:.12:1; y=(x.^2-3*x+5).*exp(-5*x).*sin(x); plot(x,y,x,y,'o')
```

It can be seen that the "curve" is formed by joining the adjacent points with straight lines. The curves around the sample points are not very smooth. One may select another vector of densely distributed $\boldsymbol{x}_1$ *data, and then, use* `interp1()` *function to calculate the interpolations to these points.*

```
>> x1=0:.02:1; y0=(x1.^2-3*x1+5).*exp(-5*x1).*sin(x1);
   y1=interp1(x,y,x1); y2=interp1(x,y,x1,'pchip');
   y3=interp1(x,y,x1,'spline'); y4=interp1(x,y,x1,'nearest');
   plot(x1,[y1',y2',y3',y4'],x,y,'o',x1,y0), L=49
   e1=max(abs(y0(1:L)-y2(1:L))), e2=max(abs(y0-y3)), e3=max(abs(y0-y4))
```

Then, with different methods, the interpolation results are compared with the theoretical results from the given function obtained as shown in Figure 8.1 (b), with maximum absolute errors $e_1 = 0.0177$, $e_2 = 0.0086$, $e_3 = 0.1598$.

It can be seen that the linear interpolation is exactly the same as the one in Figure 8.1 (a) and the quality of `'nearest'` *option is very poor. The interpolation results under*

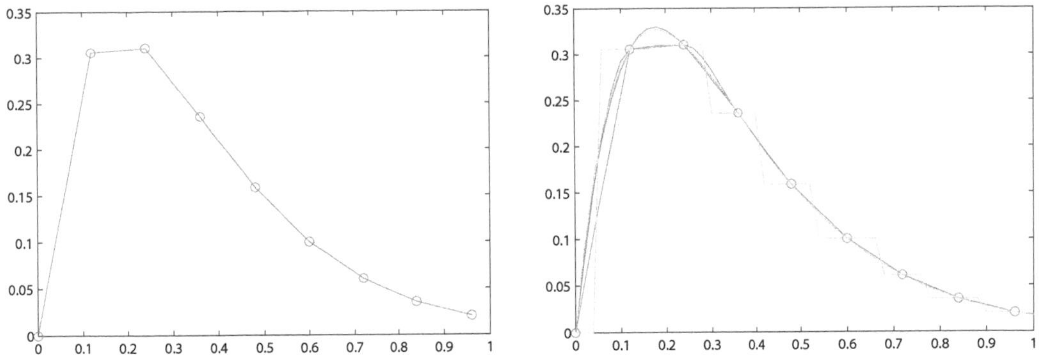

(a) sample points (b) comparison of different methods

FIGURE 8.1: Results of one-dimensional interpolation methods.

the `'pchip'` *and* `'spline'` *options are much smoother. In fact, the interpolation using the* `'spline'` *method gives much better interpolation to the original function. Thus, it is recommended to use the* `'spline'` *option in solving one-dimensional interpolation problems.*

Example 8.2 Write a piece of code, which allows the user to draw manually a smooth curve, interpolated by splines.

Solution *In applications, the user can pick a few points with the* `ginput()` *function. Then, interpolation can be made to obtain a smooth curve. The above idea can be implemented in the MATLAB function shown below*

```
function sketcher(vis)
x=[]; y=[]; i=1; h=[]; axes(gca);
while 1, [x0,y0,but]=ginput(1);
   if but==1, x=[x,x0]; y=[y,y0];
      h(i)=line(x0,y0); set(h(i),'Marker','o'); i=i+1; else, break
end, end
if nargin==0, delete(h); end
xx=[x(1):(x(end)-x(1))/100: x(end)];
yy=interp1(x,y,xx,'spline'); line(xx,yy)
```

II. Lagrange interpolation algorithm and its application

The Lagrange interpolation algorithm is the most presented interpolation method in the interpolation part of numerical analysis textbooks, e.g. [3]. For known sample points x_i, y_i, the interpolation on $\boldsymbol{x}$ vector can be obtained as

$$\phi(\boldsymbol{x}) = \sum_{i=1}^{N} y_i \prod_{j=1,j\neq i}^{N} \frac{\boldsymbol{x} - x_j}{(x_i - x_j)}. \tag{8-1-1}$$

Based on the above algorithm, a MATLAB function is prepared as follows:

```
function y=lagrange(x0,y0,x)
ii=1:length(x0); y=zeros(size(x));
for i=ii, ij=find(ii~=i); y1=1;
   for j=1:length(ij), y1=y1.*(x-x0(ij(j))); end
```

```
    y=y+y1*y0(i)/prod(x0(i)-x0(ij));
  end
```

Example 8.3 Consider a well-known function $f(x) = 1/(1+25x^2), -1 \leqslant x \leqslant 1$, where a set of points can be evaluated as sample points. Based on the points, the interpolation points can be obtained using Lagrange interpolation method, and the results are shown in Figure 8.2 (a).

```
>> x0=-1+2*[0:10]/10; y0=1./(1+25*x0.^2);
   x=-1:.01:1; y=lagrange(x0,y0,x);  % Lagrange interpolation
   ya=1./(1+25*x.^2); plot(x,ya,x,y,':',x0,y0,'o')
```

From the interpolation results, it can be seen that the interpolation is far from the theoretic curve. The higher the degree of the polynomials, the more serious the divergence. This phenomenon is referred to as the *Runge phenomenon.* Thus, for this example, the Lagrange interpolation method failed. The function `interp1()` can be used instead to solve the problem. The spline interpolation can be obtained with the following statements, and the fitting results are shown in Figure 8.2 (b). Thus, the function provided in MATLAB does not have the Runge phenomenon any more, so it can be safely used.

```
>> y1=interp1(x0,y0,x,'pchip'); y2=interp1(x0,y0,x,'spline');
   plot(x,ya,x,y1,':',x,y2,'--')
```

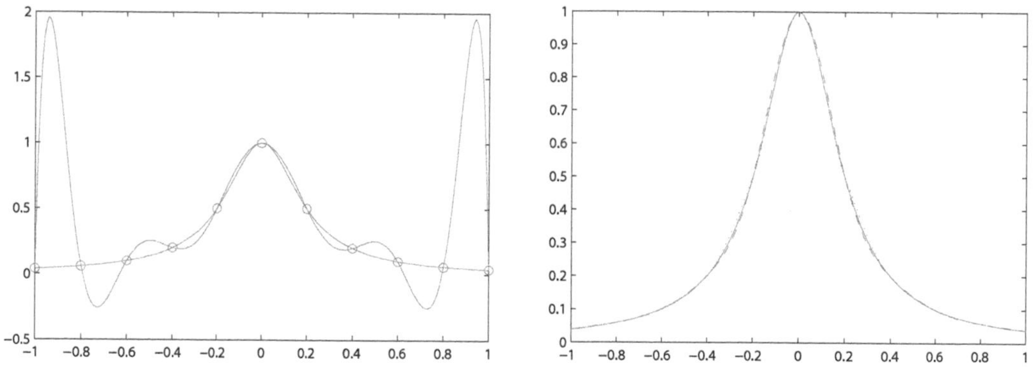

(a) failure of Lagrange interpolation (b) `interp1()` interpolation

FIGURE 8.2: Interpolation results comparison.

III. Forecast problems in one-dimensional interpolations

The so-called *forecast problem* is to predict the data in the future based on the existing ones. Practical examples of forecast technique are population forecast and weather forecast. As mentioned earlier, forecast problems are essentially the extrapolation problems, and can be implemented in `interp1()` function, with the option `'extrap'`. If the option `'spline'` is used, `'extrap'` option cannot be used, since extrapolation problems can be handled automatically with the spline interpolation computations.

Example 8.4 In Example 2.31, the population information was stored in an Excel file.

The samples can be extracted from it on the basis of 5-year interval. Please interpolate from the sample data the number of population in the years 1949–2015.

Solution *Since the population information in the Excel contains the data in the period of 1949–2011, and the samples are made as 1949, 1954, $\cdots$, 2004, 2009, the data before 2009 can be regarded as interpolation, and after that, it can be regarded as extrapolation problems. The data obtained, as well as the original data, are shown in Figure 8.3.*

```
>> X=xlsread('census.xls','B5:C67'); t=X(:,1); p=X(:,2);
   t0=t(1:5:end); p0=p(1:5:end); t1=1949:2015;
   y=interp1(t0,p0,t1,'spline','extrap'); plot(t,p,t1,y,t0,p0,'o')
```

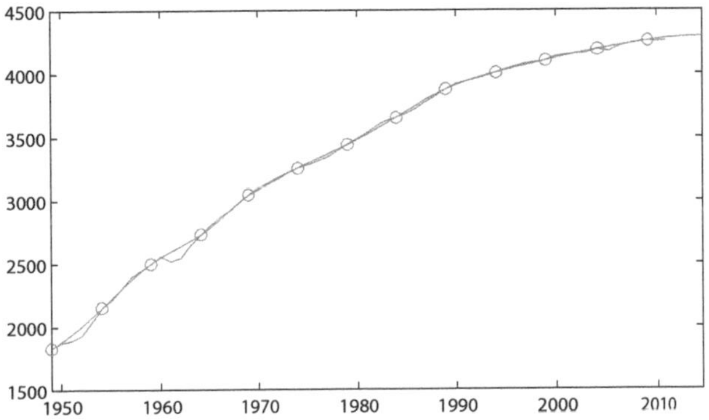

FIGURE 8.3: Population computation with interpolation and extrapolation approaches.

It should be noted that, population forecast is a complicated problem, and should be studied with specific models, such as population dynamics. The population are affected by other factors, such as population policies, natural disasters and others. The data-based results are in fact unreliable.

8.1.2 Definite integral evaluation from given samples

The definite integral evaluation based on sample points has been discussed and the trapezoidal approach `trapz()` has been presented in Section 3.9.1. From Example 3.58, it is seen that if the sample points are sparsely distributed, there exist large errors in the results. If interpolation is used in the evaluation of the intermediate points, the interpolation-based numerical integration function can be written as

```
function y=quadspln(x0,y0,a,b)
f=@(x)interp1(x0,y0,x,'spline'); y=integral(f,a,b);
```

whose syntax is $I=$ `quadspln(`x_0`,`y_0`,`a`,`b`)`, where x_0 and y_0 are vectors composed of sample points, $[a,b]$ is the integration interval. With `quadspln()` function, the definite integral can be obtained as illustrated in the following examples.

Example 8.5 Consider again the problem in Example 3.58. Use the interpolation-based method to evaluate the definite integral.

Solution *From the theoretical method it is known that the integral is 2. The trapezoidal method can be applied to evaluate the integral. If the step-size is too large, the approximation* *accuracy is not satisfactory. Here the interpolation-based method is applied to the same sample points*

```
>> x0=linspace(0,pi,30); y0=sin(x0); I1=trapz(x0,y0) % trapezoidal method
   I2=quadspln(x0,y0,0,pi) % evaluate the integral with interpolation method
```

The trapezoidal result is $I_1 = 1.9980$, *and with interpolation* $I_2 = 2$, *with absolute error of* -3.36×10^{-7}. *It can be seen that the interpolation-based integral algorithm yields more accurate results.*

An even more exaggerated example is as follows. Suppose that there are only 5 unevenly distributed sample points measured, the results by the use of interpolation method and trapezoidal methods can be compared

```
>> x0=[0,0.4,1 2,pi]; y0=sin(x0);  % sample points generation
   plot(x0,y0,x0,y0,'o')    % the sample points are shown in Figure 8.4 (a)
   I=quadspln(x0,y0,0,pi)   % only 1% relative error obtained
   I1=trapz(x0,y0)    % with the trapz() function, relative error reaches 7.9%!
```

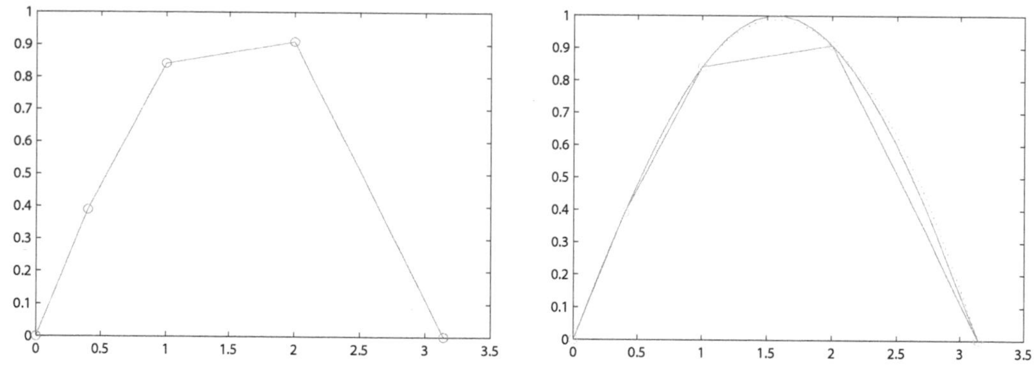

(a) distribution of sample points (b) comparison of different methods

FIGURE 8.4: The integration results with only five sample points.

and the results with these two methods are $I = 2.0191$, $I_1 = 1.8416$, *respectively. In fact, even with such sparsely distributed sample points, the fitting results are still very satisfactory. The interpolation function together with the sample points are shown in Figure 8.4 (b), which are quite close to the original true function.*

```
>> x=linspace(0,pi,30); y0a=sin(x); y=interp1(x0,y0,x,'spline');
   plot(x0,y0,x,y,':',x,y0a,x0,y0,'o')
```

Example 8.6 Consider again the oscillatory function in Example 3.59. Assume that 150 sample points are measured. Evaluate the definite integral with the `quadspln()` function and verify the accuracy of the results.

Solution *The numerical solutions to the definite integral can be obtained from the generated sample points*

```
>> x=linspace(0,3*pi/2,200); y=cos(15*x); I=quadspln(x,y,0,3*pi/2)
```

and the result obtained is $I = 0.066672375$. *Note that the true integral is 1/15.*

Clearly, the interpolation-based method achieves quite high accuracy. The following statements can be used to draw the original and interpolated curves as shown in Figure 8.5. The fitting results are also quite satisfactory. The difference between the original and interpolated curves are hardly seen from the figure.

```
>> x0=[0:3*pi/2/1000:3*pi/2]; y0=cos(15*x0); % theoretical value
   y1=interp1(x,y,x0,'spline'); plot(x,y,x0,y1,':')
```

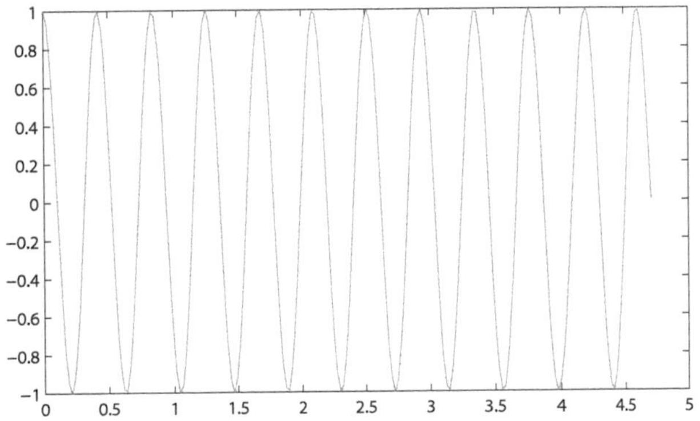

FIGURE 8.5: Original function and interpolation curves.

From this example, it can be seen that if there are significant oscillations in the integrand, the accuracy of integral evaluations cannot be guaranteed if the number of sample points is large enough.

8.1.3 Two-dimensional grid data interpolation

The two-dimensional (2D) interpolation function `interp2()` is provided in MATLAB with the syntax z_1=`interp2(`x_0,y_0,z_0,x_1,y_1`,method)`, where x_0, y_0, z_0 are the measured sample points in mesh grid form. The arguments x_1, y_1 are the points to be interpolated. They are not necessarily given in mesh grid form. They can be in any form, scalars, vectors, matrices or even multi-dimensional arrays. The returned argument z_1 is the interpolation results, which is exactly the same data type as x_1 or y_1. The `method` options are `'linear'`, `'pchip'` and `'spline'`. Similar to the one-dimensional interpolation function, the method `'spline'` is recommended. Examples will be given in the following to illustrate the 2D grid data interpolation.

Example 8.7 From the given function $z = (x^2-2x)e^{-x^2-y^2-xy}$, generate a set of relatively sparsely distributed mesh grid sample data. From the data, interpolate the whole surface, and compare the results with the exact surface.

Solution *The grid data can be obtained as shown in Figure 8.6 (a) by the following MATLAB scripts. It can be seen that the surface obtained is not very smooth.*

```
>> [x,y]=meshgrid(-3:.6:3,-2:.4:2); z=(x.^2-2*x).*exp(-x.^2-y.^2-x.*y);
   surf(x,y,z), axis([-3,3,-2,2,-0.7,1.5])
```

Now, select more densely distributed interpolation points in mesh grid form. The following statements can be used to evaluate the interpolation points with the interpolated surface shown in Figure 8.6 (b).

```
>> [x1,y1]=meshgrid(-3:.2:3, -2:.2:2); % generate sparse meshgrid
   z1=interp2(x,y,z,x1,y1); surf(x1,y1,z1), axis([-3,3,-2,2,-0.7,1.5])
```

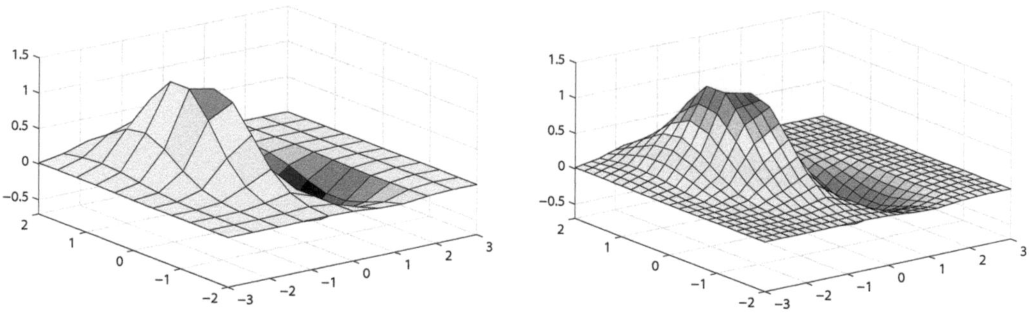

(a) surface from the sample points (b) interpolation with linear method

FIGURE 8.6: Comparison of two-dimensional interpolations.

It can be seen that the interpolation results using the linear method is still rather rough. Let us try `'pchip'` *and* `'spline'` *options and the obtained results are compared in Figure 8.7.*

```
>> z1=interp2(x,y,z,x1,y1,'pchip'); z2=interp2(x,y,z,x1,y1,'spline');
   subplot(121), surf(x1,y1,z1), axis([-3,3,-2,2,-0.7,1.5])
   subplot(122), surf(x1,y1,z2), axis([-3,3,-2,2,-0.7,1.5])
```

It can be seen that the interpolation results are all satisfactory, especially with the spline interpolation option. Thus, the `'spline'` *option is recommended.*

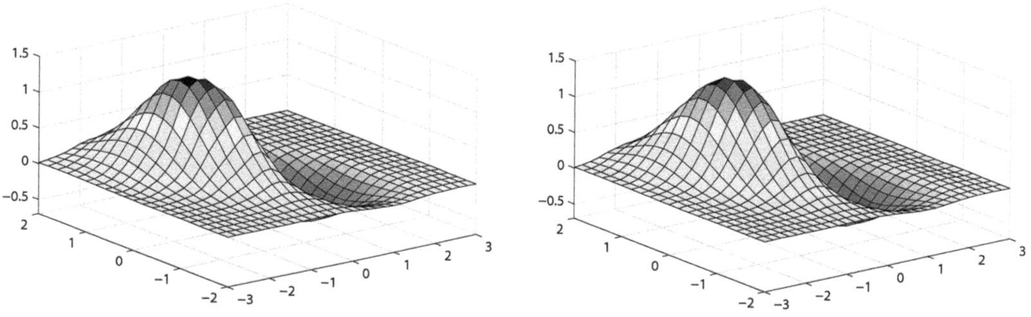

(a) cubic interpolation (b) spline interpolation

FIGURE 8.7: Comparisons of interpolation with other methods.

Furthermore, since the original function is known, the exact solutions can be obtained. The following statements can be used to compute the error surface between the two interpolated matrices z_1 *and* z_2 *and the exact* z*, respectively, as shown in Figure 8.8 (a), (b). It can be seen that the spline interpolation results are much more accurate than the ones by the cubic interpolation. The error surface comparison suggests again that the* `'spline'` *option is recommended.*

```
>> z=(x1.^2-2*x1).*exp(-x1.^2-y1.^2-x1.*y1);  % exact functions
   subplot(121), surf(x1,y1,abs(z-z1)), axis([-3,3,-2,2,0,0.08])
   subplot(122), surf(x1,y1,abs(z-z2)), axis([-3,3,-2,2,0,0.025])
```

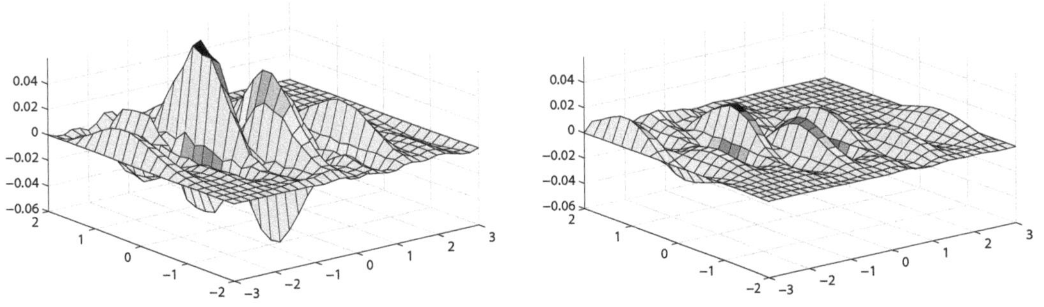

(a) cubic algorithm (b) spline algorithm

FIGURE 8.8: Error surfaces of the interpolation results.

8.1.4 Two-dimensional scattered data interpolation

Through the above examples, it can be seen that the two-dimensional interpolation problems can easily be solved with the function `interp2()`. However, it should be noted that there are some restrictions. For example, only the data given in grid format can be handled by the function `interp2()`. If a scattered sample data set is provided, the `interp2()` function cannot be used. However, in many practical problems, the scattered data set (x_i, y_i, z_i) is usually provided, rather than the grid data. The general function `griddata()` can then be used instead, with the syntax $\boldsymbol{z}$=`griddata(`$\boldsymbol{x}_0$,$\boldsymbol{y}_0$,$\boldsymbol{z}_0$,$\boldsymbol{x}$,$\boldsymbol{y}$,`'v4')`, where $\boldsymbol{x}_0$, $\boldsymbol{y}_0$, $\boldsymbol{z}_0$ are the vectors composed of the sample points. They can be the data vectors of arbitrarily distributed sample points. The arguments $\boldsymbol{x}, \boldsymbol{y}$ are the expected interpolation positions, which can be described as single point, vectors or mesh grid matrices. The returned argument $\boldsymbol{z}$ is in the same format as $\boldsymbol{x}$, representing the interpolation results. The option `'v4'` is the unnamed interpolation algorithm used in MATLAB version 4.0, which has many advantages. Apart from the `'v4'` option, other options such as `'linear'`, `'pchip'` and `'nearest'` can also be used. However, the option `'v4'` is recommended.

Example 8.8 Consider again the function $z = (x^2 - 2x)e^{-x^2-y^2-xy}$. In the rectangular region, $x \in [-3, 3]$, $y \in [-2, 2]$, a set of 200 sample points (x_i, y_i) can be selected randomly and the values z_i can be calculated. Based on the data, perform the interpolation using `griddata()` and visualize the error surface.

Solution *The randomly selected 200 can be generated with the following statements, and the vectors* $\boldsymbol{x}$, $\boldsymbol{y}$ *and bmz can be established. Since the data set is not given in the grid format, the surface plot cannot be drawn directly from the data. Scattered points can be displayed instead as shown in Figure 8.9 (a), with the sample points distribution on the x-y plane shown in Figure 8.9 (b). It can be seen that the sample points are evenly scattered.*

```
>> x=-3+6*rand(200,1); y=-2+4*rand(200,1);
   z=(x.^2-2*x).*exp(-x.^2-y.^2-x.*y);    % data generation
   subplot(121), plot3(x,y,z,'x'), axis([-3,3,-2,2,-0.7,1.5]), grid
   subplot(122), plot(x,y,'x')    % two-dimensional distribution
```

The points to be interpolated can still be selected as grid format data using the method shown in Example 8.7. The interpolated data can then be obtained with the `'cubic'`

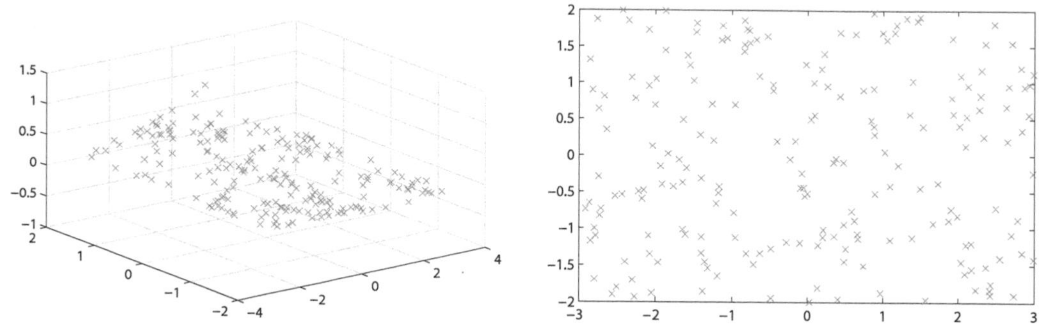

(a) scattered three dimension plot (b) projection in the x-y plane

FIGURE 8.9: Visualization of the known sample points.

and 'v4' *options, and the interpolated surfaces are shown in Figures 8.10 (a) and (b), respectively. It can be seen that the surface interpolation using the* 'v4' *algorithm is much better. Some of the points with the* 'cubic' *options are actually missing.*

```
>> [x1,y1]=meshgrid(-3:.2:3, -2:.2:2);
   z1=griddata(x,y,z,x1,y1,'pchip'); subplot(121), surf(x1,y1,z1)
   axis([-3,3,-2,2,-0.7,1.5]); z2=griddata(x,y,z,x1,y1,'v4');
   subplot(122),  surf(x1,y1,z2), axis([-3,3,-2,2,-0.7,1.5])
```

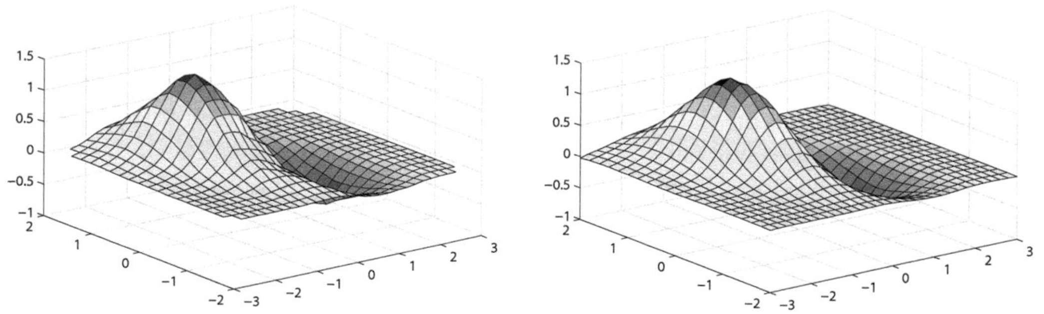

(a) cubic interpolation (b) 'v4' interpolation

FIGURE 8.10: Comparison of interpolation surfaces for a 2D function.

Next, let us check the interpolation errors. The error surfaces using the two interpolation algorithms are obtained as shown in Figures 8.11 (a) and (b), respectively. It can be seen that the interpolation quality of the 'v4' *option is much superior to the one obtained with the cubic interpolation algorithm.*

```
>> z0=(x1.^2-2*x1).*exp(-x1.^2-y1.^2-x1.*y1);
   subplot(121), surf(x1,y1,abs(z0-z1)); axis([-3,3,-2,2,0,0.1])
   subplot(122), surf(x1,y1,abs(z0-z2));  axis([-3,3,-2,2,0,0.1])
```

Example 8.9 In the previous example, the sample points in the x-y plane are evenly scattered. Now, let us remove some of the points, and then, perform interpolation and observe what happens.

Solution *If the points within the circle centered at* $(-1,-1/2)$ *with a radius of 0.5 are*

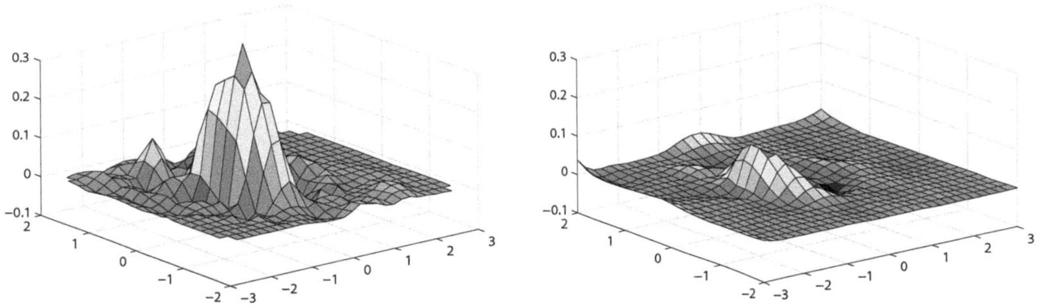

(a) cubic interpolation (b) 'v4' interpolation

FIGURE 8.11: Interpolations errors for a two-dimensional function.

removed from the vectors $\boldsymbol{x}$, $\boldsymbol{y}$ and $\boldsymbol{z}$, the following statements can be used to remove the samples.

```
>> ii=find((x+1).^2+(y+0.5).^2>0.5^2); % find the points within the circle
   x=x(ii); y=y(ii); z=z(ii); plot(x,y,'x') % display the samples
   t=[0:.1:2*pi,2*pi]; x0=-1+0.5*cos(t); y0=-0.5+0.5*sin(t);
   line(x0,y0)    % superimpose the circle on the samples
```

The new samples are distributed as shown in Figure 8.12 (a) with the circle superimposed. The samples within the circle have been removed. With the new set of samples, the interpolated surface is obtained as shown in Figure 8.12 (b). The quality of fittings looks satisfactory.

```
>> [x1,y1]=meshgrid(-3:.2:3, -2:.2:2); z1=griddata(x,y,z,x1,y1,'v4');
   surf(x1,y1,z1), axis([-3,3,-2,2,-0.7,1.5]) % draw interpolated surface
```

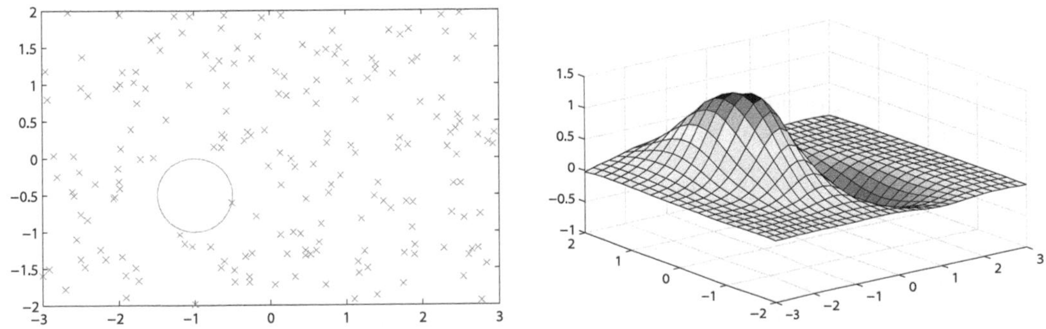

(a) sample data distribution (b) Surface interpolation

FIGURE 8.12: The new samples distribution and interpolated surface.

Error surface of the interpolation is visualized in Figure 8.13 (a) where it can be observed that although some samples are removed, the fitting result is still satisfactory.

```
>> z0=(x1.^2-2*x1).*exp(-x1.^2-y1.^2-x1.*y1); % compute theoretical data
   surf(x1,y1,abs(z0-z1)), axis([-3,3,-2,2,0,0.15]) % draw error surface
```

The error contours are also shown in Figure 8.13 (b), superimposed with the circle. It can be seen that the fitting results are satisfactory, apart from the regions where samples are removed.

```
>> contour(x1,y1,abs(z0-z1),30); hold on, plot(x,y,'x'); line(x0,y0)
```

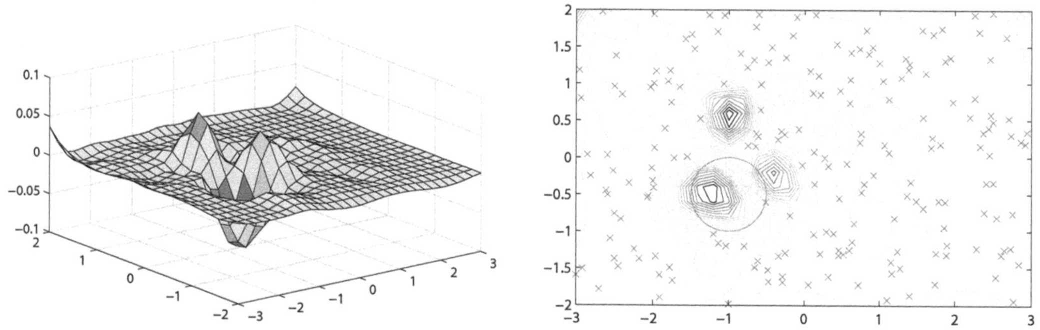

(a) error surface (b) two-dimensional contour

FIGURE 8.13: Error surface and error contour for the 2D interpolation.

It can be concluded that the quality of interpolation depends mainly on the distribution of measured samples. If the samples within a certain area are not sufficient, the quality of interpolation in that area cannot be obtained satisfactorily. Clearly, the more sample data, the better interpolation accuracy.

8.1.5 Optimization problems based on scattered sample data

In practical applications, sometimes the mathematical form of the objective function is not known, instead, a set of scattered sample data are known. In this case, spline interpolation method can be used to approximate the objective function, such that the optimization problems can be solved.

Example 8.10 Consider again the function given in Example 8.7. Generate a set of sample data, and based on the scattered sample data, find and validate the minimum of the function.

Solution *A set of scattered sample data can be generated first, and an anonymous function for the objective function can be constructed, within it, spline interpolation function can be embedded. With the following statement, the interpolation-based optimization results are* $x = 0.6069, y = -0.3085$. *The contour plot of the objective function can also be obtained as shown in Figure 8.14. It can be seen that the results are correct.*

```
>> x=-3+6*rand(200,1); y=-2+4*rand(200,1); % generate random data
   z=(x.^2-2*x).*exp(-x.^2-y.^2-x.*y);      % generate scattered sample data
   f=@(p)griddata(x,y,z,p(1),p(2),'v4'); x=fminunc(f,[0,0])
   [x0,y0]=meshgrid(-3:0.1:2,-2:0.1:2);
   z0=(x0.^2-2*x0).*exp(-x0.^2-y0.^2-x0.*y0); contour(x0,y0,z0,30)
```

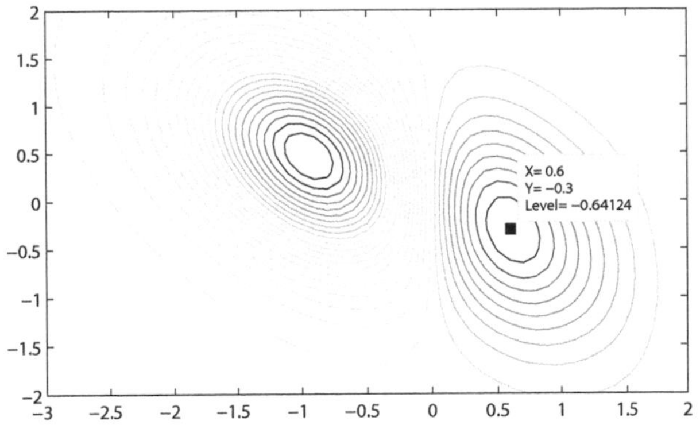

FIGURE 8.14: Contour plots with the solutions.

8.1.6 High-dimensional data interpolations

Three-dimensional mesh grid can also be generated with the previously discussed function `meshgrid()`, with $[\boldsymbol{x},\boldsymbol{y},\boldsymbol{z}]=$`meshgrid`$(\boldsymbol{x}_1,\boldsymbol{y}_1,\boldsymbol{z}_1)$, where $\boldsymbol{x}_1$, $\boldsymbol{y}_1,\boldsymbol{z}_1$ are the necessary segmentation vectors. The returned arguments $\boldsymbol{x},\boldsymbol{y},\boldsymbol{z}$ are in three-dimensional arrays, in grid format.

The n-dimensional grid data can be generated with the function `ndgrid()`, and the syntax of the function is $[\boldsymbol{x}_1,\cdots,\boldsymbol{x}_n]=$`ndgrid`$(\boldsymbol{v}_1,\cdots,\boldsymbol{v}_n)$, where $\boldsymbol{v}_1$, $\cdots,\boldsymbol{v}_n$ are the segmentation of the n-dimensional data, which are given in vectors. The returned argument $\boldsymbol{x}_1,\cdots,\boldsymbol{x}_n$ are the n-dimensional arrays in grid format.

It should be noted that the data structures generated by functions `ndgrid()` and `meshgrid()` are different. Function `meshgrid()` can only be used in two- and three-dimensional data generation, while `ndgrid()` can be used in the generation of higher-dimensional data. For two-dimensional data, the mesh grid data generated by $[\boldsymbol{x},\boldsymbol{y}]=$`meshgrid`$(\cdots)$ and $[\boldsymbol{x}_1,\boldsymbol{y}_1]=$`ndgrid`$(\cdots)$ satisfy $\boldsymbol{x}=\boldsymbol{x}_1^{\mathrm{T}}$, and $\boldsymbol{y}=\boldsymbol{y}_1^{\mathrm{T}}$, while in three-dimensional data generations, the data generated by $[\boldsymbol{x},\boldsymbol{y},\boldsymbol{z}]=$`meshgrid`$(\cdots)$ and $[\boldsymbol{x}_1,\boldsymbol{y}_1,\boldsymbol{z}_1]=$`ndgrid`$(\cdots)$ have the same $\boldsymbol{z}$ and $\boldsymbol{z}_1$, and $\boldsymbol{x}(:,:,i)$ array is a fixed matrix for each i. Similar things happen to the arrays $\boldsymbol{x}_1^{\mathrm{T}}$, $\boldsymbol{y}$ and $\boldsymbol{y}_1^{\mathrm{T}}$, and besides, $\boldsymbol{x}(:,:,i)=\boldsymbol{x}_1(:,:,i)^{\mathrm{T}}$. A function `mesh2nd()` can be written to implement conversion between the two types of grid data.

```
function [x1,y1,z1,v1]=mesh2nd(x,y,z,v)
if nargin==3, x1=x.'; y1=y.'; z1=z.';
elseif nargin==4, z1=z;
   for i=1:size(x,3), x1(:,:,i)=x(:,:,i).';
      y1(:,:,i)=y(:,:,i).'; v1(:,:,i)=v(:,:,i).';
   end
else, error('Error in input arguments'), end
```

For the samples obtained in grid format, the functions `interp3()` and `interpn()` can be used to solve interpolation problems. The syntaxes of these functions are similar to the `interp2()` function. If the scattered three-dimensional or n-dimensional data are given, the functions `griddata3()` and `griddatan()` can be used, and these functions are similar to the `griddata()` function.

Example 8.11 Consider again the function $V(x,y,z)$ in Example 2.42.

$$V(x,y,z)=\sqrt{x^x+y^{(x+y)/2}+z^{(x+y+z)/3}}.$$

Generate a set of data and show the volume visualization representation of the interpolated data.

Solution *Function* `meshgrid()` *can be used to generate a set of sparsely distributed sample data. Function* `vol_visual4d()` *can be used to compare the original data and interpolated data, as shown in Figure 8.15. It can be seen that the interpolated data is quite accurate.*

```
>> [x,y,z]=meshgrid(0:0.3:2); [x0 y0 z0]=meshgrid(0:0.1:2);
   V=sqrt(x.^x+y.^((x+y)/2)+z.^((x+y+z)/3));
   V0=sqrt(x0.^x0+y0.^((x0+y0)/2)+z0.^((x0+y0+z0)/3));
   V1=interp3(x,y,z,V,x0,y0,z0,'spline'); vol_visual4d(x0,y0,z0,V1)
```

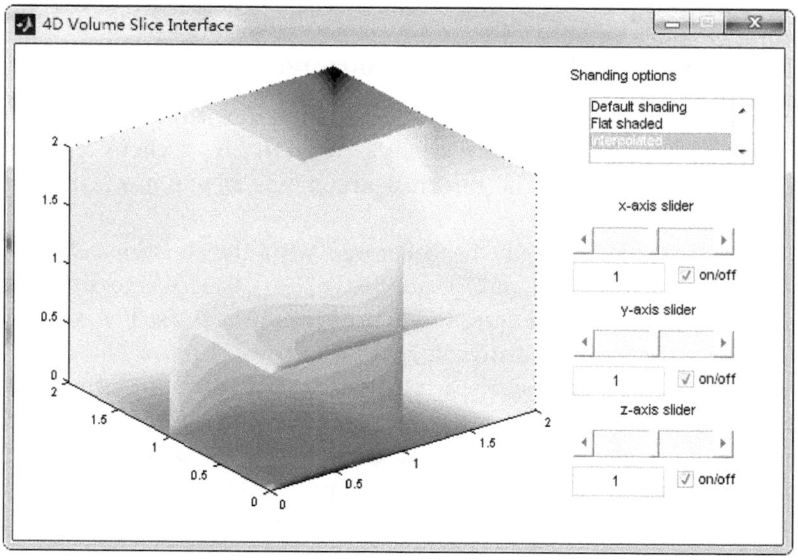

FIGURE 8.15: Slice views of the original and interpolation data.

8.2 Spline Interpolation and Numerical Calculus

The Spline Toolbox provided in MATLAB can be used to better solve the interpolation problems. Moreover, numerical differentiation and integration problems can be solved easily using this toolbox. This section can be regarded as an extension to Sections 8.1, 3.8 and 3.9.

Spline functions can be regarded as an effective approximation method. The most widely used spline functions are the cubic splines and B-splines. The creation of these splines will be shown in this section, and then, numerical differentiation and integration can be solved using the Spline Toolbox.

8.2.1 Spline interpolation in MATLAB

I. Cubic splines and MATLAB solutions

The definition of cubic spline function is that for n sample points (x_i, y_i) $(i = 1, 2, \cdots, n)$ where $x_1 < x_2 < \cdots < x_n$, the following three conditions are to be satisfied. The function $S(x)$ is referred to as *cubic spline* function with n nodes.

(i) $S(x_i)=y_i$, $(i=1,2,\cdots,n)$, i.e., the curve passes through all the samples.

(ii) In each interval $[x_i, x_{i+1}]$, $S(x)$ is given as a cubic polynomial

$$S(x) = c_{i1}(x - x_i)^3 + c_{i2}(x - x_i)^2 + c_{i3}(x - x_i) + c_{i4}. \tag{8-2-1}$$

(iii) $S'(x)$ and $S''(x)$ are continuous in the interval $[x_1, x_n]$.

A cubic spline object can be established with the function `csapi()` provided in the Spline Toolbox. The syntax of the function is simply $S=$`csapi(`$\boldsymbol{x}$`,`$\boldsymbol{y}$`)`, where $\boldsymbol{x} = [x_1, x_2, \cdots, x_n]$, $\boldsymbol{y} = [y_1, y_2, \cdots, y_n]$ are the sample points. Therefore, the returned argument S is a cubic spline object, whose fields include the sub-intervals, cubic function coefficients of each piecewise cubic function.

The interpolation curves can be drawn with the function `fnplt()`. For given independent variable vector $\boldsymbol{x}_p$, the interpolation results can be evaluated from the function `fnval()` with the syntax `fnplt(`S`)`, $\boldsymbol{y}_p=$`fnval(`S`,`$\boldsymbol{x}_p$`)`, where the vector $\boldsymbol{y}_p$ contains the interpolation results for the given vector $\boldsymbol{x}_p$.

Example 8.12 Compute the cubic spline interpolation results for the sparsely distributed sample points in Example 8.6.

Solution *The cubic spline interpolation can be obtained and compared with the theoretical data as shown in Figure 8.16.*

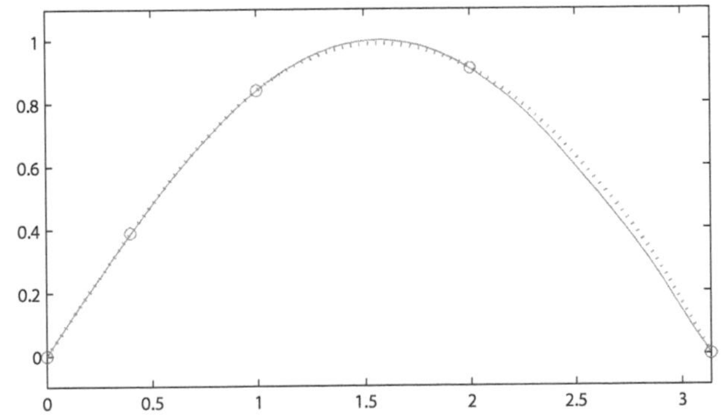

FIGURE 8.16: Interpolation with cubic splines.

```
>> x0=[0,0.4,1 2,pi]; y0=sin(x0); sp=csapi(x0,y0), fnplt(sp,':');
   hold on, ezplot('sin(t)',[0,pi]); plot(x0,y0,'o'); sp.coefs
```

The coefficients of the piecewise polynomials are given in Table 8.1. For instance, in the interval $(0.4, 1)$*, the interpolation polynomial can be expressed as*

$$S_2(x) = -0.1627(x - 0.4)^3 - 0.1876(x - 0.4)^2 + 0.9245(x - 0.4) + 0.3894.$$

TABLE 8.1: The coefficients of the piecewise cubic functions.

interval	c_0	c_1	c_2	c_3
$(0, 0.4)$	-0.16265031	0.007585654	0.99653564	0
$(0.4, 1)$	-0.16265031	-0.18759472	0.92453202	0.38941834
$(1, 2)$	0.024435717	-0.48036529	0.52375601	0.84147098
$(2, \pi)$	0.024435717	-0.40705814	-0.36366741	0.90929743

Example 8.13 Interpolate the data given in Example 8.1 using cubic splines.

Solution *The following statements can be used to establish a cubic spline object-based on the given data. The obtained piecewise cubic function coefficients are shown in Table 8.2. The cubic spline object can then be used in the interpolation method.*

```
>> x=0:.12:1; y=(x.^2-3*x+5).*exp(-5*x).*sin(x); % generate samples
   sp=csapi(x,y); fnplt(sp), sp.breaks, sp.coefs % build a spline object
```

TABLE 8.2: Coefficients of piecewise cubic spline interpolation.

piecewise interval	coefficients of cubic polynomials				piecewise interval	coefficients of cubic polynomials			
	c_1	c_2	c_3	c_4		c_1	c_2	c_3	c_4
$(0, 0.12)$	24.7396	-19.359	4.5151	0	$(0.48, 0.6)$	-0.2404	0.7652	-0.5776	0.1588
$(0.12, 0.24)$	24.7396	-10.4526	0.9377	0.3058	$(0.6, 0.72)$	-0.4774	0.6787	-0.4043	0.1001
$(0.24, 0.36)$	4.5071	-1.5463	-0.5022	0.3105	$(0.72, 0.84)$	-0.4559	0.5068	-0.2621	0.0605
$(0.36, 0.48)$	1.9139	0.07623	-0.6786	0.2358	$(0.84, 0.96)$	-0.4559	0.3427	-0.1601	0.03557

The function `csapi()` can be used to establish the cubic spline object for multivariate data in the grid format, with $S=$`csapi(`$\{\boldsymbol{x}_1,\boldsymbol{x}_2,\cdots,\boldsymbol{x}_n\},\boldsymbol{z}$`)`, where the variables $\boldsymbol{x}_i$ and $\boldsymbol{z}$ are the grid points, and the cubic spline object is returned in S.

Example 8.14 Interpolate the grid data in Example 8.8 with cubic splines and draw the interpolation surface.

Solution *The following statements can be used to establish the cubic spline object* `sp`. *The surface obtained is shown in Figure 8.17, which is the same as the one obtained with the* `interp2()` *function.*

```
>> x0=-3:.6:3; y0=-2:.4:2; [x,y]=ndgrid(x0,y0);
   z=(x.^2-2*x).*exp(-x.^2-y.^2-x.*y); sp=csapi({x0,y0},z); fnplt(sp);
```

It should be noted that the matrices $\boldsymbol{x}$ *and* $\boldsymbol{y}$ *are obtained with* `ndgrid()` *rather than* `meshgrid()`. *Be careful that the arrangement of the matrices is different in this example.*

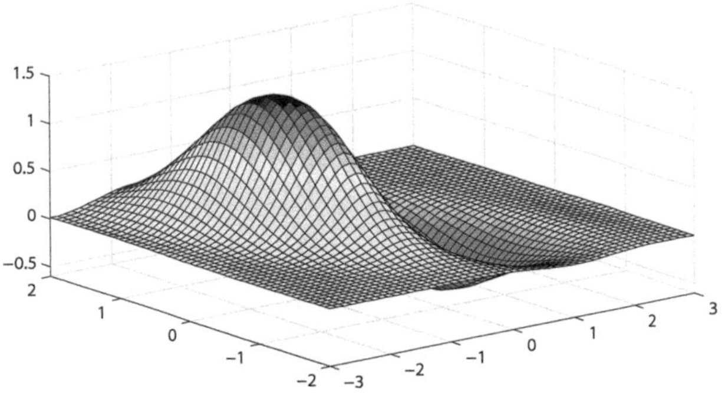

FIGURE 8.17: Interpolation results of two-dimensional function.

II. B-spline and its MATLAB functions

B-spline is another type of commonly used splines. Suppose the interested interval (a, b) is divided into several intervals, $a = t_0 < t_1 < t_2 < \cdots < t_m = b$, where t_i is referred to as a *knot*, the piecewise approximate function can be written as

$$F(t) = \sum_{i=0}^{m} p_i B_{i,k}(t), \tag{8-2-2}$$

where p_i are the coefficients, k is the order, with $k \leqslant m$. $B_{i,k}(x)$ is the kth order B-spline basis, which can be computed recursively with

$$B_{i,0}(t) = \begin{cases} 1, & \text{if } t_i < t < t_{i+1} \\ 0, & \text{otherwise,} \end{cases} \tag{8-2-3}$$

and for $j = 1, 2, \cdots, k$, $i = 0, 1, 2, \cdots, m$

$$B_{i,j}(t) = \frac{t - t_i}{t_{i+j} - t_i} B_{i,j-1}(t) + \frac{t_{i+j+1} - t}{t_{i+j+1} - t_{i+1}} B_{i+1,j-1}(t). \tag{8-2-4}$$

The function `spapi()` for defining the B-spline object is introduced. If the samples are given in vectors $\boldsymbol{x}$ and $\boldsymbol{y}$, the following statements can be used to define a B-spline object S with the syntax $S=$ `spapi(`k`,`$\boldsymbol{x}$`,`$\boldsymbol{y}$`)`, where k is the order of B-spline. It can be seen that kth order B-spline is essentially kth order piecewise polynomials. Normally one may select $k = 4$ or 5 to ensure good interpolation results. For specific problems, a suitable increase in k may improve the quality of interpolation results.

Example 8.15 Interpolate the functions given in Examples 8.12 and 8.13 with B-splines. Compare with the results obtained using cubic splines.

Solution *For the Example 8.12, the following statements can be used to interpolate the data with the results shown in Figure 8.18 (a). It can be seen that one can hardly distinguish the interpolation curve from the theoretical one.*

```
>> x0=[0,0.4,1 2,pi]; y0=sin(x0); ezplot('sin(t)',[0,pi]); hold on
   sp1=csapi(x0,y0); fnplt(sp1,'--'); % cubic spline interpolation
   sp2=spapi(5,x0,y0); fnplt(sp2,':') % B-spline with order k = 5
```

From the above observations, the B-spline interpolation is far better than the cubic spline for this example. For the data in Example 8.13, the B-spline interpolation results are shown in Figure 8.18 (b). Again, here, the B-spline interpolation is much better than the cubic spline interpolation. In practical applications, if k selected is too large, a suitable one will be assigned instead. Please try an order of 100, and see what happens.

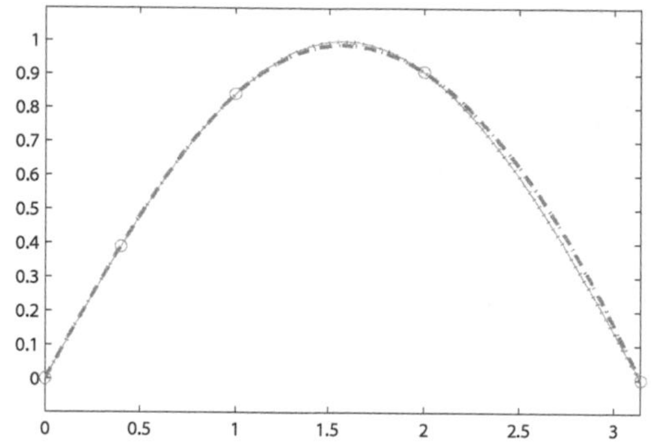

(a) data interpolation for Example 8.12

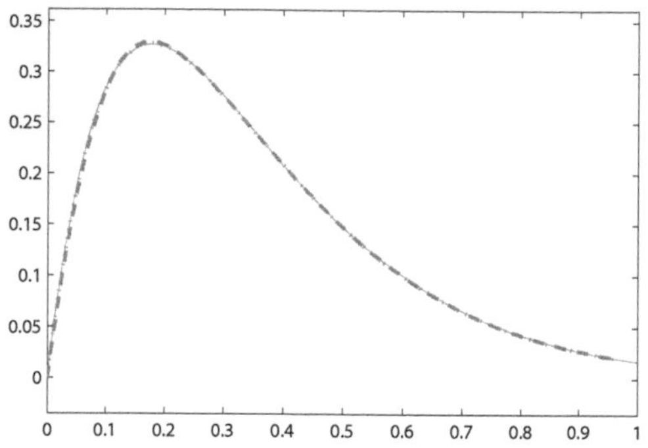

(b) data interpolation for Example 8.13

FIGURE 8.18: The interpolation curves with spline interpolations.

```
>> x=0:.12:1; y=(x.^2-3*x+5).*exp(-5*x).*sin(x); % generate samples
   ezplot('(x^2-3*x+5)*exp(-5*x)*sin(x)',[0,1]), hold on % draw curve
   sp1=csapi(x,y); fnplt(sp1,'--'); sp2=spapi(5,x,y); fnplt(sp2,':')
```

8.2.2 Numerical differentiation and integration with splines

It has been shown that splines can be used to evaluate numerical integrals and even when the samples are sparsely distributed, good results can still be obtained. Compared with the algorithms in Sections 3.8 and 3.9, the spline-based algorithm has its own advantages. The spline-based integration is defined as $F(x) = \int_{x_0}^{x} f(t)\,dt$, where x_0 is the pre-specified boundary. Clearly, the definite integral can be evaluated from $I = F(b) - F(a)$.

I. Numerical differentiation

The spline function-based numerical differentiation to the given sample points can be calculated from the function `fnder()`

S_d = `fnder(`S`,`k`)` % kth order derivative of S

S_d = `fnder(`S`,[`$k_1,\cdots,k_n$`])` % partial derivatives for multivariate functions

Example 8.16 Consider the sample points in Example 8.13. Compute the numerical differentiations to the samples with cubic function or B-spline functions and compare the results with the theoretical ones.

Solution *From the generated samples, the cubic and B-spline data objects can be established. Then, the derivatives can be obtained with* `fnder()` *function, and the curves can be shown in Figure 8.19.*

```
>> syms x; f=(x^2-3*x+5)*exp(-5*x)*sin(x); ezplot(diff(f),[0,1]),
   hold on, x0=0:.12:1; y0=double(subs(f,x,x0)); % generate samples
   sp1=csapi(x0,y0); dsp1=fnder(sp1,1); fnplt(dsp1,'--') % derivative
```

```
sp2=spapi(5,x0,y0); dsp2=fnder(sp2,1); % derivative with B-spline
fnplt(dsp2,':'); axis([0,1,-0.8,5])    % compare the results
```

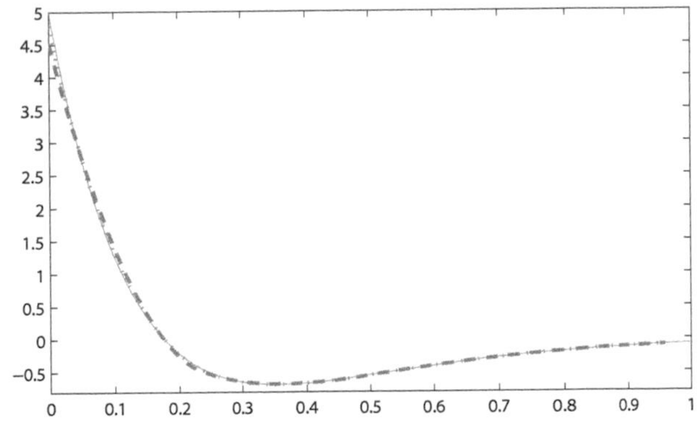

FIGURE 8.19: Numerical differentiations based on spline interpolations.

From Figure 8.19, many other theoretical curves can also be displayed. It can be seen that the numerical differentiations with B-spline is very satisfactory. Piecewise cubic polynomials may also give very good results. Since the samples are extremely sparsely distributed, the method in Section 3.8 cannot yield good results.

Example 8.17 Fit the $\partial^2 z/(\partial x \partial y)$ surface with the data obtained in Example 8.14. Compare the surface with the exact results.

Solution *The following statements can be used to generate the data. With the B-spline fitting, the numerical differentiation can be evaluated and the surface is then shown in Figure 8.20 (a).*

```
>> x0=-3:.3:3; y0=-2:.2:2; [x,y]=ndgrid(x0,y0);
   z=(x.^2-2*x).*exp(-x.^2-y.^2-x.*y); % generate samples
   sp=spapi({5,5},{x0,y0},z); dspxy=fnder(sp,[1,1]); fnplt(dspxy)
```

The following statements can be used to evaluate the exact partial derivatives theoretically and the surface is obtained as shown in Figure 8.20 (b). It can be seen that the results are almost the same, which indicates that the numerical method is reliable.

```
>> syms x y; z=(x^2-2*x)*exp(-x^2-y^2-x*y); % original function
   ezsurf(diff(diff(z,x),y),[-3 3],[-2 2]) % draw partial derivative surface
```

II. Numerical integration

The cubic and B-splines can be used to approximate the integrand that is defined by a given data set. From the spline interpolation, the numerical integration can be obtained. The method to be introduced here is different from the `quadspln()` function discussed in Section 8.1.2. The function `quadspln()` can only be used to evaluate the definite integral, while the function `fnint()` can be used to obtain an approximate integration function. Of course, it can also be used to evaluate definite integrals, $\boldsymbol{f}_i$ = `fnint(`S`)`, where $\boldsymbol{f}_i$ are vectors which return the integration at each $\boldsymbol{x}$ point. The indefinite integral can be obtained by adding a constant to the results. If the definite integral over the $[a,b]$ interval is required, we can use I = `diff(fnint(`S`,`$[a,b]$`))`.

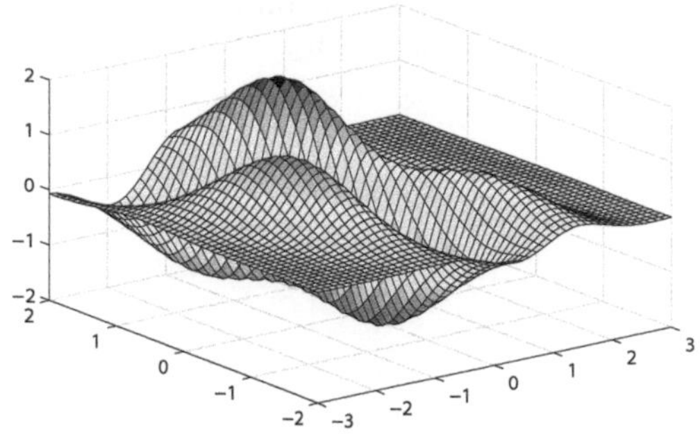

(a) spline interpolation results

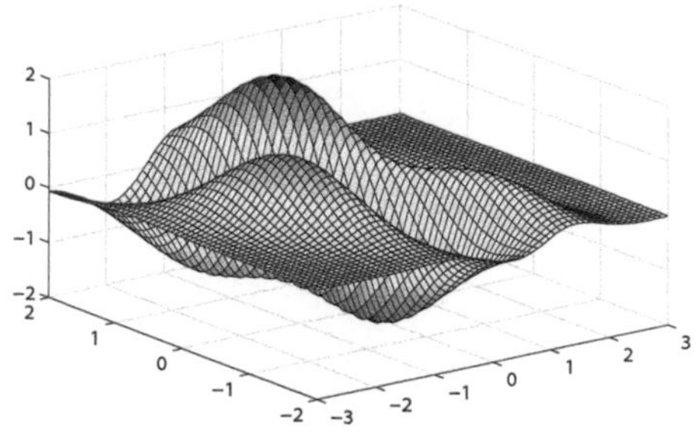

(b) theoretical results

FIGURE 8.20: The second-order partial derivative surface.

Example 8.18 Consider again the sparsely distributed samples in Example 8.5. Compute the definite and indefinite integrals using spline interpolations.

Solution *The two spline objects can be established and with the function* `fnint()`*, the integral function can be evaluated and also the definite integral can be obtained*

```
>> x=[0,0.4,1 2,pi]; y=sin(x); % generate sparsely distributed samples
   sp1=csapi(x,y); a=fnint(sp1,1); xx=fnval(a,[0,pi]); p1=xx(2)-xx(1)
   sp2=spapi(5,x,y); b=fnint(sp2,1); xx=fnval(b,[0,pi]); p2=xx(2)-xx(1)
```

and the definite integrals by the two spline functions are obtained as $p_1 = 2.01905235$*, and* $p_2 = 1.99994177102$*, respectively. It can be seen that the results with cubic spline are the same as the results obtained in Example 8.5. With the B-splines, much better results can be obtained.*

The approximate primitive function can also be obtained from sample data. For instance, the approximate primitive function can be obtained with B-splines and it is displayed together with the true one as shown in Figure 8.21.

```
>> ezplot('-cos(t)+2',[0,pi]); hold on, fnplt(a,'--'); fnplt(b,':')
```

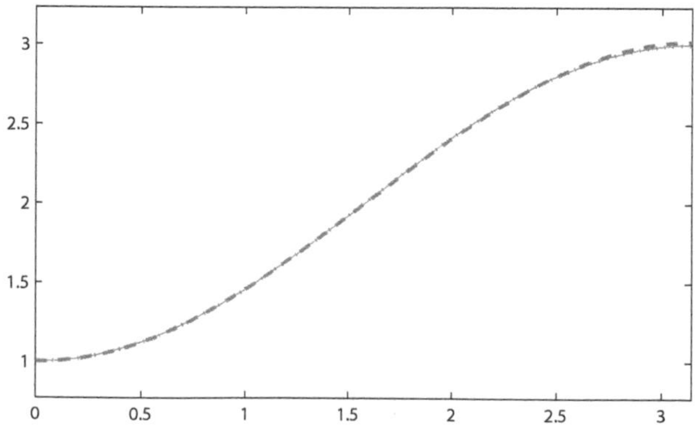

FIGURE 8.21: The approximate primitive function from spline interpolation.

It should be noted that `fnint()` function can only be used to evaluate integral functions for univariate functions, and for multivariate functions, integral functions can be obtained with `fnder()` function, with negative orders.

Example 8.19 Draw the integral surface for the function studied in Example 3.69.

$$J = \int_{-1}^{1}\int_{-2}^{2} e^{-x^2/2}\sin(x^2+y)\,dx\,dy.$$

Solution *The B-spline object can be created first, and the integral object can be constructed with* `fnder()` *function. The three-dimensional surface of the integral can be evaluated and the surface can be obtained, which is exactly the same as the one obtained in Example 3.70.*

```
>> x0=-2:0.1:2; y0=-1:0.1:1; [x y]=ndgrid(x0,y0);
   z=exp(-x.^2/2).*sin(x.^2+y); S=spapi({5,5},{x0,y0},z);
   S1=fnder(S,[-1 -1]); S2=fnval(S1,{x0,y0}); surf(y0,x0,S2)
```

8.3 Fitting Mathematical Models from Data

The interpolation methods discussed earlier can be used to evaluate the values of unknown points, however, mathematical expressions cannot be obtained. In practical applications, mathematical expressions are needed. In this section, some methods, such as polynomial fitting, multivariate linear regression and least squares nonlinear curve fittings are presented.

8.3.1 Polynomial fitting

The Lagrange interpolation is a type of polynomial fitting. The objective of the polynomial fitting is to find a set of coefficients $a_i, i = 1, 2, \cdots, n+1$, such that the polynomial

$$\varphi(x) = a_1x^n + a_2x^{n-1} + \cdots + a_nx + a_{n+1} \tag{8-3-1}$$

can be used to fit the original data in least squares sense. The difference between the polynomial fitting and the interpolation is that, in polynomial fitting, the samples are not necessarily on the fitting curve. A MATLAB function `polyfit()` can be used to solve polynomial fitting problems, with the syntax $\boldsymbol{p}$=`polyfit(`$\boldsymbol{x}$`,`$\boldsymbol{y}$`,`n`)`, where the arguments $\boldsymbol{x}$ and $\boldsymbol{y}$ are the vectors of sample points. The argument n is the selected degree of polynomial fitting. The returned argument $\boldsymbol{p}$ stores the coefficients of the polynomial in descending order. The function `poly2sym()` can be used to convert the results into symbolic polynomials. Also, the function `polyval()` can be used to evaluate the values of polynomials. The following examples illustrate the use of polynomial fitting.

Example 8.20 Consider the sample points in Example 8.1. Solve the polynomial fitting problems for different degrees and find a suitable degree.

Solution *Using the following statements, the fitting curve is shown in Figure 8.22 (a)*

```
>> x0=0:.1:1; y0=(x0.^2-3*x0+5).*exp(-5*x0).*sin(x0); % generate samples
   p3=polyfit(x0,y0,3); vpa(poly2sym(p3),10)  % get third-order polynomial
   x=0:.01:1; ya=(x.^2-3*x+5).*exp(-5*x).*sin(x); % generate denser points
   y1=polyval(p3,x); plot(x,y1,x,ya,x0,y0,'o')     % fitting quality
```

where the cubic function is $p_3(x) = 2.84x^3 - 4.7898x^2 + 1.9432x + 0.0598$. *From the fitting curve, it can be seen that the fitting quality is very poor. Thus, to address the fitting accuracy problem, it is natural to increase the degree of the polynomial. For different degrees such as 4, 5 and 6, the fitting curves can be obtained as shown in Figure 8.22 (b).*

```
>> p4=polyfit(x0,y0,4); y2=polyval(p4,x); p5=polyfit(x0,y0,5);
   y3=polyval(p5,x); p6=polyfit(x0,y0,6); y4=polyval(p6,x);
   plot(x,ya,x0,y0,'o',x,y2,x,y3,x,y4) % fitting of different polynomials
```

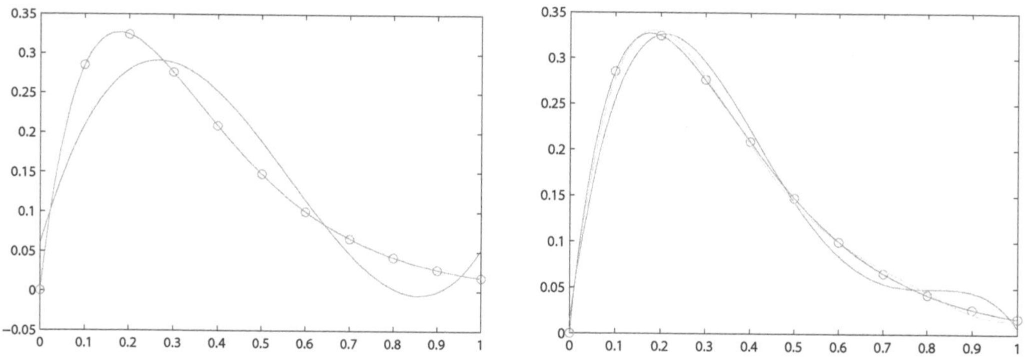

(a) cubic polynomial fittings (b) the fitting for other polynomials

FIGURE 8.22: Fitting results of polynomials of different degrees.

From the fitting results, it can be seen that when $n \geqslant 6$*, the fitting quality is satisfactory. The sixth degree polynomial obtained can be obtained from*

```
>> vpa(poly2sym(p6),4) % display 6th degree polynomial
```

that is, $p_6(x) = -11.24x^6 + 43.44x^5 - 68.6x^4 + 56.46x^3 - 24.88x^2 + 4.828x + 0.0001977$.

Polynomial fitting to a given function is in fact related to the Taylor series expansion. However, to evaluate the Taylor series, it is required that the original function be given. This is not realistic for many practical problems. For this example, since the original function is known, the Taylor series can be used to get a truncated polynomial

```
>> syms x; y=(x^2-3*x+5)*exp(-5*x)*sin(x); P=vpa(taylor(y,'Order',7),5)
```

which reads as follows

$$P = -205.0x^6 + 192.2x^5 - 142.0x^4 + 77.67x^3 - 28.0x^2 + 5.0x.$$

Comparing the results of truncated Taylor series expansion with the polynomial fitting, it can be seen that the two polynomials are significantly different. This means that the polynomial approximation to a given function may not be unique. Although the mathematical forms of the two polynomials are completely different, the fitting curves could be very close within a specific interval.

Example 8.21 Consider again the function in Example 8.3. Observe the polynomial fittings to the original function.

Solution *Polynomial fittings are not always accurate. Consider the samples in Example 8.3. Using different degrees n, the polynomial fitting can be obtained, with fitting results shown in Figure 8.23 (a).*

```
>> x0=-1+2*[0:10]/10;y0=1./(1+25*x0.^2); x=-1:.01:1; ya=1./(1+25*x.^2);
   p3=polyfit(x0,y0,3); p5=polyfit(x0,y0,5); p8=polyfit(x0,y0,8);
   y1=polyval(p3,x); y2=polyval(p5,x); y3=polyval(p8,x);
   p10=polyfit(x0,y0,10); y4=polyval(p10,x);     % different orders
   plot(x,ya,x,y1,x,y2,'-.',x,y3,'--',x,y4,':') % fitting comparisons
```

In fact, truncated Taylor series expansion to this function is even poorer. The following statements can be used to find the Taylor series for the original function, and the fitting is shown in Figure 8.23 (b). It can be seen that the Taylor series fitting is erroneous for the example, and the polynomial obtained is $p(x) = 1 - 25x^2 + 625x^4 - 15625x^6 + 390625x^8$.

```
>> syms x; y=1/(1+25*x^2); p=taylor(y,x,'Order',10), ezplot(p,[-1,1])
```

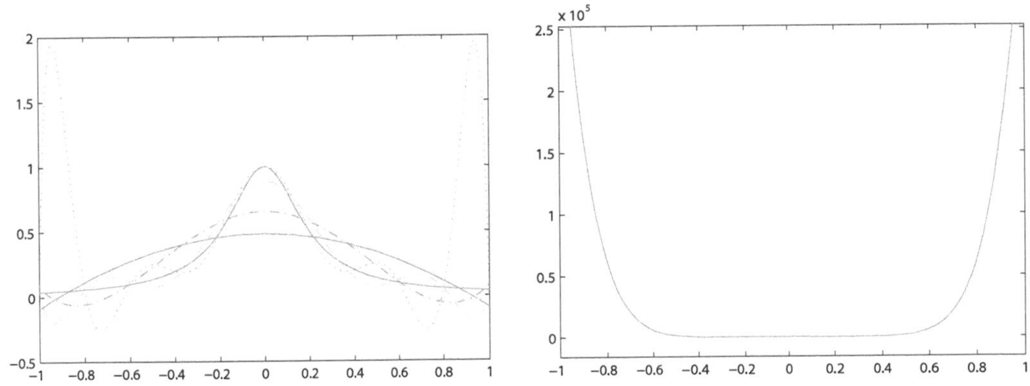

(a) fitting of derivatives (b) truncated Taylor series expansion

FIGURE 8.23: Polynomial fitting and the truncated Taylor series.

8.3.2 Curve fitting by linear combination of basis functions

Assume that the original function is composed of linear combination of a set of known basis functions $f_1(x), f_2(x), \cdots, f_n(x)$

$$g(x) = c_1 f_1(x) + c_2 f_2(x) + c_3 f_3(x) + \cdots + c_n f_n(x), \tag{8-3-2}$$

where $c_1, c_2, \cdots, c_n$ are undetermined constants. The measured data points obtained are $(x_1, y_1), (x_2, y_2), \cdots, (x_M, y_M)$. The following linear equations can be established

$$\boldsymbol{Ac} = \boldsymbol{y}, \tag{8-3-3}$$

where

$$\boldsymbol{A} = \begin{bmatrix} f_1(x_1) & f_2(x_1) & \cdots & f_n(x_1) \\ f_1(x_2) & f_2(x_2) & \cdots & f_n(x_2) \\ \vdots & \vdots & \ddots & \vdots \\ f_1(x_M) & f_2(x_M) & \cdots & f_n(x_M) \end{bmatrix}, \quad \boldsymbol{y} = \begin{bmatrix} y_1 \\ y_2 \\ \vdots \\ y_M \end{bmatrix}, \tag{8-3-4}$$

and $\boldsymbol{c} = [c_1, c_2, \cdots, c_n]^{\mathrm{T}}$. The least squares solution of the problem is $\boldsymbol{c} = \boldsymbol{A} \backslash \boldsymbol{y}$.

Example 8.22 The measured data points (x_i, y_i) are given in Table 8.3. Assume that the original function is given in the form

$$y(x) = c_1 + c_2 \mathrm{e}^{-3x} + c_3 \cos(-2x) \mathrm{e}^{-4x} + c_4 x^2.$$

Compute the undetermined constants c_i with least squares method.

Solution *The undetermined constants* c_i *can be obtained from the following:*

TABLE 8.3: Measured data.

x_i	0	0.2	0.4	0.7	0.9	0.92	0.99	1.2	1.4	1.48	1.5
y_i	2.88	2.2576	1.9683	1.9258	2.0862	2.109	2.1979	2.5409	2.9627	3.155	3.2052

```
>> x=[0,0.2,0.4,0.7,0.9,0.92,0.99,1.2,1.4,1.48,1.5]';
   y=[2.88;2.2576;1.9683;1.9258;2.0862;2.109;2.1979;2.5409;...
      2.9627;3.155;3.2052];            % input the given samples
   A=[ones(size(x)) exp(-3*x), cos(-2*x).*exp(-4*x) x.^2];
   c=A\y; c1=c', x0=[0:0.01:1.5]'; % find least squares solution
   A1=[ones(size(x0)),exp(-3*x0),cos(-2*x0).*exp(-4*x0),x0.^2];
   y1=A1*c; plot(x0,y1,x,y,'x')      % fitting results
```

and the result is $\mathbf{c}^{\mathrm{T}} = [1.2200208134, 2.33972067466, -0.6797329188, 0.8699983522]$. *The fitting curve is shown in Figure 8.24 with the given samples. It can be seen that the fitting is successful.*

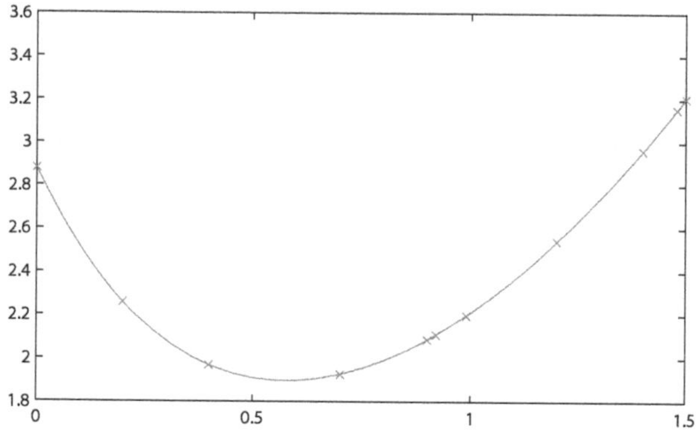

FIGURE 8.24: Original data and the fitting curve.

Example 8.23 Given the measured data in Table 8.4, find a fitting function.

TABLE 8.4: Measured data.

x_i	1.1052	1.2214	1.3499	1.4918	1.6487	1.8221	2.0138	2.2255	2.4596	2.7183	3.6693
y_i	0.6795	0.6006	0.5309	0.4693	0.4148	0.3666	0.3241	0.2865	0.2532	0.2238	0.1546

Solution *The measured data can be shown in Figure 8.25 (a).*

```
>> x=[1.1052,1.2214,1.3499,1.4918,1.6487,1.8221,2.0138,...
      2.2255,2.4596,2.7183,3.6693];
   y=[0.6795,0.6006,0.5309,0.4693,0.4148,0.3666,0.3241,...
      0.2864,0.2532,0.2238,0.1546];  % input the given samples
   plot(x,y,x,y,'*') % direct plot, nothing special observed from plots
```

By inspection, it is hard to determine a possible form of the fitting function. One may perform a nonlinear transformation to the original data, and see whether they can be described by linear functions. For instance, the logarithmic transformation to both $\boldsymbol{x}, \boldsymbol{y}$ *can be attempted as shown in Figure 8.25 (b), and it can be observed that it is linear.*

```
>> x1=log(x); y1=log(y);  plot(x1,y1,x1,y1,'*') % logarithmic fitting
```

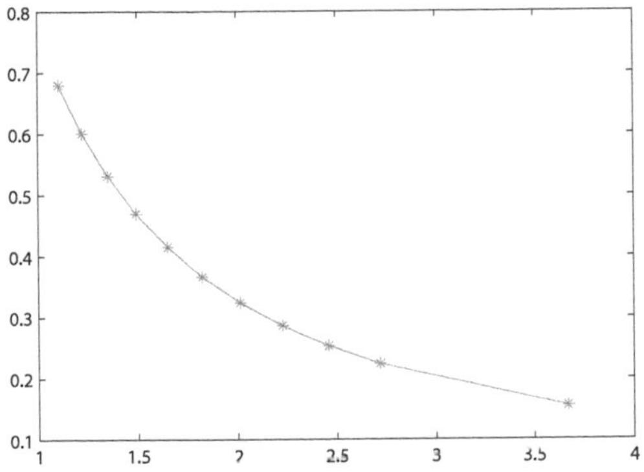

(a) original data

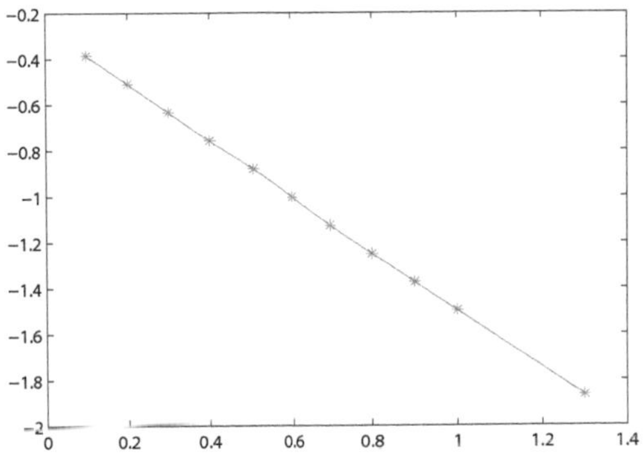

(b) fitting of logarithmic transformed data

FIGURE 8.25: Original data and fitting results for the transformed data.

The linear fitting function can be tested where $\ln y = a\ln x + b$*, i.e.,* $y = \mathrm{e}^b x^a$*. The coefficients* a, b *and* e^b *can be obtained with the following statements*

```
>> A=[x1' ones(size(x1'))]; c=[A\y1']', d=exp(c(2)) % fitting parameters
```

then, $\boldsymbol{c} = [-1.23389448522593, -0.26303708610220]^{\mathrm{T}}$, $\mathrm{e}^b = 0.76871338819924$. *The fitting function is then* $y(x) = 0.76871338819924x^{-1.23389448522593}$.

Example 8.24 Polynomial fitting can be regarded as a special example of the linear combination of basis function fittings. In this case, the basis functions can be written as $f_i(x) = x^{n+1-i}$, $i = 1, 2, \cdots, n+1$. Solve the polynomial fitting problem in Example 8.20 and observe the fitting results.

Solution *From the above algorithm, the sample data can be generated and the polynomial fitting can be performed which yields exactly the same results as in Example 8.20.*

```
>> x=[0:.1:1]'; y=(x.^2-3*x+5).*exp(-5*x).*sin(x); n=6; A=[];
   for i=1:n+1, A(:,i)=x.^(n+1-i); end, c=A\y % least squares solution
```

which means that the polynomial is

$$p_6(x) = -11.24x^6 + 43.44x^5 - 68.60x^4 + 56.46x^3 - 24.88x^2 - 4.8x + 0.0002.$$

8.3.3 Least squares curve fitting

Given a set of data $x_i, y_i, i = 1, 2, \cdots, N$, and given the original function, referred to as the *prototype function*, $\hat{y}(x) = f(\boldsymbol{a}, x)$, where $\boldsymbol{a}$ is the vector of undetermined constants, the objective of the least squares approximation is to find the undetermined constants which minimize the objective function

$$J = \min_{\boldsymbol{a}} \sum_{i=1}^{N} [y_i - \hat{y}(x_i)]^2 = \min_{\boldsymbol{a}} \sum_{i=1}^{N} [y_i - f(\boldsymbol{a}, x_i)]^2. \tag{8-3-5}$$

The `lsqcurvefit()` function provided in the Optimization Toolbox can be used to solve the least squares curve fitting problems. The syntax of the function is `[`$\boldsymbol{a}$`,`$J_{\rm m}$`] = lsqcurvefit(`*fun*`,`$\boldsymbol{a}_0$`,`$\boldsymbol{x}$`,`$\boldsymbol{y}$`,`$\boldsymbol{a}_{\rm m}$`,`$\boldsymbol{a}_{\rm M}$`,`*opts*`)`, where *fun* is the MATLAB description to the prototype function. It can either be an M-function, an anonymous or an inline function. The argument $\boldsymbol{a}_0$ is a vector containing the initial guess of $\boldsymbol{a}$. The vectors $\boldsymbol{x}$ and $\boldsymbol{y}$ store respectively the input and output data. The undetermined constants are returned in the $\boldsymbol{a}$ vector, and the objective function is in $J_{\rm m}$.

Example 8.25 The sample data set is generated with the following statements

```
>> x=0:.1:10; y=0.12*exp(-0.213*x)+0.54*exp(-0.17*x).*sin(1.23*x);
```

and assume that the prototype function satisfies $y(x) = a_1 {\rm e}^{-a_2 x} + a_3 {\rm e}^{-a_4 x}\sin(a_5 x)$, where a_i are the undetermined constants. The least squares curve fitting algorithm can be used to find the coefficients. Compute the coefficients which minimize the objective function.

Solution *The prototype function can first be expressed by an anonymous function, then, the following statements can be used to evaluate the undetermined constants*

```
>> f=@(a,x)a(1)*exp(-a(2)*x)+a(3)*exp(-a(4)*x).*sin(a(5)*x); % prototype
   [a,res]=lsqcurvefit(f,[1,1,1,1,1],x,y) % least squares fitting
   x1=0:0.02:10; y1=f(a,x1); plot(x1,y1,x,y,'o') % fitting results
```

and the estimated vector $\boldsymbol{a} = [0.1200, 0.2130, 0.5400, 0.1700, 1.2300]^{\rm T}$, *with the minimized objective function* $1.7927{\times}10^{-16}$. *It can be seen that the fitting results are extremely accurate. The sample points and the fitting curves are shown in Figure 8.26.*

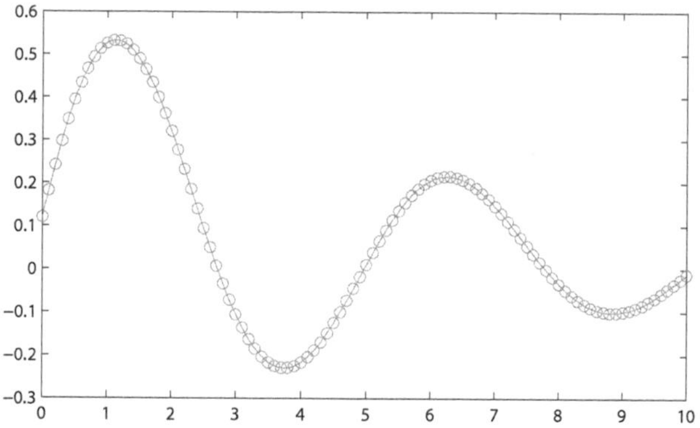

FIGURE 8.26: Comparisons of fitting results.

Example 8.26 The measured data points are given in Table 8.5, and also the prototype function is known to be $y(x) = ax + bx^2{\rm e}^{-cx} + d$. Compute using least squares method the undetermined constants a, b, c, d.

TABLE 8.5: Measured data.

x_i	0.1	0.2	0.3	0.4	0.5	0.6	0.7	0.8	0.9	1
y_i	2.3201	2.6470	2.9707	3.2885	3.6008	3.9090	4.2147	4.5191	4.8232	5.1275

Solution *The following statements can be entered in MATLAB workspace*

```
>> x=0.1:0.1:1; % generate samples
   y=[2.3201,2.6470,2.9707,3.2885,3.6008,3.9090,...
      4.2147,4.5191,4.8232,5.1275];
```

Let $a_1 = a, a_2 = b, a_3 = c, a_4 = d$, the prototype function can be rewritten as

$$y(x) = a_1 x + a_2 x^2 \mathrm{e}^{-a_3 x} + a_4.$$

Thus, an anonymous function can be written for the prototype function, and the following statement can be used to evaluate the coefficients

```
>> f=@(a,x)a(1)*x+a(2)*x.^2 .*exp(-a(3)*x)+a(4); % prototype function
   a=lsqcurvefit(f,[1;2;2;3],x,y), y1=f(a,x); plot(x,y,x,y1,'o')
```

with $\boldsymbol{a} = [3.1001, 1.5027, 4.0046, 2.0000]^{\mathrm{T}}$. *Change the initial search point to* $\boldsymbol{x}_0 = [1,0,0,0]$,

```
>> a=lsqcurvefit(f,[1;0;0;0],x,y), y1=f(a,x); plot(x,y,x,y1,'o')
```

$\boldsymbol{a} = [3.0746, 0.0000, -15.5811, 2.0513]^{\mathrm{T}}$. *If initial point is* $\boldsymbol{x}_0 = [1,0,0,1]^{\mathrm{T}}$, $\boldsymbol{a} = [3.0826, 0.0000, -11.2685, 2.0462]^{\mathrm{T}}$. *It can be seen that for this example, the fitting model is not unique, however, they all give good fitting results. The sample points and the fitted curve are shown in Figure 8.27, and it can be seen that the fitting is satisfactory.*

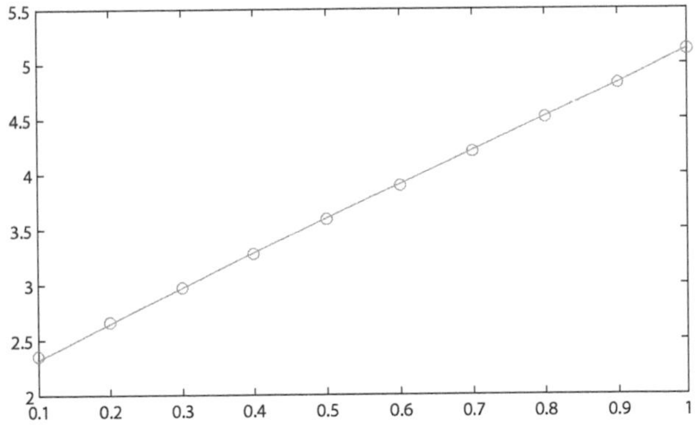

FIGURE 8.27: Comparison of fitting results.

8.3.4 Least squares fitting of multivariate functions

If a function has several independent variables, with a prototype function

$$z = f(\boldsymbol{a}, \boldsymbol{x}),$$

function `lsqcurvefit()` can still be used in fitting the function to find the vector $\boldsymbol{a}$, where $\boldsymbol{a} = [a_1, a_2, \cdots, a_n]$, $\boldsymbol{x} = [x_1,\ x_2, \cdots, x_m]$. The prototype function should be described by M-functions or anonymous functions, and the function `lsqcurvefit()` can be used to evaluate the undetermined constant vector $\boldsymbol{a}$. In describing the x_i variable in the objective function, the whole column should be extracted with $\boldsymbol{x}$`(:,`i`)`.

Example 8.27 Assume that the prototype function is given by

$$v = a_1 x^{a_2 x} + a_3 y^{a_4 (x+y)} + a_5 z^{a_6 (x+y+z)},$$

where, there are three independent variables. If the data are given in the data file c8data1.dat, where the first three columns are the independent variables x, y, z, and the 4th column is the output variable v, please find the undetermined constants a_i using least squares method.

Solution *To solve the problem, a vector form of the independent variables is described in* $\boldsymbol{x}$, *with* $x_1 = x$, $x_2 = y$, $x_3 = z$, *and prototype function is rewritten as*

$$v = a_1 x_1^{a_2 x_1} + a_3 x_2^{a_4(x_1+x_2)} + a_5 x_3^{a_6(x_1+x_2+x_3)},$$

and it can be described by an anonymous function. Also, since the data is given in a data file, function `load()` *can be used to input the data into MATLAB workspace. Sub-matrix extraction approach can be used to create matrix* $\boldsymbol{X}$ *and vector* $\boldsymbol{v}$, *so that the undetermined constants* $\boldsymbol{a} = [0.1, 0.2, 0.3, 0.4, 0.5, 0.6]$ *can be found with the following statements, to minimize the sum of squared errors, to* $1.0904{\times}10^{-7}$. *In fact, the original data were generated by the function with* $\boldsymbol{a} = [0.1, 0.2, 0.3, 0.4, 0.5, 0.6]$, *and it can be seen that the undetermined constants obtained are accurate.*

```
>> f=@(a,x)a(1)*x(:,1).^(a(2)*x(:,1))+...
          a(3)*x(:,2).^(a(4)*(x(:,1)+x(:,2)))+...
          a(5)*x(:,3).^(a(6)*(x(:,1)+x(:,2)+x(:,3)));
   XX=load('c8data1.dat'); X=XX(:,1:3); v=XX(:,4);
   a0=[2 3 2 1 2 3]; [a,f,err,key]=lsqcurvefit(f,a0,X,v)
```

8.4 Rational Function Approximations

8.4.1 Approximation by continued fraction expansions

Continued fraction is often regarded as an effective way in approximating certain functions. The typical form of a continued fraction expansion to a given function $f(x)$ can be expressed by

$$f(x) = b_1 + \cfrac{(x-a)^{c_1}}{b_2 + \cfrac{(x-a)^{c_2}}{b_3 + \cfrac{(x-a)^{c_3}}{b_4 + \cfrac{(x-a)^{c_4}}{b_5 + \cfrac{(x-a)^{c_5}}{\cdots}}}}}, \tag{8-4-1}$$

where, b_i are constants, c_i are rational numbers and a is the reference point for continued fraction expansion.

There is no continued fraction expansion functions provided in MATLAB, while low-level support of MuPAD has a set of functions. Therefore, an interface `contfrac()` is written to directly find the continued fraction expansion of a given function, with the syntaxes

cf = `contfrac(`f`,`n`)`, [*cf*,r] = `contfrac(`f`,`n`,`a`)`

where, f is the symbolic expression of the original function, a is the reference point (with a default value of 0), n is the expected level of the expansion. The returned variable *cf* is the expansion of MuPAD expression. The returned variable r is the rational approximation of the function. If f is a constant, the returned variable is returned in *cf*. The listing of the function is as follows

```
function [cf,r]=contfrac(f,varargin)
[n,a]=default_vals({6,0},varargin{:});
if isanumber(f), cf=feval(symengine,'contfrac',f,n);
   p1=char(cf); k=strfind(p1,','); k1=strfind(p1,'/');
   if nargout>1, r=sym(p1(k(end)+1:k1-1))/sym(p1(k1+1:end-1)); end
 else, if isfinite(a), str=num2str(a); else, str='infinity'; end
   cf=feval(symengine,'contfrac',f,['x=' str],n);
   if nargout>1, r=feval(symengine,'contfrac::rational',cf); end
end
```

A supporting function isanumber() is written below, and it is used to detect whether a is a number, either in double-precision or symbolic forms

```
function key=isanumber(a) % returns 1 if argument is a number
key=0; if length(a)~=1, return; end
switch class(a)
   case 'double', key=1;                        % if a double precision number
   case 'sym', try, double(a); key=1; catch, end % or a symbolic number
end
```

Example 8.28 Let us first observe the continued fraction expansion-based approximation to a constant. The irrational number π can be approximated with a 20-level continued fraction. Find a suitable degree such that good approximation can be obtained.

Solution *The 20-level continued fractions to π can be directly obtained with*

```
>> [cf,r]=contfrac(pi,20), latex(cf) % write continued fraction of π
```

The returned variable cf *is a continued fraction object with coefficient vector*

$$c = [3, 7, 15, 1, 292, 1, 1, 1, 2, 1, 3, 1, 14, 2, 1, 1, 2, 2, 2, 2],$$

and the irrational number π is approximated as 14885392687/4738167652, (with accurate 20 decimal digits). From the coefficient vector, the continued fraction of π can be written as

$$\pi \approx 3 + \cfrac{1}{7 + \cfrac{1}{15 + \cfrac{1}{1 + \cfrac{1}{292 + \cfrac{1}{1 + \cfrac{1}{1 + \cfrac{1}{1 + \cfrac{1}{2 + \cfrac{1}{1 + \cfrac{1}{3 + \cfrac{1}{1 + \cfrac{1}{14 + \cfrac{1}{2 + \cfrac{1}{1 + \cfrac{1}{1 + \cfrac{1}{2 + \cdots}}}}}}}}}}}}}}}}.$$

It can be seen from the continuous fraction expression that, the value of 292 is relatively large compared with other entities, so that the roundoff error can be very accurate, if the rational expansion is rounded up to this level, with $103993/33102 \approx 3.141592653012$. *If the digit 20 in the command is substituted to other digits, the fitting results and accuracies are given in Table 8.6, such that continued fractions yield accurate approximation.*

TABLE 8.6: Approximation of π, with different levels of continued fractions.

level	rational	approximations
0	π	3.14159265358979323846
8, 9	103993/33102	3.141592653
10	208341/66317	3.1415926535
11	312689/99532	3.1415926536
12	1146408/364913	3.14159265359
13, 14	5419351/1725033	3.14159265358981
15	80143857/25510582	3.1415926535897926
16	165707065/52746197	3.1415926535897934
17	411557987/131002976	3.14159265358979326
18	1068966896/340262731	3.141592653589793235
19	6167950454/1963319607	3.14159265358979323838
20	14885392687/4738167652	3.14159265358979323849

If command `[cf,r]=contfrac(pi,120)` *is given, the rational approximation of* π *is expressed as*

$$\pi \approx \frac{1244969988745789040106366155256015114976454553337906033890313}{396286255419907262612262286579004636343912658949984351074158}.$$

To assess the accuracy of the above rational approximation, the following command can be given, and the accuracy is around 2.975×10^{-120}.

```
>> 10^log10(abs(vpa(sym(['1244969988745789040106366155256015114976454555',...
   '3337906033890313/396286255419907262612262286579004636343912658949',...
   '9984351074158-pi']),200))) % estimate the error, thanks John D'Errico
```

Example 8.29 Establish the first 10 levels of continued fractions for the function $f(x) = \mathrm{e}^{-x}\sin x/(x+1)^3$ and obtain the rational approximation.

Solution *The first 10 levels of continued fractions can be obtained with*

```
>> syms x; f=sin(x)*exp(-x)/(x+1)^3; [cf,r]=contfrac(f,10)
```

and the continued fraction `cf` *can be written as*

$$f(x) \approx \cfrac{x}{1+\cfrac{x}{\cfrac{1}{4}+\cfrac{x}{-\cfrac{12}{5}+\cfrac{x}{-\cfrac{25}{43}+\cfrac{x}{\cfrac{7396}{1685}+\cfrac{x}{\cfrac{2839225}{4863128}+\cfrac{x}{-\cfrac{44767468256}{2592461805}+\cdots}}}}}}},$$

and the rational approximation r can be written as

$$r(x) = \frac{\begin{array}{c}-170455846739x^5 + 472453225650x^4 + 3615529382220x^3 \\ -20275122684600x^2 + 28175852788020x\end{array}}{\begin{array}{c}2071713977216x^5 + 14187032489655x^4 + 58214153847990x^3 \\ +110354057230620x^2 + 92428288467480x + 28175852788020\end{array}}.$$

The curves of the original function $f(x)$ and its rational approximation $r(x)$ in the interval $[0, 2]$ are shown in Figure 8.28 (a). It can be seen that the fitting is very satisfactory, in fact, the two curves cannot be distinguished from the figure, unless a series of zooming actions are made.

```
>> ezplot(f,[0,2]); hold on; ezplot(r,[0,2]) % compare the two functions
```

If the interval is increased to $(0, 5)$, the fitting results shown in Figure 8.28 (b) suggest that more levels are needed to improve the fitting quality.

```
>> ezplot(f,[0,5]), hold on; ezplot(r,[0,5]) % compare over lager interval
```

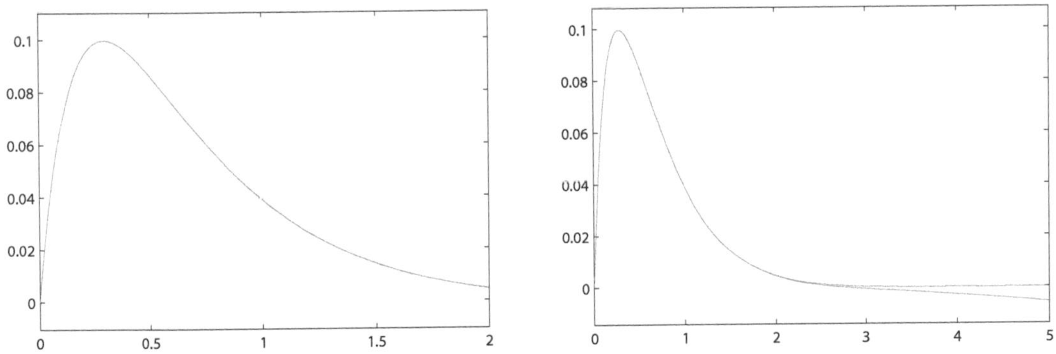

(a) fitting results in $(0, 2)$ interval

(b) fitting of the enlarged region

FIGURE 8.28: Comparison of continued fraction expansion approximation.

Alternatively, the center point of expansion can be shifted to the right, for instance, let $x = 0.9$, and increase the level number to 11, the new continued fraction expansion can be obtained with the following statements and the fitting result is shown in Figure 8.29. It can be seen that there is no difference visible between the two curves.

```
>> [cf,r1]=contfrac(f,11,0.9) % compute the continued fraction expansion
   ezplot(f,[0,5]), hold on, ezplot(r1,[0,5]) % compare the two functions
```

The approximated rational expression can be written manually as

$$r_1(x) \approx \frac{3.11x^5 - 49.44x^4 + 309.9x^3 - 892.75x^2 + 976.28x + 0.000806}{48.34x^5 + 264.81x^4 + 1420.84x^3 + 3247.73x^2 + 3012.28x + 976.29}.$$

The continued fraction expansion can be written as

$$f(x) \approx 0.04643 + \cfrac{x-0.9}{-12.0628 + \cfrac{x-0.9}{-0.1239 + \cfrac{x-0.9}{11.3557 + \cfrac{x-0.9}{0.07631 + \cfrac{x-0.9}{44.90233 + \cdots}}}}},$$

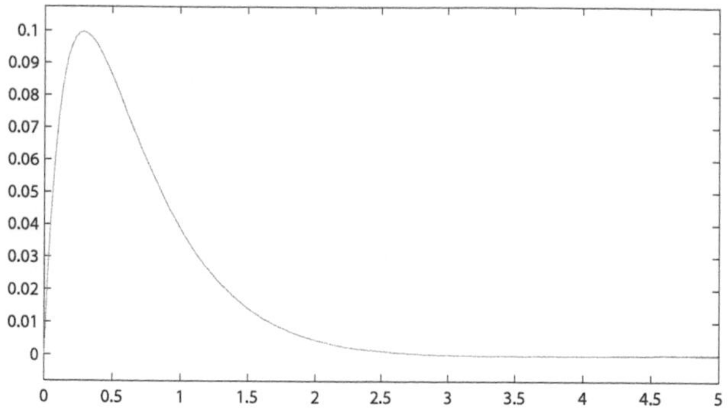

FIGURE 8.29: Improved continued fraction approximation over $[0,5]$.

It should be noted that, in this example, the center point should not exceed $x=1$. Also, the coefficients in the rational approximation are not sensitive. If the above finite digit approximation is used, the fitting quality is still satisfactory.

```
>> r2=(3.11*x^5-49.44*x^4+309.9*x^3-892.75*x^2+976.28*x+0.000806)/...
     (48.34*x^5+264.81*x^4+1420.84*x^3+3247.73*x^2+3012.28*x+976.29);
   ezplot(r2,[0,5])
```

8.4.2 Padé rational approximations

Assume that the power series expansion of a given function $f(s)$ is

$$f(s)=c_1+c_2s+c_3s^2+c_4s^3+\cdots=\sum_{i=1}^{\infty}c_is^{i-1}, \tag{8-4-2}$$

and assume that the r/mth degree Padé approximation is expressed by

$$G_m^r(s)=\frac{\beta_{r+1}s^r+\beta_rs^{r-1}+\cdots+\beta_1}{\alpha_{m+1}s^m+\alpha_ms^{m-1}+\cdots+\alpha_1}=\frac{\sum\limits_{i=1}^{r+1}\beta_is^{i-1}}{\sum\limits_{i=1}^{m+1}\alpha_is^{i-1}}, \tag{8-4-3}$$

where $\alpha_1=1,\ \beta_1=c_1$. Let $\sum\limits_{i=1}^{\infty}c_is^{i-1}=G_m^r(s)$. The following equation can be obtained:

$$\sum_{i=1}^{m+1}\alpha_is^{i-1}\sum_{i=1}^{\infty}c_is^{i-1}=\sum_{i=1}^{r+1}\beta_is^{i-1}. \tag{8-4-4}$$

Equating the terms with the same power of s, the coefficients $\alpha_i,\ i=2,\cdots,m+1$ and $\beta_i,\ i=2,\cdots,k+1$ can be obtained from

$$\boldsymbol{W}\boldsymbol{x}=\boldsymbol{w},\quad \boldsymbol{v}=\boldsymbol{V}\boldsymbol{y}, \tag{8-4-5}$$

where

$$\boldsymbol{x} = [\alpha_2, \alpha_3, \cdots, \alpha_{m+1}]^{\mathrm{T}}, \quad \boldsymbol{w} = [-c_{r+2}, -c_{r+3}, \cdots, -c_{m+r+1}]^{\mathrm{T}}$$
$$\boldsymbol{v} = [\beta_2 - c_2, \beta_3 - c_3, \cdots, \beta_{r+1} - c_{r+1}]^{\mathrm{T}}, \quad \boldsymbol{y} = [\alpha_2, \alpha_3, \cdots, \alpha_{r+1}]^{\mathrm{T}}, \tag{8-4-6}$$

and

$$\boldsymbol{W} = \begin{bmatrix} c_{r+1} & c_r & \cdots & 0 & \cdots & 0 \\ c_{r+2} & c_{r+1} & \cdots & c_1 & \cdots & 0 \\ \vdots & \vdots & \cdots & \vdots & \ddots & \vdots \\ c_{r+m} & c_{r+m-1} & \cdots & c_{m-1} & \cdots & c_{r+1} \end{bmatrix}, \tag{8-4-7}$$

$$\boldsymbol{V} = \begin{bmatrix} c_1 & 0 & 0 & \cdots & 0 \\ c_2 & c_1 & 0 & \cdots & 0 \\ \vdots & \vdots & \vdots & \ddots & \vdots \\ c_r & c_{r-1} & c_{r-2} & \cdots & c_1 \end{bmatrix}. \tag{8-4-8}$$

It can be shown [4] that, if the degree of numerator is one less than that of the denominator, the Padé approximation is equivalent to the Cauer II form of the continued fraction expansion. A MATLAB function `padefcn()` can be prepared to compute the Padé rational approximation to a given $f(x)$ function. The function is

```
function [nP,dP]=padefcn(c,r,m)
w=-c(r+2:m+r+1)'; vv=[c(r+1:-1:1)'; zeros(m-1-r,1)];
W=rot90(hankel(c(m+r:-1:r+1),vv)); V=rot90(hankel(c(r:-1:1)));
x=[1 (W\w)']; y=[1 x(2:r+1)*V'+c(2:r+1)];
dP=x(m+1:-1:1)/x(m+1); nP=y(r+1:-1:1)/x(m+1);
```

Example 8.30 Find the rational approximation to the function $f(x) = \mathrm{e}^{-2x}$.

Solution *Let us select the degree of the numerator as 0, and select different degrees for the denominator. The Padé rational approximation can be obtained by the following MATLAB scripts, and the fitting results are given in Figure 8.30.*

```
>> syms x; c=taylor(exp(-2*x),'Order',10); c=sym2poly(c);
   c=c(end:-1:1); x=0:0.01:8; nd=[3:7]; plot(x,exp(-2*x));
   for i=1:length(nd) % try different orders
      [n,d]=padefcn(c,0,nd(i)); y=polyval(n,x)./polyval(d,x); line(x,y)
   end
```

It can be seen from Figure 8.30 that third-degree Padé approximation gives a reasonably good fitting. If the degree is increased, the quality of fitting may also increase. The eighth-degree fitting to the original function is actually very satisfactory, with

$$P_8(s) = \frac{157.5}{x^8 + 4x^7 + 14x^6 + 42x^5 + 105x^4 + 210x^3 + 315x^2 + 315x + 157.5}.$$

Based on the above algorithm, the symbolic version of the `padefcn()` functions can be rewritten as

```
function G=padefcnsym(f,r,m)
c=taylor(f,'Order',r+m+1); c=polycoef(c); c=c(end:-1:1);
w=-c(r+2:m+r+1)'; vv=[c(r+1:-1:1)'; zeros(m-1-r,1)];
```

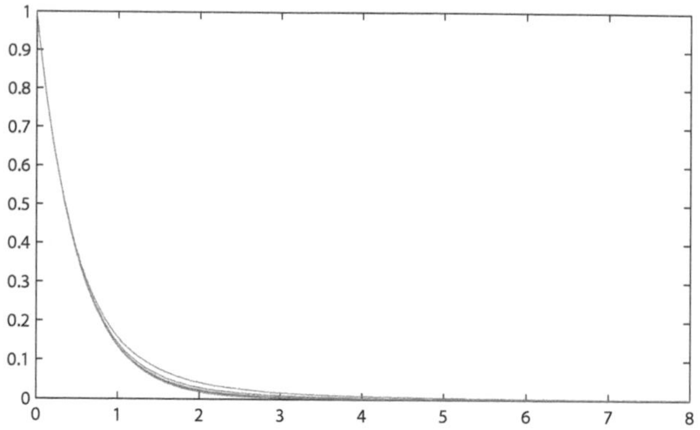

FIGURE 8.30: Original curve and fitting curves.

```
W=rot90(hankel(c(m+r:-1:r+1),vv)); V=rot90(hankel(c(r:-1:1)));
X=[1 (W\w)']; y=[1 X(2:r+1)*V'+c(2:r+1)]; dP=X(m+1:-1:1)/X(m+1);
nP=y(r+1:-1:1)/X(m+1); syms x; G=poly2sym(nP,x)/poly2sym(dP,x);
```

Example 8.31 Solve the Padé approximation problem in Example 8.30 with symbolic computation.

Solution *The Padé approximation can be obtained with*

```
>> syms x; f=exp(-2*x); G=padefcnsym(f,0,8) % symbolic computations
```

The Padé approximation is given below, and it can be seen that the rational approximation is exactly the same as the one obtained in the previous example.

$$G(x) = \frac{315}{2\left(x^8+4x^7+14x^6+42x^5+105x^4+210x^3+315x^2+315x+315/2\right)}.$$

8.4.3 Special approximation polynomials

Apart from the Padé approximation functions, many polynomials, such as Legendre, Chebyshev, Laguerre and Hermite polynomials can be used in function approximation. The polynomials are defined as

(i) **Legendre polynomials**, whose mathematical descriptions are

$$P_n(x) = \frac{1}{2^{n-1}(n-1)!}\frac{\mathrm{d}^{n-1}}{\mathrm{d}x^{n-1}}(x^2-1)^{n-1},\ n=1,2,3,\cdots. \tag{8-4-9}$$

The polynomials can be constructed recursively for $n=2,3,\cdots$ with

$$P_1(x)=1,\ P_2(x)=x,\ P_{n+1}(x)=\frac{2n-1}{n}xP_n(x)-\frac{n-1}{n}P_{n-1}(x). \tag{8-4-10}$$

(ii) **Chebyshev polynomials**, with mathematical form of

$$T_n(x)=\cos\Big[(n-1)\arccos x\Big],\ \text{with } |x|\leqslant 1,\ n=1,2,3,\cdots, \tag{8-4-11}$$

and the recursive implementations are

$$T_1(x)=1,\ T_2(x)=x,\ T_{n+1}=2xT_n(x)-T_{n-1}(x),\ n=2,3,\cdots. \tag{8-4-12}$$

(iii) **Laguarre polynomials**, whose mathematical form are

$$L_n(x) = \frac{e^x}{(n-1)!} \frac{d^{n-1}}{dx^{n-1}} x^{n-1} e^{-x}, \text{ with } x \geqslant 0, \ n = 1, 2, 3, \cdots, \tag{8-4-13}$$

and the recursive formula for $n = 2, 3, \cdots$ are

$$L_1(x) = 1, \ L_2(x) = 1 - x, \ L_{n+1}(x) = \frac{2n-1-x}{n} L_n(x) - \frac{n-1}{n} L_{n-1}(x). \tag{8-4-14}$$

(iv) **Hermite polynomials**, with mathematical form

$$H_n(x) = (-1)^{n-1} e^{x^2} \frac{d^{n-1}}{dx^{n-1}} e^{-x^2}, \ |x| < \infty, \ n = 1, 2, 3, \cdots, \tag{8-4-15}$$

and the recursive formula for $n = 2, 3, \cdots$ are

$$H_1(x) = 1, \ H_2(x) = 2x, \ H_{n+1} = 2xH_n(x) - 2(n-1)H_{n-1}(x). \tag{8-4-16}$$

These polynomials can be generated with the following MATLAB function, with the syntax $\boldsymbol{P}$=`fitting_poly(`*type*`,`n`,`x`)`, where the argument *type* can be used to indicate the type of polynomials. The vector of the polynomials is returned in argument $\boldsymbol{P}$.

```
function P=fitting_poly(type,N,x)
switch type % handle different types of fitting polynomials
   case {'P','Legendre'}, P=[1,x];     % Legendre polynomials
      for n=2:N, P(n+1)=(2*n-1)/n*x*P(n)-(n-1)/n*P(n-1); end
   case {'T','Chebyshev'}              % Chebyshev polynomials
      P=[1,x]; for n=2:N, P(n+1)=2*x*P(n)-P(n-1); end
   case {'L','Laguerre'}, P=[1,1-x]; % Laguerre polynomials
      for n=2:N, P(n+1)=(2*n-1-x)/n*P(n)-(n-1)/n*P(n-1); end
   case {'H','Hermite'},               % Hermite polynomials
      P=[1,2*x]; for n=2:N, P(n+1)=2*x*P(n)-2*(n-1)*P(n-1); end
end
```

Example 8.32 Please generate the 10th terms of the four polynomials.

Solution *With the following function calls, the expected polynomials can easily be obtained, and the last terms in the vectors are expanded*

```
>> syms x; P=fitting_poly('P',10,x); P=expand(P(end))
   L=fitting_poly('L',10,x); L=expand(L(end))
   T=fitting_poly('T',10,x); T=expand(T(end))
   H=fitting_poly('H',10,x); H=expand(H(end))
```

The 10th terms in the four polynomials can easily be obtained

$$\begin{cases} P(x) = \dfrac{46189x^{10}}{256} - \dfrac{109395x^8}{256} + \dfrac{45045x^6}{128} - \dfrac{15015x^4}{128} + \dfrac{3465x^2}{256} - \dfrac{63}{256}, \\ T(x) = 512x^{10} - 1280x^8 + 1120x^6 - 400x^4 + 50x^2 - 1, \\ L(x) = \dfrac{x^{10}}{3628800} - \dfrac{x^9}{36288} + \dfrac{x^8}{896} - \dfrac{x^7}{42} + \dfrac{7x^6}{24} - \dfrac{21x^5}{10} + \dfrac{35x^4}{4} - 20x^3 + \dfrac{45x^2}{2} - 10x + 1, \\ H(x) = 1024x^{10} - 23040x^8 + 161280x^6 - 403200x^4 + 302400x^2 - 30240. \end{cases}$$

The polynomials can be used as the basis functions so that the least squares methods can be used in approximating functions. For instance, if Chebyshev polynomials are used, the prototype function with m terms can be expressed mathematically as

$$\boldsymbol{y}=\boldsymbol{f}(\boldsymbol{a},\boldsymbol{x})=a_1T_1(\boldsymbol{x})+a_2T_2(\boldsymbol{x})+\cdots+a_mT_m(\boldsymbol{x}), \tag{8-4-17}$$

and `lsqcurvefit()` function can be used to find the undetermined parameters a_i. The function `fitting_poly()` is not suitable for use in function approximation problems. An alternative function can be written to describe the prototype function with Chebyshev polynomials

```
function y=cheby_poly(a,x) % a and x are both column vectors
a=a(:); x=x(:); n=length(a); X=[ones(size(x)) x]; % the first two columns
for i=2:n-1, X(:,i+1)=2*x.*X(:,i)-X(:,i-1); end, y=X*a;
```

Example 8.33 Consider the polynomial fitting problem studied in Example 8.24. Please solve the fitting problem with Chebyshev polynomials.

Solution *The Chebyshev polynomials fitting can be obtained with the following statements, and eventually the polynomial can be obtained*

```
>> x0=[0:.1:1]'; y0=(x0.^2-3*x0+5).*exp(-5*x0).*sin(x0); % generate samples
   a0=ones(7,1); a=lsqcurvefit(@cheby_poly,a0,x0,y0)      % least squares
   syms x; T=fitting_poly('T',6,x); P=vpa(expand(T*a),4) % get polynomial
```

with vector $\boldsymbol{a}=[-41.6760, 74.3281, -52.0080, 27.6921, -10.6832, 2.7153, -0.3514]^{\mathrm{T}}$, *while the polynomial obtained is*

$$P(x)=-11.24x^6+43.44x^5-68.6x^4+56.46x^3-24.88x^2+4.828x+0.0001975,$$

which is almost the same as the one obtained in Example 8.24. It should be noted that the least squares fitting established in this way is essentially optimal polynomial fitting, and the fitting quality should be similar with `polyfit()` *function results.*

8.5 Special Functions and Their Plots

In definite integral computations, it is often found that some functions cannot be integrated, for instance the integrand e^{-x^2}. Special functions $\mathrm{erf}(\cdot)$ can be introduced to denote the integral. Also, in solving some particular forms of nonlinear ordinary differential equations, other special functions are usually defined. In practical applications, a great amount of special functions are being used, among them, the commonly used ones are Gamma functions and Beta functions. For differential equations, Bessel, Legendre, Mittag–Leffler functions are often used. In this section, introductions to these functions are presented, and their plots are given.

8.5.1 Gamma functions

I. Ordinary Gamma function

Gamma function is the analytical description of the following infinite integral

$$\Gamma(\alpha)=\int_0^\infty \mathrm{e}^{-t}t^{\alpha-1}\,\mathrm{d}t. \tag{8-5-1}$$

It can be validated through integration by parts that $\Gamma(\alpha+1)=\alpha\Gamma(\alpha)$, and $\Gamma(1)=1$. It can be seen that, if α is a non-negative integer, $\Gamma(\alpha+1)=\alpha!$. Therefore, Gamma function is an extension of factorials in the real domain. If α is a negative integer, $\Gamma(\alpha+1)$ tends to $\pm\infty$. Gamma functions can be evaluated directly with `y=gamma(x)`, where $\boldsymbol{x}$ can be a real vector, matrix, symbolic variable or other data types, and $\boldsymbol{y}$ evaluates Gamma function to all the entities in $\boldsymbol{x}$.

Example 8.34 Please show from definition that, for any nonnegative integer k, there exists $\Gamma(k)=(k-1)!$.

Solution *The property can be shown directly with the statements*

```
>> syms t k; assume(k,'integer'); assumeAlso(k>=0); % non-negative integer
   I=int(exp(-t)*t^(k-1),t,0,inf) % compute from definition
```

Example 8.35 Please validate some of the properties of Gamma functions

$$\Gamma\left(\frac{1}{2}\right)=\sqrt{\pi},\ \Gamma(\alpha)\Gamma(1-\alpha)=\frac{\alpha}{\sin\pi\alpha},\ \Gamma(\alpha)\Gamma(-\alpha)=\frac{-\pi}{\alpha\sin\pi\alpha}, \tag{8-5-2}$$

$$\Gamma\left(\frac{1}{2}+\alpha\right)\Gamma\left(\frac{1}{2}-\alpha\right)=\frac{\pi}{\cos\pi\alpha},\quad \lim_{\alpha\to\infty}\frac{\alpha^z\Gamma(\alpha)}{\Gamma(\alpha+z)}=1, \mathrm{Re}z>0. \tag{8-5-3}$$

Solution *With symbolic* `gamma()` *function, the properties can be validated directly*

```
>> syms a z; I1=gamma(sym(1/2)), % compute with gamma()
   I2=simplify(gamma(a)*gamma(1-a)), I3=simplify(gamma(a)*gamma(-a))
   I4=simplify(gamma(1/2+a)*gamma(1/2-a))
   I5=limit(a^z*gamma(a)/gamma(a+z),a,Inf)
```

Example 8.36 Please draw Gamma function in the interval $(-5,5)$.

Solution *The following statements can be used to draw Gamma function, as shown in Figure 8.31. Since at points $\alpha=0,-1,-2,\cdots$, the values of $\Gamma(\alpha)$ tends to infinity. In order to have a better view of the plot, the y-axis is zoomed manually to make it more informative.*

```
>> x=-5:0.002:5; plot(x,gamma(x)), ylim([-15,15]) % draw Gamma function curve
```

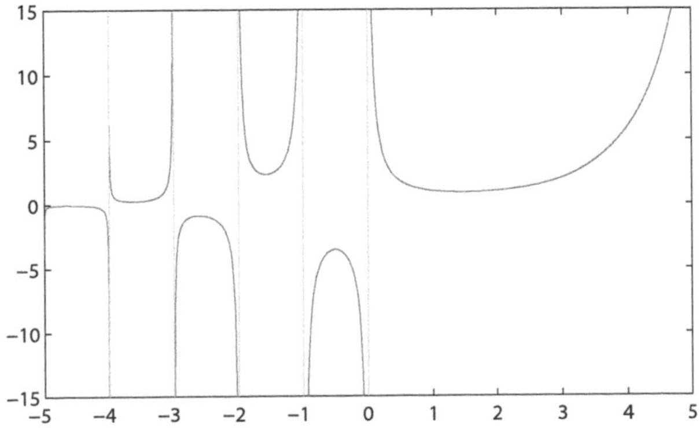

FIGURE 8.31: Gamma function plot.

Example 8.37 Some integral problems can be solved using the concept of Gamma function, and the results can be regarded as "analytical solutions" of the definite integrals. Please solve the following indefinite integral

$$I=\int_0^\infty t^{x-1}\cos t\ \mathrm{d}t, x>0.$$

Solution *With the following statements the problem can be solved*

```
>> syms t; syms x positive; I=simplify(int(t^(x-1)*cos(t),t,0,inf))
```

and the result obtained is a piecewise function mathematically interpreted as

$$I=\begin{cases}\dfrac{2^x\sqrt{\pi}\,\Gamma\left(x/2\right)}{2\Gamma\left((1-x)/2\right)}, & x<1\\ \text{no analytical solution}, & x=1 \text{ or } 2\leqslant x\\ \cos(\pi x/2)\,\Gamma\left(x\right), & x<1 \text{ or } x\in(1,2).\end{cases}$$

The values of Gamma functions can also be evaluated through numerical integrals, also it can be evaluated through the following infinite series

$$\Gamma(x)=\frac{1}{x}\mathrm{e}^{-\gamma x}\prod_{n=1}^{\infty}\left(\frac{n}{n+x}\right)\mathrm{e}^{x/n}, \tag{8-5-4}$$

where, $\gamma\approx 0.57721566490153286$ is the Euler γ constant.

II. Incomplete Gamma function

If the upper bound of the Gamma integral is finite, then, Gamma function is referred to as *incomplete Gamma function*, defined as

$$\Gamma(x,\alpha)=\frac{1}{\Gamma(\alpha)}\int_0^x \mathrm{e}^{-t}t^{\alpha-1}\,\mathrm{d}t,\quad \alpha\geqslant 0. \tag{8-5-5}$$

The incomplete Gamma function can be evaluated with $\boldsymbol{y}$ = `gammainc(`$\boldsymbol{x}$`,`α`)`.

III. Gamma function with complex arguments

It is also noted that the `gamma()` function in MATLAB applies only to real arguments. If the arguments are complex, a MATLAB function below can be written, based on the definition in (8-5-1)

```
function y=gamma_c(z), f=@(t)exp(-t).*t.^(z-1);
if isreal(z), y=gamma(z);
else, y=integral(f,0,inf,'ArrayValued',true); end
```

8.5.2 Beta functions

The Beta function is defined as

$$B(m,\alpha)=\int_0^1 t^{m-1}(1-t)^{\alpha-1}\,\mathrm{d}t=\frac{\Gamma(m)\Gamma(\alpha)}{\Gamma(m+\alpha)},\quad m,\alpha>0. \tag{8-5-6}$$

Beta function can be evaluated directly with $\boldsymbol{y}$ = `beta(`$\boldsymbol{m}$`,`$\boldsymbol{x}$`)`.

Example 8.38 Please draw the functions of Beta functions for different values of m.

Solution *If* $m = 1$, *the Beta function plot can be obtained immediately, as shown in Figure 8.32 (a). For different other values of* m, *the surface of Beta function can be obtained. Through proper adjustment of the viewpoint, the surface can be shown in Figure 8.32 (b).*

```
>> m=1; x=0.1:0.1:3; y=beta(m,x); subplot(121), plot(x,y); subplot(122)
   m=1:10; Z=[]; for i=m, Z=[Z; beta(i,x)]; end; surf(x,m,Z)
```

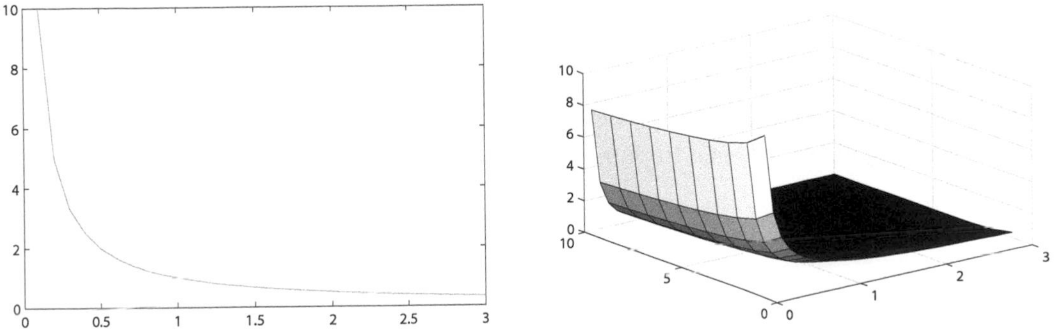

(a) Beta function when $m = 1$ (b) surface of Beta function for different m

FIGURE 8.32: Beta function plots.

Similar to incomplete Gamma functions, incomplete Beta function is defined as

$$B_x(z,m) = \frac{1}{B(z,w)} \int_0^x t^{m-1}(1-t)^{z-1}\,\mathrm{d}t, \tag{8-5-7}$$

with $\mathscr{R}(m) > 0$, $\mathscr{R}(z) > 0$, and $0 \leqslant x \leqslant 1$. Incomplete Beta function can be evaluated directly with $\boldsymbol{y}$=`betainc(`x,z,m`)`.

8.5.3 Legendre functions

Consider the Legendre differential equation

$$(1-t^2)\frac{\mathrm{d}^2x}{\mathrm{d}t^2} - 2t\frac{\mathrm{d}x}{\mathrm{d}t} + n(n+1)x = 0, \tag{8-5-8}$$

and there is no analytical solution, therefore, the general solution of the equation can be written as

$$x(t) = \mathrm{C}_1 P_n(t) + \mathrm{C}_2 Q_n(t), \tag{8-5-9}$$

where, $\mathrm{C}_1, \mathrm{C}_2$ are arbitrary constants, $P_n(t)$ and $Q_n(t)$ are, respectively, first and second type Legendre functions, defined as

$$P_n(t) = \sum_{k=0}^{\infty}(-1)^k \frac{\Gamma(k+n+1)}{(k!)^2\Gamma(n-k+1)}\left(\frac{1-t}{2}\right)^k, \quad |1-t| < 2, \tag{8-5-10}$$

$$Q_n(t) = \frac{1}{2}P_n(t)\ln\left(\frac{t+1}{t-1}\right) - \sum_{k=1}^{n}\frac{1}{k}P_{k-1}(t)P_{n-k}(t). \tag{8-5-11}$$

Extending (8-5-8), a series of Legendre differential equations can be constructed as

$$(1-t^2)\frac{\mathrm{d}^2x}{\mathrm{d}t^2} - 2t\frac{\mathrm{d}x}{\mathrm{d}t} + \left[n(n+1) - \frac{m^2}{1-t^2}\right]x = 0, \tag{8-5-12}$$

where, the Legendre functions can be denoted as $P_n^m(t)$, satisfying

$$P_n^m(t) = (-1)^m(1-t^2)^{m/2}\frac{\mathrm{d}^m}{\mathrm{d}t^m}P_n(t). \tag{8-5-13}$$

Function `legendre()` can be used to evaluate Legendre functions, $P_n^m(t)$, with the syntax $\boldsymbol{Y}=$`legendre(`n`,`$\boldsymbol{x}$`)`. Here $\boldsymbol{Y}$ is a matrix, whose rows are $P_n^0(\boldsymbol{x})$, $P_n^1(\boldsymbol{x})$, $\cdots$, $P_n^n(\boldsymbol{x})$. Also, it is required that $-1 < \boldsymbol{x} < 1$.

Example 8.39 Legendre functions can be obtained directly with the following statements, as shown in Figure 8.33. Legendre functions of other orders can be drawn similarly.

```
>> x=-1:0.04:1; Y=legendre(2,x); plot(x,Y) % draw Legendre plots
```

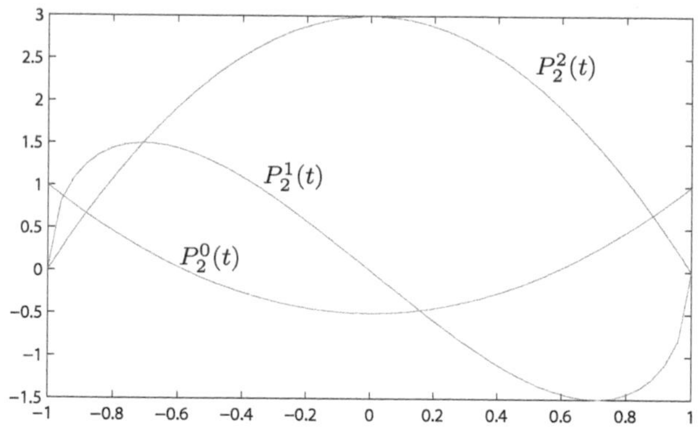

FIGURE 8.33: Legendre function plots.

8.5.4 Bessel functions

Consider the following Bessel differential equation

$$t^2\frac{\mathrm{d}^2x}{\mathrm{d}t^2} + t\frac{\mathrm{d}x}{\mathrm{d}t} + (t^2-\lambda^2)x = 0. \tag{8-5-14}$$

If λ is not integer, power series approximation can be used, and the general solution of the differential equation can be written as

$$x(t) = \mathrm{C}_1 J_\lambda(t) + \mathrm{C}_2 J_{-\lambda}(t), \tag{8-5-15}$$

where, C_1 and C_2 are arbitrary constants, $J_\lambda(t)$ is the first type λth order Bessel function defined as

$$J_\lambda(t) = \sum_{m=0}^{\infty}(-1)^m\frac{t^{\lambda+2m}}{2^{\lambda+2m}m!\,\Gamma(\lambda+m+1)}, \tag{8-5-16}$$

and it applies to the cases when λ is not an integer. If $\lambda = n$ is a positive integer, the first type Bessel has the following properties

$$J_n(t) = (-1)^n J_{-n}(t), \quad \frac{J_n(x)}{dt} = \frac{n}{t}J_n(t) - J_{n+1}(t), \quad \int t^n J_{n-1}(t)\,dt = t^n J_n(t). \quad (8\text{-}5\text{-}17)$$

If $\lambda = n$ is an integer, $J_n(t)$ and $J_{-n}(t)$ are linearly independent, thus, the general solution cannot be described with (8-5-15). The second type nth order Bessel function should be defined, and it is also known as *Neumann function*.

$$N_\lambda(t) = \frac{J_\lambda(t)\cos\lambda t - J_{-\lambda}(t)}{\sin\lambda t}, \quad (8\text{-}5\text{-}18)$$

and when $\lambda = n$, the solution in (8-5-14) can be rewritten as

$$x(t) = C_1 J_n(t) + C_2 H_n(t). \quad (8\text{-}5\text{-}19)$$

Function `besselj()` can be used to evaluate the first type Bessel functions, with the syntax `y=besselj(λ,x)`, where λ is the order. The second type Bessel function can be evaluated with `bessely()` function, and the syntax of the function is exactly the same as `besselj()` function. The third type Bessel function, also known as *Hankel function*, can be evaluated with `besselh()`.

Example 8.40 Draw the plot of the first type Bessel functions.

Solution *The plots of Bessel functions with various parameters can be obtained as shown in Figures 8.34 (a) and (b).*

```
>> lam=0; x=-10:0.1:10; y=besselj(lam,x); subplot(121), plot(x,y);
   lam=-2:2; subplot(122), for i=lam, plot(x,besselj(i,x)); hold on; end
```

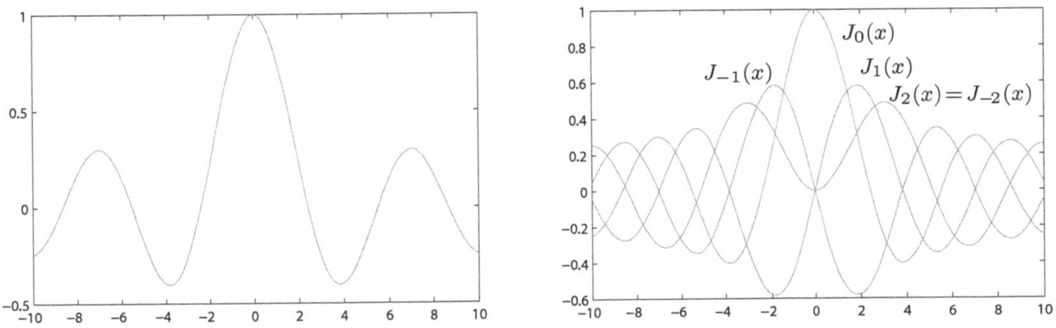

(a) $\lambda = 0$ (b) for various λ values

FIGURE 8.34: First type Bessel function curves.

8.5.5 Mittag–Leffler functions

Mittag–Leffler function is the extension of simple exponential functions. The simplest form of Mittag–Leffler function was introduced by Swedish mathematician Magnus Gustaf Mittag–Leffler in 1903 [5], it is also known as Mittag–Leffler function with one variable.

Later, Mittag–Leffler functions with two and more variables are introduced. The importance of Mittag–Leffler function in fractional-order calculus is similar to the exponential functions in traditional calculus. In this section, definitions and computations of Mittag–Leffler functions are presented.

The Mittag–Leffler function with one variable is defined as

$$\mathscr{E}_\alpha(z)=\sum_{k=0}^{\infty}\frac{z^k}{\Gamma(\alpha k+1)}, \tag{8-5-20}$$

where, α is a complex number, and for $\mathscr{R}(\alpha)>0$, the infinite series is convergent.

It is obvious that, exponential function e^z is a special case of Mittag–Leffler function, since it can be seen that when $\alpha=1$

$$\mathscr{E}_1(z)=\sum_{k=0}^{\infty}\frac{z^k}{\Gamma(k+1)}=\sum_{k=0}^{\infty}\frac{z^k}{k!}=\mathrm{e}^z. \tag{8-5-21}$$

Besides, the functions for $\alpha=2,\alpha=1/2$ can also be derived as

$$\mathscr{E}_2(z)=\sum_{k=0}^{\infty}\frac{z^k}{\Gamma(2k+1)}=\sum_{k=0}^{\infty}\frac{\left(\sqrt{z}\right)^{2k}}{(2k)!}=\cosh\sqrt{z}, \tag{8-5-22}$$

$$\mathscr{E}_{1/2}(z)=\sum_{k=0}^{\infty}\frac{z^k}{\Gamma\left(k/2+1\right)}=\mathrm{e}^{z^2}\left(1+\mathrm{erf}(z)\right)=\mathrm{e}^{z^2}\mathrm{erfc}(-z). \tag{8-5-23}$$

Consider again the Mittag–Leffler function with one variable, if the value of 1 in the Gamma function is substituted with another free variable β, the Mittag–Leffler function with two variables is defined as

$$\mathscr{E}_{\alpha,\beta}(z)=\sum_{k=0}^{\infty}\frac{z^k}{\Gamma(\alpha k+\beta)}, \tag{8-5-24}$$

where, α,β are complex, and convergent conditions for any complex variable z are that $\mathscr{R}(\alpha)>0$, $\mathscr{R}(\beta)>0$. If $\beta=1$, the Mittag–Leffler is reduced to a univariate Mittag–Leffler function, i.e.,

$$\mathscr{E}_{\alpha,1}(z)=\mathscr{E}_\alpha(z). \tag{8-5-25}$$

Other special cases of Mittag–Leffler functions can be derived

$$\mathscr{E}_{1,2}(z)=\sum_{k=0}^{\infty}\frac{z^k}{\Gamma(k+2)}=\frac{1}{z}\sum_{k=0}^{\infty}\frac{z^{k+1}}{(k+1)!}=\frac{\mathrm{e}^z-1}{z}, \tag{8-5-26}$$

$$\mathscr{E}_{1,3}(z)=\sum_{k=0}^{\infty}\frac{z^k}{\Gamma(k+3)}=\sum_{k=0}^{\infty}\frac{z^k}{(k+2)!}=\frac{1}{z^2}\sum_{k=0}^{\infty}\frac{z^{k+2}}{(k+2)!}=\frac{\mathrm{e}^z-1-z}{z^2}. \tag{8-5-27}$$

More generally, we have [6]

$$\mathscr{E}_{1,m}(z)=\sum_{k=0}^{\infty}\frac{z^k}{\Gamma(k+m)}=\frac{1}{z^{m-1}}\sum_{k=0}^{\infty}\frac{z^{k+m-1}}{(k+m-1)!}=\frac{1}{z^{m-1}}\left(\mathrm{e}^z-\sum_{k=0}^{m-2}\frac{z^k}{k!}\right), \tag{8-5-28}$$

and it can be found that

$$\mathscr{E}_{2,2}(z)=\sum_{k=0}^{\infty}\frac{z^k}{\Gamma(2k+2)}=\frac{1}{\sqrt{z}}\sum_{k=0}^{\infty}\frac{(\sqrt{z})^{2k+1}}{(2k+1)!}=\frac{\sinh\sqrt{z}}{\sqrt{z}}, \tag{8-5-29}$$

$$\mathscr{E}_{2,1}(z^2) = \sum_{k=0}^{\infty} \frac{z^{2k}}{\Gamma(2k+1)} = \sum_{k=0}^{\infty} \frac{z^{2k}}{(2k)!} = \cosh z, \tag{8-5-30}$$

$$\mathscr{E}_{2,2}(z^2) = \sum_{k=0}^{\infty} \frac{z^{2k}}{\Gamma(2k+2)} = \frac{1}{z}\sum_{k=0}^{\infty} \frac{z^{2k+1}}{(2k+1)!} = \frac{\sinh z}{z}. \tag{8-5-31}$$

The Mittag–Leffler function with one and two variables can be evaluated with MATLAB function, with the following syntaxes

F_1 = `mittag_leffler(`α`,`z`)`, and F_1 = `mittag_leffler([`α`,`β`],`z`)`

```
function f=mittag_leffler(aa,z)
aa=[aa 1]; a=aa(1); b=aa(2); % please use MATLAB R2008a or earlier
syms k; f=simplify(symsum(z^k/gamma(a*k+b),k,0,inf));
```

Please note that due to the limitations of `symsum()` function in the new versions, the `mittag_leffler()` function sometimes may not yield correct results. Therefore, MATLAB R2008a or earlier versions are recommended.

Example 8.41 Find the analytic expression of Mittag–Leffler functions with one variable α, for $\alpha = 1/3, 3, 4, 5, \cdots$.

Solution *With the direct use of* `mittag_leffler()` *function, the analytical solutions of the necessary Mittag–Leffler functions can be obtained*

```
>> syms z; I1=mittag_leffler(1/sym(3),z), I2=mittag_leffler(3,z)
   I3=mittag_leffler(4,z), I4=mittag_leffler(5,z)
```

and the following results can be obtained

$$\mathscr{E}_{1/3}(z) = -\frac{\mathrm{e}^{z^3}\left(-6\pi\Gamma(2/3) + \sqrt{3}\Gamma^2(2/3)\Gamma\left(1/3, z^3\right) + 2\Gamma\left(2/3, z^3\right)\pi\right)}{2\pi\Gamma(2/3)},$$

$$\mathscr{E}_3(z) = \frac{1}{3}\mathrm{e}^{\sqrt[3]{z}} + \frac{2}{3}\mathrm{e}^{-\sqrt[3]{z}/2}\cos\left(\frac{\sqrt{3}}{2}\sqrt[3]{z}\right),$$

$$\mathscr{E}_4(z) = \frac{1}{4}\mathrm{e}^{\sqrt[4]{z}} + \frac{1}{4}\mathrm{e}^{-\sqrt[4]{z}} + \frac{1}{2}\cos\left(\sqrt[4]{z}\right),$$

$$\mathscr{E}_5(z) = \frac{\mathrm{e}^{\sqrt[5]{z}}}{5} + \frac{2}{5}\mathrm{e}^{\cos(2\pi/5)\sqrt[5]{z}}\cos\left(\sin\left(\frac{2\pi}{5}\right)\sqrt[5]{z}\right) + \frac{2}{5}\mathrm{e}^{-\cos(\pi/5)\sqrt[5]{z}}\cos\left(\sin\left(\frac{\pi}{5}\right)\sqrt[5]{z}\right).$$

Example 8.42 Please find the Mittag–Leffler functions with two variables, such as $\mathscr{E}_{4,1}(z)$, $\mathscr{E}_{4,5}(z)$, $\mathscr{E}_{5,6}(z)$ and $\mathscr{E}_{1/2,4}(z)$.

Solution *With proper parameters α and β, the Mittag–Leffler functions can be obtained directly with*

```
>> syms z, I5=mittag_leffler(4,1,z), I6=mittag_leffler(4,5,z)
   I7=mittag_leffler(5,6,z), I8=mittag_leffler(1/sym(2),4,z)
```

and the results obtained are

$$\mathscr{E}_{4,1}(z) = \frac{1}{4}\mathrm{e}^{\sqrt[4]{z}} + \frac{1}{4}\mathrm{e}^{-\sqrt[4]{z}} + \frac{1}{2}\cos\sqrt[4]{z},$$

$$\mathscr{E}_{4,5}(z) = -\frac{1}{4} + \frac{1}{4z}\left(\mathrm{e}^{\sqrt[4]{z}} + \mathrm{e}^{-\sqrt[4]{z}} + \mathrm{e}^{\mathrm{j}\sqrt[4]{z}} + \mathrm{e}^{-\mathrm{j}\sqrt[4]{z}}\right),$$

$$\mathscr{E}_{5,6}(z) = -\frac{1}{z} + \frac{e^{\sqrt[5]{z}}}{5z}\left[1 + e^{(-1)^{2/5}} + e^{(-1)^{4/5}} + e^{-(-1)^{1/5}} + e^{-(-1)^{3/5}}\right],$$

$$\mathscr{E}_{1/2,4}(z) = \frac{e^{z^2}}{z^6} - \frac{1}{z^6} - \frac{1}{z^4} - \frac{1}{2z^2} + \frac{ze^{z^2}\operatorname{erf}(z)}{z^7} - \frac{8}{15\sqrt{\pi}z} - \frac{4}{3\sqrt{\pi}z^3} - \frac{2}{\sqrt{\pi}z^5}.$$

The nth order derivative of Mittag–Leffler functions with two variables, $\mathscr{E}_{\alpha,\beta}(z)$ can be obtained with

$$\frac{d^n}{dz^n}\mathscr{E}_{\alpha,\beta}(z) = \sum_{k=0}^{\infty} \frac{(k+n)!}{k!\,\Gamma(\alpha k + \alpha n + \beta)} z^k \tag{8-5-32}$$

In some particular fields, Mittag–Leffler functions with three or four parameters are used, and the definitions of these functions are

$$\mathscr{E}_{\alpha,\beta}^{\gamma}(z) = \sum_{k=0}^{\infty} \frac{\Gamma(k+\gamma)}{\Gamma(\alpha k+\beta)\Gamma(\gamma)} \frac{z^k}{k!}, \tag{8-5-33}$$

and

$$\mathscr{E}_{\alpha,\beta}^{\gamma,q}(z) = \sum_{k=0}^{\infty} \frac{\Gamma(kq+\gamma)}{\Gamma(\alpha k+\beta)\Gamma(\gamma)} \frac{z^k}{k!}, \tag{8-5-34}$$

where α, β, γ are complex numbers, q is an integer, and the convergent conditions for any z are $\mathscr{R}(\alpha) > 0$, $\mathscr{R}(\beta) > 0$, $\mathscr{R}(\gamma) > 0$. It is also known that

$$\mathscr{E}_{\alpha,\beta}^{1}(z) = \mathscr{E}_{\alpha,\beta}(z), \quad \mathscr{E}_{\alpha,\beta}^{\gamma,1}(z) = \mathscr{E}_{\alpha,\beta}^{\gamma}(z). \tag{8-5-35}$$

In general, for any positive integer n, one has

$$\frac{d^n}{dz^n}\mathscr{E}_{\alpha,\beta}^{\gamma,q}(z) = \frac{\Gamma(qn+\gamma)}{\Gamma(\gamma)}\mathscr{E}_{\alpha,\beta+n\alpha}^{\gamma+qn,q}(z). \tag{8-5-36}$$

Based on the previous expression, a MATLAB can be written to evaluate numerically the derivatives of Mittag–Leffler functions. Truncation algorithm is used in the function. Sometimes the function may fail to converge, and in this case, the function `MLF()` written by Slovakia scholar, Professor Igor Podlubny [7], can be used instead. The function is reliable, but the speed of it is extremely slow. The reliable function is embedded in the function, so as to ensure both the speed and the reliability of the function.

```
function f=ml_func(aa,z,varargin)
aa=[aa,1,1,1]; a=aa(1); b=aa(2); c=aa(3); q=aa(4); f=0; k=0; fa=1;
[n,eps0]=default_vals({0,eps},varargin{:});
if n==0
   while norm(fa,1)>=eps0 % check truncation error
      fa=gamma(k*q+c)/gamma(c)/gamma(k+1)/gamma(a*k+b) *z.^k;
      f=f+fa; k=k+1;          % accumulation and continue
   end
   if any(~isfinite(f)), eps1=round(-log10(eps0)); % if not converge
      if c*q==1, f=MLF(a,b,z,eps1); f=reshape(f,size(z)); % use MLF
      else, error('Error: truncation method failed'); end, end
else, aa(2)=aa(2)+n*aa(1); aa(3)=aa(3)+aa(4)*n;
   f=gamma(q*n+c)/gamma(c)*ml_func(aa,z,0,eps0); % compute derivative
end
```

The syntaxes of the function are

```
f =ml_func(α,z,n,ϵ0)          % nth order derivative of ℰα(z)
f =ml_func([α,β],z,n,ϵ0)      % nth order derivative of ℰα,β(z)
```

where the default value of ϵ_0 is **eps**, and the default value of n is zero, meaning to evaluate the original Mittag–Leffler function. In fact, the function can be used to deal with Mittag–Leffler functions with three or four variables.

Example 8.43 Evaluate numerically the function $\mathscr{E}_{1/2,4}(z)$ studied in Example 8.42, and compare the results with analytical results.

Solution *The following statements can be used to find the analytical and numerical results of the function, as shown in Figure 8.35. It can be seen that the two results agree well. For this example, if the time interval is too large, the truncation method may fail, and function* `MLF()` *can be called automatically. However, the speed of the evaluation process may become very slow.*

```
>> syms z, I8=mittag_leffler([1/sym(2),4],z);
   t=0:0.01:2; y=subs(I8,z,t); y1=ml_func([1/2,4],t); plot(t,y,t,y1)
```

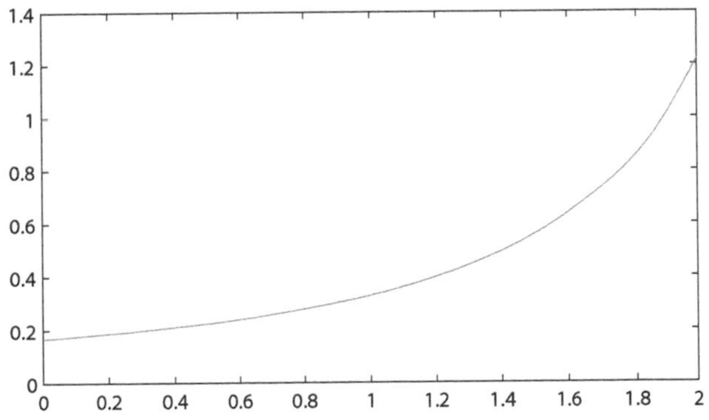

FIGURE 8.35: Mittag–Leffler function plot.

8.6 Signal Analysis and Digital Signal Processing

8.6.1 Correlation analysis

Correlation analysis can be used to characterize both stochastic and deterministic signals. Consider a deterministic signal $x(t)$. The auto-correlation function is defined as

$$R_{xx}(\tau) = \lim_{T\to\infty} \frac{1}{T} \int_0^T x(t)x(t+\tau)\,dt, \quad \tau \geqslant 0, \tag{8-6-1}$$

and the correlation function is an even function, i.e., $R_{xx}(-\tau) = R_{xx}(\tau)$. The cross-correlation function of two signals $x(t)$ and $y(t)$ is defined as

$$R_{xy}(\tau) = \lim_{T\to\infty} \frac{1}{T}\int_0^T x(t)y(t+\tau)\,\mathrm{d}t, \quad \tau \geqslant 0. \tag{8-6-2}$$

Clearly, when the functions are all given, the auto-correlation and cross-correlation functions can be evaluated by using Symbolic Math Toolbox. In some cases, it is possible to obtain analytical solutions.

Example 8.44 Consider the signal $x(t) = A_1\cos(\omega_1 t+\theta_1) + A_2\cos(\omega_2 t+\theta_2)$, and $\omega_1 \neq \pm\omega_2$. Compute its auto-correlation function.

Solution *It is known that the symbolic variables should be declared first. Then, the function $x(t)$ can be defined, and the following statements can be used to evaluate the auto-correlation function.*

```
>> syms A1 A2 t1 t2 t tau T w1 w2; assume(w1~=w2);
   assumeAlso(w1~=-w2); x=A1*cos(w1*t+t1)+A2*cos(w2*t+t2);
   Rxx=simplify(limit(int(x*subs(x,t,t+tau),t,0,T)/T,T,inf))
```

The analytical solution can be found that

$$R_{xx}(\tau) = \frac{1}{2}A_1^2\cos(\omega_1\tau) + \frac{1}{2}A_2^2\cos(\omega_2\tau).$$

Given two sets of data $x_i, y_i, (i = 1, 2, \cdots, n)$, the following formula can be used to evaluate the correlation coefficients

$$r_{xy} = \frac{\sqrt{\sum(x_i-\bar{x})(y_i-\bar{y})}}{\sqrt{\sum(x_i-\bar{x})^2}\sqrt{\sum(y_i-\bar{y})^2}}. \tag{8-6-3}$$

The MATLAB function `corrcoef()` can be used to obtain the correlation coefficients $\boldsymbol{R}$ of vectors $\boldsymbol{x}$ and $\boldsymbol{y}$ using the following syntaxes:

$\boldsymbol{R}$= `corrcoef(`$\boldsymbol{x}$`,`$\boldsymbol{y}$`)`, or $\boldsymbol{R}$= `corrcoef([`$\boldsymbol{x}$`,`$\boldsymbol{y}$`])`

Example 8.45 Generate sample points from the functions $y_1 = te^{-4t}\sin 3t$ and $y_2 = te^{-4t}\cos 3t$, and then, compute the correlation coefficient matrix.

Solution *With the following statements, the correlation matrix can be obtained*

```
>> x=0:0.01:5; y1=x.*exp(-4*x).*sin(3*x); y2=x.*exp(-4*x).*cos(3*x);
   R=corrcoef(y1,y2) % compute correlation coefficient matrix
```

and the correlation coefficients obtained are

$$\boldsymbol{R} = \begin{bmatrix} 1 & 0.477585851214039 \\ 0.477585851214039 & 1 \end{bmatrix}.$$

Consider now the sampled data $x_i, y_i, (i = 1, 2, \cdots, n)$. The auto-correlation function of the discrete signal x_i can be evaluated from

$$c_{xx}(k) = \frac{1}{N}\sum_{l=1}^{n-[k]-1} x(l)x(k+l), \quad 0 \leqslant k \leqslant m-1, \tag{8-6-4}$$

where $m < n$. The auto-correlation function obtained is also an even function. Similarly, the cross-correlation function is defined as

$$c_{xy}(k) = \frac{1}{N} \sum_{l=1}^{n-[k]-1} x(l)y(k+l), \quad 0 \leqslant k \leqslant m-1. \tag{8-6-5}$$

Similarities of two discrete signals can be studied with the help of correlation functions. The auto-correlation function and cross-correlation function can both be evaluated with the use of function `xcorr()`. The syntaxes are

C_{xx}=`xcorr(`$\boldsymbol{x}$,N`)`, and C_{xy}=`xcorr(`$\boldsymbol{x}$,$\boldsymbol{y}$,N`)`

where N is the maximum value of k and it can be omitted.

Example 8.46 Consider again the data generated from Example 8.45. Compute numerically the auto-correlation, cross-correlation functions and compare the results with theoretical curves.

Solution *Specify a time vector in the interval $t \in (0, 5)$. The output signals can be generated in vectors $\boldsymbol{x}$ and $\boldsymbol{y}$. The auto-correlation and cross-correlation functions can be obtained as shown in Figures 8.36 (a) and (b), respectively.*

```
>> t=0:0.05:5; x=t.*exp(-4*t).*sin(3*t); y=t.*exp(-4*t).*cos(3*t);
   N=30; c1=xcorr(x,N); x1=[-N:N]; subplot(121), stem(x1,c1)
   subplot(122), c1=xcorr(x,y,N); x1=[-N:N]; stem(x1,c1)
```

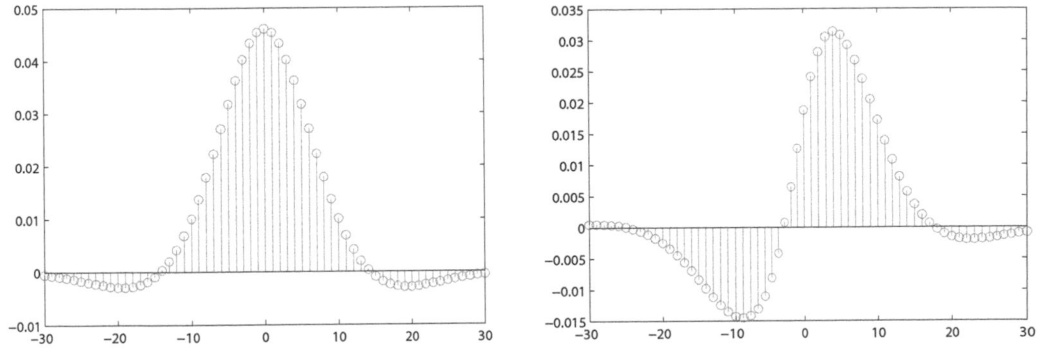

(a) auto-correlation analysis (b) cross-correlation analysis

FIGURE 8.36: Correlation function analysis.

8.6.2 Power spectral analysis

For the discrete sample data vector $\boldsymbol{y}$, the MATLAB function `psd()` can be used to evaluate its power spectral density. However, it is found that the result obtained is not satisfactory. The Welch transform based estimation algorithm is introduced here instead [8].

Assume that n is the length of the vector $\boldsymbol{y}$ to be processed. These data can be divided into $\mathcal{K} = [n/m]$ groups with a vector length of m such that

$$x^{(i)}(k) = y[k + (i-1)m], \; 0 < k \leqslant m-1, \; 1 \leqslant i \leqslant \mathcal{K}. \tag{8-6-6}$$

With the Welch algorithm, the following $\mathcal{K}$ equations can be established.

$$J_{\mathrm{m}}^{(i)}(\omega) = \frac{1}{mU}\left|\sum_{k=0}^{m-1} x^{(i)}(k)w(k)\mathrm{e}^{-\mathrm{j}\omega k}\right|^2, \tag{8-6-7}$$

where $w(k)$ is the data processing window function. For instance, it can be the Hamming window defined as

$$w(k) = a - (1-a)\cos\left(\frac{2\pi k}{m-1}\right),\ k = 0, \cdots, m-1. \tag{8-6-8}$$

If $a = 0.54$, and

$$U = \frac{1}{m}\sum_{k=0}^{m-1} w^2(k), \tag{8-6-9}$$

the power spectral density of the signal can be estimated with

$$P_{\mathrm{xx}}^{w}(\omega) = \frac{1}{\mathcal{K}}\sum_{j=1}^{\mathcal{K}} J_{\mathrm{m}}^{(i)}(\omega). \tag{8-6-10}$$

The following procedure can be used to compute the power spectral densities of the signal [9].

(i) Calculate $X_m^{(i)}(l)=\sum_{k=0}^{m-1} x^{(i)}(k)w(k)\mathrm{e}^{-\mathrm{j}[2\pi/(m\Delta t)]lk}$ in groups with `fft()`.

(ii) In group i, compute $\left|X_m^{(i)}(l)\right|^2$, the accumulative sum $Y(l)=\sum_{i=1}^{\mathcal{K}}\left|X_m^{(i)}(l)\right|^2$.

(iii) From the following formula, the power spectral density can be obtained

$$P_{\mathrm{xx}}^{w}\left(\frac{2\pi}{m\Delta t}l\right) = \frac{1}{\mathcal{K}mU}Y(l). \tag{8-6-11}$$

The function `fft()` is used in calculating $X_m^{(i)}$

(i) MATLAB function `fft()` can be used to compute discrete Fourier transforms. The resulting P_{xx}^{w} should be multiplied by Δt.

(ii) To increase the effectiveness of the algorithm, m should be chosen as $2^k - 1$ for integer k.

A MATLAB function `psd_estm()` can be written to estimate the power spectral density of a given sequence, listed as

```
function [Pxx,f]=psd_estm(y,m,T,a)
if nargin==3, a=0.54; end
k=[0:m-1]; Y=zeros(1,m); m2=floor(m/2); f=k(1:m2)*2*pi/(length(k)*T);
w=a-(1-a)*cos(2*pi*k/(m-1)); K=floor(length(y)/m); U=sum(w.^2)/m;
for i=1:K, xi=y((i-1)*m+k+1)'; Xi=fft(xi.*w); Y=Y+abs(Xi).^2; end
Pxx=Y(1:m2)*T/(K*m*U);
```

The syntax of the function is $[P_{\mathrm{xx}}, \boldsymbol{f}] = \texttt{psd\_estm}(\boldsymbol{y}, m, \Delta t, a)$. In this function, in order to avoid the aliasing phenomenon, half of the transformation data should be selected. In this function, the definitions of $\boldsymbol{y}$ and m are the same, and Δt is the sample time. The returned arguments $\boldsymbol{f}$ and P_{xx} are respectively the frequency and the power spectral density.

8.6.3 Filtering techniques and filter design

Before presenting technical details of filters, the applications of filters are illustrated through an example.

Example 8.47 To illustrate the concept of filtering techniques, the function $y(x) = \mathrm{e}^{-x}\sin(5x)$ is studied. Suppose a set of function data is generated, disturbed by the white noise signal with the zero mean and the variance of 0.05. Draw the corrupted signal first.

Solution *The following statements can be used, and the noisy signal is shown in Figure 8.37.*

```
>> x=0:0.002:2; y=exp(-x).*sin(5*x);          % generate data
   r=0.05*randn(size(x)); y1=y+r; plot(x,y1) % draw noised plot
```

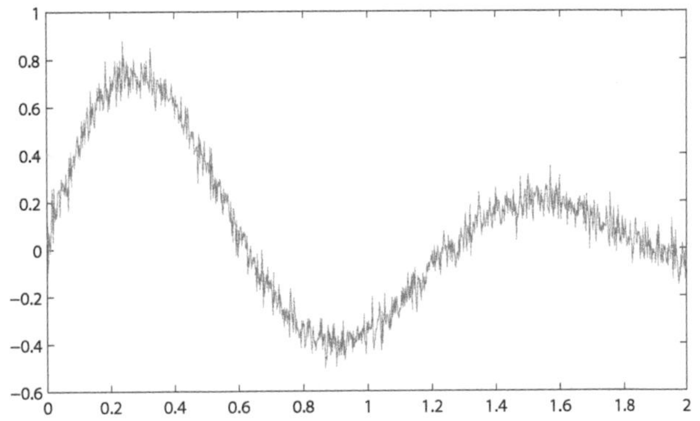

FIGURE 8.37: Data corrupted by noises.

For the corrupted signal shown in Figure 8.37, a noise elimination technique should be applied. For instance, filters can be used to filter out the noise to extract the actual signal.

I. Linear filters — general models

Linear filters can generally be written as

$$H(z) = \frac{b_1 + b_2 z^{-1} + b_3 z^{-2} + \cdots + b_{n+1} z^{-n}}{1 + a_1 z^{-1} + a_2 z^{-2} + \cdots + a_m z^{-m}}. \tag{8-6-12}$$

Assume that the input signal is $x(n)$, the filtered output signal can be expressed using the following difference equation:

$$\begin{aligned} y(k) &= -a_1 y(k-1) - a_2 y(k-2) - \cdots - a_m y(k-m) \\ &\quad + b_1 x(k) + b_2 x(k-1) + b_3 x(k-2) \cdots + b_{n+1} x(k-n). \end{aligned} \tag{8-6-13}$$

With different combinations of n and m, three types of filters are defined

(i) **FIR filter** The finite impulse response (FIR) filter requires $m = 0$ in (8-6-12). The variable $\boldsymbol{a}$ is then a scalar. This filter is also known as the *moving average* (MA) filter. The vector $\boldsymbol{b}$ can be used to define the filter.

(ii) **IIR filter** The infinite impulse response (IIR) filter, also known as *auto-regressive*

(AR) or all-pole filter required $n = 0$, i.e., $\boldsymbol{b}$ is a scalar and the vector $\boldsymbol{a}$ can be used to represent the filter. An advantage of FIR is that it is always stable [10].

(iii) **ARMA filter** The auto-regressive moving average (ARMA) filter is also known as *general IIR* filter, where n and m are not zero. The vectors $\boldsymbol{a}$ and $\boldsymbol{b}$ can be used to express the filter.

Assume that the filter is given by vectors $\boldsymbol{a}$ and $\boldsymbol{b}$, and the signal to be filtered is given by vector $\boldsymbol{x}$. The function `filter()` can be used to evaluate the filtered signal vector $\boldsymbol{y}$ where `y=filter(b,a,x)`.

Filters can be classified into low-pass filters, high-pass filters, band-stop and band-pass filters. It is known from the filter names that a low-pass filter allows the signal with low frequency to pass the filter directly with little or no attenuation while the signal's high-frequency content will be filtered out or attenuated. High-pass filters filter out the low-frequency signal components. A band-pass filter allows the signal components of certain frequency range to pass directly, while those of other frequencies are filtered out. In real applications, these filters are used depending on the purpose of the signal processing. For example, the noise in Example 8.47 is a high-frequency noise, thus, a low-pass filter is needed, i.e., the magnitude is set to 1 or nearly 1 for low-frequency signals, while to 0 for high-frequency noises. In this way, the noise may be eliminated. If the filter is known, the function `freqz()` can be used to analyze the magnitude of the filter, such that `[h,w]=freqz(b,a,N)`, where N is the number of points to be analyzed. The returned argument $\boldsymbol{h}$ is the complex gain, and $\boldsymbol{w}$ is the frequency vector. The complex gain includes the information of magnitudes and phases. If only the magnitude information is required, `plot(w,abs(h))` can be used to visualize the filter frequency response. Alternatively, Bode magnitude plot can be obtained with the command `semilogx(w,20*log10(abs(h)))`.

Example 8.48 Assume that a filter is given by

$$H(z)=\frac{1.2296\times 10^{-6}(1+z^{-1})^7}{\begin{array}{c}(1-0.7265z^{-1})(1-1.488z^{-1}+0.5644z^{-2})\\(1-1.595z^{-1}+0.6769z^{-2})(1-1.78z^{-1}+0.8713z^{-2})\end{array}}.$$

Compute the complex gain and observe the filtered signals.

Solution *The vectors* $\boldsymbol{b}$ *and* $\boldsymbol{a}$ *of the filter can be entered into MATLAB, where the function* `conv()` *can be used to perform polynomial multiplications. The following statements can also draw the gain-frequency curve as shown in Figure 8.38 (a). It can be seen from the curve that, for low frequencies, the gain is close to 1, which means that no filtering action is made over these frequencies. For high frequencies, the gain approaches to 0, which means that the high-frequency noise can be removed or attenuated from the signal.*

```
>> b=1.2296e-6*conv([1 4 6 4 1],[1 3 3 1]); a=conv([1,-0.7265],...
     conv([1,-1.488,0.5644],conv([1,-1.595,0.6769],[1,-1.78,0.8713])));
   x=0:0.002:2; y=exp(-x).*sin(5*x); r=0.05*randn(size(x)); y1=y+r;
   [h,w]=freqz(b,a,100); subplot(121), plot(w,abs(h)) % draw magnitude
   subplot(122), y2=filter(b,a,y1);  plot(x,y1,x,y2)  % filtered signal
```

It can be seen from the results shown in Figure 8.38 (b) that, in the filtered signal, the noise is successfully removed. Note that there exist small delays in the filtered signal. In real-time or online filtering, this delay is unavoidable. However, when performing off-line filtering, it is possible to have zero delay by using the so-called "zero-phase filtering" which can be done by the MATLAB function `filtfilt()` *with the same syntax as* `filter()`.

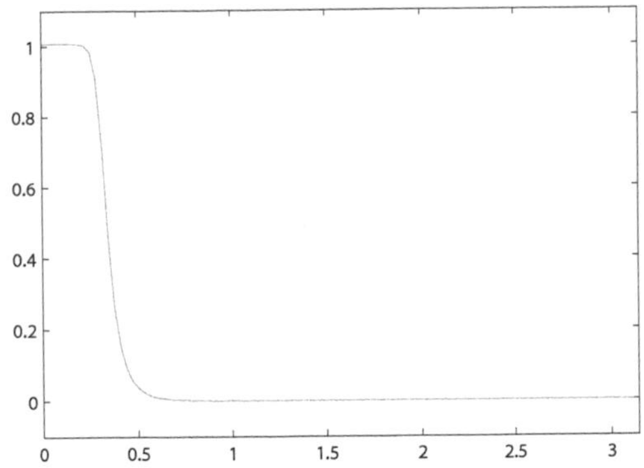

(a) the gain of the filter

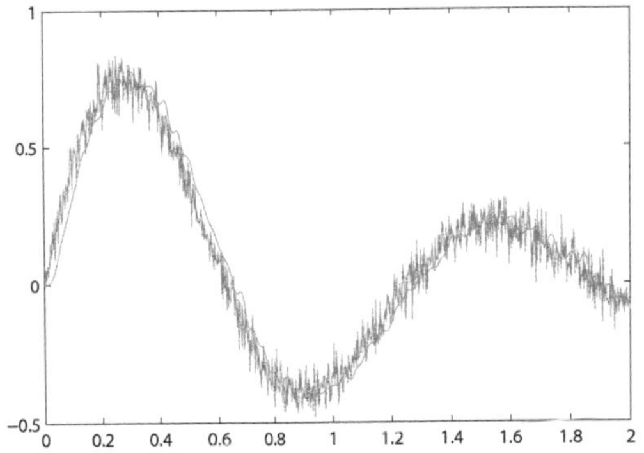

(b) original signal and the filtered one

FIGURE 8.38: Filter gain and filter results.

II. Filter design using MATLAB

It has been shown from the previous examples that the noisy signal can be filtered with a properly designed filter. There are various algorithms and relevant toolboxes in MATLAB for filter design and simulation problems. The most widely used filters are the Butterworth filters and the Chebyshev filters. They can be designed with the `butter()`, `cheby1()` (Chebyshev type I) and `cheby2()` (Chebyshev type II) functions. The syntaxes of these functions are

$$[b,a] = \texttt{butter}(n,\omega_n),\quad [b,a] = \texttt{cheby1}(n,r,\omega_n),\quad [b,a] = \texttt{cheby2}(n,r,\omega_n)$$

where n is the order of the filter, which can either be selected by the user, or be a relevant function, for instance, `buttord()`. The argument ω_n is the normalized frequency, which defines the ratio of the filter frequency and the Nyquist frequency of the signal. Assume that there are N sampling points, and step-size is Δt. The fundamental frequency can then be calculated as $f_0 = 1/\Delta t\,\text{Hz}$. Then, the Nyquist frequency is defined as $f_0/2$.

Example 8.49 Consider the signal in Example 8.47. Design Butterworth filters for different combinations of orders and the natural frequencies of w_n, and compare the filtering results.

Solution *Again $\omega_n = 0.1$ can be selected. With the following statements, the Butterworth filters of different orders can be designed, and the gain and filter results are shown in Figures 8.39 (a) and (b), respectively. It can be seen that with the increment of order n, the filter curve becomes smoother, and the delay also increases.*

```
>> f1=figure; f2=figure;
   for n=5:2:20, figure(f1); [b,a]=butter(n,0.1);
      y2=filter(b,a,y1); plot(x,y2); hold on
      figure(f2); [h,w]=freqz(b,a,100); plot(w,abs(h)); hold on
   end
```

If one selects the 7th order filter structure, different values of ω_n can be tested for Butterworth filters. The gain and filter results can be obtained as shown in Figures 8.40 (a) and (b), respectively. It can be seen that when ω_n increases, the delay will decrease and filtering results may get worse. The large value of ω_n may not provide any benefit.

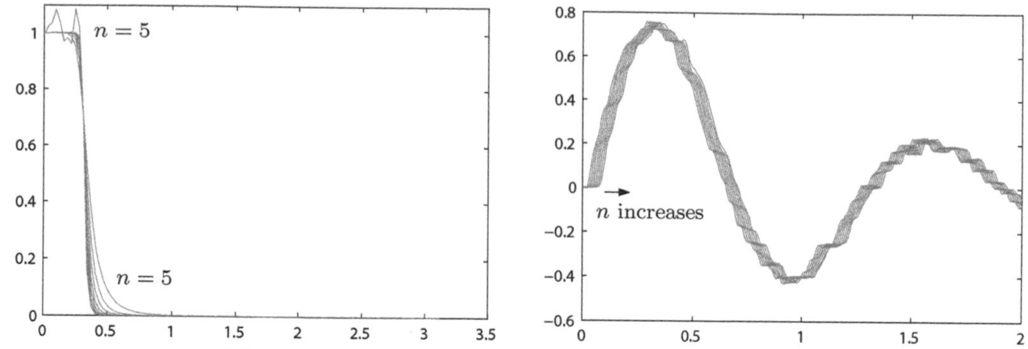

(a) gain (b) filtered curves

FIGURE 8.39: Butterworth filters with different orders.

```
>> for wn=0.1:0.1:0.7, figure(f1); [b,a]=butter(7,wn);
     y2=filter(b,a,y1); plot(x,y2); hold on
     figure(f2); [h,w]=freqz(b,a,100); plot(w,abs(h)); hold on
   end
```

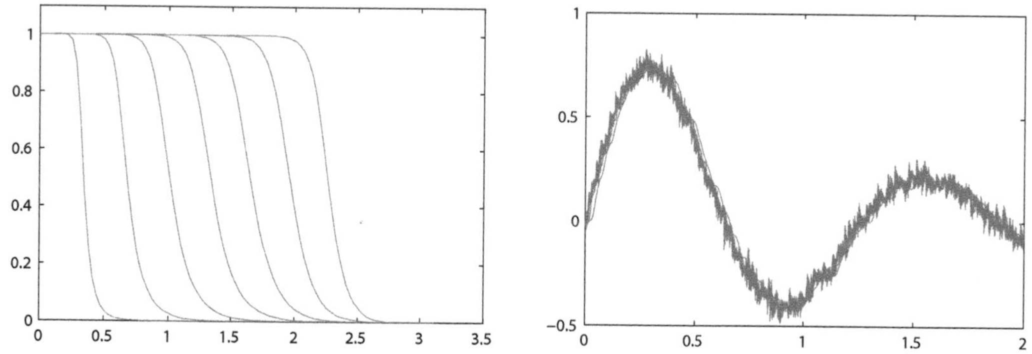

(a) gain (b) filtered curves

FIGURE 8.40: Butterworth filters with different frequencies.

If high-pass filters are required, the simplest way is to design it with the formula $1 - H(z^{-1})$, where $H(z^{-1})$ is the low-pass filter designed using the methods shown earlier. Also, the function `butter()` can be used to design high-pass and band-pass filters with the syntaxes

$[\boldsymbol{b},\boldsymbol{a}]=$ `butter(`n`,`$w_{\rm n}$`,'high')` % high-pass filter

$[\boldsymbol{b},\boldsymbol{a}]=$ `butter(`n`,`$[w_1,w_2]$`)` % band-path filter

Exercises

Exercise 8.1 *Generate a sparsely distributed data from the following functions. Use one-dimensional interpolation method to smooth the curves, with different methods. Compare the interpolation results with the theoretical curves.*

(i) $y(t) = t^2 e^{-5t} \sin t$, *where* $t \in (0, 2)$,

(ii) $y(t) = \sin(10t^2 + 3)$, *for* $t \in (0, 3)$.

Exercise 8.2 *Generate a set of mesh grid data and randomly distribute the data from the prototype function*

$$f(x, y) = \frac{1}{3x^3 + y} e^{-x^2 - y^4} \sin(xy^2 + x^2 y).$$

Fit the original 3D surface with two-dimensional interpolation methods and compare the results with the theoretical ones.

Exercise 8.3 *Assume that a set of data is given as shown in Table 8.7. Interpolate the data into a smooth curve in the interval* $x \in (-2, 4.9)$. *Compare the advantages and disadvantages of the algorithms.*

TABLE 8.7: Measured data for Problem 8.3.

x_i	−2	−1.7	−1.4	−1.1	−0.8	−0.5	−0.2	0.1	0.4	0.7	1	1.3
y_i	0.1029	0.1174	0.1316	0.1448	0.1566	0.1662	0.1733	0.1775	0.1785	0.1764	0.1711	0.1630
x_i	1.6	1.9	2.2	2.5	2.8	3.1	3.4	3.7	4	4.3	4.6	4.9
y_i	0.1526	0.1402	0.1266	0.1122	0.0977	0.0835	0.0702	0.0579	0.0469	0.0373	0.0291	0.0224

Exercise 8.4 *Assume that a set of measured data is given in a file c8pdat.dat. Draw the 3D surface using interpolation methods.*

Exercise 8.5 *Assume that a set of measured data is given in a file c8pdat3.dat, whose 1∼3 columns are the coordinates of* x, y, z, *and the fourth column saves the measured function value* $V(x, y, z)$. *Perform three-dimensional interpolation from the data.*

Exercise 8.6 *Generate a set of data from the function*

$$f(x) = \frac{\sqrt{1+x} - \sqrt{x-1}}{\sqrt{2+x} + \sqrt{x-1}},$$

for x=3 : 0.4 : 8. *The cubic splines and B-splines can be used to perform data interpolation tasks. From the fitted splines, take the second-order derivatives and compare the results with the theoretical curves.*

Exercise 8.7 *Assume that the measured data are given in Table 8.8. Draw the 3D surface plot for* (x, y) *within the rectangular region* $(0.1, 0.1) \sim (1.1, 1.1)$.

Exercise 8.8 *For the measured data samples* (x_i, y_i) *given in Table 8.9, piecewise cubic polynomial splines can be used, and find the coefficients of each polynomial.*

TABLE 8.8: Measured data for Problem 8.7.

y_i	x_1	x_2	x_3	x_4	x_5	x_6	x_7	x_8	x_9	x_{10}	x_{11}
0	0.1	0.2	0.3	0.4	0.5	0.6	0.7	0.8	0.9	1	1.1
0.1	0.8304	0.8272	0.824	0.8209	0.8182	0.8161	0.8148	0.8146	0.8157	0.8185	0.823
0.2	0.8317	0.8324	0.8358	0.842	0.8512	0.8637	0.8797	0.8993	0.9226	0.9495	0.9801
0.3	0.8358	0.8434	0.8563	0.8746	0.8986	0.9284	0.9637	1.0045	1.0502	1.1	1.1529
0.4	0.8428	0.8601	0.8853	0.9186	0.9598	1.0086	1.0642	1.1253	1.1903	1.2569	1.3222
0.5	0.8526	0.8825	0.9228	0.9734	1.0336	1.1019	1.1763	1.254	1.3308	1.4017	1.4605
0.6	0.8653	0.9104	0.9684	1.0383	1.118	1.2045	1.2937	1.3793	1.4539	1.5086	1.5335
0.7	0.8807	0.9439	1.0217	1.1117	1.2102	1.311	1.4063	1.4859	1.5377	1.5484	1.5052
0.8	0.899	0.9827	1.082	1.1922	1.3061	1.4138	1.5021	1.5555	1.5572	1.4915	1.346
0.9	0.92	1.0266	1.1482	1.2768	1.4005	1.5034	1.5661	1.5678	1.4888	1.3156	1.0454
1	0.9438	1.0752	1.2191	1.3624	1.4866	1.5684	1.5821	1.5032	1.315	1.0155	0.6247
1.1	0.9702	1.1278	1.2929	1.4448	1.5564	1.5964	1.5341	1.3473	1.0321	0.6126	0.1476

TABLE 8.9: Measured data for Problem 8.8.

x_i	1	2	3	4	5	6	7	8	9	10
y_i	244.0	221.0	208.0	208.0	211.5	216.0	219.0	221.0	221.5	220.0

Exercise 8.9 *The one-dimensional and two-dimensional data given in Exercises 8.3 and 8.7 can be used for cubic splines and B-splines interpolation. Find the derivatives of the related interpolated functions.*

Exercise 8.10 *Consider again the data in Exercise 8.3. Polynomial fitting can be used to model the data. Select a suitable degree such that good approximation by polynomials can be achieved. Compare the results with interpolation methods.*

Exercise 8.11 *Consider again the data in Exercise 8.3. Assume that the prototype of the function for the data is*

$$y(x) = \frac{1}{\sqrt{2\pi}\sigma}\mathrm{e}^{-(x-\mu)^2/2\sigma^2}.$$

The values of the parameters μ and σ are not known. Use least squares curve fitting methods to see whether suitable μ and σ can be identified. Observe the fitting results.

Exercise 8.12 *The polynomial generating function* `fitting_poly()` *discussed in the text is suitable for producing polynomials in symbolic form, while it is not suitable to polynomial evaluation needed in curve fitting. Please write numerical versions useful in the* `lsqcurvefit()` *function call.*

Exercise 8.13 *Consider the polynomial fitting problem studied in Example 8.20. If Legendre, Chebyshev, Laguerre and Hermit polynomials are used, respectively, please find the suitable orders and the fitting performances.*

Exercise 8.14 *Express the irrational constants* e, $\sqrt{19}$, lg2, $\sin 1°$, *Euler* γ *in terms of continued fractions. Observe how many continued fraction levels are expected to get a suitable approximation.*

Exercise 8.15 *Find good approximations to the functions given below using continued fraction expansions and Padé approximations. Observe the fitting results obtained and find suitable degrees of the rational functions.*

$$\text{(i) } f(x) = \mathrm{e}^{-2x}\sin 5x, \qquad \text{(ii) } f(x) = \frac{x^3+7x^2+24x+24}{x^4+10x^3+35x^2+50x+24}\mathrm{e}^{-3x}.$$

Exercise 8.16 *Assume that the data in Exercise 8.7 satisfies a prototype function of* $z(x,y) = a\sin(x^2y) + b\cos(y^2x) + cx^2 + dxy + e$. *Identify the values of* a, b, c, d, e *with least squares method. Verify the identification results.*

Exercise 8.17 *Please extend the function* `gamma(x)` *such that the Gamma function with complex argument* $\boldsymbol{x}$ *can be evaluated.*

Exercise 8.18 *Evaluate and draw the following Mittag–Leffler functions and verify* (8-5-30).

$$\text{(i) } \mathscr{E}_{1,1}(z), \quad \text{(ii) } \mathscr{E}_{2,1}(z), \quad \text{(iii) } \mathscr{E}_{1,2}(z), \quad \text{(iv) } \mathscr{E}_{2,2}(z)$$

Exercise 8.19 *Prove the following identities involving Mittag–Leffler functions and graphically visualize both sides of each identity for verification purposes.*

$$\text{(i) } \mathscr{E}_{\alpha,\beta}(x) + \mathscr{E}_{\alpha,\beta}(-x) = 2\mathscr{E}_{\alpha,\beta}(x^2), \quad \text{(ii) } \mathscr{E}_{\alpha,\beta}(x) - \mathscr{E}_{\alpha,\beta}(-x) = 2x\mathscr{E}_{\alpha,\alpha+\beta}(x^2),$$

$$\text{(iii) } \mathscr{E}_{\alpha,\beta}(x) = \frac{1}{\Gamma(\beta)} + \mathscr{E}_{\alpha,\alpha+\beta}(x), \quad \text{(iv) } \mathscr{E}_{\alpha,\beta}(x) = \beta\mathscr{E}_{\alpha,\beta+1}(x) + \alpha x\frac{\mathrm{d}}{\mathrm{d}x}\mathscr{E}_{\alpha,\beta+1}(x).$$

Exercise 8.20 *Given a signal* $f(t) = \mathrm{e}^{-3t}\cos(2t+\pi/3) + \mathrm{e}^{-2t}\cos(t+\pi/4)$, *evaluate the formula of the auto-correlation function of the signal. Generate a sequence of randomly distributed data, and verify the results in a numerical way.*

Exercise 8.21 *Evaluate the auto-correlation function for the Gaussian distribution function defined as*

$$f(t) = \frac{1}{\sqrt{2\pi}\times 3}\mathrm{e}^{-t^2/(2\times 3^2)}.$$

Generate a sequence of signals and compare them using Gaussian distributed data to check whether the resulted description is close to the theoretical result.

Exercise 8.22 *Assume that the noisy signal can be generated with the following statements*

```
>> t=0:0.005:5; y=15*exp(-t).*sin(2*t); % generate samples of a signal
   r=0.3*randn(size(y)); y1=y+r;         % add random noises
```

Find the Nyquist frequency of the signal. Based on the Nyquist frequency, design an eighth-order Butterworth filter which can be used to effectively filter out the noise while having a relatively small delay due to filtering.

Exercise 8.23 *High-pass filters can be used to filter out information with low-frequencies, and retain the high-frequency details. Design a high-pass filter for the data shown in Exercise 8.22. From the obtained high frequency noise information, compare the statistical behavior of the noise signal obtained.*

Bibliography

[1] Press W H, Flannery B P, Teukolsky S A, et al. Numerical recipes, the art of scientific computing. Cambridge: Cambridge University Press, 1986. Free textbook at `http://www.nrbook.com/a/bookcpdf.php`

[2] Wikipedia. List of numerical analysis topics

[3] Lancaster L, Šalkauskas K. Curve and surface fitting: an introduction. London: Academic Press, 1986

[4] Bosley M J, Lees F P. A survey of transfer function derivations from higher-order state-variable models. Automatica, 1972, 8:765–775

[5] Wikipedia. Mittag–Leffler function

[6] Podlubny I. Fractional differential equations. San Diego: Academic Press, 1999

[7] Podlubny I. Mittag–Leffler function. MATLAB Central File ID: #8737

[8] Oppenheim A V, Schafer R W. Digital signal processing. Englewood Cliffs: Prentice-Hall, 1975

[9] Xue D. Analysis and computer aided design of nonlinear systems with Gaussian inputs. Ph.D. thesis, Sussex University, U.K., 1992

[10] The MathWorks Inc. Signal processing toolbox user's guide, 2007

Chapter 9

Probability and Mathematical Statistics Problems

The theory of probability and mathematical statistics is a very important branch in experimental sciences. The solutions to probability and mathematical statistics problems sometimes could be quite involved. The traditional statistics usually relies heavily on lookup-tables. A very comprehensive statistics toolbox is provided in MATLAB which contains a lot of handy functions for solving related probability and mathematical statistics problems. In Section 9.1, the basic concepts of *probability density function* (PDF) and *cumulative distribution function* (CDF) are introduced. Given probability distributions, various probability related example problems are solved using Statistics Toolbox. In this section, pseudorandom number generators of different distributions, such as uniform distribution, normal distribution, Poisson distribution, etc., are introduced and demonstrated. In Section 9.2, probability problems are studied, either by theoretic computation, or with histograms, based on experimental data. A very brief introduction to Monte Carlo method is also presented in the section. Some random walk problems are studied with simulation methods. In Section 9.3, the computation of statistical quantities, such as mean, variance, moments and covariances is covered for both univariate and multivariate distributions together. Also, an introduction to outlier detection problems is given in the section. The parametric estimation and interval estimation problems, together with their MATLAB implementations, are given in Section 9.4 where multi-variable linear regression and least squares data fitting problems are also presented. In Section 9.5, the hypothesis test problems including mean value test, normality test and given distribution test, are discussed. The variance analysis problems and applications are presented in Section 9.6 with detailed example solutions in MATLAB. In Section 9.7, principal component analysis approach is briefly introduced, and its applications in dimension reduction is demonstrated through a simple example.

For readers who wish to check the detailed explanations of probability and statistics, we recommend the open source textbook [1] for probability and the online e-textbook [2] for statistics.

9.1 Probability Distributions and Pseudorandom Numbers

9.1.1 Introduction to probability density functions and cumulative distribution functions

The *probability density function* for a continuous stochastic variable is often denoted as $p(x)$, where

$$p(x) \geqslant 0, \text{ and } \int_{-\infty}^{\infty} p(x)\,\mathrm{d}x = 1. \tag{9-1-1}$$

The *cumulative distribution function* $F(x)$ is defined as the probability of an event when $\xi \leqslant x$ happens for the stochastic variable ξ

$$F(x) = \int_{-\infty}^{x} p(\xi)\,\mathrm{d}\xi. \tag{9-1-2}$$

This function is a monotonic increasing function satisfying

$$0 \leqslant F(x) \leqslant 1, \quad \text{and } F(-\infty) = 0, \ F(\infty) = 1. \tag{9-1-3}$$

For multivariate stochastic variable problems, joint probability density function $p(x_1, x_2, \cdots, x_n) \geqslant 0$ can be used, with

$$\int_{-\infty}^{\infty} \cdots \int_{-\infty}^{\infty} p(x_1, x_2, \cdots, x_n)\,\mathrm{d}x_1\,\mathrm{d}x_2 \cdots \mathrm{d}x_n = 1, \tag{9-1-4}$$

and multivariate cumulative distribution function is defined as

$$F(x_1, x_2, \cdots, x_n) = \int_{-\infty}^{x_n} \cdots \int_{-\infty}^{x_1} p(\xi_1, \xi_2, \cdots, \xi_n)\,\mathrm{d}\xi_1\,\mathrm{d}\xi_2 \cdots \mathrm{d}\xi_n. \tag{9-1-5}$$

Example 9.1 Assume that the joint probability density function of two stochastic variables is given below, please determine the parameter c.

$$p(x,y) = \begin{cases} c\,\mathrm{e}^{-(4x+3y)}, & 0 < x < 1, y > 0 \\ 0, & \text{otherwise.} \end{cases}$$

Solution *Bearing in mind the property in* (9-1-4), *the value of c can be determined with the following statements, and it is found* $c = 12\mathrm{e}^4/(\mathrm{e}^4 - 1)$.

```
>> syms c x y; p=c*exp(-(4*x+3*y)); % define probability density function
   P=int(int(p,y,0,inf),x,0,1); c=solve(P==1) % evaluate probability
```

Given the probability $f_i = F(x_i)$, to determine the value of x_i, lookup-table methods can normally be used. Since $F(x_i)$ is monotonic, x_i can easily be found. This problem is also referred to as the *inverse distribution problem*. In fact, functions provided in the Statistics Toolbox can be used to easily, accurately and straightforwardly solve this type of inverse problem. In this section, some of the commonly encountered probability distributions will be summarized and visualized.

9.1.2 Probability density functions and cumulative distribution functions of commonly used distributions

Many similar functions are provided in Statistics Toolbox. The functions with suffixes of `pdf`, `cdf` and `inv` are used to indicate the probability density function, cumulative distribution function and inverse distribution function, respectively. The related functions are summarized in Table 9.1. Other suffixes of the function include `rnd`, `stat` and `fit`, meaning random number generator, statistics analysis and parametric estimation.

TABLE 9.1: Keywords and function names in Statistics Toolbox.

keyword	distribution	parameters	keyword	distribution	parameters
`beta`	Beta distribution	a,b	`bino`	binomial distribution	n,p
`chi2`	χ^2 distribution	k	`ev`	extreme value	μ,σ
`exp`	exponential	λ	`f`	F distribution	p,q
`gam`	Gamma distribution	a,λ	`geo`	geometric	p
`hyge`	hypergeometric	m,p,n	`logn`	lognormal distribution	μ,σ
`mvn`	multivariate normal	$\boldsymbol{\mu},\boldsymbol{\sigma}$	`nbin`	negative binomial	ν_1,ν_2,δ
`ncf`	noncentral F	k,δ	`nct`	noncentral T	k,δ
`ncx2`	noncentral χ^2	k,δ	`norm`	normal distribution	μ,σ
`poiss`	Poisson distribution	λ	`rayl`	Rayleigh distribution	b
`t`	T-distribution	k	`unif`	uniform distribution	a,b
`wbl`	Weibull distribution	a,b			

Apart from these functions, unified functions such as `pdf()` are also provided to evaluate probability density functions of a certain distribution. Similar unified functions such as `cdf()`, `icdf()`, `fitdist()` are also provided and will be presented later in the section.

I. Normal distribution

The probability density function of normal distribution is defined as

$$p(x) = \frac{1}{\sqrt{2\pi}\sigma} e^{-(x-\mu)^2/(2\sigma^2)}, \tag{9-1-6}$$

where μ and σ^2 are, respectively, the mean and variance of the normal distribution. The functions `normpdf()` and `normcdf()` provided in the Statistics Toolbox can be used to evaluate the probability density functions and cumulative distribution functions of such a distribution. Also, if the distribution value $\boldsymbol{F}$ is known, the function `norminv()` can be used to evaluate the corresponding $\boldsymbol{x}$. The syntaxes of these functions are

$\boldsymbol{p}$=`normpdf(`$\boldsymbol{x}$`,`μ`,`σ`)`, $\boldsymbol{F}$=`normcdf(`$\boldsymbol{x}$`,`μ`,`σ`)`, $\boldsymbol{x}$=`norminv(`$\boldsymbol{F}$`,`μ`,`σ`)`

where $\boldsymbol{x}$ is a vector of given points, $\boldsymbol{p}$ vector contains the values of probability density functions and at $\boldsymbol{x}$. The vector $\boldsymbol{F}$ returns the cumulative distribution functions at $\boldsymbol{x}$. Clearly, the normal distribution is a function of the mean μ and the standard deviation σ.

If the unified functions are used, the syntaxes are

$\boldsymbol{p}$=`pdf('norm',`$\boldsymbol{x}$`,`μ`,`σ`)`, $\boldsymbol{F}$=`cdf('norm',`$\boldsymbol{x}$`,`μ`,`σ`)`, $\boldsymbol{x}$=`icdf('norm',`$\boldsymbol{F}$`,`μ`,`σ`)`

Example 9.2 For different combinations of (μ, σ^2) such as $(-1,1), (0,0.1), (0,1)$, $(0,10)$, $(1,1)$, draw the probability density functions and cumulative distribution functions of normal distributions.

Solution *With MATLAB Statistics Toolbox, these density functions can be drawn easily and accurately for any combination of μ and σ^2. Therefore, the use of computer mathematics languages can help the readers to solve effectively the probability and statistics problems.*

The problem can easily be solved with MATLAB. One can assign a vector $\boldsymbol{x}$ composed of points in the interval $(-5,5)$ and two other vectors defined to represent the possible combinations of the variables μ and σ^2. Then, the functions `normpdf()` *and* `normcdf()` *can be used to draw the corresponding probability density functions and cumulative distribution functions, as shown in Figures 9.1 (a) and (b), respectively.*

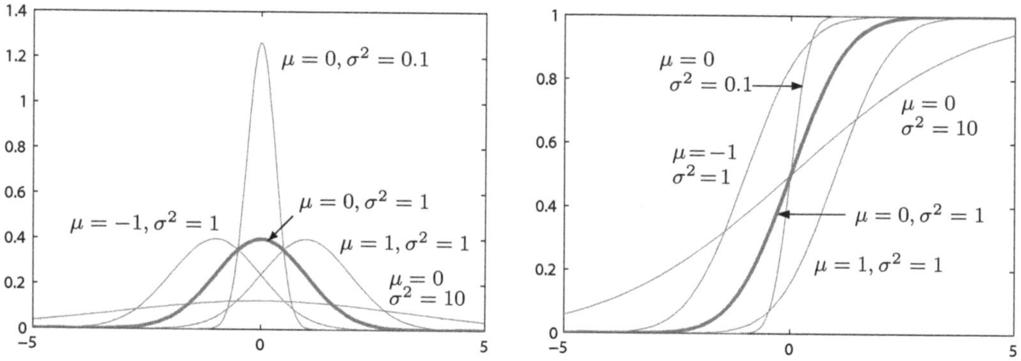

(a) probability density functions (b) cumulative distribution functions

FIGURE 9.1: PDFs and CDFs of normal distribution.

```
>> x=[-5:.02:5]'; p=[]; F=[];
   mu=[-1,0,0,0,1]; sig=[1,0.1,1,10,1]; sig=sqrt(sig);
   for i=1:length(mu) % plots for different parameters μ and σ
      p=[p,normpdf(x,mu(i),sig(i))]; F=[F,normcdf(x,mu(i),sig(i))];
   end
   subplot(121), plot(x,p), subplot(122), plot(x,F)
```

From the probability density function curves, it can be observed that if σ^2 remains the same, the shapes of the probability density functions are exactly the same as if they are obtained by translating the curves according to μ. If the values of σ are different, the shapes are also different. The smaller the value of σ^2, the narrower the curve of probability density functions.

The combination of $\mu = 0, \sigma^2 = 1$ is referred to as the standard normal distribution, *which is usually denoted by $N(0,1)$.*

II. Poisson distribution

The Poisson distribution is different from the continuous distribution functions discussed earlier. It requires that vector $\boldsymbol{x}$ is composed of positive integers. The probability density function of the Poisson distribution is

$$p(x) = \frac{\lambda^x}{x!}\mathrm{e}^{-\lambda x},\ x = 0, 1, 2, 3, \cdots, \tag{9-1-7}$$

where λ is a positive integer.

Poisson distribution is a function of positive integer parameter λ. The related functions `poisspdf()`, `poisscdf()` and `poissinv()` are provided in Statistics Toolbox to evaluate the probability density functions, cumulative distribution functions and inverse distribution function, respectively. The syntaxes of the function are

p=`poisspdf(`x,λ`)`, F=`poisscdf(`x,λ`)`, x=`poissinv(`F,λ`)`

where the vector x contains given integers, which can be specified with $x = 0\!:\!k$. The values in p are the probability density function corresponding to the points in the vector x. The vector F returns the cumulative distribution function corresponding to vector x.

The unified functions are used in the following syntaxes

p=`pdf('poiss',`x,λ`)`, F=`cdf('poiss',`x,λ`)`, x=`icdf('poiss',`F,λ`)`

Example 9.3 For the parameter λ selected as $\lambda = 1, 2, 5, 10$, draw the probability density functions and cumulative distribution functions of the Poisson distributions.

Solution *One can create a vector of* x *first, and then, for different values of* λ*, the functions* `poisspdf()` *and* `poisscdf()` *can be used in a loop structure. The probability density functions and cumulative distribution functions can be obtained as shown in Figures 9.2 (a), (b), respectively.*

```
>> x=[0:15]'; p=[]; F=[]; lam1=[1,2,5,10];
   for i=1:length(lam1) % plots for different values of parameter λ
      p=[p,poisspdf(x,lam1(i))]; F=[F,poisscdf(x,lam1(i))];
   end
   subplot(121), plot(x,p), subplot(122), plot(x,F)
```

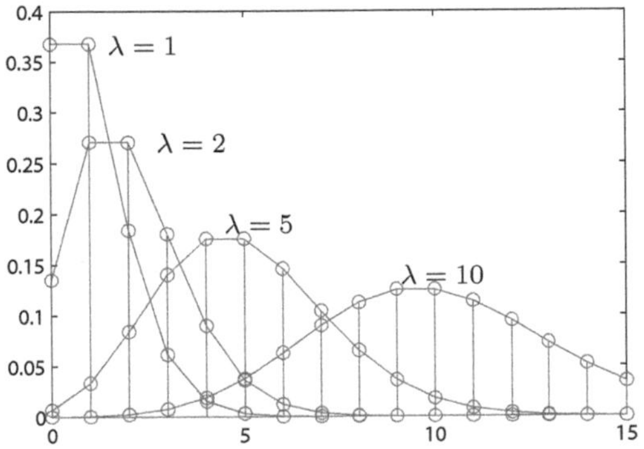

(a) probability density functions

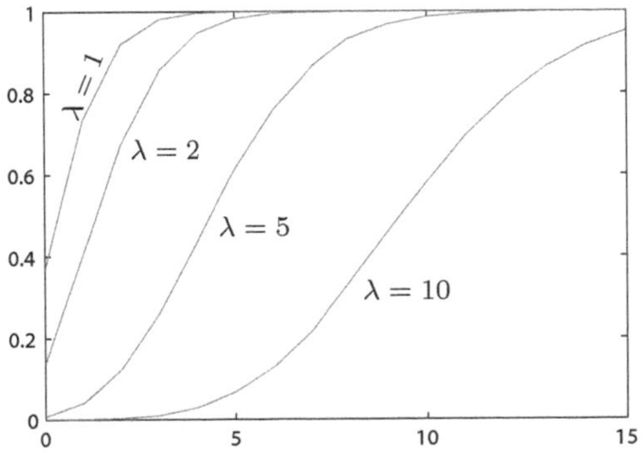

(b) cumulative distribution functions

FIGURE 9.2: PDFs and CDFs of the Poisson distribution.

III. Rayleigh distribution

The probability density function of Rayleigh distribution is defined as

$$p(x) = \begin{cases} \dfrac{x}{b^2} e^{-x^2/(2b^2)}, & x \geqslant 0 \\ 0, & x < 0, \end{cases} \tag{9-1-8}$$

which contains a parameter b. The functions `raylpdf()`, `raylcdf()` and `raylinv()` provided in the Statistics Toolbox can be used to evaluate the relevant functions. Also, the unified functions can be used. The syntaxes of these functions are

```
p=raylpdf(x,b), F=raylcdf(x,b), x=raylinv(F,b)
p=pdf('rayl',x,b), F=cdf('rayl',x,b), x=icdf('rayl',F,b)
```

Example 9.4 Draw the probability density functions and cumulative distribution functions of the Rayleigh distribution for $b = 0.5, 1, 3, 5$.

Solution *Assign a vector $\boldsymbol{x}$ in the interval $(-0.1, 5)$, and also a parameter vector $\mathbf{b}_1$, the* `raylpdf()` *and* `raylcdf()` *functions can be used to draw probability density functions and cumulative distribution functions, as shown in Figures 9.3 (a) and 9.3 (b), respectively.*

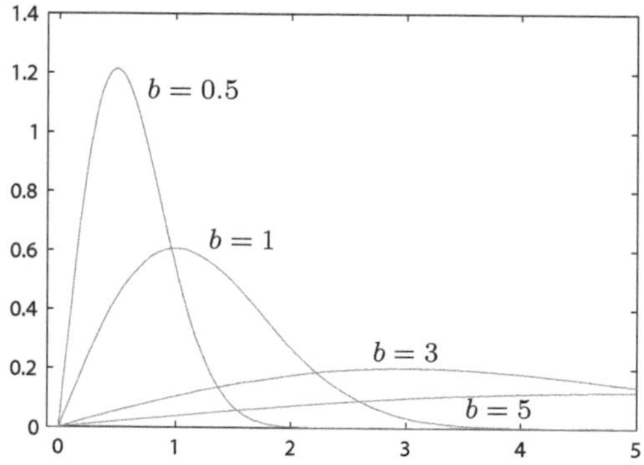

(a) probability density functions

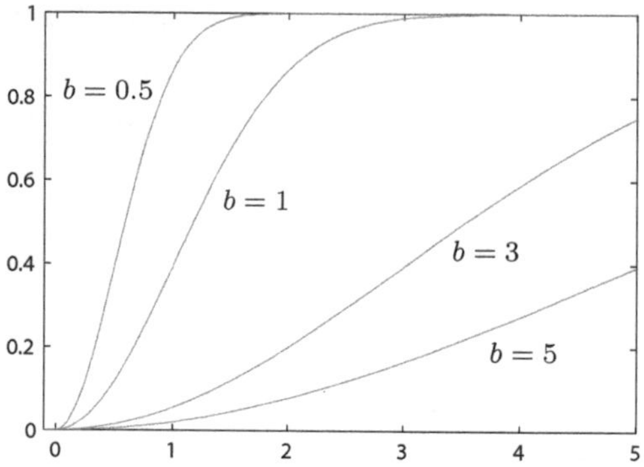

(b) cumulative distribution functions

FIGURE 9.3: PDFs and CDFs of the Rayleigh distribution.

```
>> x=[-eps:-0.02:-0.05,0:0.02:5]; x=sort(x'); b1=[.5,1,3,5]; p=[]; F=[];
   for i=1:length(b1) % PDF and CDF plots for different values of b
      p=[p,raylpdf(x,b1(i))]; F=[F,raylcdf(x,b1(i))];
   end
   subplot(121), plot(x,p), subplot(122), plot(x,F)
```

IV. Gamma distribution

The probability density function of the Gamma distribution is defined as

$$p(x) = \begin{cases} \dfrac{\lambda^a x^{a-1}}{\Gamma(a)} \mathrm{e}^{-\lambda x} & x \geqslant 0 \\ 0 & x < 0. \end{cases} \tag{9-1-9}$$

The probability density function of Gamma distribution is a function of the parameters a and λ. The functions `gampdf()`, `gamcdf()` and `gaminv()` and the unified functions can be used with the following syntaxes

```
p=gampdf(x,a,λ), F=gamcdf(x,a,λ), x=gaminv(F,a,λ)
p=pdf('gam',x,a,λ), F=cdf('gam',x,a,λ), x=icdf('gam',F,a,λ)
```

Example 9.5 For the parameter combinations (a, λ) as $(1, 1), (1, 0.5), (2, 1), (1, 2), (3, 1)$, draw the probability density functions and cumulative distribution functions of the Gamma distribution.

Solution *A vector* $\boldsymbol{x}$ *can be established over the interval* $(-0.5, 5)$*, two variable vectors are also defined as* `a1` *and* `lam1`*, then, the functions* `gampdf()` *and* `gamcdf()` *can be used to plot the probability density functions and cumulative distribution functions as shown in Figures 9.4 (a) and (b), respectively.*

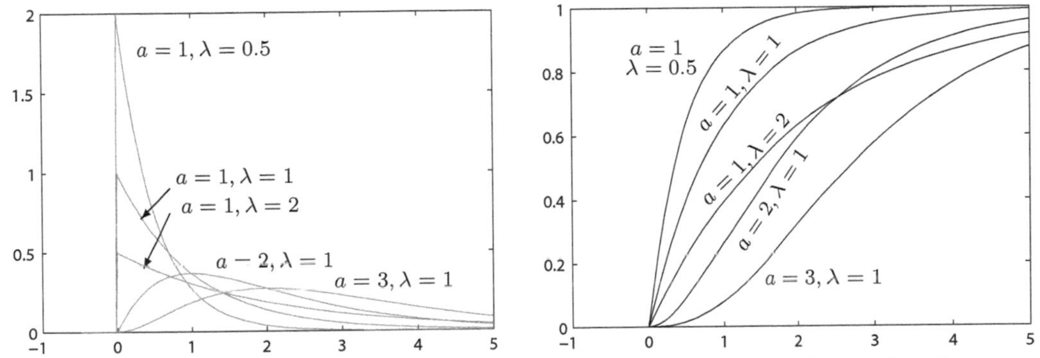

(a) probability density functions (b) cumulative distribution functions

FIGURE 9.4: PDFs and CDFs of the Gamma distribution.

```
>> x=[-0.5:.02:5]'; p=[]; F=[]; a1=[1,1,2,1,3]; lam1=[1,0.5,1,2,1];
   for i=1:length(a1)
      p=[p,gampdf(x,a1(i),lam1(i))]; F=[F,gamcdf(x,a1(i),lam1(i))];
   end
   subplot(121), plot(x,p), subplot(122), plot(x,F)
```

There is a minor problem in the probability density function curves. Since a step-size of 0.02 is selected to form the $\boldsymbol{x}$ *vector, the probability density function will jump from 0 to the maximum value. To avoid this, the* $\boldsymbol{x}$ *vector generating statement should be modified to*

```
>> x=[eps, 0:0.02:5]; x=sort(x'); % insert ε in vector
```

V. χ^2 distribution

The probability density function of the χ^2 distribution is defined as

$$p(x) = \begin{cases} \dfrac{1}{2^{k/2}\Gamma(k/2)} x^{k/2-1} \mathrm{e}^{-x/2}, & x \geqslant 0 \\ 0, & x < 0, \end{cases} \tag{9-1-10}$$

where the parameter k is a positive integer. It can also be seen that the χ^2 function is a special Gamma distribution, with $a = k/2$ and $\lambda = 1/2$. The PDF of the χ^2 distribution is also a function of k. The functions `chi2pdf()`, `chi2cdf()` and `chi2inv()` can be used to evaluate relevant functions for χ^2 distribution. The syntaxes of the functions are

```
p=chi2pdf(x,k), F=chi2cdf(x,k), x=chi2inv(F,k)
p=pdf('chi2',x,k), F=cdf('chi2',x,k), x=icdf('chi2',F,k)
```

Example 9.6 For the parameter k selected as $1, 2, 3, 4, 5$, draw the probability density functions and cumulative distribution functions of χ^2 distributions.

Solution *Select a vector $\boldsymbol{x}$ within the interval $(-0.05, 1)$, and also define a vector $\boldsymbol{k}_1$. Call the functions* `chi2pdf()` *and* `chi2cdf()` *directly. The PDFs and CDFs can be drawn as shown in Figures 9.5 (a) and (b), respectively.*

```
>> x=[-eps:-0.02:-0.05,0:0.02:2]; x=sort(x'); k1=1:5; p=[]; F=[];
   for i=1:length(k1)
      p=[p,chi2pdf(x,k1(i))]; F=[F,chi2cdf(x,k1(i))];
   end
   subplot(121), plot(x,p), subplot(122), plot(x,F)
```

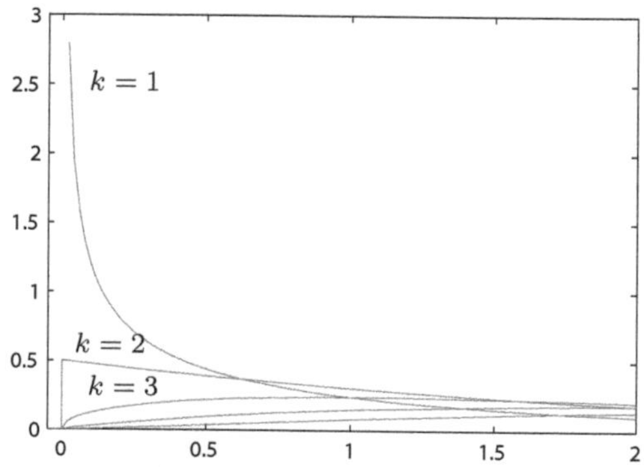

(a) probability density functions

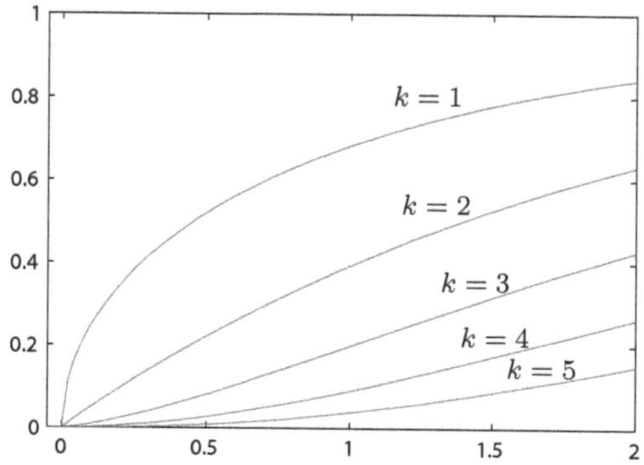

(b) cumulative distribution functions

FIGURE 9.5: PDFs and CDFs of the χ^2 distribution.

VI. T distribution

The probability density function of the T distribution is defined as

$$p(x) = \frac{\Gamma\left((k+1)/2\right)}{\sqrt{k\pi}\,\Gamma\left(k/2\right)}\left(1 + \frac{x^2}{k}\right)^{-(k+1)/2}, \tag{9-1-11}$$

with a positive integer parameter k. The functions `tpdf()`, `tcdf()` and `tinv()` can be used to evaluate relevant functions, with the syntaxes

$\boldsymbol{p}$=`tpdf(`$\boldsymbol{x}$`,`k`)`, $\boldsymbol{F}$=`tcdf(`$\boldsymbol{x}$`,`k`)`, $\boldsymbol{x}$=`tinv(`$\boldsymbol{F}$`,`k`)`

$\boldsymbol{p}$=`pdf('t',`$\boldsymbol{x}$`,`k`)`, $\boldsymbol{F}$=`cdf('t',`$\boldsymbol{x}$`,`k`)`, $\boldsymbol{x}$=`icdf('t',`$\boldsymbol{F}$`,`k`)`

Example 9.7 Draw the probability density functions and cumulative distribution functions of T distribution for $k = 1, 2, 5, 10$.

Solution *Again an $\boldsymbol{x}$ vector is defined over the interval $(-5, 5)$. Another vector $\boldsymbol{k}_1$ can also be established and with the following statements, the probability density functions and cumulative distribution functions can be obtained as shown in Figures 9.6 (a) and (b), respectively.*

```
>> x=[-5:0.02:5]'; k1=[1,2,5,10]; p=[]; F=[];
   for i=1:length(k1)
      p=[p,tpdf(x,k1(i))]; F=[F,tcdf(x,k1(i))];
   end
   subplot(121), plot(x,p), subplot(122), plot(x,F)
```

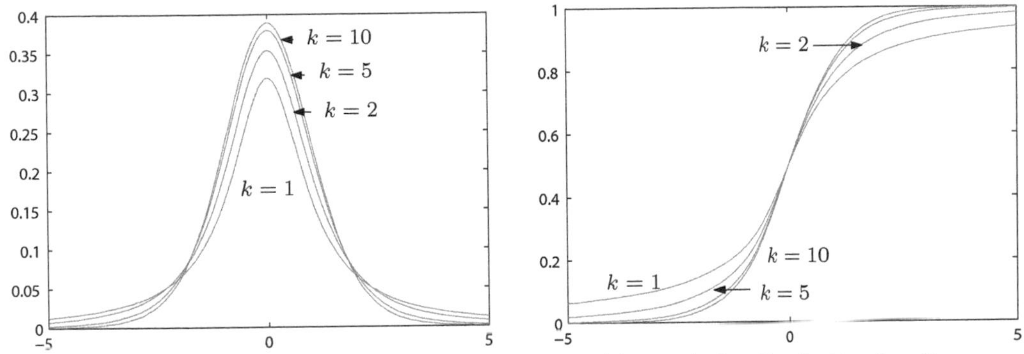

(a) probability density functions (b) cumulative distribution functions

FIGURE 9.6: PDFs and CDFs of the T distribution.

VII. F distribution

The probability density function of the F distribution is defined as

$$p(x) = \begin{cases} \dfrac{\Gamma\left((p+q)/2\right)}{\Gamma\left(p/2\right)\Gamma\left(q/2\right)} p^{p/2} q^{q/2} x^{p/2-1} (p+qx)^{-(p+q)/2}, & x \geqslant 0 \\ 0, & x < 0, \end{cases} \tag{9-1-12}$$

and the probability density function is a function of parameters p and q, with p and q positive integers. The syntaxes of the functions are

$\boldsymbol{p}$=fpdf($\boldsymbol{x}$,a,b), $\boldsymbol{F}$=fcdf($\boldsymbol{x}$,p,q), $\boldsymbol{x}$=finv($\boldsymbol{F}$,p,q)
$\boldsymbol{p}$=pdf('f',$\boldsymbol{x}$,a,b), $\boldsymbol{F}$=cdf('f',$\boldsymbol{x}$,p,q), $\boldsymbol{x}$=icdf('f',$\boldsymbol{F}$,p,q)

Example 9.8 Draw the probability density functions and cumulative distribution functions for the T distributions with the (p,q) pairs as $(1,1),(2,1),(3,1),(3,2),(4,1)$.

Solution *Define a vector $\boldsymbol{x}$ in the interval $(-0.1,1)$, and two other vectors $\boldsymbol{p}_1$ and $\boldsymbol{q}_1$ can be defined for the given pairs. The probability density functions and cumulative distribution functions can be obtained with the functions* `fpdf()` *and* `fcdf()`*, as shown in Figures 9.7 (a) and (b), respectively.*

```
>> x=[-eps:-0.02:-0.05,0:0.02:1]; x=sort(x');
   p1=[1 2 3 3 4]; q1=[1 1 1 2 1]; p=[]; F=[];
   for i=1:length(p1)
      p=[p,fpdf(x,p1(i),q1(i))]; F=[F,fcdf(x,p1(i),q1(i))];
   end
   subplot(121), plot(x,p), subplot(122), plot(x,F)
```

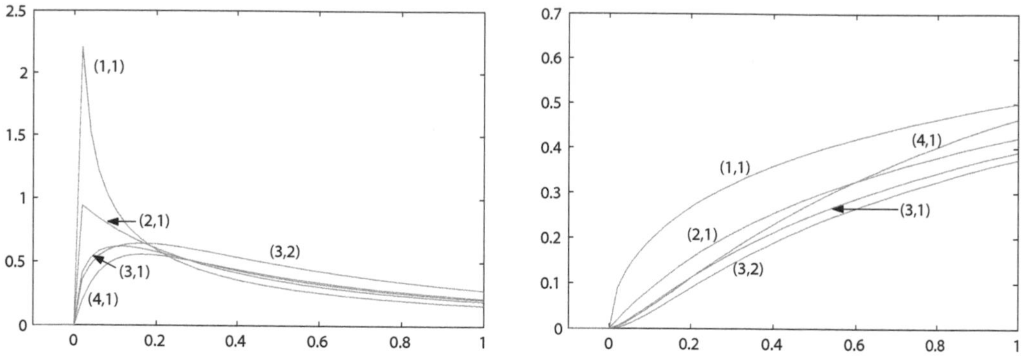

(a) probability density functions (b) cumulative distribution functions

FIGURE 9.7: PDFs and CDFs of the F distribution.

VIII. Alpha-stable distribution

The probability density function of *alpha-stable distribution* is [3]

$$p(x) = \exp\left\{-\sigma^\alpha |x|^\alpha \left[1 - \mathrm{j}\beta\,\mathrm{sign}(x)\tan\frac{\pi\alpha}{2}\right] + \mathrm{j}\mu x\right\}, \tag{9-1-13}$$

where, $0 < \alpha \leqslant 2$, also known as *index of stability*, $-1 < \beta < 1$ is known as the *skewness parameter*, $\sigma > 0$ is the *scaling factor* and μ is the *shift*, or *location parameter*. When $\beta = \mu = 0$, the distribution is symmetrical.

When $\alpha = 1$, the distribution is a symmetric Cauchy distribution, whose probability density function is represented as

$$p(x) = \exp\left\{-\sigma|x| \left[1 - \frac{2\mathrm{j}}{\pi}\beta\,\mathrm{sign}(x)\ln|x|\right] + \mathrm{j}\mu x\right\}. \tag{9-1-14}$$

A set of alpha-stable functions in MATLAB is written and released by Mark Veillette [4], the relevant names of the functions are `stblpdf()`, `stblcdf()`, `stblinv()`, `stblfit()` and `stblrnd()`. The syntaxes of the functions are

$\boldsymbol{p}$=`stblpdf(`$\boldsymbol{x}$`,`α`,`β`,`σ`,`μ`)`, $\boldsymbol{F}$=`stblcdf(`$\boldsymbol{x}$`,`α`,`β`,`σ`,`μ`)`, $\boldsymbol{x}$=`stblinv(`$\boldsymbol{F}$`,`α`,`β`,`σ`,`μ`)`

It should be noted that since the functions are not those provided in the Statistics Toolbox, the functions `pdf()`, `cdf()` cannot be used in the evaluation of the distribution. alpha-stable stable distribution is useful in modeling and simulation of stochastic processes with heavy tails, and it is useful in many fields such as finance and economics.

Example 9.9 Please observe the probability density functions of alpha-stable distribution under different combinations of the parameters.

Solution *Assume that $\beta = \mu = 0$, $\sigma = 1$, and change the values of α, the probability density function curves can be obtained as shown in Figure 9.8 (a). It can be seen that the curves obtained are symmetrical. If $\beta = 0.5$, the probability density functions can also be obtained, as shown in Figure 9.8 (b), and it can be seen that the distributions are skewed.*

```
>> x=-5:0.01:5; b1=0; b2=0.5; m=0; s=1; Y1=[]; Y2=[];
   for a=0.5:0.25:1.5, % try different values of α
```

```
   Y1=[Y1; stblpdf(x,a,b1,s,m)]; Y2=[Y2; stblpdf(x,a,b2,s,m)];
end
subplot(121), plot(x,Y1); subplot(122), plot(x,Y2)
```

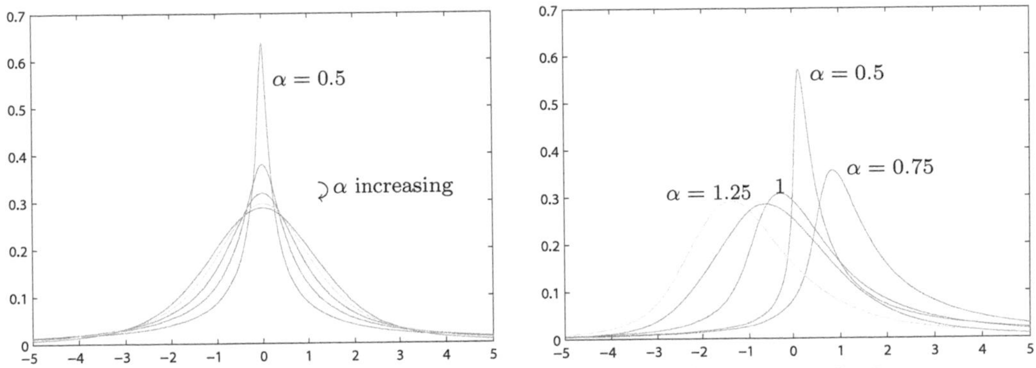

(a) symmetric distribution (b) skewed distribution

FIGURE 9.8: PDFs of alpha-stable distribution.

9.1.3 Random numbers and pseudorandom numbers

In scientific research and statistical analysis, random data are always expected. There are two ways of generating random numbers. One is to generate the random numbers by specific electronic devices, which is referred to as the *physical method*. The other method is to generate random numbers using mathematical algorithms. The random numbers generated by this method are calculated from corresponding mathematical formulae, and the random numbers generated are referred to as *pseudorandom numbers*.

There are two advantages in utilizing pseudorandom numbers. One is that the random numbers are repeatable for the same random number seed, which makes repetitive experiments possible. The other advantage is that the distributions of the random numbers generated can be specified as needed. For instance, random numbers satisfying uniform distribution, normal distribution or Poisson distribution, etc. can be selected by the user.

Two functions, `rand()` and `randn()`, have been introduced in Section 4.1 for generating uniformly distributed and normally distributed random matrices. Apart from these two types of random numbers, other types of random numbers are summarized in Table 9.1. In particular, several commonly used pseudorandom number generators are list below.

```
A=gamrnd(a,λ,n,m) % generates an n × m random matrix satisfying Γ
B=chi2rnd(k,n,m)  % generates pseudorandom numbers satisfying χ²
C=trnd(k,n,m)     % T distribution
D=frnd(p,q,n,m)   % F distribution
E=raylrnd(b,n,m)  % Rayleigh distribution
```

A unified function `random()` is used to generate pseudorandom numbers, with the syntax $\boldsymbol{N}$=`random(`*type*`,`*parameters*`,`n`,`m`)`, where, *type* and *parameters* are the same as the ones in Table 9.1. For instance, the above $\boldsymbol{A}$ and $\boldsymbol{C}$ matrices can be generated alternatively with the function

$\boldsymbol{A}$=random('gam',a,λ,n,m), $\boldsymbol{C}$=random('t',k,n,m)

Function rng() can be used to control the generation of pseudorandom numbers, and the control variables can be obtained with c=rng, where c is a structured variable, with many useful fields. For instance, the field Seed is useful to select the seed in pseudorandom number generation. With such a member, the same set of random numbers can be generated if necessary, so that repetitive experiments are possible.

Example 9.10 Please generate two sets of identical 1000 T distribution pseudorandom numbers.

Solution *If the function* random() *is called twice, with the same commands, the random numbers generated are definitely different. This is not good in doing repetitive experiments. If two sets of identical random numbers are expected, it is useful to save the data. Alternatively, it is better to have the same type of settings in the random number generators. The following commands can be used to generate the expected identical pseudorandom numbers.*

```
>> c=rng; A1=random('t',1,1,1000); rng(c); % use the same setting
   A2=random('t',1,1,1000); norm(A1-A2)    % the difference should be zero
```

9.2 Solving Probability Problems

Probability is the measure of the likeliness that an event will occur. The computation of probability is illustrated in this section for discrete numbers and continuous functions. Different kinds of graphical representation approaches are demonstrated in the section. Also, Monte Carlo algorithm and random walk simulations are presented.

9.2.1 Histogram and pie representation of discrete numbers

Assume that a set of data $x_1, x_2, \cdots, x_n$ are measured, and it is known that the numbers are in the interval (a, b). Sometimes, it is useful to divide the interval into m bins, such that $b_1 = a$, $b_{m+1} = b$. The number k_j, $j = 1, 2, \cdots, m$ of the samples of x_i falling in each bin (b_j, b_{j+1}) can be counted. The entities $f_j = k_j/n$ are referred to as the *frequency*.

With MATLAB function hist(), the frequencies can be obtained with

$\boldsymbol{k}$=hist($\boldsymbol{x}$,$\boldsymbol{b}$); $\boldsymbol{f}$=$\boldsymbol{k}$/n; bar($\boldsymbol{b}$,$\boldsymbol{f}$); or pie($\boldsymbol{f}$)

With the vectors $\boldsymbol{b}$ and $\boldsymbol{f}$, the histogram and pie chart can also be obtained. These kinds of graphical interpretations will be demonstrated through the following examples.

Example 9.11 Generate a 30000×1 vector with numbers satisfying Rayleigh distribution for $b = 1$. Verify the probability density function of the random numbers with histograms.

Solution *From the above presentations, the function* raylrnd() *can be used to generate a vector containing* 30000×1 *random values satisfying Rayleigh distribution. Assigning a vector* xx, *the function* hist() *can be used to evaluate the numbers in each bin defined in vector* xx. *In fact, the actual bin should be generated by adding half the width of bin. The probability density function can be approximated with the* bar() *function. Also, the exact result can be obtained for comparison as shown in Figure 9.9. It can be seen that the generated data correspond to the Rayleigh distribution very well.*

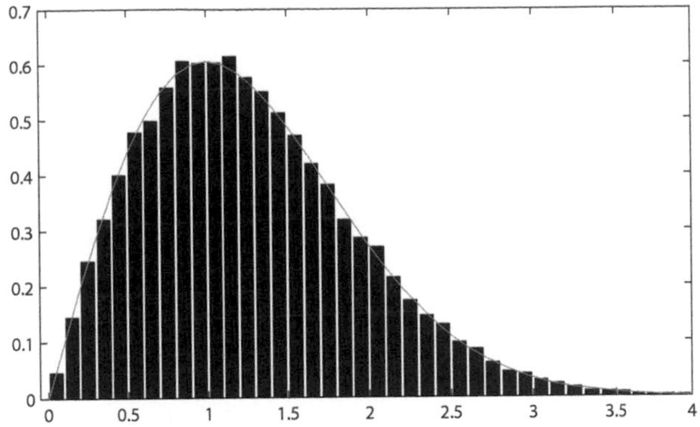

FIGURE 9.9: Distribution evaluation with Rayleigh functions.

```
>> b=1; p=raylrnd(1,30000,1); xx=0:.1:4; x1=xx+0.05; yy=hist(p,x1);
   yy=yy/(30000*0.1); bar(x1,yy), y=raylpdf(xx,1); line(xx,y)
```

Example 9.12 Assume that a set of 200 data are measured to the lifetime of incandescent lamps in given in Table 9.2 [5]. It can be seen that the data are distributed in the intervals $(500, 1500)$. Please create suitable bins and represent the frequencies with histogram and pie charts.

TABLE 9.2: Life in hours of 200 incandescent lamps (Data from [5]).

1067	919	1196	785	1126	936	918	1156	920	948	855	1092	1162	1170	929
950	905	972	1035	1045	1157	1195	1195	1340	1122	938	970	1237	956	1102
1022	978	832	1009	1157	1151	1009	765	958	902	923	1333	811	1217	1085
896	958	1311	1037	702	521	933	928	1153	946	858	1071	1069	830	1063
930	807	954	1063	1002	909	1077	1021	1062	1157	999	932	1035	944	1049
940	1122	1115	833	1320	901	1324	818	1250	1203	1078	890	1303	1011	1102
996	780	900	1106	704	621	854	1178	1138	951	1187	1067	1118	1037	958
760	1101	949	992	966	824	653	980	935	878	934	910	1058	730	980
844	814	1103	1000	788	1143	935	1069	1170	1067	1037	1151	863	990	1035
1112	931	970	932	904	1026	1147	883	867	990	1258	1192	922	1150	1091
1039	1083	1040	1289	699	1083	880	1029	658	912	1023	984	856	924	801
1122	1292	1116	880	1173	1134	932	938	1078	1180	1106	1184	954	824	529
998	996	1133	765	775	1105	1081	1171	705	1425	610	916	1001	895	709
610	916	1001	895	709	860	1110	1149	972	1002					

Solution *For clarity in presentation, the data are stored in a data file, c9dlamp.dat. The data can be imported into MATLAB workspace with* `load()` *function. The bins can be selected as* $[500, 600, 700, \cdots, 1500]$. *In fact, since the width of each bin is 100, half of the width should be added to the bins to compute their centers before using* `hist()` *function, to count the frequency. The histogram and pie chart of the data can be obtained as shown in Figure 9.10.*

```
>> A=load('c9dlamp.dat'); bins=[500:100:1500]+50;     % load data from file
```

```
f=hist(A,bins)/length(A); subplot(121), bar(bins,f) % draw histogram
subplot(122), pie(f)                                % draw pie chart
```

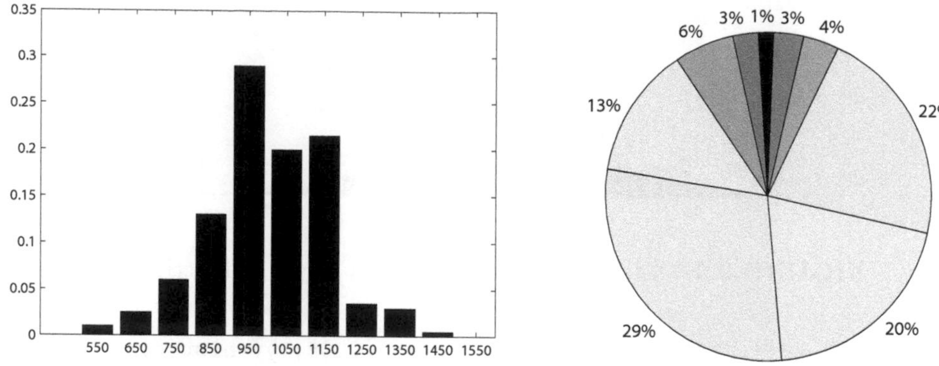

(a) histogram representation

(b) pie chart

FIGURE 9.10: Histogram and pie chart of the lamp lifetime.

9.2.2 Probability computation of continuous functions

It has been shown earlier that the definition of the cumulative distribution function $F(x)$ is the probability of a stochastic variable ξ falling in the interval $(-\infty, x)$. For instance, the probability of the random number ξ falling in the interval $[x_1, x_2]$ can be represented by $P[x_1 \leqslant \xi \leqslant x_2]$. The following formulas can be used to evaluate the probabilities

$$\begin{cases} P[\xi \leqslant x] = F(x), & \text{\% probability of } \xi \leqslant x \\ P[x_1 \leqslant \xi \leqslant x_2] = F(x_2) - F(x_1), & \text{\% probability of } x_1 \leqslant \xi \leqslant x_2 \\ P[\xi \geqslant x] = 1 - F(x), & \text{\% probability of } \xi \geqslant x, \end{cases} \tag{9-2-1}$$

where the cumulative distribution functions are given.

Example 9.13 Assume that a stochastic variable x satisfies Rayleigh distribution, with $b = 1$. Compute the probabilities when the variable x falls in the intervals $[0.2, 2]$ and $[1, \infty)$.

Solution *The probabilities of the stochastic variable x falling in the intervals $(-\infty, 0.2]$ and $(-\infty, 2]$ can easily be found. Therefore, the probability of variable x falling into the interval $[0.2, 2]$ can be evaluated from*

```
>> b=1; p1=raylcdf(0.2,b); p2=raylcdf(2,b); P1=p2-p1
```

and the probability is $P_1 = 0.844863$. Also, the probability of the variable x falling in the interval $(1, \infty]$ can alternatively be evaluated from

```
>> p1=raylcdf(1,b); P2=1-p1
```

and the probability can be found as $P_2 = 0.606531$.

For multivariate cases, joint probability density function $p(x_1, x_2, \cdots, x_n)$ is used. The probability of the stochastic variables ξ_i falling in the area $\xi_i \leqslant a_i$ can be evaluated from

$$P(\boldsymbol{\xi} \leqslant \boldsymbol{a}) = \int_{-\infty}^{a_1} \int_{-\infty}^{a_2} \cdots \int_{-\infty}^{a_n} p(\xi_1, \xi_2, \cdots, \xi_n) \, d\xi_n \cdots d\xi_2 \, d\xi_1. \tag{9-2-2}$$

Example 9.14 Assume that the joint probability density function of the two stochastic variables ξ and η is given by

$$p(x,y) = \begin{cases} x^2 + \dfrac{xy}{3}, & 0 \leqslant x \leqslant 1,\ 0 \leqslant y \leqslant 2 \\ 0, & \text{otherwise.} \end{cases}$$

Determine the probability $P(\xi < 1/2, \eta < 1/2)$.

Solution *From the given joint probability density function $p(x,y)$, the probability $P(\xi < x_0, \eta < y_0)$ can be evaluated with the following statements*

```
>> syms x y; f=x^2+x*y/3; P=int(int(f,x,0,1/2),y,0,1/2)
```

and it can be found that the probability is $P = 5/192$. In the original problem, $p(x,y)$ is zero when the independent variables x and/or y are zero. Therefore, in the integrals, only the first quadrant is considered.

Example 9.15 Assume that between locations A and B, there are six sets of traffic lights. The probability of a red light at each set of traffic lights is the same, with $p = 1/3$. Suppose that the number of red traffic lights on the road for locations A to B satisfies a binomial distribution $B(6,p)$. Find the probability of the event that one meets only once the red traffic light from A to B. Also, varying the value of p to plot the probability functions.

Solution *Since it is known that at each traffic light, the probability of meeting a red light satisfies a binomial distribution, and it can be calculated with functions* `binopdf()` *or* `pdf()`. *Assume that the number of meeting red traffic light is x, then, the values of x can be $0, 1, 2, \cdots, 6$, and the corresponding probability density can be evaluated with*

```
>> x=0:6; y=binopdf(x,6,1/3) % evaluate PDF for binomial distribution
```

and it can be found that $\boldsymbol{y} = [0.0878, 0.2634, 0.3292, 0.2195, 0.0823, 0.0165, 0.0014]$. It can be seen that, the probability of meeting a red light once is 0.2634, and the probability of meeting a red light at least once can be evaluated in either of the two ways, one is subtract from 1 the probability when no red light is met, while the other is to sum up all but the first entity of $\boldsymbol{y}$ vector. It can be found that the probability is $p = 91.22\%$.

```
>> P=1-y(1)  % or to sum up P=sum(y(2:end))
```

If the value of p in the binomial distribution changes, loop structure can be used to calculate the probability of meeting at least once, as shown in Figure 9.11.

```
>> p0=0.05:0.05:0.95; y=[];
   for p=p0, y=[y 1-binopdf(0,6,p)]; end, plot(p0,y,1/3,P,'o')
```

9.2.3 Monte Carlo solutions to mathematical problems

Buffon's needle is an old example of the statistical experiment, carried out by a French mathematician Georges-Louis Leclerc, Comte de Buffon (1707–1788) in the 18th century. He dropped a needle onto a floor, made of parallel strips of wood, and counted the number of times the needle lie across a line between two strips.

The so-called *Monte Carlo method* is another interesting statistical experiment to find the values of uncertain variables using a large number of random quantities, based on random distributions.

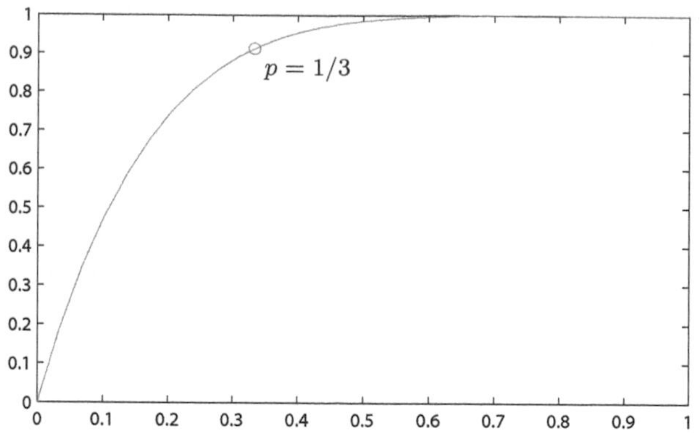

FIGURE 9.11: Probability of having at least one red light with different p.

Consider the plot shown in Figure 9.12 (a). Generate N sets of two random numbers x_i and y_i satisfying uniform distribution in the interval $[0, 1]$. Assume that the number of points falling in the quarter circle, i.e., $x_i^2+y_i^2 \leqslant 1$, is N_1. Since the area of the quarter circle is $\pi/4$, while the area of the square is 1, It is known from probability theory that the probability of a random point falls in the quarter circle is $\pi/4$. If the number of random points is large enough, it can be seen that $N_1/N \approx \pi/4$, from which it is found that $\pi \approx 4N_1/N$. The value of π can be approximated using such a formula.

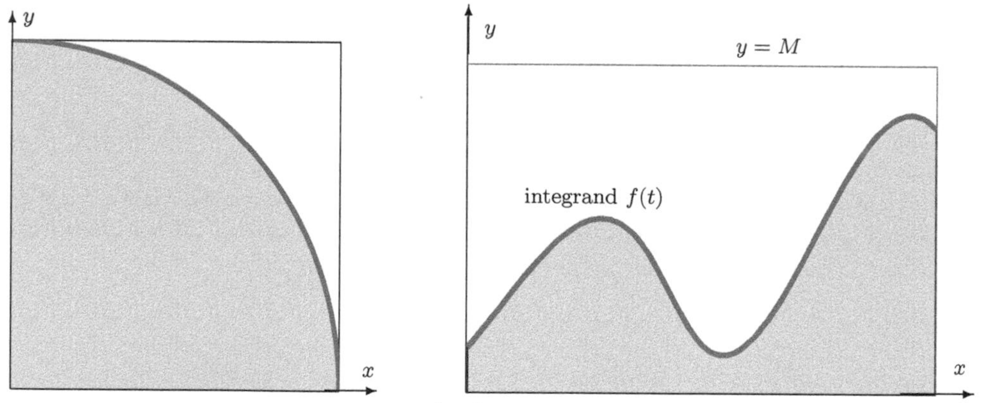

FIGURE 9.12: Applications of Monte Carlo method.

Example 9.16 Use Monte Carlo method to evaluate approximately the value π.

Solution *The following MATLAB statements can be used*

```
>> N=100000; x=rand(1,N); y=rand(1,N); i=(x.^2+y.^2)<=1; % random points
   N1=sum(i); p=N1/N*4 % approximate the value of π with Monte Carlo method
```

and it is found that $\pi \approx 3.1418$. *If the value of* N *further increases, the accuracy may or may not increase. More importantly, never expect to find the exact value of* π *in this way.*

Consider also the definite integral problem $\displaystyle\int_a^b f(x)\,dx$ shown in Figure 9.12 (b). Again generate N sets of two random numbers x_i and y_i satisfying the uniform distribution in the intervals $[a, b]$ and $(0, M)$, respectively. For a value of x_i, the numbers of points satisfying $f(x_i) \leqslant y_i$ is counted as N_1. It is found that $\dfrac{N_1}{N} \approx \dfrac{1}{M(b-a)} \displaystyle\int_a^b f(x)\,\mathrm{d}x$, from which the integral can be approximated with [6]

$$\int_a^b f(x)\,\mathrm{d}x \approx \frac{M(b-a)N_1}{N}. \tag{9-2-3}$$

It should be noted that the evaluation is only possible when $M \geqslant f(x) \geqslant 0$, otherwise, modified representations should be used.

Example 9.17 Evaluate $\displaystyle\int_1^3 \left[1 + \mathrm{e}^{-0.2x}\sin(x+0.5)\right]\mathrm{d}x$ using Monte Carlo method.

Solution *The following statements can be used, and the integral can be evaluated numerically and analytically such that* $p = 2.7395$, $I = 2.74393442001018$.

```
>> f=@(x)1+exp(-0.2*x).*sin(x+0.5); a=1; b=3; M=2; N=100000;
   x=a+(b-a)*rand(N,1); y=M*rand(N,1); i=y<=f(x); N1=sum(i);
   p=M*N1*(b-a)/N  % approximately compute integral with Monte Carlo method
   syms x; I=vpa(int(1+exp(-0.2*x)*sin(x+0.5),x,a,b)) % direct integral
```

9.2.4 Simulation of random walk processes

Brownian motion is an interesting stochastic process. Suppose there is a particle in a two-dimensional plane, it moves randomly in speed and direction. For simplicity, the position of it can be evaluated recursively from

$$x_{i+1} = x_i + \sigma\Delta x_i,\ y_{i+1} = y_i + \sigma\Delta y_i, \tag{9-2-4}$$

where σ is the scaling factor related to the diffusion coefficient. The increments Δx_i and Δy_i satisfies standard normal distribution. Simulation is an effective approach in studying this kind of process.

Example 9.18 Please study with simulation approach for Brownian motion. If the increments are replaced by alpha-stable distribution process, please study the motion.

Solution *For simplicity, select $\sigma = 1$ in both the two processes. The following statements can be used to compute Brownian motion and alpha-stable motion, with $\alpha = 1.5$, as shown in Figures 9.13 (a) and (b), respectively. The alpha-stable motion is also known as* Lévy flight.

```
>> n=1000; x=zeros(2,n); y=zeros(2,n); s=1; r1=randn(2,n); a=1.5;
   r2=stblrnd(a,0,1,0,2,n); % generate alpha-stable random numbers
   for i=2:n, % compute recursively with loop structure
      x(1,i)=x(1,i-1)+s*r1(1,i); y(1,i)=y(1,i-1)+s*r1(2,i);
      x(2,i)=x(2,i-1)+s*r2(1,i); y(2,i)=y(2,i-1)+s*r2(2,i);
   end
   subplot(121), plot(x(1,:),y(1,:),'-o') % draw Brownian motion
   subplot(122), plot(x(2,:),y(2,:),'-o') % draw alpha-stable motion
```

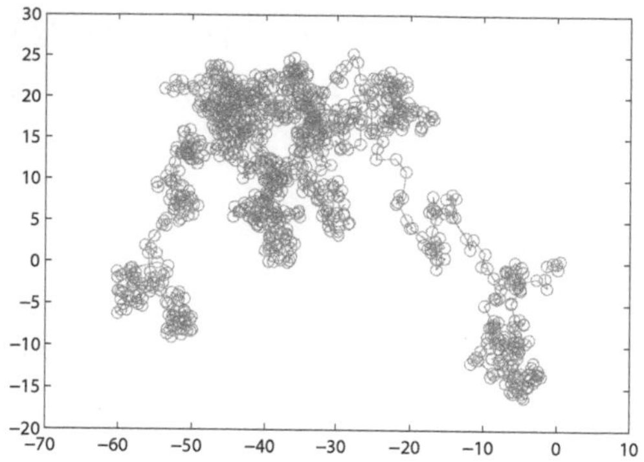

(a) Brownian motion

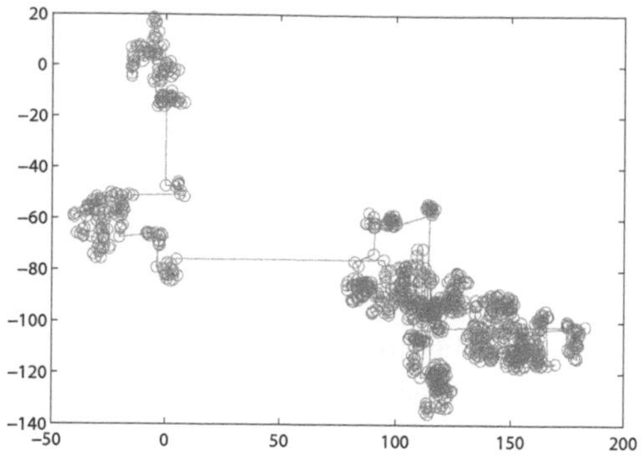

(b) alpha-stable motion

FIGURE 9.13: Random walk plots.

9.3 Fundamental Statistical Analysis

9.3.1 Mean and variance of stochastic variables

Assume that the probability density function of a continuous stochastic variable x is given by $p(x)$. The following formulae can be used to evaluate the *mathematical expectation* $\mathrm{E}[x]$ and the *variance* $\mathrm{D}[x]$, respectively:

$$\mathrm{E}[x] = \int_{-\infty}^{\infty} xp(x)\,\mathrm{d}x,\ \ \mathrm{D}[x] = \int_{-\infty}^{\infty} (x - \mathrm{E}[x])^2\, p(x)\,\mathrm{d}x, \tag{9-3-1}$$

and mathematical expectation is also known as the *mean* of the function.

Clearly, these two important statistical quantities or statistics can be evaluated through the direct integral approach.

Example 9.19 Calculate the mean and variance of Gamma distribution with $a > 0, \lambda > 0$ using direct integration method.

Solution *The following statements can be used to evaluate the mean and variance of Gamma distribution analytically:*

```
>> syms x; syms a lam positive
   p=lam^a*x^(a-1)/gamma(a)*exp(-lam*x); m=int(x*p,x,0,inf)
   s=simplify(int((x-1/lam*a)^2*p,x,0,inf))
```

The results obtained are $m = a/\lambda$ and $s = a/\lambda^2$, while unfortunately, the latter one can only be obtained with earlier versions of MATLAB.

For a set of measured data $x_1, x_2, x_3, \cdots, x_n$, the *sample mean* and *sample variance* are alternatively defined as

$$\bar{x} = \frac{1}{n}\sum_{i=1}^{n} x_i,\ \ \hat{s}_x^2 = \frac{1}{n-1}\sum_{i=1}^{n}(x_i - \bar{x})^2, \tag{9-3-2}$$

with $\hat{s}_x$ referred to as the *standard deviation.*

Another important quantity is the *median value*. For a given set of sorted data $x_1 \leqslant x_2 \leqslant \cdots \leqslant x_n$, if n is odd, the median value is defined as $x_{(n+1)/2}$; while if n is even, the median value is defined as $(x_{n/2-1} + x_{n/2+1})/2$.

Like sample mean values, median values are sometime important in statistics. Consider the case, there is a set of normally distributed data, and the range of all the numbers are around $(-5, 5)$, but there are a couple of points in the set located far away. For instance, their values are about 30. These points are considered as *outliers*, and due to the existence of these points, the sample mean value may be seriously affected, while the median values are unchanged.

Let a set of measured data be expressed by vector $\boldsymbol{x} = [x_1, x_2, x_3, \cdots, x_n]^{\mathrm{T}}$. The MATLAB functions `mean()`, `var()`, `std()` and `median()` can be used to evaluate the sample mean, sample variance, standard deviation and median value, respectively, with

m=`mean(`$\boldsymbol{x}$`)`, $s2$=`var(`$\boldsymbol{x}$`)`, s=`std(`$\boldsymbol{x}$`)`, m_1=`median(`$\boldsymbol{x}$`)`

These four functions can also be used to process the data given in matrices on a column basis. Also, these functions can be used to measure the properties of all the elements in a matrix or multi-dimensional array with the command m_1=`mean(`$\boldsymbol{x}$`(:))`.

Example 9.20 Generate a set of 30000 normally distributed random numbers with the mean and variance given by 0.5 and 1.5^2, respectively. Compute the sample mean and variance of the generated data. If the total number of points is significantly reduced, what may happen?

Solution *The following statements can be used to generate the random numbers. The sample mean, variance, standard deviation and median value can be evaluated directly from*

```
>> p=random('norm',0.5,1.5,30000,1); mean(p), var(p), std(p), median(p)
```

and it can be seen that $\mu = 0.4990$, $\sigma^2 = 2.2188$ *and* $\bar{s} = 1.4896$, $m_1 = 0.5066$, *pretty close to the original exact values. If one reduces the number of points, for instance to 300 points, the following statements can be used, and it can be seen that the statistical quantities become* $\mu = 0.4698$, $\sigma^2 = 2.5523$, $\bar{s} = 1.5976$, $m_1 = 0.4557$, *which deviates far away from the expected values.*

```
>> p=random('norm',0.5,1.5,300,1); mean(p), var(p), std(p), median(p)
```

The commonly encountered distribution functions have been summarized earlier. If the distribution is given, the functions such as `normstat()` and `gamstat()` can be used directly to evaluate the mean and variance of the normal and Gamma distributions. Similar to the suffix `pdf`, the suffix `stat` is used to indicate the statistics properties. For instance, the syntax of `gamstat()` is `[`μ`,`σ^2`]`=`gamstat(`a`,`λ`)`, where the returned arguments are the mean and variance of the given distribution.

Example 9.21 Evaluate the mean and variance of Rayleigh distribution with the parameter b=0.45.

Solution *Since Rayleigh distribution is considered, the function* `raylstat()` *should be used, and the mean and variance of the distribution can be evaluated as* $m = 0.5640, s = 0.0869$.

```
>> b=0.45; [m,s]=raylstat(b) % compute the mean and variance
```

9.3.2 Moments of stochastic variables

Suppose that x is a continuous variable, and $p(x)$ is its probability density function; its rth *raw moment* and rth *central moment* are defined as

$$\nu_r = \int_{-\infty}^{\infty} x^r p(x)\,\mathrm{d}x, \quad \mu_r = \int_{-\infty}^{\infty} (x-\mu)^r p(x)\,\mathrm{d}x, \tag{9-3-3}$$

and it can be seen that $\nu_1 = \mathrm{E}[x]$, $\mu_2 = \mathrm{D}[x]$.

Example 9.22 Consider the raw moment and central moment problems for Gamma distributions in Example 9.19. Summarize the general formula from the first few quantities.

Solution *The raw moment of the signal can easily be evaluated from*

```
>> syms x; syms a lam positive; p=lam^a*x^(a-1)/gamma(a)*exp(-lam*x);
   for n=1:6, m=factor(int(x^n*p,x,0,inf)), end % compute the raw moments
```

and the first five raw moments are

$$\frac{a}{\lambda},\ \frac{a}{\lambda^2}(a+1),\ \frac{a}{\lambda^3}(a+1)(a+2),\ \frac{a}{\lambda^4}(a+1)(a+2)(s+3),$$
$$\frac{a}{\lambda^5}(a+1)(a+2)(a+3)(a+4),\ \frac{a}{\lambda^6}(a+1)(a+2)(a+3)(a+4)(a+5)$$

which can be summarized as

$$\nu_k = \frac{1}{\lambda^k}a(a+1)(a+2)\cdots(a+k-1) = \frac{1}{\lambda^k}\prod_{i=0}^{k-1}(a+k) = \frac{\lambda^{-k}\Gamma(a+k)}{\Gamma(a)}.$$

In fact these results can also be obtained from

```
>> syms k x; assume(k,'integer'); assumeAlso(k>0) % k is a positive integer
   m=simplify(int((x)^k*p,x,0,inf))
```

with $m = \lambda^{-k}\Gamma(k+a)/\Gamma(a)$, which is exactly the same as the one summarized above.

Similarly, the central moments of the stochastic variable can be evaluated from

```
>> for n=1:7, s=simplify(int((x-1/lam*a)^n*p,x,0,inf)), end
```

The first seven central moments obtained are (again the results can only be obtained with MATLAB R2008a or earlier versions)

$$0,\ \frac{a}{\lambda^2},\ \frac{2a}{\lambda^3},\ \frac{3a(a+2)}{\lambda^4},\ \frac{4a(5a+6)}{\lambda^5},\ \frac{5a(3a^2+26a+24)}{\lambda^6},\ \frac{6a(35a^2+154a+120)}{\lambda^7}.$$

No further compact formulae, however, can be summarized from the above results.

For a set of random samples $x_1, x_2, \cdots, x_n$, the rth raw moment and central moment are defined as

$$A_r = \frac{1}{n}\sum_{i=1}^{n} x_i^r, \quad B_r = \frac{1}{n}\sum_{i=1}^{n}(x_i-\bar{x})^r. \tag{9-3-4}$$

The `moment()` function provided in the Statistics Toolbox can be used to evaluate the central moments of all given order. There is no such function to evaluate the raw moment. In fact, according to the definition, the rth raw and central moments can be evaluated from

$A_r =$ `sum(x.^r)/length(x)`, and $B_r =$ `moment(x,r)`

Example 9.23 Consider again the random numbers generated earlier. Please find the moments from the generated pseudorandom numbers.

Solution *The moments of different orders can be evaluated with*

```
>> A=[]; B=[]; p=normrnd(0.5,1.5,30000,1); n=1:5;
   for r=n, A=[A, sum(p.^r)/length(p)]; B=[B,moment(p,r)]; end
```

and the moments obtained are

$$\boldsymbol{A} = [0.508053449, 2.5154715759, 3.5456642980, 18.8911497429, 40.791198267]$$

$$\boldsymbol{B} = [0, 2.257353268872, -0.026041939834, 15.381462246655, -1.208702122593].$$

Actually, the analytical method can be used to evaluate the exact values of the moments, with the following statements

```
>> syms x; A1=[]; B1=[]; p=1/(sqrt(2*pi)*1.5)*exp(-(x-0.5)^2/(2*1.5^2));
   for i=1:5
      A1=[A1,vpa(int(x^i*p,x,-inf,inf),12)];
      B1=[B1,vpa(int((x-0.5)^i*p,x,-inf,inf),12)];
   end
```

with $\boldsymbol{A}_1 = [0.5, 2.5, 3.5, 18.625, 40.8125]$, *and* $\boldsymbol{B}_1 = [0, 2.25, 0, 15.1875, 0]$, *which agree very well with the above numerical results.*

9.3.3 Covariance analysis of multivariate stochastic variables

Assume that the data pairs (x_1, y_1), (x_2, y_2), (x_3, y_3), $\cdots$, (x_n, y_n) are samples of a function with two stochastic variables (x, y). The covariance s_{xy} and correlation coefficient η of the two variables are defined as

$$s_{xy} = \frac{1}{n-1}\sum_{i=1}^{n}(x_i - \bar{x})(y_i - \bar{y}), \quad \eta = \frac{s_{xy}}{s_x s_y}, \tag{9-3-5}$$

and from the above definition, a matrix $\boldsymbol{C}$ can also be defined

$$\boldsymbol{C} = \begin{bmatrix} c_{xx} & c_{xy} \\ c_{yx} & c_{yy} \end{bmatrix}, \tag{9-3-6}$$

where $c_{xx} = \sigma_x^2$, $c_{xy} = c_{yx} = s_{xy}$ and $c_{yy} = \sigma_y^2$. The matrix $\boldsymbol{C}$ is referred to as the *covariance matrix.*

The definition of the covariance matrix with more variables can also be extended from the above definition. The function `cov()` can also be used to evaluate the covariance matrix, with the syntax $\boldsymbol{C}$=`cov(`$\boldsymbol{X}$`)`, where each column in matrix $\boldsymbol{X}$ is regarded as the samples of a random signal. If $\boldsymbol{X}$ is a vector, however, the sample variance will be returned instead.

Example 9.24 Generate four sets of standard normally distributed random data points. Evaluate the covariance matrix for the four signals.

Solution *The random signals can be generated with the* `randn()` *function, such that it has four columns, each one containing 30000 elements. This means that four signals are defined with each one containing 30000 samples. The covariance matrix can then be obtained, and theoretically, it should be an identity matrix, if the four signals generated are independent.*

```
>> p=randn(30000,4); C=cov(p) % compute covariance matrix
```

The covariance matrix can be found as follows, which is quite near to an identity matrix. Thus, the four signals generated this way are indeed independent.

$$\boldsymbol{C}=\begin{bmatrix} 1.006377779 & 0.001258868464 & 0.004726044354 & -0.000519010644 \\ 0.001258868464 & 1.003955935 & -0.000945450079 & 0.004802025815 \\ 0.004726044354 & -0.000945450079 & 1.011043483 & -0.01189535769 \\ -0.0005190106435 & 0.004802025815 & -0.01189535769 & 0.9947556372 \end{bmatrix}.$$

9.3.4 Joint PDFs and CDFs of multivariate normal distributions

Assume that a set of n stochastic variables $\xi_1, \xi_2, \cdots, \xi_n$ satisfy normal distribution. Also, it is assumed that their means are $\mu_1, \mu_2, \cdots, \mu_n$, from which a mean vector $\boldsymbol{\mu}$ can be defined. The covariance matrix can be represented as $\boldsymbol{\Sigma}^2$. Thus, the joint probability density function is defined as

$$p(x_1, x_2, \cdots, x_n) = \frac{1}{\sqrt{2\pi}}\boldsymbol{\Sigma}^{-1}\mathrm{e}^{-(\boldsymbol{x}-\boldsymbol{\mu})^{\mathrm{T}}\boldsymbol{\Sigma}^{-2}(\boldsymbol{x}-\boldsymbol{\mu})/2}, \tag{9-3-7}$$

where $\boldsymbol{x} = [x_1, x_2, \cdots, x_n]^{\mathrm{T}}$.

The function `mvnpdf()` is provided in Statistics Toolbox to evaluate the joint PDF, with the following syntax $\boldsymbol{p}$=`mvnpdf(`$\boldsymbol{X}$,$\boldsymbol{\mu}$,$\boldsymbol{\Sigma}^2$`)`, where $\boldsymbol{X}$ is a matrix with n columns, each representing a stochastic variable.

Example 9.25 If the mean vector and covariance matrix are defined as $\boldsymbol{\mu} = [-1, 2]^{\mathrm{T}}$, $\boldsymbol{\Sigma}^2 = [1, 1; 1, 3]$, draw the joint probability density function. If the covariance matrix is a diagonal matrix, draw the surface of new joint probability density function.

Solution *The random matrix with two columns can be generated from the grid data obtained with the* `meshgrid()` *function. The function* `mvnpdf()` *can be used to evaluate the joint probability density function as a column vector. The function* `reshape()` *can be used to restore the grid matrix. The surface of the joint probability density function can be obtained as shown in Figure 9.14 (a).*

```
>> mu1=[-1,2]; Sigma2=[1 1; 1 3]; % mean vector and covariance matrix
   [X,Y]=meshgrid(-3:0.1:1,-2:0.1:4); xy=[X(:) Y(:)];  % grid data
   p=mvnpdf(xy,mu1,Sigma2); P=reshape(p,size(X));        % joint pdf
   surf(X,Y,P)                        % the surface of the joint pdf
```

If diagonal covariance matrix is studied, the following statements can then be introduced, and the joint probability density function can be obtained as shown in Figure 9.14 (b).

```
>> Sigma2=diag(diag(Sigma2));  % eliminate the non-diagonal elements
   p=mvnpdf(xy,mu1,Sigma2); P=reshape(p,size(X)); surf(X,Y,P)
```

The `mvnrnd()` function provided in the Statistics Toolbox can be used to generate multivariate pseudorandom numbers under normal distribution, with the syntax $\boldsymbol{R}$=`mvnrnd(`$\boldsymbol{\mu}$,$\boldsymbol{\Sigma}^2$,m,n`)`, where the m sets of random numbers will be returned in an $m \times n$ matrix $\boldsymbol{R}$, with each column corresponding to a random signal.

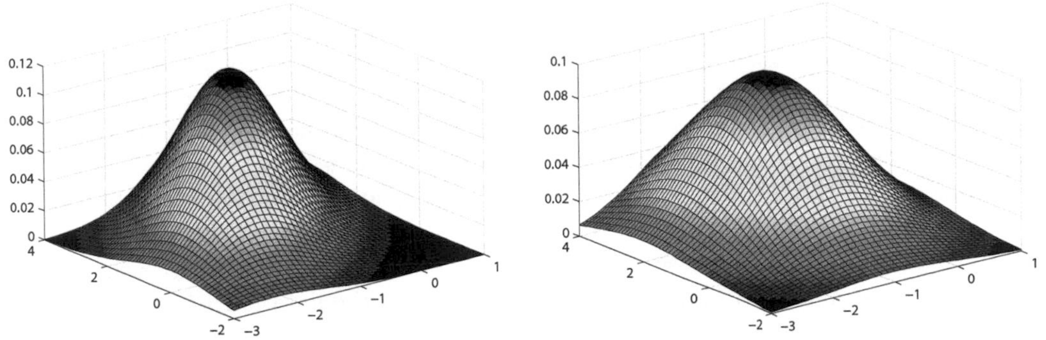

(a) original covariance matrix (b) diagonal covariance matrix

FIGURE 9.14: The joint probability density function surface.

Example 9.26 Observe the two types of normally distributed random numbers with specifications in Example 9.25.

Solution *The following statements can be used to generate the two sets of random numbers, with each set containing 2000 samples. The distribution of the samples on the x-y plane are obtained as shown in Figures 9.15 (a) and (b), respectively. It can be seen that when the covariance matrix is diagonal, there is no relationship between the two variables. Thus, the variables are independent.*

```
>> mu1=[-1,2]; Sigma2=[1 1; 1 3]; % non-diagonal covariance matrix
   R1=mvnrnd(mu1,Sigma2,2000); subplot(121), plot(R1(:,1),R1(:,2),'o')
   Sigma2=diag(diag(Sigma2)); % create a diagonalized covariance matrix
   R2=mvnrnd(mu1,Sigma2,2000); subplot(122), plot(R2(:,1),R2(:,2),'o')
```

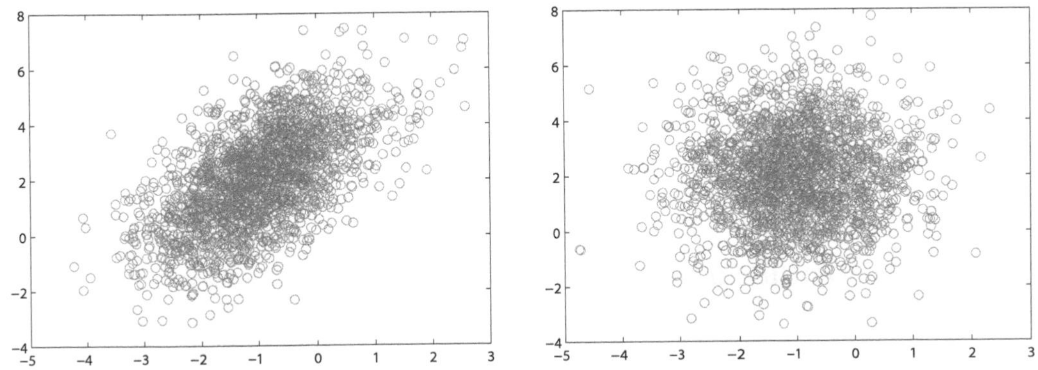

(a) covariance matrix (b) diagonal covariance matrix

FIGURE 9.15: Distribution of two-dimensional stochastic variables.

9.3.5 Outliers, quartiles and box plots

An *outlier* is an observation that lies outside the overall pattern of a distribution [7]. Some of the outliers can be spotted by direct observations from the data set, with histograms or data distribution plots. It is usually helpful to spot and remove the outliers, especially in multivariate cases.

The concept of median value, also known as the second *quartile* q_2, of the vector $\boldsymbol{v}$ was discussed earlier. With the concept median value in mind, the original data vector can further be divided into two halves, one with data smaller than the median value, denoted as $\boldsymbol{v}_1$, and the other, denoted by $\boldsymbol{v}_3$. The median values, q_1 and q_3, of the two vectors can also be obtained. The three median values are denoted as quartiles, and can be obtained with $\boldsymbol{q}$=`quantile(`$\boldsymbol{v}$`,3)`, where $\boldsymbol{q} = [q_1, q_2, q_3]$. The quartiles q_1 and q_3 are referred to the first and third quartiles of the data set. The *interquartile range* (IQR) is defined as the distance between the first and third quartiles, $\text{IQR} = q_3 - q_1$, and the data fall $1.5 \times \text{IQR}$ above q_3 or below q_1 are considered the outliers.

The box plot of the data vector $\boldsymbol{v}$ can be obtained with `boxplot(`$\boldsymbol{v}$`)`. The concept of the box plot will be presented through an example.

Example 9.27 Consider the data set studied in Example 9.12. Draw and interpret the box plot, and find the quartiles and outliers.

Solution *The box plot of the data set can be obtained directly with the following function calls, as shown in Figure 9.16.*

```
>> A=load('c9dlamp.dat'); boxplot(A), q=quantile(A,3)
```

In the box plot, we can see the three horizontal lines in the box, representing the values of $\boldsymbol{q} = [909.5, 997, 1108]$, *the quartiles. Also, it can be seen that there are crosses, marked for the outliers in the data set. The other two horizontal lines are the minimum and maximum values, after the outliers excluded.*

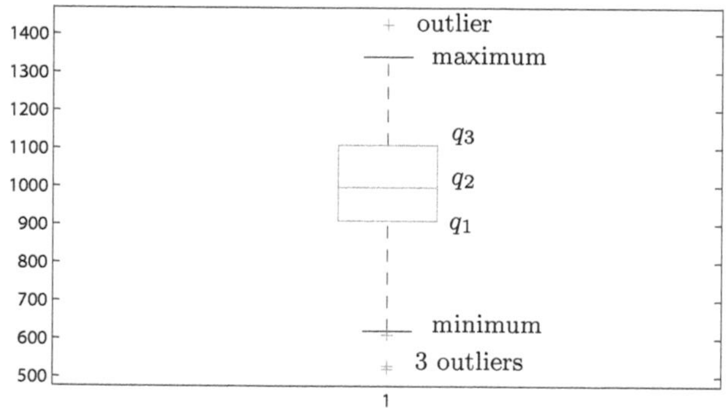

FIGURE 9.16: The box plot with quartiles and outliers.

In the syntax of `boxplot()` function, if $\boldsymbol{v}$ is a matrix with m columns, m boxes will be displayed in the plot.

An alternative function `outliers()` is written by Niccolo Battistini [8] of Rutgers University. The outliers in a vector $\boldsymbol{v}$ can be detected with interquartile range method, as well as Grubbs method. The syntax of the function is `[`$\boldsymbol{v}_1$`,`$\boldsymbol{v}_2$`]`=`outliers(`$\boldsymbol{v}$`,`*opts*`,`α`)`, where *opts* can be `'grubbs'` or `'quartile'`, however, the former one is recommended, and α is the significance level. The returned vector $\boldsymbol{v}_2$ contains the outliers, while $\boldsymbol{v}_1$ is the vector with outliers removed.

Example 9.28 Consider again the data in Example 9.27. Please find the outliers with Grubbs method.

Solution *With the following statements, the outliers can be detected and returned in vector* $\boldsymbol{v}_2$, *with* $\boldsymbol{v}_2 = [521, 529]$.

```
>> A=load('c9dlamp.dat'); [v1 v2]=outliers(A,'grubbs',0.05)
```

For multivariate problems, an outlier detection function `moutlier1()` written by Antonio Trujillo-Ortiz [9] of Universidad Autonoma de Baja California, Mexico is recommended, with `moutlier1(`$\boldsymbol{X}$`,`α`)`, with $\boldsymbol{X}$ the m column matrix, and α the significance level.

Example 9.29 The information of 29 NBA teams were given in Table 9.3 [7]. Please find the outliers in the multivariate problem.

TABLE 9.3: Some data of NBA teams (Data from [7]).

team number	value ($millions)	revenue ($millions)	income ($millions)	team number	value ($millions)	revenue ($millions)	income ($millions)
1	447	149	22.8	2	401	160	13.5
3	356	119	49	4	338	117	−17.7
5	328	109	2	6	290	97	25.6
7	284	102	23.5	8	283	105	18.5
9	282	109	21.5	10	280	94	10.1
11	278	82	15.2	12	275	102	−16.8
13	274	98	28.5	14	272	97	−85.1
15	258	72	3.8	16	249	96	10.6
17	244	94	−1.6	18	239	85	13.8
19	236	91	7.9	20	230	85	6.9
21	227	63	-19.7	22	218	75	7.9
23	216	80	21.9	24	208	72	15.9
25	202	78	-8.4	26	199	80	13.1
27	196	70	2.4	28	188	70	7.8
29	174	70	-15.1				

Solution *The data can be entered into MATLAB workspace first, and with the* `moutlier1()` *function call, the 14th team is spotted as the outlier.*

```
>> X=[447,149,22.8; 401,160,13.5; 356,119,49; 338,117,-17.7; 328,109,2;
   290,97,25.6; 284,102,23.5; 283,105,18.5; 282,109,21.5; 280,94,10.1;
   278,82,15.2; 275,102,-16.8; 274,98,28.5; 272,97,-85.1; 258,72,3.8;
   249,96,10.6; 244,94,-1.6; 239,85,13.8; 236,91,7.9; 230,85,6.9;
   227,63,-19.7; 218,75,7.9; 216,80,21.9; 208,72,15.9; 202,78,-8.4;
   199,80,13.1; 196,70,2.4; 188,70,7.8; 174,70,-15.1];
 moutlier1(X,0.05)  % find outliers from the multivariate problem
```

The result can be verified with the projections of the scatter plot on x–y, x–z *and* y–z *planes, as shown in Figure 9.17.*

```
>> subplot(131), plot(X(:,1),X(:,2),'o') % projection on x--y plane
   subplot(132), plot(X(:,1),X(:,3),'o') % projection on x--z plane
   subplot(133), plot(X(:,2),X(:,3),'o') % projection on y--z plane
```

It is obvious from Figures 9.17 (b) and (c) that, the data in the 14th team is indeed the outliers, from multivariate point of view.

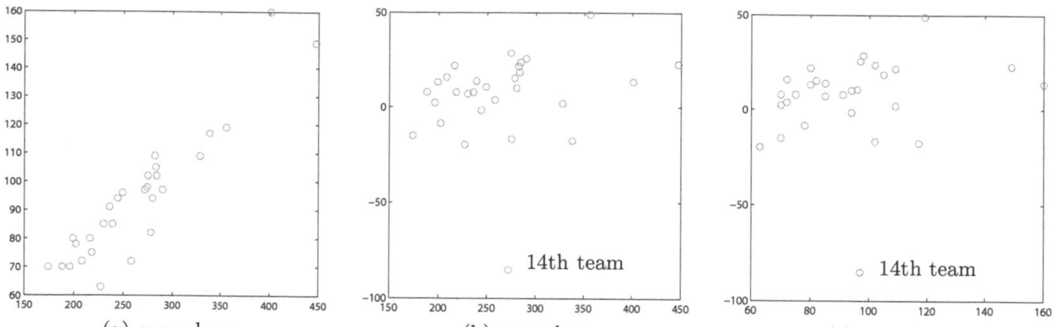

(a) x–y plane (b) x–z plane (c) y–z plane

FIGURE 9.17: Projections of 3D scatter plot on different planes.

9.4 Statistical Estimations

9.4.1 Parametric estimation and interval estimation

If a set of data $\boldsymbol{x} = [x_1, x_2, \cdots, x_n]^{\mathrm{T}}$ is measured, and it is known that they satisfy a certain distribution, for instance, normal distribution, the function `normfit()` can be used to evaluate the mean μ and variance σ^2 via maximum likelihood method. Also, the confidence intervals $\Delta\mu$ and $\Delta\sigma^2$ can be obtained, with the syntax $[\mu, \sigma^2, \Delta\mu, \Delta\sigma^2]=$`normfit`$(\boldsymbol{x}, P_{\mathrm{ci}})$, where the argument P_{ci} is the confidence level, for instance, 0.95 for 95% of confidence. The function `norminv()` can then be used to evaluate the relevant needed quantities. It can be predicted that the bigger the value of P_{ci}, the narrower the confidence interval, and the more accurate the parameters are estimated.

Similar to the PDFs, other functions for evaluating the means and variances for other distributions can also be made. For instance, the mean and variance of Gamma distribution can be estimated by using `gamfit()`, and those for Rayleigh distribution by using `raylfit()`, as summarized in Table 9.1.

Example 9.30 Generate a set of random numbers with Gamma distribution, where $a = 1.5$ and $\lambda = 3$. From the generated data, the parameters can be estimated with different confidence levels, and compare the results.

Solution *First, generate a set of 30000 random data. Select the confidence levels at 90%, 92%, 95%, 98%. The following statements can be used to estimate the parameters with the given confidence levels*

```
>> p=gamrnd(1.5,3,30000,1); Pv=[0.9,0.92,0.95,0.98]; A=[];
   for i=1:length(Pv) % try different confidence levels
      [a,b]=gamfit(p,Pv(i)); A=[A; Pv(i),a(1),b(:,1)',a(2),b(:,2)'];
   end
```

The results are obtained and shown in Table 9.4. It can be seen that the estimated parameters are not affected with the confidence levels. However, the confidence intervals can be different. Normally, a confidence level of 95% should be used.

TABLE 9.4: Parameter estimation results.

confidence	estimation results of a			estimation results of λ		
level	$\hat{a}$	$a_{\min}$	$a_{\max}$	$\hat{\lambda}$	$\lambda_{\min}$	$\lambda_{\max}$
90%	1.506500556	1.505099132	1.507901979	2.991117941	2.987797191	2.994438691
92%	1.506500556	1.505380481	1.507620631	2.991117941	2.988463861	2.993772021
95%	1.506500556	1.505801226	1.507199886	2.991117941	2.98946084	2.992775042
98%	1.506500556	1.506220978	1.506780134	2.991117941	2.990455465	2.991780417

Now consider the impact of the size of random number group on the estimation, by selecting respectively 300, 3000, 30000, 300000, 3000000 random numbers. The confidence level of 95% is assumed; therefore, the estimation intervals may also change and the variations are shown in Table 9.5.

TABLE 9.5: Parameter estimation results.

number of	estimation results of a			estimation results of λ		
values	$\hat{a}$	$a_{\min}$	$a_{\max}$	$\hat{\lambda}$	$\lambda_{\min}$	$\lambda_{\max}$
300	1.548677954	1.540991679	1.55636423	2.91172985	2.896265076	2.927194623
3000	1.476057561	1.473908973	1.47820615	3.040589493	3.035438607	3.04574038
30000	1.503327455	1.502624743	1.504030167	2.976242793	2.974591027	2.977894559
300000	1.509546583	1.509323617	1.50976955	2.984774009	2.984252596	2.985295421
3000000	1.498005677	1.497935817	1.498075536	3.006048895	3.005882725	3.006215065

```
>> num=[300,3000,30000,300000,3000000]; A=[];
   for i=1:length(num), p=gamrnd(1.5,3,num(i),1);
      [a,b]=gamfit(p,0.95); A=[A;num(i),a(1),b(:,1)',a(2),b(:,2)'];
   end
```

It can be seen from the table that, when the number of random samples is small, the estimated parameters may not be satisfactory. Normally 30,000 samples are acceptable.

9.4.2 Multivariate linear regression and interval estimation

Assume that the output signal y is the linear combination of n input signals $x_1, x_2, \cdots, x_n$ such that

$$y = a_1x_1 + a_2x_2 + a_3x_3 + \cdots + a_nx_n, \tag{9-4-1}$$

where $a_1, a_2, \cdots, a_n$ are undetermined constants. Assume that m groups of experiments are made with the measured data arranged in the following format:

$$\begin{aligned} y_1 &= x_{11}a_1 + x_{12}a_2 + \cdots + x_{1,n}a_n + \varepsilon_1 \\ y_2 &= x_{21}a_1 + x_{22}a_2 + \cdots + x_{2,n}a_n + \varepsilon_2 \\ &\vdots \\ y_m &= x_{m1}a_1 + x_{m2}a_2 + \cdots + x_{m,n}a_n + \varepsilon_m. \end{aligned} \tag{9-4-2}$$

where ε_i is the measuring error of the ith observation. The following matrix equation can then be set up

$$\boldsymbol{y} = \boldsymbol{X}\boldsymbol{a} + \boldsymbol{\varepsilon}, \tag{9-4-3}$$

where $\boldsymbol{a} = [a_1, a_2, \cdots, a_n]^{\mathrm{T}}$ is the vector with undetermined constants. If the original data were obtained in experiments, there exist errors in each equation in (9-4-1) denoted by $\boldsymbol{\varepsilon} = [\varepsilon_1, \varepsilon_2, \cdots, \varepsilon_m]^{\mathrm{T}}$. Also, the vector $\boldsymbol{y} = [y_1, y_2, \cdots, y_m]^{\mathrm{T}}$ is the observed value, and the matrix $\boldsymbol{X}$ is composed of observed independent variables such that

$$\boldsymbol{X} = \begin{bmatrix} x_{11} & x_{12} & \cdots & x_{1n} \\ x_{21} & x_{22} & \cdots & x_{2n} \\ \vdots & \vdots & \ddots & \vdots \\ x_{m1} & x_{m2} & \cdots & x_{mn} \end{bmatrix}. \tag{9-4-4}$$

Assume that the objective function for the problem is to have the minimized sum of the squared errors, i.e., $J = \min \boldsymbol{\varepsilon}^{\mathrm{T}}\boldsymbol{\varepsilon}$, then, the undetermined constant vector $\boldsymbol{a}$ of the linear regression model can be obtained from

$$\hat{\boldsymbol{a}} = (\boldsymbol{X}^{\mathrm{T}}\boldsymbol{X})^{-1}\boldsymbol{X}^{\mathrm{T}}\boldsymbol{y}. \tag{9-4-5}$$

From the knowledge of linear algebra illustrated in Chapter 4, the least squares solution to the above equation can be obtained from $\boldsymbol{a}=\boldsymbol{X}\backslash\boldsymbol{y}$ or more formally $\boldsymbol{a}=$`inv(`$\boldsymbol{X}$`'*`$\boldsymbol{X}$`)*`$\boldsymbol{X}$`'*`$\boldsymbol{y}$.

A multivariate linear regression function `regress()` for parameter estimation and confidence interval estimation is provided in MATLAB, and the syntax of the function is `[`$\hat{\boldsymbol{a}}$`,`$\boldsymbol{a}_{\mathrm{ci}}$`]=regress(`$\boldsymbol{y}$`,`$\boldsymbol{X}$`,`$\alpha$`)`, where $1-\alpha$ is the confidence level specified by the user.

Example 9.31 Assume that the linear regression model is

$$y = x_1 - 1.232x_2 + 2.23x_3 + 2x_4 + 4x_5 + 3.792x_6.$$

With 120 set of generated data points x_i, the output vector $\boldsymbol{y}$ can be computed first. Based on this information, the undetermined constants a_i can be estimated, with given confidence intervals.

Solution *Linear regression can be used to process the generated data. The following statements can be used to construct the matrix $\boldsymbol{X}$ and the vector $\boldsymbol{y}$. The least squares method can be used to estimate the undetermined constants $\boldsymbol{a}$ under 98% of confidence.*

```
>> a=[1 -1.232 2.23 2 4,3.792]'; X=randn(120,6); y=X*a; % generate data
   [a,aint]=regress(y,X,0.02) % multivariate linear regression and intervals
```

The estimated parameter and the confidence interval can be obtained as

$$\boldsymbol{a} = \begin{bmatrix} 1 \\ -1.232 \\ 2.23 \\ 2 \\ 4 \\ 3.792 \end{bmatrix}, \quad \boldsymbol{a}_{\rm int} = \begin{bmatrix} 1 & 1 \\ -1.232 & -1.232 \\ 2.23 & 2.23 \\ 2 & 2 \\ 4 & 4 \\ 3.792 & 3.792 \end{bmatrix}.$$

It can be seen that there is no error in the estimated results, and the confidence interval is 100%. If the samples are corrupted by noises, for instance, normally distributed noise $N(0,0.5)$ can be added to the samples within the interval. The following statements can then be used to perform linear regression analysis. The estimated parameters and their confidence intervals can be obtained and shown in Figure 9.18 (a), with `errorbar()` *function.*

```
>> yhat=y+sqrt(0.5)*randn(120,1); [a,aint]=regress(yhat,X,0.02)
   errorbar(1:6,a,aint(:,1)-a,aint(:,2)-a) % regression parameters
```

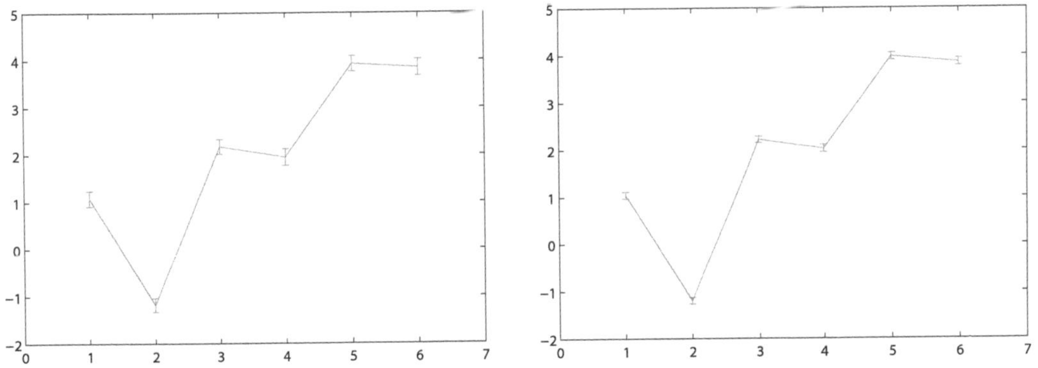

(a) with noise $\sigma^2 = 0.5$ (b) with noise $\sigma^2 = 0.1$

FIGURE 9.18: Parameter estimations with confidence intervals.

The parameters and the intervals are obtained as

$$\boldsymbol{a} = \begin{bmatrix} 1.03887888369425 \\ -1.20360949260729 \\ 2.19454416841833 \\ 1.89146235051598 \\ 34.0628445587228 \\ 8.70540411609337 \end{bmatrix}, \quad \boldsymbol{a}_{\rm int} = \begin{bmatrix} 0.92959508774995 & 1.14816267963855 \\ -1.313314026578 & -1.09390495863658 \\ 2.07413724843311 & 2.31495108840356 \\ 1.77199410730299 & 2.01093059372896 \\ 33.9418338141614 & 34.1838553032841 \\ 8.59651534574785 & 8.8142928864389 \end{bmatrix}.$$

Reducing the variance to 0.1, the estimated parameters are obtained using the following statements as shown in Figure 9.18 (b). It can be seen that the estimation results are even more accurate.

```
>> yhat=y+sqrt(0.1)*randn(120,1); [a,aint]=regress(yhat,X,0.02);
   errorbar(1:6,a,aint(:,1)-a,aint(:,2)-a) % regression parameters
```

9.4.3 Nonlinear least squares parametric and interval estimations

Assume that a set of data $x_i, y_i, i = 1, 2, \cdots, N$ is measured satisfying a given prototype functional relationship $\hat{y}(x) = f(\boldsymbol{a}, x)$, where $\boldsymbol{a}$ is the vector containing undetermined constants. Since the measured data is corrupted with noises, the original function can be written as $\hat{y}(x) = f(\boldsymbol{a}, x) + \varepsilon$, where ε represents the *residual error*. An objective function can be introduced

$$I = \min_{\boldsymbol{a}} \sum_{i=1}^{N} [y_i - \hat{y}(x_i)]^2 = \min_{\boldsymbol{a}} \sum_{i=1}^{N} [y_i - f(\boldsymbol{a}, x_i)]^2. \tag{9-4-6}$$

Minimizing the above objective function, the undetermined constants $\boldsymbol{a}$ can be estimated. Substituting the estimated $\boldsymbol{a}$ back to the prototype function, the residue error $\varepsilon_i = y_i - f(\boldsymbol{a}, x_i)$ can be obtained. Similar to the least squares estimation shown in Section 8.3.3, the least squares fitting with Levenberg–Marquardt algorithm is implemented in the function `nlinfit()`. The Jacobian vector $\boldsymbol{j}_i$ of the residual error with respect to $\boldsymbol{a}$ can also be obtained. Confidence interval estimation can also be obtained with the function `nlparci()`, and the results can be obtained with 95% confidence level, with the syntaxes

```
[a,r,J]=nlinfit(x,y,fun,a0)  % least squares estimation
c=nlparci(a,r,J)             % confidence interval with 95% of confidence
```

where $\boldsymbol{x}$ and $\boldsymbol{y}$ contain the measured data. The function *fun* represents the prototype function, which can be described either by an M-function, an anonymous function or an inline function. The initial values $\boldsymbol{a}_0$ of the estimated parameters should also be specified. It can be seen that the input arguments are the same as the `lsqcurvefit()` function. The returned argument $\boldsymbol{a}$ vector contains the estimated parameters, and $\boldsymbol{r}$ returns the residue error vector for the estimation. Matrix $\boldsymbol{J}$ is the Jacobian and based on the information, the estimation of confidence intervals $\boldsymbol{c}$ can be found. The parametric estimation and confidence interval estimation are illustrated by the following examples.

Example 9.32 The parametric estimation can be performed to the measured data in Example 8.25 with least squares fitting method. The 95% confidence level is assumed. Please solve the parametric estimation and confidence interval estimation problems.

Solution *Assume that the prototype function is in the form of* $y(x) = a_1 e^{-a_2 x} + a_3 e^{-a_4 x} \sin(a_5 x)$, *where the parameters* a_i *are the undetermined constants. An anonymous function can be used to describe the prototype function and the undetermined constants can be evaluated by using* `nlinfit()` *function*

```
>> f=@(a,x)a(1)*exp(-a(2)*x)+a(3)*exp(-a(4)*x).*sin(a(5)*x);
   x=0:0.1:10; y=f([0.12,0.213,0.54,0.17,1.23],x);
   [a,r,j]=nlinfit(x,y,f,[1;1;1;1;1]); ci=nlparci(a,r,j)
```

and it is found that

$$\boldsymbol{a} = \begin{bmatrix} 0.11999999763418 \\ 0.21299999458274 \\ 0.54000000196818 \\ 0.17000000068705 \\ 1.22999999996315 \end{bmatrix}, \quad \boldsymbol{c}_i = \begin{bmatrix} 0.11999999712512 & 0.11999999814323 \\ 0.21299999340801 & 0.21299999575747 \\ 0.54000000124534 & 0.54000000269101 \\ 0.17000000036077 & 0.17000000101332 \\ 1.22999999978603 & 1.23000000014028 \end{bmatrix}.$$

It can be seen that the results obtained are more accurate than the default results obtained with `lsqcurvefit()`*. However, more accurate results are not possible. The associated* `nlparci()` *function can be used to get the confidence intervals for the parameters, with 95% of confidence level.*

If the original samples y_i are corrupted with noises uniformly distributed in the interval $[0, 0.02]$, the following statements can be used to estimate the parameters and confidence intervals for the new samples

```
>> y=f([0.12,0.213,0.54,0.17,1.23],x)+0.02*rand(size(x));
   [a,r,j]=nlinfit(x,y,f,[1;1;1;1;1]); ci=nlparci(a,r,j)
   errorbar(1:5,a,ci(:,1)-a,ci(:,2)-a)
```

and it can be found that

$$\boldsymbol{a} = \begin{bmatrix} 0.12281531581639 \\ 0.17072641296744 \\ 0.55113088779121 \\ 0.17347639675132 \\ 1.2291686258648 \end{bmatrix}, \quad \boldsymbol{c}_i = \begin{bmatrix} 0.11857720435195 & 0.12705342728083 \\ 0.16221631527879 & 0.17923651065609 \\ 0.54465309442893 & 0.55760868115349 \\ 0.17055714192171 & 0.17639565158094 \\ 1.22755955648343 & 1.23077769524618 \end{bmatrix}.$$

The estimated parameters and their confidence intervals are shown in Figure 9.19.

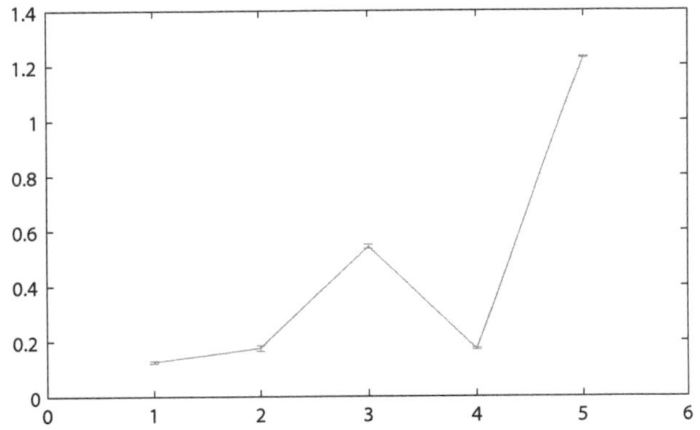

FIGURE 9.19: Estimated parameters and confidence intervals.

Example 9.33 Assume that the prototype function is

$$f(\boldsymbol{a}, \boldsymbol{x}) = (a_1 x_1^3 + a_2)\sin(a_3 x_2 x_3) + (a_4 x_2^3 + a_5 x_2 + a_6).$$

Solve the multivariate nonlinear regression problem for a set of noisy data with the function `nlinfit()`.

Solution *Assume that the initial values of a_i are all 1's. The following statements can be used to define the function f, and to generate a data set $\boldsymbol{X}$ serving as the observed data.*

```
>> a=[1;1;1;1;1;1]'; % initial search vector
   f=@(a,x)(a(1)*x(:,1).^3+a(2)).*sin(a(3)*x(:,2).*x(:,3))+...
     (a(4)*x(:,3).^3+a(5)*x(:,3)+a(6)); % prototype function description
   X=randn(120,4); y=f(a,X)+sqrt(0.2)*randn(120,1); % noised data
```

With the observed data, the following statements can be used to estimate a_i, and the results can be shown in Figure 9.20 (a) where we can observe that the fitting is satisfactory.

```
>> [ahat,r,j]=nlinfit(X,y,f,[0;2;3;2;1;2]); ci=nlparci(ahat,r,j);
   y1=f(ahat,X); plot([y y1]) % nonlinear regression parameters and intervals
```

The estimated parameters and the confidence intervals are

$$\boldsymbol{a} = \begin{bmatrix} 1.04839959073146 \\ 1.01882085899938 \\ 0.98446778587739 \\ 0.99107092667601 \\ 1.02519403669663 \\ 1.05040136072101 \end{bmatrix}, \quad \boldsymbol{c}_i = \begin{bmatrix} 0.96893545453576 & 1.12786372692717 \\ 0.88884528207121 & 1.14879643592754 \\ 0.89167073268603 & 1.07726483906874 \\ 0.96675355288178 & 1.01538830047025 \\ 0.9270517440877 & 1.12333632930555 \\ 0.99419821923406 & 1.10660450220796 \end{bmatrix}.$$

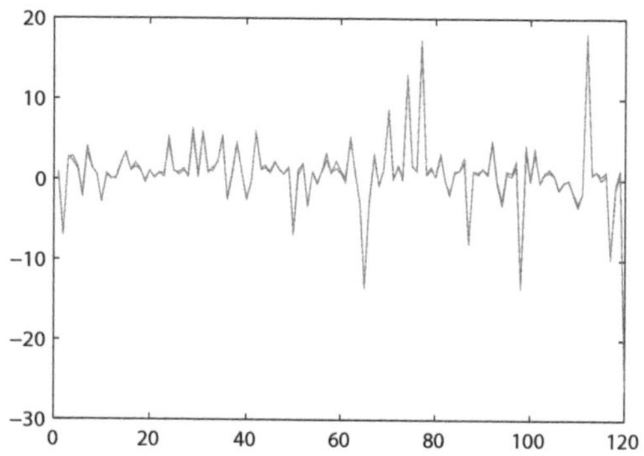

(a) nonlinear regression

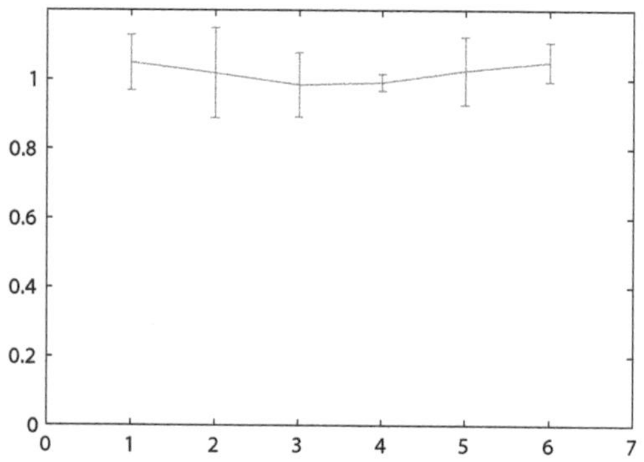

(b) parametric estimation and confidence interval

FIGURE 9.20: Nonlinear regression for multivariate functions.

The results are exactly the same as the ones obtained previously. With the function `nlparci()`*, the confidence interval can also be found. The confidence intervals with 95% confidence level are visualized in Figure 9.20 (b).*

```
>> errorbar(1:6,ahat,ci(:,1)-ahat,ci(:,2)-ahat) % plot with error bars
```

9.4.4 Maximum likelihood estimations

Maximum likelihood estimation (MLE) is a frequently used method in statistics for parameter and interval estimation. Normally when a statistical model is known, with undetermined constants, and also a set of experimental data is obtained, maximum likelihood approach can be used to estimate the undetermined constants and intervals. In MATLAB, `mle()` function can be used to carry out maximum likelihood estimation tasks. Here only one syntax of the function is presented, $[\boldsymbol{p},\boldsymbol{p}_1]$ `=mle('norm',`$\boldsymbol{X}$`,`α`)`, where $\boldsymbol{X}$ is a data vector assumed to satisfy a normal distribution, and α is the confident level. The returned variables $\boldsymbol{p}$ contained the estimated mean and variance, while $\boldsymbol{p}_1$ returns the estimated intervals.

Example 9.34 A batch of light bulbs produced by a factory may have different lumen levels. The lumen here is regarded as a stochastic variable ξ. It is assumed that ξ satisfies the normal distribution $N(\mu, \sigma^2)$. Now take 120 samples randomly from the products, and the lumen levels of the samples are given in Table 9.6. Please estimate the mean and variance using maximum likelihood method.

TABLE 9.6: The lumen levels of the test samples.

216	203	197	208	206	209	206	208	202	203	206	213	218	207	208	202	194	203	213	211
193	213	208	208	204	206	204	206	208	209	213	203	206	207	196	201	208	207	213	208
210	208	211	211	214	220	211	203	216	224	211	209	218	214	219	211	208	221	211	218
218	190	219	211	208	199	214	207	207	214	206	217	214	201	212	213	211	212	216	206
210	216	204	221	208	209	214	214	199	204	211	201	216	211	209	208	209	202	211	207
202	205	206	216	206	213	206	207	200	198	200	202	203	208	216	206	222	213	209	219

Solution *For the clarity in presentation, the data was stored in an ASCII file c9dlumen.dat. The data should be loaded into MATLAB workspace first, and the function* `mle()` *can be used to estimate the parameters*

```
>> X=load('c9dlumen.dat');      % load the data into MATLAB workspace
   [p,p1]=mle('norm',X,0.05) % estimate the mean and variance with MLE method
```

The estimated mean is 208.8167, with an interval of $[207.6737, 209.9596]$*, and the estimated variance is 6.2968, with interval* $[5.6118, 7.2428]$.

9.5 Statistical Hypothesis Tests

9.5.1 Concept and procedures for statistic hypothesis test

For a given data set, we can assume certain statistic properties, for instance, certain distribution. How to check the assumption made on the data set is also known as *hypothesis test*. Hypothesis test is very important in statistics. For instance, someone may propose a hypothesis that the average lifetime of a certain light bulb product is over 3000 hours. How to check whether this hypothesis is correct? The exact way to check it is to turn all the bulbs on, and measure the lifetime of all the bulbs. This is, of course, not a feasible solution to the test. In statistics, one may randomly select a certain number of bulbs for the hypothesis test.

I. Test of significance

One may claim that the average value of a product specification is μ_0. To test whether this claim is true, one can select randomly n samples, and find the sample mean $\bar{x}$ and sample deviation s. Now the hypothesis can be made as $\mathscr{H}_0$: $\mu = \mu_0$, i.e., the mean of the product specification is μ_0. The following procedures can be carried out to perform hypothesis test:

(i) Establish a statistical variable u as

$$u = \frac{\sqrt{n}(\bar{x} - \mu_0)}{s}, \tag{9-5-1}$$

and statistical variable u satisfies a standard normal distribution $N(0, 1)$.

(ii) Select a significance level α, for instance, $\alpha = 0.05$.

(iii) Compute the inverse normal distribution function $K_{\alpha/2}$ such that

$$\int_{-K_{\alpha/2}}^{K_{\alpha/2}} \frac{1}{\sqrt{2\pi}} \mathrm{e}^{-\frac{x^2}{2}} \mathrm{d}x < 1 - \alpha, \tag{9-5-2}$$

which can be evaluated with $K_{\alpha/2}$=`norminv(1-α/2,0,1)`. This value can alternatively be evaluated with $K_{\alpha/2}$=`icdf('norm',1-α/2,0,1)`.

(iv) Make decision: If $|u| < K_{\alpha/2}$, the hypothesis $\mathscr{H}_0$ cannot be rejected, otherwise, the hypothesis $\mathscr{H}_0$ can be rejected with $(1-\alpha) \times 100\%$ of confidence.

Example 9.35 Suppose it is known that the average strength of a product is $\mu_0 = 9.94$kg. Now a new manufacturing technique is adopted. Two hundred pieces of the product were selected randomly and the average strength is measured as $\bar{x} = 9.73$kg, with a standard deviation $s = 1.62$kg. Check whether the new manufacturing technique will affect the average strength of the product.

Solution *The hypothesis for the problem can be defined as* $\mathscr{H}_0$: $\mu = 9.94\,\mathrm{kg}$, *meaning the new manufacturing technique will not affect the average strength of the product. To solve this type of hypothesis test problem, the following statements can be issued*

```
>> n=200; mu0=9.94; xbar=9.73; s=1.62; u=sqrt(n)*(mu0-xbar)/s
   alpha=0.02; K=norminv(1-alpha/2,0,1), H=abs(u)<K
```

With the above MATLAB statements, it is found that $u = 1.8332$, $K = 2.3263$, *and most importantly,* $H = 1$, *meaning* $|u| < K$, *therefore, the hypothesis* $\mathscr{H}_0$ *cannot be rejected. In other words, the new manufacturing technique will not affect the average strength of the product.*

II. Comparing two means

Another type of typical hypothesis test is usually used to check whether two sets of data have significant difference or not.

Select randomly n_1 samples from the first set of data, and measure the sample mean $\bar{x}_1$ and sample deviation s_1. Select also randomly n_2 samples from the second set, and measure $\bar{x}_2$ and s_2. The hypothesis is defined as $\mathscr{H}_0$: $\mu_1 = \mu_2$, i.e., the mean of the two sets of data are equal. In other words, the two sets of data have no significant difference. The hypothesis test can be carried out in the following procedures:

(i) A statistical variable t can be computed

$$t = \frac{\bar{x}_1 - \bar{x}_2}{\sqrt{s_1^2/n_1 + s_2^2/n_2}}, \tag{9-5-3}$$

which satisfies T distribution.

(ii) Select a significance level α and compute T_0, with T_0=`tinv(α/2,k)` or T_0=`icdf('t',α/2,k)`, where $k = \min(n_1 - 1, n_2 - 1)$.

(iii) Make decision: If $|t| < |T_0|$, the hypothesis $\mathscr{H}_0$ cannot be rejected, otherwise, it can be rejected with $(1-\alpha) \times 100\%$ confidence.

Example 9.36 Twenty patients suffering from insomnia are divided randomly into groups A and B, with ten patients each. They were given medicines A and B, respectively. The extended sleeping hours are measured as shown in Table 9.7. Determine whether there are significant differences in the healing effect.

TABLE 9.7: Extended sleeping hours.

A	1.9	0.8	1.1	0.1	−0.1	4.4	5.5	1.6	4.6	3.4
B	0.7	−1.6	−0.2	−1.2	−0.1	3.4	3.7	0.8	0	2

Solution *The simplest way to check the difference of two sets of data is to make the hypothesis —* $\mathscr{H}_0$: $\mu_1 = \mu_2$, *i.e., the means of the two sets of data are equal. The following statements can be made to carry out the hypothesis test*

```
>> x=[1.9,0.8,1.1,0.1,-0.1,4.4,5.5,1.6,4.6,3.4];
   y=[0.7,-1.6,-0.2,-1.2,-0.1,3.4,3.7,0.8,0,2];
   n1=length(x); n2=length(y); k=min(n1-1,n2-1);
   t=(mean(x)-mean(y))/sqrt(std(x)^2/n1+std(y)^2/n2)
   a=0.05; T0=tinv(a/2,k), H=abs(t)<abs(T0)
```

with $t = 1.8608$, $k = 9$, $T_0 = -2.2622$. *Since* $H = 1$, *the hypothesis cannot be rejected. In other words, the medicines do not have significant difference in effect.*

Since the two set of samples are known, the box plots for the two sets can be obtained as shown in Figure 9.21, which also cross-validated the above conclusion.

```
>> boxplot([x.' y.']) % draw box plots for the two sets of data
```

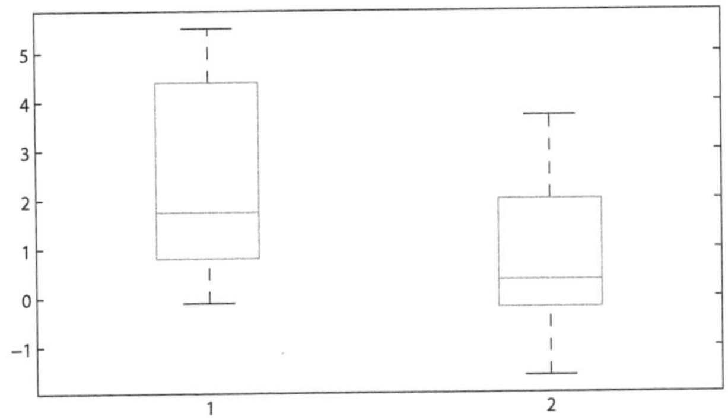

FIGURE 9.21: The box plots of the two sets.

9.5.2 Hypothesis tests for distributions

MATLAB solution to simple hypothesis test problems have been shown in the previous examples. In fact, many MATLAB functions are provided in the Statistics Toolbox to solve different hypothesis test problems, for instance, mean value test for normal distributions, normality test and arbitrary given distribution test. In this subsection, hypothesis tests based on the Statistics Toolbox are presented.

I. Hypothesis test of the mean of normal distribution

Given a set of data satisfying the normal distribution with its standard deviation σ known. The hypothesis $\mathscr{H}_0$ is made to claim that the mean of the distribution is μ. This hypothesis test can be performed through Z test with the function `ztest()`, where `[`H`,`s`,`$\mu_{\rm ci}$`]=ztest(`$\boldsymbol{X}$`,`μ`,`σ`,`α`)`, with H the test result. When $H=0$, it means that the hypothesis $\mathscr{H}_0$ cannot be rejected, otherwise, the hypothesis should be rejected, with a confidence level of α. The value of s is the significance level, and $\mu_{\rm ci}$ is the confidence interval of the mean.

If the standard deviation for the normal distribution is not known, the T test can be used to perform hypothesis test on the means. The MATLAB function `ttest()` can be used to perform such a test, with `[`H`,`s`,`$\mu_{\rm ci}$`]=ttest(`$\boldsymbol{X}$`,`μ`,`α`)`.

Example 9.37 Generate a set of random numbers satisfying the normal distribution. Perform hypothesis test on the mean of the random numbers.

Solution *Let us first generate 400 random numbers satisfying* $N(1,2^2)$. *Since the standard deviation is 2, the hypothesis* $\mathscr{H}_0: \mu=1$ *can be established. With the following MATLAB statements*

```
>> r=normrnd(1,2,400,1); [H,p,ci]=ztest(r,1,2,0.02) % carry out Z test
```

it can be found that $H=0, p=0.43594320476, \boldsymbol{c}_i=[0.8453, 1.3105]$. *Since* $H=0$, *the hypothesis cannot be rejected. In other words, the hypothesis can be accepted with 98% of confidence.*

If the hypothesis is changed to $\mathscr{H}_0: \mu=0.5$, *the following statements can be used, and the value of H is 1. In this case, the hypothesis* $\mathscr{H}_0$ *should be rejected. The confidence interval for the mean value remains the same.*

```
>> [H,p,ci]=ztest(r,0.5,2,0.02) % carry out Z test
```

It is found that $H=1, p=7.5118\times10^{-9}, \boldsymbol{c}_i=[0.845, 1.3105]$.

If the standard deviation is not known, the T test can be used to test the $\mathscr{H}_0: \mu=1$ *hypothesis by using the following MATLAB statement*

```
>> [H,p,ci]=ttest(r,1,0.02) % carry out T test
```

and the results obtained are $H=0, p=0.4517, \boldsymbol{c}_i=[0.8363534005, 1.3194589956]$, *which means that the hypothesis can be accepted with 98% confidence.*

II. Hypothesis test of normality

Testing whether a set of random numbers satisfies the normal distribution can be performed with MATLAB directly. Two functions, `jbtest()` and `lillietest()` are provided in the Statistics Toolbox, which implement the Jarque–Bera and Lilliefors hypothesis test algorithms, respectively [10]. They can be used to test whether the random samples are normal or not. The syntaxes of the two functions are

```
[H,s]=jbtest(X,α)        % Jarque–Bera test
[H,s]=lillietest(X,α)    % Lilliefors test
```

Example 9.38 Consider again the data in Example 9.34, with the data given in Table 9.6. Please verify whether the normal distribution assumption is acceptable.

Solution *The data should be entered into MATLAB workspace first, and the function* `jbtest()` *can be used*

```
>> X=load('c9dlumen.dat'); % load the data into MATLAB workspace
   [H,p]=jbtest(X,0.05)    % test whether the data satisfy normal distribution
```

The obtained results are $H = 0$, $p = 0.7281$, *which means that the experimental data satisfies a normal distribution.*

Having shown that the distribution is normal, the mean and variance with confidence intervals can be estimated with the use of the `normfit()` *function*

```
>> [mu1,sig1,mu_ci,sig_ci]=normfit(X,0.05) % parameter estimation
```

and the estimated parameters are $\mu = 208.8167, \sigma^2 = 6.3232$, *and the corresponding confidence intervals are* $(207.6737, 209.9596)$, *and* $(5.6118, 7.2428)$, *respectively. The estimation results are quite close to the one obtained in Example 9.34, using maximum likelihood method.*

Example 9.39 Generate a set of random numbers satisfying Gamma distribution. Verify whether the same random numbers also satisfy a normal distribution using the hypothesis test method. Of course, we know that it does not satisfy normal distribution, and the result of hypothesis test should be 1.

Solution *The following statements can be used to generate a set of pseudorandom numbers satisfying the Gamma distribution. The function* `jbtest()` *can then be used, and the following results can be obtained.*

```
>> r=gamrnd(1,3,400,1); [H,p,c,d]=jbtest(r,0.05) % normality test
```

It can be found that $H = 1$, $p = 0$, *i.e., the hypothesis* $\mathscr{H}_0$ *should be rejected.*

III. Kolmogorov–Smirnov test for other distributions

The Jarque–Bera and Lilliefors hypothesis test algorithms can only be used to test whether the set of data to be tested satisfies normal distribution or not. They cannot be used for testing other distributions. The Kolmogorov–Smirnov test algorithm is an effective algorithm which can be used to test whether it satisfies arbitrarily given distributions. The `kstest()` function can be used `[`H`,`s`]=kstest(`$\boldsymbol{X}$`,`*cdffun*`,`α`)`, where the matrix *cdffun* is composed of two columns, one for the independent variable, and the other for the cumulative distribution function of the target distribution. By establishing the *cdffun* matrix, the existing function or the user-defined functions can be used. Thus, this MATLAB function can be used to test arbitrary distribution function.

Example 9.40 Test whether the random numbers generated in Example 9.39 satisfy the Gamma distribution.

Solution *Assume that the data satisfy Gamma distribution. Then, the function* `gamfit()` *can be used to estimate the two parameters* a *and* λ

```
>> r=gamrnd(1,3,400,1); alam=gamfit(r) % parameter estimation
```

and $\hat{a} = 1.0456$ *and* $\hat{\lambda} = 3.2868$. *The cumulative distribution function of Gamma distribution is computed by* `gamcdf(sort(r),alam(1),alam(2))`. *The results can be given in the* `kstest()` *function so that the hypothesis test can be performed*

```
>> r=sort(r); [H,p]=kstest(r,[r gamcdf(r,alam(1),alam(2))],0.05)
```

and for this example, $H = 0$ *and* $p = 0.8772$. *Thus, the hypothesis can be accepted which means that the samples satisfy Gamma distribution, with* $\hat{a} = 1.0456$ *and* $\hat{\lambda} = 3.2868$, *which validates the distribution test problem. In fact, the samples are generated by Gamma distribution with* $a = 1$ *and* $\lambda = 3$.

9.6 Analysis of Variance

The hypothesis test discussed earlier can only be used to test the behaviors of one or two statistic variables. For problems with more than two variables, *analysis of variance* should be employed. Analysis of variance, also known as *ANOVA*, is a statistical analysis approach proposed by the British statistician Ronald Fisher. It can be used in many fields, such as in medical research, scientific tests and quality control.

Since the classification of the test samples are different, the method of variance analysis may also be different. Commonly used classification methods include one-way method, two-way method and n-way method. The analysis of variance method and its solutions in MATLAB will be demonstrated via several examples.

9.6.1 One-way ANOVA

One-way analysis of variance is to check whether the external factor has a significant impact on the observations. Assume that for a certain disease, N types of medicines can be tested. The target of the analysis is determining whether the medicines have the same healing effect. In mathematics words, whether each group has the same mean.

To do the test, the patients are divided randomly into N groups, with m patients each. The healing time can be measured, and denoted as $y_{i,j}$, where i is the group number, $i = 1, 2, \cdots, N$, and j is the patient number, $j = 1, 2, \cdots, m$. Using the colon expression in MATLAB, the measured data in the ith group for all patients is $y_{i,:}$, and the data for the jth patient for all groups are $y_{:,j}$. Also, $\bar{y}_{i,:}$ denotes the average healing time in the ith group, while $\bar{y}_{:,:}$ for the average healing time for all the patients. The standard variance analysis table can be constructed as shown in Table 9.8, where SS stands for *sum of squares*, DOF stands for *degree of freedom*. Based on the table, certain necessary rules can be extracted.

Hypothesis tests are used in the analysis of variance, that is, to propose $\mathscr{H}_0$ assuming that all the average observed measures are the same. The most important quantities are the last two columns of the table, the statistic F and the probability p, which can be evaluated from the inverse distribution function. If the value of the probability $p < \alpha$, with α the selected confidence level, the hypothesis $\mathscr{H}_0$ should be rejected. Otherwise, the hypothesis cannot be rejected.

A function `anova1()` provided in the Statistics Toolbox can be used to perform one-way analysis of variance for the test data. The syntax of the function is `[p,tab,stats]=anova1(`$\boldsymbol{X}$`)`, where $\boldsymbol{X}$ is the data set to be analyzed. The experimental

TABLE 9.8: Variance analysis table.

source	sum of squares	DOF	mean squares	F	probability p
groups	$\text{SSA}=\sum_i n_i\bar{y}_{i,:}^2-N\bar{y}_{:,:}^2$	$I-1$	$\text{MSSA}=\dfrac{\text{SSA}}{I-1}$	$\dfrac{\text{MSSA}}{\text{MSSE}}$	$p=P(F_{I-1,N-I}>c)$
error	$\text{SSE}=\sum_i\sum_k y_{i,k}^2-\sum_i n_i\bar{y}_{i,:}^2$	$N-I$	$\text{MSSE}=\dfrac{\text{SSE}}{N-I}$		
total	$\text{SST}=\sum_i\sum_k y_{i,k}^2-N\bar{y}_{:,:}^2$	$N-1$			

data should be given as an $m \times n$ matrix, with each column corresponding to the measured data of the same group. The probability p, variance analysis table `tab` in the format of Table 9.8 are returned. The other returned argument `stats` is the structured data containing the statistic quantities. Two windows will also be automatically opened, one displaying the table shown in Table 9.8 and the other showing the box plot.

Example 9.41 Suppose that there are five medicines to cure a certain disease. Assume that there are 30 patients, randomly divided in 5 test groups, with 6 patients in each test group. For each test group, only one medicine is used. The healing time for each patient is recorded as shown in Table 9.9. Conclude whether there is significant difference in the effect. (Data and example source, Reference [11].)

TABLE 9.9: Experimental data on healing time (days).

patient	medicine numbers					patient	medicine numbers				
number	1	2	3	4	5	number	1	2	3	4	5
1	5	4	6	7	9	2	8	6	4	4	3
3	7	6	4	6	5	4	7	3	5	6	7
5	10	5	4	3	7	6	8	6	3	5	6

Solution *From the given table, matrix* $\boldsymbol{A}$ *can be established in MATLAB and the mean value of each column is obtained as* $[7.5, 5, 4.3333, 5.1667, 6.1667]$. *The analysis of variance is then carried out*

```
>> A=[5,4,6,7,9; 8,6,4,4,3; 7,6,4,6,5; 7,3,5,6,7; 10,5,4,3,7; 8,6,3,5,6];
   mean(A), [p,tbl,stats]=anova1(A) % analysis of variance
```

and the probability is returned as $p = 0.01359$. *Two windows are opened automatically by function* `anova1()`, *as shown respectively in Figures 9.22 (a) and (b). The value of probability is* $p = 0.0136 < \alpha$, *where* $\alpha = 0.02$ *or* 0.05. *Thus, the hypothesis is to be rejected with 98% of confidence, i.e., it can be concluded that there is significant difference among the medicines. In fact, from the box plot obtained, the healing time of medicine 3 is significantly shorter than medicine 1.*

ANOVA Table

Source	SS	df	MS	F	Prob>F
Columns	36.4667	4	9.11667	3.9	0.0136
Error	58.5	25	2.34		
Total	94.9667	29			

(a) variance table

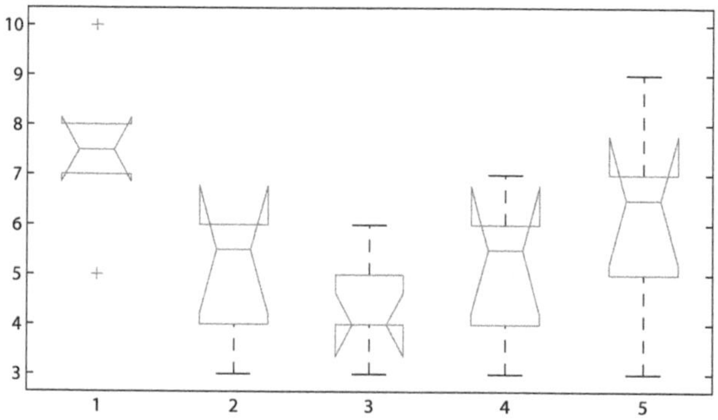

(b) box plot

FIGURE 9.22: Results of one-way analysis of variance.

9.6.2 Two-way ANOVA

If there are two factors which may affect the statistical properties of a certain phenomenon, the concept of two-way variance analysis can be used. The measured data $y_{i,j,k}$ are expressed as a three-dimensional array $y(i,j,k)$ in MATLAB.

Based on the two-way analysis, three hypotheses are introduced:

$$\begin{cases} \mathscr{H}_1 : \alpha_1 = \alpha_2 = \cdots = \alpha_I, & \alpha_i \text{ is the effect when factor 1 acts alone} \\ \mathscr{H}_2 : \beta_1 = \beta_2 = \cdots = \beta_J, & \beta_j \text{ is the effect when factor 2 acts alone} \\ \mathscr{H}_3 : \gamma_1 = \gamma_2 = \cdots = \gamma_{IJ}, & \gamma_k \text{ is the effect when both factors act.} \end{cases} \tag{9-6-1}$$

For the two-way analysis of variance, the ANOVA table constructed is shown in Table 9.10, where the interactive SSAB effect can be evaluated from (9-6-2).

$$\mathrm{SSAB} = K\sum_{ij} \bar{y}_{i,j,:}^2 - JK\sum_{i} \bar{y}_{i,:,:}^2 - IK\sum_{j} \bar{y}_{:,j,:}^2 + IJK\bar{y}_{:,:,:}^2. \tag{9-6-2}$$

The definitions of the three probabilities are

$$\begin{cases} p_A = P\left(F_{[I-1,IJ(K-1)]} > c_1\right) & \text{if } p_A < c_1 \text{ reject hypothesis } \mathscr{H}_1 \\ p_B = P\left(F_{[J-1,IJ(K-1)]} > c_2\right) & \text{if } p_B < c_2 \text{ reject hypothesis } \mathscr{H}_2 \\ p_{AB} = P\left(F_{[(I-1)(J-1),IJ(K-1)]} > c_3\right) & \text{if } p_{AB} < c_3 \text{ reject hypothesis } \mathscr{H}_3. \end{cases} \tag{9-6-3}$$

Two-way analysis of variance problems can be solved with the `anova2()` function `[`p`,tab,stats] = anova2(`$\boldsymbol{X}$`,`n`)`. The argument n is the number of subjects in each group.

TABLE 9.10: Table of two-way analysis of variance.

source	square of sums	DOF	mean squared error	F	p
factor A	$\text{SSA}=JK\sum_i \bar{y}_{i,:,:}^2 - IJK\bar{y}_{:,:,:}^2$	$I-1$	$\text{MSSA}=\dfrac{\text{SSA}}{I-1}$	$\dfrac{\text{MSSA}}{\text{MSSE}}$	p_A
factor B	$\text{SSB}=IK\sum_i \bar{y}_{:,j,:}^2 - IJK\bar{y}_{:,:,:}^2$	$J-1$	$\text{MSSB}=\dfrac{\text{SSB}}{J-1}$	$\dfrac{\text{MSSB}}{\text{MSSE}}$	p_B
interaction	SSAB see (9-6-2)	$(I-1)(J-1)$	$\text{MSSAB}=\dfrac{\text{SSAB}}{(I-1)(J-1)}$	$\dfrac{\text{MSSAB}}{\text{MSSE}}$	p_{AB}
errors	$\text{SSE}=\sum_{ijk} y_{i,j,k}^2 - K\sum_i\sum_j \bar{y}_{i,j,:}^2$	$IJ(K-1)$	$\text{MSSE}=\dfrac{\text{SSE}}{IJ(K-1)}$		
total	$\text{SST}=\sum_{ijk} y_{i,j,k}^2 - IJK\bar{y}_{:,:,:}^2$	$IJK-1$			

Example 9.42 Suppose that there are three species of pines planted in four different living conditions. Select randomly five samples in each group of subjects for measurement. The chest radii of the trees are measured, and the data are given in Table 9.11. Perform two-way analysis of variance on the data and check whether there exists significant differences (Data source, book [11]).

TABLE 9.11: Measured data for the pines.

pine	living conditions																			
species	1					2					3					4				
1	23	15	26	13	21	25	20	21	16	18	21	17	16	24	27	14	17	19	20	24
2	28	22	25	19	26	30	26	26	20	28	19	24	19	25	29	17	21	18	26	23
3	18	10	12	22	13	15	21	22	14	12	23	25	19	13	22	16	12	23	22	19

Solution *The data in the table can be entered into MATLAB workspace. Calling the function* `anova2()`*, the results are obtained as shown in Figure 9.23, which are exactly the same as the ones in Reference [11].*

```
>> B=[23,15,26,13,21,25,20,21,16,18,21,17,16,24,27,14,17,19,20,24;
      28,22,25,19,26,30,26,26,20,28,19,24,19,25,29,17,21,18,26,23;
      18,10,12,22,13,15,21,22,14,12,23,25,19,13,22,16,12,23,22,19];
   anova2(B',5); % two-way variance analysis
```

ANOVA Table

Source	SS	df	MS	F	Prob>F
Columns	355.6	2	177.8	9.68	0.0003
Rows	49.65	3	16.55	0.9	0.4478
Interaction	106.4	6	17.733	0.97	0.4588
Error	882	48	18.375		
Total	1393.65	59			

FIGURE 9.23: Results of two-way analysis of variance.

It can be seen from the results that, since p_A is very small, the hypothesis $\mathscr{H}_1$ should be rejected. It can be concluded that factor A has a significant impact upon the phenomenon observed, i.e., the tree species has significant impact on the chest radius of the trees. The other two hypothesis cannot be rejected.

9.6.3 n-way ANOVA

Similar to two-way analysis of variance presented earlier, three-way and even n-way analysis of variance are supported in the Statistics Toolbox with the function `manova1()`. Interested readers may refer to References [12, 13].

9.7 Principal Component Analysis

Principal components analysis (PCA) is an effective approach in modern statistic analysis. Assume that a phenomenon is caused simultaneously by several factors, principal component analysis method can be used to identify which of the factors are important, and the less important factors can be neglected. In this way, the dimensions of the original problem can be reduced, so as to simplify the original problems.

Assume that a phenomenon can be affected by n factors, $x_1, x_2, \cdots, x_n$, and there are M sets if measured data. An $M \times n$ matrix $\boldsymbol{X}$ can be used to represent the data, and the means of each column is denoted by $\bar{x}_i$, $i = 1, 2, \cdots, n$. The main procedures of principal component analysis are as follows:

(i) With the `corr()` function, the $n \times n$ covariance matrix $\boldsymbol{R}$ of matrix $\boldsymbol{X}$ can be established

$$r_{ij} = \frac{\sqrt{\displaystyle\sum_{k=1}^{M}(x_{ki} - \bar{x}_i)(x_{kj} - \bar{x}_j)}}{\sqrt{\displaystyle\sum_{k=1}^{M}(x_{ki} - \bar{x}_i)^2 \sum_{k=1}^{M}(x_{kj} - \bar{x}_j)^2}}. \tag{9-7-1}$$

(ii) The eigenvectors $\boldsymbol{e}_i$ of matrix $\boldsymbol{R}$ can be established, according to the sorted eigenvalues $\lambda_1 \geqslant \lambda_2 \geqslant \cdots \geqslant \lambda_n \geqslant 0$. Normalization are made to each column of the eigenvector matrix, i.e., $||\boldsymbol{e}_i|| = 1$, or $\sum_{j=1}^{n} e_{ij}^2 = 1$. The operation can be performed with `eig()` function, and since the eigenvalues were made in ascending order, it should be rearranged with `fliplr()` function.

(iii) Compute the contribution from each and the principal components

$$\text{contribution from the principal component: } \gamma_i = \frac{\lambda_i}{\displaystyle\sum_{k=1}^{n} \lambda_k}, \tag{9-7-2}$$

$$\text{accumulated contributions: } \delta_i = \frac{\displaystyle\sum_{k=1}^{i} \lambda_k}{\displaystyle\sum_{k=1}^{n} \lambda_k}. \tag{9-7-3}$$

If the contribution rate of the first s eigenvalues exceed a certain value, e.g., 85%~95%, the s factors can be regarded as the principal components, and the original n-dimensional problems can be reduced to s-dimensional problems.

(iv) Define new variables $\boldsymbol{Z} = \boldsymbol{X}\boldsymbol{L}$, such that

$$\begin{cases} z_1 = l_{11}x_1 + l_{21}x_2 + \cdots + l_{n1}x_n \\ z_2 = l_{12}x_1 + l_{22}x_2 + \cdots + l_{n2}x_n \\ \quad\vdots \\ z_n = l_{1n}x_1 + l_{2n}x_2 + \cdots + l_{nn}x_n, \end{cases} \tag{9-7-4}$$

where, the coefficients l_{ji} in the ith column of the transformation matrix can be computed with $l_{ji} = \sqrt{\lambda_i}e_{ji}$. Principal component analysis can be carried out on the obtained coefficients l_{ij}. Normally the first s components can be used as the principal components, and the entities in the columns after sth column in matrix $\boldsymbol{L}$ may be very close to zeros. The last $n-s$ variables in the z variable in (9-7-4) can be neglected, the m new variables can be used to describe the original problem.

$$\begin{cases} z_1 = l_{11}x_1 + l_{21}x_2 + \cdots + l_{n1}x_n \\ \quad\vdots \\ z_s = l_{1s}x_1 + l_{2s}x_2 + \cdots + l_{ns}x_n. \end{cases} \tag{9-7-5}$$

Under appropriate linear transformation, the original n-dimensional problem can be reduced to an s-dimensional problem.

Assume that a variable is affected by many factors, and these factors can be measured with observers in real applications. The measured data sometimes contain redundant information. Principal component method can be used to extract information, and high-dimensional problems can be reduced as low-dimensional ones. An example is given below to demonstrate the use of principal component analysis applications.

Example 9.43 Assume that a set of three-dimensional sample points are generated with the parametric equations $x = t\cos 2t, y = t\sin 2t, z = 0.2x + 0.6y$. Please reduce the dimensions with principal component analysis approach.

Solution *The data can be generated with the following MATLAB statements, and the three-dimensional curve is shown in Figure 9.24 (a).*

```
>> t=[0:0.1:3*pi]'; x=t.*cos(2*t); y=t.*sin(2*t); z=0.2*x+0.6*y;
   X=[x y z]; R=corr(X); [e,d]=eig(R), d=diag(d), plot3(x,y,z)
```

It can be seen from the figure that, the curve should be located within a certain plane. Thus, a new coordinate system can be introduced, so that the original three-dimensional curve can be compressed and shown in a two-dimensional plane.

$$\boldsymbol{R}=\begin{bmatrix} 1 & -0.0789 & 0.2536 \\ -0.0789 & 1 & 0.9443 \\ 0.2536 & 0.9443 & 1 \end{bmatrix}, \ \boldsymbol{e}=\begin{bmatrix} 0.2306 & -0.9641 & 0.1314 \\ 0.6776 & 0.256 & 0.6894 \\ -0.6983 & -0.0699 & 0.7124 \end{bmatrix}, \ \boldsymbol{d}=\begin{bmatrix} 0 \\ 1.0393 \\ 1.9607 \end{bmatrix}.$$

It can be seen that the final $\boldsymbol{L}$ matrix can be established, and since the first two eigenvalues are very large while the third one tends to zero, two factors are necessary for effectively describing the original problem.

```
>> d=d(end:-1:1); e=fliplr(e); D=[d'; d'; d'];      % sort the eigenvalues
   L=real(sqrt(D)).*e, Z=X*L; plot(Z(:,1),Z(:,2)) % draw 2D plot
```

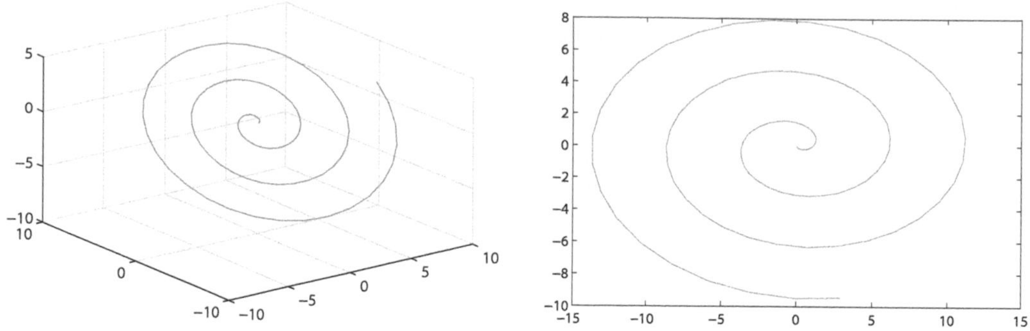

(a) original 3D curve (b) reduced 2D curve

FIGURE 9.24: Three-dimensional curve and dimension reduction effect.

Thus, the transformation matrix can be found as

$$\boldsymbol{L}=\begin{bmatrix}0.1840 & -0.9829 & 0\\ 0.9653 & 0.2610 & 0\\ 0.9975 & -0.0713 & 0\end{bmatrix},$$

i.e., a new set of coordinates can be introduced

$$\begin{cases} z_1 = 0.1840x + 0.9653y + 0.9975z \\ z_2 = -0.9829x + 0.2610y - 0.0713z. \end{cases}$$

In this way, the three-dimensional problem can be reduced to a two-dimensional problem. The reduced two-dimensional curve can be obtained as shown in Figure 9.24 (b). It can be seen that the two-dimensional curve can be extracted, and it contains all the information of the original plot.

Exercises

Exercise 9.1 *The PDF of Rayleigh distribution is given by*

$$p_{\rm r}(x)=\begin{cases}\dfrac{x}{b^2}\mathrm{e}^{-x^2/(2b^2)}, & x\geqslant 0\\ 0, & x<0.\end{cases}$$

Derive analytically the cumulative distribution function, mean, variance, central moment and raw moments of the distribution. Generate a pseudorandom sequence satisfying Rayleigh distribution and verify numerically whether the numerical generation is correct or not.

Exercise 9.2 *Assume that the number ξ of cars passing through a traffic light satisfies Poisson distribution. Also, assume that the probabilities of no car passing through and exactly one car passing through are the same. Please find out the probability of at least two cars passing through the traffic light.*

Exercise 9.3 *Assume that in a foreign language examination, the randomly selected samples indicate that the scores satisfy approximately the normal distribution, with a mean value of 72. The number of those scores higher than 96 is 2.3% of all the number of students. Find the probability of a student whose score is between 60 and 80.*

Exercise 9.4 *Generate 30000 pseudorandom numbers satisfying normal distribution of* $N(0.5, 1.4^2)$*. Find the mean and standard deviation of the data set. Observe the histogram of the data to see whether they agree with theoretical distribution. Change the width of the bins and see what may happen.*

Exercise 9.5 *Please carry out coin-tossing experiment with MATLAB. If you toss a coin 1000 times, check how many times you get heads. What happens if you toss the coin 10000, 100000 or more times? —* Hints: *You may define a standard uniformly distributed number* $\xi > 0.5$ *as heads in the computerized experiment.*

Exercise 9.6 *Assume a set of data is measured and given in Table 9.12* [7]*. Please find the three quartiles, and draw the box plot. Are there any outliers in the data set?*

TABLE 9.12: The measured data in Exercise 9.6.

0	3.9	5.64	8.22	0	5.62	3.92	6.81	30.61	0	73.2	0	46.7	0
0	26.41	22.82	0	0	3.49	0	0	4.81	9.57	5.36	0	5.66	0
59.76	12.38	15.74	0	0	0	0	9.37	20.78	7.1	7.89	5.53		

Exercise 9.7 *Assume that a set of data was measured as shown in Table 9.13. Use MATLAB to perform the following hypothesis tests:*

(i) Assume that the data satisfies normal distribution with a standard deviation of 1.5. Test whether the mean value of the data is 0.5.

(ii) If the standard deviation is not known, test whether the mean is still 0.5.

(iii) Test whether the distribution of the data is normal.

TABLE 9.13: The measured data in Exercise 9.7.

−1.7908	0.3238	4.6927	−2.3586	−0.0940	2.8943	5.3067	3.1634	−3.2812	−3.4389
0.0903	2.5006	−0.6758	−3.2431	−3.2440	−0.0521	−0.0796	0.4653	−0.7905	2.0690
3.9223	5.6959	0.7327	−1.083	−1.8152	−2.9145	−2.6714	1.7065	0.0819	2.3258
0.4135	1.6804	−1.3172	1.132	1.047	0.5219	4.4827	−1.112	0.5201	1.9318
3.2618	0.4735	2.031	−0.7177	−2.3273	0.6606	1.2325	−0.9750	2.3831	3.4477
−1.0665	2.5546	−4.8203	−2.5004	−0.2812	1.2122	−2.0178	1.2073	−1.1251	1.236
0.5169	0.6259	2.7278	2.9135	−1.6181	1.6246	1.8958	0.7403	−1.1234	−1.0142
−1.2615	−1.9909	0.9925	−1.1022	−2.1428	3.3757	3.357	4.6585	0.04734	0.1640
1.8206	1.5924	1.0887	0.47461	−1.7976	−0.7326	−1.5161	−0.1190	0.4540	−5.0103
−0.0652	0.48874	3.2303	0.49816	−0.40375	1.0868	0.80414	5.4782	1.1275	1.5649
1.5803	−0.1215	−0.118	−0.0612	0.8908	0.4704	0.1872	3.8942	2.8812	0.7631
2.0033	3.372	0.2005	1.3923	0.23873	−0.80559	−2.1176	−3.8764	1.8988	−0.8300

Exercise 9.8 *For a prototype linear function* $y = a_1x_1 + a_2x_2 + a_3x_3 + a_4x_4 + a_5x_5$*, with five independent variables* x_1, x_2, x_3, x_4, x_5 *and one output* y*, the data in Table 9.14 are obtained. Find the values* a_i *and their confidence intervals using linear regression method.*

TABLE 9.14: The measured data in Exercise 9.8.

x_1	8.11	9.25	7.63	7.89	12.94	10.11	7.57	9.92	7.74	7.3	9.48	11.91
x_2	2.13	2.66	0.83	1.54	1.74	0.79	0.68	2.93	2.01	1.35	2.81	2.23
x_3	3.98	−0.68	1.42	−0.96	−0.28	3.37	4.58	2.15	2.66	3.69	1	−0.98
x_4	6.55	6.85	6.25	5.34	6.85	7.2	6.12	6.07	5.51	6.6	6.15	6.43
x_5	5.92	7.54	5.39	4.65	6.47	5.1	6.04	5.37	6.54	6.55	5.8	3.95
y	27.676	38.774	23.314	23.828	35.154	21.779	25.516	29.845	32.642	28.443	31.5	23.554

Exercise 9.9 *Assume that a set of measured data x_i and y_i are given in Table 9.15, and the prototype function is $f(x) = a_1\mathrm{e}^{-a_2x}\cos(a_3x+\pi/3)+a_4\mathrm{e}^{-a_5x}\cos(a_6x+\pi/4)$. Estimate the values of a_i and their confidence intervals.*

TABLE 9.15: The measured data in Exercise 9.9.

x	1.027	1.319	1.204	0.684	0.984	0.864	0.795	0.753	1.058	0.914	1.011	0.926
y	8.8797	5.9644	7.1057	8.6905	9.2509	9.9224	9.8899	9.6364	8.5883	9.7277	9.023	9.6605

Exercise 9.10 *Suppose that tests have been made on a group of randomly selected fuses, and it is found that the burn-out currents of the fuses are 10.4, 10.2, 12.0, 11.3, 10.7, 10.6, 10.9, 10.8, 10.2, 12.1 A. Suppose that these random values satisfy normal distribution. Find the burn-out current and its confidence interval under the confidence level $\alpha \leqslant 0.05$.*

Exercise 9.11 *Assume that the boiling points under certain atmospheric pressures are tested with the multiple measured data 113.53, 120.25, 106.02, 101.05, 116.46, 110.33, 103.95, 109.29, 93.93, 118.67° C. Check whether they satisfy normal distribution under the confidence level of $\alpha \leqslant 0.05$.*

Exercise 9.12 *Assume the measured data given in Table 9.16 satisfies the following prototype function $y(t) = c_1\mathrm{e}^{-5t}\sin(c_2t) + (c_3t^2 + c_4t^3)\mathrm{e}^{-3t}$. Find from the data the parameters c_i's and their confidence interval.*

TABLE 9.16: The measured data in Exercise 9.12.

t	0	0.1	0.2	0.3	0.4	0.5	0.6	0.7	0.8	0.9
y	0	0.1456	0.2266	0.2796	0.3187	0.3479	0.3677	0.3777	0.3782	0.37
t	1	1.1	1.2	1.3	1.4	1.5	1.6	1.7	1.8	1.9
y	0.3546	0.3335	0.3085	0.2812	0.253	0.225	0.198	0.1726	0.1492	0.1279
t	2	2.1	2.2	2.3	2.4	2.5	2.6	2.7	2.8	2.9
y	0.109	0.0922	0.0776	0.065	0.0541	0.0449	0.0371	0.0305	0.025	0.0204

Exercise 9.13 *Assume that 12 sample plants are randomly selected from areas A and B. The iron element content in $\mu g/g$ is measured as shown in Table 9.17. Assume that the iron element content in the plant satisfies a normal distribution and the variance of the distribution is not affected by the area. Test whether the distribution of the iron element content is the same.*

TABLE 9.17: The measured data in Exercise 9.13.

area A	11.5	18.6	7.6	18.2	11.4	16.5	19.2	10.1	11.2	9	14	15.3
area B	16.2	15.2	12.3	9.7	10.2	19.5	17	12	18	9	19	10

Exercise 9.14 *Assume that 12 samples are obtained for a stochastic variable as 9.78, 9.17, 10.06, 10.14, 9.43, 10.60, 10.59, 9.98, 10.16, 10.09, 9.91, 10.36. Find the deviation of the data and its confidence interval.*

Exercise 9.15 *Assume that stochastic variables A and B are sampled in Table 9.18. Check whether they have significant statistic differences.*

TABLE 9.18: The measured data in Exercise 9.15.

A	10.42	10.48	7.98	8.52	12.16	9.74	10.78	10.18	8.73	8.88	10.89	8.1
B	12.94	12.68	11.01	11.68	10.57	9.36	13.18	11.38	12.39	12.28	12.03	10.8

Exercise 9.16 *Suppose that five different dyeing techniques are tested for the same cloth. Different dyeing techniques and different machines are tested randomly, and the percentage of washing shrinkage are given in Table 9.19. Judge whether the dyeing techniques have significant effect on the washing shrinkage.*

TABLE 9.19: The measured data in Exercise 9.16.

machine	dyeing techniques					machine	dyeing techniques				
number	1	2	3	4	5	number	1	2	3	4	5
1	4.3	6.1	6.5	9.3	9.5	2	7.8	7.3	8.3	8.7	8.8
3	3.2	4.2	8.6	7.2	11.4	4	6.5	4.2	8.2	10.1	7.8

Exercise 9.17 *Assume that the heights of randomly selected Year-5 pupils in three schools are measured in Table 9.20. Check whether there are significant differences in the heights in these three schools.* ($\alpha = 0.05$)

Exercise 9.18 *The table in Table 9.21 recorded the daily output of three operators on four different machines. Check the following*

(i) whether there are significant differences in the skill of the operators

(ii) whether there are significant differences in the machines

(iii) whether the interaction is significant ($\alpha = 0.05$)

Bibliography

[1] Grinstead C M, Snell J L. Grinstead and Snell's introduction to probability. The CHANCE Project: Open source textbook at `http://math.dartmouth.edu/~prob/prob/prob.pdf`, 2006

TABLE 9.20: The measured data in Exercise 9.17.

school	measured height data					
1	128.1	134.1	133.1	138.9	140.8	127.4
2	150.3	147.9	136.8	126	150.7	155.8
3	140.6	143.1	144.5	143.7	148.5	146.4

TABLE 9.21: The measured data in Exercise 9.18.

machine	operator number									machine	operator number								
number	1			2			3			number	1			2			3		
M_1	15	15	17	19	19	16	16	18	21	M_3	15	17	16	18	17	16	18	18	18
M_2	17	17	17	15	15	15	19	22	22	M_4	18	20	22	15	16	17	17	17	17

[2] StatSoft Inc. Electronic statistics textbook. Tulsa, OK: StatSoft. Electronics textbook at `http://www.statsoft.com/textbook/stathome.html`, 2007

[3] Janicki A, Weron A. Simulation and chaotic behavior of α-stable stochastic processes. New York: Marcel Dekker Inc, 1994

[4] Veillette M. STBL: α-stable distributions for MATLAB. MATLAB Central File ID: # 37514, 2012

[5] Ross M S. Introduction to probability and statistics for engineers and scientists (4th edition). Burlington, MA: Elsevier Academic Press, 2009

[6] Landau D P, Binder K. A guide to Monte Carlo simulations in statistical physics. Cambridge University Press, 2000

[7] Moore D S, McCabe G P, Craig B A. Introduction to the practice of statistics, the 6th Edition. New York: W H Freeman and Company, 2007

[8] Battistini N. Outliers. MATLAB Central File ID: #35048

[9] Trujillo-Ortiz A. Moutlier1. MATLAB Central File ID: #12252

[10] Conover W J. Practical nonparametric statistics. New York: Wiley, 1980

[11] Lu X. Applied statistics. Beijing: Tsinghua University Press, 1999. (in Chinese)

[12] Cody R P, Smith J K. Applied statistics and the SAS programming language. Prentice Hall, fifth edition, 2006

[13] The MathWorks Inc. Statistics toolbox user's manual, 2007

Chapter 10

Topics on Nontraditional Mathematical Branches

In the previous chapters, traditional branches of advanced applied mathematics have been summarized, and our focus was on computer aided solutions to those problems. Over the last few decades, many new applied math topics emerged which are referred to as *nontraditional mathematics* in this book. For instance, fuzzy logic and fuzzy inference are presented and used for imitating imprecise human thinking and linguistic behaviors. The artificial neural networks are established based on the mathematical model imitating the neural network of biological systems. The genetic algorithm-based optimization procedures are proposed based on the principles of survival of the fittest. These new branches of applied mathematics are promising areas of research in science and engineering offering important tools for real-life problems. In Section 10.1, classical set theory, fuzzy set and fuzzy inference are presented with their implementations in MATLAB. Fundamental introduction to rough set and rough set-based attribute reduction is given in Section 10.2 with real-life examples. Section 10.3 introduces artificial neural network in general and feedforward neural network in particular with back-propagation algorithms, where MATLAB solutions for network construction, training and generalization are presented using data fitting problems for illustration. Radial basis neural networks are also illustrated, and graphical user interface-based solution is demonstrated. In Section 10.4, evolutionary optimization algorithms, including genetic algorithms and particle swarm optimization methods, are introduced. The global solutions to optimization problems are also explored. Wavelet-based methods and solutions are given in Section 10.5 using signal de-noising as an application example. In Section 10.6, a comprehensive introduction to fractional-order calculus and numerical solutions to fractional-order ordinary differential equations is given, where from a programming point of view, the design and application of the classes and objects dedicated for fractional-order calculus are demonstrated thoroughly.

It should be noted that only very brief introductions to the mathematical background and theoretical description are given. Our focus, again, is on the solutions of the related math problems using MATLAB.

10.1 Fuzzy Logic and Fuzzy Inference

10.1.1 MATLAB solutions to classical set problems

Set theory is the foundation of modern mathematics. The so-called *set* is a collection of objects, with each object defined as a *member* of the set. If the object a is a member of set A, it is denoted as $a \in A$, and we call it "a belongs to set A." If b is not a member of the set A, it is denoted as $b \notin A$. A set is called an *enumerable set* if all its members can be enumerated. In MATLAB, enumerable sets can be represented by vectors or cell arrays.

Example 10.1 The following MATLAB statements can be used to define enumerable sets

```
>> A=[1 2 3 5 6 7 9 3 4 11]  % set of digits, repeated members are allowed
   B={1 2 3 5 6 7 9 3 4 11}  % the above set can also be described by cells
   C={'ssa','jsjhs','su','whi','kjshd','kshk'}  % set of strings
```

The set operations provided in MATLAB are summarized in Table 10.1 with brief explanations. These functions can be nested to establish complicated set operations.

TABLE 10.1: Set operations under MATLAB.

operations	MATLAB functions	descriptions to set operations
set union	A=`union(`B,C`)`	Union of sets B and C, such that $A = B \bigcup C$, where the returned results are sorted
difference	A=`setdiff(`B,C`)`	The difference of the sets B and C, denoted as $A = B \setminus C$. This operation removes the members in set C from set B. The remaining members, after sorting, are returned
intersection	A=`intersect(`B,C`)`	The sorted results of the intersect of sets B and C, where $A = B \bigcap C$
exclusive or	A=`setxor(`B,C`)`	Exclusive or operation of the sets B and C, i.e., removes set $B \bigcap C$ from set $B \bigcup C$, mathematically $A = (B \bigcup C) \setminus (B \bigcap C)$. Sorted results are returned
unique	A=`unique(`B`)`	Find the non-repeating elements in set B, i.e., to delete the extra repeated members with returned set sorted
belong to	`key`=`ismember(`a,B`)`	Check where a belongs to set B, `key`$= a \in B$, returns 1 or 0. If a is a set, `key` is a vector of 0's and 1's

Example 10.2 For the three sets $A = \{1, 4, 5, 8, 7, 3\}$, $B = \{2, 4, 6, 8, 10\}$ and $C = \{1, 7, 4, 2, 7, 9, 8\}$, perform the relevant set operations. Verify the commutative law $(A \bigcup B) \bigcap C = (A \bigcap C) \bigcup (B \bigcap C)$.

Solution *The three sets can be expressed in MATLAB easily and the related functions can be called to perform set operations*

```
>> A=[1,4,5,8,7,3]; B=[2,4,6,8,10]; C=[1,7,4,2,7,9,8]; % define sets
   D=unique(C), E=union(A,B), F=intersect(A,B)  % set operations
```

and the unique set D is found such that $D = [1, 2, 4, 7, 8, 9]$, where the extra repeated set member 7 is removed, and the result is sorted. The union and intersection are also found as $E = [1, 2, 3, 4, 5, 6, 7, 8, 10]$, and $F = [4, 8]$.

The following statement is used to verify the commutative law where the set difference is taken between the sets at the left-hand side and the right-hand side. The result is an empty matrix, indicating that the commutative law holds for the given sets.

```
>> L=intersect(union(A,B),C); R=union(intersect(A,C),intersect(B,C));
   G=setdiff(L,R) % the set difference is an empty set
```

The function `ismember()` *can be called such that*

```
>> H=ismember(A,B), I=A(ismember(A,B)) % show I = A∩B
```

and it is found that $H = [0,1,0,1,0,0]$, *indicating the second and fourth members in set* A, *i.e., members 4 and 8, belong to set* B. *These members can be extracted to form set* I, *which is the same as the intersection of sets* A *and* B, *i.e.,* $I = A\bigcap B$.

Example 10.3 Consider the sets A and B containing strings
$A = \{$`'skhsak','ssd','ssfa'`$\}$, and $B = \{$`'sdsd','ssd','sssf'`$\}$.
Find the union and intersection of A and B. If $C=\{$`'jsg','sjjfs','ssd'`$\}$, verify the *distributive law* such that $(A\bigcup C)\bigcap B = (A\bigcap B)\bigcup(C\bigcap B)$.

Solution *The set containing strings can be expressed by cell objects. The set operations can be calculated easily by the following scripts*

```
>> A={'skhsak','ssd','ssfa'}; B={'sdsd','ssd','sssf'};
   F=union(A,B), D=intersect(A,B) % simple set computation
```

and the union and intersection of sets A *and* B *are found*
$F=\{$`'sdsd','skhsak','ssd','ssfa','sssf'`$\}$, $D=\{$`'ssd'`$\}$.

When set C *is specified, the distributive law can be verified with the following statements, since, the result* R *is an empty set.*

```
>> C={'jsg','sjjfs','ssd'}; L=intersect(union(A,C),B);
   R=union(intersect(A,B),intersect(C,B)); E=setdiff(L,R)
```

Subset and set inclusion are the important concepts in set theory. The so-called *set inclusion* means that if all the members in set A belong to set B, it is said that A is included in set B, denoted by $A \subseteq B$, or if set B includes set A, A is also called a *subset* of set B, while B is referred to as a *superset* of A. If $B\setminus A$ is not empty, the set inclusion is referred to as *strict inclusion*, denoted by $A \subset B$. Set A is then referred to as a *proper subset* of B. Set inclusion functions are not directly provided in MATLAB, but the following statements can be used to check the set inclusion and strict inclusion.

```
v=all(ismember(A,B)) % v=1 for A ⊆ B, all the elements in A belong to B
v=all(ismember(A,B))&(length(setdiff(B,A))>0) % v=1 for A ⊂ B
```

Example 10.4 Consider the sets E, F in Example 10.2, check whether $F \subset E$ is true. Verify the reflexive law of set A, i.e., $A \subseteq A$.

Solution *The following statements can be used to confirm the set inclusion relationship, with* `k`$=1$, *meaning* F *is a subset of* E.

```
>> A=[1,4,5,8,7,3]; B=[2,4,6,8,10]; % input the two sets
   E=union(A,B); F=intersect(A,B); k=all(ismember(F,E)) % set operations
```

In fact, $F = A\bigcup B$ *and* $E = A\bigcap B$, *thus,* $E \subset F$. *It can also be verified that* $A \subseteq A$, *i.e., the reflexive law, with* `key`$=0$, `key1`$=1$, *meaning* A *is not a proper subset of* A. *However,* A *is a subset of set* A.

```
>> key=all(ismember(A,A)) & (length(setdiff(A,A))>0)
   key1=all(ismember(A,A)) % A ⊆ A is satisfied
```

Example 10.5 Let us consider the verification of the well-known *Goldbach's conjecture*, which states that any even integer greater than 2 can be expressed by the sum of two prime numbers. It is one of the oldest unsolved problems in number theory which dates back to the time of Euler.

Solution *For finite even integers, the full set of all the possible sums of two prime numbers within a given range can be constructed as set* c*. One can then check whether the terms in the set of even integers* c_1 *not belonging to* c *is empty. The following statements can be used to verify the conjecture for even integers in* $[4, 2000]$.

```
>> iA=primes(1040); c=[]; for i=iA, c=[c i+iA]; end,
   c=unique(c); c1=4:2:2000; c2=ismember(c1,c); C=c1(c2==0)
```

It can be seen that the set C *is an empty one, which verifies the conjecture for small integer numbers. It should be noted that a larger* c *should be constructed. With the parallel computers, the even integers* c *up to* 4×10^{18} *have been verified with no exception* [1].

10.1.2 Fuzzy sets and membership functions

I. Membership functions

From the classical set theory, it is seen that an event a either belongs to set A, or does not belong to set A. There is no other relationship. In modern science and engineering, the fuzzy concepts might be useful, which may require that the event a belongs to set A to some extent. This is the fundamental motivation of fuzzy set theory.

The concept of fuzzy set was proposed by Professor Lotfi A. Zadeh in 1965 [2]. Currently the concept of fuzzy logic has been applied to almost all research and application areas. For example, fuzzy control is an attractive and promising research topic in control engineering.

Actually, fuzzy concept has been used earlier in the book. For instance, in the variable-step computation, it has been stated that "when the error is large $\cdots$." The word "large" is a fuzzy description. However, fuzzy procedures were not used there.

In real world situations, precision is not everything. Professor Zadeh points out that "As complexity rises, precise statements lose meaning, and meaningful statements lose precision."

Example 10.6 The membership functions of fuzzy sets "old" and "young" were introduced by Professor Zadeh. Assume that the universe $U = [0, 120]$, the membership functions are defined as

$$\mu_{\rm O}(u) = \begin{cases} 0, & 0 \leqslant u \leqslant 50 \\ \left[1 + \left(\dfrac{u-50}{5}\right)^{-2}\right]^{-1}, & 50 < u \leqslant 120, \end{cases}$$

$$\mu_{\rm Y}(u) = \begin{cases} \left[1 + \left(\dfrac{u-25}{5}\right)^{-2}\right]^{-1}, & 0 \leqslant u \leqslant 25 \\ 0, & 25 < u \leqslant 120. \end{cases}$$

The plots of the two membership functions can be drawn with the following statements, as shown in Figure 10.1.

```
>> u=0:0.1:120; mu_o=1./(1+((u-50)/5).^(-2)).*(u>50);
   mu_y=1./(1+((u-25)/5).^(-2)).*(u<25); plot(u,mu_y,u,mu_o)
```

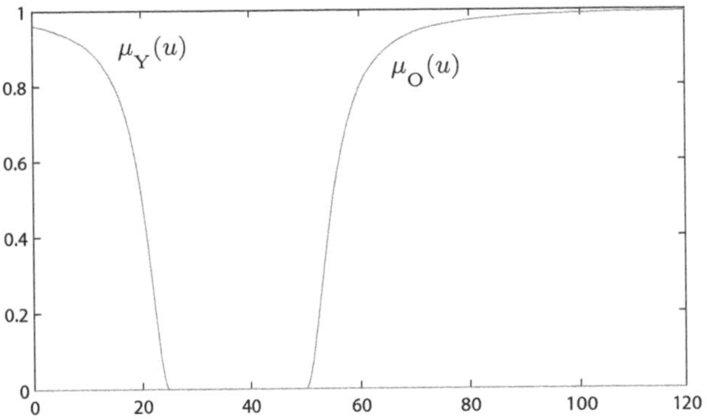

FIGURE 10.1: Membership functions of "old" and "young" fuzzy sets.

Membership function is a very important concept in fuzzy set theory. It states the "grade of membership" of an element a belonging to set A, denoted as $\mu_A(a)$. The membership function of classical set is either 1 or 0, indicating a either belongs to A or does not. However, for fuzzy sets, the value of the membership function is between 0 and 1. Some of the commonly used membership functions are summarized below.

(i) **Bell-shaped membership functions** The mathematical description to bell-shaped membership function is

$$f(x) = \frac{1}{1+\left|\dfrac{x-c}{a}\right|^{2b}}, \tag{10-1-1}$$

and function `gbellmf()` in Fuzzy Logic Toolbox can be used to evaluate the membership function with $\boldsymbol{y}$=`gbellmf(`$\boldsymbol{x}$`,[`a,b,c`])`, where $\boldsymbol{x}$ is the value of independent variable $\boldsymbol{x}$, and $\boldsymbol{y}$ contains the values of membership functions.

Example 10.7 The following statements can be used to draw the bell-shaped membership functions for different combinations of parameters a, b, c, in Figure 10.2.

```
>> x=[0:0.05:10]'; y=[]; a0=1:5; b=2; c=3;
   for a=a0, y=[y gbellmf(x,[a,b,c])]; end
   y1=[]; a=1; b0=1:4; c=3; for b=b0, y1=[y1 gbellmf(x,[a,b,c])]; end
   y2=[]; a=2; b=2; c0=1:4; for c=c0, y2=[y2 gbellmf(x,[a,b,c])]; end
   subplot(131), plot(x,y); subplot(132), plot(x,y1);
   subplot(133), plot(x,y2)
```

It can be seen from the curves that the shapes of the membership function can be modified by changing respectively the parameters a, b, c. One may use these parameters to shape the membership functions depending on the application requirements.

(ii) **Gaussian membership functions** The mathematical description to Gaussian membership function is given by

$$f(x) = \mathrm{e}^{-\frac{(x-c)^2}{2\sigma^2}}, \tag{10-1-2}$$

and function `gaussmf()` in Fuzzy Logic Toolbox can be used to evaluate the membership function with $\boldsymbol{y}$=`gaussmf(`$\boldsymbol{x}$`,[`σ,c`])`, where $\boldsymbol{x}$ is the value of independent variable $\boldsymbol{x}$, and $\boldsymbol{y}$ contains the values of membership function.

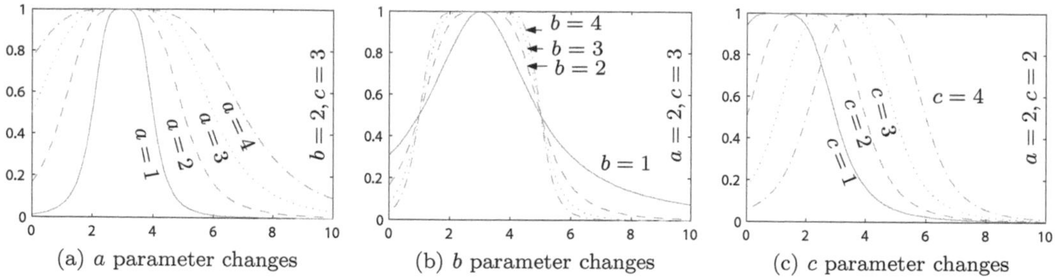

$a=1$
$a=2$
$a=3$
$a=4$
$b=2, c=3$
(a) a parameter changes
$b=4$
$b=3$
$b=2$
$b=1$
$a=2, c=3$
(b) b parameter changes
$c=1$
$c=2$
$c=3$
$c=4$
$a=2, c=2$
(c) c parameter changes

FIGURE 10.2: Bell-shaped membership function.

Example 10.8 For different values of c and σ, the Gaussian membership function can be drawn as shown in Figure 10.3. The shape of the function is the same as the normal PDF studied in Chapter 9. It can be seen that when the values of c change, the shape of the membership function is unchanged. Only translations are made.

```
>> x=[0:0.05:10]'; y=[]; c0=1:4; s=3; y1=[]; c=5; sig0=1:4;
   for c=c0, y=[y gaussmf(x,[s,c])]; end
   for sig=sig0, y1=[y1 gaussmf(x,[sig,c])]; end;
   subplot(121), plot(x,y); subplot(123), plot(x,y1)
```

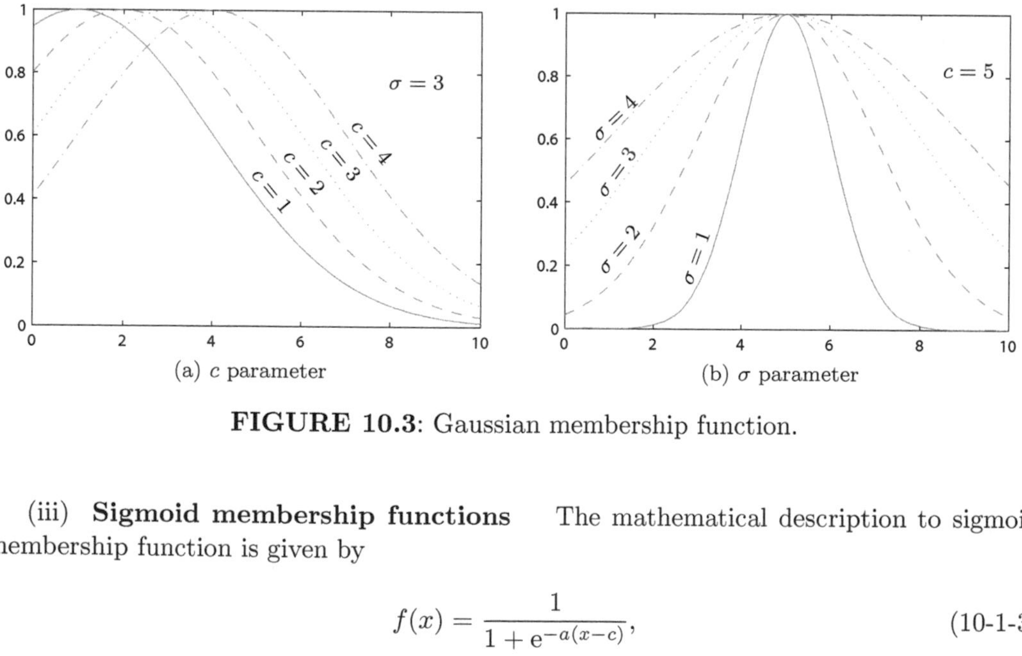

$\sigma=3$
$c=1$
$c=2$
$c=3$
$c=4$
(a) c parameter
$c=5$
$\sigma=1$
$\sigma=2$
$\sigma=3$
$\sigma=4$
(b) σ parameter

FIGURE 10.3: Gaussian membership function.

(iii) **Sigmoid membership functions** The mathematical description to sigmoid membership function is given by

$$f(x) = \frac{1}{1+\mathrm{e}^{-a(x-c)}}, \tag{10-1-3}$$

which can be evaluated with the MATLAB function $\boldsymbol{y}$=`sigmf(`$\boldsymbol{x}$`,[`a`,`c`])`.

Example 10.9 The shapes of sigmoid function for different parameters of a and c are visualized in Figure 10.4. When c varies, the curve may translate to left or right with the shape of the curve unchanged.

```
>> x=[0:0.05:10]'; y=[]; c0=1:3; a=3;
   for c=c0, y=[y sigmf(x,[a,c])]; end
   y1=[]; c=5; a0=1:2:5; for a=a0, y1=[y1 sigmf(x,[a,c])]; end;
   subplot(121), plot(x,y); subplot(122), plot(x,y1)
```

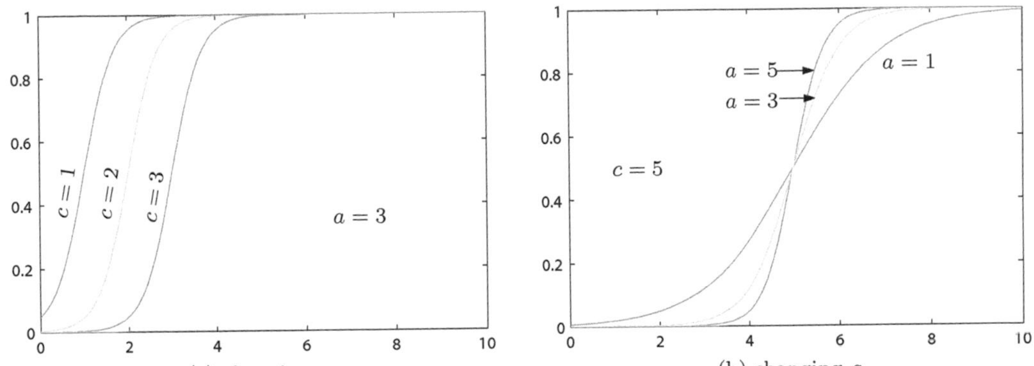

(a) changing c (b) changing a

FIGURE 10.4: Sigmoid membership function.

II. An interactive membership function editor

A graphical user interface is provided for membership function manipulations in the Fuzzy Logic Toolbox: the command `mfedit` can be used to open the interface shown in Figure 10.5. The prototypes of three membership functions are given, and one can edit the membership function with mouse drag and click.

If one more fuzzy set, or membership function is to be added, the menu item Edit → Add custom MF can be selected and a dialog box is shown in Figure 10.6 (a). The membership function can then be introduced as shown in Figure 10.6 (b).

III. Building fuzzy inference systems

The `newfis()` function provided in the Fuzzy Logic Toolbox can be used to construct the data structure of the fuzzy inference system (FIS). The syntax of the function is *fis* = `newfis(`*name*`)`, where the string variable *name* is used to describe the name of the FIS. The structured variable `fis` can be established. The properties in the FIS include the fuzzification, fuzzy inference and defuzzification, etc. These properties can also be used in the definition of `newfis()` function, or, they can be defined later. Having defined the FIS object `fis`, the function `addvar()` can be used to define input and output variables to the object, with the following syntaxes

```
fis=addvar(fis,'input',iname,vi)    % add an input variable iname
fis=addvar(fis,'output',oname,vo)   % add an output variable oname
```

where $v_{\rm i}$ and $v_{\rm o}$ are the ranges, i.e., the *universes*, of the input and output variables. They are described as row vectors of the minimum and maximum of the variable. The input and output variables can also be defined with the graphical user interface `fuzzy`. The membership functions can be specified with the `addmf()` function, or directly, with the `mfedit()` interface.

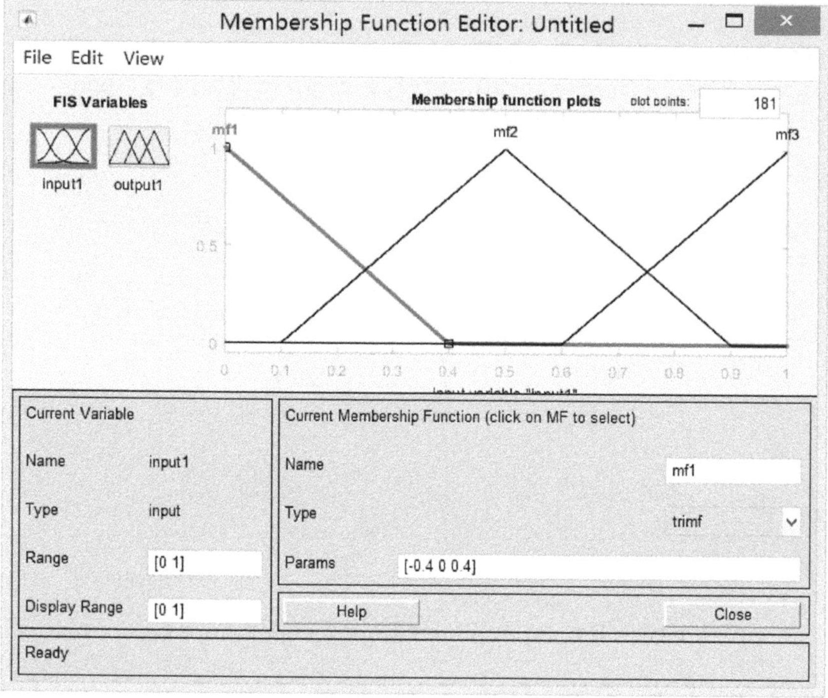

FIGURE 10.5: Editing interface for membership functions.

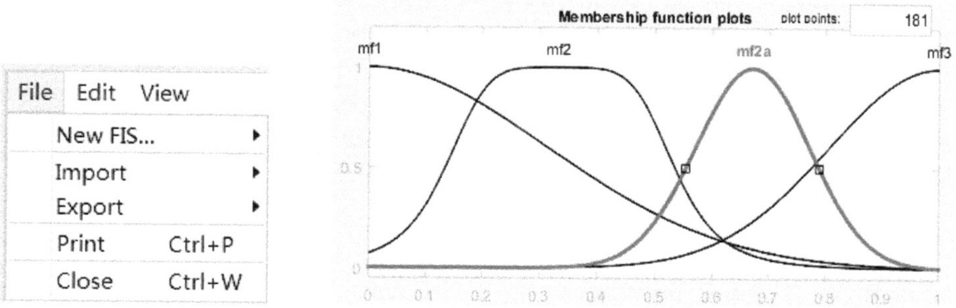

(a) Edit menu

(b) modified membership function

FIGURE 10.6: Modifications of membership functions.

Example 10.10 Suppose that there are two input variables, $\mathtt{ip}_1$ and $\mathtt{ip}_2$, and one output variable `op`. Assume that the universe of the input $\mathtt{ip}_1$ is $(-3, 3)$, with three membership functions, all selected as bell-shaped function. The universe for input signal $\mathtt{ip}_2$ is $(-5, 5)$, also three Gaussian-type membership functions are defined. The universe of the output `op` is $(-2, 2)$, whose membership function is sigmoidal functions. The framework of the fuzzy inference system can be established. Also, the graphical user interface `fuzzy()` can be used to edit the FIS.

```
>> fff=newfis('c10mfis');        % set up a new fuzzy inference model
   fff=addvar(fff,'input','ip1',[-3,3]);  % define input 1
   fff=addvar(fff,'input','ip2',[-5,5]);  % define input 1
```

```
fff=addvar(fff,'output','op',[-2,2]);  % define the output
fuzzy(fff)                % edit using graphical interface fuzzy()
```

The fuzzy inference system modification interface can be opened with the function `fuzzy()`, as shown in Figure 10.7.

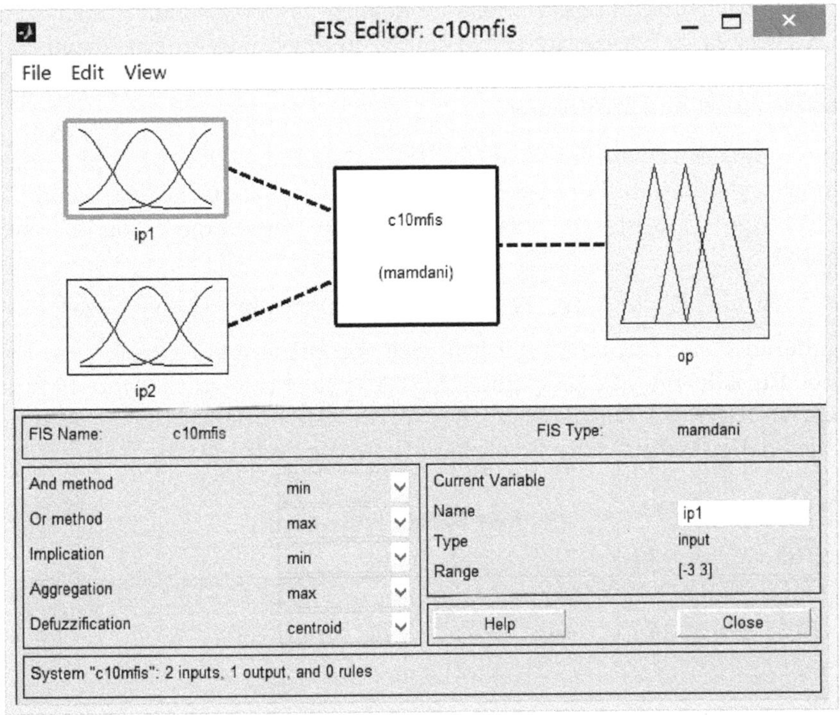

FIGURE 10.7: Graphical user interface of fuzzy inference system.

The Edit → Membership functions menu item can be selected in the interface, and the membership function editing interface will be displayed as shown in Figure 10.5. The icon labeled $\mathtt{ip}_1$ in the interface can be selected, and Edit → Add MFs menu item can be selected, a dialog box shown in Figure 10.8 (a) opens. The user can define the membership functions in the dialog box. For instance, the output membership function, after editing procedure, can be obtained as shown in Figure 10.8 (b).

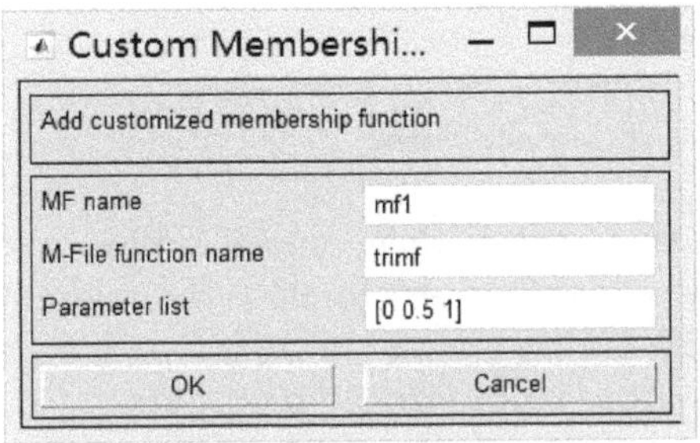

(a) membership function dialog box

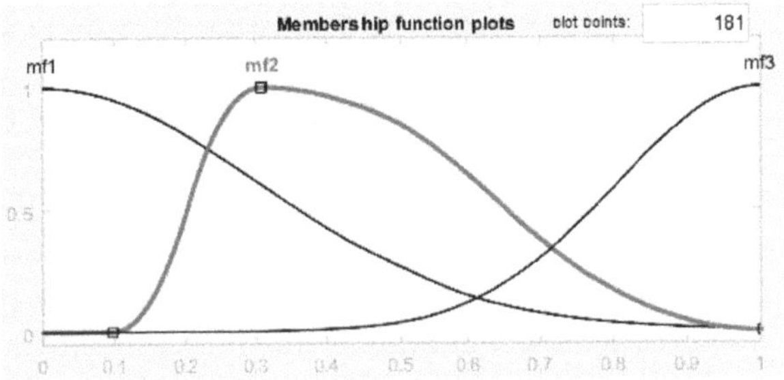

(b) modified membership function for output

FIGURE 10.8: Edited results of membership functions.

10.1.3 Fuzzy rules and fuzzy inference

I. Fuzzification

If three fuzzy sets or membership functions are defined for a certain signal, the physical meanings of the three sets may be "very small," "medium" and "very large." If five sets are used, the physical meanings can also be defined as "very small," "small," "medium," "large" and "very large." An exact signal can be fuzzified into a fuzzy signal.

II. Fuzzy rules and inference

If all the input signals are fuzzified, the `if – else` clauses can be used to represent the fuzzy inference relationship. For instance, if the input signal $\mathtt{ip}_1$ is "very small," and input $\mathtt{ip}_2$ is "very large," then, set the output signal `op` to "very large." The inference rule can be represented by

`if` $\mathtt{ip}_1$ is " very small" `and` $\mathtt{ip}_2$ is "very large," `then,` `op`= "very large"

Fuzzy inference rule can be established with the `ruleedit()` interface, or by the menu item Edit → Rules in the `mfedit()` interface. A typical rule editing interface is shown in Figure 10.9, and the user can add all rules to the rule library, with the Add rule button. One may also delete certain rules by clicking the Delete rule button.

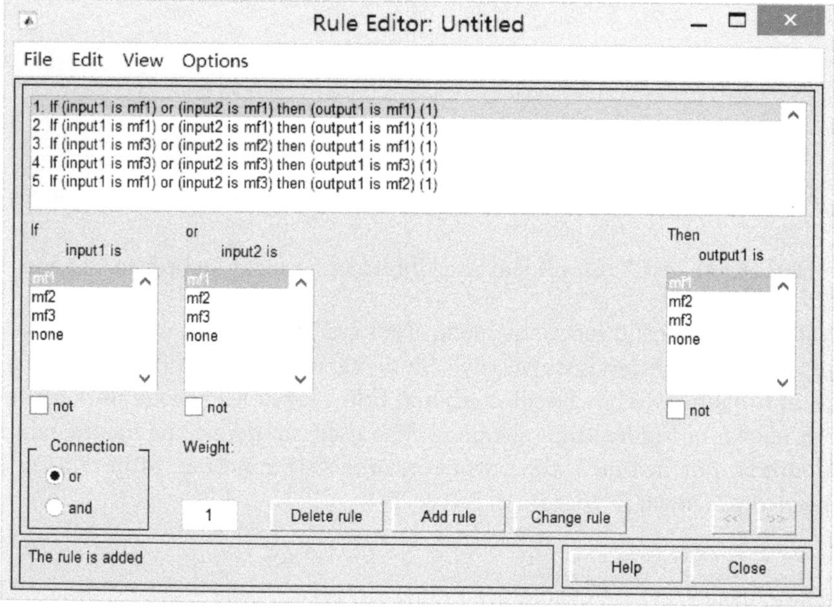

FIGURE 10.9: Edit box of fuzzy inference rules.

The rules can further be modified by clicking the Change Rule button. When the modification of the rules is completed, the Close button can be used to close the relevant windows. The fuzzy inference can be displayed as 3D surfaces with the View → Surface button, as shown in Figure 10.10, indicating the map from input signals to outputs.

Fuzzy rules can also be expressed by vectors. Multiple rules can be expressed by matrices comprising different vectors. The matrix is referred to as a *fuzzy rule matrix*. In each row of the matrix, there are $m+n+2$ elements, where m and n represent the numbers of input and output variables. The first m elements in the vector correspond to the sequence numbers

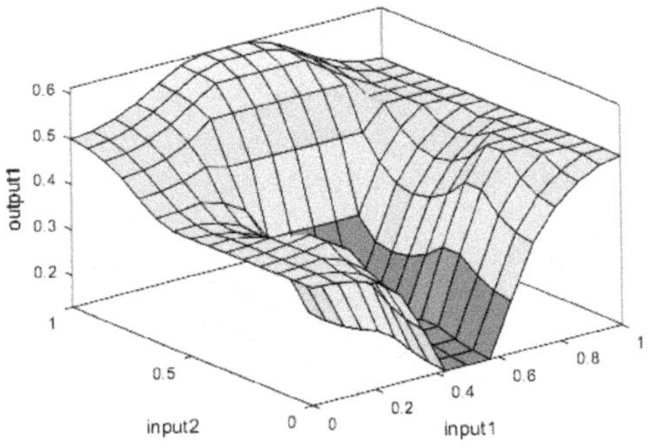

FIGURE 10.10: 3D surface.

of the input fuzzy sets, while the next n elements correspond to the numbers of the output sets. The $(m+n+1)$th element expresses the weight, while the last element represents the logic relationship, such that 1 represents "and" and 2 represents "or." For instance, the third rule in Example 10.9 can be expressed by vector $[3,2,1,1,2]$.

From these rule vectors, a matrix $\boldsymbol{R}$ consisting of all the rule vectors can be established. The following command can be used to add a rule matrix to a `fis` object: `fis=addrule(fis,R)`.

III. Defuzzification

Through fuzzy inference, a fuzzy output variable `op` can be generated. In practice, the exact or crispy value of the obtained fuzzy output signal is needed. The process of converting a fuzzy signal to the exact signal is referred to as *defuzzification*. Various defuzzification algorithms are supported in the Fuzzy Logic Toolbox, and they can be selected from the dialog box in Figure 10.7, i.e., from the list box shown in Figure 10.11.

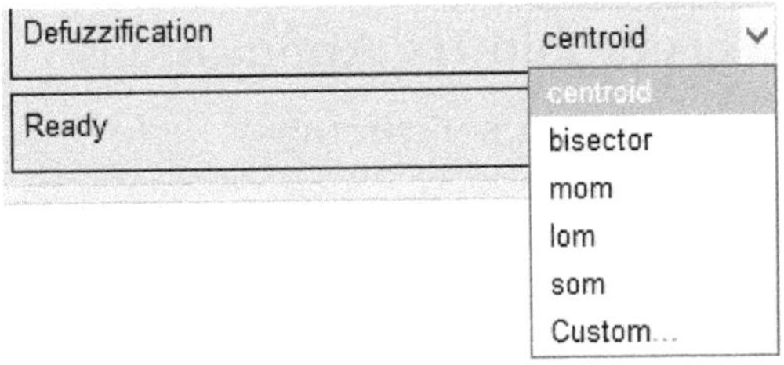

FIGURE 10.11: FIS system.

The fuzzy inference data thus created can be saved to a file by the File → Export → To Disk menu item, with the file extension of `*.fis`. For instance, the fuzzy inference system created above can be saved into file c10mfis.fis. The process can be saved too with the `writefis()` function. The menu item File → Export → To Workspace saves the FIS into MATLAB workspace.

Fuzzy inference problems can be solved by `y=evalfis(X,fis)` function where $\boldsymbol{X}$ is a matrix, whose columns contain the exact values of each input signal. The function `evalfis()` can be used to define the fuzzy inference system, to perform fuzzification first, then, to perform fuzzy inference and finally to defuzzify the inference results. The exact output is returned in argument $\boldsymbol{y}$.

Example 10.11 Assume that the fuzzy inference system is the one defined in the last example. Draw the 3D surface of the fuzzy inference results.

Solution *The following statements can be used to import the fuzzy inference system, draw mesh grids in the interested region on the x-y plane, and call* `evalfis()` *function to evaluate the z values. The 3D surface can then be obtained as shown in Figure 10.12.*

```
>> fff=readfis('c10mfis.fis');  % read in the fuzzy inference system
   [x,y]=meshgrid(-3:.2:3,-5:.2:5);                % create mesh grids
   x1=x(:); y1=y(:); z1=evalfis([x1 y1],fff); % fuzzy inference
   z=reshape(z1,size(x)); surf(x,y,z)              % surface plot
```

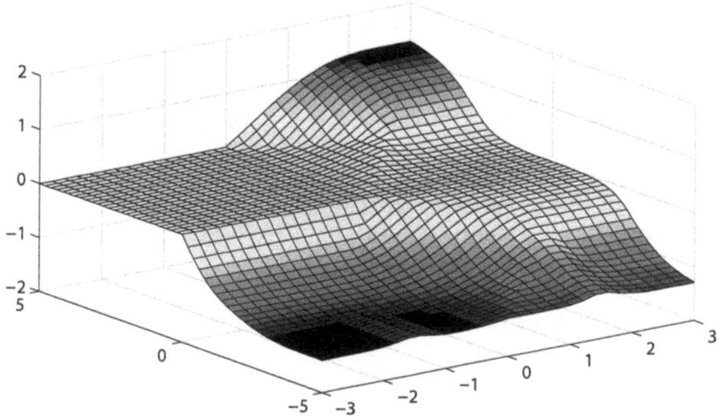

FIGURE 10.12: Output surface obtained from fuzzy inference system.

10.2 Rough Set Theory and Its Applications

10.2.1 Introduction to rough set theory

I. Rough set theory

The idea of rough set was publicized by Professor Zdzisław Pawlak, a Polish mathematician, in 1982, for the development of automatic rule generation systems and soft-computing problems. In the earlier 1990s, the researchers began to realize the importance of rough sets. In 1991, Pawlak published a research monograph, which established the mathematical basis of rough sets [3]. Rough set theory provides a new mathematical framework for processing imprecise and incomplete information, and finding hidden rules from huge amounts of data.

It is widely recognized that rough set theory is the foundation for data mining, knowledge discovery and information reduction problems.

II. Fundamental concepts in rough sets

Assume that $X, Y \in U$ and R is the equivalent relationship defined on the universe U. The *lower-approximation set* of the set X on R is defined as

$$\underline{\mathscr{R}}(X) = \bigcup\{Y \in U/R:\ Y \subseteq X\}, \tag{10-2-1}$$

where $\underline{\mathscr{R}}(X)$ is the maximum set of entities that for sure belong to the set X, also referred to as *positive region*, denoted by POS(X).

Similarly, the *upper-approximation set* of the set X on R is defined as

$$\overline{\mathscr{R}}(X) = \bigcup\{Y \in U/R:\ Y \cap R \neq \phi\}, \tag{10-2-2}$$

where ϕ is an empty set, and $\overline{\mathscr{R}}(X)$ is the minimum set of entities which possibly belongs to set X.

Based on the above definitions, the *boundary set* can be defined as $\mathrm{BND}(X) = \overline{\mathscr{R}}(X) - \underline{\mathscr{R}}(X)$. If BND($X$) is an empty set, the set X reduces to a crisp set on R. If on the other hand, BND(X) is non-empty, X is a *rough set* on R.

Suppose that the *decision table* can be represented by matrix $\boldsymbol{S}$ in MATLAB. One can extract its first m columns to represent *conditional attributes* C, and the rest of the columns for *decisional attributes* D. The upper-approximation set $\overline{\mathscr{R}}(X)$ and lower-approximation set $\underline{\mathscr{R}}(X)$ can be obtained by calling the `rslower()` and `rsupper()` functions, respectively, such that

S_l = `rslower(`X`,`$\boldsymbol{a}$`,`$\boldsymbol{S}$`)` % find the lower-approximation set S_l

S_u = `rsupper(`X`,`$\boldsymbol{a}$`,`$\boldsymbol{S}$`)` % find the upper-approximation set S_u

S_d = `setdiff(`S_u`,`S_l`)` % find the boundary set S_d

and the boundary set can be obtained by calling the `setdiff()` function in MATLAB. The contents of `rslower()` and `rsupper()` functions are listed below

```
function w=rslower(y,a,T)
z=ind(a,T); w=[]; [p,q]=size(z);
for u=1:p,
   zz=setdiff(z(u,:),0); if ismember(zz,y), w=cat(2,w,zz); end
end

function w=rsupper(y,a,T)
z=ind(a,T); w=[]; [p,q]=size(z);
for u=1:p
   zz=setdiff(z(u,:),0); zzz=intersect(zz,y);
   if length(zzz)~=0, w=cat(2,w,zz); end
end
```

and the common supporting function `ind()` is used to evaluate the indiscernibility relationship to be defined later.

Example 10.12 For the toy set $U = \{x_1, x_2, x_3, x_4, x_5, x_6, x_7\}$. Assume that they have three attributes, namely, "color R_1," "shape R_2" and "size R_3." For the color attribute, it is assumed that $R_1 = \{0, 1, 2\}$, representing "red," "green" or "blue." The shape attribute $R_2 = \{0, 1, 2\}$, denoting "square," "round" and "triangular." The size attribute $R_3 = \{0, 1\}$, indicating "large" and "small."

For the relationship in the attributes, the red toys can be described as $\{x_1, x_2, x_7\}$, green ones $\{x_3, x_4\}$ and blue ones $\{x_5, x_6\}$. It can then be written as $U|R_1 = \{\{x_1, x_2, x_7\}, \{x_3, x_4\}, \{x_5, x_6\}\}$.

III. Information decision system

The information decision system T can be expressed as a quadruple $T = (U, A, C, D)$, where U is the set of entities, i.e., the *universe.* A is the attribute set. If set A can further be divided into conditional attribute set C and decisional attribute set D, i.e., $C \cup D = A$ and $C \cap D = \phi$, the information system is referred to as a *decision system* or a *decision table.*

In rough set theory, a decision table can be used to describe the entities in a given universe. A decision table is a two-dimensional table with each row corresponding to an entity, and each column corresponding to a certain attribute. Again the attributes can be classified as conditional attributes and decisional attributes. The entities in the universe can be classified into different decisional classes due to its different conditional attributes. Table 10.2 is a typical example of a decision table. The universe $U = \{x_1, x_2, \cdots\}$ is a set of entities, $C = \{c_1, \cdots, c_m\}$ is a conditional attribute set, while $D = \{d_1, \cdots, d_k\}$ is a decisional attribute set. The term f_{ij} represents the jth conditional attribute of the ith entity, and g_{ij} represents the jth decisional attribute of the ith entity.

TABLE 10.2: Decision table.

	C				D		
U	c_1	c_2	$\cdots$	c_m	d_1	$\cdots$	d_k
x_1	f_{11}	f_{12}	$\cdots$	f_{1m}	g_{11}	$\cdots$	g_{1k}
x_2	f_{21}	f_{22}	$\cdots$	f_{2m}	g_{21}	$\cdots$	g_{2k}
$\vdots$	$\vdots$	$\vdots$	$\vdots$	$\vdots$	$\vdots$	$\vdots$	$\vdots$
x_n	f_{n1}	f_{n2}	$\cdots$	f_{nm}	g_{n1}	$\cdots$	g_{nk}

TABLE 10.3: Example 10.13.

	C properties				D
U	a	b	c	d	sales
1	1	0	1	1	1
2	1	0	0	0	1
3	0	0	1	0	0
4	1	1	0	1	0
5	1	1	1	2	2
6	2	1	0	2	2
7	2	2	0	2	2

Example 10.13 Consider the toy problem in Example 10.12. Assume that there are four relevant attributes, "a" for color, "b" for shape, "c" for size and "d" for price. It is known that $x_{1,2,4,5}$ are yellow, x_3 is red and $x_{6,7}$ are green. Also, $x_{1,2,3}$ are square, $x_{4,5,6}$ are round and x_7 is triangular; $x_{1,3,5}$ are large while the rest is small; $x_{2,3}$ are cheap ones, $x_{1,4}$ are medium priced and $x_{5,6,7}$ are expensive ones. From sales status, $x_{3,4}$ are good, $x_{1,2}$ are average and $x_{5,6,7}$ are poor. Construct the decision table.

Solution *Assume that for the color attribute, $\{0, 1, 2\}$ can be used to describe "red," "yellow" and "green;" for shape, $\{0, 1, 2\}$ for "square," "round" and "triangular;" for size, $\{0, 1\}$ for "small" and "large;" for price, $\{0, 1, 2\}$ for "low," "medium" and "high," and for sales, the decision attribute, $\{0, 1, 2\}$ for "good," "average" and "poor." The decision table can easily be constructed as shown in Table 10.3. One of the applications of rough set theory is that it can be used to check which of the conditional attributes contributes the most to the sales attribute, and which has little impact.*

Example 10.14 Let the universe $U = \{x_1, x_2, x_3, x_4, x_5, x_6, x_7, x_8, x_9, x_{10}\}$, with relations $R = \{R_1, R_2\}$, and

$U/R_1 = \{\{x_1, x_2, x_3, x_4\}, \{x_5, x_6, x_7, x_8\}, \{x_9, x_{10}\}\}$,

$U/R_2 = \{\{x_1, x_2, x_3\}, \{x_4, x_5, x_6, x_7\}, \{x_8, x_9, x_{10}\}\}$

If $X = \{x_1, x_2, x_3, x_4, x_5\}$, find the upper- and lower-approximation sets of X.

Solution *From the given conditions, the three subsets for U/R_1 and U/R_2 can be*

represented as $\{0,1,2\}$*. Thus, the decision table in Table 10.4 can be established in* $\boldsymbol{S}$*. Here the decision table is rotated for neat type-setting purposes. If* $X = \{1,2,3,4,5\}$*, the first two columns can be extracted for vector* a*. The upper- and lower-approximation sets of* X *can be obtained by using*

TABLE 10.4: Rotated decision table.

Universe X	x_1	x_2	x_3	x_4	x_5	x_6	x_7	x_8	x_9	x_{10}
U/R_1 relation	0	0	0	0	1	1	1	1	2	2
U/R_2 relation	0	0	0	1	1	1	1	2	2	2

```
>> S=[0,0; 0,0; 0,0; 0,1; 1,1; 1,1; 1,1; 1,2; 2,2; 2,2];
   X=[1,2,3,4,5]; a=[1,2]; S1=rslower(X,a,S)
   S2=rsupper(X,a,S), Sd=setdiff(S2,S1)  % compute the three sets
```

The sets are obtained as $S_1 = [1,2,3,4]$*,* $S_2 = [1,2,3,4,5,6,7]$ *and* $S_\mathrm{d} = [5,6,7]$*.*

It can be seen from the decision table that, the sets for $\{U/R_1, U/R_2\}$ *are* $\{0,0\}$*,* $\{0,1\}$*,* $\{1,1\}$*,* $\{1,2\}$ *and* $\{2,2\}$*. For the selected entities* $X = \{1,2,3,4,5\}$*, those involved with* $\{U/R_1, U/R_2\}$ *are only* $\{0,0\}$*,* $\{0,1\}$ *and* $\{1,1\}$*. Thus, those belonging to set* X *are* $\{1,2,3,4\}$*, i.e.,* $\{x_1, x_2, x_3, x_4\}$*. The lower-approximation set is then* $\{x_1, x_2, x_3, x_4\}$*. Similarly, the set with the entities which may probably belong to* X *is the upper-approximation set of* X*. Since* x_6, x_7 *and* x_5 *are all* $\{1,1\}$*, thus, apart from the entities in the lower-approximation set, the upper-approximation set also includes* x_5, x_6, x_7*. The latter three entities belong to the boundary set. Besides, since the boundary set is non-empty, it belongs to a rough set.*

In the information system, for each subset $R \subseteq A$, the indiscernibility relationship is defined as

$$\mathrm{IND}(R) = \{(x,y) \in U \times U : \ r \in R : \ r(x) = r(y)\}. \tag{10-2-3}$$

The MATLAB implementation of the function is listed below

```
function aa=ind(a,x)
[p,q]=size(x); [ap,aq]=size(a); z=1:q;
tt=setdiff(z,a); x(:,tt(size(tt,2):-1:1))=-1;
for r=q:-1:1, if x(1,r)==-1, x(:,r)=[]; end, end
for i=1:p, v(i)=x(i,:)*10.^(aq-[1:aq]'); end
y=v'; [yy,I]=sort(y); y=[yy I];
[b,k,l]=unique(yy); y=[l I]; m=max(l); aa=zeros(m,p);
for ii=1:m, for j=1:p, if l(j)==ii, aa(ii,j)=I(j); end, end, end
```

10.2.2 Data processing problem solutions using rough sets

I. Reductions in rough set

Nowadays people live in the information explosive era. It is easier and easier to get more and more information. However, a huge amount of unprocessed information causes "data disasters." To tackle this kind of problem, the so-called *data mining* techniques should be used.

The aim of the reduction of information system is simply deleting and neglecting the unnecessary and redundant information, without affecting the original decision. New decision rules can be generated with the reduced information set.

Attribute reduction means finding the minimum set of conditional attributes without affecting the decisional attributes. One decision table may simultaneously have several reductions. The set of attributes which is common to all the reductions is referred to as the *core set* of R, denoted as Core(R).

Details of the attribute reduction algorithms are not given here. Only a few MATLAB-based functions, such as `redu()` and `core()`, are introduced with their usage explained. The Rough Set Data Analysis (RSDA) Toolbox with the related functions can be obtained from the aforementioned book. For details of the algorithms and implementations, please refer to Reference [4]. The updated version of the Toolbox is provided with the package of the book.

Assume again that the decision table is described by matrix $\boldsymbol{S}$, whose $\boldsymbol{c}$ columns are stored in matrix $\boldsymbol{C}$ for conditional attributes, and the $\boldsymbol{d}$ columns are the decisional attributes in matrix $\boldsymbol{D}$. Thus, the minimum set of attributes can be extracted by the `redu()` function, and the core set can be extracted by using the function `core()`, with the following syntaxes

```
y=redu(c,d,S)  % attribute reduction, the minimum set from C to D
y=core(c,d,S)  % extract the core set from C to D
```

II. Application examples of rough set in information reduction

With the use of the attribute reduction concepts, and in particular, the relevant MATLAB functions, two illustrative examples are demonstrated in the following [4].

Example 10.15 LCDs are often used to show the 10 digits, from 0~9, with a seven-segment display unit. The seven-segment display units are shown in Figure 10.13 (a). If a

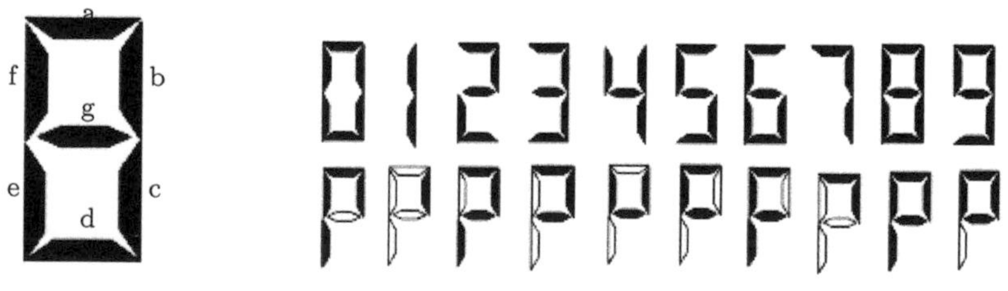

(a) seven-segment display (b) reduced display

FIGURE 10.13: LCD displays and reduction results.

segment is lighted, it is denoted as 1, otherwise, 0. The truth table for the encoding system is given in Table 10.5. Find the unnecessary segments by attribute reduction using rough set theory.

Solution *For digit recognition by human vision, it is necessary to use a seven-segment LCD display. If one segment is missing, the recognition of digits may become very difficult. However, sometimes it may not be necessary to have all the seven segments in place in order to recognize digit by computer vision. If one does find a certain segment which is not necessary, the LCD display may be reduced or further simplified. With the use of rough set*

TABLE 10.5: Truth table of LCD.

code	C attributes							D	code	C attributes							D
X	a	b	c	d	e	f	g	value	X	a	b	c	d	e	f	g	value
0	1	1	1	1	1	1	0	0	5	1	0	1	1	0	1	1	5
1	0	1	1	0	0	0	0	1	6	1	0	1	1	1	1	1	6
2	1	1	0	1	1	0	1	2	7	1	1	1	0	0	0	0	7
3	1	1	1	1	0	0	1	3	8	1	1	1	1	1	1	1	8
4	0	1	1	0	0	1	1	4	9	1	1	1	1	0	1	1	9

techniques, reduction can be tried to check whether certain segments can be reduced, without affecting the recognition of digits. The following commands can be given

```
>> C=[1,1,1,1,1,1,0; 0,1,1,0,0,0,0; 1,1,0,1,1,0,1; 1,1,1,1,0,0,1;
      0,1,1,0,0,1,1; 1,0,1,1,0,1,1; 1,0,1,1,1,1,1; 1,1,1,0,0,0,0;
      1,1,1,1,1,1,1; 1,1,1,1,0,1,1];
   D=[0:9]'; X=[C D]; c=1:7; d=8;
   Y=core(c,d,X) % where columns 1-7 are C properties, 8 for D
```

with the reduction result $\boldsymbol{Y} = [1, 2, 5, 6, 7]$, *which means that the segments a, b, e, f and g in the LCD are irreducible. The segments 3 and 4, i.e., the segments c and d in Figure 10.13 (a) can be removed without affecting the recognition of digits by computers. The corresponding mapping patterns are shown in Figure 10.13 (b). However, for human visual recognition, the above reduction may be useless. On the other hand, for the recognition by machines, this kind of reduction is no doubt useful, since less information is needed. The reduction idea shown in this example is promising for character recognition via machine vision.*

Example 10.16 The outbreak of severe acute respiratory syndrome (SARS) in 2003 caused terrors globally. The accurate diagnosis of SARS is very difficult. Some data collected from newspapers and magazines are given in Table 10.6. All the 12 attributes are analyzed with rough set theory to see whether some of the attributes are redundant for the diagnosis purpose. The major attributes can be found from the 12 attributes. *It should be noted that some data may not be accurate, and the entries are far from complete. Thus, the conclusion here is not usable for clinical diagnosis.*

Solution *From the data given in the table, the reduction of attributes with rough sets can be used to find which attributes contribute much more towards the diagnosis of SARS*

```
>> D=[1; 0; 0; 0; 0; 0; 0; 1; 1; 1; 1; 1];
   C=[1,1,1,1,0,0,0,0,1,1,0,1; 0,0,0,0,0,0,0,0,0,0,0,0;
      1,0,1,0,0,0,0,0,0,1,0,0; 0,0,0,1,1,1,1,0,1,0,1,1;
      1,0,0,1,1,1,1,1,0,1,1,0; 0,1,0,1,1,1,1,1,1,0,0,1;
      1,0,0,0,1,1,1,0,0,1,1,1; 1,1,1,1,0,0,0,0,1,1,0,1;
      1,0,1,1,1,0,0,0,1,1,0,1; 1,1,1,1,0,0,0,0,1,1,0,1;
      1,0,1,1,1,0,0,0,1,1,0,1; 1,0,1,1,1,0,0,0,1,1,0,1];
   Y=redu(1:12,13,[C D])
```

with $\boldsymbol{Y} = [3, 4]$, *indicating that attributes 3 and 4 are irreducible, which means that "blood test" and "fever with temperature higher than* $38^\circ C$*" are the irreducible attributes.*

TABLE 10.6: Collected incomplete data for diagnosing SARS.

universe	*C* properties												*D*
U	c_1	c_2	c_3	c_4	c_5	c_6	c_7	c_8	c_9	c_{10}	c_{11}	c_{12}	SARS
1	1	1	1	1	0	0	0	0	1	1	0	1	1
2	0	0	0	0	0	0	0	0	0	0	0	0	0
3	1	0	1	0	0	0	0	0	0	1	0	0	0
4	0	0	0	1	1	1	1	0	1	0	1	1	0
5	1	0	0	1	1	1	1	1	0	1	1	0	0
6	0	1	0	1	1	1	1	1	1	0	0	1	0
7	1	0	0	0	1	1	1	0	0	1	1	1	0
8	1	1	1	1	0	0	0	0	1	1	0	1	1
9	1	0	1	1	1	0	0	0	1	1	0	1	1
10	1	1	1	1	0	0	0	0	1	1	0	1	1
11	1	0	1	1	1	0	0	0	1	1	0	1	1
12	1	0	1	1	1	0	0	0	1	1	0	1	1

c_1 — hacking cough, c_2 — breath difficulties, c_3 — blood test, c_4 — temperature $\geqslant 38°$C

c_5 — X-ray test abnormal, c_6 — with phlegm, c_7 — high white blood cells, c_8 — chill

c_9 — ache in muscles, c_{10} — hypodynamia, c_{11} — pleurodynia, c_{12} — headache

III. MATLAB graphical user interface for rough set reduction

Based on the rough set theory and its reduction methods, a MATLAB interface is written and included in the RSDA Toolbox. The command `rsdav3` can be given to start the interface, as shown in Figure 10.14. The Browse button can be used to import the decision table, and the column numbers for attributes in C and D can also be specified. Then, the Redu button can be used to perform attribute reduction tasks, and the results are shown in the Results box. The free toolbox is provided in the book package.

10.3 Neural Network and Applications in Data Fitting Problems

Artificial neural networks (ANN) were originated from studying and understanding the behavior of complicated neural networks of living creatures. In human brains, there are about 10^{11} interlinked units, known as *neurons*. Each neuron has about 10^4 links with other neurons [5]. When the mathematical representations of artificial neurons are established, the interconnected neurons can be used to construct artificial neural networks. However, due to the limitations of the status of computers today, neural networks as complicated as human brains cannot be built so far.

In this section, an introduction to mathematical descriptions of artificial neurons and neural networks are given first, followed by the descriptions of MATLAB solutions to the neural network-based problems. The neural networks graphical user interface is demonstrated in detail.

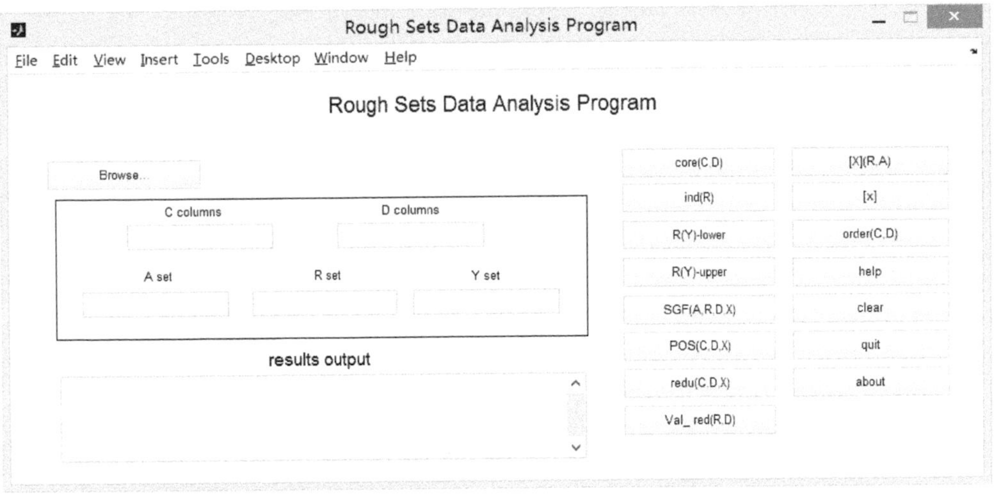

FIGURE 10.14: Graphical user interface for data analysis with rough sets.

10.3.1 Fundamentals of neural networks

I. Concept and structure of an artificial neuron

The structure of an artificial neuron is shown in Figure 10.15, where $x_1, x_2, \cdots, x_n$ are input signals. The weighted sum of them, plus the threshold b, forms linear function u_i. The signal is further processed by the nonlinear transfer function $f(u_i)$ to generate the output signal y.

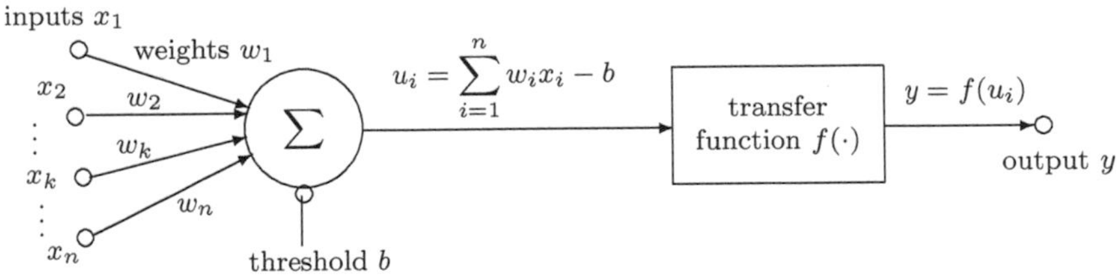

FIGURE 10.15: Basic structure of an artificial neuron.

In this artificial neuron model, the weights w_i and the *transfer function* or activation function $f(\cdot)$ are the two important elements. The weights can be considered as the intensities of the input signals which can be determined by repeated training from the samples. Normally the transfer function should be selected as a monotonic function such that the inverse function uniquely exists. Commonly used transfer functions are sigmoidal functions and logarithmic sigmoidal functions, expressed respectively by

$$\text{sigmoid function} \quad f(x) = \frac{2}{1+\mathrm{e}^{-2x}} - 1 = \frac{1-\mathrm{e}^{-2x}}{1+\mathrm{e}^{-2x}}, \tag{10-3-1}$$

$$\text{logarithmic sigmoid function} \quad f(x) = \frac{1}{1+\mathrm{e}^{-x}}. \tag{10-3-2}$$

The simple saturation function and step function can also be used as transfer functions. The shapes of some typical transfer functions are shown through the following examples.

Example 10.17 Draw some of the commonly used activation functions.

Solution *The following MATLAB statements can be used to draw the sigmoid functions, as shown in Figure 10.16.*

```
>> x=-2:0.01:2; y=tansig(x); plot(x,y)
```

The `logsig()` *function can be used to replace the* `tansig()` *function to draw the logarithmic sigmoid function. Also, other MATLAB functions can be used to draw the commonly used transfer functions, as shown in Figure 10.16.*

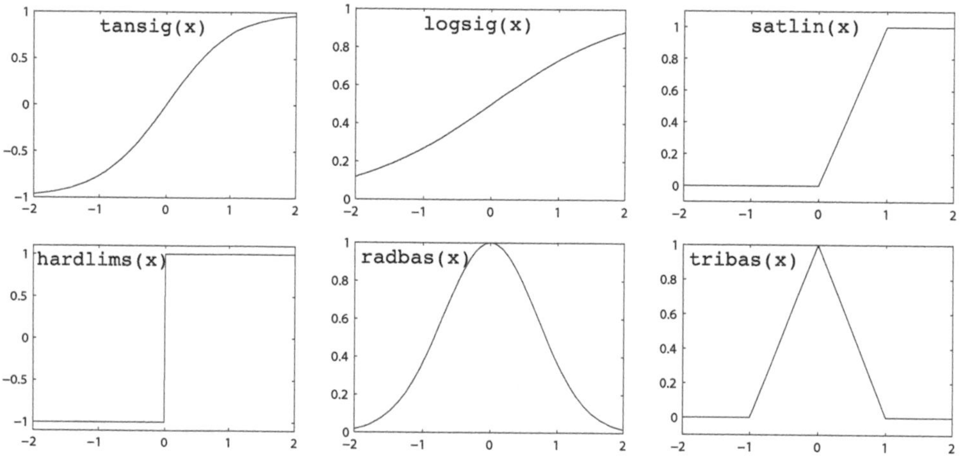

FIGURE 10.16: Different curves of the commonly used transfer functions.

II. Artificial neural networks

The artificial neurons can be connected together to form a network, known as the *artificial neural network*. The word "artificial" is usually dropped off and *neural network* is commonly used. With different ways of connections, different types of neural networks can be established. In this section, only the feedforward structure will be presented. Since in the training procedure, the error is propagated in the reversed direction, this network type is often referred to as the *back-propagation* (BP) neural network. The typical structure of a BP neural network is shown in Figure 10.17. In the network, there are the input layer, several intermediate layers known as *hidden layers*, and the output layer. In this book, the last hidden layer is in fact the output layer.

There are significant amounts of neural networks in literature, such as feedforward networks, Hopfield networks, self-organizing networks. In recent versions of Neural Network Toolbox of MATLAB, several typical network structures are recommended. For instance, for function fitting and approximation purposes, the `fitnet()` function can be used to create the neural network. Usually a two-layer feedforward neural network is adopted, and the default training algorithm is Levenberg–Marquardt back-propagation algorithm.

For pattern recognition purposes, `patternnet()` is recommended.

10.3.2 Feedforward neural network

The use of a neural network is a three-step procedure. In the first step, the neural network structure should be defined. In the second step, the network must be trained with samples,

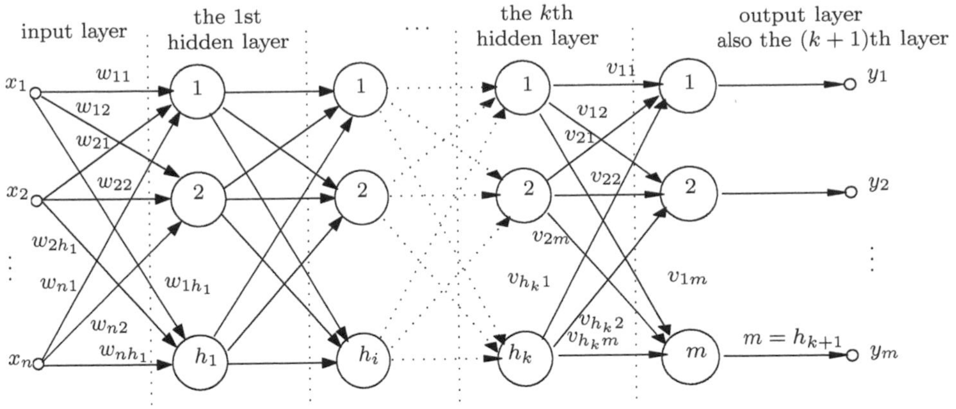

FIGURE 10.17: Basic structure of neural network.

and proper weights in the neurons should be found. In the third step, the generalization can be made. In other words, simulation and validation of the network should be performed.

I. Network structure

Consider a feedforward neural network with two layers, i.e., $k = 2$, with m and n the numbers of output and input ports, respectively. Denote p as the number of nodes in the hidden layer. The illustration of the neural network is given in Figure 10.18. The signals before and after the transfer functions of the hidden layer nodes are

$$u_j = \sum_{i=1}^{n} w_{ij}x_i + b_{1j}, \quad u_j' = F_1(u_j), \quad j = 1, \cdots, p, \tag{10-3-3}$$

where, b_{1j} is the threshold of hidden layer nodes, while the signals before and after the transfer functions at the output layer are respectively

$$y_j' = \sum_{i=1}^{p} v_{ji}u_i' + b_{2j}, \quad y_j = F_2(y_j'), \quad j = 1, \cdots, m. \tag{10-3-4}$$

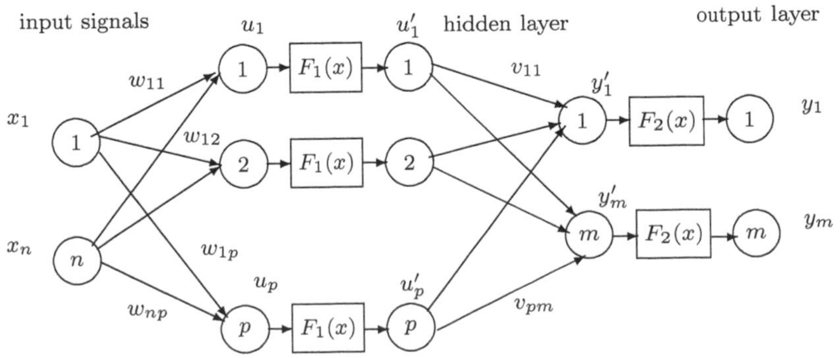

FIGURE 10.18: Structure of the feedforward neural network with one hidden layer.

If there are N sets of practical sample data for training, the relationships between the inputs and outputs are as follows

$$\begin{array}{cccccccc} x_{11} & x_{12} & \cdots & x_{1n} & \Rightarrow & \hat{y}_{11} & \cdots & \hat{y}_{1m} \\ x_{21} & x_{22} & \cdots & x_{2n} & \Rightarrow & \hat{y}_{21} & \cdots & \hat{y}_{2m} \\ \vdots & \vdots & \vdots & \vdots & \vdots & \vdots & \vdots & \\ x_{N1} & x_{N2} & \cdots & x_{Nn} & \Rightarrow & \hat{y}_{N1} & \cdots & \hat{y}_{Nm}. \end{array} \tag{10-3-5}$$

The problem in neural network training is to train the network repeatedly with the weighting quantities w_{ij}, v_{ij} and threshold b_{ij}, such that the error between the computed output y_i and the real output signal $\hat{y}_i$ is minimized. An appropriate criterion is to have the sum of the squared errors minimized

$$\min_{\boldsymbol{W},\boldsymbol{V},\boldsymbol{B}} \sum_{l=1}^{N}\sum_{i=1}^{m}(y_{li}-\hat{y}_{li})^2. \tag{10-3-6}$$

For this kind of unconstrained optimization problem, the target is to solve equations, and find the final weighting matrices $\boldsymbol{W}$ and $\boldsymbol{V}$, to minimize the criterion. Various algorithms can be used to select the initial values of weighting quantities $\boldsymbol{W}_{ij}$, $\boldsymbol{V}_{ij}$, and assume the initial values are assigned as $\boldsymbol{W}_{ij}^0$, $\boldsymbol{V}_{ij}^0$, the following recursive algorithm can be used [6]

$$\boldsymbol{W}_{ij}^{l+1} = \boldsymbol{W}_{ij}^{l} + \beta e_j^l a_i^l, \quad \boldsymbol{V}_{jt}^{l+1} = \boldsymbol{V}_{jt}^{l} + \alpha d_t^l \gamma_j^l, \tag{10-3-7}$$

where, $i = 1,\cdots,n$,$j = 1,\cdots,p$, $t = 1,\cdots,m$, and α and β are the velocity constants. Intermediate variables a_i^l, γ_j^l, e_j^l and d_t^l can be evaluated recursively.

It is quite easy to construct a feedforward network under BP algorithm with the Neural Network Toolbox of MATLAB. The function `fitnet()` can be used, with the syntax `net=fitnet([`h_1`,`h_2`,`$\cdots$`,`h_k`])`, where, h_i are number of nodes in the ith hidden layer. It is worth mentioning that the actual number of layers is $k+1$, with the last layer the output layer with linear transfer functions. The neural network object `net` is then created, whose important properties are listed in Table 10.7. Examples are given below to show the creation of neural network objects.

Example 10.18 Assume that there are two inputs and one output signal. Establish a feedforward neural network object with the `fitnet()` function.

Solution *Consider a feedforward network, with one hidden layer composed of 8 nodes. The following statements can be used to establish the neural network object in MATLAB workspace*

```
>> net=fitnet(8); % one hidden layer with 8 nodes
```

In fact, a two-layer network is constructed, with the second layer the output layer, and the number of nodes are in fact the number of output signals.

Now assume that in a new network, there are two hidden layers. In the first hidden layer there are 4 nodes, and in the second layer, there are 8 nodes. The following statements can be used to establish the neural network model.

```
>> net=fitnet([4 6]); % two hidden layers, with 4 and 6 nodes in each layer
```

TABLE 10.7: Common properties of neural network.

property names	data type	descriptions	defaults
`net.IW`	cell	input and hidden layer weights, where `net.IW{1}` stores the weights of the first hidden layer, while `net.IW{`$i+1$`}` for the ith hidden layer	random
`net.numInputs`	integer	number of inputs, it can be calculated automatically from the sizes of $\boldsymbol{x}_{\rm m}$ or $\boldsymbol{x}_{\rm M}$	
`net.numLayers`	integer	number of layers	
`net.LW`	cell	output layer weighting matrix	random
`net.trainParam.epochs`	integer	maximum training steps, when the error criterion is met, the training stops	100
`net.trainParam.lr`	double	learning rate	0.01
`net.trainParam.goal`	double	training error criterion, if the error is smaller than this value, the training stops	0
`net.trainFcn`	string	training algorithms, the selectable options are `'traincgf'` (conjugate gradient with Fletcher–Reeves updates), `'train'` (batch training), `'traingdm'` (gradient descent with momentum), `'trainlm'` (Levenberg-Marquardt algorithm)	`'train'`

II. Training of neural networks

If the neural network model `net` has been established, the function `train()` can be used to train the parameters in the network

`[net,tr,`$\boldsymbol{Y}_1$`,`$\boldsymbol{E}$`] = train(net,`$\boldsymbol{X}$`,`$\boldsymbol{Y}$`)`

where the variable $\boldsymbol{X}$ is an $n \times M$ matrix, with n the number of input signals, and M the number of samples for training. The variable $\boldsymbol{Y}$ is an $m \times M$ matrix, with m the number of outputs. The variables $\boldsymbol{X}$ and $\boldsymbol{Y}$ store respectively the inputs and outputs data of the samples. From the function call, the network can be trained, and the returned variable `net` is the trained object. The argument `tr` is a structured variable with training information, where `tr.epochs` returns the number of *epochs*, meaning steps in the training process of an artificial neural network, and `tr.perf` returns the objective function values in each training step. The arguments $\boldsymbol{Y}_1$ and $\boldsymbol{E}$ matrices are the output and model errors of the network, respectively, and the training error can be visualized by `plotperform(tr)`.

If the training up to the maximum epochs cannot find a satisfactory network, a warning message will be given. The training results can be used as the initial weights, and one can continue the training process until a satisfactory network is obtained. If the satisfactory network still cannot be obtained, there might be problems in the network structure and a new structure should be tested.

III. Generalization of neural networks

After training, the neural network can be tested and validated, then, it can be used as a computation unit. In practical applications, the samples can be divided randomly into two groups, with one group used in training, and the other group used in validation.

When the input signals other than the ones in the samples are provided, the output can be evaluated from the unit. This process is also known as the *simulation* or *generalization* of neural networks. This process can be used to solve data fitting problems. For the input signals in matrix $\boldsymbol{X}_1$, the output can be evaluated using the function `sim()` where $\boldsymbol{Y}_1$ = `sim(net,`$\boldsymbol{X}_1$`)`. Alternatively, if `net` is the name of the neural network object, the function $\boldsymbol{Y}_1$ = `net(`$\boldsymbol{X}_1$`)` can be used directly to compute the output signals.

Example 10.19 For the problem in Example 8.25, use neural network to fit the data. If the input signal is a sinusoidal function, please compute the output signal through the neural network.

Solution *The following statements can be given to compute the sample data. A two-layer feedforward network structure can be selected. Assume that the hidden layer comes with 5 nodes. The network parameters are trained for generalization, with the training error shown in Figure 10.19 (a) and the generalization results shown in Figure 10.19 (b). It can be observed that the training is satisfactory, and almost no difference can be observed from the true results.*

```
>> x=0:.5:10; x0=[0:0.1:10];
   y=0.12*exp(-0.213*x)+0.54*exp(-0.17*x).*sin(1.23*x);
   y0=0.12*exp(-0.213*x0)+0.54*exp(-0.17*x0).*sin(1.23*x0);
   net=fitnet(5); [net,tr]=train(net,x,y); subplot(121), plotperform(tr)
   y1=net(x0); subplot(122), plot(x,y,'o',x0,y0,x0,y1,':');
```

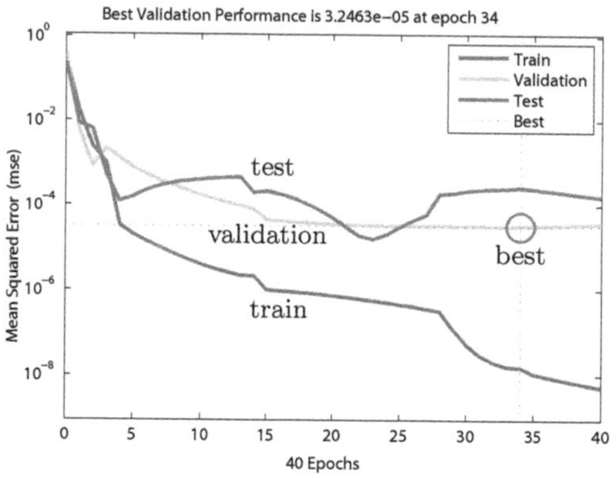

(a) training error

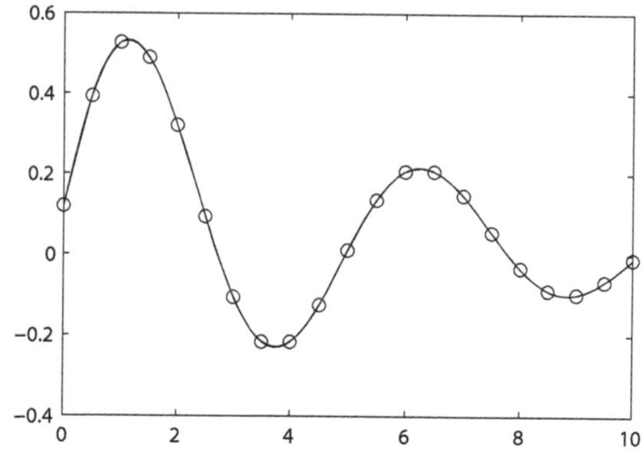
(b) fitting by neural network

FIGURE 10.19: Data fitting with neural network.

The following statements can be used to extract the weights of the trained neural network

```
>> w1=net.IW{1}, w2=net.LW{2,1} % weights for hidden layer and output
```

and the weights obtained are

from input to hidden layer: $\boldsymbol{w}_1^{\mathrm{T}} = [7.4494, -3.8019, -3.9805, 4.5079, 7.6978]$,

from hidden layer to output: $\boldsymbol{w}_2 = [0.1988, 0.7808, -1.0970, -1.4687, 1.0793]$

and each time, the trained weights may be completely different. The weights of the neural network have no physical meaning; slight change in the weights may lead to different computation results.

The structure of the neural network can be obtained with command `view(net)`*, as shown in Figure 10.20.*

The number of hidden layer nodes can further be increased, for instance, to 20. The

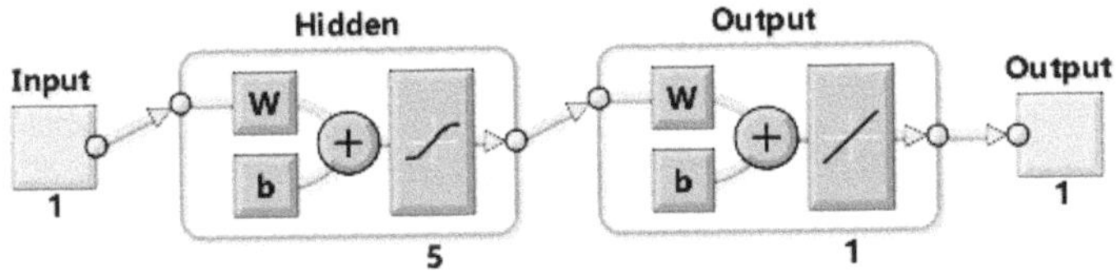

FIGURE 10.20: Structure of feedforward neural network.

new network can be created and trained, and the training error reaches an extremely small value for only 4 epochs, as shown in Figure 10.21 (a). The curve fitting, i.e., generalization, results are shown in Figure 10.21 (b).

```
>> net1=fitnet(20); [net1,tr]=train(net1,x,y);
   subplot(121), plotperform(tr)
   subplot(122), y1=net1(x0); plot(x0,y0,x0,y1,x,y,'o')
```

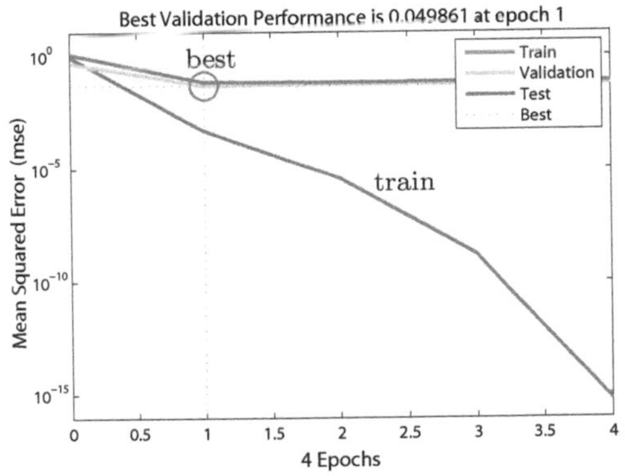

(a) training error

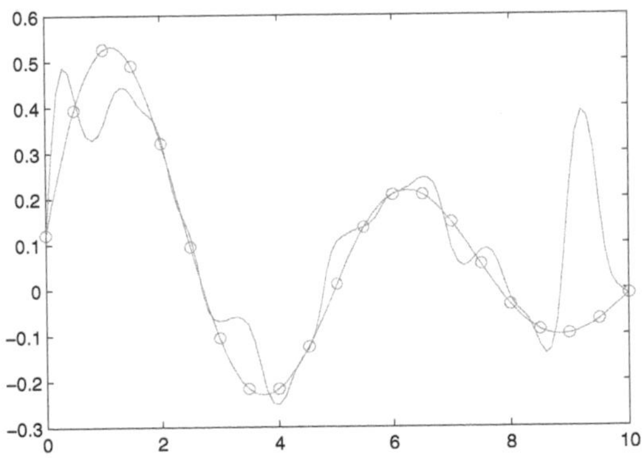

(b) fitting results with neural network

FIGURE 10.21: Fitting results with 20 nodes in the hidden layer.

It can be seen from the generalization results that the curve fitting results are very poor, albeit on the samples, the fittings are good. This means that due to the increase of hidden layer nodes, the generalization process goes wrong. However, to date there is no universally accepted method on how to assign reasonable numbers of nodes. The node numbers and number of layers can only be assigned by trial-and-error methods.

If a neural network object `net` is created, function `gensim(net)` can be used to generate a Simulink model for the network. Therefore, the neural network block generated can be directly used in simulation.

Example 10.20 For the neural network constructed in the previous example, please evaluate the output of the system under Simulink environment, if the input signal is a sinusoidal signal.

Solution *A Simulink block can be generated with* `gensim()` *function, and the sinusoidal signal block is used as the input of the system, shown in Figure 10.22 (a). With the excitation of the sinusoidal input, the output of the system can be obtained with the Simulink model. Alternatively, the same output signal can be obtained with the commands, as shown in Figure 10.22 (b).*

```
>> t=0:0.01:2*pi; y=net(sin(t)); plot(tout,yout,t,y,'--')
```

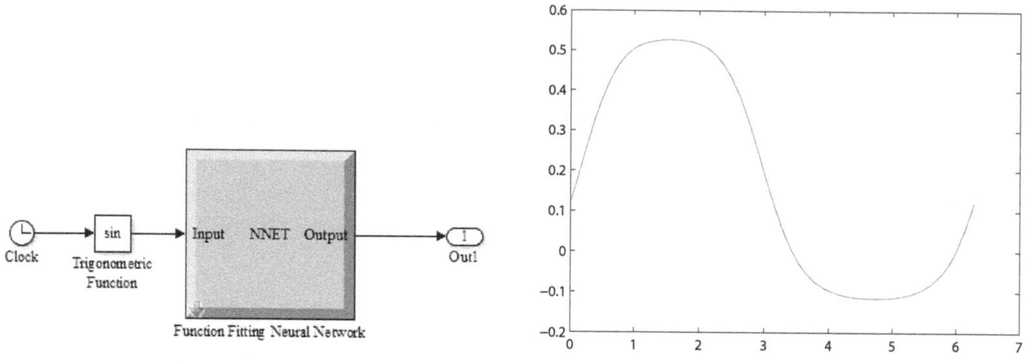

(a) Simulink model (c10mnn1a.mdl) (b) output signal

FIGURE 10.22: Simulink model and output signal.

Example 10.21 Fitting the two-dimensional data with neural network for the data given in Example 8.7.

Solution *The data for neural network training and generalization can be generated first, and a two-layer neural network, with 20 nodes in the hidden layer can be established and trained, and the generalized surface obtained is shown in Figure 10.23 (a). It can be seen that the fitting is not good.*

```
>> N=200; x=-3+6*rand(1,N); y=-2+4*rand(1,N);
   z=(x.^2-2*x).*exp(-x.^2-y.^2-x.*y);
   [x2,y2]=meshgrid(-3:.1:3, -2:.1:2); x1=x2(:)'; y1=y2(:)';
   net=fitnet(20); [net,b]=train(net,[x; y],z);
   z1=net([x1; y1]); z2=reshape(z1,size(x2)); surf(x2,y2,z2)
```

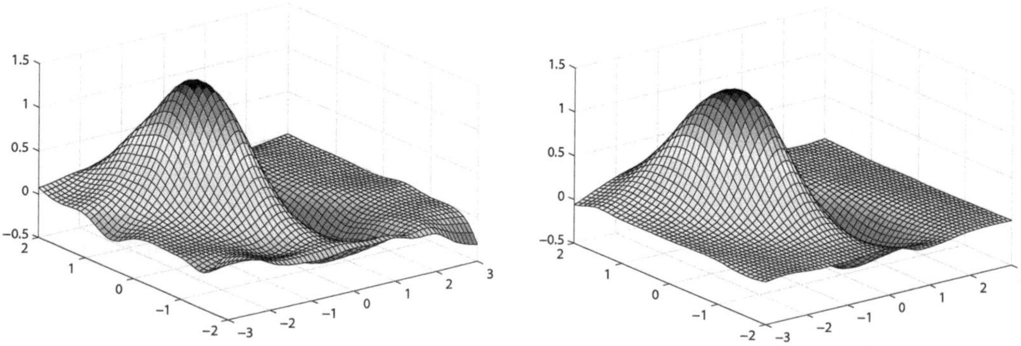

(a) two-layer network (b) three-layer network

FIGURE 10.23: Fitting results with different structures.

We can now try a three-layer network structure. Assume that there are 10 nodes each in hidden layers 1 and 2. In the third layer, of course, one should use one node, the same as the number of outputs. The generalization result, represented as a 3D surface, is shown in Figure 10.23 (b). The fitting quality is significantly improved with the new neural network.

```
>> net=fitnet([10,10]); [net,b]=train(net,[x; y],z);
   z1=net([x1; y1]); z2=reshape(z1,size(x2)); surf(x2,y2,z2)
```

Now select 20 nodes in both hidden layers. From the new generalization results shown in Figure 10.24 (a), it can be seen that even though the number of nodes increased, the generalization becomes even worse.

```
>> net=fitnet([20,20]); [net,b]=train(net,[x; y],z);
   z1=net([x1; y1]); z2=reshape(z1,size(x2)); surf(x2,y2,z2)
```

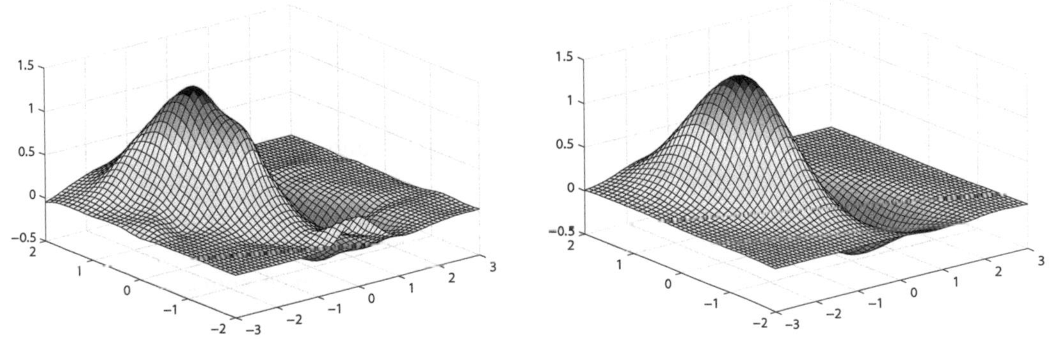

(a) 20 nodes in both layers (b) four-layer network

FIGURE 10.24: Fitting results with different structures.

A four-layer neural network can also be tried, with 7 nodes each on the three hidden layers. The fitting result is shown in Figure 10.24 (b). It can be seen that the fitting quality is even better than the one in Figure 10.23 (b).

```
>> net=fitnet([7 7 7]); [net,b]=train(net,[x; y],z);
   z1=net([x1; y1]); z2=reshape(z1,size(x2)); surf(x2,y2,z2)
```

Example 10.22 Consider the exaggerated interpolation problem studied in Example 8.5, where only five samples were used to restore the sinusoidal curve. Please check whether it is possible to construct a neural network to do the same job.

Solution *Different number of nodes can be tried for the problem, however, unfortunately, the sinusoidal curve cannot be restored with only five samples, an example of the fitting result is obtained in Figure 10.25. In the example in Chapter 8, due to the assumptions in spline function, the sinusoidal curve can be restored easily.*

```
>> x=[0,0.4,1 2,pi]; y=sin(x); x1=0:0.01:pi;
   net=fitnet(10); net=train(net,x,y); y1=net(x1); plot(x1,y1,x,y,'o')
```

10.3.3 Radial basis neural networks and applications

Radial basis function (RBF) is a type of exponential function defined as

$$\varphi(\boldsymbol{x}) = \mathrm{e}^{-b||\boldsymbol{x}-\boldsymbol{c}||} = \mathrm{e}^{-b(\boldsymbol{x}-\boldsymbol{c})^{\mathrm{T}}(\boldsymbol{x}-\boldsymbol{c})}, \tag{10-3-8}$$

where, $\boldsymbol{c}$ is the center, and $b > 0$ is an adjustable parameter. Function `radbas()` is used

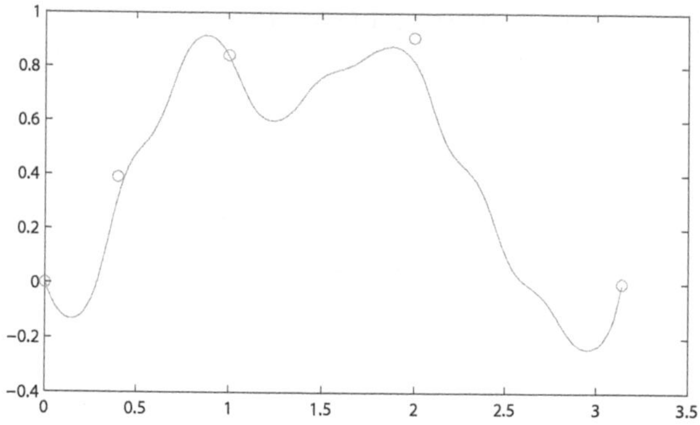

FIGURE 10.25: Failure in fitting of a sinusoidal curve.

to compute the standard radial basis function $y_i = \mathrm{e}^{-x_i^2}$, and the function in (10-3-8) can easily be found through simple computation.

Radial basis neural network is a special neural network structure. Consider the typical two-layer feedforward neural network shown in Figure 10.18. If the transfer function of $F_1(x)$ is radial basis function, while $F_2(x)$ in the output layer is a linear function, the network structure is referred to as *radial basis network.*

Example 10.23 Draw the RBF curves for different (c, b) parameters.

Solution *Selecting the centers at* $c = -2, 0, 2$, *and* $b = 1$, *the curves of the radial basis functions can be obtained, as shown in Figure 10.26 (a). It can be seen that the shapes of the curves are exactly the same. Only translations of* c *units are made.*

```
>> x=-4:0.1:4; cc=[-2,0,2]; b=1;
   for c=cc, y=exp(-b*(x-c).^2); plot(x,y); hold on; end
```

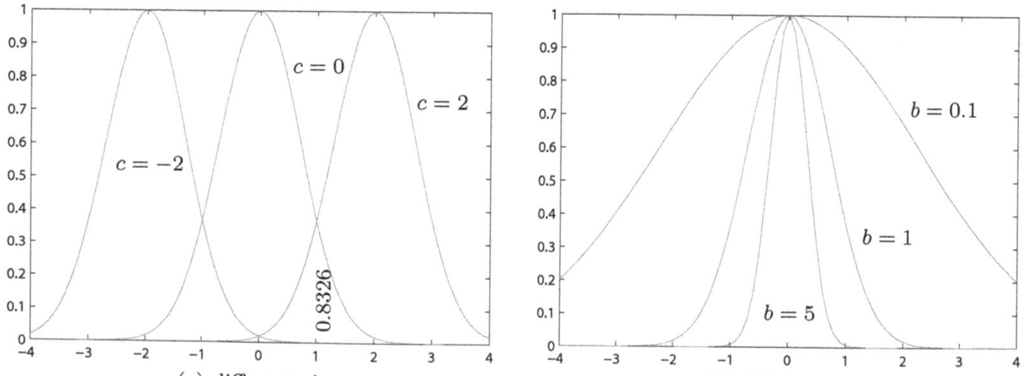

(a) different c's (b) different b's

FIGURE 10.26: Radial basis functions under different parameters.

Selecting the center at $c = 0$, *and assume* $b = [0.1, 1, 5]$, *the curves of radial basis function can be obtained as shown in Figure 10.26 (b).*

```
>> x=-4:0.1:4; bb=[0.1,1,5]; c=0;
   for b=bb, y=exp(-b*(x-c).^2); plot(x,y); hold on; end
```

Although the structure of radial basis network is of feedforward type, it is not BP network, since the training is not completed with back-propogation algorithms. With the Neural Network Toolbox of MATLAB, it can be seen that the use of radial basis network is easier, since two functions `newrbe()` and `sim()` can be used in the establishment and training process. The number of hidden layer nodes is selected automatically with the function call. Examples are given below to demonstrate the use of these functions.

Example 10.24 Consider the data in Example 10.19. Please fit the curve using a radial basis neural network.

Solution *The samples can be generated first, and the RBF neural can be created, trained and generalized, and the fitting results are shown in Figure 10.27. It can also be seen that the fitting quality is slightly better than the one with* `fitnet()`*. The default number of hidden layer nodes is 21, and the weights can be retrieved with* `net.IW{1}` *and* `net.LW{2,1}` *commands.*

```
>> x=0:.5:10; y=0.12*exp(-0.213*x)+0.54*exp(-0.17*x).*sin(1.23*x);
   u=[0:0.1:10]; v=0.12*exp(-0.213*u)+0.54*exp(-0.17*u).*sin(1.23*u);
   net=newrbe(x,y); y1=net(u); plot(x,y,'o',u,v,u,y1,':');
```

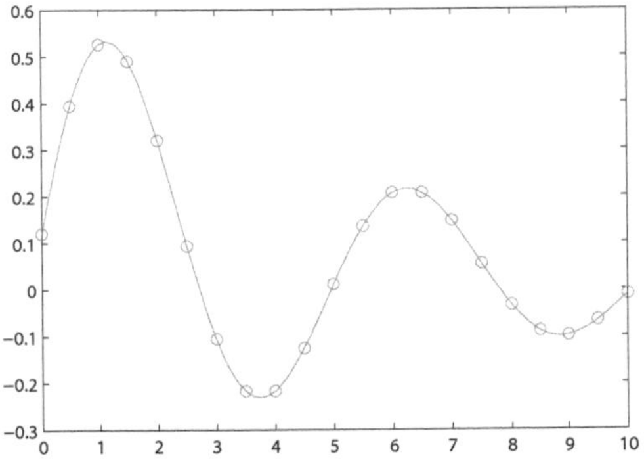

FIGURE 10.27: One-dimensional fitting.

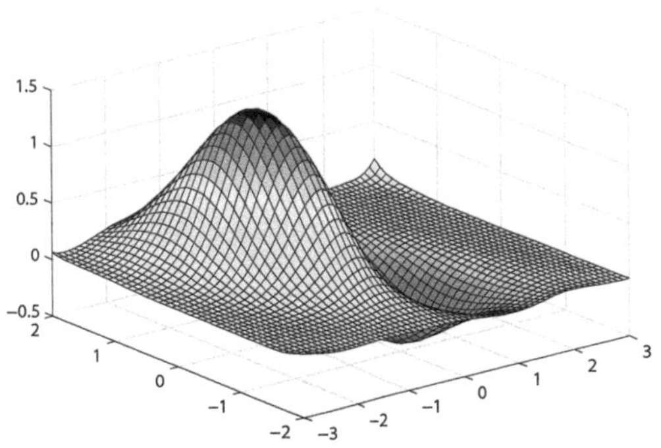

FIGURE 10.28: Two-dimensional fitting.

Example 10.25 Please fit the 2D surface of Example 10.21 with RBF network.

Solution *With RBF network, the fitting surface can be obtained as shown in Figure 10.28, and the number of hidden layer nodes is automatically selected as 121.*

```
>> N=200; x=-3+6*rand(1,N); y=-2+4*rand(1,N);
   z=(x.^2-2*x).*exp(-x.^2-y.^2-x.*y); net=newrbe([x; y],z);
   [x2,y2]=meshgrid(-3:.1:3, -2:.1:2); x1=x2(:)'; y1=y2(:)';
   z1=sim(net,[x1; y1]); z2=reshape(z1,size(x2)); surf(x2,y2,z2)
```

Example 10.26 Let us revisit the exaggerated problem in Example 10.22. Please assess the fitting quality using RBF network.

Solution *With the following statements, the RBF network can be established, and the fitting result is obtained as shown in Figure 10.29, with 5 hidden layer nodes. It can be seen that the fitting quality is very high, and the numerical integral obtained is* $I = 1.9963$*. It is more accurate than the cubic spline result in Example 8.18, yet less accurate than the B-spline result.*

```
>> x=[0,0.4,1 2,pi]; y=sin(x); x1=0:0.01:pi; y0=sin(x1);
   net=newrbe(x,y); y1=net(x1); plot(x1,y1,x1,y0,x,y,'o')
   f=@(x)net(x); I=integral(f,0,pi)
```

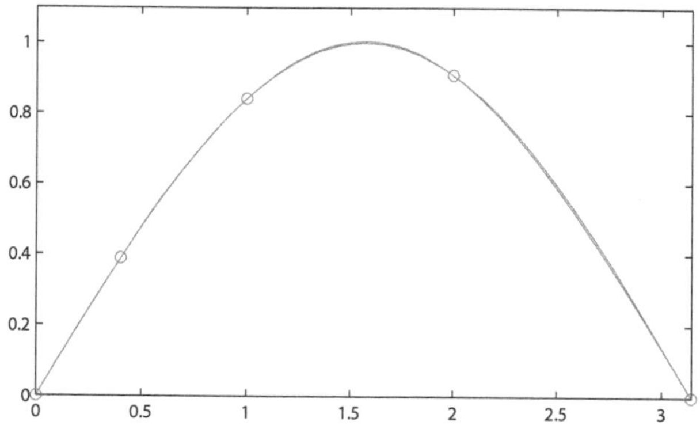

FIGURE 10.29: Fitting result of sinusoidal curve.

With `view(net)` *command, the structure of the RBF network is obtained as shown in Figure 10.30.*

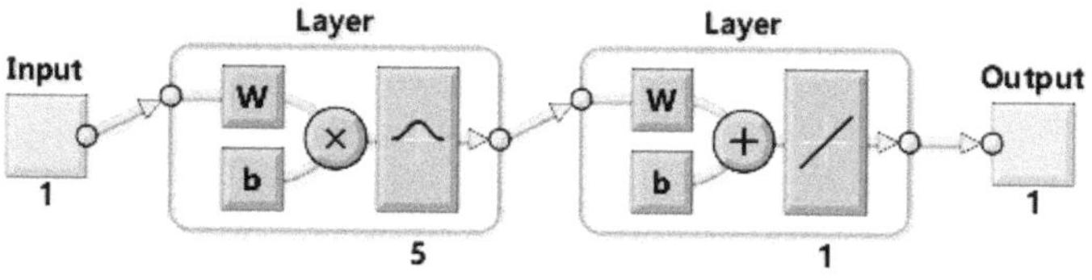

FIGURE 10.30: Structure of RBF network.

10.3.4 Graphical user interface for neural networks

A graphical user interface is provided in the Neural Network Toolbox. The command `nntool` can be used to start the GUI shown in Figure 10.31. The interface can be used to create neural network, then, train and simulate the network. The example below is given to show the use of the interface.

Example 10.27 Consider the problem in Example 10.19. Use the `nntool` interface to solve the same problem.

Solution *The following statements can be used to import data into MATLAB workspace. Then,* `nntool` *can be used to load the interface shown in Figure 10.31. The interface can be used in neural network data fitting.*

```
>> x=0:.5:10; x0=[0:0.1:10];
   y=0.12*exp(-0.213*x)+0.54*exp(-0.17*x).*sin(1.23*x);
   y0=0.12*exp(-0.213*x0)+0.54*exp(-0.17*x0).*sin(1.23*x0);
   nntool     % start the user interface of neural networks
```

Click the Import *button to import the samples and the dialog box shown in Figure 10.32*

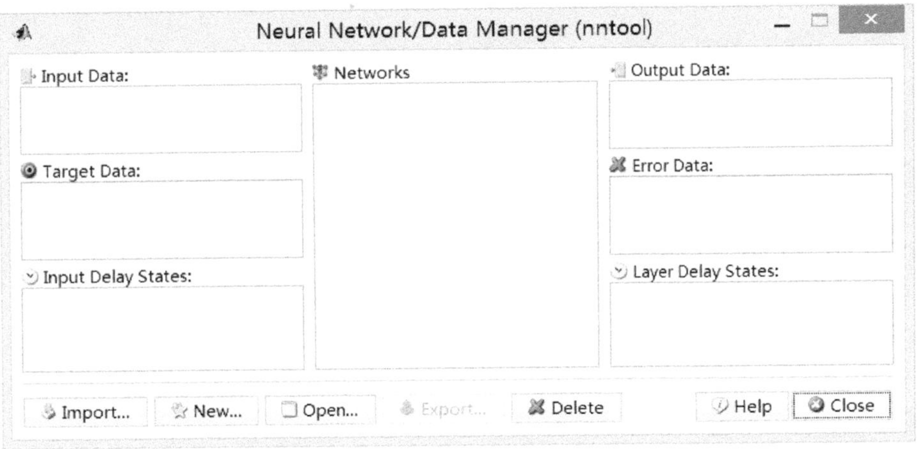

FIGURE 10.31: Graphical user interface for neural network.

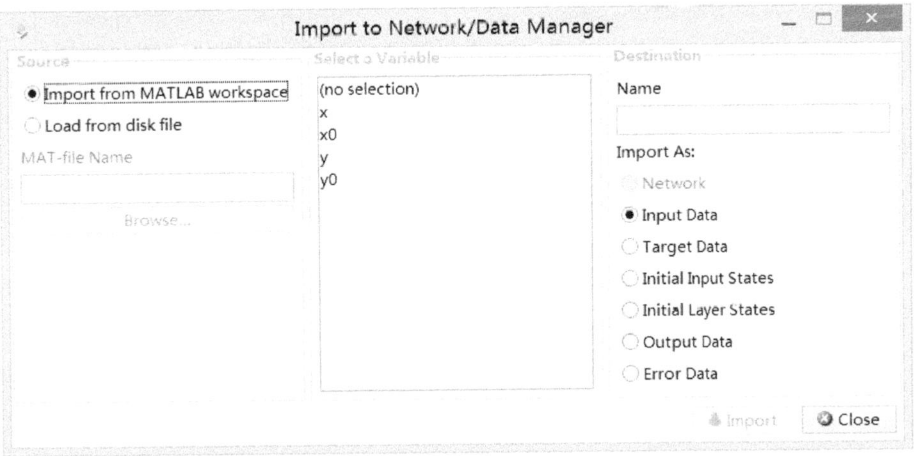

FIGURE 10.32: Data input interface.

appears. Select the variables `x` *and* `xx` *as input variables, and* `y` *as target variables, by selecting the combinations in the interface.*

Click the New Network *button to select the structure of neural network, with the graphical user interface shown in Figure 10.33 (a). The default (*Feedforward Backprop*) network structure can be established, with (*Number of layers*) set to 2. The node numbers in each layer can also be set, where the number for layer 1 is 8. One may also set the transfer functions by selecting the* Transfer Function *list box to* Logsig*. Click the* Create *button, and the neural network structure can be created.*

The network structure can be displayed by clicking the View *button, as shown in Figure 10.33 (b).*

The network can be trained by clicking the Train *button, and the dialog box shown in Figure 10.34 is displayed. The training data can be specified first and training parameters can be set by clicking the* Training Parameters *tag to show the dialog box shown in Figure 10.35. The network can then be trained by clicking the* Train *button, and the training curve is shown in Figure 10.19 (a). If the termination conditions are satisfied, the training stops*

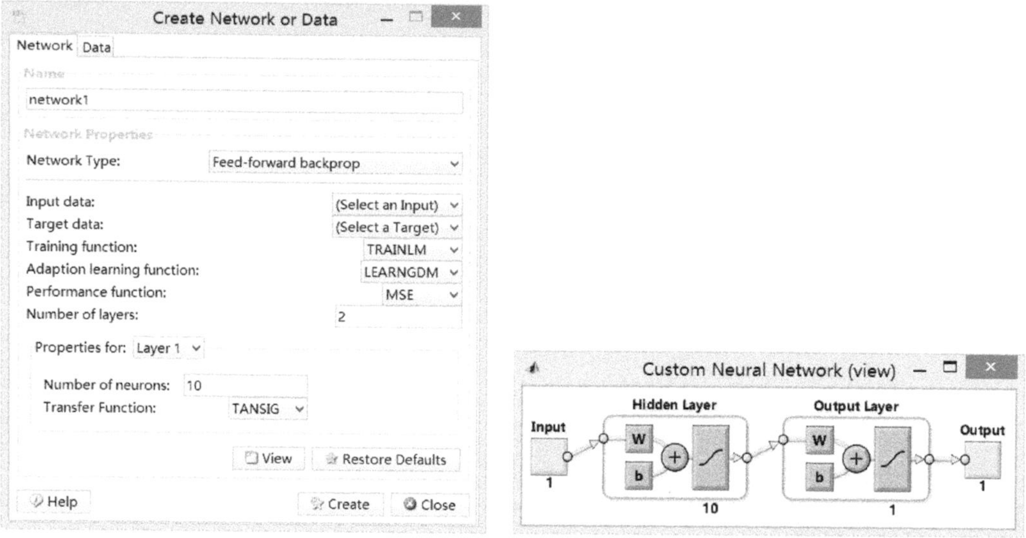

(a) neural network structure setting (b) display of neural network

FIGURE 10.33: Structure setting and display of neural network.

and the required parameters can be obtained. The trained network can be exported back to the environment with the Export *button. The generalization results are obtained with the following statements, as shown in Figure 10.19 (b).*

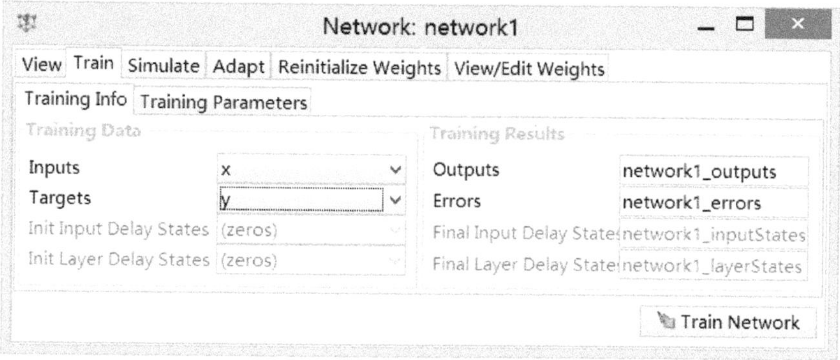

FIGURE 10.34: Dialog box for neural network training.

```
>> y1=sim(network1,x0); plot(x,y,'o',x0,y0,x0,y1,':')
```

10.4 Evolutionary Computing and Global Optimization Problem Solutions

It has been pointed out in Chapter 6 that the conventional searching methods in optimization may lead to local optimum points. Thus, various parallel searching algorithms

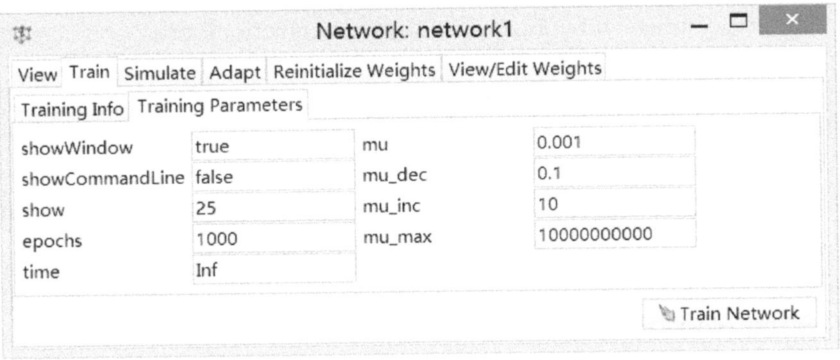

FIGURE 10.35: Training parameter setting dialog box.

are proposed, aimed at finding the global optimum solutions. Evolutionary computing is attractive for the innovation of ideas, and its capabilities in finding the global optimum solutions. Among the evolution methods, genetic algorithm based approaches and particle swarm optimization algorithms are the most widely used ones. In this section, global optimization algorithms and their application with MATLAB functions are presented.

10.4.1 Basic idea of genetic algorithms

Genetic algorithm (GA) is a class of evolutionary computing methods following the law of "survival of the fittest." [7] The method was first proposed by Professor John Holland of Michigan University in 1975. The main idea of the method is to search from a population consisting of randomly distributed individuals. The individuals are encoded, regarded as genes with chromosomes, in a certain way. The population evolves generation by generation through reproduction, crossover and mutation, until individuals with the best fitness function are found. Another similar method is the particle swarm optimization (PSO) method, which imitates the coordinated motion in flocks of birds searching for food. In this section, MATLAB solutions to optimization problems using genetic algorithms and particle swarm optimization methods are introduced.

The general procedures of a simple genetic algorithm are as follows:

(i) Select an initial population $\boldsymbol{P}_0$ with N individuals. Evaluate the objective functions for all the individuals. The initial population $\boldsymbol{P}_0$ can be established randomly.

(ii) Set the generation to $i = 1$, which means the first generation.

(iii) Compute the values of selective functions, i.e., select some individuals in a probabilistic way from the current population.

(iv) Create the population of the next generation $\boldsymbol{P}_{i+1}$, by reproduction, crossover and mutation.

(v) Set $i = i + 1$. If the termination conditions are not satisfied, go to (iii) to continue evolution.

Compared with traditional optimization methods, the genetic algorithms have mainly the following differences [8]:

(i) In searching the optimum points, the genetic algorithms allow searches from many initial points in a parallel way. Thus, it is more likely to find global optimum points than with the traditional methods, which initiate searches from a single point.

(ii) Genetic algorithms do not depend on the gradient information of the objective functions. Only the fitness functions, i.e., objective functions are necessary in optimum points search.

(iii) Genetic algorithms evaluate and select the objective function in a probabilistic way rather than a deterministic way. Thus, there are slight differences among each run.

10.4.2 Solutions to optimization problems with genetic algorithms

There are several genetic algorithm toolboxes under MATLAB. The Global Optimization Toolbox (former name is Genetic Algorithm and Direct Search Toolbox, GADT) by MathWorks is the official toolbox, and its functions are updated in each release of MATLAB. Apart from that, the Genetic Algorithm Optimization Toolbox (GAOT) [9] developed by Christopher Houck, Jeffery Joines and Michael Kay, North Carolina State University and the GA Toolbox [8] written by Peter Fleming and Andrew Chipperfield of Sheffield University are among the most commonly used free toolboxes.

For instance, in the GAOT, the main function `ga()` was renamed as `gaopt()` to avoid the conflict with `ga()` function in Global Optimization Toolbox. The new file name is used throughout this book. The function can be used directly to solve optimization problems. The benefit of such a function is that the intermediate search results are also available. In GAOT Toolbox, three selective functions are provided, such as `roulette()`, `normGeomSelect()` and `tournSelect()`, with `normGeomSelect()` the default.

In this section, the two toolboxes are explored for optimization problems.

I. Applications of GAOT Toolbox in optimization

The `gaopt()` function in GAOT Toolbox is illustrated first. `gaopt()` can be called in the following formats:

```
[a,b,c]=gaopt(bound,fun)                  % the simplest form
[x,b,c]=gaopt(bound,fun,p,v,P0,fun1,n)   % with more arguments
```

where `bound=[`$\boldsymbol{x}_{\rm m}$`,`$\boldsymbol{x}_{\rm M}$`]` stores the lower-bound $\boldsymbol{x}_{\rm m}$ and upper-bound $\boldsymbol{x}_{\rm M}$ of the decision variables $\boldsymbol{x}$. The argument *fun* is the file name string of M-function describing the objective function. It should be noted that the description of objective function is different from the other optimization routines. Examples will be given later to demonstrate the syntax of the objective function. The returned argument $\boldsymbol{a}$ is composed of the solution $\boldsymbol{x}$ and the value of the objective function $f_{\rm opt}$. The argument `b` stores the information of the final population. The argument `c` returns the intermediate search results.

In the second syntax, the argument $\boldsymbol{p}$ is the additional variables in the objective function, $\boldsymbol{v}$ is the display precision control vector, $\boldsymbol{P}_0$ is the initial population, and *fun1* is the extra function name, with default of `'maxGenTerm'`, the maximum allowed generation, and n is the number of generations. Of course, other functions are allowed, for instance, selective function, and mutation function, etc. Details can be found from Reference [9].

Example 10.28 Now consider a simple function $f(x) = x\sin(10\pi x)+2$, $x \in (-1,2)$. Find the maximum value of $f(x)$, and the value of x.

Solution *The objective function within the specified interval is drawn as shown in Figure 10.36, with the following simple MATLAB statements. Clearly, $f(x)$ is oscillatory with many extreme points in the interval.*

```
>> ezplot('x*sin(10*pi*x)+2',[-1,2])
```

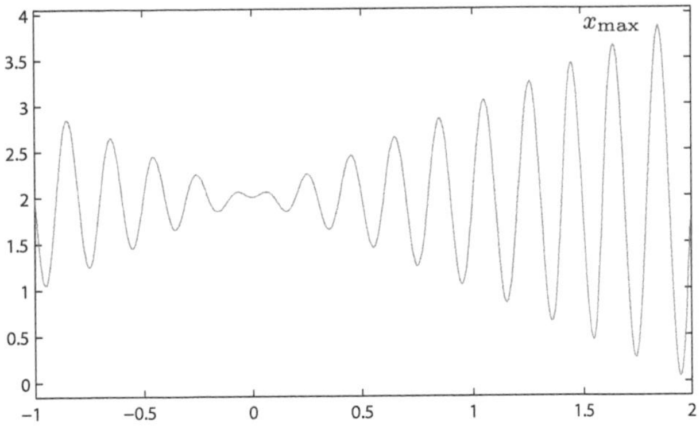

FIGURE 10.36: Objective function curve.

Optimization functions provided in the Optimization Toolbox, like all traditional optimization approaches, need the initial values for searching the optimum. Different initial values may yield different optimization results. For this example, the following statements are tested and the search results for different initial values are given in Table 10.8.

```
>> f=@(x)-x.*sin(10*pi*x)-2; v=[];
   for x0=[-1:0.8:1.5,1.5:0.1:2]
      x1=fmincon(f,x0,[],[],[],[],-1,2); v=[v; x0,x1,f(x1)];
   end
```

So, from this example, when the initial value is selected randomly, it may be very hard to find the global optimum point, unless the `fmincon_global()` *developed in Chapter 6 is used.*

TABLE 10.8: Optimal solutions from different search points x_0.

x_0	x_1	$f(x_1)$	x_0	x_1	$f(x_1)$	x_0	x_1	$f(x_1)$
−1	−1	−2	1.4	1.4506983	−3.4503492	1.7	1.2508097	−3.2504050
−0.2	−0.65155382	−2.6507777	1.5	0.25396846	−2.2519973	1.8	1.8505475	−3.8502738
0.6	0.65155379	−2.6507777	1.6	1.6506138	−3.6503069	1.9	0.4522327	−2.4511207

To solve the problem with a genetic algorithm, an objective function should be expressed by an M-function

```
function [sol,y]=c10mga1(sol,options)
x=sol(1); y=x.*sin(10*pi*x)+2;
```

It can be seen that the objective function description is different from the ones required by traditional optimization algorithms in the following sense: (i) the maximum function is specified; (ii) the input and returned arguments are differently defined. With the use of `gaopt()` *function, the following statements can be given.*

```
>> [a,b,c,d]=gaopt([-1,2],'c10mga1'); x=a(1), f=a(2)
```

where the global result returned in a is $x^* = 1.8505$*, with the optimum value* $f(x^*) = 3.8503$.

The intermediate results are given in Table 10.9, with the abbreviation gen. *for the generation number. It can be seen that from generation 12, the results obtained are acceptable for this example.*

TABLE 10.9: Intermediate search results using genetic algorithm.

generations	x	$f(x)$	generations	x	$f(x)$	generations	x	$f(x)$
1	1.833411	3.590014	6	1.85168	3.849101	11	1.850281	3.850209
2	1.647955	3.644557	8	1.851659	3.849144	12	1.85054	3.850274
3	1.858199	3.796899	9	1.851091	3.850004	100	1.850547	3.850274

Example 10.29 Find the minimum value of the function $f(x) = (x_1+x_2)^2+5(x_3-x_4)^2+(x_2-2x_3)^4+10(x_1-x_4)^4$ using genetic algorithm.

Solution *It is obvious that the global optimum solution to the problem is* $x_1 = x_2 = x_3 = x_4 = 0$. *In order to use a genetic algorithm, the M-function below should be established. It should be noted that here, the maximum objective function is expressed.*

```
function [S,f]=c10mga3(S,options), x=S(1:3);
f=-(x(1)+x(2))^2-5*(x(3)-x(4))^2-(x(2)-2*x(3))^4-10*(x(1)-x(4))^4;
```

Selecting the ranges for x_i *such that* $-1 \leqslant x_i \leqslant 1$, $i = 1, 2, 3, 4$, *the function* gaopt() *can be called to solve this optimization problem, with the intermediate results shown in Table 10.10.*

TABLE 10.10: Intermediate search results using genetic algorithm.

generations	x_1	x_2	x_3	x_4	$f(x_1, x_2, x_3, x_4)$
1	0.053034683	0.40724952	0.13840848	−0.01682375	−0.33287436
5	0.063661288	0.084587914	0.042225122	0.096858012	−0.036913734
10	0.061948198	−0.00062694769	0.089349763	0.074719112	−0.0058649774
14	0.061948198	−0.035582875	0.089349763	0.074719112	−0.003874042
21	0.061948198	−0.039947967	0.089349763	0.087316197	−0.0027943115
28	0.030899005	−0.028464643	0.089349763	0.087316197	−0.0019697798
32	0.022521945	−0.0085565429	0.089313262	0.087316197	−0.0016188571
37	0.016334305	−0.0076107093	0.089291541	0.087316197	−0.0015513478
41	0.016334305	−0.0082057068	−0.02253066	−0.02328211	-9.5374641×10^{-5}
44	0.014834811	−0.0082619779	−0.016819586	−0.017633545	-5.8042763×10^{-5}
70	0.014761402	−0.013351969	−0.017506533	−0.017633545	-1.3300401×10^{-5}
77	0.014761402	−0.013966665	−0.01184689	−0.01104842	-8.2658502×10^{-6}
85	0.013772082	−0.013965442	−0.011333	−0.01107414	-4.1891828×10^{-6}
100	0.013121641	−0.01320957	−0.010778171	−0.010696633	-3.264236×10^{-6}

```
>> [a,b,c,d]=gaopt([-1,1; -1 1; -1 1; -1 1],'c10mga3'); x=a(1:4)
```

It is found from the solutions that $x_1 = 0.0131$, $x_2 = -0.0132$, $x_3 = -0.0108$, $x_4 = -0.0107$, *with some errors in the solutions.*

Similar to the procedures used in the previous example, here 2000 generations are tested. The intermediate results are shown in Table 10.11. It can be found that the results are $\boldsymbol{x}^* = [-0.0032, 0.0032, -0.0014, -0.0014]$, *which are more accurate than the ones obtained in the previous example.*

```
>> xmM=[-ones(4,1),ones(4,1)];
   [a,b,c,d]=gaopt(xmM,'c10mga3',[],[],[],'maxGenTerm',2000);
```

TABLE 10.11: Intermediate search results for more generations.

generations	x_1	x_2	x_3	x_4	$f(x_1, x_2, x_3, x_4)$
1	0.52943125	−0.18554265	0.17102598	0.42105685	−0.50969811
4	0.30026929	−0.15235023	0.0046551376	0.073501054	−0.072705948
9	0.25969559	−0.24700356	0.0046551376	0.073501054	−0.040194938
21	0.18429634	−0.17440885	0.051175939	0.073501054	−0.009963743
32	0.18448868	−0.17115196	0.055131714	0.062716218	−0.0089361333
40	0.15139802	−0.17115196	0.055131714	0.060727809	−0.0074944505
60	0.12157312	−0.14269825	0.0069953616	0.011040366	−0.0026235247
127	0.11836763	−0.11881963	0.0087307561	0.01355787	−0.0016683703
147	−0.055877859	0.060775121	0.0097285242	0.014063368	−0.00036014666
252	−0.044906673	0.04604043	0.0078889251	0.0078620251	-7.9664135×10^{-5}
562	−0.025653189	0.024757092	0.0070596265	0.007048969	-1.2253193×10^{-5}
765	−0.024463789	0.024672404	0.0070538013	0.007039069	-9.9062381×10^{-6}
841	−0.0026816486	0.003508872	−0.001446297	−0.0014440959	-6.8602555×10^{-7}
2000	−0.0031643672	0.0031641119	−0.0013999645	−0.0014000418	-1.3622025×10^{-9}

Now consider again the unconstrained optimization problem in Chapter 6. The solution to such a problem can be solved with the following statements

```
>> f=@(x)(x(1)+x(2))^2+5*(x(3)-x(4))^2+(x(2)-2*x(3))^4+10*(x(1)-x(4))^4;
   ff=optimset; ff.MaxIter=10000; ff.TolX=1e-7;
   x=fminsearch(f,10*ones(4,1),ff)
```

with the solution $\boldsymbol{x}^{\mathrm{T}} = [0.0304, -0.0304, -0.7534, -0.7534]\times10^{-6}$. *It can be seen that the accuracy of traditional methods is much higher than that of the genetic algorithm-based methods.*

Example 10.30 Consider again the revised Rastrigin function in Example 6.26

$$f(x_1, x_2) = 20 + (x_1/30-1)^2 + (x_2/20-1)^2 - 10[\cos(x_1/30-1)\pi + \cos(x_2/20-1)\pi],$$

Run 100 times of the `gaopt()` function, and compare the time and successful rate in finding the global optimal solutions, with the `fmincon_global()` function.

Solution *The objective for* `gaopt()` *can be written as*

```
function [S,f]=c10mga5(S,options), x=S(1:2);
f=-20-(x(1)/30-1)^2-(x(2)/20-1)^2+10*(cos(pi*(x(1)/30-1))+...
    cos(pi*(x(2)/20-1)));
```

If the gaopt() *function is called 100 times, 60.35 seconds are needed.*

```
>> X=[]; range=[-100 100; -100 100];
   tic, for i=1:100, x=gaopt(range,'c10mga5'); X=[X; x]; end, toc
```

Analyzing the results in matrix $\boldsymbol{X}$*, it is found that all the 100 calls reached the global optimal point, although the results are not very accurate. With the code in Chapter 6, the successful rate is not thus high, and the time elapsed is longer, about 150 seconds. It can be seen that GAOT has obvious advantages in unconstrained optimization problems.*

10.4.3 Solving constrained problems

It has been shown that the ordinary genetic algorithm-based methods can only be used in solving unconstrained problems. For constrained optimization problems, it is sometimes possible to convert it into unconstrained problems.

Example 10.31 Solve the following linear programming problem with genetic algorithm.

$$\min_{\boldsymbol{x} \text{ s.t.} \begin{cases} -2x_1+x_2+x_3\leqslant 9 \\ -x_1+x_2\geqslant -4 \\ 4x_1-2x_2-3x_3=-6 \\ x_{1,2}\leqslant 0, x_3\geqslant 0 \end{cases}} (x_1 + 2x_2 + 3x_3).$$

Solution *From the equality constraint, it is found that* $x_3 = (6+4x_1-2x_2)/3$*. Substituting* x_3 *into the original problem, problems with two variables can be reformulated. The following M-function can be written to describe the objective function*

```
function [sol,y]=c10mga4(sol,options)
x=sol(1:2); x=x(:); x(3)=(6+4*x(1)-2*x(2))/3;
y1=[-2 1 1]*x; y2=[-1 1 0]*x;
if (y1>9 | y2<-4 | x(3)<0), y=-100; else, y=-[1 2 3]*x; end
```

where x_3 *can be calculated first from* x_1 *and* x_2*. The constraints are checked with* $\boldsymbol{x}$*. For the points where the constraints are not satisfied, the objective functions are set to* -100 *as penalties. The constrained optimization problem can now be solved using the genetic algorithm. The intermediate results are listed in Table 10.12.*

```
>> [a,b,c]=gaopt([-200 0; -200 0],'c10mga4',[],[],[],'maxGenTerm',1000)
```

The obtained results are $x_1 = -6.9983, x_2 = -10.9967$*, and it is found from* $x_3 = (6+4x_1-2x_2)/3$ *that* $x_3 = 3.2583\times 10^{-5}$.

In fact, the linear programming function can be used to find a more precise solution, where $\boldsymbol{x}^{\mathrm{T}} = [-6.99999999999967, -10.99999999999935, 0]$.

```
>> f=[1 2 3]; A=[-2 1 1; 1 -1 0]; B=[9; 4]; Aeq=[4 -2 -3]; Beq=-6;
   x=linprog(f,A,B,Aeq,Beq,[-inf;-inf;0],[0;0;inf]);
```

10.4.4 Solving optimization problems with Global Optimization Toolbox

There are several global optimization problem solvers provided in the new Global Optimization Toolbox. Specifically, particle swarm optimization (PSO) algorithm, simulated

TABLE 10.12: Intermediate search results using genetic algorithm.

generations	x_1	x_2	$f(\boldsymbol{x})$	generations	x_1	x_2	$f(\boldsymbol{x})$
1	−186.2892	−196.6231	−100	596	−6.9076	−10.8243	28.5382
58	−1.0380	0	−0.8100	797	−6.9372	−10.9094	28.6862
63	−1.2029	0	0.0143	841	−6.9797	−10.9683	28.8983
114	−1.3579	−2.0200	0.7895	921	−6.9909	−10.9825	28.9547
141	−6.1889	−9.9597	24.9444	995	−6.9983	−10.9966	28.9914
230	−6.6159	−10.2321	27.0796	999	−6.9983	−10.9967	28.9915
481	−6.7766	−10.5617	27.8832	1000	−6.9983	−10.9967	28.9915

annealing algorithm, genetic algorithm, as well as pattern search algorithm are implemented, with the latter two targeting at solving constrained optimization problems.

I. Particle swarm optimizations

Particle swarm optimization (PSO) is a class of evolutionary computing optimization algorithm initially proposed in Reference [10]. The algorithm is motivated by the phenomenon in nature where birds are seeking food. It is a useful algorithm in finding global solutions to optimization problems.

Assume that within a certain area, there is a piece of food (global optimum point), and there are a flock of randomly distributed birds (or particles). Each particle has its personal best value $p_{i,\mathrm{b}}$, and the swarm has its best value g_b up to now. The position and speed of each particle can be updated with the formula

$$\begin{cases} v_i(k+1) = \phi(k)v_i(k) + \alpha_1\gamma_{1i}(k)[p_{i,\mathrm{b}} - x_i(k)] + \alpha_2\gamma_{2i}(k)[g_\mathrm{b} - x_i(k)] \\ x_i(k+1) = x_i(k) + v_i(k+1), \end{cases} \tag{10-4-1}$$

where γ_{1i} and γ_{2i} are uniformly distributed random numbers in the interval $[0,1]$. The argument $\phi(k)$ is the momentum function, while α_1 and α_2 are acceleration constants.

In the new Global Optimization Toolbox for MATLAB R2014b, `particleswarm()` function is provided to solve unconstrained optimization problems with particle swarm optimization algorithms, with the syntaxes

$[\boldsymbol{x}, f_\mathrm{m}, \mathtt{key}] = \mathtt{particleswarm}(\mathit{problem})$,
$[\boldsymbol{x}, f_\mathrm{m}, \mathtt{key}] = \mathtt{particleswarm}(f, n, \boldsymbol{x}_\mathrm{m}, \boldsymbol{x}_\mathrm{M}, \mathit{opts})$

while in the structured variable *problem*, the fields `solver`, `objective`, `nvars` are essential.

Example 10.32 Please solve the revised Rastrigin function problem with particle swarm optimization approach.

Solution *The same* `particleswarm` *function can be called 100 times with the following statements*

```
>> f=@(x)20+(x(1)/30-1)^2+(x(2)/20-1)^2-...
       10*(cos(pi*(x(1)/30-1))+cos(pi*(x(2)/20-1))); tic
   X=[]; for i=1:100, [x g]=particleswarm(f,2); X=[X; x g]; end, toc
```

The 100 runs of `particleswarm()` *function takes only about 5.3 seconds, and the successful rate for finding the global optimum point is 98%, and the accuracy is also much higher than that of the* `ga()` *function.*

Example 10.33 Solve the unconstrained optimization problem in Example 10.29 by the PSO method.

Solution *The objective function for the PSO solver can be in an anonymous function, and the solutions can be found within 0.15 seconds, and the norm of the decision variable* $\boldsymbol{x}$ *is* 5.3×10^{-5}.

```
>> f=@(x)(x(1)+x(2))^2+5*(x(3)-x(4))^2+(x(2)-2*x(3))^2+10*(x(1)-x(4))^2;
   tic, [x g]=particleswarm(f,4), norm(x), toc
```

Example 10.34 Consider again the constrained optimization problem in Example 10.31. Find the solution with the particle swarm optimization method.

Solution *Similar to the M-function for the genetic algorithm, another M-function can be written for the objective function with the constraints. It should be noted that vectorized format should be used.*

```
function y=c10mpso4(x)
x3=(6+4*x(1)-2*x(2))/3; x=[x x3]'; y1=[-2 1 1]*x; y2=[-1 1 0]*x;
y=[1 2 3]*x; if y1>9|y2<-4|x3<0|x1>0|x2>0, y=100; end
```

With the above MATLAB statements, the accurate solution $\boldsymbol{x}^{\rm T}=[-7,-11,0]$ *can be obtained.*

```
>> x=particleswarm(@c10mpso4,2,[-50,-50],[0,0]); % call particle swarms
   x=[x(1:2) (6+4*x(1)-2*x(2))/3]                % find the 3rd argument
```

II. Genetic algorithm based solver

The `ga()` function provided in the Global Optimization Toolbox offers an alternative way in solving optimization problems using genetic algorithms. The function is quite similar to other traditional optimization functions, and it can be used to solve constrained problems as well. The syntaxes of the function are

$[\boldsymbol{x},f,\texttt{flag},\texttt{out}]=\texttt{ga}(\mathit{fun},n,\mathit{opts})$ % unconstrained problem

$[\boldsymbol{x},f,\texttt{flag},\texttt{out}]=\texttt{ga}(\mathit{fun},n,\boldsymbol{A},\boldsymbol{B},\boldsymbol{A}_{\rm eq},\boldsymbol{B}_{\rm eq},\boldsymbol{x}_{\rm m},\boldsymbol{x}_{\rm M},\mathit{CFun},\texttt{intcon},\mathit{opts})$

where *fun* is a MATLAB description of the objective function, whose format is consistent to the ones in the Optimization Toolbox. The argument n is the number of variables to be optimized. The *opts* argument contains the control properties which can be set by the function `gaoptimset()`, similar to `optimset()` function in Chapter 6. The commonly used control properties are listed in Table 10.13. It seems that the function can be used in solving mixed integer programming problems, with `intcon` arguments.

The search results are returned in the vector $\boldsymbol{x}$. The other input and returned arguments are the same as the functions in the Optimization Toolbox. In particular, if `flag` is larger than 0, successful solutions to the problem are achieved. However, even if `flag` is positive, the solution found is not necessarily a global optimum solution.

TABLE 10.13: Commonly used control properties in GADS Toolbox.

property name	explanation to the options
`Generations`	maximum allowed generations, with a default 100
`InitialPopulation`	initial population matrix, the default population is created by random numbers
`PopulationSize`	the number of individuals in the population, with default 20
`SelectionFcn`	setting of the selective function, and the default `@selectionstochunif`, and other functions `@selectionremainder`, `@selectionuniform`, `@selection roulette`, `@selectiontournament`
`TolFun`	similar to `optimset` properties, the termination condition for the objective function. Also, `TolX`, `TolCon` can be set

Example 10.35 Let us consider the revised Rastrigin function studied earlier. Please run `ga()` function 100 times and see the successful rate in finding the global optimal solutions.

Solution *With default settings of the* `ga()` *function, the function is called 100 times with the following loop structure*

```
>> f=@(x)20+(x(1)/30-1)^2+(x(2)/20-1)^2-...
         10*(cos(pi*(x(1)/30-1))+cos(pi*(x(2)/20-1)));
   X=[]; tic, for i=1:100, [x,ff]=ga(f,2); X=[X; x ff]; end, toc
```

About 100 seconds are needed, and it is happy to see that all the solutions are global optimum ones, although the accuracy is not very high.

Example 10.36 Consider again the linear programming problem studied in Example 10.31, please solve again with `ga()` function.

Solution *Since* `ga()` *function can be used to solve constrained optimization problems, there is no need to express the constraints as penalties in the objective function. The constrained optimization problem can be solved directly with*

```
>> f=@(x)[1 2 3]*x(:); A=[-2 1 1; 1 -1 0]; B=[9; 4];
   Aeq=[4 -2 -3]; Beq=-6; xm=[-inf;-inf;0]; xM=[0;0;inf];
   [x a k]=ga(f,3,A,B,Aeq,Beq,xm,xM)
```

and sometimes the global optimal solution can be found, while sometimes the solution cannot be found, even though linear programming problem is a convex one.

Example 10.37 Solve again the nonlinear integer programming problem studied in Example 6.45, with genetic algorithm.

Solution *The mathematical form of the problem is given by*

$$\min \quad 2y_1^2/16 + y_2^2/100 - 4y_1 - y_2,$$
$$\boldsymbol{y} \text{ s.t. } \begin{cases} y_1^2/16-6y_1/4+y_2/10-11\leqslant 0 \\ -y_1y_2/40+3y_2/10+e^{y_1/4-3}-1\leqslant 0 \\ y_2\geqslant 30 \end{cases}$$

with the nonlinear constraints given in file c6mdisp.m. With `ga()` *function, the* `IntCon` *field should be assigned, and the problem can be solved with genetic algorithm, with the following statements, with the result* $\boldsymbol{y} = [16, 50]$. *It can be seen that the result is the same as the one obtained in the original example.*

```
>> clear P; P.fitnessfcn=@(y)2*y(1)^2/16+y(2)^2/100-4*y(1)-y(2);
   P.nonlcon=@c6mdisp; P.IntCon=[1,2]; P.lb=[-200; 30]; P.ub=[200; 200];
   P.solver='ga'; P.options=gaoptimset; P.nvars=2; [y,a key c]=ga(P)
```

Example 10.38 Consider the nonlinear programming problem studied in Example 6.36, with its standard form given by

$$\min_{\boldsymbol{x}\ \text{s.t.}\ \begin{cases} x_3+9.625x_1x_4+16x_2x_4+16x_4^2+12-4x_1-x_2-78x_4=0\\ 16x_1x_4+44-19x_1-8x_2-x_3-24x_4=0\\ -0.25x_5-x_1\leqslant -2.25\\ x_1-0.25x_5\leqslant 2.25\\ -0.5x_5-x_2\leqslant -1.5\\ x_2-0.5x_5\leqslant 1.5\\ -1.5x_5-x_3\leqslant -1.5\\ x_3-1.5x_5\leqslant 1.5\end{cases}} x_5.$$

Please solve the problem with `ga()` function.

Solution *Since there are two equation constraints, for genetic algorithm solvers which does not supporting constraints, the nonlinear equation solver should be embedded in the objective function, and it can be seen that the solution process would be extremely complicated. With* `fmincon_global()` *function in Chapter 6, the global optimal solution can be found easily.*

Now let us try the new `ga()` *function, which allows the handling of constraints. The following statements can be used to solve the problem. Unfortunately, the result obtained is* $\boldsymbol{x} = [1.7650, 2.1223, 0.1556, 1.5729, 2.3657]$, *far away from the global optimal solution, with* $x_5 = 0.8175$.

```
>> f=@(x)x(5); fnl=@c6exnls;
   A=[-1 0 0 0 -0.25; 1 0 0 0 -0.25; 0 -1 0 0 -0.5;
      0 1 0 0 -0.5; 0 0 -1 0 -1.5; 0 0 1 0 -1.5];
   B=[-2.25; 2.25; -1.5; 1.5; -1.5; 1.5]; Aeq=[]; Beq=[];
   xm=[]; xM=[]; x=ga(f,5,A,B,Aeq,Beq,xm,xM,fnl)
```

The time consuming function call was tested 30 times, which takes about 350 seconds, and none of the results are near the global optimal point. While with the use of `fmincon_global()` *function, each time the global optimal solution can be found.*

Summarizing the above results, it can be seen that, to the opinion of the authors, the `ga()` function is not recommended in real applications, unless it is significantly improved in later releases.

III. Other optimization problem solvers

Two other functions `simulannealbnd()` and `patternsearch()` are also provided in the Global Optimization Toolbox. Function `simulannealbnd()` can be used to solve unconstrained optimization problem with bounds on the decision variable, the syntaxes of the function are

$\boldsymbol{x}=$ `simulannealbnd(`f`,`$\boldsymbol{x}_0$`,`$\boldsymbol{x}_\mathrm{m}$`,`$\boldsymbol{x}_\mathrm{M}$`,`*opts*`)`, $\boldsymbol{x}=$ `simulannealbnd(`*problem*`)`

The pattern search based function `patternsearch()` can be used to solve nonlinear programming problems, and its syntax is the same as `fmincon()` function. It should be noted that the two functions all need an initially selected search point.

Example 10.39 Solve again with the revised Rastrigin function problem with the two solvers.

Solution *It is found that in the two functions, initial search points should be specified. 100 runs of each function can be carried out with the following statements, and the initial search point is generated randomly.*

```
>> f=@(x)20+(x(1)/30-1)^2+(x(2)/20-1)^2-...
         10*(cos(pi*(x(1)/30-1))+cos(pi*(x(2)/20-1)));
   xm=[-100; -100]; xM=[100; 100]; tic, X=[]; Y=[];
   for i=1:100, x0=-100+200*rand(2,1);
      [x g]=simulannealbnd(f,x0,xm,xM); X=[X; x' g];
   end, toc
   tic, for i=1:100, x0=-100+200*rand(2,1);
      [x g]=patternsearch(f,x0); Y=[Y; x' g];
   end, toc
```

For the simulated annealing algorithm, the successful rate for finding the global optimal point is 44%, and the time elapsed is 118 seconds. For the pattern search algorithm, the successful rate is similar, and the time elapsed is only 4.5 seconds.

A Lévy flight based PSO solver is developed by Zhuo Li [11]*, where vectorized form is used in describing the objective function. With such a tool, 100 runs are executed, and all the runs find the global optimal solution, with rather high precision. The time elapsed for the 100 runs is about 42 seconds.*

```
>> f=@(x)20+(x(:,1)/30-1).^2+(x(:,2)/20-1).^2-...
         10*(cos(pi*(x(:,1)/30-1))+cos(pi*(x(:,2)/20-1)));
   tic, X=[]; for i=1:100
      [g x]=levyPSO(f,xm',xM',100,600); X=[X; x g];
   end, toc
```

Summarizing all the functions tried up to now for the same problem, comparisons are given in Table 10.14.

TABLE 10.14: Comparisons of revised Rastrigin function solutions.

solver function	toolbox	successful rate	time	norm error
`fminunc_global()`	author developed function	98%	153	3.1×10^{-12}
`ga()`	Global Optimization Toolbox	100%	21	0.039
`gaopt()`	GAOT Toolbox (free)	100%	60	3.28×10^{-8}
`particleswarm()`	Global Optimization Toolbox	97%	5.3	4.54×10^{-8}
`patternsearch()`	Global Optimization Toolbox	29%	4.5	1.4×10^{-9}
`simulannealbnd()`	Global Optimization Toolbox	44%	118	0.14
`levyPSO()`	Free Lévy-flight based PSO Solver	100%	42	3.71×10^{-10}

Example 10.40 Solve again the problem in Example 6.36, with pattern search algorithm.

Solution *The genetic algorithm function* `ga()` *was tried in Example 10.38, but failed. The following statements can be used with the pattern search algorithm, and the solution* *found is* $\boldsymbol{x} = [1.9085, 0.8169, 2.0208, 0.1250, 1.3662]$. *Clearly it is again not the global optimal solution.*

```
>> clear P; P.objective=@(x)x(5); P.nonlcon=@c6exnls;
   P.Aineq=[-1 0 0 0 -0.25; 1 0 0 0 -0.25; 0 -1 0 0 -0.5;
            0 1 0 0 -0.5; 0 0 -1 0 -1.5; 0 0 1 0 -1.5];
   P.bineq=[-2.25; 2.25; -1.5; 1.5; -1.5; 1.5]; P.options=gaoptimset;
   P.solver='patternsearch'; P.X0=rand(5,1); patternsearch(P)
```

Running the `patternsearch()` *function 100 times, the time elapsed is 355 seconds, much slower than the other functions. The successful rate for finding the global solution is extremely low, about 2%. Again this algorithm is not recommended to use in real applications with constraints.*

With `fmincon_global()` *function written in Chapter 6, the global optimal solution can be found easily, within 2.54 seconds, the global solution is found*

$$\boldsymbol{x} = [2.4544, 1.9088, 2.7263, 1.3510, 0.8175].$$

```
>> P.solver='fmincon'; P.options=optimset;
   tic, fmincon_global(P,-100,100,5,10), toc %
```

10.4.5 Towards accurate global minimum solutions

In general, it can be concluded that the traditional optimization algorithms may find a solution with good accuracy, however, the global optimality cannot be ensured. For oscillatory surfaces, local minima are inevitable. The evolutionary computing algorithms can likely find the global optimal solutions, since multiple initial points are used. However, the accuracy of the solution may not be satisfactorily high. A combination of the two types of optimization algorithms can be made, for instance, with the use of evolutionary algorithms, a less accurate global solution can be found first. This solution can then be used as an initial search point for the traditional optimization algorithms in order to find the accurate global optimum solutions.

Example 10.41 Solve accurately the optimization problem

$$\min_{x,y \text{ s.t. } \begin{cases} -1\leqslant x\leqslant 3 \\ -3\leqslant y\leqslant 3 \end{cases}} \sin(3xy) + (x-0.1)(y-1) + x^2 + y^2.$$

Solution *The surface plot of the objective function can be obtained as shown in Figure 10.37. It can be seen that the surface is oscillatory, which makes the global optimization a difficult task.*

```
>> [x,y]=meshgrid(-1:0.1:3,-3:0.1:3);
   z=sin(3*x.*y+2)+(x-0.1).*(y-1)+x.^2+y.^2; surf(x,y,z);
```

To solve numerically the original problem, new variables $x_1 = x, x_2 = y$ *can be selected, and the original problem can be rewritten as*

$$\min_{\boldsymbol{x} \text{ s.t. } \begin{cases} -1\leqslant x_1\leqslant 3 \\ -3\leqslant x_2\leqslant 3 \end{cases}} \sin(3x_1x_2) + (x_1-0.1)(x_2-1) + x_1^2 + x_2^2.$$

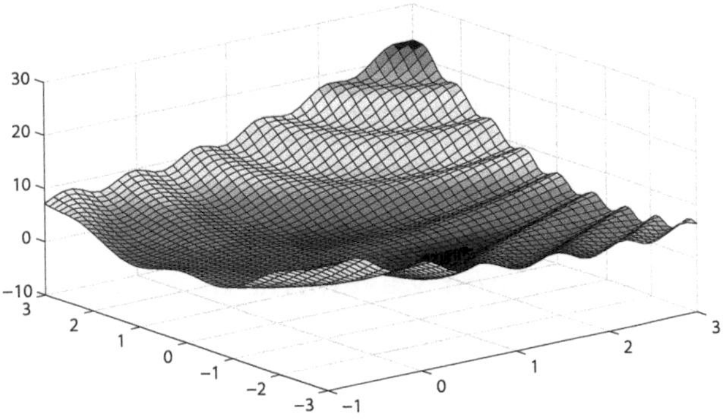

FIGURE 10.37: Surface of the objective function.

If the traditional nonlinear programming technique is used, using an anonymous function to describe the objective function, the following statements can be used

```
>> f=@(x)sin(3*x(1)*x(2)+2)+(x(1)-0.1)*(x(2)-1)+x(1)^2+x(2)^2;
   x0=rand(1,2); x=fmincon(f,x0,[],[],[],[],[-1;-3],[3;3]), f(x)
```

and the optimum point is $\boldsymbol{x}^{\mathrm{T}} = [0.9299, 0.6577]$, *with the objective function value* 0.3742.

Using GAOT, the objective function can be written as

```
function [sol,y]=c10mga6(sol,options)
x=sol(1:2); y=-sin(3*x(1)*x(2)+2)-(x(1)-0.1)*(x(2)-1)-x(1)^2-x(2)^2;
```

Within the selected search boundaries, the problem can be solved again to find the solution, which is $x_1 = 1.225588050491944$, $x_2 = -0.918235615244950$. *The objective function in this case is* -0.795033762096730. *It is clear that the solution by traditional optimization method is a poor local minimum.*

```
>> xmM=[-1 3; -3 3]; % set the boundaries
   [a,b,c,d]=gaopt(xmM,'c10mga6'); x=a(1:2), f0=-a(3)
```

Using the results from applying the genetic algorithm as the initial search points, the following statements can be used to find a more accurate solution which is $x_1 = 1.225533324173843$ *and* $x_2 = -0.918286942316160$, *and the objective function becomes* -0.795033772873523, *which is slightly better than the genetic algorithm result.*

```
>> ff=optimset; ff.TolX=1e-10; ff.TolFun=1e-20;
   x=fmincon(f,a(1:2),[],[],[],[],[-1;-3],[3;3],[],ff), f(x)
```

10.5 Wavelet Transform and Its Applications in Data Processing

Fourier transform is a very important technique in signal processing. However, the ordinary Fourier transform has certain limitations. For instance, it maps the time-domain representation into the frequency-domain representation, where features in the original domain may get lost. Therefore, for many types of signals, the processed results may not be satisfactory. For stationary signals, Fourier transform may be very useful. However, for non-stationary and transient signals with sudden changes, the Fourier transform may not be suitable, since the features of the sudden changes may be neglected. Thus, other types of transforms, for instance, short-time Fourier transform should be used. Moreover, the wavelet transform technique emerged in the 1980s is a very powerful tool used widely, in particular, in signal and image processing.

10.5.1 Wavelet transform and waveforms of wavelet bases

A wavelet is a function with zero mean which can be translated and scaled to form a family of functions. In this section, fundamental concepts of continuous and discrete wavelet transforms are introduced. Some of the commonly used wavelet bases are introduced.

I. Continuous wavelet transform

The continuous wavelet transformation formula is given by

$$\mathscr{W}_{a,b}^{\varphi}[f(t)] = \frac{1}{\sqrt{|a|}}\int_{-\infty}^{\infty} f(t)\overline{\varphi_{a,b}(t)}\,\mathrm{d}t = W_{\varphi}(a,b), \tag{10-5-1}$$

where

$$\varphi_{a,b}(t) = \varphi\left(\frac{t-b}{a}\right), \quad \text{and} \quad \int_{-\infty}^{\infty} \varphi(t)\,\mathrm{d}t = 0, \tag{10-5-2}$$

and the function $\varphi(t)$ is referred to as *wavelet basis*, and $\varphi_{a,b}(t)$ is generated by translation and scaling from the wavelet basis.

Example 10.42 Consider the "Mexican hat" wavelet basis given by $\varphi(t) = \dfrac{1-t^2}{\sqrt{2\pi}}\mathrm{e}^{-\frac{t^2}{2}}$. Draw the wavelet bases for different values of a and b.

Solution *From the given mathematical formula of the wavelet basis, the function* `ezplot()` *can be used to draw the relevant curves. The translation and scaling operations can be carried out with the* `subs()` *function. The curves for different values of a and b can be obtained as shown in Figures 10.38 (a) and (b).*

```
>> syms t; f=(1-t^2)*exp(-t^2/2)/sqrt(2*pi);
   subplot(121), ezplot(f,-4,4), hold on;  % wavelet basis plot
   ezplot(subs(f,t,t-1),-4,4); ezplot(subs(f,t,t+1),-4,4) % translation
   subplot(122), ezplot(f,-4,4), hold on;
   ezplot(subs(f,t,t/2),-4,4); ezplot(subs(f,t,2*t),-4,4) % scaling
```

From the above example, it is seen that the parameter b can translate the wavelet basis to the right or left while a expands or shrinks the waveform accordingly. If $a < 1$, the waveform of the basis will be shrunken to form the new wavelet signal. Wavelet analysis is performed by computing the coefficients for different combinations of a and b. The wavelet signals multiplied by the corresponding coefficients sum up to reconstruct the original signal. Similar to other integral transform problems, the inverse wavelet transforms are also defined:

$$f(t) = \frac{1}{C_{\varphi}}\int_{-\infty}^{\infty}\int_{-\infty}^{\infty} W_{\varphi}(a,b)\varphi_{a,b}(t)\,\mathrm{d}a\,\mathrm{d}b, \tag{10-5-3}$$

where

$$C_{\varphi} = \int_{-\infty}^{\infty} \frac{|\hat{\varphi}(\omega)|^2}{|\omega|}\,\mathrm{d}\omega. \tag{10-5-4}$$

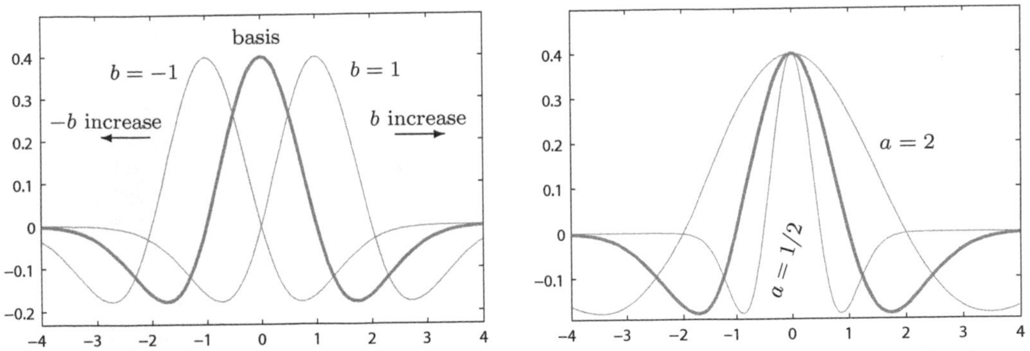

(a) translation for different b's when $a=1$ (b) scaling for different a's when $b=0$

FIGURE 10.38: Waveforms of Mexican hat basis for different a and b.

The coefficients of continuous wavelet transforms can be evaluated with the `cwt()` function, with the following syntaxes

```
Z=cwt(y,a,fun)            % wavelet coefficients matrix Z
Z=cwt(y,a,fun,'plot')     % absolute value plots
```

where the *fun* is the name of wavelet basis to be discussed more later. Here, the wavelet basis name for Mexican hat wavelet basis is denoted by `'mexh'`.

Example 10.43 Perform continuous wavelet transform for the function $f(t)=\sin t^2$, and visualize its coefficients.

Solution *The data in the interval $t\in[0,2\pi]$ can be generated directly with the following statements, and the time domain function is shown in Figure 10.39 (a).*

```
>> t=0:0.03:2*pi; y=sin(t.^2); plot(t,y)
```

Taking the option `'mexh'` wavelet basis as the template, the wavelet coefficients $W_\varphi(a,b)$ can be obtained as shown in Figure 10.39 (b).

```
>> a=1:32; Z=cwt(y,a,'mexh','plot');
```

The following statements can be used to draw the surface plot of the wavelet coefficients, as shown in Figure 10.40.

```
>> surf(t,a,Z); shading flat; axis([0 2*pi,0,32,min(Z(:)) max(Z(:))])
```

II. Discrete wavelet transform

If the signal $f(t)$ is expressed by a discrete sequence $f(k)$, and the wavelet basis $\varphi(t)$ is selected, the translated and scaled wavelet function can be written as $\varphi_{a,b}(t)=\sqrt{2}\varphi\left(2^m t-n\right)$, whose discrete form can be written as $\varphi_{a,b}(k)=\sqrt{2}\varphi\left(2^m k-n\right)$. The wavelet transform of the discrete signal is written as

$$\mathscr{W}_{n,m}^{\varphi}[f(k)]=\sqrt{2}\sum_k f(k)\overline{\varphi\left(2^m k-n\right)}\,\mathrm{dt}=W_\varphi(m,n), \tag{10-5-5}$$

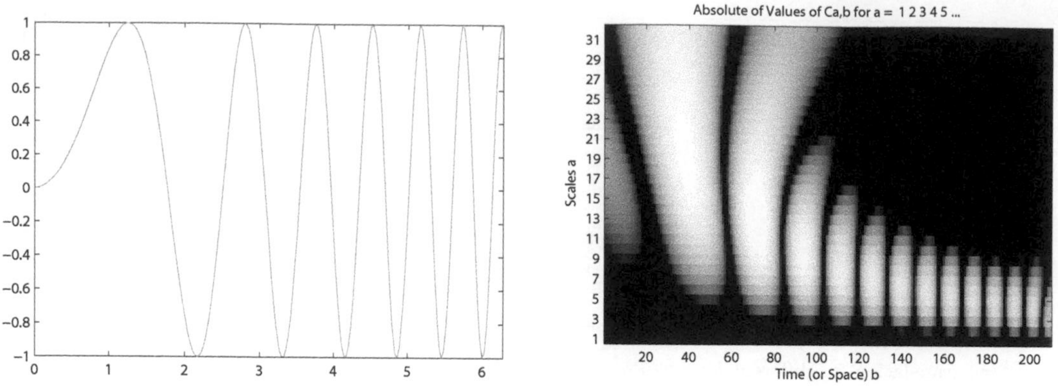

(a) time domain function $f(t)$ (b) continuous wavelet transform coefficients

FIGURE 10.39: Continuous wavelet transform.

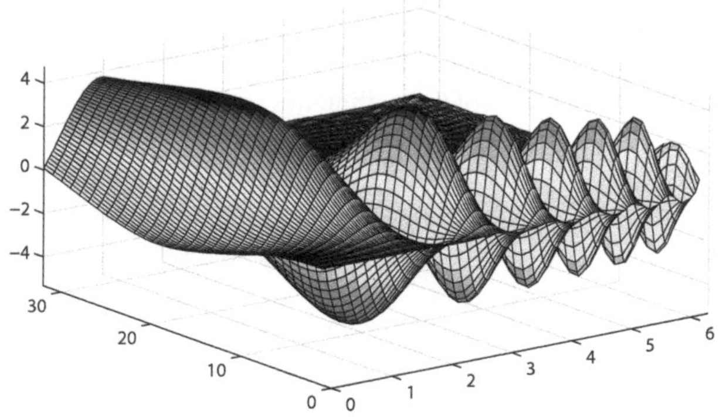

FIGURE 10.40: Surface plot of continuous wavelet transform coefficients.

and the inverse discrete wavelet transform is expressed as

$$f(k) = \sum_m \sum_n W_{n,m}(k)\varphi_{m,n}(k). \tag{10-5-6}$$

The function `dwt()` provided in the Wavelet Toolbox can be used for discrete wavelet transform, with the syntax $[\boldsymbol{C}_{\rm a},\boldsymbol{C}_{\rm d}]=$`dwt`$(\boldsymbol{x},$*fun*$)$, where $\boldsymbol{x}$ is the original data, *fun* is the selected wavelet basis, which can be `'mexh'` or other functions to be discussed more later. The returned argument $\boldsymbol{C}_{\rm a}$ can be used to describe the approximated coefficients of the original function, while the argument $\boldsymbol{C}_{\rm d}$ returns the detailed coefficients. Normally the argument $\boldsymbol{C}_{\rm a}$ corresponds to the low- and mid-frequency information, while $\boldsymbol{C}_{\rm d}$ corresponds to the high-frequency information. The lengths of vectors $\boldsymbol{C}_{\rm a}$ and $\boldsymbol{C}_{\rm d}$ are half of the length of the original vector $\boldsymbol{x}$.

The vectors $\boldsymbol{C}_{\rm a}$ and $\boldsymbol{C}_{\rm d}$ can be used to perform inverse wavelet transform with the function $\hat{\boldsymbol{x}}=$`idwt`$(\boldsymbol{C}_{\rm a},\boldsymbol{C}_{\rm d},$*fun*$)$ to restore the original vector $\hat{\boldsymbol{x}}$.

Example 10.44 Generate a signal sequence corrupted by noises. Perform discrete wavelet transform to the sequence. Perform also inverse wavelet transform to the sequence, and see whether the original function can be restored.

Solution *The white noise with a standard deviation of 0.1 is added to the original signal*

$f(t) = \sin t^2$ given in Example 10.43. The corrupted data are shown in Figure 10.41. It can be seen that the noise level is quite visible. Using discrete wavelet transform, the waveforms of approximated and detailed coefficients are shown in the same window. It can be seen that the resulted signal removes noises to some extent. For better filtering results, more steps in wavelet transform should be used.

```
>> x=0:0.002:2*pi; y=sin(x.^2); r=0.1*randn(size(x));
   y1=y+r; subplot(211); plot(x,y1)   % draw the original corrupted signal
   [cA,cD]=dwt(y1,'db4');             % discrete wavelet transform
   subplot(223), plot(cA); subplot(224), plot(cD)
```

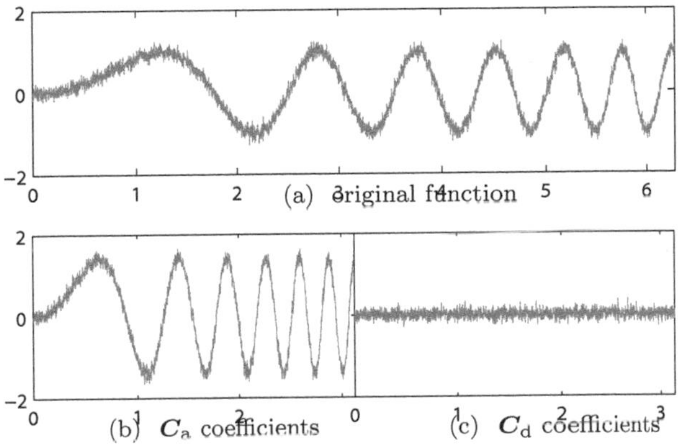

FIGURE 10.41: Discrete wavelet transform of a given function.

The obtained C_a and C_d vectors can be used to perform the inverse discrete wavelet transform. Comparing the transformed sequence and the original sequence, it can be seen that the error is very small, with an error norm of 6.16×10^{-12}.

```
>> y2=idwt(cA,cD,'db4'); norm(y1-y2)
```

III. Wavelet bases provided in the Wavelet Toolbox

A large amount of wavelet bases such as Haar basis, Daubechies basis, Mexican hat basis, Bior basis, etc. can be generated and tested by `wavemngr()` function. The syntax of the function is `wavemngr('read',1)`. For instance, Haar wavelet basis is denoted as `'haar'`. Daubechies families are denoted as `'db1'`, `'db2'`, etc. Bior families are `'bior1.3'`, `'bior2.4'`, etc. Mexican hat wavelet basis can be denoted as `'mexh'`. The wavelet basis function can be evaluated with the `wavefun()` function. The function can be called as

$[\varphi,x]=$ `wavefun`(fun,n) % Gaussian, Mexican hat wavelet bases

$[\phi,\varphi,x]=$ `wavefun`(fun,n) % Daubechies, Symlets orthogonal families

$[\phi_1,\varphi_1,\phi_2,\varphi_2,x]=$ `wavefun`(fun,n) % Bior basis

where n is the number of iterations computed, default is 8. The argument φ is the wavelet basis, and ϕ is the wavelet derivatives. In Bior wavelet bases, the vectors ϕ_1 and φ_1 are used for wavelet computation, while the vectors ϕ_2 and φ_2 are used for wavelet reconstruction.

Example 10.45 Select Daubechies 6 wavelet basis (`'db6'`) to draw the wavelet waveforms of different iterations.

Solution *The iteration numbers 2, 4, 6, 8 can be used in the function call to draw the wavelets as shown in Figure 10.42. It can be seen that when the iteration is 8, smooth waveform can be achieved. Thus, normally, one should select the iteration number to 8.*

```
>> [a,y,x]=wavefun('db6',2); subplot(141), plot(x,y)
   [a,y,x]=wavefun('db6',4); subplot(142), plot(x,y)
   [a,y,x]=wavefun('db6',6); subplot(143), plot(x,y)
   [a,y,x]=wavefun('db6',8); subplot(144), plot(x,y)
```

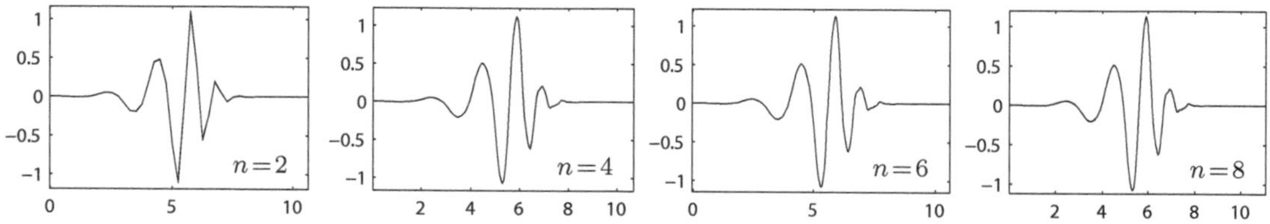

FIGURE 10.42: The waveforms of Daubechies 6 for different iterations.

Example 10.46 Show the waveforms of some commonly used wavelet bases.

Solution *The following statements can be used to draw wavelets of some selected wavelet bases, as shown in Figure 10.43, where* `'db1'` *is in fact the Haar basis. Daubechies wavelet families are smooth when* `'db6'` *is used. For Symlets wavelet families, the function* `'sym6'` *is smooth.*

```
>> subplot(5,4,1), [a,y,x]=wavefun('db1'); plot(x,y), % other db bases
   subplot(5,4,9), [a,y,x]=wavefun('sym2'); plot(x,y),% other sym bases
   subplot(5,4,13), [a,y,x]=wavefun('coif2'); plot(x,y)
   subplot(5,4,15), [y,x]=wavefun('gaus2'); plot(x,y)
   subplot(5,4,17), [a,y,b,c,x]=wavefun('bior1.3'); plot(x,y)
```

10.5.2 Wavelet transform in signal processing problems

Similar to the Fourier transform technique, wavelet transform is also very useful in signal processing and image processing problems. Some unique characteristics of wavelet transform are not offered in conventional frequency-domain analysis. Here, wavelet transform-based signal decomposition and reconstruction in MATLAB are demonstrated by a de-noising example in this section.

Wavelet transform can be used to perform signal decomposition, that is, to express a given signal S by two parts, $\boldsymbol{C}_{\mathrm{a}_1}$ and $\boldsymbol{C}_{\mathrm{d}_1}$. The lengths of the two vectors are half the length of S. The vector $\boldsymbol{C}_{\mathrm{a}_1}$ preserves the low- and mid-frequency information or "approximate" information, while $\boldsymbol{C}_{\mathrm{d}_1}$ retains the high-frequency, or "detailed" information. From the viewpoint of de-noising signal processing task, the latter high-frequency content can be regarded as noise information. The vector $\boldsymbol{C}_{\mathrm{a}_1}$ can further be decomposed with wavelet transform into $\boldsymbol{C}_{\mathrm{a}_2}$ and $\boldsymbol{C}_{\mathrm{d}_2}$. Further decomposition leads to $\boldsymbol{C}_{\mathrm{a}_3}$ and $\boldsymbol{C}_{\mathrm{d}_3}$, and so on. The decomposition process is shown in Figure 10.44 (a). The vectors $\boldsymbol{C}_{\mathrm{a}_i}$ are referred to as the *approximate coefficients*, and $\boldsymbol{C}_{\mathrm{d}_i}$ is the *detailed coefficients*.

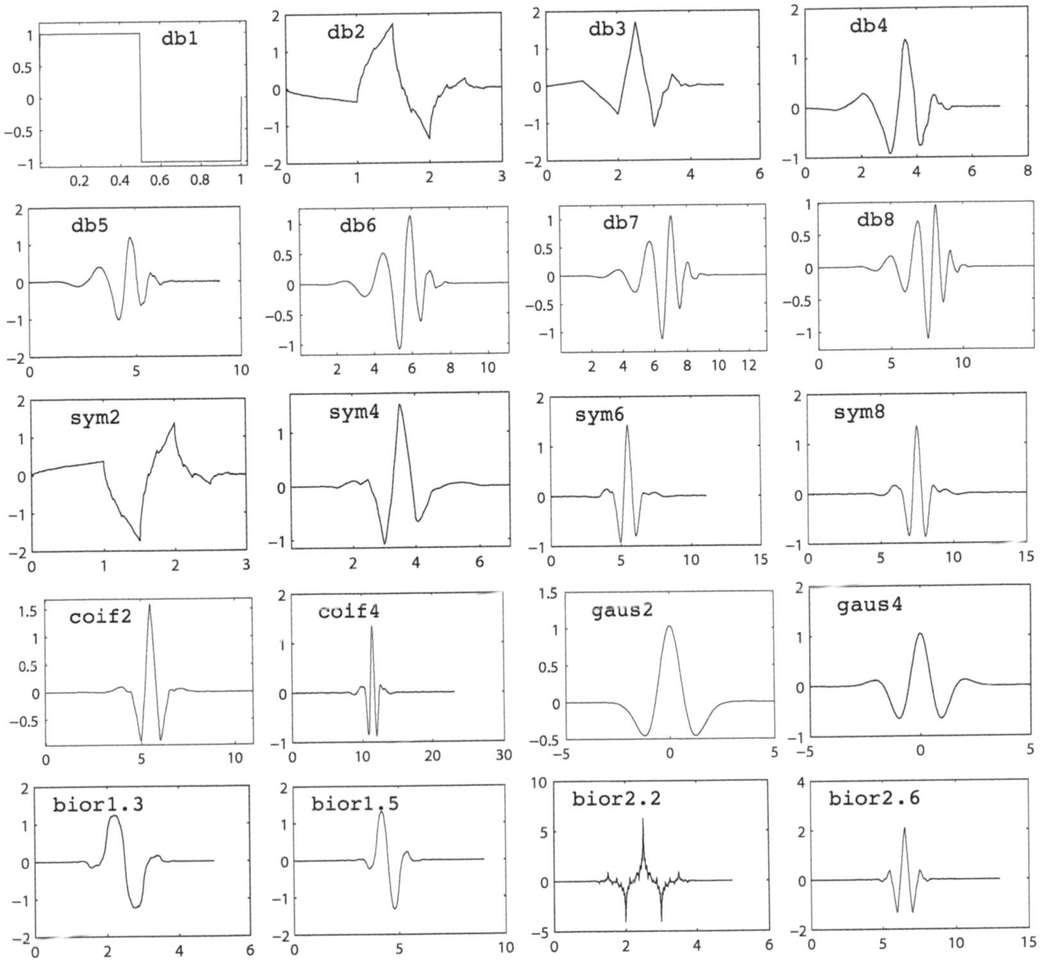

FIGURE 10.43: Waveforms of commonly used wavelet bases.

The function `wavedec()` provided in Wavelet Toolbox can be used in one-dimensional wavelet decomposition, with $[\boldsymbol{C},\boldsymbol{L}]=\texttt{wavedec}(\boldsymbol{x},n,\mathit{fun})$, where $\boldsymbol{x}$ is the original signal, n is the level of decomposition. *fun* is the name of the wavelet basis, for instance `'db6'`. After the decomposition, the two vectors $\boldsymbol{C}$ and $\boldsymbol{L}$ are returned, as shown in Figure 10.44 (b). The original vector $\boldsymbol{C}$ is decomposed into a vector whose length is the same as that of $\boldsymbol{x}$. The short sub-vectors are joined together, and the lengths of the sub-vectors are indexed in the vector $\boldsymbol{L}$.

The approximate coefficients $\boldsymbol{C}_{\rm a}$ and detailed coefficients $\boldsymbol{C}_{\rm d}$ after decomposition can be extracted from the vectors $\boldsymbol{C}$ and $\boldsymbol{L}$, with the functions `appcoef()` and `detcoef()`, respectively, by the syntaxes

$\boldsymbol{C}_{{\rm a}_i}=\texttt{appcoef}(\boldsymbol{C},\boldsymbol{L},\mathit{fun},n)$; % extract approximate coefficients

$\boldsymbol{C}_{{\rm d}_i}=\texttt{detcoef}(\boldsymbol{C},\boldsymbol{L},i)$; % extract detailed coefficients

where n is the level of extraction.

Some of the noises can be filtered out from the approximate coefficients and detailed coefficients, by reconstructing the signal with the function `wrcoef()`, whose syntax

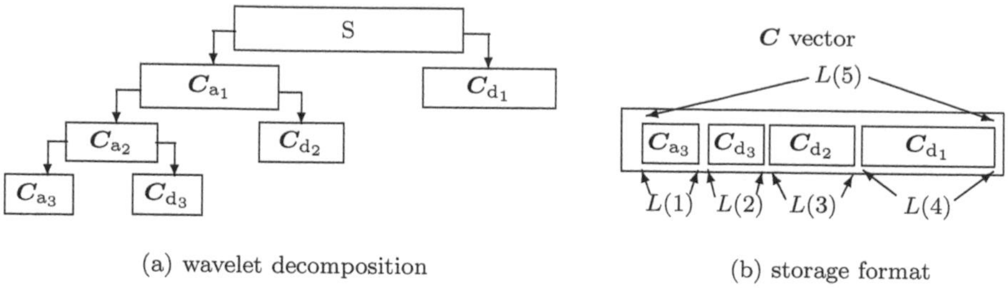

FIGURE 10.44: Illustrations of wavelet decomposition.

is $\hat{x}$=`wrcoef(`*type*`,`C`,`L`,`*fun*`,`n`)`, where *type* can be set to `'a'` or `'d'`, representing approximate coefficients and detailed coefficients, respectively. If `'a'` is selected, the reconstructed signal may filter out certain high-frequency noises.

Example 10.47 For the data given in Example 10.44, three-level wavelet decomposition can be performed. Compare the de-noise effects for different wavelet bases.

Solution *The noisy signal in Example 10.44 shown in Figure 10.41 (a) can still be used for de-noising tests.*

```
>> x=0:0.002:2*pi; y=sin(x.^2); r=0.1*randn(size(x)); y1=y+r; plot(x,y1)
```

For three-level wavelet decomposition, the following statements can be used to get the waveforms of the sub-vectors as shown in Figure 10.45, where the widths of the axes have been rearranged manually. It can be seen that part of the noise is filtered out in each decomposition level. Thus, the final C_{a_3} contains less noise information.

```
>> [C,L]=wavedec(y1,3,'db6'); cA3=C(1:L(1)); subplot(141), plot(cA3)
   dA3=C(L(1)+1:sum(L([1 2]))); subplot(142), plot(dA3)
   dA2=C(sum(L(1:2))+1:sum(L(1:3))); subplot(143), plot(dA2)
   dA1=C(sum(L(1:3))+1:sum(L(1:4))); subplot(144), plot(dA1)
```

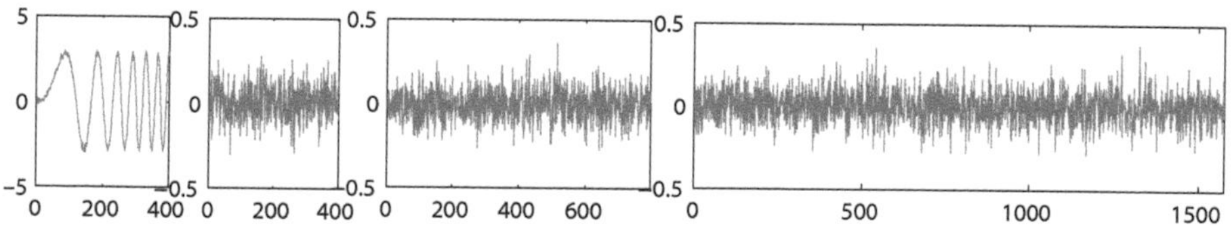

FIGURE 10.45: Wavelet decomposition illustrations.

The wavelet basis `'db6'` *can be used, and the approximate coefficients can be obtained as shown in Figure 10.46 (a). If* `'db2'` *is used instead, the de-noising result shown in Figure 10.46 (b) can be obtained. The two wavelet bases do not have significant differences in de-noising effect for this example.*

```
>> A3=wrcoef('a',C,L,'db6',3); subplot(121), plot(A3); subplot(122)
   [C,L]=wavedec(y1,3,'db2'); A3=wrcoef('a',C,L,'db2',3); plot(A3)
```

Two commonly used wavelet bases, `'bior2.6'` *and* `'coif4'` *can also be tried for the de-noising purposes, and the results are still the same as the one shown in Figure 10.46 (a). It is seen that for this example, there is no significant difference in the de-noising effect.*

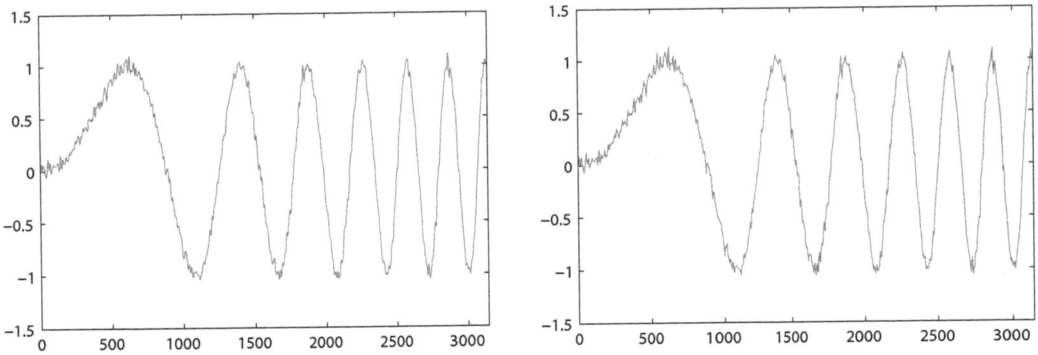

(a) 'db6' basis (b) 'db2' basis

FIGURE 10.46: De-noising effects under different wavelet bases.

```
>> [C,L]=wavedec(y1,3,'bior2.6'); A3=wrcoef('a',C,L,'bior2.6',3);
   [C,L]=wavedec(y1,3,'coif4'); A3a=wrcoef('a',C,L,'coif4',3);
   subplot(121), plot(A3), subplot(122), plot(A3a)
```

Example 10.48 Reconsider the digital filtering problem in Example 8.48. Filter the corrupted signal using wavelet method and compare the results.

Solution *The filter in Example 8.48 can be used again for the noisy signal. The 4-level wavelet filter with* 'db6' *basis is used for the same signal. The de-noising effects are compared in Figure 10.47. It can be seen that there exists delay in the filtering result in Chapter 8 while the wavelet-based filtering contributes no delay and the de-noising effect is much better.*

```
>> b=1.2296e-6*conv([1 4 6 4 1],[1 3 3 1]); a=conv([1,-0.7265],...
      conv([1,-1.488,0.5644],conv([1,-1.595,0.6769],[1,-1.78,0.8713])));
   x=0:0.002:2; y=exp(-x).*sin(5*x); r=0.05*randn(size(x)); y1=y+r;
   y2=filter(b,a,y1);  % filter used in Example 8.48
   [C,L]=wavedec(y1,4,'db6'); A4=wrcoef('a',C,L,'db6',4);
   plot(x,y,x,y2,x,A4)  % comparisons of the two filters
```

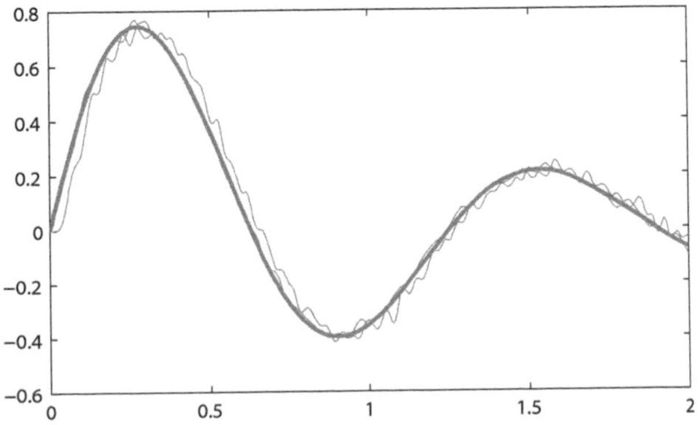

FIGURE 10.47: De-noising effect comparisons for given signal.

10.5.3 Graphical user interface in wavelets

An easy-to-use graphical user interface is provided in the Wavelet Toolbox, which can be used for solving one-dimensional and two-dimensional wavelet transform problems. The command `wavemenu` is used to start the interface, and the window shown in Figure 10.48 appears. If a one-dimensional wavelet transform problem is to be solved, the Wavelet 1-D button should be clicked. The interface guides the user to import data file, select wavelet basis and perform wavelet analysis for the data. Details of using the interface will not be given in this book. The interested readers are suggested to consult the users' manual of Wavelet Toolbox and play with the demos offered in the toolbox.

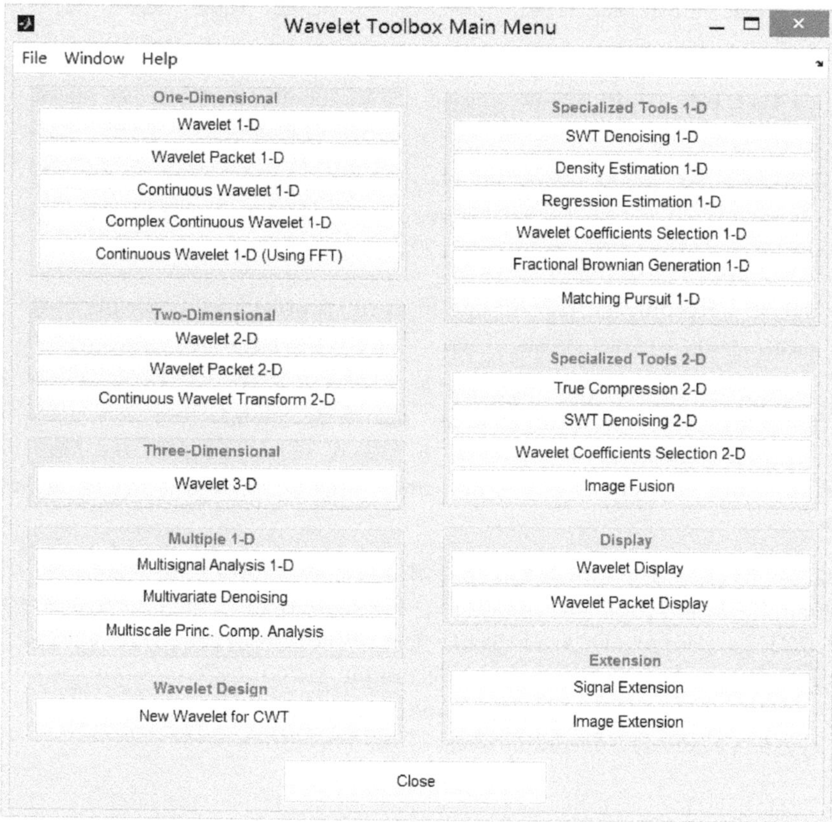

FIGURE 10.48: Graphical user interface for wavelet analysis.

10.6 Fractional-order Calculus

In Chapter 3, the calculus problems have been discussed, and the related materials are also shown in other chapters. It is quite interesting to note that before the invention of calculus, only algebra, geometry, and statics could be subjects of study. With calculus, lots of new subjects emerge such as dynamics, modeling, etc., which enable us to better characterize the world around us. It is known that the nth order derivative of a function

$f(t)$ can be mathematically described by $\mathrm{d}^n y/\,\mathrm{d}x^n$, a notation invented by Leibniz not Newton who suggested dot notation. With Leibniz's notation, one may ask "What does $n=1/2$ mean to the notation?" Actually, this is the question asked in a letter by the French mathematician Guillaume François Antoine L'Hôpital to one of the inventors of calculus, German mathematician Gottfried Wilhelm Leibniz more than 300 years ago. In answering the letter, Leibniz said, "It will lead to a paradox, from which one day useful consequences will be drawn." This marks the beginning of fractional-order calculus. However, earlier research concentrated on theoretical math issues. It is now being widely used in many areas. For instance, in the discipline of automatic control, fractional-order controller is a promising new topic. In this section, an easy-to-follow introduction to fractional-order calculus will be presented with how to get started using MATLAB in mind. Our treatment in this section, although not very mathematically rigorous, is practically useful since we presented a comprehensive introduction to numerical solutions to fractional-order filtering and fractional-order ordinary differential equations. Furthermore, from a programming point of view, the design and application of MATLAB classes and objects dedicated for fractional-order calculus demonstrated thoroughly in this section are of great value.

10.6.1 Definitions of fractional-order calculus

Various definitions appeared in the development and studies of fractional-order calculus. Some of the definitions are directly extended from the conventional integer-order calculus. The commonly used definitions are summarized as follows:

I. Fractional-order Cauchy integral formula

The formula is extended from integer-order calculus

$$\mathscr{D}^\alpha f(t)=\frac{\Gamma(\alpha+1)}{\mathrm{j}2\pi}\int_{\mathrm{C}}\frac{f(\tau)}{(\tau-t)^{\alpha+1}}\,\mathrm{d}\tau, \tag{10-6-1}$$

where C is the closed-path that encircles the poles of the function $f(t)$.

The integrals and derivatives for sinusoidal and cosine functions can be expressed by

$$\frac{\mathrm{d}^k}{\mathrm{d}t^k}\Big[\sin at\Big]=a^k\sin\left(at+\frac{k\pi}{2}\right),\ \frac{\mathrm{d}^k}{\mathrm{d}t^k}\Big[\cos at\Big]=a^k\cos\left(at+\frac{k\pi}{2}\right). \tag{10-6-2}$$

It can also be shown with Cauchy's formula that if k is not an integer, the above formula is still valid.

II. Grünwald–Letnikov definition

The fractional-order differentiation and integral can be defined in a unified way such that

$${}_{t_0}\mathscr{D}_t^\alpha f(t)=\lim_{h\to 0}\frac{1}{h^\alpha}\sum_{j=0}^{[(t-t_0)/h]}(-1)^j\binom{\alpha}{j}f(t-jh), \tag{10-6-3}$$

where $\binom{\alpha}{j}$ are the binomial coefficients; the subscripts to the left and right of $\mathscr{D}$ are the lower- and upper-bounds in the integral. The value α can be positive or negative, corresponding to differentiation and integration, respectively.

III. Riemann–Liouville definition

The fractional-order integral is defined as

$$ {}_{t_0}\mathscr{D}_t^{-\alpha} f(t) = \frac{1}{\Gamma(\alpha)} \int_{t_0}^{t} (t-\tau)^{\alpha-1} f(\tau)\,\mathrm{d}\tau, \tag{10-6-4}$$

where $0 < \alpha < 1$, and t_0 is the initial time. Let $t_0 = 0$, the notation of integral can be simplified to $\mathscr{D}_t^{-\alpha} f(t)$. The Riemann–Liouville definition is a widely used definition for fractional-order differentiation and integral [12]. Similarly, fractional-order differentiation is defined as

$$ {}_{t_0}\mathscr{D}_t^{\beta} f(t) = \frac{\mathrm{d}^n}{\mathrm{d}t^n}\left[{}_{t_0}\mathscr{D}_t^{-(n-\beta)} f(t)\right] = \frac{1}{\Gamma(n-\beta)} \frac{\mathrm{d}^n}{\mathrm{d}t^n}\left[\int_{t_0}^{t} \frac{f(\tau)}{(t-\tau)^{\beta-n+1}}\,\mathrm{d}\tau\right], \tag{10-6-5}$$

where $n-1 < \beta \leqslant n$.

IV. Caputo definition

The Caputo fractional-order differentiation is defined by

$$ {}_{t_0}\mathscr{D}_t^{\alpha} f(t) = \frac{1}{\Gamma(1-\alpha)} \int_{t_0}^{t} \frac{f^{(m+1)}(\tau)}{(t-\tau)^{\alpha}}\,\mathrm{d}\tau, \tag{10-6-6}$$

where $\alpha = m+\gamma$, m is an integer and $0 < \gamma \leqslant 1$. Similarly, by Caputo's definition, the integral is described by

$$ {}_{t_0}\mathscr{D}_t^{-\gamma} f(t) = \frac{1}{\Gamma(\gamma)} \int_{t_0}^{t} \frac{f(\tau)}{(t-\tau)^{1-\gamma}}\,\mathrm{d}\tau,\quad \gamma > 0. \tag{10-6-7}$$

10.6.2 Properties and relationship of various fractional-order differentiation definitions

I. Properties of fractional-order derivatives

The properties of fractional-order calculus are summarized below [14]:

(i) The fractional-order differentiation ${}_{t_0}\mathscr{D}_t^{\alpha} f(t)$ of an analytic function $f(t)$ with respect to t is also analytic.

(ii) If $\alpha = n$, the fractional-order derivative is identical to integer-order derivative, and also ${}_{t_0}\mathscr{D}_t^{0} f(t) = f(t)$.

(iii) The fractional-order differentiation is linear, i.e., for any constants c, d

$$ {}_{t_0}\mathscr{D}_t^{\alpha}\left[cf(t) + dg(t)\right] = c\,{}_{t_0}\mathscr{D}_t^{\alpha} f(t) + d\,{}_{t_0}\mathscr{D}_t^{\alpha} g(t). \tag{10-6-8}$$

(iv) Fractional-order differentiation operators satisfy commutative law, i.e.,

$$ {}_{t_0}\mathscr{D}_t^{\alpha}\left[{}_{t_0}\mathscr{D}_t^{\beta} f(t)\right] = {}_{t_0}\mathscr{D}_t^{\beta}\left[{}_{t_0}\mathscr{D}_t^{\alpha} f(t)\right] = {}_{t_0}\mathscr{D}_t^{\alpha+\beta} f(t). \tag{10-6-9}$$

II. Laplace transforms of fractional-order derivaatives

The Laplace transform of a fractional-order integral can be expressed by

$$ \mathscr{L}\left[{}_{t_0}\mathscr{D}_t^{-\gamma} f(t)\right] = s^{-\gamma}\mathscr{L}[f(t)]. \tag{10-6-10}$$

Under Riemann–Liouville definition, Laplace transform of the differentiation of a function can be written as

$$\mathscr{L}\left[{}^{\rm RL}_{t_0}\mathscr{D}^{\alpha}_t f(t)\right] = s^{\alpha}\mathscr{L}[f(t)] - \sum_{k=0}^{n-1} s^k \left[\, {}_{t_0}\mathscr{D}^{\alpha-k-1}_t f(t)\right]_{t=t_0}. \qquad (10\text{-}6\text{-}11)$$

The Laplace transform of integrals under Caputo definition is identical to those under Riemann–Louiville definition, and the Laplace transform to differentiations under Caputo definition satisfies

$$\mathscr{L}\left[{}^{\rm C}_{t_0}\mathscr{D}^{\gamma}_t f(t)\right] = s^{\gamma}F(s) - \sum_{k=0}^{n-1} s^{\gamma-k-1} f^{(k)}(t_0). \qquad (10\text{-}6\text{-}12)$$

Especially, when the initial values of $f(t)$ and their integer-order derivatives are all zero, $\mathscr{L}[{}^{\rm C}_{t_0}\mathscr{D}^{\alpha}_t f(t)] = s^{\alpha}\mathscr{L}[f(t)]$. It is also seen that Caputo definition is more suitable in describing a class of fractional-order differential equations with nonzero initial conditions.

III. Relationships among different definitions

It can be shown that for great varieties of functions, the Grünwald–Letnikov and Riemann–Liouville definitions are equivalent [13]. The major difference between Caputo and Riemann–Liouville definitions are the differentiations to a constant, with the former, the differentiation is bounded, while in the later, it is unbounded.

If function $y(t)$ has nonzero initial conditions, and the order $\alpha \in (0,1)$, it can be found by comparing the Caputo and Riemann–Liouville definitions that

$${}^{\rm C}_{t_0}\mathscr{D}^{\alpha}_t y(t) = {}^{\rm RL}_{t_0}\mathscr{D}^{\alpha}_t (y(t) - y(t_0)), \qquad (10\text{-}6\text{-}13)$$

where, the αth order derivative of constant $y(t_0)$ is

$${}^{\rm RL}_{t_0}\mathscr{D}^{\alpha}_t y(t_0) = y(t_0)(t-t_0)^{-\alpha}/\Gamma(1-\alpha), \qquad (10\text{-}6\text{-}14)$$

and the relationship between the two definitions can be derived as

$${}^{\rm C}_{t_0}\mathscr{D}^{\alpha}_t y(t) = {}^{\rm RL}_{t_0}\mathscr{D}^{\alpha}_t y(t) - \frac{y(t_0)(t-t_0)^{-\alpha}}{\Gamma(1-\alpha)}. \qquad (10\text{-}6\text{-}15)$$

More generally, if the order $\alpha > 1$, denote $m = \lceil \alpha \rceil$, then

$${}^{\rm C}_{t_0}\mathscr{D}^{\alpha}_t y(t) = {}^{\rm RL}_{t_0}\mathscr{D}^{\alpha}_t y(t) - \sum_{k=0}^{m-1} \frac{y^{(k)}(t_0)}{\Gamma(k-\alpha+1)}(t-t_0)^{k-\alpha}, \qquad (10\text{-}6\text{-}16)$$

and the $0 \leqslant \alpha \leqslant 1$ case is only a special one in the formula.

If $\alpha < 0$, it has been pointed out earlier that, Riemann–Liouville and Caputo definitions are identical.

10.6.3 Evaluating fractional-order differentiation

I. Computation via Grünwald–Letnikov definitions

The most straightforward way in evaluating the fractional-order derivatives numerically is to use the Grünwald–Letnikov definition which is given again

$${}_{t_0}\mathscr{D}^{\alpha}_t f(t) = \lim_{h\to 0} \frac{1}{h^{\alpha}} \sum_{j=0}^{[(t-t_0)/h]} (-1)^j \binom{\alpha}{j} f(t-jh) \approx \frac{1}{h^{\alpha}} \sum_{j=0}^{[(t-t_0)/h]} w_j^{(\alpha)} f(t-jh), \quad (10\text{-}6\text{-}17)$$

where $w_j^{(\alpha)} = (-1)^j \binom{\alpha}{j}$ are the coefficients of binomial expression $(1-z)^\alpha$, and can be evaluated recursively from

$$w_0^{(\alpha)} = 1, \quad w_j^{(\alpha)} = \left(1 - \frac{\alpha+1}{j}\right) w_{j-1}^{(\alpha)}, \; j = 1, 2, \cdots. \tag{10-6-18}$$

If the step-size h is small enough, (10-6-17) can be used directly to calculate approximately the values of the fractional-order derivatives, with an accuracy of $o(h)$ [13]. The following MATLAB function can be written, with $\boldsymbol{y}_1$ = `glfdiff(`$\boldsymbol{y},\boldsymbol{t},\gamma$`)`, where $\boldsymbol{y}, \boldsymbol{t}$ describe the original function using evenly distributed samples. The returned vector $\boldsymbol{y}_1$ is the γth order derivative.

```
function dy=glfdiff(y,t,gam)
h=t(2)-t(1); y=y(:); t=t(:);
w=1; for j=2:length(t), w(j)=w(j-1)*(1-(gam+1)/(j-1)); end
for i=1:length(t), dy(i)=w(1:i)*[y(i:-1:1)]/h^gam; end
```

Example 10.49 It is well known in the traditional calculus framework that, the derivatives of constants are zeros, and the first-order integral is straight lines. Please see what happens with fractional-order differentiation and integrals.

Solution *A vector* $\boldsymbol{y}$ *of constants can be generated first. Function* `glfdiff()` *can be used to evaluate fractional-order derivatives and integrals, as shown in Figure 10.49. It can be seen that the behaviors of fractional-order calculus are different from the traditional ones.*

```
>> t=0:0.01:1.5; gam=[-1 -0.5 0.3 0.5 0.7]; y=ones(size(t)); dy=[];
   for a=gam, dy=[dy; glfdiff(y,t,a)]; end, plot(t,dy)
```

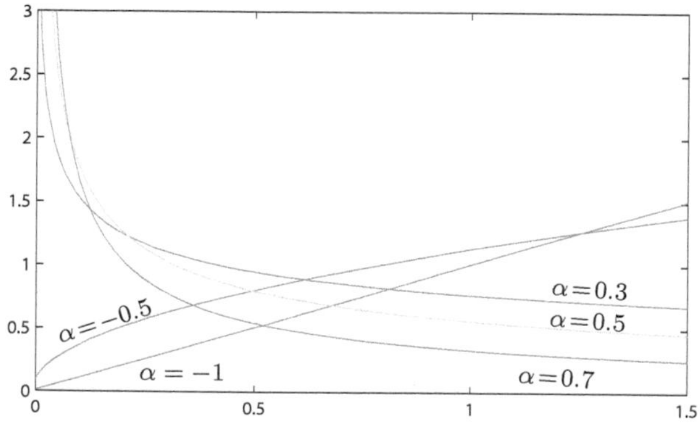

FIGURE 10.49: Fractional-order derivatives and integrals of a constant.

Example 10.50 For function $f(t) = \mathrm{e}^{-t}\sin(3t+1)$, $t \in (0,\pi)$, study the behaviors of the fractional-order derivatives.

Solution *Select the step-sizes,* $T = 0.01$ *and* 0.001*, the 0.5th order derivative can be evaluated as shown in Figure 10.50 (a). It can be seen that, when t is close to* 0*, there are some differences. For larger t, the two solutions are very close. Normally speaking, selecting* $T = 0.01$ *may give accurate results.*

```
>> t=0:0.001:pi; y=exp(-t).*sin(3*t+1); dy=glfdiff(y,t,0.5); plot(t,dy);
   t=0:0.01:pi; y=exp(-t).*sin(3*t+1); dy=glfdiff(y,t,0.5); line(t,dy)
```

For different selections of γ, the 3D surface representation of the fractional-order derivative can be obtained as shown in Figure 10.50 (b).

```
>> Z=[]; t=0:0.01:pi; y=exp(-t).*sin(3*t+1); % describe the original function
   for gam=0:0.1:1, Z=[Z; glfdiff(y,t,gam)]; end % compute FO derivatives
   surf(t,0:0.1:1,Z); axis([0,pi,0,1,-1.2,6]) % draw the surface
```

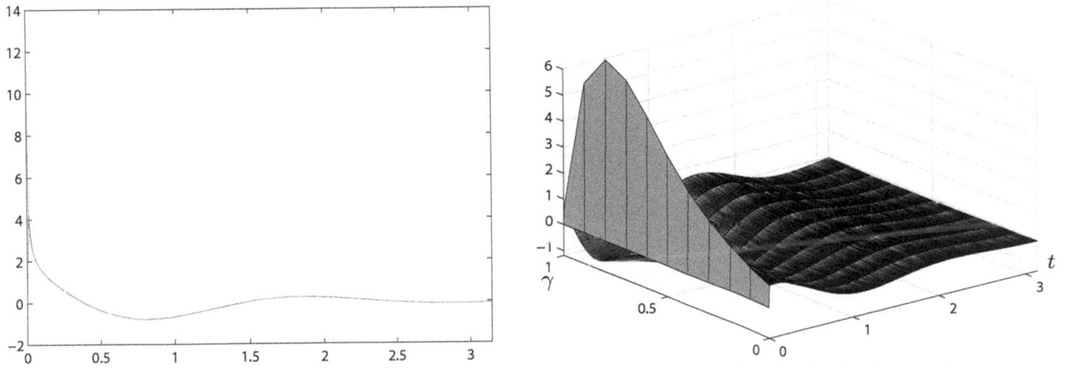

(a) comparisons under different step-sizes (b) fractional-order derivative surface

FIGURE 10.50: Fractional-order differentiation of the function.

Example 10.51 Consider a sinusoidal function $f(t) = \sin(3t+1)$. Find its 0.75th order derivative with Grünwald–Letnikov definition and Cauchy's formula, and compare the results.

Solution *According to Cauchy's formula and* (10-6-2), *the 0.75th order derivative can be obtained as* ${}_0\mathscr{D}_t^{0.75} f(t) = 3^{0.75}\sin(3t+1+0.75\pi/2)$. *With the Grünwald–Letnikov definition, the derivative can be evaluated with* `glfdiff()` *function. The two derivative curves are compared in Figure 10.51 (a).*

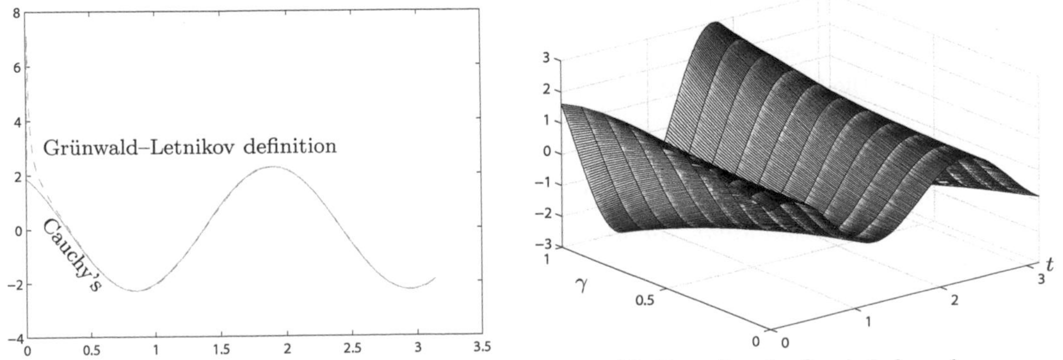

(a) comparisons of different definitions (b) 3D surface by Cauchy's formula

FIGURE 10.51: Comparisons and 3D surfaces.

```
>> t=0:0.01:pi; y=sin(3*t+1); y1=3^0.75*sin(3*t+1+0.75*pi/2);
   y2=glfdiff(y,t,0.75); plot(t,y1,t,y2,'--')
```

It can be seen that when t is small, there exists significant difference between the two definitions. This is because, in these two definitions, the assumptions of the function at $t < 0$ are considered differently. In Cauchy's formula, it is assumed that the function can also be described by $f(t) = \sin(3t+1)$ when $t < 0$, however, for Grünwald–Letnikov definition, it is assumed that $f(t) = 0$ for $t < 0$. Therefore, in the later case, there is a jump at $t = 0$, which causes the difference.

Again, the 3D surface can also be found for different orders γ, with the following statements, as shown in Figure 10.51 (b). It can be seen that the fractional-order derivatives provide information between the original function and its first-order derivative.

```
>> gam=[0:0.1:1]; Y=[]; t=0:0.01:pi; y=sin(3*t+1);
   for a=gam, Y=[Y; 3^a*sin(3*t+1+a*pi/2)]; end, surf(t,gam,Y)
```

II. Numerical computation of Caputo derivatives

It was pointed out that the fractional-order integrals under Caputo and Grünwald–Letnikov definitions are equivalent, function `glfdiff()` can be used directly in evaluating integrals. For fractional-order derivatives, if $\alpha > 0$, the fractional-order derivative under Caputo definition can be evaluated directly with (10-6-16), and the MATLAB implementation is given as

```
function dy=caputo(y,t,gam,vec,L)
t0=t(1); dy=glfdiff(y,t,gam); if nargin<=4, L=10; end
if gam>0, m=ceil(gam); if gam<=1,vec=y(1); end
   for k=0:m-1, dy=dy-vec(k+1)*(t-t0).^(k-gam)./gamma(k+1-gam); end
   yy1=interp1(t(L+1:end),dy(L+1:end),t(1:L),'spline'); dy(1:L)=yy1;
end
```

The syntax of the function is $\boldsymbol{y}_1$ = `caputo(`$\boldsymbol{y}$,$\boldsymbol{t}$,α,$\boldsymbol{y}_0$,L`)`, with, $\alpha \leqslant 0$, the Grünwald–Letnikov integral is returned. If $\alpha < 1$, the initial vector $y(t_0)$ is extracted directly with $\boldsymbol{y}$, while if $\alpha > 1$, the initial vector $\boldsymbol{y}_0$ should be supplied for function $y(t)$, with $\boldsymbol{y}_0 = [y(t_0), y'(t_0), \cdots, y^{(m-1)}(t_0)]$, and $m = \lceil \alpha \rceil$. In practical computations, there might exist large errors in the first few terms of the Caputo derivatives, therefore, interpolation can be performed for the first L terms. The default value of L is 10, and when the order increases, the value of L should also be increased.

Example 10.52 Consider again the sinusoidal function in Example 10.51, $f(t) = \sin(3t+1)$. Please find the 0.3th, 1.3th and 2.3th order derivatives.

Solution *It can be seen that the value of the function at $t = 0$ is $\sin 1$, thus, the difference of the two definitions is $d(t) = t^{-0.3}\sin 1/\Gamma(0.7)$. The two curves under Grünwald–Letnikov and Caputo definitions can be obtained as shown in Figure 10.52 (a). It can be seen that with nonzero initial conditions, the difference of the two definitions are rather large.*

```
>> t=0:0.01:pi; y=sin(3*t+1); d=t.^(-0.3)*sin(1)/gamma(0.7);
   y1=glfdiff(y,t,0.3); y2=caputo(y,t,0.3,0); plot(t,y1,t,y2,'--',t,d,':')
```

Since the curve of ${}_0^{\rm C}\mathscr{D}_t^{2.3}y(t)$ is expected, the initial values $y'(0)$ and $y''(0)$ are required. These values can be obtained symbolically and converted to double-precision ones. The 1.3th and 2.3th order derivatives can be obtained. To have good approximations in the first few terms, interpolation is performed, and the results are shown in Figure 10.52 (b).

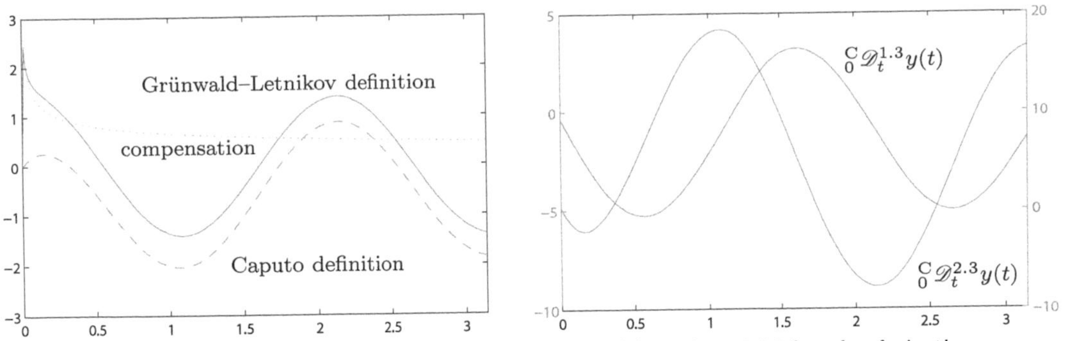

(a) 0.3th order derivative (b) 1.3th and 2.3th order derivatives

FIGURE 10.52: Fractional-order derivatives under different orders.

```
>> syms t; y=sin(3*t+1); y00=sin(1); y10=double(subs(diff(y,t),t,0));
   y20=double(subs(diff(y,t,2),t,0)); t=0:0.01:pi; y=sin(3*t+1);
   y1=caputo(y,t,1.3,[y00 y10],10); y2=caputo(y,t,2.3,[y00,y10,y20],30);
   plotyy(t,y1,t,y2)
```

III. Using filtering algorithm to compute the fractional-order calculus

In the fractional-order derivative evaluation method discussed above, the function $f(t)$ or its samples are known. In many other applications, the signal $f(t)$ is generated dynamically. In this case, a filter can be designed to evaluate fractional-order derivative of $f(t)$ in real-time.

For the fractional-order derivative, its Laplace representation is s^γ, which exhibits straight lines in both Bode magnitude and phase plots. Thus, it is not possible to find a finite order filter to fit the straight lines for all the frequencies. However, it is useful to fit the frequency responses over a frequency range of interest.

Different continuous filters have been studied in Reference [14], among which the Oustaloup's filter [15] has certain advantages. For the selected frequency range of interest, $(\omega_{\rm b}, \omega_{\rm h})$, the continuous filter can be written as

$$G_{\rm f}(s) = K\prod_{k=1}^{N}\frac{s+\omega_k'}{s+\omega_k}, \tag{10-6-19}$$

where, the poles, zeros and gain can be evaluated from

$$\omega_k' = \omega_{\rm b}\omega_{\rm u}^{(2k-1-\gamma)/N},\ \ \omega_k = \omega_{\rm b}\omega_{\rm u}^{(2k-1+\gamma)/N},\ \ K = \omega_{\rm h}^{\gamma}, \tag{10-6-20}$$

with $\omega_{\rm u} = \sqrt{\omega_{\rm h}/\omega_{\rm b}}$. Based on the above algorithm, the following function can be written

```
function G=ousta_fod(gam,N,wb,wh)
k=1:N; wu=sqrt(wh/wb);
wkp=wb*wu.^((2*k-1-gam)/N); wk=wb*wu.^((2*k-1+gam)/N);
G=zpk(-wkp,-wk,wh^gam); G=tf(G);
```

The continuous filter can be designed as $G=$ `ousta_fod(`γ`,`N`,`$\omega_{\rm b}$`,`$\omega_{\rm h}$`)`, where γ is the order of derivative, and N is the order of the filter.

Example 10.53 Select the frequency range of interest as $\omega_{\rm b} = 0.01, \omega_{\rm h} = 1000\,\text{rad/sec}$, and design the continuous-time approximate fractional-order filters. For the function $f(t) = e^{-t}\sin(3t+1)$, calculate the 0.5th order derivative and verify the obtained results.

Solution *The 5th order filter can be designed*

```
>> G5=ousta_fod(0.5,5,0.01,1000), bode(G5)
```

and the filter designed is

$$G(s) = \frac{31.62s^5 + 6248s^4 + 1.122\times10^5s^3 + 1.996\times10^5s^2 + 3.514\times10^4s + 562.3}{s^5 + 624.8s^4 + 3.549\times10^4s^3 + 1.996\times10^5s^2 + 1.111\times10^5s + 5623}$$

The Bode diagrams of the above approximate filter are shown in Figure 10.53 (a), superimposed by the theoretical straight lines. The filter output is shown in Figure 10.53 (b). Moreover, the 0.5th order derivative obtained through Grünwald–Letnikov definition can be obtained using the following MATLAB scripts and the obtained derivative curve is also shown in Figure 10.53 (b). Clearly, the filter output of this example is fairly accurate.

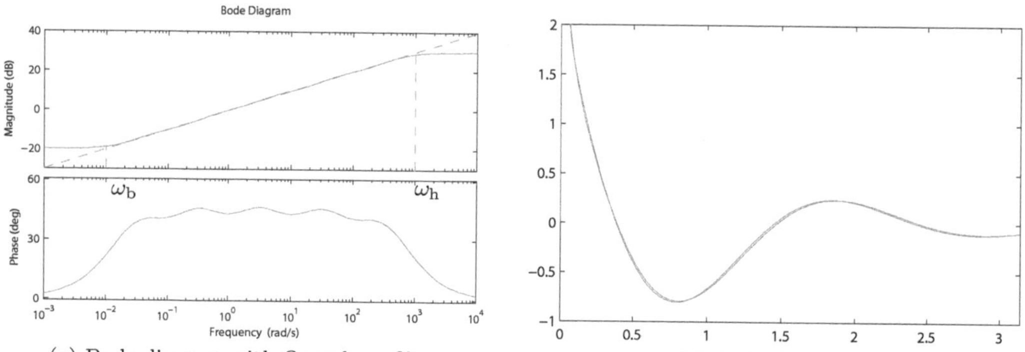

(a) Bode diagram with Oustaloup filter (b) derivative curves

FIGURE 10.53: Comparisons of approximate fractional-order filters.

```
>> t=0:0.001:pi; y=exp(-t).*sin(3*t+1);
   y1=lsim(G,y,t); y2=glfdiff(y,t,0.5); plot(t,y1,t,y2)
```

Of course, if one is not satisfied with the filters, the frequency range of interest and the order of the filter can both be increased. The fitting result in the frequency interval $(10^{-4}, 10^4)$ *and fitting results are shown in Figure 10.54. It can be seen that the order* $N = 5$ *is not suitable, higher order should be tried.*

```
>> G=ousta_fod(0.5,5,1e-4,1e4); G1=ousta_fod(0.5,7,1e-4,1e4);
   G2=ousta_fod(0.5,9,1e-4,1e4); G3=ousta_fod(0.5,11,1e-4,1e4);
   bode(G,'-',G1,'--',G2,':',G3,'-.')
```

IV. A modified Oustaloup filter

In practical applications, it is frequently found that the filter from using the `ousta_fod()` function cannot exactly fit the whole expected frequency range of interest. A new improved filter for a fractional-order derivative in the frequency range of interest $[\omega_b, \omega_h]$, which is shown to perform better, is introduced in this subsection. The modified filter is [16]

$$s^\gamma \approx \left(\frac{d\omega_h}{b}\right)^\gamma \left(\frac{ds^2 + b\omega_h s}{d(1-\gamma)s^2 + b\omega_h s + d\gamma}\right) \prod_{k=-N}^{N} \frac{s+\omega_k'}{s+\omega_k}, \tag{10-6-21}$$

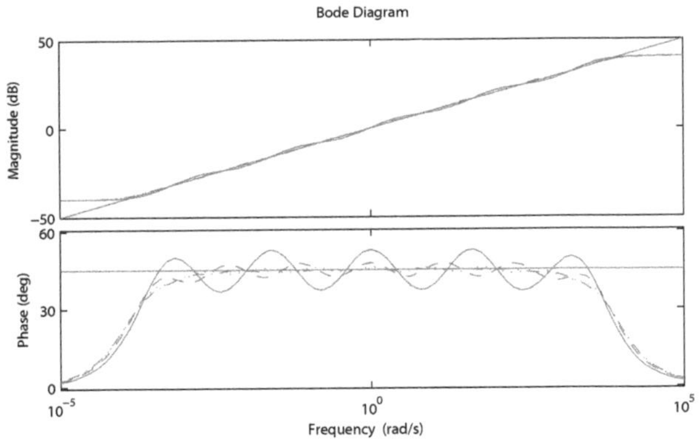

FIGURE 10.54: Fitting results under different orders in filters.

where

$$\omega_k' = (d\omega_{\rm b}/b)^{(\gamma-2k)/(2N+1)}, \quad \omega_k = (b\omega_{\rm h}/d)^{(\gamma+2k)/(2N+1)}. \tag{10-6-22}$$

Through a number of experimentation confirmation and theoretic analyses, the modified filter achieves good approximation when $b = 10$ and $d = 9$. With the above algorithm, a MATLAB function `new_fod()` is written

```
function G=new_fod(r,N,wb,wh,b,d)
if nargin==4, b=10; d=9; end
mu=wh/wb; k=-N:N; w_kp=(mu).^((k+N+0.5-0.5*r)/(2*N+1))*wb;
w_k=(mu).^((k+N+0.5+0.5*r)/(2*N+1))*wb; K=(d*wh/b)^r;
G=zpk(-w_kp',-w_k',K)*tf([d,b*wh,0],[d*(1-r),b*wh,d*r]);
```

with the syntax $G_{\rm f}$=`new_fod`(γ,N,$\omega_{\rm b}$,$\omega_{\rm h}$,b,d), and due to the limitations of the algorithm, $\omega_h\omega_b = 1$ is required.

Example 10.54 Consider again the problem in Example 10.53, and select $\omega_{\rm b} = 0.001$, $\omega_{\rm h} = 1000$. Please observe the behaviors of the new filter, and the fractional-order derivatives.

Solution *With the two filters designed, Bode diagrams can be obtained as shown in Figure 10.55 (a). The fractional-order derivative curve can be obtained as shown in Figure 10.55 (b). It can be seen from the requency response fitting that, the modified filter is better than the Oustaloup filter.*

```
>> G1=ousta_fod(0.5,2,0.001,1000); G2=new_fod(0.5,2,0.001,1000);
   subplot(121), bode(G1,'-',G2,'--'), t=0:0.001:pi;
   y=exp(-t).*sin(3*t+1); y1=lsim(G1,y,t); y2=lsim(G2,y,t);
   y0=glfdiff(y,t,0.5); subplot(122), plot(t,y1,t,y2,t,y0)
```

10.6.4 Solving fractional-order differential equations

A class of linear time-invariant (LTI) fractional-order differential equations (FODE) can be written as [13]

$$a_n \mathscr{D}_t^{\beta_n} y(t) + a_{n-1}\mathscr{D}_t^{\beta_{n-1}} y(t) + \cdots + a_1 \mathscr{D}_t^{\beta_1} y(t) + a_0 \mathscr{D}_t^{\beta_0} y(t) = u(t), \tag{10-6-23}$$

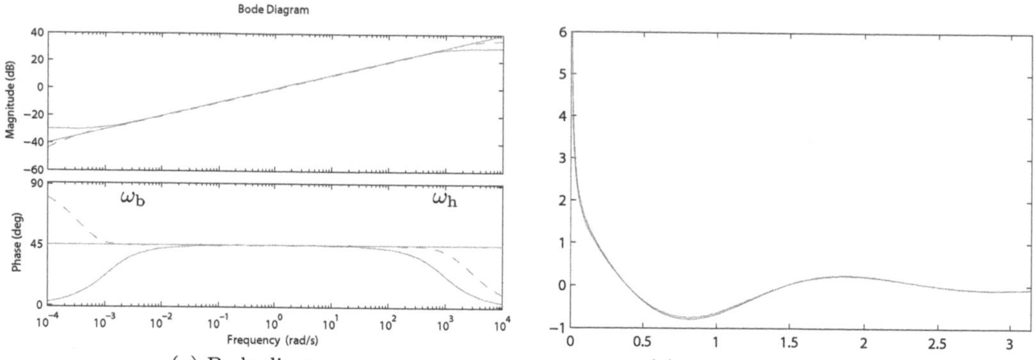

(a) Bode diagram (b) fractional-order derivative

FIGURE 10.55: Fractional-order derivatives with different filters.

where $u(t)$ is the input signal. If the initial conditions are zero, the equation is Riemann–Liouville differential equation, while the ones with nonzero initial conditions, are referred to as *Caputo differential equations.* In this section, numerical solutions of the two types of differential equations will be introduced. If the initial conditions are zero, the simplified transfer function can be written as

$$G(s)=\frac{Y(s)}{U(s)}=\frac{1}{a_n s^{\beta_n}+a_{n-1}s^{\beta_{n-1}}+\cdots+a_1 s^{\beta_1}+a_0 s^{\beta_0}}. \tag{10-6-24}$$

I. Analytical solutions of linear fractional-order equations

With Mittag–Leffler function presented in Chapter 8, the analytical solution of the n-term fractional-order differential equation is given in general form [17] by

$$\begin{aligned} y(t) &= \frac{1}{a_n}\sum_{m=0}^{\infty}\frac{(-1)^m}{m!}\sum_{\substack{k_0+k_1+\cdots+k_{n-2}=m\\ k_0\geqslant 0;\cdots,k_{n-2}\geqslant 0}}(m;k_0,k_1,\cdots,k_{n-2})\\ &\quad \prod_{i=0}^{n-2}\left(\frac{a_i}{a_n}\right)^{k_i} t^{(\beta_n-\beta_{n-1})m+\beta_n+\sum_{j=0}^{n-2}(\beta_{n-1}-\beta_j)k_j-1}\\ &\quad \mathscr{E}^{(m)}_{\beta_n-\beta_{n-1},\beta_n+\sum_{j=0}^{n-2}(\beta_{n-1}-\beta_j)k_j}\left(-\frac{a_{n-1}}{a_n}t^{\beta_n-\beta_{n-1}}\right), \end{aligned} \tag{10-6-25}$$

where $\mathscr{E}_{\lambda,\mu}(z)$ is the Mittag–Leffler function defined in (8-5-24), m is an integer.

The system response expression (10-6-25) sometimes is too complicated to use. Here a special form of the model $G(s)=1/(a_2 s^{\beta_2}+a_1 s^{\beta_1}+a_0)$ is studied. The step response of the system is written as [18]

$$y(t)=\frac{1}{a_2}\sum_{k=0}^{\infty}\frac{(-1)^k\hat{a}_0^k t^{-\hat{a}_1+(k+1)\beta_2}}{k!}\,\mathscr{E}^{(k)}_{\beta_2-\beta_1,\beta_2+\beta_1 k+1}\left(-\hat{a}_1 t^{\beta_2-\beta_1}\right), \tag{10-6-26}$$

where $\hat{a}_0=a_0/a_2, \hat{a}_1=a_1/a_2$. Similar to the `ml_func()` function, the step response solution function can be written as

```
function y=ml_step(a0,a1,a2,b1,b2,t,eps0)
y=0; k=0; ya=1; a0=a0/a2; a1=a1/a2; if nargin==6, eps0=eps; end
```

```
while max(abs(ya))>=eps0 % if error exists, continue add terms
   ya=(-1)^k/gamma(k+1)*a0^k*t.^((k+1)*b2).*... % compute the next term
       ml_func([b2-b1,b2+b1*k+1],-a1*t.^(b2-b1),k,eps0);
   y=y+ya; k=k+1;
end
y=y/a2;
```

whose syntax is `y=ml_step(`a_0`,`a_1`,`a_2`,`β_1`,`β_2`,t,`ϵ_0`)`.

Example 10.55 Please solve numerically the fractional-order differential equation $\mathscr{D}^{0.8}y(t) + 0.75\mathscr{D}^{0.4}y(t) + 0.9y(t) = u(t)$, with zero initial conditions, and the input is unit step signal.

Solution *It is obvious that* $a_0 = 0.9, a_1 = 0.75, a_2 = 1, \beta_1 = 0.4, \beta_2 = 0.8$. *The numerical solution of the original differential equation can be obtained, as shown in Figure 10.56.*

```
>> t=0:0.001:5; y=ml_step(0.9,0.75,1,0.4,0.8,t); plot(t,y)
```

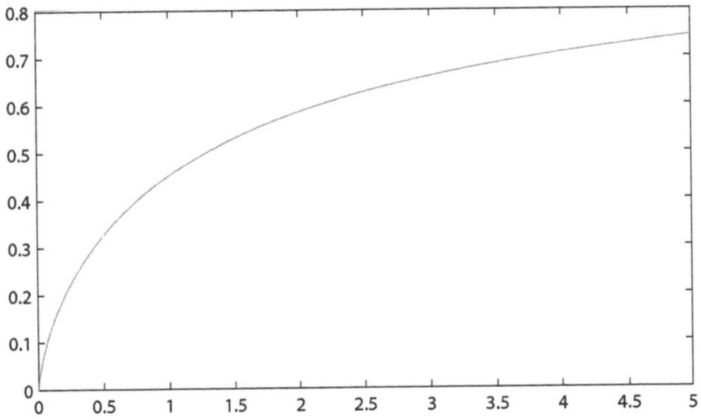

FIGURE 10.56: Solution of the fractional-order differential equation.

From the above simulation example, it can be seen that both methods give correct results. However, the computation based on Grünwald–Letnikov's definition is much simpler than the Mittag–Leffler function-based method. Thus, it will be used later in the book.

II. A closed-form solution to linear fractional-order differential equations with zero initial conditions

If the initial values of the output signal $y(t)$, input signal $u(t)$ and their derivatives at $t = 0$ are all zeros, and the right hand side of the equation contains $\hat{u}(t)$ alone, the original differential equation can be simplified as

$$a_n\mathscr{D}_t^{\beta_n}y(t) + a_{n-1}\mathscr{D}_t^{\beta_{n-1}}y(t) + \cdots + a_1\mathscr{D}_t^{\beta_1}y(t) + a_0\mathscr{D}_t^{\beta_0}y(t) = \hat{u}(t), \tag{10-6-27}$$

where, $\hat{u}(t)$ is composed of the linear combination of a signal $u(t)$ and its fractional-order derivatives, and can be evaluated independently

$$\hat{u}(t) = b_1\mathscr{D}_t^{\gamma_1}u(t) + b_2\mathscr{D}_t^{\gamma_2}u(t) + \cdots + b_m\mathscr{D}_t^{\gamma_m}u(t). \tag{10-6-28}$$

For simplicity in the presentation, assume that $\beta_n > \beta_{n-1} > \cdots > \beta_1 > \beta_0 > 0$. If the following two special cases emerge, conversions should be made first, then, the solutions can be found.

(i) If the order of the original equation does not satisfy the inequalities, the terms in the original equation should be sorted first.

(ii) If there exists negative β_i's, fractional-order integral-differential equation is involved. In this case, new variable $z(t) = \mathscr{D}_t^{\beta_0} y(t)$ should be introduced, such that the original equation can be converted to the fractional-order differential equation og signal $z(t)$.

Consider the Grünwald–Letnikov definition in (10-6-17), the modified discrete version can be written as

$$ {}_{t_0}\mathscr{D}_t^{\beta_i} y(t) \approx \frac{1}{h^{\beta_i}} \sum_{j=0}^{[(t-t_0)/h]} w_j^{(\beta_i)} y_{t-jh} = \frac{1}{h^{\beta_i}} \left[y_t + \sum_{j=1}^{[(t-t_0)/h]} w_j^{(\beta_i)} y_{t-jh} \right] \tag{10-6-29} $$

where $w_0^{(\beta_i)}$ can still be evaluated recursively

$$ w_0^{(\beta_i)} = 1, \quad w_j^{(\beta_i)} = \left(1 - \frac{\beta_i + 1}{j} \right) w_{j-1}^{(\beta_i)}, \; j = 1, 2, \cdots. \tag{10-6-30} $$

Substituting the formula to (10-6-23), the closed-form numerical solution to the fractional-order differential equation can be obtained as

$$ y_t = \frac{1}{\displaystyle\sum_{i=0}^{n} \frac{a_i}{h^{\beta_i}}} \left[u_t - \sum_{i=0}^{n} \frac{a_i}{h^{\beta_i}} \sum_{j=1}^{[(t-t_0)/h]} w_j^{(\beta_i)} y_{t-jh} \right]. \tag{10-6-31} $$

Now consider the general form of the fractional-order differential equation in (10-6-38), the right-hand side of the function can be evaluated, and the equation can be converted into the form in (10-6-23). In practical programming, the original linear problem can be equivalently converted to evaluate the $\hat{y}(t)$ signal under the excitation of $u(t)$, then, take fractional-order derivatives to $\hat{y}(t)$. Based on the algorithm, a MATLAB function `fode_sol()` can be written to solve numerically linear fractional-order differential equations with zero initial conditions.

```
function y=fode_sol(a,na,b,nb,u,t) % fractional-order differential equation
h=t(2)-t(1); D=sum(a./[h.^na]); nT=length(t); D1=b(:)./h.^nb(:);
nA=length(a); vec=[na nb]; y1=zeros(nT,1); W=ones(nT,length(vec));
for j=2:nT, W(j,:)=W(j-1,:).*(1-(vec+1)/(j-1)); end
for i=2:nT,
   A=[y1(i-1:-1:1)]'*W(2:i,1:nA); y1(i)=(u(i)-sum(A.*a./[h.^na]))/D;
end
for i=2:nT, y(i)=(W(1:i,nA+1:end)*D1)'*[y1(i:-1:1)]; end
```

The function can be called with $\boldsymbol{y}=$`fode_sol(`$\boldsymbol{a}$`,`$\boldsymbol{n}_{\rm a}$`,`$\boldsymbol{b}$`,`$\boldsymbol{n}_{\rm b}$`,`$\boldsymbol{u}$`,`$\boldsymbol{t}$`)`, where, the time and input vectors can be obtained in vectors in $\boldsymbol{t}$ and $\boldsymbol{u}$.

Example 10.56 Solve numerically the following fractional-order differential equation

$$ \mathscr{D}_t^{3.5} y(t) + 8\mathscr{D}_t^{3.1} y(t) + 26\mathscr{D}_t^{2.3} y(t) + 73\mathscr{D}_t^{1.2} y(t) + 90\mathscr{D}_t^{0.5} y(t) = 90 \sin t^2. $$

Solution *The vectors* $\boldsymbol{a}$ *and* $\boldsymbol{n}$ *can be extracted from the original equation first, and then,* *function* `fode_sol()` *can be used to solve the original equations. The curves of input and output signals can be obtained as shown in Figure 10.57. So, one should verify the numerical results obtained. The simplest way is to modify the control parameters in the solution. For instance, one can change the step-size from 0.002 to 0.001, and check whether they give the same results. For this example, it can be seen that the results are the same, as shown in Figure 10.57.*

```
>> a=[1,8,26,73,90]; n=[3.5,3.1,2.3,1.2,0.5];
   G=fotf(a,n,1,0); t=0:0.002:10; u=90*sin(t.^2); y=lsim(G,u,t);
   subplot(211), plot(t,y); subplot(212), plot(t,u)
```

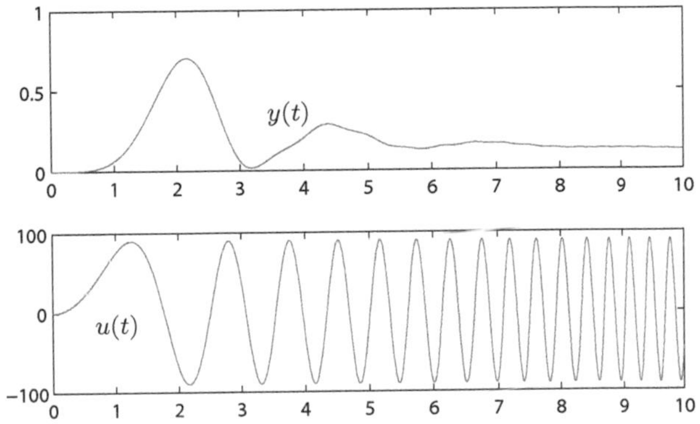

FIGURE 10.57: Input signal and solution of the equation.

III. Solutions of Caputo differential equations with nonzero initial conditions

If there exists nonzero initial values in the input, output and their derivatives, the above algorithm cannot be used to find the numerical solutions. Numerical solution algorithms are needed to solve Caputo fractional-order differential equations.

Consider the linear Caputo fractional-order differential equation given by

$$a_n\, {}_{t_0}^{\rm C}\mathscr{D}_t^{\beta_n} y(t) + a_{n-1}\, {}_{t_0}^{\rm C}\mathscr{D}_t^{\beta_{n-1}} y(t) + \cdots + a_1\, {}_{t_0}^{\rm C}\mathscr{D}_t^{\beta_1} y(t) + a_0\, {}_{t_0}^{\rm C}\mathscr{D}_t^{\beta_0} y(t) = \hat{u}(t). \tag{10-6-32}$$

For convenience, assume that $\beta_n > \beta_{n-1} > \cdots > \beta_1 > \beta_0 \geqslant 0$. Also, assume that only $\hat{u}(t)$ function appears in the right hand side of the equation. If there are fractional-order derivatives of $u(t)$ in the right hand side of the equation, the method used earlier can be applied to evaluate $\hat{u}(t)$ first the linear combinations.

If $m = \lceil \beta_n \rceil$, m initial conditions, $y(t_0)$, $y'(t_0)$, $\cdots$, $y^{(m-1)}(t_0)$, are needed to uniquely solve the original fractional-order differential equation. Thus, an auxiliary signal $z(t)$ is introduced such that

$$z(t) = y(t) - y(t_0) - y'(t_0)t - \cdots - \frac{1}{(m-1)!} y^{(m-1)}(t_0)t^{m-1}, \tag{10-6-33}$$

where the initial values of $z(t)$ and its first $(m-1)$th order derivatives are all zeros. The above expression can also be written as

$$y(t) = z(t) + y(t_0) + y'(t_0)t + \cdots + \frac{1}{(m-1)!} y^{(m-1)}(t_0)t^{m-1}. \tag{10-6-34}$$

Since the initial values of $z(t)$ and its derivatives are all zeros, ${}^{\rm C}\mathscr{D}_t^{\beta_i}z(t) = {}^{\rm RL}\mathscr{D}_t^{\beta_i}z(t)$, the β_ith order Caputo derivative of $y(t_0)+y'(t_0)t+\cdots+y^{(m-1)}(t_0)t^{m-1}/(m-1)!$ can be evaluated directly with

```
function s=poly2caputo(a,r,t0), syms v tau; if nargin==2, t0=0; end
if r==ceil(r), s=diff(poly2sym(a,u),r);
else, s=int(diff(poly2sym(a,'tau'),ceil(r))/((u-tau)^(r-ceil(r)+1))...
               /gamma(ceil(r)-r),tau,t0,u);
end
```

The syntax of the function is s=`poly2caputo(`a`,`β`,`t_0`)`, where,

$$\boldsymbol{a} = [y^{(m-1)}(t_0), y^{(m-2)}(t_0), \cdots, y'(t_0), y(t_0)] \tag{10-6-35}$$

is the vector of initial conditions, β is the order. The returned s is the symbolic expression of the derivative of the polynomial.

With the compensation function given above, the original differential equation can be converted to

$$\begin{aligned} &a_n\,{}^{\rm RL}_{t_0}\mathscr{D}_t^{\beta_n}z(t) + a_{n-1}\,{}^{\rm RL}_{t_0}\mathscr{D}_t^{\beta_{n-1}}z(t) + \cdots + a_1\,{}^{\rm RL}_{t_0}\mathscr{D}_t^{\beta_1}z(t) + a_0\,{}^{\rm RL}_{t_0}\mathscr{D}_t^{\beta_0}z(t) \\ &\quad = \hat{u}(t) - \sum_{i=0}^{n} a_i\,{}^{\rm C}\mathscr{D}^{\beta_i}\left[y(t_0)+y'(t_0)t+\cdots+\frac{1}{(m-1)!}y^{(m-1)}(t_0)t^{m-1}\right]. \end{aligned} \tag{10-6-36}$$

Similar to `fode_sol()` function, a Caputo differential equation solver can be written, with the syntax $\boldsymbol{y}$=`fode_caputo(`$\boldsymbol{a}$`,`$\boldsymbol{n}_{\rm a}$`,`$\boldsymbol{y}_0$`,`$\boldsymbol{u}$`,`$\boldsymbol{t}$`)`, where, $\boldsymbol{u}$ is the samples of the input signal, and $\boldsymbol{t}$ is the vector of time.

```
function [y,z]=fode_caputo(a,na,y0,u,t), h=t(2)-t(1);
D=sum(a./[h.^na]); nT=length(t); nb=0; b=1; D1=b(:)./h.^nb(:);
nA=length(a); vec=[na nb]; y1=zeros(nT,1); W=ones(nT,length(vec));
for i=1:length(a), u=u-a(i)*subs(poly2caputo(y0,na(i),t(1)),'v',t); end
for j=2:nT, W(j,:)=W(j-1,:).*(1-(vec+1)/(j-1)); end
for i=2:nT,
   A=[y1(i-1:-1:1)]'*W(2:i,1:nA); y1(i)=(u(i)-sum(A.*a./[h.^na]))/D;
end
z=y1'; y=z+polyval(y0,t);
```

Example 10.57 Consider again the linear equation studied in Example 10.56. If the equation is a Caputo equation, with initial conditions $y(0)=1$, $y'(0)=-1$, $y''(0)=2$, and $y'''(0)=3$, please solve the equation again.

Solution *An initial condition vector can be constructed based on the initial conditions, and then, the following statements can be used to solve numerically the Caputo equation, and the result obtained is shown in Figure 10.58.*

```
>> a=[1,8,26,73,90]; n=[3.5,3.1,2.3,1.2,0.5];
   t=0:0.001:10; u=90*sin(t.^2); y0=[3 2 -1 1]; % compute the input
   y=fode_caputo(a,n,y0,u,t); plot(t,y) % solve Caputo equation
```

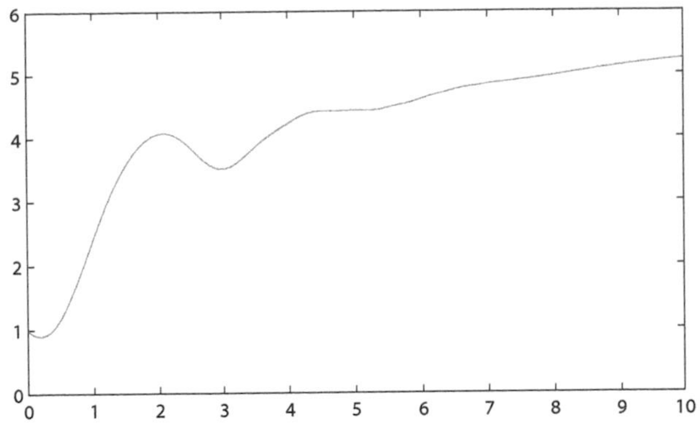

FIGURE 10.58: Numerical solutions of Caputo equations.

10.6.5 Block diagram based solutions of nonlinear fractional-order ordinary differential equations

If nonlinear fractional-order differential equation is given, and in particular, the nonlinear differential equation is only part of the whole system model, conventional solvers are not suitable in finding the numerical solutions, and a block diagram based algorithm is needed. In this section, the Oustaloup filter or the modified filter are used as the kernel in constructing nonlinear fractional-order differential equations. Zero and nonzero initial condition problems are both discussed in the section.

I. Nonlinear equations with zero initial conditions

It can be seen that the Oustaloup filters given in the previous subsection are effective ways for evaluating fractional-order differentiations. Since the orders of the numerator and denominator in the ordinary Oustaloup filter are the same, it is likely to cause algebraic loops. Thus, a low-pass filter should be appended to the filter. The block in Figure 10.59 (a) can be used for modeling fractional-order differentiators.

With the use of masking technique in Simulink [19], the designed block can be masked as shown in Figure 10.59 (b). Double click such a block and a dialog box appears as in Figure 10.59 (c). The corresponding parameters can be filled into the dialog box to complete the fractional-order differentiator block. The following code can be attached to the masked block.

```
wb=ww(1); wh=ww(2); G=ousta_fod(gam,n,wb,wh);
num=G.num{1}; den=G.den{1}; str='Fractional\n';
if isnumeric(gam)
   if gam>0, str=[str, 'Der  s^' num2str(gam)];
   else, str=[str, 'Int  s^{' num2str(gam) '}']; end
else, str=[str, 'Der  s^gam']; end
```

Example 10.58 Solve the linear fractional-order differential equations in Example 10.56, and compare the results with other methods.

Solution *For linear fractional-order differential equations, the block diagram-based method is not as straightforward as the method used in Example 10.56. An auxiliary variable* $z(t) = \mathscr{D}_t^{0.5}y(t)$ *can be introduced, and the original differential equation can be rewritten as*

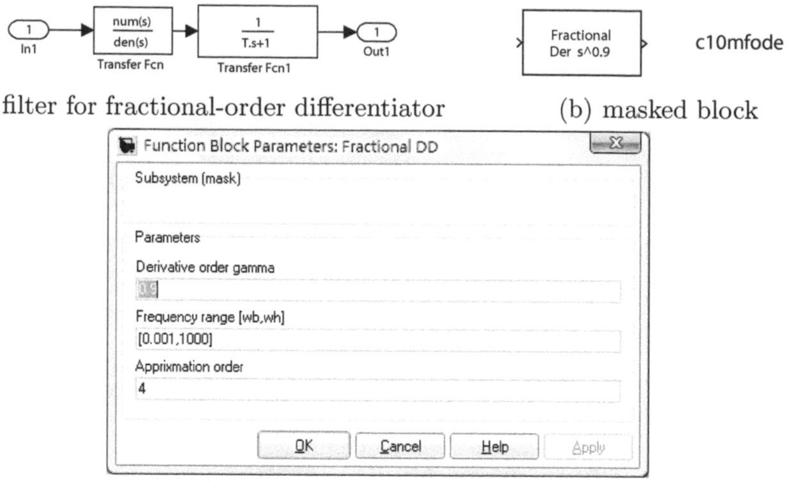

(a) filter for fractional-order differentiator (b) masked block

(c) parameter dialog box

FIGURE 10.59: Fractional-order differentiator block design.

$$z(t) = \sin(t^2) - \frac{1}{90}\left[\mathscr{D}_t^3 z(t) + 8\mathscr{D}_t^{2.6} z(t) + 26\mathscr{D}_t^{1.8} z(t) + 73\mathscr{D}_t^{0.7} z(t)\right].$$

The Simulink block diagram shown in Figure 10.60 can be established based on the new equation. With stiff ODE solvers, the numerical solution to the problem can be found and the results are exactly the same as the one in Figure 10.57.

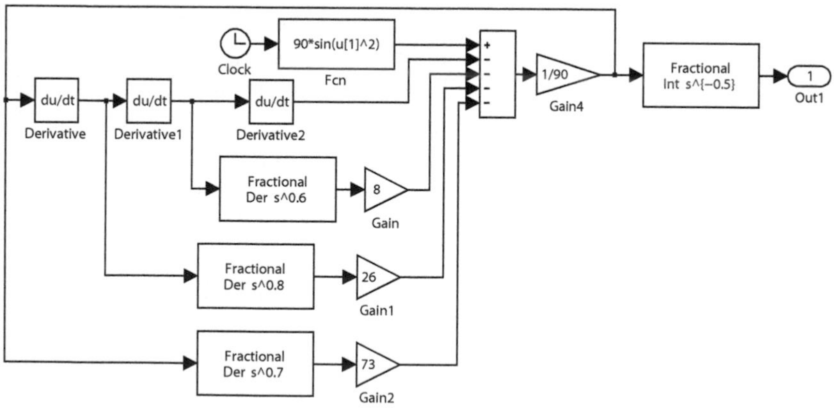

FIGURE 10.60: Simulink block diagram (file: c10mfod1.mdl).

Example 10.59 Solve the nonlinear fractional-order differential equation

$$\frac{3\mathscr{D}^{0.9}y(t)}{3 + 0.2\mathscr{D}^{0.8}y(t) + 0.9\mathscr{D}^{0.2}y(t)} + \left|2\mathscr{D}^{0.7}y(t)\right|^{1.5} + \frac{4}{3}y(t) = 5\sin 10t.$$

Solution *From the given equation, the explicit form of $y(t)$ can be written as*

$$y(t) = \frac{3}{4}\left[5\sin 10t - \frac{3\mathscr{D}^{0.9}y(t)}{3 + 0.2\mathscr{D}^{0.8}y(t) + 0.9\mathscr{D}^{0.2}y(t)} - \left|2\mathscr{D}^{0.7}y(t)\right|^{1.5}\right]$$

and from the explicit expression of $y(t)$, the block diagram in Simulink can be established in Figure 10.61 (a). From the simulation model, the simulation results can be obtained as shown in Figure 10.61 (b). The results are verified by different control parameters in the filter, and they give consistent results.

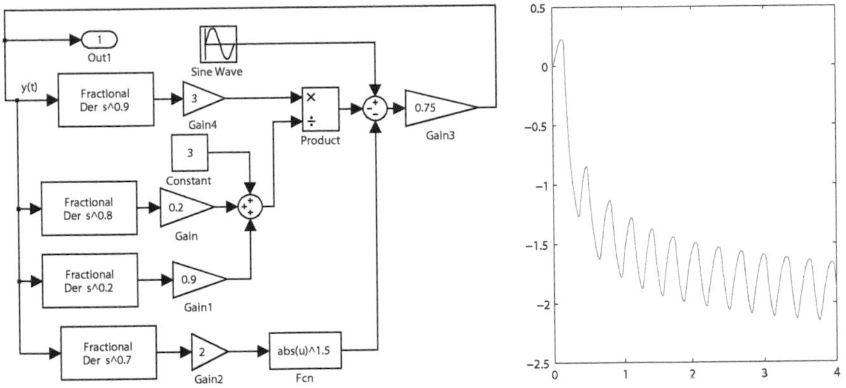

(a) Simulink model (file: c10mfod2.mdl) (b) simulation results

FIGURE 10.61: Simulink description and simulation results.

II. Numerical solution of Caputo differential equations with nonzero initial conditions

Similar to the masking process of the Riemann–Liouville operator discussed earlier, a masked Caputo operator is shown in Figure 10.62.

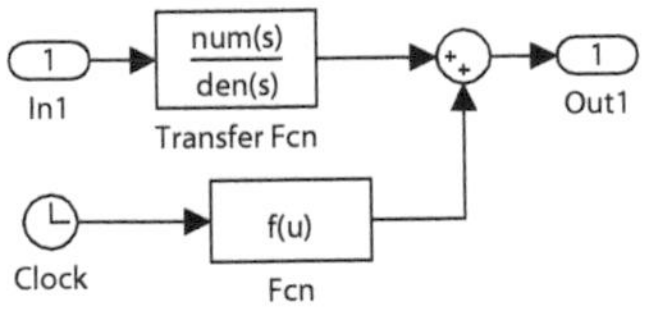

(a) Simulink filter

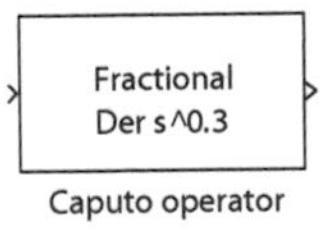

(b) masked block

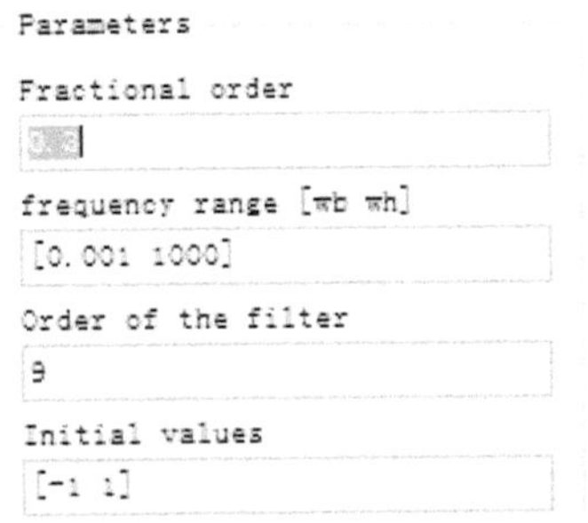

(c) parameter dialog box of the operator

FIGURE 10.62: Caputo fractional-order differentiator block.

The initialization code in the block can be modified as

```
wb=ww(1); wh=ww(2); G=ousta_fod(gam,n,wb,wh); G1=tf(G);
num=G1.num{1}; den=G1.den{1}; str='Fractional\n';
strmodel=[get_param(gcs,'Name'),'/' get_param(gcb,'Name') '/Fcn'];
set_param(strmodel,'Expr',char(poly2caputo(a,gam)));
if isnumeric(gam)
```

```
    if gam>0, str=[str, 'Der s^' num2str(gam) ];
    else, str=[str, 'Int s^{' num2str(gam) '}']; end
  else, str=[str, 'Der s^gam']; end
```

where, function poly2caputo() is the compensation function expressed in a function block Fcn, discussed earlier.

Assume that the highest order in the equation is α, and $m = \lceil \alpha \rceil$, m initial conditions, $y(t_0), y'(t_0), y''(t_0), \cdots, y^{(m-1)}(t_0)$ are needed in the numerical solution. An auxiliary signal $z(t)$ can be introduced

$$z(t) = y(t) - y(t_0) - y'(t_0)t - \frac{1}{2!}y''(t_0)t^2 - \cdots - \frac{1}{(m-1)!}y^{(m-1)}(t_0)t^{m-1}. \tag{10-6-37}$$

Thus, the signal $y(t)$ in the Caputo equation can be converted to the Riemann–Liouville equation of signal $z(t)$. Thus, the nonlinear fractional-order differential equations can be solved directly with block diagram methods.

Example 10.60 Solve the Caputo differential equation given by [20]

$${}_0^{\mathrm{C}}\mathscr{D}_t^{1.455}y(t) = -t^{0.1}\frac{\mathscr{E}_{1,1.545}(-t)}{\mathscr{E}_{1,1.445}(-t)}\mathrm{e}^t y(t){}_0^{\mathrm{C}}\mathscr{D}_t^{0.555}y(t) + \mathrm{e}^{-2t} - \left[\,{}_0^{\mathrm{C}}\mathscr{D}_t^{1}y(t)\right]^2,$$

where, $y(0) = 1, y'(0) = -1$. The analytical solution $y = \mathrm{e}^{-t}$ can be used to validate the solution. It should be noted that there were errors in the original equation in [20] and are modified here, where the Mittag–Leffler functions in two parameters are used instead.

Solution *Introducing the auxiliary signal $z(t) = y(t) - y(0) - y'(0)t = y(t) - 1 + t$, the variable $y(t) = z(t) + 1 - t$ can be substituted back to the original equation to convert it to an equation of signal $z(t)$, with zero initial conditions.*

$$\begin{aligned}{}_0^{\mathrm{C}}\mathscr{D}_t^{1.455}[z(t)+1-t] = &-t^{0.1}\mathrm{e}^t\frac{\mathscr{E}_{1,1.545}(-t)}{\mathscr{E}_{1,1.445}(-t)}y(t)\ {}_0^{\mathrm{C}}\mathscr{D}_t^{0.555}[z(t)+1-t]\\ &+\mathrm{e}^{-2t} - \left[\,{}_0^{\mathrm{C}}\mathscr{D}_t^{1}[z(t)+1-t]\right]^2.\end{aligned}$$

It is known in (10-6-16) that, for the signal $z(t)$ with zero initial conditions, ${}_0^{\mathrm{C}}\mathscr{D}^{\alpha}z(t) = {}_0^{\mathrm{RL}}\mathscr{D}^{\alpha}z(t)$. Besides, it is known from the Caputo definition, since ${}_0^{\mathrm{C}}\mathscr{D}^{1.455}[z(t)+1-t]$ takes the second order derivative with respect to t, the augmented $1-t$ term vanishes. Therefore, ${}_0^{\mathrm{C}}\mathscr{D}^{1.455}[z(t)+1-t] = {}_0^{\mathrm{RL}}\mathscr{D}^{1.455}z(t)$. The original Caputo equation can be converted to

$${}_0^{\mathrm{RL}}\mathscr{D}_t^{1.455}z(t) = -t^{0.1}\mathrm{e}^t\frac{\mathscr{E}_{1,1.545}(-t)}{\mathscr{E}_{1,1.445}(-t)}y(t){}_0^{\mathrm{C}}\mathscr{D}_t^{0.555}y(t) + \mathrm{e}^{-2t} - [\,z'(t)-1]^2.$$

The term ${}_0^{\mathrm{C}}\mathscr{D}^{0.555}y(t)$ can be modeled directly with the Caputo operator *block. Therefore, the Simulink model shown in Figure 10.63 can be established. Besides, since Mittag–Leffler is used, its S-function implementation can be constructed*

```
function [sys,x0,str,ts]=sfun_mls(t,x,u,flag,a)
switch flag
   case 0, sizes=simsizes;
      sizes.NumContStates=0; sizes.NumDiscStates=0; sizes.NumOutputs=1;
      sizes.NumInputs=1; sizes.DirFeedthrough=1; sizes.NumSampleTimes=1;
      sys=simsizes(sizes); x0=[]; str=[]; ts=[-1 0];
   case 3, sys=ml_func([1,a],u);
   case {1,2,4,9}, sys=[];
```

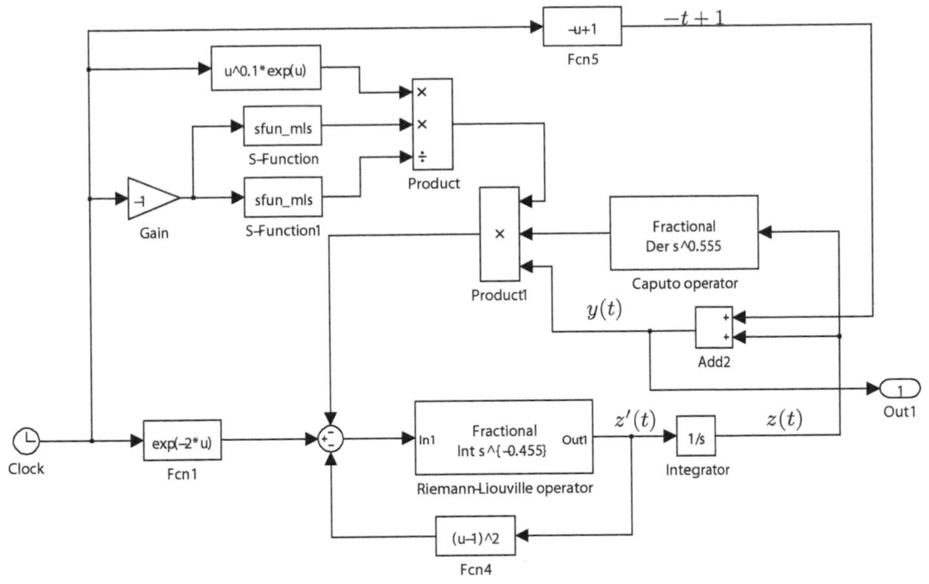

FIGURE 10.63: Simulink description of the nonlinear equation (c10mcaputo).

```
    otherwise, error(['Unhandled flag=',num2str(flag)]);
end
```

The Simulink model can be solved numerically, and the result of the nonlinear Caputo solution can be obtained as shown in Figure 10.64. It can be seen that the result is quite close to its theoretical one of e^{-t}.

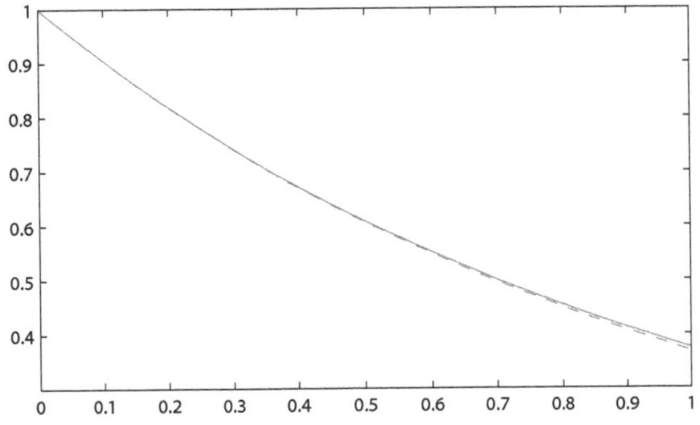

FIGURE 10.64: Solution of the nonlinear equation.

10.6.6 Object-oriented modeling and analysis of linear fractional-order systems

Consider the linear fractional-order differential equation in (10-6-23). If the initial values of the input signal $\boldsymbol{u}(t)$ and output signal $y(t)$, and their derivatives are all zeros, with the property in Laplace transform, the fractional-order transfer function can be established

$$G(s) = \frac{b_1 s^{\gamma_1} + b_2 s^{\gamma_2} + \cdots + b_m s^{\gamma_m}}{a_1 s^{\eta_1} + a_2 s^{\eta_2} + \cdots + a_{n-1} s^{\eta_{n-1}} + a_n s^{\eta_n}} \mathrm{e}^{-Ts}, \tag{10-6-38}$$

and in the transfer function model, a delay term with T seconds of time delay is introduced. The numerator and denominator are referred to as *pseudo-polynomials.*

Compared with the well-known integer-order transfer functions, it can be seen that in the fractional-order transfer function models, apart from the coefficients in numerator and denominator pseudo-polynomials, the orders are also needed. Therefore, four vectors and a vector can be used to describe uniquely the fractional-order transfer function in (10-6-38). An FOTF (fractional-order transfer function) class can be established in MATLAB to handle this type of model. Also, according to the facilities provided in the Control System Toolbox, a series of overload functions should also be written, such that the modeling and analysis of fractional-order systems are in fact as easy and straightforward as in integer-order ones.

I. Creation of FOTF class

To establish a MATLAB class, a name must be assigned. In our case, the name is selected as FOTF. Therefore, a folder `@fotf` in MATLAB's search path should be created. Two essential functions `fotf.m` and `display.m` must be written in the folder, with the first one for defining the class, while the latter for displaying the class. Also, supporting facilities are introduced.

(i) **Programming with the `fotf.m` function**. A MATLAB function `fotf.m` should be created to declare the FOTF class, as follows

```
function G=fotf(a,na,b,nb,T)
if nargin==0,                                                % an empty FOTF
   G.a=[]; G.na=[]; G.b=[]; G.nb=[]; G.ioDelay=0; G=class(G,'fotf');
elseif isa(a,'fotf'), G=a;                                   % existing FOTF
elseif nargin==1 & isa(a,'double'), G=fotf(1,0,a,0,0); % a gain block
elseif isa(a,'tf') | isa(a,'ss'),                            % integer-order one
   [n,d]=tfdata(tf(a),'v'); nn=length(n)-1:-1:0;
   nd=length(d)-1:-1:0; G=fotf(d,nd,n,nn,a.ioDelay);
elseif nargin==1 & a=='s', G=fotf(1,0,1,1,0);                % an operator
else, ii=find(abs(a)<eps); a(ii)=[]; na(ii)=[];              % ordinary FOTF
   ii=find(abs(b)<eps); b(ii)=[]; nb(ii)=[];
   if nargin==5, G.ioDelay=T; else, G.ioDelay=0; end
   G.a=a; G.na=na; G.b=b; G.nb=nb; G=class(G,'fotf');
end
```

Now in MATLAB command window, a command $G=$`fotf(`$\boldsymbol{a}$`,`$\boldsymbol{n}_\mathrm{a}$`,`$\boldsymbol{b}$`,`$\boldsymbol{n}_\mathrm{b}$`,`$T$`)` can be used to create a fractional-order transfer function object, with

$$\boldsymbol{a} = [a_1, a_2, \cdots, a_n],\ \boldsymbol{b} = [b_1, b_2, \cdots, b_m],$$

$$\boldsymbol{n}_\mathrm{a} = [\beta_1, \beta_2, \cdots, \beta_n],\ \boldsymbol{n}_\mathrm{b} = [\gamma_1, \gamma_2, \cdots, \gamma_m].$$

These vectors are used to represent the coefficients and orders of denominators and numerators of the fractional-order transfer functions. The argument T is the delay constant. It can be omitted if there is no delay in the model.

Similar to the definitions of integer-order transfer functions, $s=$`fotf('s')` command can also be used to declare a fractional-order differentiation operator s. Also, $G=$`fotf(`k`)` can be used to convert a constant into an FOTF object. If G is an LTI object in Control System Toolbox, command $G=$`fotf(`G`)` can be used to convert it into a fractional-order one.

(ii) **Display function** Another essential function is the display function, named as `display.m`. The listing of the function is given below, and the function can be used to display the FOTF object automatically, once created. The function `simplify()` is an overload function used to simplify the object and will be presented later.

```
function display(G)
G=simplify(G); strN=polydisp(G.b,G.nb); strD=polydisp(G.a,G.na);
nn=length(strN); nd=length(strD); nm=max([nn,nd]);
disp([char(' '*ones(1,floor((nm-nn)/2))) strN]), ss=[];
T=G.ioDelay; if T>0, ss=[' exp(-' num2str(T) 's)']; end
disp([char('-'*ones(1,nm)), ss]);
disp([char(' '*ones(1,floor((nm-nd)/2))) strD])
function strP=polydisp(p,np) % subfunction to display pseudo-polynomial
if length(np)==0, p=0; np=0; end,
P=''; [np,ii]=sort(np,'descend'); p=p(ii); L=length(p);
for i=1:L, P=[P,'+',num2str(p(i)),'s^{',num2str(np(i)),'}']; end
P=P(2:end); P=strrep(P,'s^{0}',''); P=strrep(P,'+-','-');
P=strrep(P,'^{1}',''); P=strrep(P,'+1s','+s');
strP=strrep(P,'-1s','-s'); nP=length(strP);
if nP>=2 & strP(1:2)=='1s', strP=strP(2:end); end
```

Example 10.61 Please enter the following FOTF model into MATLAB workspace.

$$G(s)=\frac{0.8s^{1.2}+2}{1.1s^{1.8}+1.9s^{0.5}+0.4}\,\mathrm{e}^{-0.5s}.$$

Solution *The following MATLAB statements can be issued, and an FOTF object G can be constructed in MATLAB workspace.*

```
>> G=fotf([1.1,1.9,0.4],[1.8,0.5,0],[0.8,2],[1.2,0],0.5) % FOTF object
```

(iii) **Other facilities**. Further, apart from the two essential functions fotf.m and display.m, if we want to access the fields in the FOTF object directly with MATLAB, the following two files are written. With these functions, commands like G`.nb` and G`.na`$=$ `[0.1,0.2]` are supported

```
function A=subsasgn(G,index,InputVal)
switch index.subs
   case {'a','na','b','nb','ioDelay'},
      eval(['G.' index.subs,'=InputVal;']);
      if length(G.a)~=length(G.na) | length(G.b)~=length(G.nb)
         error('Error: field pairs (na,a) or (nb,b) mismatched.')
      else, A=fotf(G.a,G.na,G.b,G.nb,G.ioDelay); end
   otherwise,
      error('Error: Available fields are a, na, b, na, ioDelay.');
end
```

```
function A=subsref(G,index)
switch index.subs
   case {'a','na','b','nb','ioDelay'}, A=eval(['G.' index.subs]);
   otherwise,
      error('Error: Available fields are a, na, b, na, ioDelay.');
end
```

It should be noted that all the files should be placed in the `@fotf` folder, and nowhere else, otherwise, existing MATLAB functions will be affected.

II. Interconnections of FOTF objects

With the use of Control System Toolbox of MATLAB, the integer-order LTI models can be evaluated with sum, product and `feedback()` operations, when the parallel, series or feedback connections are involved. Following the idea used in the Control System Toolbox, the following overload functions can be written, and these functions should be placed again in the `@fotf` folder [21]. The functions with some modifications and extensions are listed here.

(i) **Product of two FOTF objects**, with $G=G_1*G_2$, computing the overall model, when two FOTFs $G_1(s)$ and $G_2(s)$ are in series connection. The formula for finding series connected system is

$$G(s)=G_1(s)G_2(s)=\frac{N_1(s)N_2(s)}{D_1(s)D_2(s)}\mathrm{e}^{-(\tau_1+\tau_2)s}. \tag{10-6-39}$$

```
function G=mtimes(G1,G2) % handle series connected FOTF blocks
G1=fotf(G1); G2=fotf(G2); a=kron(G1.a,G2.a); b=kron(G1.b,G2.b);
na=kronsum(G1.na,G2.na); nb=kronsum(G1.nb,G2.nb);
G=simplify(fotf(a,na,b,nb,G1.ioDelay+G2.ioDelay));
```

For simplicity, Kronecker sum $\boldsymbol{A}\oplus\boldsymbol{B}$ is introduced

$$\boldsymbol{A}\oplus\boldsymbol{B}=\begin{bmatrix} a_{11}+\boldsymbol{B} & \cdots & a_{1m}+\boldsymbol{B}\\ \vdots & \ddots & \vdots\\ a_{n1}+\boldsymbol{B} & \cdots & a_{nm}+\boldsymbol{B}\end{bmatrix}, \tag{10-6-40}$$

in contrast to Kronecker product $\boldsymbol{A}\otimes\boldsymbol{B}$, and similar to `kron()` function, a Kronecker sum function `kronsum()` is written

```
function C=kronsum(A,B)
[ma,na]=size(A); [mb,nb]=size(B);
A=reshape(A,[1 ma 1 na]); B=reshape(B,[mb 1 nb 1]);
C=reshape(bsxfun(@plus,A,B),[ma*mb na*nb]); % structure copied from kron
```

(ii) **Sum function**, with $G=G_1+G_2$, computing the overall model, when two FOTFs, $G_1(s)$ and $G_2(s)$, are connected in parallel. If the delays of the two blocks are the same, where $\tau_1=\tau_2$, denoted by τ, the sum action can be carried out, and the formula for the evaluation of parallel connection is

$$G(s)=G_1(s)+G_2(s)=\frac{N_1(s)D_2(s)+N_2(s)D_1(s)}{D_1(s)D_2(s)}\mathrm{e}^{-\tau s}. \tag{10-6-41}$$

```
function G=plus(G1,G2)          % handle parallel connected FOTF blocks
G1=fotf(G1); G2=fotf(G2);       % unify to FOTF objects
if G1.ioDelay==G2.ioDelay       % if delays are the same, problem solvable
   a=kron(G1.a,G2.a); na=kronsum(G1.na,G2.na);
   b=[kron(G1.a,G2.b),kron(G1.b,G2.a)];
   nb=[kronsum(G1.na,G2.nb),kronsum(G1.nb,G2.na)];
   G=simplify(fotf(a,na,b,nb,G1.ioDelay));
else, error('cannot handle different delays'); end
```

(iii) **Negative feedback function**, with G=`feedback(`G_1,G_2`)`. The overall model of two FOTFs in negative feedback connection can be obtained, if there are no delays of the two blocks. The feedback connection of the two blocks can be processed. If positive feedback is involved, G_2 should be replaced by $-G_2$. The formula for finding the overall model is given by

$$G(s) = \frac{G_1(s)}{1+G_1(s)G_2(s)} = \frac{N_1(s)D_2(s)}{D_1(s)D_2(s)+N_1(s)N_2(s)}. \tag{10-6-42}$$

```
function G=feedback(G1,G2)           % handle feedback connected FOTF blocks
G1=fotf(G1); G2=fotf(G2);            % unify to FOTF objects
if G1.ioDelay==0 & G2.ioDelay==0     % blocks without delays, problem solvable
   b=kron(G1.b,G2.a); nb=kronsum(G1.nb,G2.na);
   na=[kronsum(G1.nb,G2.nb), kronsum(G1.na,G2.na)];
   a=[kron(G1.b,G2.b), kron(G1.a,G2.a)]; G=simplify(fotf(a,na,b,nb,0));
else, error('cannot handle blocks with delays'); end
```

(iv) **Other simple supporting functions**, with `uminus()` for $G_1(s) = -G(s)$, to enable the use of $G_1 = -G$ command. Function G=`inv(`G_1`)` can be used to compute $G(s) = 1/G_1(s)$; function `minus()` is used to compute $G(s) = G_1(s) - G_2(s)$, allowing $G = G_1 - G_2$; Function `eq()` is used to check whether the two FOTFs G_1 and G_2 are equal. If `key`$=G_1$`==`G_2 returns 1, the two FOTFs are equal.

```
function G=uminus(G1), G=G1; G.b=-G.b;
function G=inv(G1), G=fotf(G1.b,G1.nb,G1.a,G1.na);
function G=minus(G1,G2), G=G1+(-G2);
function key=eq(G1,G2), G=G1-G2; key=(length(G.nb)==0|norm(G.b)<1e-10);
```

(v) **Right division function**, with the syntax $G = G_1/G_2$, compute $G(s) = G_1(s)/G_2(s)$.

```
function G=mrdivide(G1,G2)
G1=fotf(G1); G2=fotf(G2); G=G1*inv(G2); G.ioDelay=G1.ioDelay-G2.ioDelay;
if G.ioDelay<0, warning('block with positive delay'); end
```

(vi) **Power function**, with the syntax $G = G_1$`^`n. If G_1 is an FOTF, then, n must be an integer. Otherwise, G_1 must be an s operator.

```
function G1=mpower(G,n)
if n==fix(n), % find out whether power is an integer
   if n>=0, G1=1; for i=1:n, G1=G1*G; end  % integer power of FOTF
   else, G1=inv(G^(-n)); end                % handling negative power
elseif G==fotf(1,0,1,1), G1=fotf(1,0,1,n); % power of operator
else, error('mpower: power must be an integer.'); end
```

(vii) **Simplification function**, with the syntax G=`simplify(G)`, used to simplify the pseudo-polynomials in the numerator and denominator. The subfunction `polyuniq()` is used to collect the terms in the pseudo-polynomials. The subfunction is a low-level function of `simplify()` and cannot be called directly by other functions.

```
function G=simplify(G1)
[a,n]=polyuniq(G1.a,G1.na); G1.a=a; G1.na=n; na=G1.na;
[a,n]=polyuniq(G1.b,G1.nb); G1.b=a; G1.nb=n; nb=G1.nb;
if length(nb)==0, nb=0; G1.nb=0; G1.b=0; end
nn=min(na(end),nb(end)); nb=nb-nn; na=na-nn;
G=fotf(G1.a,na,G1.b,nb,G1.ioDelay);
function [a,an]=polyuniq(a,an) % collect terms in the pseudo-polynomial
[an,ii]=sort(an,'descend'); a=a(ii); ax=diff(an); key=1;
for i=1:length(ax) % here ax is the order difference vector
   if ax(i)==0,     % collect terms of the same order
      a(key)=a(key)+a(key+1); a(key+1)=[]; an(key+1)=[];
   else, key=key+1; end
end
```

Example 10.62 Please input the fractional-order PID controller $G_c(s) = 5+2s^{-0.2}+3s^{0.6}$ into MATLAB workspace

Solution *An FOTF operator s can be specified first, then, the following commands can be used to input the fractional-order PID controller model*

```
>> s=fotf('s'); Gc=5+2*s^(-0.2)+3*s^0.6 % declare operator then add up
```

An FOTF model is returned as $G_c(s) = (3s^{0.8} + 5s^{0.2} + 2)/s^{0.2}$.

Example 10.63 Please enter the following complicated fractional-order transfer function model into MATLAB workspace

$$G(s) = \frac{(s^{0.3}+3)^2}{(s^{0.2}+2)(s^{0.4}+4)(s^{0.4}+3)}.$$

Solution *For complicated models, an FOTF operator s can be defined first, then, simple commands can be used to input the system into MATLAB workspace.*

```
>> s=fotf('s'); G=(s^0.3+3)^2/(s^0.2+2)/(s^0.4+4)/(s^0.4+3)
```

The result is expressed in FOTF format as

$$G(s) = \frac{s^{0.6}+6s^{0.3}+9}{s+2s^{0.8}+7s^{0.6}+14s^{0.4}+12s^{0.2}+24}.$$

Example 10.64 Assume that in a typical unity negative feedback system, the two transfer functions in the forward path are given by

$$G(s) = \frac{0.8s^{1.2}+2}{1.1s^{1.8}+0.8s^{1.3}+1.9s^{0.5}+0.4},\quad G_c(s) = \frac{1.2s^{0.72}+1.5s^{0.33}}{3s^{0.8}},$$

please find the closed-loop fractional-order system.

Solution *The two components can be entered into MATLAB environments first as FOTF objects, then, function* `feedback()` *can be called to get the closed-loop model*

```
>> G=fotf([1.1,0.8 1.9 0.4],[1.8 1.3 0.5 0],[0.8 2],[1.2 0]);
   Gc=fotf([3],[0.8],[1.2 1.5],[0.72 0.33]); G0=feedback(G*Gc,1)
```

The closed-loop model can be obtained

$$G_0(s)=\frac{0.96s^{1.59}+1.2s^{1.2}+2.4s^{0.39}+3}{3.3s^{2.27}+2.4s^{1.77}+0.96s^{1.59}+1.2s^{1.2}+5.7s^{0.97}+1.2s^{0.47}+2.4s^{0.39}+3}.$$

With the FOTF class established, further analysis and design functions can be written for linear fractional-order transfer functions, and the developed functions are listed in Table 10.15. For details, please check [22].

TABLE 10.15: Analysis functions of linear fractional-order systems.

syntax	explanation of the function
y = `step(`G`,`t`)`	compute the unit step response of G, if no argument is returned, draw automatically the step response
y = `lsim(`G`,`u`,`t`)`	compute the time response G to arbitrary input u
`key` = `isstable(`G`)`	assess the stability of fractional-order system G
H = `rlocus(`G`)`	draw the root locus for the system G
H = `bode(`G`,`w`)`	compute/draw Bode diagram of the system G, over frequency vector w. Similar functions are `nyquist()` and `nichols()`
n = `norm(`G`,`k`)`	compute the norms of the system G, with default $k=2$, for $\mathcal{H}_2$ norm. For k = `inf`, compute $\mathcal{H}_\infty$

Example 10.65 Consider again the fractional-order PID control problem, with

$$G(s)=\frac{(s^{0.3}+3)^2}{(s^{0.2}+2)(s^{0.4}+4)(s^{0.4}+3)},\quad G_c(s)=0.1s^{1.3}+0.02+\frac{10}{s^{0.95}}.$$

Please find the norms of the plant model, and draw the open-loop Bode diagram and closed-loop step response of the system, with FOTF tools.

Solution *The two blocks can be entered as FOTF blocks, the open-loop and closed-loop models can be obtained, respectively. Based on the models, the Bode diagram and step response can be obtained easily as shown in Figures 10.65 (a) and (b). It can be seen that, with the use of the FOTF tools, the analysis of linear fractional-order systems is as simple as it is in dealing with integer-order linear systems.*

```
>> s=fotf('s'); G=(s^0.3+3)^2/(s^0.2+2)/(s^0.4+4)/(s^0.4+3) % plant
   Gc=0.1*s^1.3+0.02+10/s^0.95; % enter the controller model
   w=logspace(-1,3); subplot(121), bode(G*Gc,w) % open-loop Bode diagram
   t=0:0.01:10; subplot(122), step(feedback(G*Gc,1),t) % closed-loop
```

The norms of the plant model can be evaluated directly with the overload `norm()` *function, with $n_1=2.7167$ and $n_2=8.6115$. The following statements can be used to assess the stability of the closed-loop system, from which it can be seen that the system is stable.*

```
>> n1=norm(G), n2=norm(G,inf) % compute the norms of the plant model
   G1=feedback(G*Gc,1); key=isstable(G1) % key is 1 for stable
```

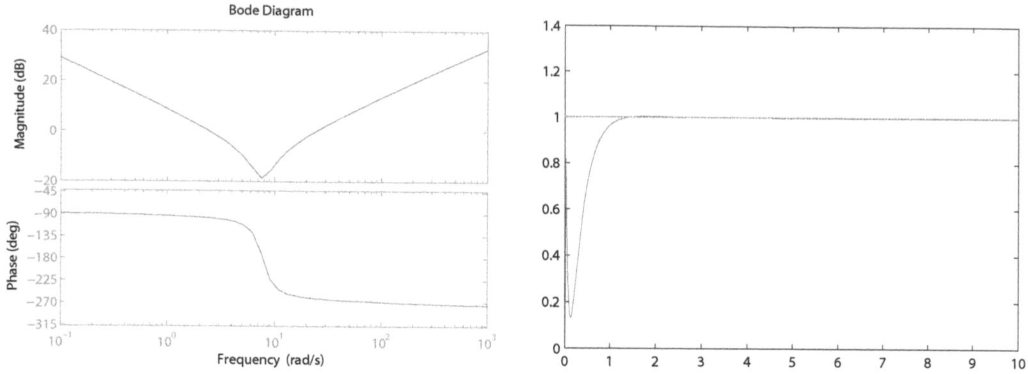

(a) open-loop Bode diagram (b) closed-loop step response

FIGURE 10.65: System analysis plots.

Exercises

Exercise 10.1 *Consider a tipping problem in a restaurant*[23]. *Assume that the average rate for the tips is 15% of the consumption. The service level and food quality are used to calculate the tip. The service level can be written as "good," "average" and "poor," and the food quality can also be expressed as other fuzzy descriptions. Establish a fuzzy inference system for determining the tips.*

Exercise 10.2 *Consider the sampled data set* (x_i, y_i) *given in Table 10.16. Construct a neural network model and plot the training curve in the interval* $x \in (1, 10)$. *Test different neural network structures and training algorithms. Compare the fitting results under different structures.*

TABLE 10.16: Measured data of Exercise 10.2.

x_i	1	2	3	4	5	6	7	8	9	10
y_i	244.0	221.0	208.0	208.0	211.5	216.0	219.0	221.0	221.5	220.0

Exercise 10.3 *For the actual measured data given in Table 10.17, construct a neural network to fit the surface in the rectangular region* $(0.1, 0.1) \sim (1.1, 1.1)$. *Compare the results with data interpolation algorithms.*

Exercise 10.4 *Solve the benchmark problems in Exercise 6.9 using genetic algorithms and PSO methods.*

Exercise 10.5 *Solve the constrained optimization problem with genetic algorithms and PSO methods and compare the results with traditional optimization algorithms.*

TABLE 10.17: Measured data of Exercise 10.3.

y_i	x_1	x_2	x_3	x_4	x_5	x_6	x_7	x_8	x_9	x_{10}	x_{11}
0	0.1	0.2	0.3	0.4	0.5	0.6	0.7	0.8	0.9	1	1.1
0.1	0.8304	0.8273	0.8241	0.8210	0.8182	0.8161	0.8148	0.8146	0.8158	0.8185	0.8230
0.2	0.8317	0.8325	0.8358	0.8420	0.8513	0.8638	0.8798	0.8994	0.9226	0.9496	0.9801
0.3	0.8359	0.8435	0.8563	0.8747	0.8987	0.9284	0.9638	1.0045	1.0502	1.1	1.1529
0.4	0.8429	0.8601	0.8854	0.9187	0.9599	1.0086	1.0642	1.1253	1.1904	1.257	1.3222
0.5	0.8527	0.8825	0.9229	0.9735	1.0336	1.1019	1.1764	1.254	1.3308	1.4017	1.4605
0.6	0.8653	0.9105	0.9685	1.0383	1.118	1.2046	1.2937	1.3793	1.4539	1.5086	1.5335
0.7	0.8808	0.9440	1.0217	1.1118	1.2102	1.311	1.4063	1.4859	1.5377	1.5484	1.5052
0.8	0.8990	0.9828	1.082	1.1922	1.3061	1.4138	1.5021	1.5555	1.5573	1.4915	1.346
0.9	0.9201	1.0266	1.1482	1.2768	1.4005	1.5034	1.5661	1.5678	1.4889	1.3156	1.0454
1	0.9438	1.0752	1.2191	1.3624	1.4866	1.5684	1.5821	1.5032	1.315	1.0155	0.6248
1.1	0.9702	1.1279	1.2929	1.4448	1.5564	1.5964	1.5341	1.3473	1.0321	0.6127	0.1476

$$\min \quad \frac{1}{2\cos x_6}\left[x_1x_2(1+x_5)+x_3x_4\left(1+\frac{31.5}{x_5}\right)\right].$$

$$\boldsymbol{x} \text{ s.t.} \begin{cases} 0.003079x_1^3x_2^3x_5-\cos^3 x_6\geqslant 0\\ 0.1017x_3^3x_4^3-x_5^2\cos^3 x_6\geqslant 0\\ 0.09939(1+x_5)x_1^3x_2^2-\cos^2 x_6\geqslant 0\\ 0.1076(31.5+x_5)x_3^3x_4^2-x_5^2\cos^2 x_6\geqslant 0\\ x_3x_4(x_5+31.5)-x_5[2(x_1+5)\cos x_6+x_1x_2x_5]\geqslant 0\\ 0.2\leqslant x_1\leqslant 0.5, 14<\leqslant x_2\leqslant 22, 0.35\leqslant x_3\leqslant 0.6\\ 16\leqslant x_4\leqslant 22, 5.8\leqslant x_5\leqslant 6.5, 0.14\leqslant x_6\leqslant 0.2618 \end{cases}$$

Exercise 10.6 *Assume that the corrupted signal is generated from*

```
>> t=0:0.005:5; y=15*exp(-t).*sin(2*t); r=0.3*randn(size(y)); y1=y+r;
```

Perform de-noising tasks with wavelet transforms and compare the results with the filtering techniques in Exercise 8.22.

Exercise 10.7 *A series of experimental data is given in file c10rsdat.txt which is made up as a* 60×13 *table. Each column corresponds to an attribute and the last column is the decision attribute. Use rough set reduction technique to check which attribute is most important to the event in the decision.*

Exercise 10.8 *For the signal* $f(t) = \mathrm{e}^{-3t}\sin(t+\pi/3)+t^2+3t+2$, *find the 0.2th order derivative and 0.7th order integral. Draw the relevant curves.*

Exercise 10.9 *Design a filter for Exercise 10.8. The fractional-order derivatives and integrals can be obtained with the filter. Compare the results with the ones obtained with Grünwald–Letnikov method.*

Exercise 10.10 *Consider a fractional-order linear differential equation described by* [13]

$$0.8\mathscr{D}_t^{2.2}y(t)+0.5\mathscr{D}_t^{0.9}y(t)+y(t)=1,\ y(0)=y'(0)=y''(0)=0.$$

Solve the solution using numerical method. If one changes the orders of 2.2 and 0.9 respectively to 2 and 1, an approximate integer-order differential equation can be obtained. Compare the accuracy of the integer-order approximation.

Exercise 10.11 *Find the closed-loop model from the typical negative feedback structure.*

(i) $G(s)=\dfrac{211.87s+317.64}{(s+20)(s+94.34)(s+0.17)},\ G_c(s)=\dfrac{169.6s+400}{s(s+4)},\ H(s)=\dfrac{1}{0.01s+1},$

(ii) $G(s)=\dfrac{s^{0.4}+5}{s^{3.1}+2.8s^{2.2}+1.5s^{0.8}+4},\ G_c(s)=3+2.5s^{-0.5}+1.4s^{0.8},\ H(s)=1.$

Exercise 10.12 *Consider the linear fractional-order differential equation given by*

$$\mathscr{D}x(t)+\left(\frac{9}{1+2\lambda}\right)^{\alpha}\mathscr{D}^{\alpha}x(t)+x(t)=1,\ 0<\alpha<1,$$

where $\lambda=0.5$, $\alpha=0.25$. *Solve the equation numerically.*

Exercise 10.13 *Find a good approximation for the modified Oustaloup's filter, to* $s^{0.7}$ *and see which* N *can best fit the fractional-order differentiator.*

Exercise 10.14 *Two filter approximation approaches are proposed in the chapter on fractional-order derivatives. Please compare the two filters for the following fractional-order system, in frequency and step response fitting.*

$$G(s)=\frac{s+1}{10s^{3.2}+185s^{2.5}+288s^{0.7}+1}.$$

Exercise 10.15 *Solve the following nonlinear fractional-order differential equation with the block diagram-based algorithm*

$$\mathscr{D}^2x(t)+\mathscr{D}^{1.455}x(t)+\left[\mathscr{D}^{0.555}x(t)\right]^2+x^3(t)=f(t),$$

where $f(t)$ *is a unit square wave input signal of 1 Hz and the system has zero initial conditions.*

Exercise 10.16 *Consider the fractional-order nonlinear differential equation expressed by the Simulink model shown in Figure 10.66. Write out the mathematical expression of the equation and draw the output signal* $y(t)$.

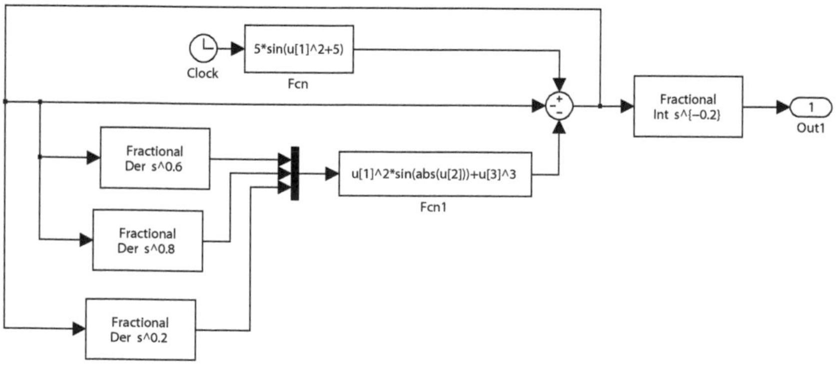

FIGURE 10.66: Simulink model (file: c10mfode4.mdl).

Exercise 10.17 *Solve the well-known Bagley–Torvik fractional-order differential equation* [20]

$$Ay''(t) + B\mathscr{D}^{3/2}y(t) + Cy(t) = C(t+1), \text{ with } y(0) = y'(0) = 1,$$

and show that the solutions are independent of the constants A, B and C.

Exercise 10.18 *Analyze the stability of the closed-loop system, and draw Bode diagram and closed-loop step response for the following system.*

$$G(s) = \frac{s^{1.2}+4s^{0.8}+7}{8s^{3.2}+9s^{2.8}+9s^2+6s^{1.6}+5s^{0.4}+9}, \quad G_c(s) = 10 + \frac{9}{s^{0.97}} + 10s^{0.98}.$$

Bibliography

[1] Weisstein E W. Goldbach conjecture. From MathWorld — A Wolfram Web Resource. `http://mathworld.wolfram.com/GoldbachConjecture.html`

[2] Zadeh L A. Fuzzy sets. Information and Control, 1965, 8:338–353

[3] Pawlak Z. Rough sets — theoretical aspects of reasoning about data. Boston, USA: Kluwer Academic Pub., 1991

[4] Zhang X F. Research and program development of rough set data analysis system. Master's thesis, Northeastern University, 2004. (in Chinese)

[5] Hagan M T, Demuth H B, Beale M H. Neural network design. PWS Publishing Company, 1995

[6] Wang X, Wang H, Wang W H. Principles and applications of artificial neural networks. Shenyang: Northeastern University Press, 2000. In Chinese

[7] Goldberg D E. Genetic algorithms in search, optimization and machine learning. Reading, MA: Addison-Wesley, 1989

[8] Chipperfield A, Fleming P. Genetic algorithm toolbox user's guide. Department of Automatic Control and Systems Engineering, University of Sheffield, 1994

[9] Houck C R, Joines J A, Kay M G. A genetic algorithm for function optimization: a MATLAB implementation, 1995

[10] Kennedy J, Eberhart R. Particle swarm optimization. Proceedings of IEEE International Conference on Neural Networks. Perth, Australia, 1995, 1942∼1948

[11] Li Z. Lévy PSO. MATLAB Central File ID: #50277

[12] Hilfer R. Applications of fractional calculus in physics. Singapore: World Scientific, 2000

[13] Podlubny I. Fractional differential equations. San Diego: Academic Press, 1999

[14] Petráš I, Podlubny I, O'Leary P. Analogue realization of fractional order controllers. Fakulta BERG, TU Košice, 2002

[15] Oustaloup A, Levron F, Nanot F, et al. Frequency band complex non integer differentiator: characterization and synthesis. IEEE Transactions on Circuits and Systems I: Fundamental Theory and Applications, 2000, 47(1):25–40

[16] Xue D, Zhao C N, Chen Y Q. A modified approximation method of fractional order system. Proceedings of IEEE Conference on Mechatronics and Automation. Luoyang, China, 2006, 1043–1048

[17] Podlubny I. The Laplace transform method for linear differential equations of the fractional order. Proc. of the $9^{\rm th}$ International BERG Conference. Kosice, Slovak Republic (in Slovak), 1997, 119–119

[18] Podlubny I. Fractional-order systems and $PI^{\lambda}D^{\mu}$-controllers. IEEE Transactions on Automatic Control, 1999, 44(1):208–214

[19] Xue D, Chen Y Q. System simulation techniques with MATLAB/Simulink. London: Wiley, 2013

[20] Cachan J, Groningen F, Paris B. The analysis of fractional differential equations. New York: Springer, 2010

[21] Monje C A, Chen Y Q, Vinagre B M, Xue D, Feliu V. Fractional-order systems and controls — fundamentals and applications. London: Springer, 2010

[22] Xue D Y, Chen Y Q. Modeling, analysis and design of control systems in MATLAB and Simulink. Singapore: World Scientific Press, 2014

[23] The MathWorks Inc. Fuzzy logic toolbox user's manual, 2007

MATLAB Functions Index

Bold page numbers indicate where to find the syntax explanation of the function. The function or model name marked by * are the ones developed by the authors, and the ones with † are overload functions. The items marked with ‡ are those downloadable freely from internet.

Index

T

U